AF342285

New Therapeutic Strategies for Type 2 Diabetes
Small Molecule Approaches

RSC Drug Discovery Series

Editor-in-Chief
Professor David Thurston, *London School of Pharmacy, UK*

Series Editors:
Dr David Fox, *Pfizer Global Research and Development, Sandwich, UK*
Professor Salvatore Guccione, *University of Catania, Italy*
Professor Ana Martinez, *Instituto de Quimica Medica-CSIC, Spain*
Professor David Rotella, *Montclair State University, USA*

Advisor to the Board:
Professor Robin Ganellin, *University College London, UK*

Titles in the Series:
1: Metabolism, Pharmacokinetics and Toxicity of Functional Groups
2: Emerging Drugs and Targets for Alzheimer's Disease; Volume 1
3: Emerging Drugs and Targets for Alzheimer's Disease; Volume 2
4: Accounts in Drug Discovery
5: New Frontiers in Chemical Biology
6: Animal Models for Neurodegenerative Disease
7: Neurodegeneration
8: G Protein-Coupled Receptors
9: Pharmaceutical Process Development
10: Extracellular and Intracellular Signaling
11: New Synthetic Technologies in Medicinal Chemistry
12: New Horizons in Predictive Toxicology
13: Drug Design Strategies: Quantitative Approaches
14: Neglected Diseases and Drug Discovery
15: Biomedical Imaging
16: Pharmaceutical Salts and Cocrystals
17: Polyamine Drug Discovery
18: Proteinases as Drug Targets
19: Kinase Drug Discovery
20: Drug Design Strategies: Computational Techniques and Applications
21: Designing Multi-Target Drugs
22: Nanostructured Biomaterials for Overcoming Biological Barriers
23: Physico-Chemical and Computational Approaches to Drug Discovery
24: Biomarkers for Traumatic Brain Injury
25: Drug Discovery from Natural Products
26: Anti-Inflammatory Drug Discovery
27: New Therapeutic Strategies for Type 2 Diabetes: Small Molecule Approaches

How to obtain future titles on publication:
A standing order plan is available for this series. A standing order will bring
delivery of each new volume immediately on publication.

For further information please contact:
Book Sales Department, Royal Society of Chemistry, Thomas Graham House,
Science Park, Milton Road, Cambridge, CB4 0WF, UK
Telephone: +44 (0)1223 420066, Fax: +44 (0)1223 420247,
Email: booksales@rsc.org
Visit our website at http://www.rsc.org/Shop/Books/

New Therapeutic Strategies for Type 2 Diabetes

Small Molecule Approaches

Edited by

Robert M. Jones
Arena Pharmaceuticals, San Diego, California, USA
Email: rjones@arenapharm.com

RSC Publishing

RSC Drug Discovery Series No. 27

ISBN: 978-1-84973-414-1
ISSN: 2041-3203

A catalogue record for this book is available from the British Library

© The Royal Society of Chemistry 2012

Published by The Royal Society of Chemistry,
Thomas Graham House, Science Park, Milton Road,
Cambridge CB4 0WF, UK

Registered Charity Number 207890

For further information see our web site at www.rsc.org

Printed in the United Kingdom by Henry Ling Limited, at the Dorset Press, Dorchester, DT1 1HD

Preface

The intent of *Therapeutic Strategies for Type 2 Diabetes* is to provide a comprehensive overview of recent advances in small molecule discovery campaigns aimed at developing new oral medications for treating this burgeoning worldwide epidemic.

I would like to take this opportunity to thank all of the authors for their timely and relevant contributions to this book, Dr. David Rotella of Montclair State University, New Jersey for "persuading" me to put this text together during cocktail hour at the 2010 NMCS in Minneapolis, Gwen Jones of RSC Publishing for her continued helpful insights and discussions during the incubation phase of this text, Dr. Dominic Behan and Dr. Graeme Semple of Arena Pharmaceuticals for allowing me to take on the project, Dr. Carleton Sage for the cover art (the first in class agonist APD668 docked into a model of GPR119), and finally to Shelley and Finley for their support over the course of this almost two-year project. From planning the list of 16 chapters and selection of key thought leaders to write on each respective topic to the realization of this text, it has been a thoroughly rewarding experience. I sincerely hope that the book will not only fill a perceived void in the scientific literature but will also provide a "go to" guide for both medicinal chemists and biologists alike working on new therapies and mechanisms in this important disease arena. Finally I hope that you the reader will enjoy and profit from the contents.

Rob Jones, PhD, San Diego

RSC Drug Discovery Series No. 27
New Therapeutic Strategies for Type 2 Diabetes: Small Molecule Approaches
Edited by Robert M. Jones
© The Royal Society of Chemistry 2012
Published by the Royal Society of Chemistry, www.rsc.org

Contents

RSC Drug Discovery Series No. 27
New Therapeutic Strategies for Type 2 Diabetes: Small Molecule Approaches
Edited by Robert M. Jones
© The Royal Society of Chemistry 2012
Published by the Royal Society of Chemistry, www.rsc.org

CHAPTER 1

Type 2 Diabetes: Disease Overview

DANIEL M. KEMP

Diabetes & Endocrinology, Merck Research Laboratories, 126 East Lincoln Avenue, Rahway, NJ 07065, USA
E-mail: daniel.kemp@merck.com

1.1 Type 2 Diabetes

1.1.1 Societal and Economic Effects

Type 2 diabetes is a metabolic disease, and is characterized by elevated circulating glucose, otherwise known as hyperglycemia. The specific molecular cause of type 2 diabetes remains unknown, though considerable progress has been made to define the metabolic characteristics of people who have, or later acquire, the disease. What we are certain of is that two essential components define the overall metabolic dysfunction in type 2 diabetes; firstly, a relative insensitivity of glucose-utilizing tissues to insulin (i.e., skeletal muscle, liver, and adipose tissue), subsequently compounded by a relative insufficiency of insulin production from the pancreas, leading to whole body glucose intolerance. This progressive phenotype of insulin resistance and glucose intolerance contrasts with that of type 1 diabetes, which results from autoimmune destruction of insulin-producing β-cells of the pancreas, and thus is predominantly characterized by insulin insufficiency alone, and is treated specifically by injection of exogenous insulin.

In adults, type 2 diabetes accounts for about 90–95% of all diagnosed cases of diabetes, and develops most often in middle-aged and older adults. Among

RSC Drug Discovery Series No. 27
New Therapeutic Strategies for Type 2 Diabetes: Small Molecule Approaches
Edited by Robert M. Jones
© The Royal Society of Chemistry 2012
Published by the Royal Society of Chemistry, www.rsc.org

U.S. residents aged 65 years and older, 10.9 million, or 26.9%, had diabetes in 2010, according to the NIDDK.[1] The total number of cases in the US alone is forecast to double from 24 million in 2009 to 44 million in 2034. Not surprisingly, in association with the growing diabetic population, spending on diabetes and related complications are projected to triple in the same period, from $100 billion in 2009 to around $300 billion in 2034. A report published in 2011 that included a dataset of 2.7 million individuals across the world concluded that diabetes prevalence is projected to be in the region of 347 million people.[2] These staggering statistics outline the emerging epidemic of type 2 diabetes, and strongly indicate the need for new and better therapies.

1.1.2   Epidemiology

Though still poorly understood, the root cause of type 2 diabetes clearly results from interplay between genetic and environmental factors. The importance of genetics for the development of type 2 diabetes has long been recognized, both at the individual level (family history) and at the population level (ethnic background). For example, the most convincing evidence of genetic predisposition at the level of the individual comes from twin studies. Concordance rates for identical twins range from 70% up to close to 90%, with lifelong follow-up, and are higher than non-identical twins, siblings, or other family members.[3] With respect to population genetics, strong evidence comes from studies like the San Antonio Heart Study, which focused on studying populations with different genetic backgrounds living in the same environment. In that study, the prevalence of type 2 diabetes was higher in Mexican Americans than in non-Hispanic whites at each level of obesity.[4] Despite the overwhelming evidence that susceptibility for type 2 diabetes is inherited, the specific susceptibility genes and their mode of inheritance have yet to be determined, and this has been an area of intense research over the past decade. Though some monogenic traits have been firmly associated with a subset of type 2 diabetes patients, annotated as MODY genes (maturity onset diabetes of the young), the advent of genome-wide association studies, enabled by the human genome project, heralded the potential to identify disease-causing genes via the association of specific genetic mutations with incidences of type 2 diabetes. To date more than 50 genetic loci have been discovered, and these loci appear to associate predominantly with genes involved in pancreatic islet function, with few if any involved in insulin resistance-related pathways.[5] Though initially surprising to some, this observation makes sense, as the primary cause of hyperglycemia is the inability of the pancreas to maintain sufficient insulin levels to drive glucose uptake into peripheral tissues. As mentioned earlier, although insulin resistance is a primary cause of the disease, type 2 diabetes only manifests when the β-cells of the pancreas fail to keep pace with demand. Mutations in genes that result in functional impairment of pancreatic β-cells are therefore most apparent in population genetics studies with hyperglycemia as the primary clinical endpoint. More recent genetic studies that focus on markers of insulin resistance are currently ongoing, and should identify additional genes involved more specifically in the function of insulin action and glucose utilization. It is highly anticipated that the results of these genetic studies should identify putative drug targets for the treatment of both insulin resistance and type 2 diabetes.

Environmental factors that influence the prevalence of type 2 diabetes can be easily exposed by studies focused on migrant populations. For example, Japanese migrants in Hawaii and Los Angeles are two to three times more likely to suffer type 2 diabetes than Japanese living in Japan.[6] Such a shift in environmental influences over just one or two generations can impart surprisingly rapid changes in prevalence and incidence of type 2 diabetes. Factors such as birth weight, in utero exposure to diabetes, diet, obesity, and physical activity can expose an underlying genetic susceptibility within specific ethnic populations. The Pima Indians of Arizona are particularly notable with respect to their genetic predisposition to type 2 diabetes, clearly exposed by environmental factors of "western lifestyle".[7] All of these specific examples speak to the broader context of how exposure to an ever evolving global economy and cultural environment has unveiled the apparent fragility of the human genome, and underscores how habitat is just as important as evolution in defining what species thrive and perish.

1.1.3 Pathophysiology

The ability of insulin to stimulate glucose disposal has been extensively studied, and abundant evidence exists to confirm that insulin action is markedly decreased in patients with type 2 diabetes. The major route of glucose disposal, demonstrated by infusion studies, is uptake into the skeletal muscle, and it is reasonable to conclude that the majority of patients with type 2 diabetes have a defect in insulin-stimulated glucose disposal into muscle. However, it is also clear that impaired insulin-dependent glucose uptake per se cannot account for the development of hyperglycemia in patients with type 2 diabetes, because relatively normal fasting plasma glucose levels are often observed in individuals who are equally insulin resistant as patients with frank diabetes. Indeed, type 2 diabetes will develop only when insulin-resistant subjects are incapable of secreting sufficient insulin to compensate for the defect in skeletal muscle insulin action. To further elaborate, as fasting hyperglycemia develops only when the pancreas fails, the consequence of this decline in insulin secretory capacity must be defined further in order to understand the pathophysiology of type 2 diabetes. We need to consider the involvement of adipose tissue as an important player in the overall characterization of the disease. This is because resistance to insulin regulation at the level of the adipose tissue, or as a result of decreased insulin secretion, leads to elevated plasma free fatty acid (FFA) concentrations. Indeed, ambient plasma FFA concentrations are elevated in type 2 diabetes, and the greater the increase in FFA concentration, the higher the plasma glucose concentration.[8]

Exacerbation of the dysmetabolic state then ensues because elevated FFA levels decrease insulin-stimulated glucose uptake, cause lipotoxicity at the level of the pancreatic β-cell, and further compromises insulin secretory function. Furthermore, elevated FFA levels stimulate gluconeogenesis in the liver, decreasing the ability of hyperglycemia to suppress hepatic glucose production, further compounding the problem. To summarize, when insulin resistance is compensated by hyperinsulinemia, whole body glucose homeostasis can be preserved. But, when the insulin secretory response declines to a point where circulating plasma FFA levels become significantly elevated, the plasma glucose concentration increases precipitously due to unsuppressed hepatic glucose output, exacerbation of pancreatic β-cell failure, and impaired insulin-dependent glucose disposal into muscle.

Although far from complete, this high level perspective of the pathophysiology of type 2 diabetes serves to highlight complex interactions between carbohydrate and fat metabolism, and between the function of multiple metabolically active tissues of the body. This is underscored by the fact that many therapeutic approaches, including those discussed in this book, target diverse mechanisms within different tissues. The repertoire of drug targets presented here aim to regulate glucose metabolism either directly or indirectly via various compensatory mechanisms.

1.1.4 Etiology

As inferred already, there are many factors that can potentially give rise to, or exacerbate, type 2 diabetes, including obesity, hypertension, and elevated cholesterol. Others include aging, high-fat diets, and an inactive lifestyle. All of these causal factors are to some degree a result of our evolving environment. The onset of type 2 diabetes has traditionally been most common in middle age and later life, although it is now being more frequently seen in adolescents and young adults due primarily to the increase in child obesity and inactivity, and this aspect is worth further consideration as a defining component of type 2 diabetes, along with implications to therapeutic intervention.

A key etiological factor linking obesity to type 2 diabetes is insulin resistance, characterized by an impaired ability of insulin to inhibit glucose output from the liver and to promote glucose uptake in fat and muscle. The physiological mechanisms connecting obesity to insulin resistance have received intense attention in recent years resulting in the emergence of several hypotheses to explain this link, such as (1) ectopic lipid accumulation in liver and muscle secondary to obesity-associated increase in serum free fatty acids, (2) altered production of various adipocyte-derived factors (collectively known as adipokines), and (3) low-grade inflammation of white adipose tissue (WAT) resulting from chronic activation of the innate immune system.[9] However, not all obese individuals are insulin resistant, and in fact insulin sensitivity has been shown to vary up to six fold in this population, highlighting the

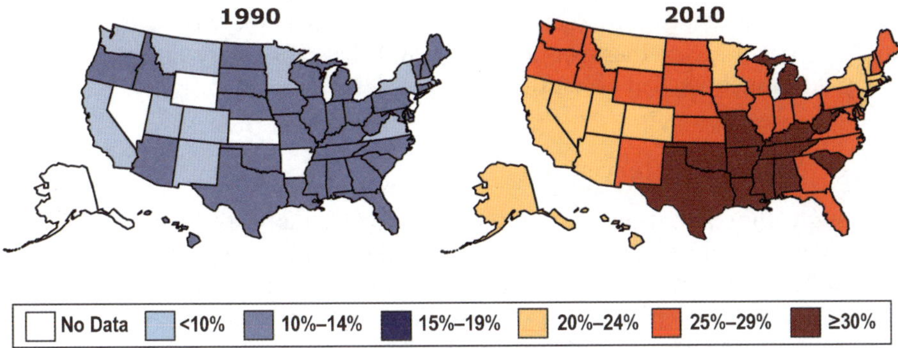

Figure 1.1   Obesity Trends Among U.S. Adults between 1990 and 2010. Source: Behavioral Risk Factor Surveillance System, CDC.

importance of identifying genetic and environmental factors that place obese individuals at the greatest risk of obesity-related complications.[10]

The degree to which obesity is affecting westernized society is worth noting, as it underlies the prevalence of type 2 diabetes, and serves as a leading indicator for metabolic dysfunction. In 2010, no U.S. state had a prevalence of obesity less than 20%. Thirty-six states had a prevalence of 25% or more; 12 of these states (Alabama, Arkansas, Kentucky, Louisiana, Michigan, Mississippi, Missouri, Oklahoma, South Carolina, Tennessee, Texas, and West Virginia) had a prevalence of 30% or more.[11] By comparing these statistics with those from 1990, the explosion in obesity rate is striking, and is depicted in Figure 1.1. The data strongly suggest that prevention (or reversal) of obesity would have a profound effect on the prevalence of type 2 diabetes. As such, many companies have focused their research on anti-obesity programs as a means to treating obesity-related metabolic diseases such as diabetes.

1.2   Treatment of Type 2 Diabetes

1.2.1   Lifestyle Management

Diet and exercise are the most powerful means to lower blood glucose levels in type 2 diabetes patients and are the foundation of effective treatment and disease management. Patient education and self-care practices are important aspects of disease management that help people with diabetes lead normal lives. In fact, the Diabetes Prevention Program (DPP), a large prevention study of people at high risk for diabetes, showed that lifestyle intervention to lose weight and increase physical activity reduced the development of type 2 diabetes by 58% during a 3-year period.[12] The reduction was even greater, 71%, among adults aged 60 years or older. However, due mainly to socio-economic factors that have become a global concern, and inadequate compliance, diet and exercise alone appear insufficient to stem the epidemic of type 2 diabetes. As such, therapeutic drug intervention is a key component of any strategy to treat diabetes patients, and this book serves to introduce some of the primary therapeutic targets and drug discovery efforts that will likely add much needed strength to the physician's armory.

1.2.2 Surgical Intervention

Although the management and prevention of diabetes through lifestyle modifications and weight loss represents the ideal therapy in appropriate candidates, supported by results from the DPP and the Finnish Prevention Study, over 95% of patients not participating in a prevention research study are unable to achieve and maintain any significant weight loss over time.[13] Robust and sustained weight loss following bariatric surgery is an emerging therapeutic option for severely obese subjects, especially when obesity is complicated by diabetes or other co-morbidities. The two most common types of procedures currently used in the U.S. are adjustable gastric bands and Roux-en-Y gastric bypass (RYGB), and these procedures can be performed laparoscopically, further reducing the perioperative morbidity and mortality associated with the surgery. While the RYGB procedure usually results in greater sustained weight loss (40–50%) than adjustable gastric banding (20–30%), it also carries greater morbidity and nutritional/metabolic issues, such as deficiencies in iron, B12, calcium, and vitamin D.[14] Following RYGB, most subjects experience improvements in diabetes control, hypertension, dyslipidemia, and other obesity-related conditions. In patients with impaired glucose tolerance most studies report 99–100% prevention of progression to diabetes, while in subjects with diabetes prior to surgery, resolution of the disease is reported in 65–90% of the cases. While improvements in insulin resistance and β-cell function are related to surgically induced weight loss, the rapid postoperative improvement in glycemia is possibly due to a combination of decreased nutrient intake and changes in gut hormones independent of weight loss.

Regulatory authorities around the world are now seriously discussing whether gastric bypass surgery should be used specifically to treat type 2 diabetes in non-obese patients, and concerted research efforts are being pursued to test less-invasive procedures that may mimic the profound glucose regulatory effects of RYGB.[15] Time will tell whether the efficacy and long-term cost-effectiveness of bariatric surgery will offset the safety and tolerability issues that exist currently.

1.2.3 Current Drug Therapy Options

The relevance of this book is shown by the fact that the standard of care for treatment of type 2 diabetes in the US has not changed meaningfully in over 15 years. One reason for this is the entrenched first- and second-line add-on options that are generic and therefore inexpensive, i.e., metformin and the sulfonylureas. However, despite the established landscape, there remains significant unmet need in glycemic control, and especially so now that several branded drug classes have failed to gain sustained acceptance due to a variety of issues, including tolerability, dosing, and safety. That said, the newest classes of approved drugs, the incretin (GLP-1) based therapies, have made significant inroads into the second-line and 3rd add-on positions. However, these drugs are yet to take a majority share in the market and are somewhat limited in the degree of efficacy attained in the clinic. The portfolio of currently approved small molecule drug classes is presented in Table 1.1.

The first-in line therapy for newly diagnosed type 2 diabetes patients is metformin, an oral drug in the biguanide class. Metformin has been shown to be particularly effective in overweight and obese people, and works by suppressing glucose production by the liver.[16] Moreover, metformin has been shown to positively impact the cardiovascular complications of diabetes, reduce LDL cholesterol and triglyceride levels, and is neutral with respect to body weight. Metformin causes few adverse effects when prescribed appropriately, with the most common side effect being gastrointestinal upset. Lactic acidosis (a buildup of lactate in the blood) can be a serious concern in overdosed individuals or when prescribed to people with contraindications. The overall risk of treatment with metformin is therefore low, and decades of use have confirmed a robustly positive efficacy/safety profile that supports its primary use in newly diagnosed patients. The issue of safety profiles for novel type 2 diabetes drugs has become paramount within the industry over the past decade, mainly due to the availability of relatively safe options currently on the market, and also because of the recent spate of high profile cases involving drugs with significant associations with heart disease and cancer risk. As a testament of its effectiveness and safety profile, in the U.S. alone, more than 48 million prescriptions for metformin were filled in 2010.

To date, therapeutic intervention for type 2 diabetes has centered on the two primary mechanisms that underlie the pathophysiology of the disease, insulin secretion and insulin action. Administration of drugs that augment secretion of endogenous insulin have proven to be excellent therapeutics, and the first generation of insulin secretagogues were the K-ATP channel blockers of the sulfonylurea chemical class. These drugs work by inhibiting ATP-sensitive

Table 1.1 Major existing drug classes approved for therapy of hyperglycemia in patients with type 2 diabetes (excluding insulin products).

Drug Class	*Target*	*Mode Of Action*
Sulfonylurea	K-ATP channel	Insulin secretion
Metformin	Unknown	Hepatic glucose output
Thiazoladinediones	PPARγ	Insulin sensitization
Acarbose	α-glucosidase	Intestinal glucose absorption
Incretin mimetics	GLP-1 receptor	GDIS
DPP4 inhibitors	DPP4 / GLP-1	GDIS

potassium channels in pancreatic β-cells. This inhibition causes cell membrane depolarization and opening of the voltage-dependent calcium channel, resulting in increased intracellular calcium in the β-cell and subsequent stimulation of insulin release.[17] Importantly, this mechanism is not responsive to glucose levels, leading to the increased risk of hypoglycemia in patients due to overwhelming levels of insulin in the circulation.

The role of the peptide GLP-1 (glucagon-like peptide 1) in augmenting insulin secretion in a glucose-dependent manner emerged over the past decade or so as a far more reliable way to regulate glucose metabolism with reduced risk of hypoglycemia. GLP-1 possesses several physiological properties that make it (and its analogs) the subject of intensive investigation for type 2 diabetes treatment. Firstly, its secretion by specific enteroendocrine cells in the gut (predominantly the ileum and colon) is dependent on the presence of nutrients in the lumen of the small intestine, and agents that cause or stimulate secretion of GLP-1 include nutrients like carbohydrate, protein, and lipid. Once in the circulation, the active hormone has a half-life of less than 2 minutes, due to rapid degradation by the enzyme dipeptidyl peptidase-4 (DPP4). GLP-1 is a potent anti-hyperglycemic hormone, induces glucose-dependent stimulation of insulin secretion while suppressing glucagon secretion from pancreatic α-cells, and the overall glucose-dependent action is particularly attractive because, when plasma glucose concentration is in the normal fasting range, GLP-1 no longer stimulates insulin that would otherwise lead to hypoglycemia.[18] As a proof of the physiological role of GLP-1 in post-prandial insulin secretion, it has been shown that an oral dose of glucose triggers a much higher peak in plasma insulin concentration compared to an intravenous dose. Within the last 5 years, several novel drugs have emerged on

Table 1.2 Emerging drug targets described in this book, all of which have entered clinical testing.

Drug Target	Mode Of Action	Primary Target Tissues
DPP4	GDIS	Gut, blood
GPR119	GDIS	Gut, pancreas
GPR40	GDIS	Pancreas, gut
TGR5	GDIS / energy expenditure	Gut, skeletal muscle
SGLT2	Glucose absorption	Kidney
SIRT1	Energy expenditure	Skeletal muscle, liver
Glucokinase	Hepatic glucose uptake	Liver, pancreas
DGAT1	GDIS	Gut, liver
PTP1B	Insulin sensitization	Liver, skeletal muscle
11b-HSD1	Anti inflammatory	Liver, adipose
SCD1	Glucose utilization	Liver
FBPase	Glucose utilization	Liver
Glycogen Phosphorylase	Glucose utilization	Liver
Glucagon Receptor	Hepatic glucose output	Liver
ACC	Glucose utilization	Liver

the market that either mimic the function of GLP-1, or alter the stability of GLP-1 via inhibition of DPP4. The efficacy and safety profile of this class of drugs has been outstanding, and as such continues to take a significant portion of the market share from other drugs with less impressive safety profiles, such as sulfonylureas and thiazoladinediones.

Throughout this book, the authors describe exciting new drug targets and mechanisms that may impact insulin release, glucose utilization, or glucose production via various distinct mechanisms, and in some cases via multiple mechanisms, as overviewed in Table 1.2. The eventual success of any of these programs will ultimately depend on how well the development teams can balance robust and sustainable glucose-lowering efficacy with an excellent safety profile, ideally with additional benefits in, for example, cardiovascular endpoints or body weight loss.

1.2.4 Emerging Mechanisms

1.2.4.1 Pancreatic Targets

Most individuals with insulin resistance don't progress to developing type 2 diabetes, as only those individuals whose β-cells fail to compensate for this insulin resistance go on to develop the disease. Therefore, even though hyperinsulinemia is almost always found in the pre-diabetic period, either absolute or relative insulin deficiency is requisite for type 2 diabetes to become established. Moreover, as shown by the United Kingdom Prospective Diabetes Study (UKPDS) and the Belfast diet intervention study, β-cell function progressively declines while the degree of insulin resistance changes little.[19] That said, agents that enhance insulin action, insulin sensitizers, or agents that inhibit gluconeogenesis could have a significant beneficial impact because a substantial amount of β-cell capacity can still be present after many years of type 2 diabetes, and interestingly, it is not uncommon to see patients with long-standing type 2 diabetes who lose large amounts of weight, either voluntarily, due to illness, or following bariatric surgery, no longer need treatment for glucose control. Drugs that enhance insulin secretion through pathways distinct from those affected by sulfonylureas may have significant value, and studies with GLP-1 and exendin-4 provide strong evidence for such approaches. Emerging research in the area of islet function and pancreas development suggests that improvement of glycemic control by enhancing insulin secretion may be complemented by mechanisms that stimulate β-cell replication and/or neogenesis. Some of the more validated islet targets represented within this book include the G-protein coupled receptors GPR119 and GPR40, both of which have demonstrated glucose-lowering efficacy in the clinic. Glucokinase (GK) is a glucose-sensing enzyme that plays an important role in the regulation of carbohydrate metabolism not only in the pancreas, but also in the liver, gut, and brain. GK activators have been shown to drive robust improvements in glucose metabolism in patients, though sustainability of these effects remains to be proven.[20]

1.2.4.2   Liver Targets

Several currently available and emerging drug classes are targeted to the liver because maintenance of normal blood glucose over acute (hours) and chronic (days to weeks) periods of time is a particularly important function of the liver. Hepatocytes are highly metabolically active cells and many well-defined signaling pathways employing dozens of enzymes are alternatively regulated, depending on whether blood glucose levels are rising or falling out of the normal range. For example, when excess glucose enters the blood after a meal, it is rapidly taken up by the liver and sequestered as glycogen via a process called glycogenesis. When blood concentrations of glucose later begin to decline, the liver activates other pathways which lead to depolymerization of glycogen, or glycogenolysis, and export of glucose back into the blood for utilization by other tissues. When hepatic glycogen reserves are exhausted, glucose synthesis, or gluconeogenesis, utilizes other cellular substrates including amino acids and non-hexose carbohydrates as fuel for the body. Potential drug targets such as GK, glycogen phosphorylase, and the glucagon receptor have been selected due to their regulatory importance in these signaling pathways. The liver also regulates bile acid and fat metabolism, regulates blood cholesterol levels, and balances the storage and utilization of these energy sources throughout the day. Targets such as TGR5, ACC, SCD1, and DGAT1 that appear not to regulate glucose directly may impact the glucoregulatory pathways in the liver via indirect routes. In summary, the liver is a predominant tissue target for type 2 diabetes drug discovery, and is very well represented in the current portfolio of new targets being critically tested in the clinic.

1.2.4.3   Targets in the Gastro-Intestinal Tract

Importance of the gut for regulation of whole body glucose metabolism has emerged over the past 5–10 years from many post-operative observations of gastric bypass surgery, an intervention primarily designed for and performed on morbidly obese patients as a life-saving intervention therapy. Besides the impressive weight loss efficacy of this and other forms of bariatric surgery, profound metabolic improvements have been observed in the majority of patients that coincidentally exhibited type 2 diabetes as a co-morbidity.[21] Importantly, the metabolic improvements occur independently of weight loss, determined by the fact that hyperglycemia is often corrected just hours to days after surgery, prior to any significant body weight effects. One of the most notable molecular responses to gastric bypass is the amplification of the GLP-1 secretory response following ingestion of a meal. Several other (and possibly many) gut-derived hormones are also augmented after gastric bypass, including PYY, oxyntomodulin, and CCK, all factors that regulate the metabolic equilibrium in different ways, and in combination are thought to comprise the overall metabolic improvement of the surgery in dysmetabolic patients.[22] Several putative drug targets discussed in this book are localized in the specialized enteroendocrine (EE) cells of the gut that regulate secretion of these metabolically active hormones, such as GPR119, TGR5, and GPR40. Other targets localized in the gastro-intestinal tract, such as DGAT1, are involved in lipid metabolism that may indirectly impact the release of GLP-1 and other hormones from the EE cells of the gut. In the case of DGAT1, the specific mechanism by which lipid metabolism within enterocytes triggers increased GLP-1 levels from EE cells is not understood.

1.2.4.4   Skeletal Muscle Targets

The acute stimulatory action of insulin on glucose uptake into skeletal muscle cells is a key requirement for the maintenance of normal glucose homeostasis. Transport of glucose across the cell surface membrane via Glut4 transporters is essentially rate limiting for glucose metabolism, and thus elevated blood levels of glucose result in part from inappropriately low rates of sugar transport across cell surface membrane due to deficient levels of insulin, or suppressed sensitivity of muscle cells to insulin. The central importance of glucose metabolism in energy production and as a provider of precursor compounds for macromolecule biosynthesis in skeletal muscle further reinforces the deleterious physiologic impact of impaired glucose transport in this disease.

While insulin secretagogues show promise for the management of glycemia in the short term and with very well tolerated compounds currently on the market, this class of drug doesn't stem the deterioration in peripheral insulin sensitivity, and so many believe the "holy grail" for type 2 diabetes therapy will emerge from mechanisms that reverse insulin resistance in skeletal muscle. To this end, the elucidation of metabolic pathways responsive to exercise in various tissues, most particularly skeletal muscle, has been an important antecedent to some of the most promising concepts of exercise mimetic drugs. From the perspective of obesity-related insulin resistance, a pivotal goal is to develop agents that increase energy expenditure while concomitantly reducing body fat and improving metabolic homeostasis. Two drug targets that may improve glucose metabolism via skeletal muscle insulin sensitization, and are currently being tested in clinical development, are PTP1B, which a negative regulator of the insulin signaling pathway, and SIRT1, an evolutionarily conserved sensor of metabolic stress.

1.2.4.5   Targeting the Kidney

As discussed throughout this chapter, there are numerous putative targets for treating type 2 diabetes that function in various tissues and/or systems of the body as a result of the complexity of whole body glucose homeostasis. Another example, the kidney, contributes to glucose homeostasis by reabsorbing approximately 180 g of glucose from the glomerular filtrate each day. Because of the activity of glucose transporters in the renal proximal tubule, <0.5 g/d is excreted in the urine of healthy adults.[23] Therefore, a new strategy to reduce hyperglycemia is to target renal glucose excretion by inhibiting SGLT2, a sodium-dependent glucose co-transporter that mediates glucose reabsorption in the kidney. SGLT2 inhibitors are currently in advanced stages of clinical development and may represent the next significant addition to the physician's armory.

1.2.5 Summary of Oral Diabetes Medications

The first oral type 2 diabetes medications were sulfonylureas, which were introduced into the market in 1955. The second-generation sulfolylureas, which are used today, were introduced in 1984. Metformin (a biguanide) was introduced in 1995, meglitinides in 1997, α-glucosidase inhibitors in 1998, and thiazolidinediones in 1999. The most recent class of oral type 2 diabetes drug introduced to the market were dipeptidyl peptidase 4 (DPP4) inhibitors in 2006. The comparative effectiveness and safety profile of these oral therapeutic options, presented in the appropriate context of the patient's co-morbidities, defines the optimal course of therapy, which varies significantly from individual to individual. The next wave of anti-diabetic small molecule drugs is described in this book. Each candidate drug target is currently undergoing extensive characterization in various phases of clinical development, as

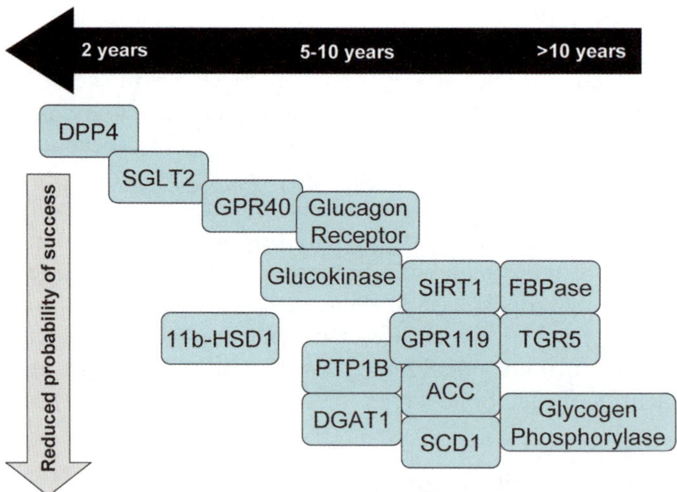

Figure 1.2 Author's opinion on relative timelines to regulatory approval and chance of success through clinical development.

reflected in Figure 1.2, and are being pursued with an equally sharp focus on effectiveness and sustainability of glucose-lowering efficacy in addition to its relative safety profile. Never has the bar been set so high for drug safety by regulatory authorities, which in turn has driven the cost (and also the risk) of drug development to new heights. The result of this and other economic business drivers has lead to a particularly crowded landscape around a handful of the most validated targets, manifesting in a highly competitive best-in-class race for most of the emerging drug classes.

References

1. Centers for Disease Control and Prevention. National diabetes fact sheet: national estimates and general information on diabetes and prediabetes in the United States, 2011.
2. G. Danaei, M. M. Finucane, Y. Lu, G. M. Singh, M. J. Cowan, C. J. Paciorek, J. K. Lin, F. Farzadfar, Y. H. Khang, G. A. Stevens, M. Rao, M. K. Ali, L. M. Riley, C. A. Robinson and M. Ezzati; Global Burden of Metabolic Risk Factors of Chronic Diseases Collaborating Group (Blood Glucose), *Lancet*, 2011 **378**, 31.
3. S. S. S Lo, R. Y. M. Tun, M. Hawa and R. D. G. Leslie, *Diabetes/ Metabolism Reviews*, 1991, **7**, 223.
4. S. Haffner, M. P. Stern, H. P. Hazuda, M. Rosenthal, J. A. Knapp and R. M. Malina, *Diabetes Care*, 1986, **9**, 153.
5. E. E. Ntzani and F. K. Kavvoura, *Curr. Vasc. Pharmacol.*, 2012, **10**, 147.
6. H. Hara, G. Egusa, M. Yamakido and R. Kawate. *Diabetes Res. Clin. Pract.*, 1994, **24**, S37.
7. W. C. Knowler, P. H. Bennet, R. F. Hamman and M. Miller, *Am. J. Epidemiol.*, 1978, **108**, 497.
8. R. Belfort, L. Mandarino, S. Kashyap, K. Wirfel, T. Pratipanawatr, R. Berria, R. A. DeFronzo and K. Cusi, *Diabetes*, 2005, **54**, 1640.
9. M. Qatanani and M Kazar, *Genes and Development*, 2007, **21**, 1443.
10. G. Reaven, *Diab. Vasc. Dis. Res.*, 2005, **2**, 105.
11. Data from the Office of Surveillance, Epidemiology, and Laboratory Services, Behavioral Risk Factor Surveillance System, CDC.
12. Diabetes Prevention Program Research Group, *N. Engl. J. Med.*, 2002, **346**, 393.
13. L. F. Meneghini, *Cell. Biochem. Biophys.*, 2007, **48**, 97.
14. F. Rubino and M. Gagner, *Ann. Surg.*, 2002, **236**, 554.
15. F. Rubino and J. Marescaux, *Ann. Surg.*, 2004, **239**, 1.
16. B. Viollet, B. Guigas, N. Sanz Garcia, J. Leclerc, M. Foretz and F. Andreelli, *Clin. Sci.*, 2012, **122**, 253.
17. S. Seino, H. Takahashi, T. Takahashi and T. Shibasaki, *Diabetes Obes. Metab.*, 2012, **14**, S1.
18. J. Gromada, B. Brock, O. Schmitz and P Rorsman, *Clin. Pharmacol. Toxicol.*, 2004, **95**, 252.

19. S. E. Kahn, *J. Clin. Endocrin. Metab.*, 2001, **86**, 4047.
20. G. E. Meininger, R. Scott, M. Alba, Y. Shentu, E. Luo, H. Amin, M. J. Davies, K. D. Kaufman and B. J. Goldstein, *Diabetes Care*, 2011, **34**, 2560.
21. L.F. Meneghini, *Cell. Biochem. Biophys.*, 2007, **48**, 97.
22. G. Mingrone and L. Castagneto-Gissey, *Diabetes Metab.*, 2009 **35**, 518.
23. E. M. Wright, B.A. Hirayama and D.F. Loo, *J. Intern. Med.*, 2007, **261**, 32.

Marketed Small Molecule Dipeptidyl Peptidase IV (DPP4) Inhibitors as a New Class of Oral Anti-Diabetics

ZHONGHUA PEI

Department of Discovery Chemistry, Small Molecule Drug Discovery, Genentech, Inc., A Member of the Roche Group, South San Francisco, CA 94080, USA
E-mail: pei.zhonghua@gene.com

2.1   Introduction

More than 47 years ago, Elrich and colleagues documented that oral glucose administration into human subjects resulted in a significant and sustained increase in plasma insulin compared to intravenous glucose administration.[1] The authors interpreted the difference "...as evidence for an additional stimulus to insulin secretion, possibly a gastrointestinal or liver factor triggered by alimentary glucose." Extensive research by several teams over the years led to the concept of incretin effect, namely the existence of gastrointestinal hormones that enhance glucose-stimulated insulin secretion from the islet β-cell.[2,3] This has sparked intensive interest in identifying the incretins and their functions. Coupled with the finding that the incretin effect is reduced in type 2 diabetics compared to healthy controls,[4] modulating the incretin signal

RSC Drug Discovery Series No. 27
New Therapeutic Strategies for Type 2 Diabetes: Small Molecule Approaches
Edited by Robert M. Jones

Published by the Royal Society of Chemistry, www.rsc.org

pathway has become an attractive potential approach for the treatment of diabetes.

Two incretins have been identified that play an important role in glucose homeostasis: glucose-dependent insulinotropic peptide (GIP, also known as gastric inhibitory peptide) and glucagon-like peptide-1 (GLP-1).[5] GLP-1 (7–36) amide is processed from proglucagon and released from enteroendocrine L cells in the distal small intestine and colon in response to oral ingestion of nutrients. Binding of GLP-1 (7–36) amide to its G-protein-coupled receptor on pancreatic β-cells increases glucose-stimulated insulin secretion (Figure 2.1).[6] GLP-1 receptor mRNA is expressed in small and large intestine, pancreas, liver, lung, kidney, and the hypothalamic nuclei which are responsible for modulating feeding behavior in rodents.[7,8] In addition, GLP-1 (7–36) amide stimulates insulin gene expression[9] and inhibits glucagon secretion from islet cells.[10] GLP-1 also slows gastric emptying thereby reducing the rate that nutrients are absorbed into the circulation.[11] Peripheral administration of GLP-1 promotes satiety and inhibits food intake in humans.[12,13] Consequently GLP-1 (7–36) amide has multiple biological effects that contribute to glucose homeostasis and promotes normalization of blood glucose levels.

GIP is a 42-amino acid peptide secreted by endocrine K cells of the duodenum in response to ingestion of nutrients (Figure 2.1).[14] The physiological actions of GIP include glucose-dependent potentiation of insulin secretion and regulation of insulin gene transcription. In contrast to GLP-1, GIP does not inhibit glucagon secretion or influence gastric emptying in humans.[15,16] In addition to the effects on β-cells, GIP promotes energy storage and reduces insulin action in adipocytes.[17]

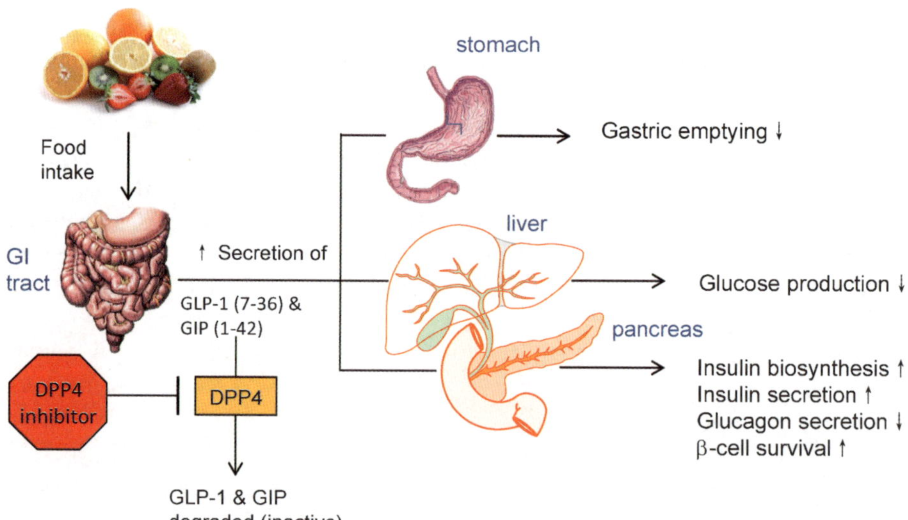

Figure 2.1   Incretin action and inactivation of incretins by DPP4.

Upon food ingestion, plasma levels of GLP-1 and GIP increase rapidly. However, both GLP-1 (7–36) and GIP (1–42) are inactivated by dipeptidyl peptidase IV (DPP-IV or DPP4) rapidly by cleavage of two amino acid residues from the N-terminus *in vivo*.[18–20] The truncated peptides lose most (if not all) of their glucose-regulating properties. DPP4 (EC 3.4.14.5), also known as lymphocyte cell surface protein CD26 or adenosine deaminase (ADA)-binding protein, was first identified by Hopsu-Havu and Glenner[21] in 1966 as an enzyme that possesses glycylproline β-naphthylamidase activity. This protein was later identified as a serine protease that preferably cleaves prolyl and alanyl peptide bonds at the penultimate position from the N-terminus.[22–24] DPP4 is expressed as a glycoprotein on the surface of cells of most tissues, including kidney, liver, intestine, placenta, prostate, skin, lymphocytes, and endothelial cells. DPP4 is catalytically active as a dimer. Proteolytic cleavage of DPP4 from cell surfaces results in a soluble circulating form with a monomeric mass of approximately 100 kDa.

DPP4 is a serine protease with the catalytic triad of Ser630-Asp708-His740 (all residue numberings are from human DPP4) oriented in a nonclassical amino acid sequence order and with significant homology to other α,β-hydroxylases (e.g., prolyl oligopeptidase, POP). The X-ray crystal structures of rat and human DPP4 have been solved.[25] Each monomer of the dimeric DPP4 consists of two domains comprised of an N-terminal eight-blade β-propeller and a C-terminal catalytic domain that adopts a α,β-hydrolase fold. Similar to POP, the active site is inside a large solvent-filled cavity surrounded by the propeller domain. The propeller domain forms two channels for substrates to access the active site with a central propeller pore of ~ 13 Å and a larger pore of ~ 20 Å located on the side. These two pores limit the size of potential peptide substrates that can approach the catalytic site of DPP4.

In addition to cleaving GLP-1 and GIP, DPP4 may play a role in the cleavage of other substrates with accessible preferred amino-terminal dipeptide sequences of Xaa-Pro- or Xaa-Ala-, resulting in inactivation or alteration of their biological activities. One such substrate is GLP-2, a 33-amino acid peptide, co-secreted along with GLP-1 from intestinal endocrine cells in the small and large intestine. Administration of GLP-2 in humans increases glucagon secretion, which may counteract the glucagonostatic effect of GLP-1.[26] Other potential DPP4 substrates include growth hormone-releasing hormone (GHRH), substance P, bradykinin, gastrin-releasing peptide, neuropeptide Y (NPY), peptide YY (PYY), and certain chemokines such as RANTES (regulated on activation normal T-cell expressed and secreted), stromal cell-derived factor, eotaxin, and macrophage-derived chemokine which may modulate immune function.[27] Besides GLP-1 and GIP, little is known about whether inhibition of DPP4 activity will increase the endogenous circulating levels of intact peptides and whether increased levels of these intact peptides will have any pharmacological consequence.[28]

Two different approaches to enhancing the beneficial effects of GLP-1 and GIP have emerged: the first is the administration of GLP-1 mimetics (or GLP-

1 receptor agonists) that are resistant to DPP4 degradation, and the other is the inhibition of DPP4 enzymatic activities. Each approach has its own advantages and limitations, as will be discussed later.[29,30] DPP4 inhibition has been shown as a viable approach for the treatment of diabetes through years of intensive research including GLP-1 infusion studies,[31] DPP4 knockout studies,[32] and small-molecule inhibitors.

2.2   Marketed DPP4 Inhibitors

Since the medicinal chemistry of DPP4 inhibitors has been reviewed extensively,[33–39] an exhaustive review is not intended here. This chapter will rather provide an overview on the various aspects of the marketed small-molecule DPP4 inhibitors: potency, selectivity, binding mode, clinical efficacy, and safety, with the emphasis on comparing and contrasting the marketed drugs to each other.

Extensive drug discovery efforts over the past two decades have resulted in the approval of five small-molecule DPP4 inhibitors by various regulatory agencies so far. The compounds, in the order of their respective first-time approval, are sitagliptin, vildagliptin, saxagliptin, alogliptin, and linagliptin (Table 2.1), although vildagliptin was only approved in Europe and alogliptin only in Japan at the time of writing. There are many other DPP4 inhibitors in clinical trials at various stages.[40] The marketed DPP4 inhibitors can be used either as a single agent or as part of a combination therapy. The discovery and evolution of these drugs are not discussed here. Rather, the reader is strongly encouraged to read the original publications on their discovery in peer-reviewed journals: vildagliptin by Novartis in 2003,[41] saxagliptin by BMS[42] and sitagliptin by Merck[43] in 2005, linagliptin by Boehringer-Ingelheim in 2007,[44] and alogliptin by Syrrx (now Takeda) in 2008.[45]

The five DPP4 inhibitors possess four different pharmacophore structures (Figure 2.2):

Table 2.1   Marketed DPP4 inhibitors.

Genetic	Trade Name[a]	Originator	Initial Approval	Annual Sales[b]
sitagliptin	Januvia	Merck	Oct 2006, FDA	$ 1.92B (2009) $ 2.38B (2010)
vildagliptin	Galvus	Novartis	Feb 2008, EMA (EU)	$ 0.181B (2009) $ 0.391B (2010)
saxagliptin	Onglyza	BMS	Jul 2009, FDA	$ 0.158B (2010)
alogliptin	Nesina	Syrrx/Takeda	Apr 2010, NDA (Jpn)	na
linagliptin	Tradjenta	Boehringer Ingelheim	May 2011, FDA	na

[a]There may be different trade names in different countries or regions; [b]Source: knowledge-express.com

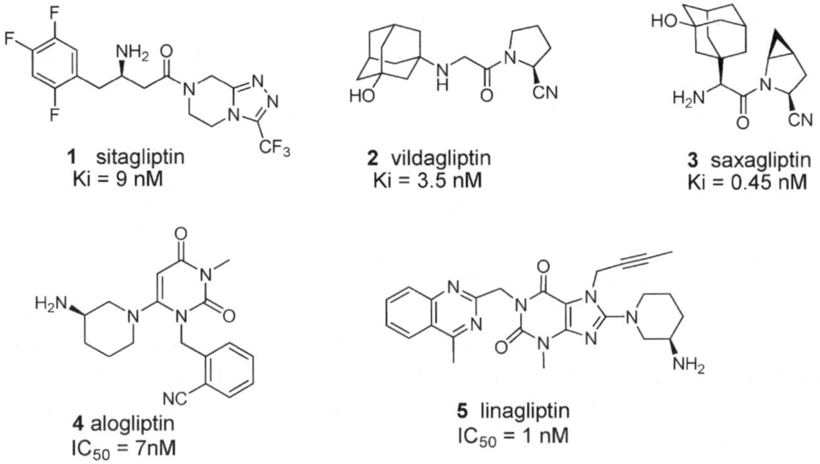

1 sitagliptin
Ki = 9 nM

2 vildagliptin
Ki = 3.5 nM

3 saxagliptin
Ki = 0.45 nM

4 alogliptin
IC_{50} = 7nM

5 linagliptin
IC_{50} = 1 nM

Figure 2.2　　Chemical structures of marketed DPP4 inhibitors.

- Both vildagliptin and saxagliptin share the cyanopyrrolidine group, which occupies the S1 pocket of the DPP4 enzyme, and a substituted glycine. In this sense, both are substrate-like dipeptide mimetics.
- Sitagliptin has a β-amino acid core, with the trifluorophenyl occupying the S1 pocket.
- Linagliptin is a xanthine derivative, with the but-2-ynyl group (a substituted alkyne group) occupying the S1 pocket.
- Alogliptin has a uracil core, with the 2-cyanophenyl group filling the S1 pocket.

The binding modes of these DPP4 inhibitors will be discussed in detail later.

2.3　Potency and Selectivity of Marketed DPP4 Inhibitors

All five inhibitors are quite potent, with saxagliptin reaching below nanomolar (Table 2.2). The selectivity over other peptidases range from several tens to greater than one thousand fold.

In 2005, the research team at Merck reported toxicities such as lung histiocytosis and thrombocytopenia when thiazolidine analog **6**, which lacks DPP8/9 selectivity over DPP4 (Figure 2.2), was administered at relatively high doses to rats over a period of four weeks.[46] Upon 5–6-week treatment of **6** in dogs, more toxicity (anemia, thrombocytopenia, splenomegaly, and multiple organ pathology) and mortality were observed. In a 4-week rat toxicity study, allo analog **7**, which has similar DPP4 potency but increased DPP8 and 9 potency compared to **6**, produced similar toxicity but at approximately one-tenth dose level or plasma exposure. Remarkably similar toxicity (alopecia, thrombocytopenia, anemia, enlarged spleen, multiple histological pathologies,

Table 2.2 Potency, selectivity, and dosage of marketed DPP4 inhibitors.

Generic Name	Compound Number	Potency (nM)	DPP8	DPP9	Dosage	Comment
Sitagliptin	1	$K_i = 9$	>2700×	>5500×	100 mg qd	
Vildagliptin	2	$K_i = 3$	270×	32×	50 mg bid	
saxagliptin	3	$K_i = 0.6$	400×	75×	5 mg qd	Slow off-rate
alogliptin	4	$IC_{50} = 7$	>10 000×	>10 000×	25 mg qd	
linagliptin	5	$IC_{50} = 1$	>10 000×	>10 000×	5 mg qd	Slow off-rate

and death) was observed in rats with a DPP8/9-selective inhibitor, **8**. Neither DPP4-selective nor DPP2 (aka QPP)-selective inhibitors produced such toxicities in rats or dogs. The authors concluded that "these results strongly suggest that inhibition of DPP8/9 produces profound toxicity" and "assessment of selectivity of potential clinical candidates may be important to an optimal safety profile for this new class of antihyperglycemic agents." Since then, most DPP4 drug discovery programs worldwide have emphasized and adopted screening paradigms for DPP8/9 selectivity in order to avoid potential toxicity.

Recently, the notion that DPP8/9 may cause *in vivo* toxicity and thus should be avoided was challenged. Firstly, both vildagliptin and saxagliptin have modest *in vitro* selectivity over DPP8 or 9, but neither drug demonstrated an inferior safety profile compared with the more selective inhibitors in human clinical trials. Secondly, in a 13-week study of vildagliptin in rats at doses that provided tissue and plasma concentrations well above the K_i values for DPP8 and 9 over a 24-hour period, toxicities reported in the Merck study were not observed.[47] Thirdly, when a potent and cell permeable DPP8/9 inhibitor **9** (Figure 2.3) was administered in rats for 2 weeks either intravenously or orally (achieving high drug concentration), no severe toxicity was observed.[48] Based on these results, it is likely that the *in vivo* toxicity observed with compounds **6–8** by the Merck team are due to modulation of unknown off-target(s) associated with these specific compounds.

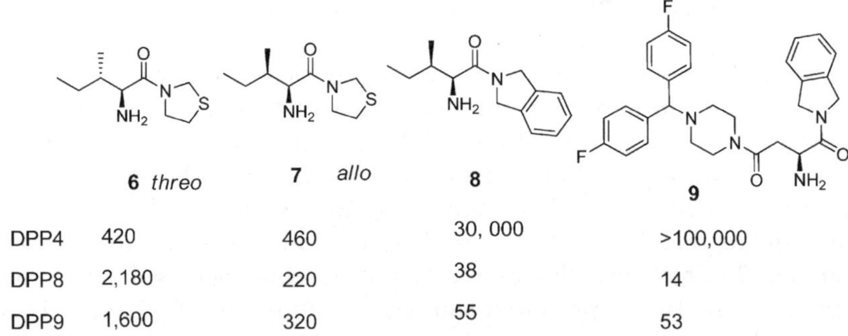

Figure 2.3 Structure and potency of DPP8 and 9 inhibitors.

2.4 Binding Mode of DPP4 Inhibitors

All DPP4 inhibitors reported so far have a basic primary or secondary amine group. This amino group (presumably ionized at physiological pH) forms electrostatic interactions with both Glu205 and Glu206 and hydrogen bonds with the hydroxyl group of Tyr662 of DPP4. These same interactions are used for endogenous substrate recognition (Figure 2.4). This is where the similarity in binding modes among the five inhibitors ends. Based on crystal structures it is clear that different inhibitors achieve potency in their unique ways. Sitagliptin forms extensive polar interactions with various residues of DPP4: one of the fluorine atoms of the CF_3 group on the heterocyclic ring interacts with Arg358, another fluorine atom interacts with Ser209 and Val207. The side chain of Arg358 is flexible and rotates away from the active site to accommodate the CF_3 group. The same movement is also observed with a conformationally constrained phenethylamine DPP4 inhibitor.[49] The triazolo ring of sitagliptin forms a π-stacking with the phenyl ring of Phe357. A water molecule (not shown in Figure 2.4a) bridges the carbonyl oxygen atom and the hydroxyl of Tyr547. The trifluorophenyl group occupies the S1 pocket and one of the fluorine atoms of the trifluorophenyl group forms polar interactions with side chains of Asn710 and Arg125.

Compared to sitagliptin, saxagliptin is a smaller molecule and occupies much less space of DPP4 yet achieving high potency (Figure 2.4b). The cyano group of both vildagliptin and saxagliptin forms a covalent bond with the catalytic serine 630, which is evident from the crystal structure of the complex. This is true for all inhibitors containing a cyanopyrolidine as the warhead. However, this covalent adduct is reversible, as demonstrated by complete recovery of enzyme activity upon dialysis[50] or replacement of a covalent inhibitor bound to DPP4 by soaking with a more potent non-covalent inhibitor.[51] This covalent bond contributes significantly to the exquisite potency of saxagliptin and presumably to the slow off-rate as well. The cyclopropyl ring fits snugly into the hydrophobic back pocket. While the adamantyl group efficiently fills the hydrophobic space, the hydroxyl group on the adamantyl ring forms an H-bond with the side chain of Tyr547.

Linagliptin occupies a very different region of DPP4 and thus interacts with quite different residues of DPP4 compared to sitagliptin (Figure 2.4c). The but-2-ynyl group occupies the hydrophobic S1 pocket. The xanthine ring interacts with the phenyl ring of Tyr547 in a face-to-face fashion, as does the methylquinazoline with the indole ring of Trp629. The phenyl group of Tyr547 is flexible and movement, relative to where it sits in the sitagliptin crystal structure, has been observed to accommodate linagliptin and other inhibitors.[51]

The cyanophenyl ring of alogliptin occupies the S1 pocket with the cyano group forming an H-bond with the side chain of Arg125 (note that the cyano group does NOT form a covalent bond with Ser630). The uracil ring is involved in a π-stacking interaction with the phenyl ring of Tyr547. The oxygen atom of one of the carbonyl groups of the uracil forms an H-bond with the main chain NH of Tyr 631. The amino group on the piperidyl group adopts

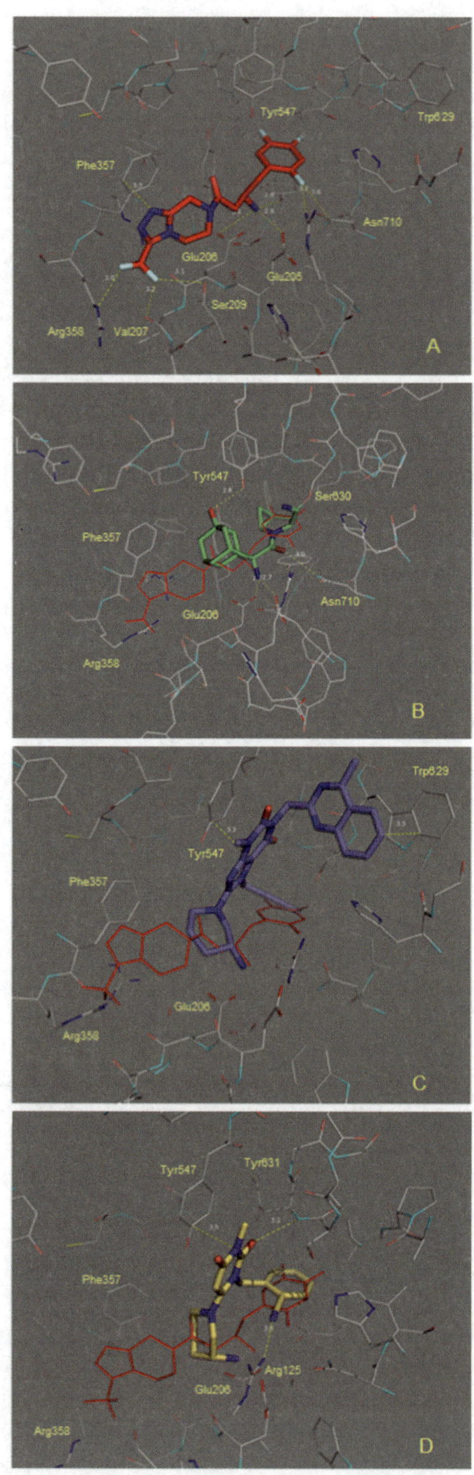

Figure 2.4

Figure 2.4　X-ray crystal structures of complexes of inhibitors bound to DPP4. (a) sitagliptin, (b) saxagliptin, (c) linagliptin, and (d) alogliptin. Selected interactions are highlighted by dashed lines with distances indicated in angstroms. Sitagliptin is drawn as a stick in (a) but as lines in (b) to (d) as an orientation marker since the enzyme in the four pictures is oriented differently in order to better show the various interactions with different inhibitors. Color coding: oxygen atoms red, nitrogen blue, carbon varies from (a) to (d). The figures are generated with PyMol from data deposited at RSCB with PDB codes 1X70, 3BJM, 2RGU, and 3G0B, respectively.

an axial (rather than equatorial) conformation in order to better interact with Glu205 and Glu206. This axial conformation is postulated to be stabilized by the intramolecular nitrile–amine interaction.[52]

2.5　Pharmacokinetics, Efficacy, and Safety of Marketed DPP4 inhibitors

All five marketed DPP4 inhibitors have reasonably good drug properties: the molecular weights are less than 500 (with linagliptin having the highest MW of 472), $c \log D$ ranges from −0.7 to 0.26, the topological polar surface area (TPSA) from 76 to 116 $Å^2$, and plasma protein binding is generally low (with the exception of linagliptin, which is highly protein bound). These properties contribute to their favorable pharmacokinetics (PK) profile: low to moderate clearance, reasonable half-lives, and good oral bioavailability.[53] All five marketed DPP4 inhibitors are dosed once daily except for vildagliptin (which is dosed twice a day) and at relatively low doses of 5–25 mg per day; with sitagliptin and vildagliptin having the highest dose of 100 mg per day (Table 2.2). The use of these drugs in moderate to severe renal impaired patients is either not recommended (sitagliptin, saxagliptin, and vildagliptin) or the dose should be reduced (alogliptin) as a significant portion of these inhibitors are eliminated through kidney. Linagliptin, being much more lipophilic than the other four drugs, is the only approved DPP4 inhibitor that does not need dose adjustment for renally impaired patients, which may find its advantage in the renally impaired patient population.[54]

At pharmacological doses, these drugs cause >80% inhibition of DPP4 enzymatic activity 12 hours after dosing and a 2∼3-fold increase of active GLP-1 levels[55] despite some reduction of GLP-1 secretion following DPP4 inhibition (presumably by some feedback mechanism).[56] All five drugs show similar extent of efficacy in diabetics in terms of lowering fasting plasma glucose level, with HbA1c (a reflection of average blood glucose levels over the past ∼120 days) reduction of ∼0.6% as a mono therapy.[57,58] The magnitude of reduction observed is dependent on the baseline: patients with higher baseline HbA1c usually achieving greater reduction upon treatment. About 40% of the patients in the treatment group achieved the HbA1c <7%, a glycemic control target set by the American Diabetes Association guidelines. The efficacy achieved by these DPP4 inhibitors is very comparable to other oral anti-diabetic agents, including metformin, sulfonylureas, alpha-glucosidase inhibitors and thiazolidindiones (glitazones). Additionally, patients taking vildagliptin and sitagliptin have reduced triglyceride, total cholesterol, and low-density lipoprotein. When the DPP4 inhibitors are combined with other anti-diabetic agents (metformin being the most common), the reduction of HbA1c is significantly greater (usually −1.0% to −1.8%) than achieved in mono therapy.[59,60]

The marketed DPP4 inhibitor drugs are generally well tolerated and their safety profiles are very similar despite their difference in chemical structure and clinical dose: two of the five approved DPP4 inhibitors are prescribed at low dose (5 mg per day), one at a medium dose (25 mg per day) and two at high dose (100 mg per day). It has been documented that there is a correlation between daily doses of oral prescription medications and idiosyncratic drug-induced liver injury; specifically, low-dose medications are associated with lower risk of drug-associated liver toxicities.[61] Since the insulinotropic activity of GLP-1 and GIP is strictly glucose dependent, all of the inhibitors have a low risk of causing hypoglycemia, which is a concern with other oral anti-diabetic agents (e.g. sulfonylureas). There is no increased cardiovascular risk compared to glitazones. There is no known risk of drug–drug interaction of these DPP4 inhibitors as they do not have significant interactions (either inhibition or induction) with drug-metabolizing enzymes CYPs except for saxagliptin. The dose of saxagliptin should be reduced when it is co-administered with any of the potent CYP3A4/5 inhibitors since saxagliptin is mainly metabolized by CYP3A4/5.

Compared to the marketed GLP-1 agonists (exenatide and liraglutide), the DPP4 inhibitors are less efficacious as measured by HbA1c reduction and have no additional benefit of causing significant body weight loss. On the other hand, DPP4 inhibitors offer the more convenient route of drug administration (oral versus injection) and a superior tolerability profile (e.g., less gastro-intestinal adverse effects such as nausea and vomiting).[62]

One of the fundamental problems that diabetics have is the gradual loss of β-cell mass and function. While numerous animal studies suggest that DPP4 inhibitors could potentially delay or stop the loss of functional β-cell mass as diabetes progresses,[63–69] and thus have the potential of being disease-modifying, any long-term clinical benefit on β-cell mass in human diabetics remains to be elucidated. While clinical measurement of β-cell mass in humans is technically difficult, treatment with DPP4 inhibitors does improve indices of β-cell function (such as HOMA-b and proinsulin insulin ratio) in clinical trials of sitagliptin and vildagliptin.[70,71] Whether such improvement of function will translate into delay of disease progression remains a question.

2.6 Summary

Incretins such as GLP-1 and GIP play an important role in glucose homeostasis. Both GLP-1 and GIP are rapidly cleaved and inactivated by the serine protease DPP4. It has been demonstrated that inhibition of DPP4 enzymatic activity could lead to increased levels of active GLP-1 and GIP thus restore the glucose homeostasis in type 2 diabetes. Five small molecule inhibitors have been approved by regulatory agencies as oral anti-diabetics. They have diverse chemical structures and achieve inhibitory potency in their unique ways. The approved DPP4 inhibitors are safe and efficacious and represent a novel and valuable therapeutic option in combating the ever prevalent type 2 diabetes.

References

1. (a) H. Elrich, L. Stimmler, C. J. Hlad and Y. Arai, *J. Clin. Endocrinol. Metab.*, 1964, **24**, 1076; (b) for a history on the incretin concept, see W. Creutzfeldt, *Reg. Peptides*, 2005, **128**, 87.
2. M. Perley and D. M. Kipnis, *J. Clin. Invest.*, 1967, **46**, 1954.
3. M. A. Nauck, E. Homberger, E. G. Siegel, R. C. Allen, R. P. Eaton, R. Ebert and W. Creutzfeldt, *J. Clin. Endocrinol. Metab.*, 1986, **63**, 492.
4. M. Nauck, F. StOckmann, R. Ebert and W. Creutzfeldt, *Diabetologia*, 1986, **29**, 46.
5. L. L. Baggio and D. J. Drucker, *Gastroenterology*, 2007, **132**, 2131.
6. S. Mojsov, G. C. Weir and J. F. Habener, *J. Clin. Invest.*, 1987, **79**, 616.
7. R. V. Campos, Y. C. Lee and D. J. Drucker, *Endocrinology*, 1994, **134**, 2156.
8. P. J. Shughrue, M. V. Lane and I. Merchenthaler, *Endocrinology*, 1996, **137**, 5159.
9. H. C. Fehmann and J. F. Habener, *Endocrinology*, 1992, **130**, 159.
10. H. Larsson, J. J. Holst and B. Ahren, *Acta Physiol. Scand.*, 1997, **160**, 413.
11. J. Schirra, M. Katschinski, C. Weidmann, T. Schafer, U. Wank, A. Arnold and B. Goke, *J. Clin. Invest.*, 1996, **97**, 92.
12. M. D. Turton, D. O'Shea, I. Gunn, S. A. Beak, C. M. Edwards, K. Meeran, S. J. Choi, G. M. Taylor, M. M. Heath, P. D. Lambert, J. P. Wilding, D. M. Smith, M. A. Ghatei, J. Herbert, S. R. Bloom, *Nature*, 1996, **379**, 69.
13. M. Zander, S. Madsbad, J. L. Madsen and J. J. Holst, *Lancet*, 2002, **359**, 824.
14. J. J. Meier, M. A. Nauck, W. E. Schmidt and B. Gallwitz, *Regul. Pept.*, 2002, **107**, 1.
15. J. J. Holst, *Horm. Metab. Res.*, 2004, **36**, 747.
16. J. J. Meier, *Best Practice & Res. Clin. Endocrin. Metab.*, 2004, **18**, 587.
17. K. Miyawaki, Y. Yamada, N. Ban, Y. Ihara, K. Tsukiyama, H. Zhou, S. Fujimoto, A. Oku, K. Tsuda, S. Toyokuni, H. Hiai, W. Mizunoya, T. Fushiki, J. J. Holst, M. Makino, A. Tashita, Y. Kobara, Y. Tsubamoto, T. Jinnouchi, T. Jomori and Y. Seino, *Nat. Med.*, 2002, **8**, 738.
18. R. Mentlein, B. Ballwitz and W. E. Schmidt, *Eur. J. Biochem.*, 1993, **214**, 829.

19. T. J. Kieffer, C. H. S. McIntosh and T. A. Pederson, *Endocrinology*, 1995, **136**, 3585.
20. C. F. Deacon, *Horm. Metab. Res.*, 2004, **36**, 761.
21. V. K. Hopsu-Havu and G. G. Glenner, *Histochemie*, 1966, **7**, 197.
22. A. J. Kenny, A. G. Booth, S. G. George, J. Ingram, D. Kershaw, E. J. Wood and A. R. Young, *Biochem. J.*, 1976, **157**, 169.
23. A. J. Kenny, A. G. Booth, E. J. Wood and A. R. Young, *Biochem. Soc. Trans.*, 1976, **4**, 347.
24. G. Puschel, R. Mentlein and E. Heymann, *Eur. J. Biochem.*, 1982, **126**, 359.
25. (a) H. B. Rasmussen, S. Branner, F. C. Wiberg and N. Wagtmann, *Nat. Struct. Biol.*, 2003, **10**, 19; (b) R. Thoma, B. Loffler, M. Stihle, W. Huber, A. Ruf and M. Hennig, *Structure*, 2003, **11**, 947; (c) K. Aertgeerts, S. Ye, M. G. Tennant, M. L. Kraus, J. Rogers and B.-C. Sang, *Protein Sci.*, 2004, **13**, 412.
26. J. J. Meier, M. A. Nauck, A. Pott, K. Heinze, O. Goetze, K. Bulut, W. E. Schmidt, B. Gallwitz and J. J. Holst, *Gastroenterology*, 2006, **130**, 44.
27. R. Mentlein, *Regul. Pept.*, 1999, **85**, 9.
28. D. L. Drucker, *Diabetes Care*, 2007, **30**, 1335.
29. P. L. Brubaker, *Trends Endocrinol. Metab.*, 2007, **18**, 240.
30. K. Raun, P. von Voss, C. F. Gotfredsen, V. Golozoubova, B. Rolin and L. B. Knudsen, *Diabetes*, 2007, **56**, 8.
31. A. Flint, A. Raben, A. Astrup and J. J. Holst, *J. Clin. Invest.*, 1998, **101**, 515.
32. D. Marguet, L. Baggio, T. Kobayashi, A. M. Bernard, M. Pierres, P. F. Nielsen, U. Ribel, T. Watanabe and D. J. Drucker, *Proc. Natl. Acad. Sci. USA*, 2000, **97**, 6874.
33. E. B. Villhauer, G. M. Coppola and T. E. Hughes, *Ann. Reports Med. Chem.*, 2001, **36**, 191.
34. S. L. Gwaltney II and J. A. Stafford, *Ann. Reports Med. Chem.*, 2005, **40**, 149.
35. A. E. Weber, *J. Med. Chem.*, 2004, **47**, 4135.
36. B. D. Green, P. R. Flatt and C. J. Bailey, *Expert. Opin. Emerging Drugs*, 2006, **11**, 525.
37. P. Wideman, *Prog. Med. Chem.*, 2007, **45**, 63.
38. Z. Pei, *Curr. Opin. Drug Discv. Dev.*, 2008, **14**, 512.
39. J.-U. Peters and P. Mattei, in *Analogue-based Drug Discovery II*, ed. J. Fischer and C. R. Ganellin, 2010, Wiley-VCH Verlag GmbH & Co., p. 109.
40. For examples, see: (a) P. Mattei, M. Boehringer, P. Di Giorgio, H. Fischer, M. Hennig, J. Huwyler, B. Kocer, B. Kuhn, B. M. Loeffler, A. McDonald, R. Narquizian, E. Rauber, E. Sebokova and U. Sprecher, *Bioorg. Med. Chem. Lett.*, 2010, **20**, 1109; (b) J. Rosenstock, A. J. Lewin, P. Norwood, V. Somayaji, T. T. Nguyen, J. G. Teeter, S. L. Johnson, H. Dai and S. G. Terra, *Diabetic Med.*, 2011, **28**, 464.

41. E. B. Villhauer, J. A. Brinkman, G. B. Naderi, B. F. Burkey, B. E. Dunning, K. Prasad, B. L. Mangold, M. E. Russell and T. E. Hughes, *J. Med. Chem.*, 2003, **46**, 2774.

42. (a) D. J. Augeri, J. A. Robl, D. A. Betebenner, D. R. Magnin, A. Khanna, J. G. Robertson, A. Wang, L. M. Simpkins, P. Taunk, Q. Huang, S.-P. Han, B. Abboa-Offei, M. Cap, L. Xin, L. Tao, E. Tozzo, G. E. Welzel, D. M. Egan, J. Marcinkeviciene, S. Y. Chang, S. A. Biller, M. S. Kirby, R. A. Parker and L. G. Hamann, *J. Med. Chem.*, 2005, **48**, 5025; (b) J. Robl and L. G. Hamann, in *RSC Drug Discovery Series No. 4*, ed. J. C. Barrish, P. H. Carter, P. T. W. Cheng and R. Zahler, Royal Society of Chemistry, 2011, p. 1.

43. (a) D. Kim, L. Wang, M. Beconi, G. J. Eiermann, M. H. Fisher, H. He, G. J. Hickey, J. E. Kowalchick, B. Leiting, K. Lyons, F. Marsilio, M. E. McCann, R. A. Patel, A. Petrov, G. Scapin, S. B. Patel, R. S. Roy, J. K. Wu, M. J. Wyvratt, B. B. Zhang, L. Zhu, N. A. Thornberry and A. E. Weber, *J. Med. Chem.*, 2005, **48**, 141; (b) A. E. Weber and N. A. Thornberry, *Ann. Report Med. Chem.*, 2007, **42**, 95.

44. M. Eckhardt, E. Langkopf, M. Mark, M. Tadayyon, L. Thomas, H. Nar, W. Pfrengle, B. Guth, R. Lotz, P. Sieger, H. Fuchs and F. Himmelsbach, *J. Med. Chem.*, 2007, **50**, 6450.

45. J. Feng, Z. Zhang, M. B. Wallace, J. A. Stafford, S. W. Kaldor, D. B. Kassel, M. Navre, L. Shi, R. J. Skene, T. Asakawa, K. Takeuchi, R. Xu, R. A. Webb and S. L. Gwaltney II, *J. Med. Chem.*, 2008, **50**, 2297.

46. G. R. Lankas, B. Leiting, R. S. Roy, G. J. Eiermann, M. G. Beconi, T. Biftu. C.-C. Chan, S. Edmondson, W. P. Feeney, H. He, D. E. Ippolito, D. Kim, K. A. Lyons, H. O. Ok, R. A. Patel, A. N. Petrov, K. A. Pryor, X. Qian, L. Reigle, A. Woods, J. K. Wu, D. Zaller, Z. Zhang, L. Zhu, A. E. Weber and N. A. Thornberry, *Diabetes*, 2005, **54**, 2988.

47. B. F. Burkey, P. K. Hoffmann, U. Hassiepen, J. Trappe, M. Juedes and J. E. Foley. *Diabetes Obesity Metab.*, 2008, **10**, 1057.

48. J.-J. Wu, H.-K. Tang, T.-K.Yeh, C.-M. Chen and H.-S. Shy, *Biochem. Pharmcol.*, 2009, **78**, 203.

49. Z. Pei, X. Li, T. W. von Geldern, D. J. Madar, K. Longenecker, H. Yong, T. H. Lubben, K. D. Stewart, B. A. Zinker, B. J. Backes, A. S. Judd, M. Mulhern, S. J. Ballaron, M. A. Stashko, A. K. Mika, W. D. A. Beno, G. A. Reinhart, R. M. Fryer, L. C. Preusser, A. J. Kempf-Grote, H. L. Sham and J. M. Trevillyan, *J. Med. Chem.*, 2006, **49**, 6439.

50. Y. B. Kim, L. M. Kopcho, M. S. Kirby, L. G. Hamann, C. A. Weigelt, W. J. Metzler and J. Marcinkeviciene, *Arch. Biochem. Biophys.*, 2006, **445**, 9.

51. K. L. Longenecker, K. D. Stewart, D. J. Madar, C. G. Jakob, E. H. Fry, S. Wilk, C. W. Lin, S. J. Ballaron, M. A. Stashko, T. H. Lubben, H. Yong, D. Pireh, Z. Pei, F. Basha, P. E. Wiedeman, T. W. von Geldern, J. M. Trevillyan and V. S. Stoll, *Biochemistry*, 2006, **45**, 7474.

52. A. Zhang, M. B. Wallace, J. Feng, J. A. Stafford, R. J. Skene, L. Shi and B. Lee, *J. Med. Chem.*, 2011, **54**, 510.

53. A. J. Scheen, *Diabetes Obes. Metab.*, 2010, **12**, 648.

54. A. H. Barnett, *Adv. Ther.*, 2011, **28**, 447.

55. B. O. Ahre'n, M. Landin-Olsson, P. A. Jansson, P.-A. Jansson, M. Svensson, D. Holmes and A. Schweizer, *J. Clin. Endocrinol. Metab.*, 2004, **89**, 2078.

56. A. El-Ouaghlidi, E. Rehring and J. J. Holst, *J. Clin. Endocrinol. Metab.*, 2007, **92**, 4165.

57. (a) K. Esposito, D. Cozzolino, G. Bellastella, M. I. Maiorino, P. Chiodini, A. Ceriello and D. Giugliano, *Diabetes, Obesity and Metabolism*, 2011, **13**, 594; (b) R. E. Amori, J. Lau and A. G. Pittas, *J. Am. Med. Assoc.*, 2008, **298**, 194.

58. M. Monami, F. Cremasco, C. Lamanna, N. Marchionni and E. Mannucci, *Diabetes Metab. Res. Rev.*, 2011, **27**, 362.

59. M. E. Cox, J. Rowell, L. Corsino and J. B. Green, *Drug Healthcare Patient Safety*, 2010, **2**, 7.

60. N. A. Nauck and I. Vardarli, *J. Diabetes Invest.*, 2010, **1**, 24.

61. G. Lammert, S. Einarsson, C. Saha, A. Niklasson, E. Bjornsson and N. Chalasani, *Hepatology*, 2008, **47**, 2003.

62. R. E. Pratley, M. Nauck, T. Bailey, E. Montanya, R. Cuddihy, S. Filetti, A. B. Thomsen, R. E. Sondergaard and M. Davies, for the 1860-LIRA-DPP-4 Study Group, *Lancet*, 2010, **375**, 1447.

63. L. Farilla, H. Hui, C. Bertolotto, E. Kang, A. Bulotta, U. Di Mario and R. Perfetti, *Endocrinology*, 2002, **143**, 4397.

64. M. K. Reimer, J. J. Holst and B. Ahren, *Eur. J. Endocrinol.*, 2002, **146**, 717.

65. A. Pospisilik, J. Martin, T. Doty, J. A. Ehses, N. Pamir, F. C. Lynn, S. Piteau, H.-U. Demuth, C. H. S. McIntosh and R. A. Pederson, *Diabetes*, 2003, **52**, 741.

66. J. Mu, J. Woods, Y. P. Zhou, R. S. Roy, Z. Li, E. Zycband, Y. Feng, L. Zhu, C. Li, A. D. Howard, D. E. Moller, N. A. Thornberry and B. B. Zhang, *Diabetes*, 2006, **55**, 1695.

67. A. Dutteroy, F. Voelker, X. Zhang, X. Ren, K. Merriam, J. Qui, S. Knight, H. Chen, T. Hughes and B. Burkey, *Diabetologia*, 2005, **48**(Suppl. 1), A178 (Abstract).

68. G. Xu, D. A. Stoffers, J. F. Habener and S. Bonner-Weir, *Diabetes*, 1999, **48**, 2270.

69. J. Zhou, X. Wang, M. A. Pineyro and J. M. Egan, *Diabetes*, 1999, **48**, 2358.

70. R. E. Pratley, A. Schweizer, J. Rosenstock, J. E. Foley, M. A. Banerji, F. X. Pi-Sunyer, D. Mills and S. Dejager, *Diabetes Obes. Metab.*, 2008, **10**, 931.

71. L. Xu, C. D. Man, B. Charbonnel, G. Meninger, M. J. Davies, D. Williams-Herman, C. Cobelli and P. P. Stein, *Diabetes Obes. Metab.*, 2008, **10**, 1212.

SGLT2 Inhibitors in Development

WILLIAM N. WASHBURN

Metabolic Disease Chemistry, Bristol-Myers Squibb Research and
Development, PO Box 5400, Princeton, NJ 08543, USA
E-mail: William.Washburn@bms.com

3.1   Introduction

Diabetes mellitus type 2 (T2DM) is a major growing health problem
throughout the world. The worldwide incidence of people with T2DM is 366
million, a number projected to soar to more than 552 million by 2030 as the
Western diet and lifestyle become more prevalent in China and India.[1] In 2011
the number of diabetics in the United States is estimated to be 25.8 million,
11.3% of the adult population aged 20 and older. This incidence represents an
increase of approximately 2 million from the 2008 estimate of 23.8 million.
Even more alarming is the finding that increasing numbers of teenagers are
succumbing to type 2 diabetes. The economic burden of diabetes in the United
States in 2007 was estimated to be $116 billion in direct medical costs and $58
billion in disability, work loss, and premature death.[2]

The hallmarks of the onset of type 2 diabetes are hyperglycemia, peripheral
insulin resistance, and β-cell dysfunction, manifested initially as delayed first-
phase insulin release. Over time insulin resistance continues to increase
ultimately causing apoptosis of increasing numbers of the insulin producing β-
cells which have become exhausted due to the ever increasing demands for
insulin needed to maintain normal glycemic levels.[3] Eventually, when adequate
insulin levels cannot be maintained, progressively increasing hyperglycemia
ensues. Hyperglycemia, abetted by other metabolic derangements, is the major
contributor to the onset of the microvascular and macrovascular complications

RSC Drug Discovery Series No. 27
New Therapeutic Strategies for Type 2 Diabetes: Small Molecule Approaches
Edited by Robert M. Jones

Published by the Royal Society of Chemistry, www.rsc.org

associated with type 2 diabetes.[4,5] Chronically elevated glycemic levels can result in higher protein glycation, reduced insulin secretion, β-cell exhaustion resulting in apoptosis, increased oxidative stress, and heightened insulin resistance.[3] The consequences of these metabolic changes are manifested by diminished wound healing as well as tissue damage of the retina, nerves, and kidney that give rise to increased incidence of gangrene, retinopathy, neuropathy, and nephropathy, resulting in amputations of extremities, blindness, renal failure, cardiovascular disease, and stroke. The resulting economic burden imposes severe and eventually unsustainable demands on the country's healthcare system.

Results from the United Kingdom Prevention of Diabetes Study (UKPDS) showed that incremental reductions in glycosylated hemoglobin (HbA1c), a marker of protein glycation, lower the risk of diabetes-related events, including myocardial infarction and microvascular complications.[4] Consequently, the focus for management of diabetic patients has become tight glycemic control with the goal of reducing glycosylated hemoglobin (HbA1c) to less than 6.5–7%. Since most patients cannot achieve this goal by adopting behavior modifications to promote weight loss and increase exercise, they consequently elect to begin a noninsulin therapy. These medications target the liver to reduce glucose output, the small intestine to decrease glucose absorption, adipose deposits or muscle to elevate glucose cellular uptake or promote glucose metabolism, serum proteases to prolong incretin action, and the pancreas to enhance insulin release.[6] With exception of the α-glucosidase inhibitors, all depend on the availability of insulin to regulate blood glucose levels. As insulin secretion diminishes due to continuing β-cell death, the patient will ultimately fail the current medication, necessitating combinations of two or more anti-diabetic agents to control glycemic levels. Ultimately insulin production becomes so low that insulin therapy is required for the patient. For many of these therapies hypoglycemia is a risk, especially if the patient has progressed to combinational therapy.

3.2   Renal Recovery of Glucose

For a healthy individual, two organs – liver and kidney – are responsible for supplying adequate amounts of glucose to maintain glucose homeostasis.[7] Depending on the nutritional status, glycogenolysis or gluconeogenesis is the major source for hepatic glucose output; moreover, in the postprandial state the liver contributes to glycemic decrease by glucose uptake and storage as glycogen. Renal glucose output, which is solely due to gluconeogenesis, is a significant contributor as it can be as much as 15–55 g per day or 20–25% of the total post-absorptive release of glucose. As important as renal gluconeogenesis may be to maintenance of glucose homeostasis, renal recovery of glucose is even more important.[8] Each day healthy kidneys filter ~ 180 liters of blood to remove metabolic waste products. As the resultant glomerular filtrate descends the proximal tubules, essentially all of the ~180 g of glucose present in the filtrate is recovered by two specialized sodium dependent glucose co-transporters (SGLT1 and SGLT2).

SGLT1 and SGLT2 are both members of the SLC5 gene family comprising twelve co-transporters.[9] The protein structure of all six sodium co-transporters SGLT1–6 is organized similarly with fourteen helical trans-membrane spanning domains. X-ray crystallography of vSGLT (a bacterial SGLT homolog obtained from *Vibrio parahaemolyticis*) revealed the existence of two inverted repeats comprising trans-membrane helices (TM1–TM5 and TM6–TM10). Compilation of a number of kinetic and structural studies is supportive of the following proposed account of SGLT-mediated sugar translocation across the cell membrane. The interface of the repeats forms a hydrophilic cavity accessible from the extracellular face. Binding of sodium ion(s) initiates a conformational change that allows access to a sugar binding site midway across the membrane. Following binding of the monosaccharide, a second conformational change closes the original cavity while opening a hydrophilic channel extending to the cytoplasmic face. Following dissociation of the sugar and sodium ion(s), a conformational change restores the original extracellular facing cavity.

Approximately 90% of glucose reabsorbed in the kidney of rats was shown to occur in S1 segments of the proximal tubules – the site of the low-affinity, high-capacity SGLT2.[10,11] The remaining 10% of glucose not reabsorbed in the S1 segment was thought to have been recovered during passage of the filtrate through the S3 segment – the site of the high-affinity, low-capacity SGLT1; however, a recent study with SGLT1 ko mice suggests that under normal glycemic conditions SGLT1 may not contribute more than 3% to renal glucose recovery.[12] Glucose recovery is achieved through a combination of active transport by SGLTs on the luminal tubule surface and facilitated diffusion by sodium-independent glucose transporter 2 (GLUT2) in the basolateral membrane to return glucose to the plasma. Coupling of glucose transport against the concentration gradient with transport of Na^+ down a concentration gradient renders the process energetically favorable. The prerequisite low Na^+ concentration inside the endothelial cell is maintained by Na/K pumps.[8] SGLT1-mediated glucose transport differs from that by SGLT2 only in that the transport ratio of glucose to Na^+ is 1:2 rather than to 1:1.

For a healthy individual the renal reabsorptive capacity for glucose significantly exceeds the glomerular glucose concentrations that might be incurred during normal glycemic excursions. For these individuals before the onset of glucosuria can occur, blood glycemic levels are required to nearly double to 200 mg/dL or 11 mmol/L in order to exceed the recovery capacity maximum or renal threshold (T_m). Due to enhanced expression of the SGLT2 transporter, the renal thresholds of T2DM diabetics are elevated even higher such that T_m is increased from approximately 350 mg/min for healthy individuals to $\sim$420 mg/min for T2DM.[7]

3.3 SGLT2 Inhibitors

3.3.1 Target Validation

The identification of the SGLT1 and SGLT2 transporters and establishment of the predominant role of the SGLT2 transporter for renal glucose recovery provided a new anti-diabetic target for medicinal chemists.[13] Confidence in this target was bolstered by reports that the rare individuals lacking a functional SGLT2 gene suffered no ill effects despite massive glucosuria resulting in a daily loss of as much as 140 g of glucose.[14,15] It was particular noteworthy that no electrolyte imbalances were noted and that the incidence of renal complications or urinary tract infections was unchanged from that of normal individuals. Moreover, the case for selective inhibition of SGLT2 was bolstered by the fact that it appears to be expressed only in the kidney whereas SGLT1 is present in heart (unknown function) and small intestine (transporter responsible for absorption of both glucose and galactose).[16] The physiological response to a defective SGLT1 gene is quite different than that for SGLT2. Depending on the extent of loss of function, individuals expressing defective SGLT1 present with glucose and galactose malsorption resulting in diarrhea of varying degrees of severity due to the inability to absorb glucose and galactose from the GI tract.[17] Due to concerns regarding the potential for GI disturbances to become manifested if SGLT1 mediated transport was reduced, most research groups elected to pursue SGLT2 inhibitors that were highly selective against SGLT1.[18]

Reduction of glycemic levels by this mechanism is dependent on the mass of glucose excreted daily in urine, which will be proportional to the volume of glomerular filtrate, the glucose concentration in the filtrate, and both the extent and duration of inhibition over the 24-hour period. Efficacy will be decreased for renal compromised T2DM patients with diminished GFR. More slowly cleared inhibitors will induce a greater response than rapidly cleared inhibitors exhibiting comparable potency unless the dose of the latter or frequency of administration is increased to compensate.

3.3.2 *O*-Glucosides

The discovery that SGLT transporters were the mediators for renal glucose recovery provided a mechanistic explanation for onset of glucosuria following administration of phlorizin to mammals.[19] The natural product phlorizin **1** is an phenolic *O*-glucoside containing a dihydrochalcone moiety that was first isolated from bark of apple trees more than 150 years ago. Prior studies had demonstrated that following ingestion of sufficient quantities of phlorizin, glucosuria ensued accompanied by transient lowering of glycemic levels for rats and human volunteers. Phlorizin was subsequently shown to be a modestly potent weakly selective SGLT2 inhibitor (EC_{50} = 33 nM with 7-fold selectivity versus SGLT1).[25] Poor oral bioavailability and rapid clearance was ascribed in part to α-glucosidase mediated cleavage of the *O*-glucoside bond generating the aglycone phloretin which is devoid of SGLT activity.

The Tanabe research group, recognizing the potential of SGLT inhibitors, modified the phlorizin structure by removing polar structural features that adversely impacted bioavailability to generate compounds such as **2**.[136] In addition, methyl carbonate pro-drugs, prepared by acylation of the C6 glucose hydroxyl, were employed to mitigate glucosidase mitigated cleavage in the gut prior to absorption. The disclosure of selective agents such as **2** and **3a** and more importantly the ability of pro-drugs of these agents to treat diabetes or prevent progression of diabetes in diabetic rodent models following sub-chronic oral administration induced other groups to pursue this approach.[20–22] These efforts generated a number of active structures (Figure 3.1), exemplified by **2**, **3a** (T1095A), **4a** (sergliflozin A), **5** (remogliflozin) and **6a**; all were based on a common pharmacophore comprising an *O*-glucoside of an phenol or heterocyclic equivalent for which the vicinal carbon was attached via a one or three atom spacer to a distal aryl or heteroaryl ring.[23,24] These compounds exhibited a consistent SAR pattern in which non-polar small substituents at the *para* position of the distal ring conferred *in vitro* potency.

At least six methyl or ethyl carbonate pro-drugs of *O*-glucoside- containing SGLT2 inhibitors entered clinical trials as T-1095 **3b**, sergliglozin[25] **4b**, remogliflozin etabonate[26] **5b**, AVE-2268,[18] BI-44847[18] (structure presumed to be **6b**), and TS-033 (presumed to be a thiopyranoside *O*-glucoside).[18] None progressed beyond early phase 2 trials. In general the glucosuric effect was immediate although 24-hour duration was not achieved unless multi-gram doses were administered qd or bid. Notably the postprandial glucose peak of an individual meal could be greatly suppressed by prior administration of these agents. Clinical progression was impeded by the rapid elimination of these agents, exemplified by $t_{1/2}$ of 1.5 hour and 30–60 min respectively for remogliflozin and sergliflozin A.[27,28] A major contributor to the rapid

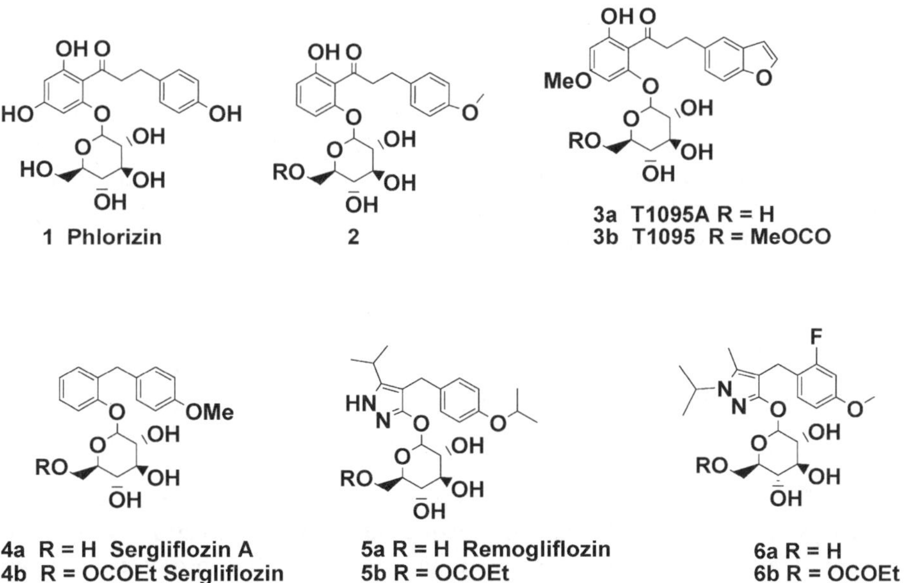

Figure 3.1   *O*-Glucoside containing SGLT2 inhibitors.

clearance for all these *O*-glucosides was hydrolytic cleavage of the active agent by the α-glucosidases present in tissues such as liver, kidney, lung, etc.

3.3.3 Biological Assays

3.3.3.1 SGLT Binding Assays

For most discovery programs, Chinese hamster ovary (CHO) cells stably expressing human SGLT2 (hSGLT2) and human SGLT1 (hSGLT1) (Genbank accession numbers M95549 and M24847, respectively) were utilized for the development of transport assays using the selective SGLT substrate α-methyl-*D*-glucopyranoside (AMG).[29] Inhibitors were assayed for the ability to inhibit [^{14}C]AMG uptake in a protein-free buffer over a 2-hour incubation period. The response curve was fitted to an empirical model to determine the inhibitor concentration at half maximal response, reported as EC_{50}. In most instances **17** or phlorizin was included as a standard. For most discovery groups the SGLT2 EC_{50} variability for these standards was typically 2-fold or less between groups; selectivity versus SGLT1, when reported, tended to exhibit wider variations. To allow the reader to assess SAR progress, whenever possible, outlier values will be cited in the text. For some programs EC_{50} values were also determined in a similar fashion for rat or mouse SGLT1 and SGLT2 depending on the *in vivo* model employed. Typically rat SGLT2 EC_{50} values were within a factor of 2–3 of the human SGLT2 EC_{50}; whereas, the greater variability regarding SGLT1 affinity between the two species meant that rat selectivity could significantly differ from that determined for human.

*3.3.3.2 Evaluation of i*n vivo *Pharmacology*

Although inherent SGLT2 affinity as reflected by EC_{50} for inhibition of SGLT2-mediated transport of AMG using a CHO cell assay was essential, a more rigorous indicator of the potential of the compound was the normalized urinary glucose excretion (UGE) induced in normal rats over 24 hours following oral administration of the test compound. Normalization of the 24-hour glucosuria to glucose output per 200 g body weight enabled comparison of these normalized UGE values within and across chemotypes. For a 200 g normal Sprague Dawley rat maintained on typical rat chow, the maximum rate of glucosuria arising from inhibition of renal recovery by potent selective SGLT2 inhibitors appears to be ∼1.4 mg/min. If the blockade is maintained over 24 hours, the UGE dose response plateaus at ∼2–2.5 g following an oral dose of approximately 1 mg/kg of a potent SGLT2 inhibitor exhibiting a highly favorable ADME rat profile. Production of a 24-hour UGE significantly less than this experimentally determined value following administration of an 1 mg/kg oral dose of a potent SGLT2 inhibitor indicated that the compound was deficient with respect to some ADME property encompassing metabolic stability, clearance, bioavailability, free fraction, etc. This assay provided the most reliable means to compare the glucosuric potential for compounds reported by different groups. Although the doses utilized varied from one group to the next, the UGE response obtained from a 1–10 mg/kg dose enabled a rough assessment of potency/duration of the agent.

A variety of rodent models have been employed to ascertain the anti-hyperglycemic potential in either acute studies of a few hours or sub-chronic studies with duration as great as 4 weeks. Male Sprague Dawley rats (Charles River) made diabetic by a single intraperitoneal injection of streptozotocin were sometimes utilized for acute initial *in vivo* screening. More rigorous studies utilized rodents genetically disposed to become diabetic with age such as Zucker diabetic fatty rats, db/db mice, and KK^{AY} mice since they more closely emulated the human condition.

3.3.4 *C*-Glucosides

Many groups sought to avoid the impediments imposed by *O*-glucoside metabolic instability by preparation of the corresponding *C*-glucoside either by replacement of the anomeric oxygen with a methylene or by excising it.[30] Initial attempts (Figure 3.2) sought to replace the glycosidic oxygen of dihydrochalcone **7** with a methylene; however, this structural change

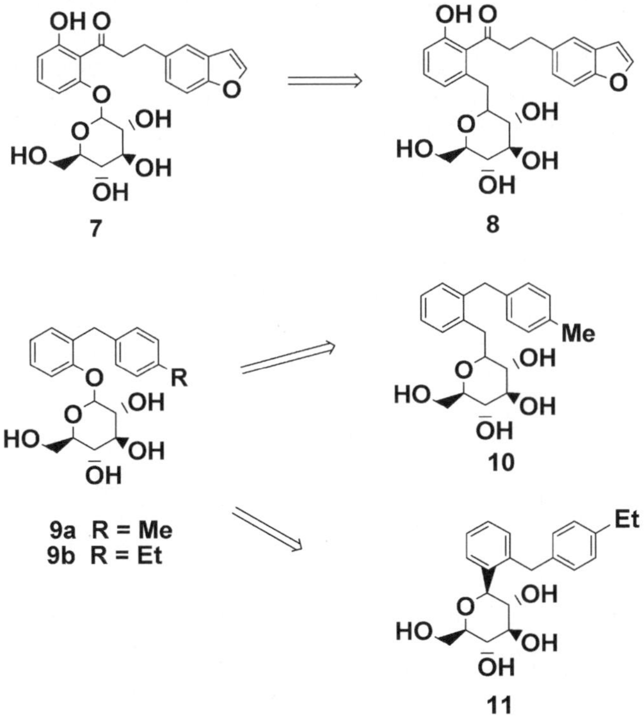

Figure 3.2 *C*-Glucoside counterparts of potent *O*-glucoside SGLT2 inhibitors.

diminished the SGLT2 affinity of **8** more than 10-fold.[31] During the pursuit of *O*-glucosides of *o*-benzylphenols, the Bristol-Myers Squibb (BMS) group subsequently reported that replacement of the glycosidic oxygen of **9a** with a methylene to generate **10** increased the EC_{50} from 8 to 500 nM. Moreover, excising the glycosidic oxygen of **9b** to directly link the aryl moiety to the glucose moiety generating **11** also increased the EC_{50} from to 2 to >4000 nM.

A fortuitous set of circumstances expedited the BMS discovery of the proper spatial presentation conducive to potent *C*-glucoside-based SGLT2 inhibitors.[30] Early in their program while evaluating the potential of *O*-glucosides of dihyroxybenzamides, attempted glucosylation of **12** with bromoacetoglucose failed to generate any desired product due to the steric encumbrance of the tertiary amide. Structural determination of the four side products, each generated in ~1% yield, revealed that decomposition of the thermally labile bromoacetoglucose had produced sufficient quantities of HBr to promote conversion of **12** to a diarylmethane **13**. Subsequent *O*-glucosylation of this aglycone to generate two isomeric mono *O*-glucosides and a bis-*O*-glucoside was not surprising; however, *C*-glucosylation also occurred to form a meta benzylated *C*-aryl glucoside **14**. Characterization revealed **14** to be a modest SGLT2 inhibitor; EC_{50} = 1300 nM (Figure 3.3). Efforts to devise a better synthetic route were unsuccessful, so **14** remained a curiosity until the group was exploring *O*-glucosides of *o*-benzylphenols. The findings that *ortho O*-glucosides of diarylmethane aglycone appropriately substituted with non-polar substituents such as **9a** (10 nM EC_{50}) were potent SGLT2 inhibitors whereas incorporation of *para* polar groups such as amides reduced potency 100-fold suggested that a similar SAR pattern might hold for *meta C*-glucosides of a diarylmethane aglycone. This merger of *ortho* benzyl phenolic *O*-glucosides

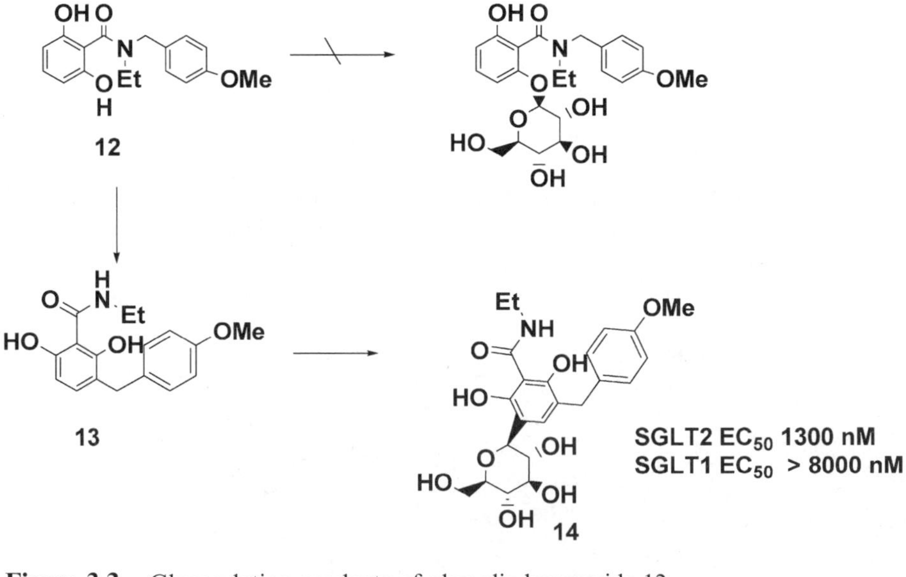

Figure 3.3 Glucosylation products of phenolic benzamide 12.

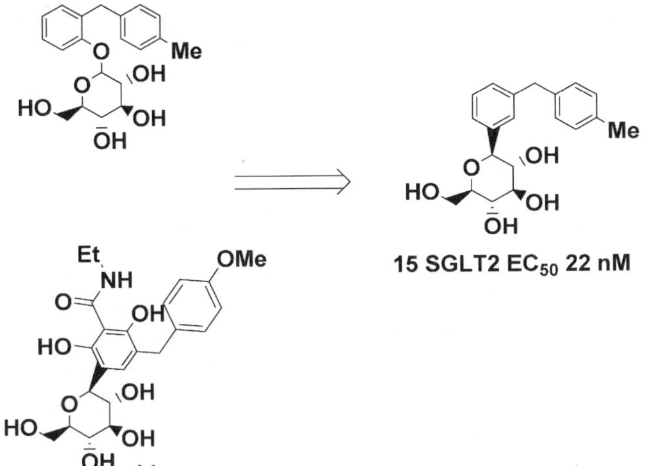

Figure 3.4 SAR Progression from O- to C-glucosides of diarylmethane as potent SGLT2 inhibitors.

with *C*-glucoside **14** led to synthesis of **15** (Figure 3.4), the first potent *C*-glucoside inhibitor of SGLT2 for which hSGLT2 EC_{50} was 22 nM.[30]

3.3.4.1 Discovery of Dapagliflozin

SAR analysis of these hydrolytically stable inhibitors revealed that *meta*-substituted diarylmethanes were more potent inhibitors of SGLT2 than the corresponding biphenyl or 1,2-diarylethane structures.[32] Moreover, the diarylmethane structure conferred greater inhibitory activity than that obtained with a diaryl sulfide or especially more than that found for a diaryl ether. Methyl substitution of the methylene bridge or replacement with a carbonyl virtually abolished SGLT activity. Small *para* lipophilic substitutents at C4′ of the distal ring increased SGLT2 affinity 10-fold, resulting in EC_{50} values of ~10 nM; whereas, substitution at C2′ or C3′ decreased affinity by a factor of 20 and 3 respectively.[30,33] Compounds in this series exhibited >1000-fold selectivity for SGLT2 than SGLT1.

These properties, in addition to the hydrolytic resistance of *C*-aryl glucosides to glucosidases, shifted the focus of the Bristol-Myers Squibb group to *C*-glucosides. Exploration of the SAR for a methyl group attached to the central aryl ring revealed that a methyl substituent at the C4 position of the central aryl ring increased SGLT2 affinity 15-fold but only 2- to 3-fold if attached to C5 or C6. In contrast, methylation of C2 was deleterious, decreasing affinity 7-fold.[30] Subsequently all further SAR efforts focused on substitution at C4 and C4′ especially since the effect of substituents at C4 and C4′ were additive. It quickly became apparent that small lipophilic substituents were preferred at the C4 position of the central aryl ring; rank ordering was Me ~ Cl > F > Et

$\sim$ H $>$ *i*-Pr. SGLT1 affinity increased as the C4 substituent increased in size; Consequently maximum selectivity of $\sim$15 000 was obtained upon incorporation of hydrogen or fluorogen at C4; selectivity versus SGLT1 progressively diminished as the group became larger – such that selectivity decreased to 4 if an *i*-propyl group was attached to C4.

Evaluation of the various combinations of favorable groups at C4 and C4′ yielded more than 20 highly selective potent SGLT2 inhibitors with EC_{50} values of 1–5 nM. Since the functional CHO cell based assay was not capable of distinguishing among these, these compounds were ranked in terms of ability to lower blood glucose at 5 h after p.o. administration at 0.1 mg/kg levels to SD rats previously rendered diabetic by streptozotocin. Upon completion of this initial screen, the dose-dependent response in this model was determined for the leading candidates when administered at 0.01, 0.03, and 0.1 mg/kg. Subsequently, the anti-diabetic potential for the top four compounds was confirmed using Zucker diabetic rats in a sub-chronic 15-day study in which declines for both fasting and postprandial glycemic levels monitored.[29] Compound **17**, now known as dapagliflozin, emerged as the clinical candidate since it consistently was judged to be superior (Figure 3.5).[33]

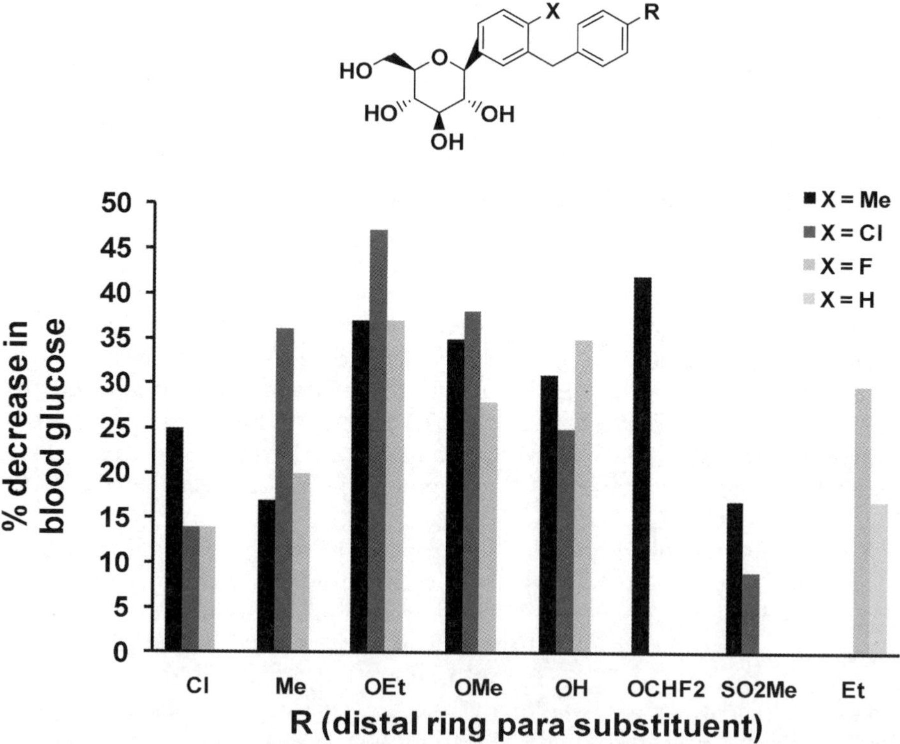

Figure 3.5   SAR for % decrease in blood glucose at 5 hours after oral administration of 0.1 mg/kg of diarylmethane *C*-glucosides to Sprague Dawley rats made diabetic by prior streptozotocin injection.

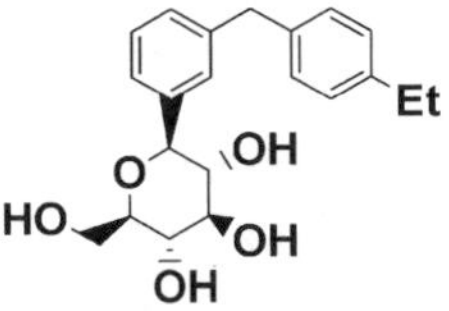

16
SGLT2 EC$_{50}$ 10 nM
SGLT1 EC$_{50}$ 18,500 nM

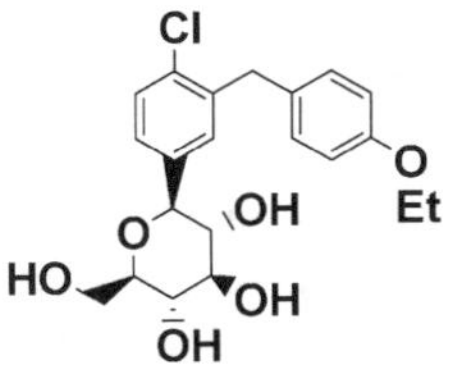

17 Dapagliflozin
SGLT2 EC$_{50}$ 1.1 nM
SGLT1 EC$_{50}$ 1300 nM

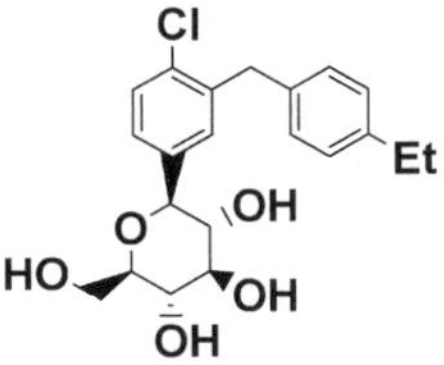

18
SGLT2 EC$_{50}$ 0.9 nM
SGLT1 EC$_{50}$ 785 nM

Figure 3.6   Dapagliflozin and closely related SGLT2 inhibitors.

3.3.4.2   Structural Modifications of Dapagliflozin

The disclosures of *C*-aryl glucoside structures exhibiting potent and selective SGLT2 inhibitory activity and subsequent clinical findings with SGLT2 inhibitors in general stimulated pursuit of SGLT2 inhibitors by a number of research groups (Figure 3.6). **16**, featured in the first Bristol-Myers Squibb disclosure,[34] especially **17**, the only example in the second disclosure,[35] and **18** in a third disclosure[36] served as leads to expedite the discovery efforts of medicinal chemistry groups. All adopted a fast follow-on strategy employing **17** as a benchmark as they sought to identify non-inferior proprietary structures that could be progressed into the clinic. These efforts generally utilized one of two strategies – modification of the aglycone component or alteration of a structural element of the glucose moiety. No portion of **17** was left unchanged as both approaches sought to maintain the favorable interactions of **17** with the SGLT2 transporter by incorporation of an isosteric counterpart or by introduction of new structural elements that did not disrupt these favorable interactions. Identification of compounds that exhibited high affinity and selectivity was not as challenging as finding candidates for which the PK profiles predicted adequate coverage would be achieved by a low daily dose comparable to that required by dapagliflozin. These efforts, entailing replacement of the distal aryl ring, incorporation of alternative substituents to either aryl ring and modification of the glucose moiety or combinations thereof identified eight *C*-glucosides that have progressed at least to phase 2.

3.3.4.2.1   Replacement of the Distal Aryl Ring

Modification of the distal aryl ring has been a fruitful exercise for six groups. The range of replacements investigated as a surrogate for the distal aryl ring encompassed monocyclic heterocycles, bicyclic carbocycles, and bicyclic heterocycles. Three groups at Tanabe, Green Cross, and Astellas explored heterocyclic replacements of the distal *p*-ethoxy substituted phenyl ring of dapagliflozin. Two groups maintained the benzenoid ring while incorporating novel substituents. In one instance the two aryl rings were incorporated into 13–15 membered macrocycles, utilizing an appropriate aliphatic chain to link

C4 and C4′. These modifications of the distal ring identified five compounds that entered clinical trials; four of which have progressed to late phase 2 or phase 3.

A retrospective analysis summarized in Table 3.1 reveals that utilization of non-polar heterocyclic isosteres of the distal aryl ring enhanced the probability of success. Utilizing **16** as a lead, the Tanabe group systematically evaluated replacement of 4′-Et-phenyl moiety of **16** with appropriately substituted heterocycles exemplified by 5-ethylfuran **1-1** (EC_{50} = 920 nM) and 5-ethylthiophene **1-2** (EC_{50} = 17 nM).[37] Encouraged by the finding that **16** and its isosteric counterpart **1-2** exhibited comparable EC_{50} values, the group incorporated substituents at C4 of the central ring to further improve potency in expectation that the SAR would emulate that previously disclosed by Bristol-Myers for diarylmethane glucosides. The 7-fold reduction in EC_{50} for **1-3** relative to **1-2** confirmed that both series followed the same SAR. Incorporation of more polar heterocycles as a distal ring replacement diminished potency as exemplified by the at least 10-fold decrease observed for pyridine (compare **1-3** and **1-5**) or pyrrazole (compare **1-4** and **1-11**) or the 4-fold reduction produced by thiazole **1-6** relative to **1-8**. The group generated several hundred examples by systematically mixing and matching of substituents on the two rings with particular focus on 5-aryl substituted thiophenes.[38] Note that although use of a polar heterocycle as distal ring is unfavorable, potent inhibitors can be obtained following incorporation of these heterocycles as a distal ring substituent. No real pattern was apparent regarding the preferred point of substitution of the pendent aryl ring attached to C5 of the thiophene moiety although *meta* substitution generally appeared to confer to greater potency; compare structures **1-7** to **1-11**. For the ∼100 compounds for which SGLT2 EC_{50} values were disclosed, the SAR was rather flat with EC_{50} values ranging from 2 to 10 nM; however, the 24-h normalized UGE measured following p.o. administration of these compounds to Sprague Dawley rats exhibited greater variability. Further characterization of **1-8** revealed it to be 410-fold selective versus SGLT1. When administered to hyperglycemic high fat KK mice at 3 mg/kg, **1-8** produced reductions in blood glycemic AUC of 48% and 25% at 6 h and 24 h respectively. Subsequently **1-8** (now canagliflozin) was selected for evaluation in clinical trials which have progressed to phase 3.[39]

More recently a group at Green Cross reported that their extensive efforts to utilize thiazoles as an alternative for the distal benzene ring of **17** revealed SGLT2 potency to be dependent upon the thiazole C5 substituent.[40] The SAR for thiazoles bearing 5-alkyl, 5-aryl, 5-carbocycles, and 5-alkoxy substituents was rather flat with a representative EC_{50} of 17 nM for the ethyl derivative **1-12**. However, substitution of the thiazole at C5 with small non-polar electron-rich heterocycles such as a 5-furanyl **1-13** or 4-thienyl **1-14** reduced EC_{50} ∼ 20-fold to 0.7–0.8 nM, a value that approached the 0.49 nM measured for **17** in the Green Cross assay. Presumably this increase in potency reflects the ability of these properly oriented electron-rich heterocyclic appendages to engage in a favorable pi interaction with residues of SGLT2. The 15-fold decrease in

Table 3.1 Potencies of diarylmethane *C*-glucosides with alterative distal rings.

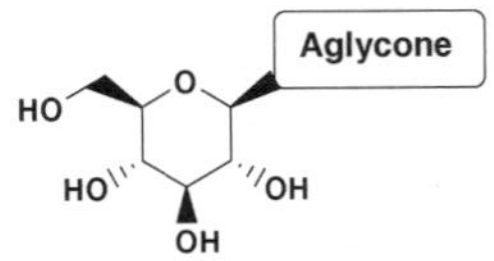

Compound ID	Aglycone	SGLT2 EC$_{50}$ (nM)	Selectivity versus SGLT1	24 h Rat UGE mg (Dose)	% Glucose Decrease (p.o. dose)	Ref.
1-1	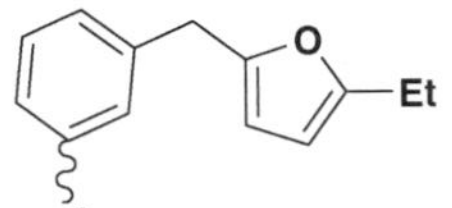	920				37
1-2	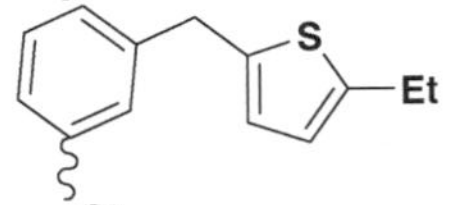	17		373 (30)		37
1-3	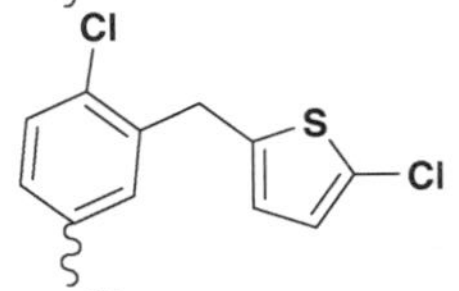	2.4		2495 (30)		37
1-4	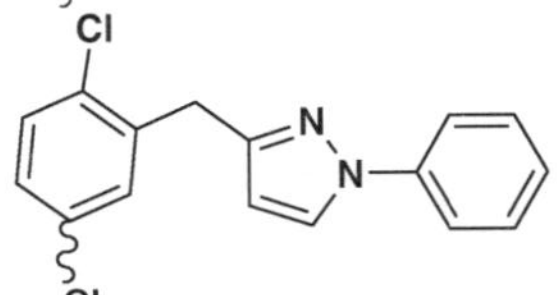	32				37
1-5	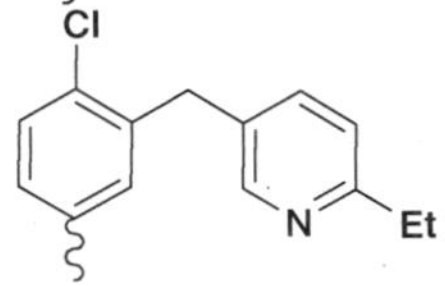	28				37

Table 3.1 (*Continued*)

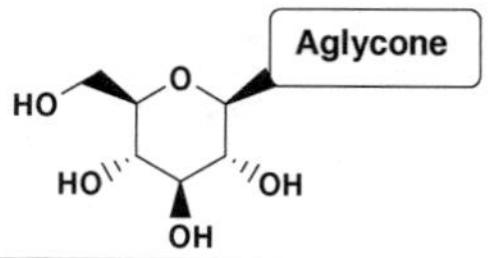

Compound ID	Aglycone	SGLT2 EC$_{50}$ (nM)	Selectivity versus SGLT1	24 h Rat UGE mg (Dose)	% Glucose Decrease (p.o. dose)	Ref.
1-6	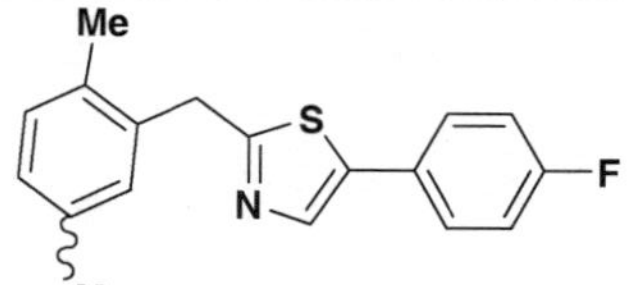	8.1		1277 (30)		37
1-7	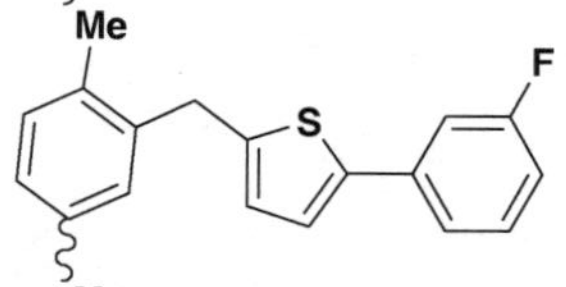	1.3		>2000 (30)		38
1-8	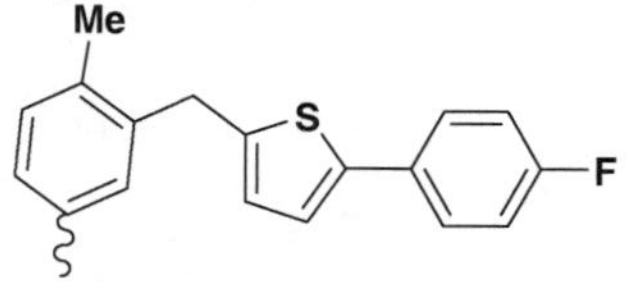 (canagliflozin)	2.2	410	3696 (30)	48% at 6 h (3 mg/kg)	37,39
1-9	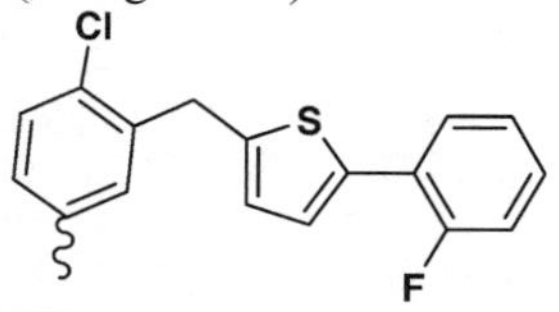	7.9		1000–2000 (30)		38

Table 3.1   (*Continued*)

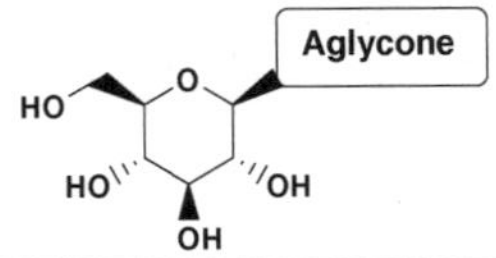

Compound ID	Aglycone	SGLT2 EC$_{50}$ (nM)	Selectivity versus SGLT1	24 h Rat UGE mg (Dose)	% Glucose Decrease (p.o. dose)	Ref.
1-10	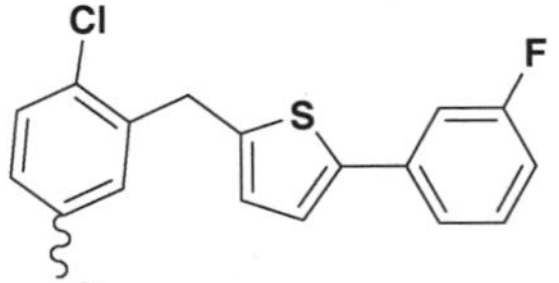	1.7				38
1-11	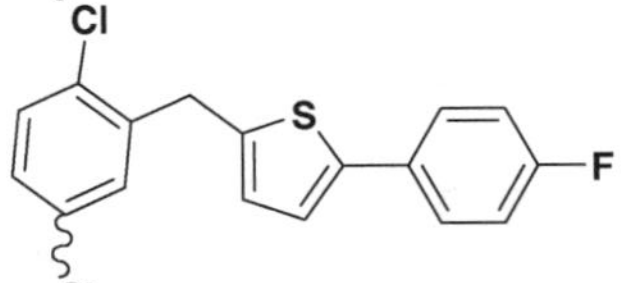	7.0		>2000 (30)		38
1-12	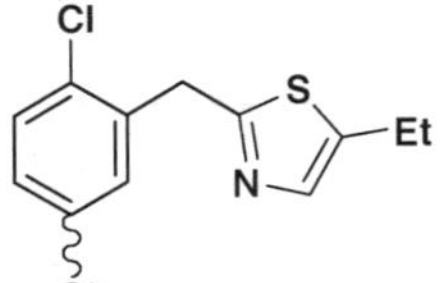	17				40
1-13	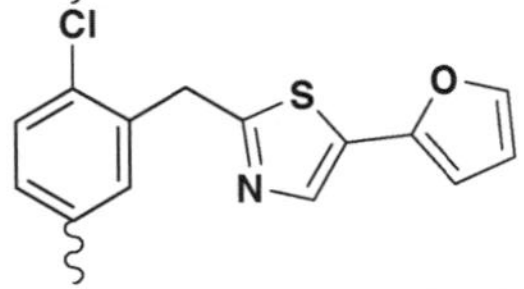	0.7		~450 (1); ~1300 (10)		40

Table 3.1   (*Continued*)

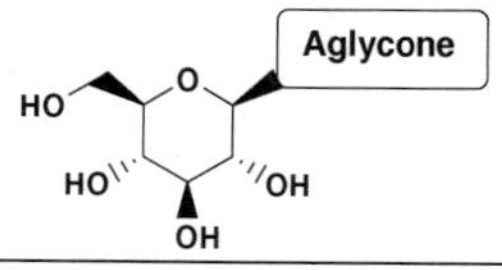

Compound ID	Aglycone	SGLT2 EC$_{50}$ (nM)	Selectivity versus SGLT1	24 h Rat UGE mg (Dose)	% Glucose Decrease (p.o. dose)	Ref.
1-14	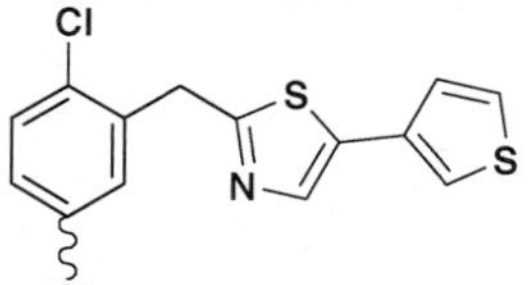	0.8		~320 (1); ~1400 (10)		40 40
1-15	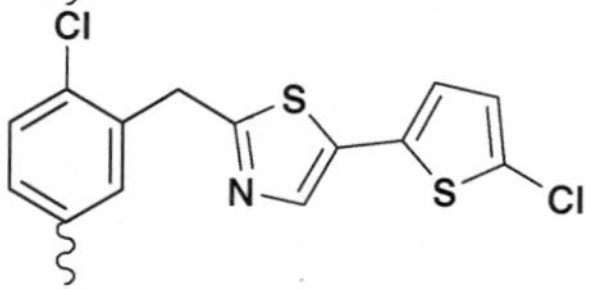	13				40
2-16	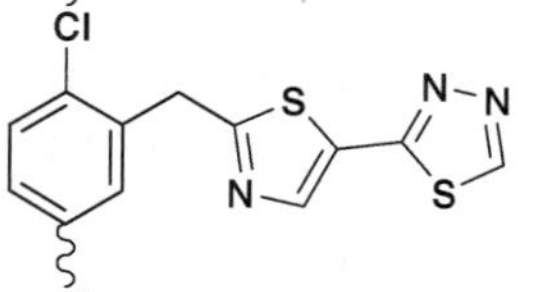	14				40
1-17	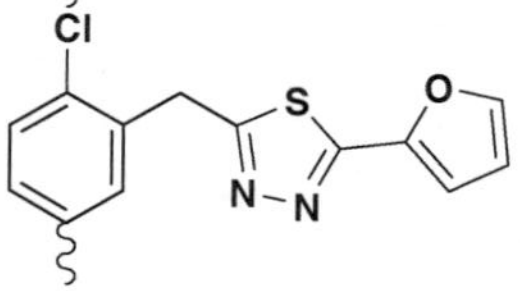	7		195 (10)	65% at 5 h (10 mg/kg)	41

Table 3.1 (*Continued*)

Compound ID	Aglycone	SGLT2 EC$_{50}$ (nM)	Selectivity versus SGLT1	24 h Rat UGE mg (Dose)	% Glucose Decrease (p.o. dose)	Ref.
1-18		3.5				41
1-19		13		344 (10)		42
1-20		64				43
1-21		22	590			45,47
1-22		99	140			45,47

Table 3.1   (*Continued*)

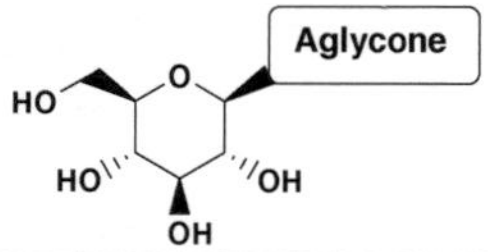

Compound ID	Aglycone	SGLT2 EC_{50} (nM)	Selectivity versus SGLT1	24 h Rat UGE mg (Dose)	% Glucose Decrease (p.o. dose)	Ref.
1-23	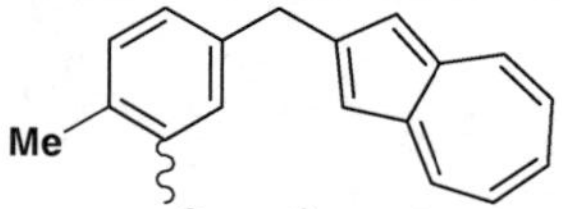 Me	29	240			45,47
1-24	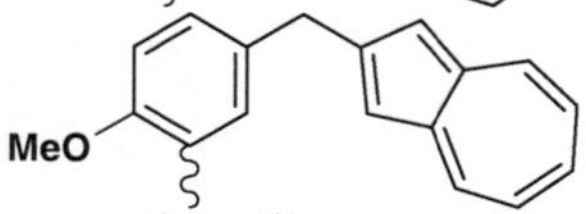 MeO	16	2100		8 h AUC 45% (3 mg/kg)	45,47
1-25	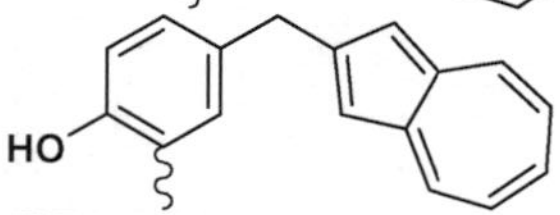 HO (YM-543)	8.9	280			45,47
1-26	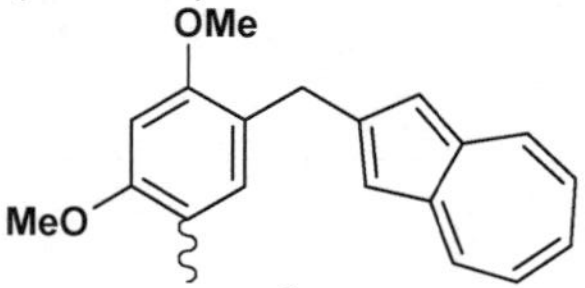 OMe MeO	5.7	100		8 h AUC (46% 3 mg/kg)	45,47
1-27	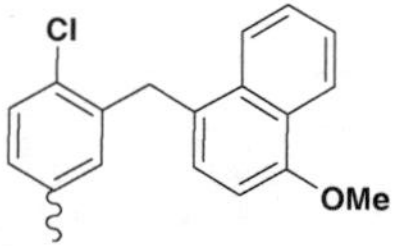 Cl OMe	6% inhibition @ 10 nM				48

Table 3.1   (*Continued*)

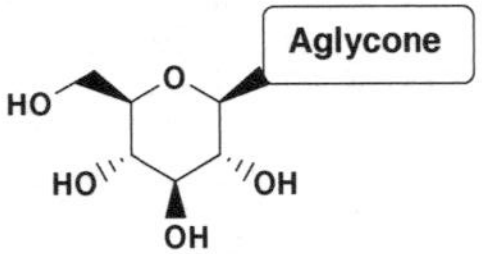

Compound ID	Aglycone	SGLT2 EC$_{50}$ (nM)	Selectivity versus SGLT1	24 h Rat UGE mg (Dose)	% Glucose Decrease (p.o. dose)	Ref.
1-28	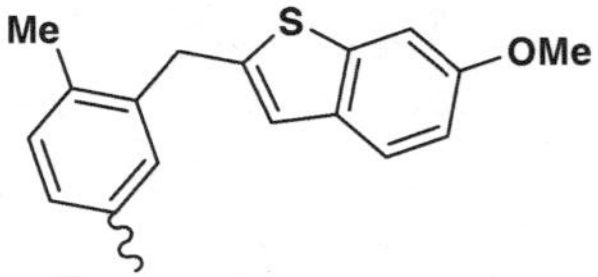	1.6		>2000	8 h AUC 46% (30 mg/kg)	38
1-29	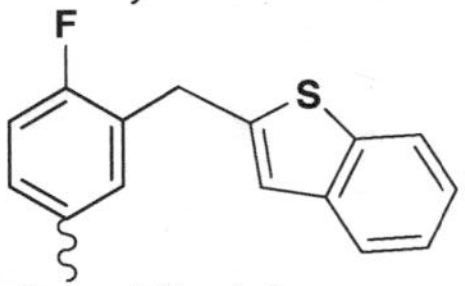	7.4			255	50
1-30	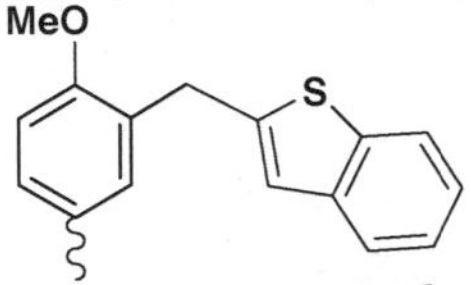 (ipragliflozin)	14			8 h AUC 39% (1 mg/kg)	49
1-31	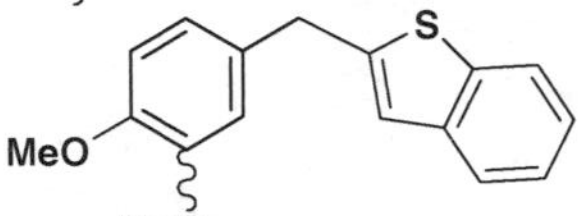	13				49
1-32	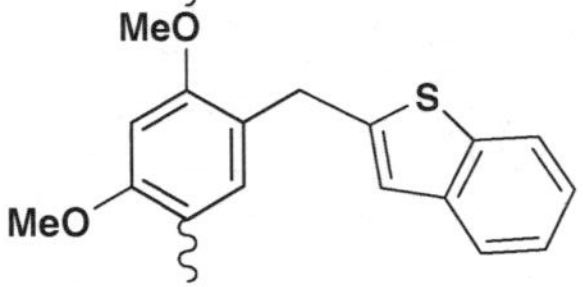	3.8			8 h AUC 34% (1 mg/kg)	49

potency observed for **1-15** and **1-16** versus **1-14** was accordance with this proposed pi interaction since both chlorination of the thiophene or conversion to a thiadiazole would diminish the electron-donating capacity of the pendent heterocycle. Despite the high potency of **1-13** and **1-14**, interest ceased once it became apparent that the normalized 24 h UGE following oral administration to SD rats was 28% and 20% respectively of that induced by **17** in the same assay. Subsequent rat PK studies suggested that the problem was an unfavorable rat PK profile: both low bioavailability (20 and 15% respectively for **1-13** and **1-14** versus 88% for **17**) and rapid elimination ($t_{1/2}$ of 2.5 h) contributed to the poor *in vivo* efficacy of **1-14**.

Efforts to replace the distal benzene ring of **17** with more polar aromatic heterocycles – thiadiazole and pyridazine – were not encouraging. When a thiadiazole was utilized as the distal ring, the potency of the most active derivatives, 5-furanyl **1-17** or 4-thiophenyl **1-18**, was 5–10-fold poorer than that obtained with the thiazole counterparts **1-13** and **1-14**.[41] Moreover, due to an even less favorable rat PK profile (8% bioavailability and a 3-fold shorter half-life of 0.86 h), the 24-h normalized UGE following administration of **1-17** at 10 mg/kg was 195 mg of glucose, i.e. ~ one-sixth of that observed for the thiazole **1-13**. Characterization of even the most active C6 substituted pyridazines also proved disappointing.[42] The EC_{50} for **1-19** was 13 nM; the low 24-h UGE of only 344 mg, obtained after a 10 mg/kg dose, was again attributed to a poor PK profile (26% bioavailability and a half-life of 1.9 h). Additional substituents at C4 or C5 of the pyridazine ring diminished activity analogous to the loss reported for *ortho* or *meta* substituents of the distal ring of the diarylmethane series. More discouraging were SGLT2 inhibitors containing a pyrimidine distal ring such as **1-20** which exhibited such weak *in vitro* potency (SGLT2 $EC_{50} \geq 64$ nM) that UGE values were not measured for this series.[43]

Two groups, Astellas and Egret/Theracos, explored bicyclic aromatic carbocycles as a distal ring replacement. Initially the Kotobuki (now Astellas) group had been pursuing *C*-glucosides containing a phlorizin-like three atom spacer joining the two aryl rings for which the central ring was oxygenated at C4 and/or C6.[44] After the first BMS disclosure, the focus appears to have shifted to chemotypes utilizing a methylene spacer to link the two aryl rings. This effort led to an application providing little biological data while disclosing a series for which the distal ring was an azulene and the central ring optionally was substituted at C6 and/or C4.[45] Orientation of the azulene moiety impacted affinity as the C2 linked **1-21** was ~ 5-fold more potent and selective for SGLT2 than the C6 linked **1-22** possibly reflecting a more favorable dipole orientation. The limited data did not reveal any benefit relative to the parent **1-21** regarding SGLT2 potency arising from substitution of the central ring at C6 with small groups such as methyl (**1-23**) or methoxyl (**1-24**) but hydroxylation (**1-25**) enhanced affinity 2- to 3-fold. A subsequent disclosure revealed that the high SGLT2 selectivity for this series was especially enhanced when C6 was substituted with methoxyl (**1-24**).[47] The 3-fold increase in affinity accompanied by decreased SGLT1 selectivity upon C4 methoxylation of **1-24** to generate **1-26** was in accord with the SAR for C4 substitution of the previously discussed diarylmethane-based *C*-glucosides. When administered at 3 mg/kg to fed KK-A^y mice, both **1-24** and **1-26** produced $\sim 45\%$ reduction in the glucose AUC over 8 h. Compound **1-25** subsequently entered clinical development as the crystalline 1:1 choline complex YM-543; however, development was halted in phase 2 by Astellas.[46] Subsequently the groups at Chengdu Institute and Egret Pharmaceuticals evaluated naphthalene as a distal ring replacement; however, the 6% inhibition observed at 10 nM for **1-27** discouraged further pursuit.[48]

Benzothiophenes attached at C2 were the main focus of both the Astellus and Tanabe groups' efforts to identify bicyclic heterocyclic replacements of the distal ring. Despite the Tanabe group's primary interest in thiophenes, a number of benzothiophene such as **1-28** containing substituents at C4 of the central benzene ring and C5′ or C6′ were disclosed to possess potencies ranging from 1 to 10 nM.[39] However, the failure to obtain maximal UGE values following administration of some potent analogs at 30 mg/kg to Sprague Dawley rats suggested that there were shortcomings in the ADME profile.

Based on the limited biological data provided by Astellas, the benzothiophene SAR appeared flat as SGLT2 potency remained unchanged at ~ 10 nM regardless whether the central ring was substituted at C4 with fluorine **1-29** or methoxyl **1-30** or methoxylated at C6 **1-31**.[49,50] However, the 3-fold potency increase upon introduction of methoxyls at both C4 and C6 **1-32** was in accord with the Kotobuki group's finding that synergy could arise following substitution at both sites of the central ring.[44] The benzothiophene series appears to exhibit greater anti-hyperglycemic activity than the previously discussed azulene-containing series since oral administration of 1 mg/kg of **1-29**, **1-30**, and **1-32** to fed KK-A^y mice reduced glucose AUC at 8 h by 54%, 39%, and 34% respectively versus an $\sim 45\%$ reduction by 3 mg/kg of the azulene counterparts **1-24** and **1-26**. Of particular interest is **1-29** (SGLT2 EC$_{50}$ = 7.4 nM with 255-fold SGLT1 selectivity) which began clinical development as the crystalline 1:1 proline complex ASP1941 but now having progressed to phase 3 trials has acquired the trade name ipragliflozin.[50,51]

3.3.4.2.2   Alternative Distal Aryl Ring Substitution

Groups at Boehringer Ingelheim, Taisho, Green Cross, and Theracos investigated replacement of the *p*-ethoxy moiety of **17** with novel substituents. (Table 3.2). Patent disclosures reveal that the Boehringer Ingelheim group focused on close analogs of **17** for which the C4 substituents of the central aryl ring and/or the C4′ substituents of the distal ring were altered to provide sufficient novelty to be outside the scope of the BMS disclosures. Assessment of the merits of these perturbations of the dapagliflozin structure is not possible since no specific biological data has been disclosed. This effort eventually led to the clinical candidate **2-1** for which the SGLT2 EC$_{50}$ is 3.1

nM and the selectivity versus SGLT1 is 2700.[52,53] **2-1** entered clinical trials as BI 10773 and now having progressed to phase 3 has acquired the trade name empagliflozin.

Disclosures from both the Taisho and Egret/Chengdu/Theracos groups reveal both SGLT2 affinity and selectivity versus SGLT1 to be responsive to the C4′ substituent. The Egret/Theracos group extensively evaluated analogs of **17** for which the *p*-ethoxy group of the distal ring had been replaced with a variety of functionalized substituents comprising mainly ethylene glycol ethers and unsaturated ethers and alkanes.[48] Inclusion of hydrophilic functionality reduced SGLT2 affinity with little impact on SGLT1. For example, oxidation of the *p*-ethoxy ether of **17** to generate an oxyacetic acid **2-2** or hydroxylation to form an ethylene glycol ether **2-3** decreased SGLT2 affinity 140- and 30-fold respectively and reduced selectivity $\sim$50- and 10-fold relative to that of **17** (EC_{50} = 3.2 nM with 970-fold SGLT1 selectivity in the Egret assay). Conversion of the hydroxyl of **2-3** to an allyl ether **2-4** essentially restored potency and selectivity within 2-fold of that of **17**. Capping the ethylene glycol moiety of **2-3** with progressively larger cycloalkyls to generate tethered cyclic ethers did not significantly increase potency relative to **2-4** but did markedly enhance selectivity $\sim$4-fold for the cyclopropyl **2-5** and cyclopentyl **2-7** ethers and 20-fold for the cyclobutyl ether **2-6**. SGLT2 affinities less than 100 nM and selectivities greater than 100-fold were obtained if C4′ was substituted with small carbon linked substituents such as ethinyl **2-8**, cyclopropanol **2-9**, or even alkyl amides **2-10**. *In vivo* studies with Sprague Dawley rats revealed compounds **2-4** and **2-5** to be promising glucosuric agents which induced normalized 24-h UGE values of $\sim$1800 and $\sim$1700 mg of glucose respectively following administration of 1 mg/kg. Subsequent characterization using db/db mice led to **2-5** being selected for clinical studies.[54] **2-5** (EGT1442) has progressed to phase 3 clinical trials as the bis-proline complex of **2-5**, now known as EGT1474.

In contrast, the Taisho group reported that appropriately tethered polar groups would so markedly increase SGLT1 affinity such that non-selectiveSGLT1/2 inhibitors were formed.[55] For example, incorporation of ethano- or propane-linked amides and ureas at C4 exemplified by **2-11**, **2-12**, and **2-13** generated non-selective SGLT2 inhibitors with EC_{50} values 10–50 nM. For these more extended C4′ substituents (compare **2-12** to **2-13**), hydroxylation of the terminal carbons did not reduce SGLT2 affinity, unlike **2-3**. When administered orally at 1 mg/kg to streptozotocin-induced diabetic Sprague Dawley rats 5 min prior to an OGTT, the dual SGLT inhibitors **2-11** and **2-13** reduced the glucose AUC at 1 h by 42% and 69 % respectively.

Following a recent report regarding weak SGLT inhibitory activity of selected members of the acerogenin family of natural product macrocycles,[56] the Green Cross group incorporated the dapagliflozin structure into 13–15 member macrocycles that linked C4 and C4′ of **16**.[57] Progressively decreasing the macrocycle size from a 15-membered to a 13-membered ring (see **2-14** and **2-15**) increased SGLT2 inhibitory activity $\sim$2-fold to 60 nM;

Table 3.2 SAR of distal ring substituents on potency of diarylmethane *C*-glucosides.

Compound ID	X	SGLT2 EC_{50} (nM)	Selectivity versus SGLT1	24-h rat UGE mg (Dose)	Ref.
2-1	(empagliflozin)	3.1	2700		53
2-2		428	>20		48
2-3		93	97		48
2-4		6	600	~1800 (1)	48
2-5	(EGT 1442)	2.3	2400	~1700 (1)	48
2-6		3	12 600		48
2-7		4	1700		48
2-8		10	140	~3000 (1)	48
2-9		3	830		48
2-10		9% at10 nM			48

Table 3.2 (*Continued*)

Compound ID	*X*	*SGLT2 EC_{50} (nM)*	*Selectivity versus SGLT1*	*24-h rat UGE mg (Dose)*	*Ref.*
		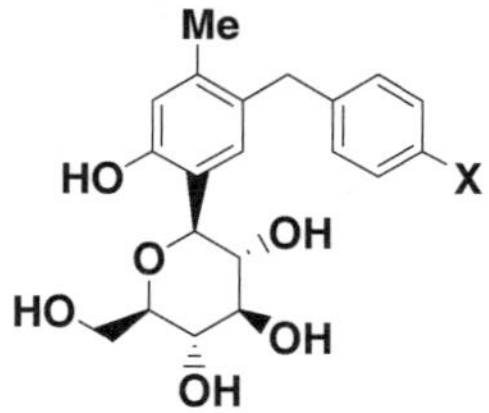			
2-11	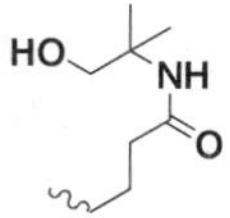	17	0.65		55
2-12	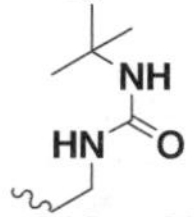	~70	1		55
2-13	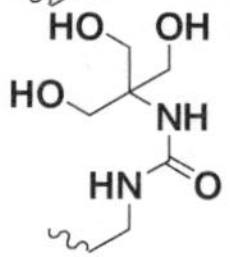	34	2		55
		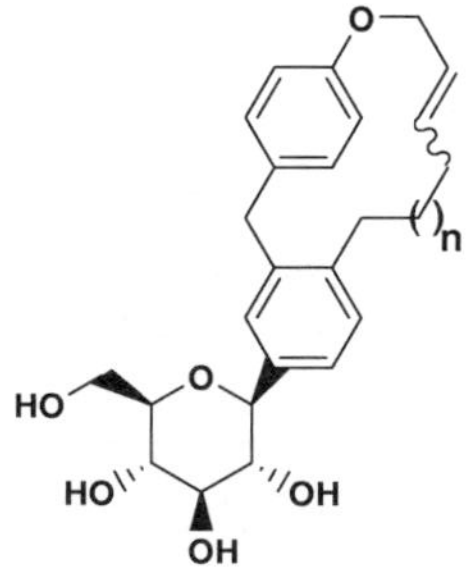			
2-14	$n = 0$	60			57
2-15	$n = 2$	103			57

however, when compared to **16**, the potency for these compounds was only 10–16% of **16**.

3.3.4.2.3 Replacement of the Central Aryl Ring

Alternative carbocylic and heterocyclic replacements of the central benzenoid aryl ring of **16** and **17** have been explored (Table 3.3). For example, Chugai disclosed a series of naphthalene *C*-glucosides for which the distal ring was either thiophene or benzthiophene. Incorporation of the naphthyl moiety provided no advantage over that conferred by a benzenoid central ring given the similar SGLT2 potencies of the two representative examples **3-1** and **3-2** to that of their respective counterparts **1-2** and **1-30**.[58]

Replacement of the central benzene ring of **16** with a bicyclic heterocycle has yielded potent SGLT2 inhibitors. Although disclosures from the Kissei group exemplified different points of attachment of a benzothiophene, the limited biological data suggests that only two of these orientations of the benzothiophene generated potent low nanomolar inhibitors. In one instance attachment of the glucose moiety at C2 and a *p*-methylbenzyl moiety at C7 yielded potent inhibitors exemplified by **3-3** (SGLT2 EC_{50} = 2 nM);[59] likewise, as exemplified by **3-4** (SGLT2 EC_{50} = 1.4 nM) glucosylation of C5 and benzylation at C3 generated an alternative series exhibiting comparable potency.[60] The 30-fold potency loss observed for **3-5** relative to **3-3** is consistent with the preference for a methano over an ethano spacer linking the central and distal rings first observed for diarylmethanes **16** and **17**. The failure to further develop these initial findings suggests that some property – ADME, SGLT1 selectivity, etc. – was lacking.

Indoles have also been shown to be acceptable as a central ring. The Kissei group disclosed that proper spatial presentation conducive to potent *in vitro* SGLT2 affinity was achieved with C3 glucosides of *N*-benzyl indoles.[61] For example, SGLT2 EC_{50} of **3-6** was 6 nM with 14-fold selectivity versus SGLT1. For this series normalized UGE values ranging from 50 to 1900 mg produced by a 1 mg/kg oral dose were highly dependent on the indole substitution. Only glucosides of 7-chloro-5-substituted *N*-benzylated indoles such as **3-7** generated a glucosuric response that exceeded 1000 mg. The Tanabe group reported that the reverse presentation entailing *N*-glucosides of C3 benzylated indoles such as **3-8** generated potent SGLT2 inhibitors.[62] No indication of SGLT1 selectivity was provided; however, when administered at 30 mg/kg, **3-8** induced a normalized UGE of >2400 mg. The Kissei group's disclosure confirmed the high SGLT2 affinity of *N*-glucosides of C3 benzylated indoles; moreover, the limited data suggests SGLT1 selectivity is promoted by substitution at C4 and reduced by C6 substituents (compare **3-9** and **3-10**).[63] These indole *N*-glucosides were not further pursued given the discouraging weak UGE of 112 mg obtained following oral administration of **3-9** at 1 mg/kg to rats.

Table 3.3 Potencies of diarylmethane *C*-glucosides with alterative central rings.

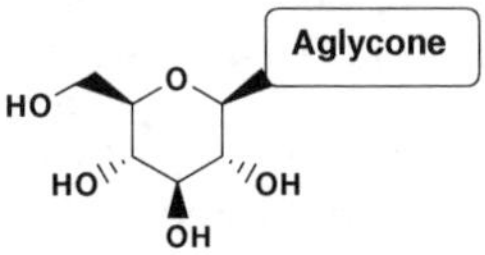

Compound ID	Aglycone	SGLT2 EC$_{50}$ (nM)	Selectivity versus SGLT1	24-h Rat UGE mg (Dose)	Ref.
3-1	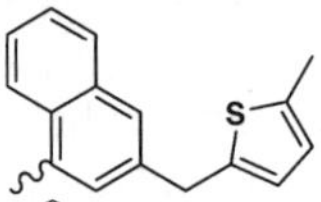	18			58
3-2	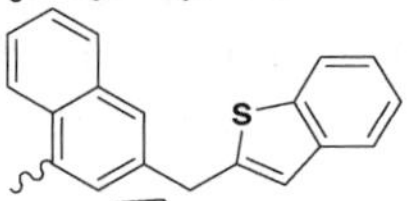	18			58
3-3	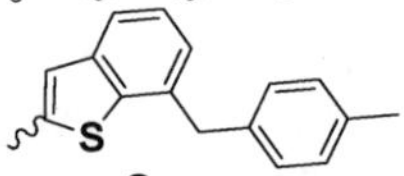	2			59
3-4	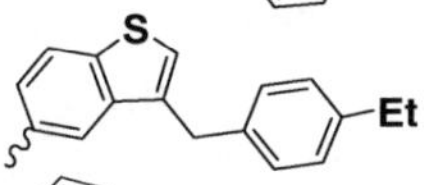	1.4			60
3-5	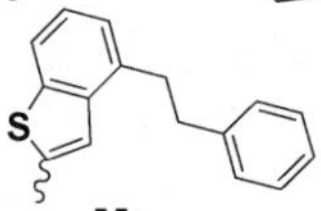	58	4		59
3-6	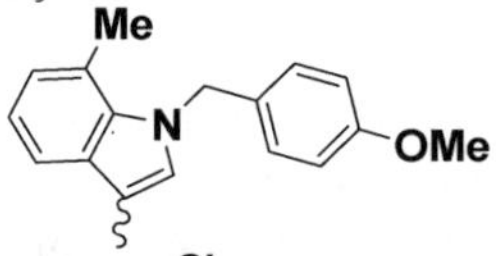	6	14		61
3-7	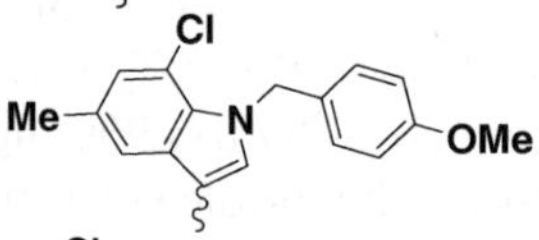			1693 (1)	61
3-8	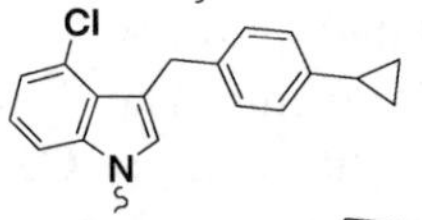	2.3		>2400 (30)	62
3-9	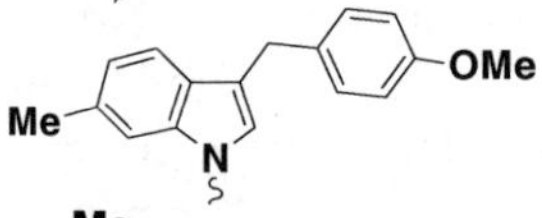	0.9	3500	112 (1)	63
3-10	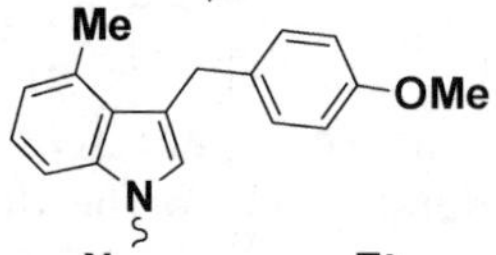	0.6	17		63
3-11	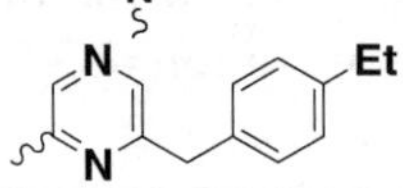				49

Table 3.3 (*Continued*)

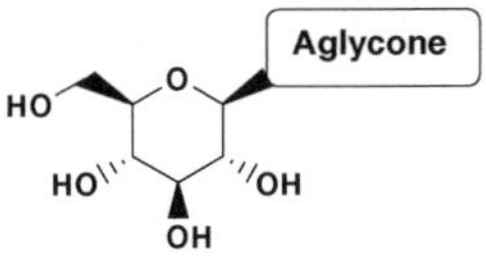

Compound ID	Aglycone	SGLT2 EC$_{50}$ (nM)	Selectivity versus SGLT1	24-h Rat UGE mg (Dose)	Ref.
3-12	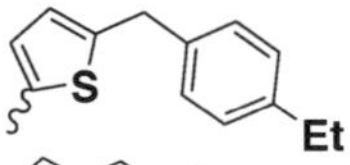 *(structure: thiophene–CH₂–phenyl–Et)*				38
3-13	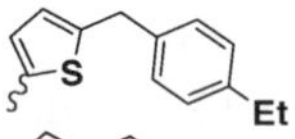 *(structure: thiophene–CH₂–phenyl–Et)*				38
3-14	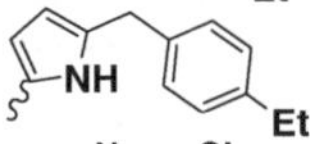 *(structure: pyrrole (NH)–CH₂–phenyl–Et)*				49
3-15	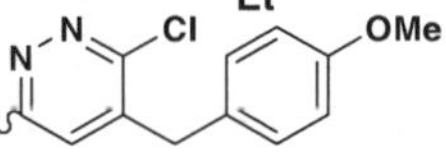 *(structure: chloropyridazine–CH₂–phenyl–OMe)*	610			64
1-16	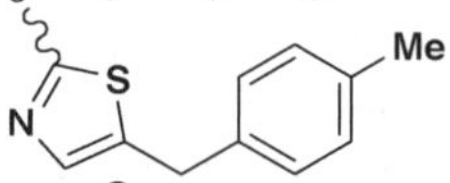 *(structure: thiazole–CH₂–phenyl–Me)*	4630			64
3-17	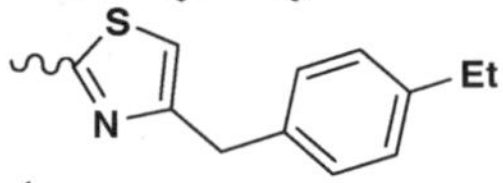 *(structure: thiazole–CH₂–phenyl–Et)*	>10 000			64
3-18	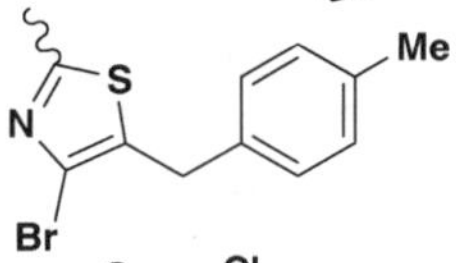 *(structure: bromothiazole–CH₂–phenyl–Me)*	1050			64
3-19	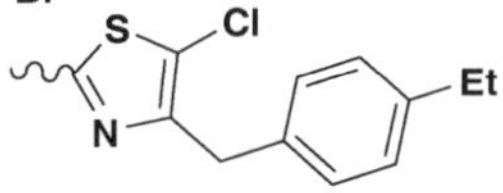 *(structure: chlorothiazole–CH₂–phenyl–Et)*	121			64

Several groups examined monocyclic heterocyclic replacements of the central aryl ring to no avail as judged by the failure to identify any clinical candidates. SGLT2 patents from several groups contain a low percentage of examples exemplified by **3-11**–**3-14** for which the monocyclic heterocyclic central ring is either an electron-deficient pyrazine[49] or an electron-rich thiophene[38] or pyrrole.[49] Although no biological data was provided for these particular examples, the paucity of such examples suggests that the *in vitro* profile for this structural modification was not encouraging.

Incorporation of polar monocyclic heterocycles consistently failed to generate potent viable SGLT2 inhibitors. For example, the Green Cross group reported that replacement of the 4-chlorophenyl moiety of **17** with a 4-chloropyridazinyl ring (**3-15**) decreased SGLT2 potency 1000-fold.[64] Likewise the SAR for thiazoles as a surrogate central ring was not encouraging.[64] C2 glycosylated thiazoles benzylated at C5, represented by **3-16**, exhibited weak SGLT activity whereas the isomeric C4 benzylated counterpart **3-17** were devoid of SGLT2 activity. Surprisingly C5 chlorination of thiazole **3-17** to generate **3-19** increased SGLT2 potency >80-fold, producing an ~120 nM inhibitor. In contrast, only a 4-fold potency increase ensued following C4 bromination of **3-16** to generate **3-18** (EC_{50} = 1050 nM). The authors postulated that the unfavorable dipolar or electronic effects of the polar pyridazine and thiazole rings were responsible for the diminution in SGLT2 potency compared to **17**.

3.3.4.2.4 Modification of the Glucoside Ring

Although many groups briefly explored modification of the glucose component of **17**, Taisho's, Pfizer's, and Lexicon's focus on this approach resulted in the identification of three clinical candidates. Alterations of the glucose moiety included either replacement of the pyranosyl oxygen or modification/replacement of a glycoside hydroxyl. Employing **17** and two close analogs **19** and **20**, for which the SGLT2 EC_{50} was also 1–2 nM, as reference compounds, the Pfizer group systematically evaluated the impact on SGLT2 affinity following changes in substitution for each of the five pyranosyl carbons of glucose.[65] The results summarized in Table 3.4 revealed that modifications of C5 produced relatively minor reductions in SGLT2 affinity.

Table 3.4 SAR for selected sugar modifications of diarylmethane *C*-glucosides.[65]

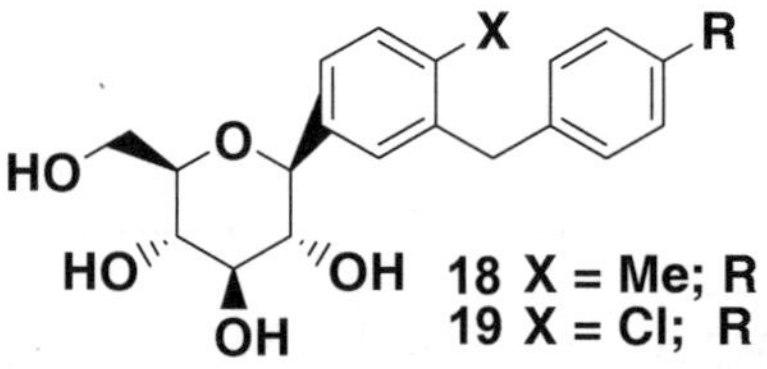

Glucose Carbon Modified	Structural Change	Fold Decrease in SGLT2 Potency	Fold Decrease in SGLT1 Potency
C2	CHOH → CH$_2$	>10 000	
C3	CHOH → CH$_2$	1300	
C4	CHOH → CH$_2$	21	
C3	(*R*) CHOH → (*S*) CHOH	2400	
C4	(*S*) CHOH → (*R*) CHOH	90	
C2	(*R*) CHOH → (*R*) CHOMe	140	
C4	(*S*) CHOH → (*S*) CHOMe	180	
C5	(*R*) CHCH$_2$OH → (*R*) CHMe	2	3 × increase
C5	(*R*) CHCH$_2$OH → (*R*) CHCH$_2$F	5	
C5	(*R*) CHCH$_2$OH → C(CH$_2$OH)$_2$	59	28
C1	(*S*) CHAr → (*R*) C(OMe)Ar	670	

In contrast, changes at C4 reduced potency $\sim$100-fold, whereas changes at C2 and C3 were especially detrimental. The finding that introduction of a second hydroxymethyl substituent at C5 reduced affinity only 59-fold prompted an exploration of various spirocycles attached at C5.

The discovery that the spiro cyclopropyl derivative **5-1** was a 3 nM SGLT2 inhibitor with 63-fold selectivity versus SGLT1 encouraged the preparation and evaluation of additional four- and five-membered spirocycles.[66] Oxetanes **5-2** and **5-3** exhibited comparable potency to **5-1** with higher selectivity. SGLT2 affinity was only slightly diminished for the cyclic sulfone **5-4** but was essentially abolished for azetidine **5-5**. Ring expansion of the oxetane moiety to a tetrahydrofurane **5-6** resulted in 15–20-fold loss of potency and selectivity. Rat PK studies with **5-1** revealed that the $t_{1/2}$ was $\sim$8 times shorter than that of **17** reflecting the facile phase 1 metabolic conversion of the ethyl side chain to the corresponding methyl ketone. Most unexpectedly, despite this rapid clearance (112 mL/min/kg), administration of 10 mg/kg of **5-1** to normal Sprague Dawley rats resulted in 24-h UGE of $\sim$2500 mg of glucose. Since a glucosuric response of this duration could not have been caused by the rapidly cleared **5-1**, the result was attributed to formation of an active ketone metabolite.

To avoid metabolism potentially confounding the subsequent UGE studies, oxetane **5-3** rather than **5-2** was utilized in a subsequent UGE. A dose-dependent increase in 24-h UGE was observed which did not exceed 80% of the $\sim$2500 mg obtained with 10 mg/kg of **17** even when **5-3** was administered at 60 mg/kg. PK studies subsequently established that, even though clearance (49 min/min/kg) was improved relative to that of **5-1**, the half-life in rat of **5-3** was 1.2 h versus 4.4 h for **17**. Enhanced phase 2 metabolism of **5-3** resulting in glucuronidation of a sugar hydroxyl (presumably C3 by analogy to the glucuronide generated from **17**) led to rapid clearance of **5-3**. As a consequence sufficiently high plasma concentrations of **5-3** could not be maintained to insure maximum SGLT2 inhibition over 24 h even when administered at 60 mg/kg.

Subsequent *in vitro* investigations with **5-1** and **5-3** revealed that the rapid clearance was due to phase 2 metabolism; however, clearance by hepatocytes was significantly slower for azetidine **5-5** than for **5-1** and **5-3** despite all three compounds exhibiting comparable lipophilicity. The failure of lipophilicity to correlate with clearance rate was underscored by low hepatocyte clearance rates observed for **17** and the two close structural analogs despite their lipophilicity being similar to that of **5-1** and **5-3**. The Pfizer group attributed these differences in clearance to a structural dependent modulation of the rate of glucuronidation arising from the presence or absence of an H-bond donor or polar functionality (CH_2OH or CH_2NH) attached to C5.[67]

To maintain a hydroxymethyl attached to C5 and capitalize on the tolerance for an additional C5 substituent, a methano-oxo bridge was introduced linking C1 and C5, thereby generating a new series of potent selective bicyclic SGLT2 inhibitors represented by structures **5-7**–**5-11**. **5-7**, the bridged counterpart of **17**, exhibited comparable SGLT2 potency ($EC_{50} = 0.9$ nM) and 2200-fold selectivity versus SGLT1. Efforts to improve upon **5-7** proved to be

Table 3.5 SAR for glucose modification on potency of diarylmethane-containing glycosides.

Compound ID	*X*	*SGLT2 EC_{50} (nM)*	*Selectivity versus SGLT1*	*24-h Rat UGE mg (Dose)*	*Ref.*
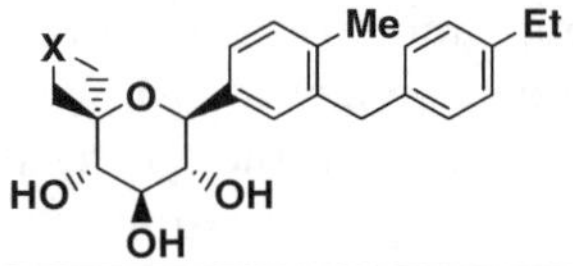					
5-1	Bond	3	63	~2400 (10)	65
5-2	O	3.4	440		66
5-3	O; Et = MeO	6.6	230	~1000 (10)	66
5-4	SO_2	14	>700		66
5-5	NH	5100	>2		66
5-6	CH_2O	53	20		65
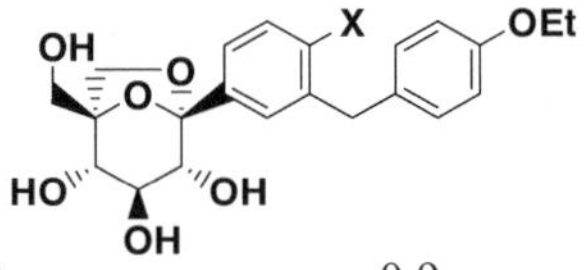					
5-7	Cl (PF-04971729)	0.9	2200	2145 (10)	67
5-8	Me	1.1	860	2045 (10)	67
5-9	F	2.9	~1000		68
5-10	CN	7.3	>1400		68
5-11	Cl; Et = 3-oxetane	16.3	>4000		68
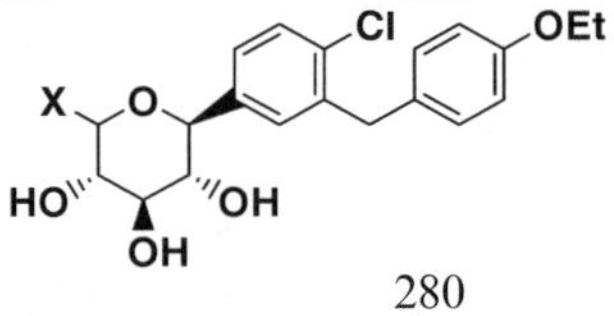					
5-12	X = 1:1 α/β OH	280			69
5-13	X = 1:20 α/β OMe	14	134		69
5-14	X = 20:1 α/β OMe	4800			69
5-15	X = 1:20 α/β OBn	620			69
5-16	X = $O(CH_2)_2OH$	36			69
5-17	X = $O(CH_2)_2NMe_2$	3600			69
5-18	X = β SMe (LX4211)	1.8	20		70,71
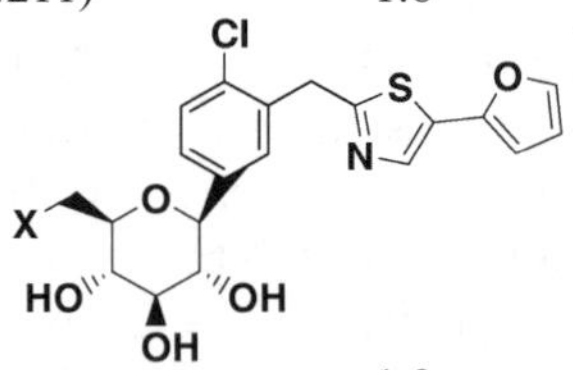					
2-13	X = CH_2OH	1.8			72
5-19	X = CH(Me)OH	13.5			72
5-20	X = CH_2OEt	69			72
5-21	X = CH_2SMe	19			72

Table 3.5 (*Continued*)

Compound ID	*X*	*SGLT2 EC_{50} (nM)*	*Selectivity versus SGLT1*	*24-h Rat UGE mg (Dose)*	*Ref.*
	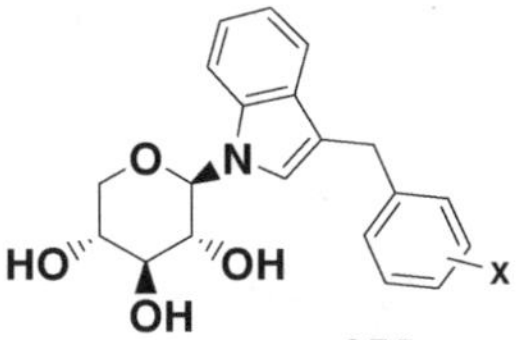				
5-22	X = 4-OMe	275	0.8		73
5-23	X = H	865			73
5-24	X = 2-OMe	12,915			73
5-25	X = 3-OMe	893			73
5-26	X = 4-cyclopropyl	161	1.3	19 (10 mg/kg)	73
	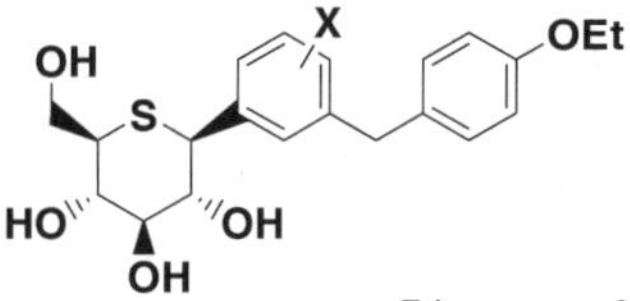				
5-27	H	74	355		74
5-28	4-F	9	847		74
5-29	4-Me	2.3	293		74
5-30	4-Cl	1.8	684		74
5-31	4-MeO	13	42		74
5-32	6-OMe	38	2640		74
5-33	6-OH	17	232		74
5-34	4,6-OMe_2	11	395		74
5-35	4-Me-6-OMe (luseogliflozin)	2.3	1770		74
5-36	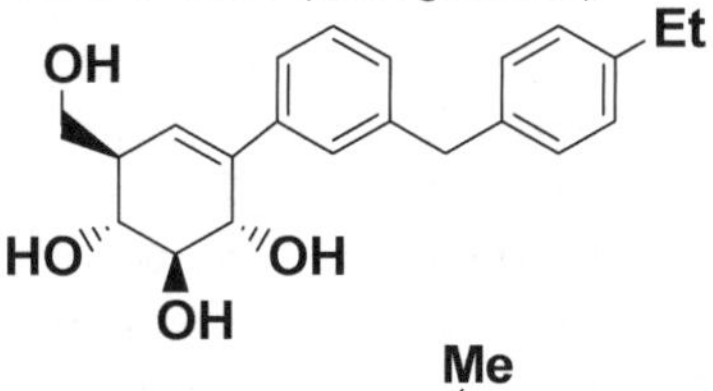	6.6			75
5-37	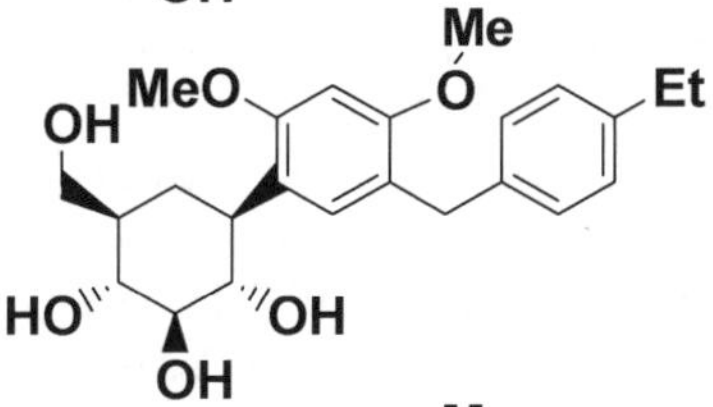	7.4			75
5-38	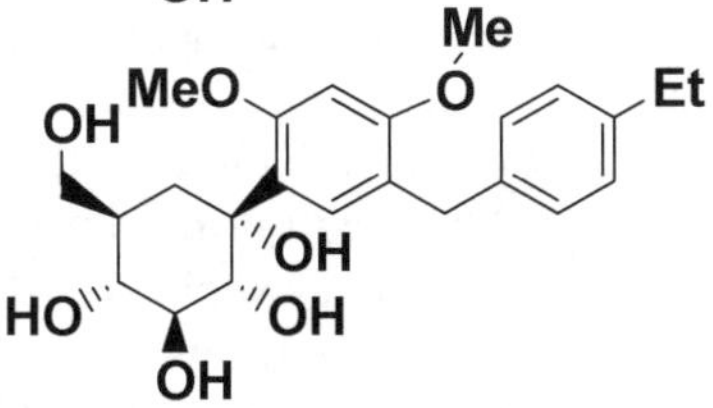	5			75

Table 3.5 (*Continued*)

Compound ID	X	SGLT2 EC$_{50}$ (nM)	Selectivity versus SGLT1	24-h Rat UGE mg (Dose)	Ref.
5-39	H	690	19		76
5-40	Cyclopropyl	11	510		76
5-41	OMe	42	220		76

unsuccessful. The progressive decrease in potency upon replacement of 4-chloro substituent with methyl (**5-8**; 1.1 nM), fluorine (**5-9**; 2.9 nM), or cyano (**5-10**; 7.3 nM) was consistent with prior SAR patterns for these C4 substituents on the central aryl ring.[67,68] It would appear that incorporation of the glucose moiety in the dioxabicyclo[3.2.1]octane framework greatly increases sensitivity to the steric demands of *para* substituents of the distal aryl ring. For example, replacement of the *p*-ethoxy group of **17** with a 3-oxytetrahydrofuran moiety generating **2-1** produced a 3-fold decrease in SGLT2 potency whereas nearly a 20-fold loss of potency resulted from conversion of **5-7** to **5-11**. **5-7** completed Phase 1 and 2 clinical trials as PF-04971729, before entering Phase 3 as ertugliflozin.

The Lexicon group also exploited the tolerance of the SGLT2 transporter for modifications of the C5 pyranosyl carbon of **17**.[69] As shown in Table 3.5, replacement of the C5 pyranosyl hydroxymethyl substituent of **17** with hydroxyl **5-12** reduced SGLT2 affinity ~250-fold; however, the potency loss was greatly attenuated upon conversion of **5-12** to the β-methyl ether **5-13** (EC$_{50}$ = 14 nM). **5-13** was 340-fold more potent than the corresponding α-isomer **5-14** (EC$_{50}$ = 4800 nM). The SAR for the more active β-alkoxy anomers of these C5 L-xylose ethers was relatively flat until the alkyl moiety became benzyl **5-15** or larger whereupon affinity substantially diminished. Potency was diminished 2–3-fold upon incorporating alkyl ethers containing a H-bond donor such as 2-hydroxyethoxy **5-16**; whereas, incorporation of a basic H-bond acceptor such as 2-diethylaminoethoxy **5-17** reduced potency >3000-fold. Subsequent SAR exploration of the aglycone confirmed that the same pattern reported for *C*-glucosides of diarylmethanes held for L-xylosides of diarylmethanes. The preferred spacer joining the two aryl rings was a methylene; a C4 substituent such as methyl or chlorine enhanced affinity 20–40-fold; and *para* substitution of the distal aryl by small nonpolar substituents such as ethoxy was preferred over polar groups such as thiazole or thiocarbamates. The group also confirmed the findings of others that modification or inversion of the remaining sugar hydroxyls would diminish affinity 40–1000-fold. When administered p.o. to DIO C-57 mice at 10–100 mg/kg, **5-13** induced dose-dependent glucosuria over the subsequent 17 h. Although **5-13** was extensively featured in a communication, a 2009 patent application, describing the preparation of crystalline **5-18** on a kg scale, indicated **5-18** (SGLT2 EC_{50} 1.8 nM) was a compound of considerable interest.[71] Subsequently the Lexicon group selected **5-18** to be the clinical candidate LX4211 despite the fact that LX4211 exhibited only 20-fold selectivity versus SGLT1.[70] Clinical trials with LX4211 have progressed to Phase 2B.

The Green Cross group attempted unsuccessfully to enhance potency of **2-13** by modification or replacement of the glucose C6 hydroxymethyl.[72] As exemplified by **5-19**, potency decreased 10–30-fold depending on the additional alkyl group attached to C6 and the chirality of the resulting secondary alcohol; the corresponding tertiary alcohols also exhibited similar reduced potency. Conversion of the hydroxymethyl to a series of alkyl ethers such as **5-20** diminished potency 10–40-fold. Similar reductions in potency ensued upon conversion to thioethers such as **5-21** or fluoro methyl. Replacement of the C6 hydroxyl with an alkyl sulfone or triazole was especially unfavorable causing >100-fold diminished potency.

Although many groups reported that the hydroxymethyl substituent of the C5 pyranosyl carbon can be replaced with alkyl or alkoxy substituents with varying degrees of potency loss, a group in Taiwan revealed that replacement with hydrogen thereby converting C5 to a methylene, as exemplified by indole N-xylosides **5-22** – **5-26**, is especially unfavorable. Under assay conditions in which **17** exhibited an EC50 of 3 nM and 161 fold selectivity versus SGLT1, comparison of **5-22** (SGLT2 EC_{50} = 275 nM) to its close *N*-glucoside counterpart **3-9** revealed that deletion of the C6 hydroxymethyl not only reduced potency ∼300-fold but also abolished selectivity versus SGLT1.[73] Although non-selective, these indole *N*-xylosides appear to bind in the same fashion as do diarylmethane *C*-glucosides since the SAR for the distal ring substitution is similar. Relative to the unsubstituted **5-23**, potency was diminished 15 times for the *ortho* chloro isomer **5-24**, was unchanged for the *meta* isomer **5-25** and was enhanced 3 times for the *para* isomer **5-22**. The glucosuric capability of this series appears greatly diminished as exemplified by the meager response induced by the slightly more potent cyclopropyl counterpart **5-26** which, following oral administration at 10 and 50 mg/kg to Sprague Dawley rats, produced respective 24-h UGE responses of only 19 and 700 mg.

Replacements of the pyranose oxygen of **16** and **17** have also been explored. Following the Bristol-Myers Squibb disclosures of *C*-glucosides, the Taisho group shifted their focus from thiapyranoside *O*-glucosides to thiapyranoside *C*-glucoside counterparts of **17**.[74] Comparison of the thiapyranoside **5-27** to its *O*-pyranoside analog **16** reveals less than a 7-fold difference in SGLT2 potency. Not surprisingly, aspects of the SAR for the thiapyranosides were similar to the *O*-pyranoside counterparts. Modulation of SGLT2 affinity was best achieved by appropriate substitution of C4 and C4′. SAR exploration quickly established that *para* substituents larger than *i*-Pr on the distal ring adversely impacted potency. For the central aryl ring the C4 substituent effect was steric not electronic in origin The SAR response was bell-shaped: the comparably sized methyl **5-29** and chlorine **5-30** substituents produced a maximum 30–40-fold increase in SGLT2 potency over that of the unsubstituted reference **5-27**; whereas a 6–8-fold enhancement was conferred by the less optimally sized C4 substituents fluorine **5-28** or methoxyl **5-31**. SGLT1 affinity increased as the C4 substituent progressively increased in size. As a consequence of the differing responses of SGLT1 and SGLT2 affinity to the substituent size, the correlation of substituent with selectivity was more complicated. Selectivity versus SGLT1 was a maximum with fluorine **5-28** at C4 and progressively diminished for larger C4 substituents. Alkoxy substituents at C6 significantly enhanced selectivity primarily by decreasing SGLT1 affinity as exemplified by the 7-fold selectivity increase provided by a C6 methoxyl without appreciable alteration of SGLT2 potency (compare **5-32** to **5-27**). In contrast, a C6 hydroxyl increased both SGLT1 and SGLT2 affinity nearly commensurately and consequently did not significantly alter selectivity (compare **5-33** to **5-27**).

Incorporation of methoxyls at both C4 and C6 (**5-34**) did not produce an additive effect but rather maintained the potency effect of C4 substitution and selectivity conferred by C6 substituent. The above SAR for thiapyranoside *C*-glucosides substituted at C4 and C6 of the central ring and C4′ of the distal ring enabled modulation of both SGLT1 and SGLT2 potencies by appropriate mixing and matching of substituents. Having established the benefits of the central aryl substituted with methoxy at C6, methyl or chlorine at C4, and the distal ring substituted with methyl, ethyl, propyl, methoxy, or ethoxy, the various combinations were evaluated. Ultimately, **5-35** was selected over other promising analogs for clinical trials based on the glucose AUC reduction achieved following oral administration to diabetic streptozotocin rats, greater metabolic stability when incubated with human hepatacytes, and a free fraction in human serum such that adequate drug levels could access the proximal tubules via glomerular filtration. Initially referred to as TS-071 in early clinical trials but more recently as luseogliflozin, **5-35** has progressed to phase 3.

Replacement of the pyranose oxygen with a methylene was investigated by the Chugai group.[75] The finding that SGLT2 affinities disclosed for **5-36** (6.6 nM) and **5-37** (7.4 nM) approximated the values reported for structurally similar oxa- and thia-pyranose analogs **16** (10 nM) and **5-34** (11 nM) suggests that the neither pyranose heteroatom engages in a critical interaction with the SGLT2 transporter. Also the observation that no significant change in affinity resulted upon hydroxylation of **5-36** to generate **5-38** (5 nM) is not consistent with the findings from Bristol-Myers Squibb and Pfizer that SGLT2 affinity is modulated by alpha C1 substituents.

In addition, the corresponding series of 5a-carba-glucopyranosides **5-39**–**5-41** was explored as a means to enhance metabolic instability by blocking the α-glucosidase cleavage observed with *O*-aryl glucosides such as phlorizin **1** and *o*-benzylphenol *O*-glucosides such as **4**.[75,76] SGLT2 potency and selectivity of the unsubstituted parent carba-glucopyranose **5-39** (EC_{50} = 690 nM ; 19-fold selective) was enhanced upon C4 substitution of the distal ring with small lipophilic groups; no SAR exploration has been reported for substitution of the central ring.[76] The highest disclosed SGLT2 affinity (11 nM) and selectivity (510-fold) was obtained with a cyclopropyl substituent **5-40**. Both potency and selectivity of the sergliflozin counterpart **5-41** were less than that measured for sergliflozin A **4a** (EC_{50} of 42 nM versus 4.8; 220-fold selective versus 480-fold), suggesting that replacement of the pyranose oxygen with a methylene reduced SGLT2 binding more than SGLT1 binding. *In vivo* studies conducted with hyperglycemic db/db mice following oral administration of **5-41** and the sergliflozin ethyl carbonate pro-drug **4b** at 100 mg/kg revealed that comparable ~50% glycemic reductions were achieved at 2 h post dose, despite the ~10-fold *in vitro* potency differential. Moreover, as a consequence of a 3-fold longer half-life (2.7 h versus 0.8 h), **5-41** maintained glycemic levels below 200 mg/dL for 6 h versus 2 h for serglifozin **4a** in this study. PK studies revealed that the longer half-life was not due to enhanced metabolic stability of **5-41** since the clearance rates of **4a** and **5-41** were 2.8 and 3.3 mL/min/kg but rather reflected the greater volume of distribution (12.6 versus 3.3 L/kg) of the more lipophilic **5-41**.

3.3.4.2.5 Structurally Constrained C-Glucosides

Three groups investigated the potential of analogs of **16** for which the central aryl ring is linked by an additional bond to the glucose moiety. Crystallographic data of complexes of **16** and **17** revealed that the central aryl ring was nearly orthogonal to the plane of the glucose moiety.[77] This realization prompted the Chugai group and the Egret/Chengdu group to synthesize and evaluate the 1,1-anhydro-hydroxymethyl -5-benzylphenylgluco-pyranoses. Note: The SARs from the two groups contain some compounds in common, for which the *in vitro* profiling data on occasion does differ by a factor of five. The Egret/Chengdu group reported **6-1** to be inactive, a finding that was consistent with prior BMS disclosures that substitution of C2 of the central aryl ring greatly reduced SGLT2 potency.[78] Consequently, no additional analogs of **6-1** were pursued. Very limited data has been disclosed regarding the Chugai group's SAR efforts pertaining to the series of 1,1-anhydro-hydroxymethyl-5-benzylphenylgluco-pyranose constrained analogs of **16** such as **6-2**. a 4 nM inhibitor that is a 2-fold more potent counterpart of **16**.[79] Phase2 clinical trials were conducted with **6-2** as RG7201; phase 3 are have begun with **6-2** as CSG 452 or most recently as its trade name tofogliflozin.[80] The second disclosure from Chugai concerned a set of thirty similarly constrained analogs of dapagliflozin **17** such as **6-4**–**6-7** for which SGLT2 EC_{50} values ranged from 1-4 nM depending on the C4 and C4′ small lipophilic substituents of the two aryl rings.[81] Oral administration of **6-4** at 0.3 mg/kg to nine to eleven week-old db/db mice reduced blood glucose levels by 52% and 25% at 6 and 24 h post dose respectively relative to that of controls.

Table 3.6   Potencies of conformationally constrained diarylmethane *C*-glucosides.

Compound ID	X	Y	SGLT2 EC$_{50}$ (nM)	Selectivity versus SGLT1	24-h Rat UGE mg (Dose)	Ref.
6-1		0% at 1 μM				78
6-2	H	4-Et (CSG 452; tofoglifozin)	4			79
6-3	H	4-OEt	71	141		78
6-4	Cl	4-OEt	1.5	283		81
			6.5	200		82
6-5	Cl	4-Et	1.3	155		81
			6.6	91		78
6-6	Cl	4-OCF$_3$	1.9			81
			0.3	10 333		82
6-7	Cl	4-*i*-Pr	1.7	596		81
			7.1	352		82
6-8	Cl	4-*t*-Bu	13	742		82

Table 3.6 (*Continued*)

Compound ID	X	Y	SGLT2 EC$_{50}$ (nM)	Selectivity versus SGLT1	24-h Rat UGE mg (Dose)	Ref.
6-9	Cl	CF$_2$	7	229		82
6-10	Cl		1.3	4615		82
6-11	F	3.8	184	1220 (1)		82
6-12	MeO	5.8	466			82
6-13	CH$_2$	0.3	18 667	370 (1 mg/kg) 1470 (9 mg/kg)		83
6-14	CHOH	40	3450			83
6-15	CH$_2$CH$_2$	33	>6000			83

Table 3.6 (*Continued*)

Compound ID	X	Y	SGLT2 EC$_{50}$ (nM)	Selectivity versus SGLT1	24-h Rat UGE mg (Dose)	Ref.
6-16	2		1.3			84
6-17	3		3			84
6-18	4		2.3			84
6-19	5		0.9	226 (1 mg/kg)		84
6-20	6		3.6			84
6-21	1	allyl	52	1700	830 (1 mg/kg)	85
6-22	1	propargyl	0.3	>3300		85
6-23	2	allyl	12	2500	1170	85
6-24	~O-allyl		1.2			

Table 3.7 Development status and profile of SGLT2 inhibitors that entered clinical trials.

SGLT2 Inhibitor	Alternative Name	SGLT2 EC_{50} (nM)	Selectivity versus SGLT1	Clinical Half-life	Preferred Clinical dose(s)	Development Status
Dapagliflozin **17**	BMS-512548	1.2*	1200*	15–17 h	5; 10 mg	FDA Review
Canagliflozin **1-8**		2.7*	262*	~ 10 h	100; 300 mg	Phase 3
Ipragliflozin **1-29**	ASP1941	5.3*	566*	12–15 h	100; 300 mg	Phase 3
Luseogliflozin **5-35**	TS 071	2.3*	3990*	9–13 h	2.5; 5 mg	Phase 3
Empagliflozin **2-1**	BI 10773	3.1*	2700*		1–50 mg	Phase 3
Tofogliflozin **6-2**	CSG 452; RO 07201	6*	1900*		Not disclosed	Phase 3
EGT 1474 **2-5**	EGT 1442	2.3	2400		5–50 mg	Phase 2
LX4211		1.8	20		300 mg	Phase 2
PF-04971729 **5-7**		0.9	2200		5–25 mg	Phase 2
T-1095 **3b**		4.4*	59*		Not disclosed	Halted phase2
Serglifozin **4b**		7.5*	280*	0.5–1 h	500:1000 mg tid	Halted phase2
Remogliflozin etabonate **5b**		12	542*	1.5 h	100; 1000 mg bid	Halted phase2

*Values from Reference 53

The SGLT2 EC_{50} SAR was similar to that of analogs of **17**; however, the SGLT1 selectivities, which ranged from 100 to 500, were somewhat reduced suggesting that the constraints imposed by the anhydro pyranose structure did not impact SGLT2 affinity but enhanced SGLT1 affinity with the result being diminished SGLT1 selectivity.

The Egret/Chengdu group, having been encouraged by the activity of **6-3**, also synthesized **6-4** to better assess the potential of this series.[82] (In the Egret/Chengdu assay SGLT2 EC_{50} and selectivity for **17** were 6.7 nM and 132-fold rather than 1.1 nM and 1200-fold reported by Bristol-Myers and others.) This promising similarity in affinity for **6-4** and **17** prompted an extensive SAR evaluation of the distal ring. Despite the constraints imposed by the anhydro pyranose structure, the SAR appeared to be similar to that previously reported for other series of SGLT2 inhibitors. Substituents at C3′ and especially C2′ of the distal ring were unfavorable; C4′alkyl substituents such as ethyl **6-5** provided no *in vitro* potency advantage over **6-4**; however, stericly more demanding alkyl substituents such as *t*-butyl **6-8** increased selectivity $\sim$5-fold over that of **6-5**. The profile for *para*-trifluoromethoxyl analog **6-6** differed for the two groups; however, the expectation is that the Egret/Chengdu values are in error given the divergence of their values for **6-6** from other analogs.

Having established that the spiro[isobenzofuan]pyranose framework maintained SGLT2 potency, further modifications encompassing the C6 hydroxyl, the bridging diaryl methylene, and the oxa methano ketal bridge were explored by the Egret/Chengdu group in hopes of achieving greater potency.[82] Contrary to the Bristol-Myers Squibb findings, substitution of the diaryl methylene bridge was tolerated. Potency was unchanged upon replacement of the methylene bridge linking the two aryls with CF_2 **6-9**. More surprisingly, incorporation of the bridging carbon into a 1,1-disubstituted cyclopropane **6-10** enhanced potency 6-fold. SGLT2 potency was also essentially unchanged upon conversion of the C6 hydroxyl to fluorine **6-11** or methyl ether **6-12** (EC_{50} = 3.8 and 5.8 nM respectively) suggesting H-bonding donor capacity of this substituent is not essential. Activity was significantly reduced upon conversion to an ester or amine.

The *in vitro* profile for this constrained series suggested that restriction of the conformational flexibility of the acyclic diarylmethane was advantageous; however, the *in vivo* characterization regarding the glucosuric potential of this series disclosed by Egret/Chengdu group was disappointing. For example, an UGE of only 1220 mg of glucose was obtained with **6-11** administered orally to Sprague Dawley rats at 1 mg/kg dose.[82] Severe diarrhea, consistent with SGLT1 inhibition, ensued 6 h after oral administration of a single 25 mg/kg dose of **6-5** to CD1 mice. Given the similarities of the human SGLT1/2 *in vitro* profiles of **6-5** and **17** and the absence of similar GI disturbances being reported with **17**, this finding raised concerns that a metabolite of **6-5** might be responsible. However, the potency and selectivity cited above are for human SGLT2 and SGLT1, not for the rodent transporters. If the species effect for

SGLT1 selectivity of **6-5** was as pronounced as was observed for **18**, the observed diarrhea could have been due to the SGLT1 mediated activity of **6-5**.

Due to concerns regarding potential metabolite liabilities arising from the ketal moiety of **6-5**, alternative means to link C1 of glucose with C6′ of the central ring were pursued.[83] Replacement of the oxomethylene bridge with an ethano group generated **6-13**, a highly potent SGLT2 inhibitor (EC$_{50}$ = 0.3 nM) with essentially no SGLT1 activity. This high potency was diminished ~100-fold upon homologation of the ethano bridge of **6-13** to the propano bridge of **6-15**. Likewise, hydroxylation of the ethano bridge at the α-benzyl carbon to generate **6-14** reduced SGLT2 potency 100-fold. When orally administered to normal Sprague Dawley rats at doses from 0.33 to 27 mg/kg, **6-13** produced dose-dependent 24 h UGE ranging from 110 to 1470 mg after normalization to 200 g body weight. Surprisingly despite the *in vitro* potency of **6-13**, doses of 9 mg/kg or higher were required to reach the UGE response plateau. Rat hepatocyte clearance studies revealed clearance of **6-13** was 33 ml/min/kg as compared to 2 ml/min/kg for **17** thereby explaining why 3 mg/kg of **6-13** were required to produce a glucosuric response comparable to that induced by 0.1 mg/kg of **17**.

The Green Cross group explored an alternative means to incorporate **17** into a macrocycle. In this instance a series of 10–14 membered macrocyclic ethers **6-16**–**6-20** of **17** was constructed connecting the glucose C6 hydroxyl to C6 of the central aryl ring of **17**.[84] Under assay conditions for which the EC$_{50}$ of **17** was 0.5 nM, SGLT2 EC$_{50}$ ranged from 0.9 to 3.6 nM for this set. However, despite the *in vitro* potency similarity for **6-19** and **17**, the *in vivo* potency of two compounds greatly differed. Oral administration of 1 mg/kg of **17** and **6-19** to healthy Sprague Dawley rats induced normalized 24 h UGE values of 1648 and 226 mg respectively. This difference probably is due to a much less favorable PK profile of **6-19**. If, as postulated by the Pfizer group, the H-bond donor interaction involving the C5 hydroxymethyl of glucose is a critical determinate preventing extensive phase 2 metabolism resulting in rapid clearance, then clearance of **6-19** and analogs would be expected to be rapid since this hydroxymethyl had been incorporated into the macrocycle. Furthermore, this analysis predicts that introduction of a second hydroxymethyl at C5 should rectify the clearance issue analogous to the Pfizer approach leading to **5-7**.

Two groups explored whether non-constrained acyclic C6 substituents attached to the central ring of analogs of **17** would emulate the advantages conferred by the anhydropyranose framework for **6-2**. The Chengdu/Egret groups evaluated a number of allyl and alkyl ethers of **18**.[85] The results illustrate the unpredictability of this approach as minor steric differences can produce markedly differences in affinity. In general, this approach yielded potent selective SGLT2 inhibitors as exemplified by **6-21**, **6-22**, and **6-23**. The 170-fold increase in potency as a result of replacing an extended allyl ether **6-21** with a propargyl ether **6-22** attests to the steric sensitivity for C6 substitution. Comparison to **18** (EC$_{50}$ = 1 nM) revealed only **6-22** enhanced potency. When administered at 1 mg/kg orally to Sprague Dawley rats, the resulting normalized UGE produced by **6-22** and **6-23** was respectively 830 and 1170 mg of glucose, responses that were 42% and 56% respectively of that generated by **18**. Efforts by the Green Cross group to improve upon the activity of **1-13** by introduction of a C6 fluorine, hydroxyl, or corresponding ethers such as **6-24** failed to increase potency beyond the 0.7 nM achieved with **1-13**.[86]

3.3.5 Non-Glycoside Containing SGLT2 Inhibitors

In addition to the above discussed glucoside-based SGLT2 inhibitors, three groups have reported that screening of compound collections identified diverse non-glycoside SGLT2 inhibitors (Figure 3.7). The Merck GMBH inhibitors **7-1**–**7-4** were the first examples of a successful screen; however, no subsequent reports have issued following the initial disclosure.[87–90] In all instances the pharmacophore and the spatial presentation of the appendages remained the same; however, for each series a different bicyclic heterocycle core – indolizine, benzoimidazole, imidazo[1,2]pyridine, and imidazo[1,2]pyrimidine – was employed. No specific SGLT1 and SGLT2 IC_{50} values were provided but instead a range of 10 to 1000 nM for SGLT2 and >10 000 nM for SGLT1. No *in vivo* data was provided.

A group at Amylin described their efforts to increase potency and selectivity of a benzooxazinone screening hit which culminated in the identification of compounds such as **7-5**.[91] Compound **7-5** was reported to be a selective competitive SGLT2 inhibitor exhibiting SGLT1 and SGLT2 respective potencies of 9140 nM and 9 nM. Under the Amgen assay conditions **7-5** exhibited comparable SGLT2 potency and greater SGLT1 selectivity than **17**. Unfortunately, compound **7-5** lacked adequate solubility and metabolic stability to undertake *in vivo* studies. Further

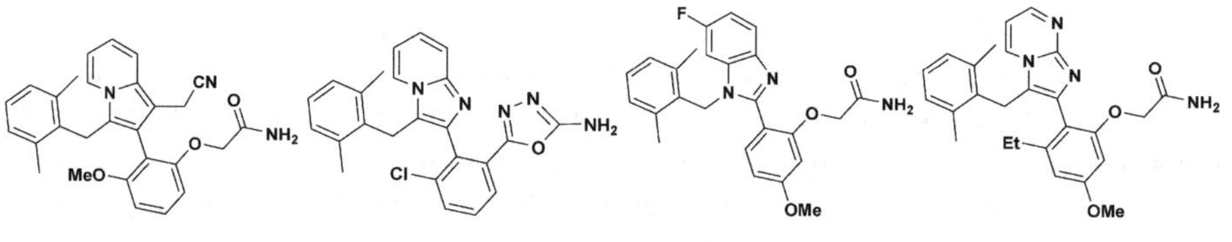

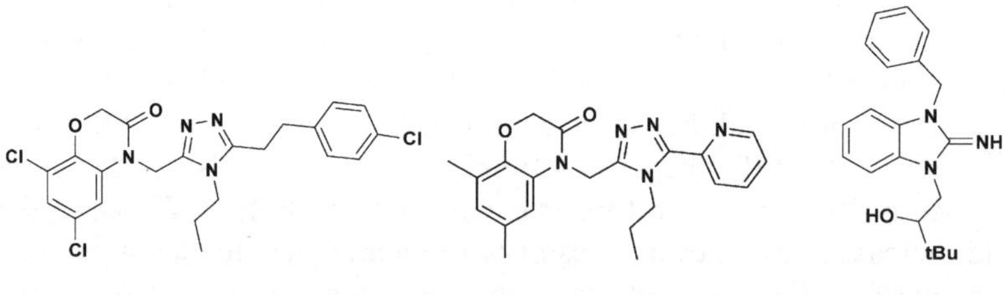

Figure 3.7   Representative examples of non-glucoside containing SGLT2 inhibitors.

SAR exploration led to somewhat metabolically more stable SGLT2 inhibitors such as **7-6** exhibiting SGLT2 EC_{50} of 12 nM with 2500-fold selectivity versus SGLT1.[92] Solubility and metabolic stability were both sufficiently increased that low micromolar rat plasma concentrations could be detected following subcutaneous administration of **7-6** at 100 mg/kg; however, the glucosuric response induced by **7-6** has yet to be disclosed.

Ligand-based virtual screening has been successfully implemented to identify novel non-glycoside SGLT2 inhibitors.[93] This approach utilized a consensus pharmacophore generated from a set of potent SGLT2 inhibitors to screen three commercial databases. Subsequent *in vitro* evaluation of the top hits revealed that compound **7-7** inhibited SGLT2 with an EC_{50} of 3800 nM. An SAR study based on **7-7** provided additional validation.

3.4 Clinical Studies with SGLT2 Inhibitors

A number of reviews have been published concerning clinical studies with SGLT2 studies.[94–100] Since dapagliflozin has progressed so much further than the other candidates, these reviews focused primarily on the phase 2/3 results obtained with dapagliflozin. In addition to reviewing the major studies completed with dapagliflozin, selected clinical results disclosed at the 2011 annual meetings of the American Diabetes Association and European Association for the Study of Diabetes for each of the other eight *C*-glucosides will be summarized. The nine SGLT2 inhibitors currently progressing in active clinical trials are all *C*-glucosides containing a diarylmethane pharmacophore. In 2011 the regulatory review process began for dapagliflozin; the remaining eight candidates are in various stages of phase 2 or phase 3 clinical trials.

Insufficient data exists to discern whether structure-specific safety issues will emerge due to the differing aglycone structures. Since maximum efficacy requires continual maintenance of SGLT2 inhibition, structural features, such as *O*-glycosides, conducive to enhanced clearance will necessitate higher or multi-day doses. Tables 3.6 and 3.7 provide a summary of *in vitro* potency and selectivity, human half-life, and apparent preferred clinical dose and clinical status for these nine *C*-glucosides as well as three *O*-glucosides – T-1095, sergliflozin, and remogliflozin. To facilitate comparison, where possible the values generated by the Boehringer-Ingelheim group for SGLT2 EC_{50} and selectivity versus SGLT1 were utilized.[53] Preferred clinical doses for these *C*-glucoside containing candidates range from 5 to 300 mg/day reflecting differences in SGLT2 affinity and PK profile. In principle dual SGLT1/2 inhibitors should be more efficacious; however, due to concern regarding the potential of SGLT1-mediated GI disturbances, most of the field elected to pursue SGLT2 inhibitors for which the selectivity versus SGLT1 ranges from 200- to 3000-fold. To date the only *C*-glucoside dual inhibitor in the clinic is Lexicon's LX4211 for which the selectivity versus SGLT1 is 20.

Given the differences in the stage of development and number of patients, adverse events will not be compared. In particular it is difficult to assess if the variations in the incidence of genital and urinary tract infections incurred during the phase 2 studies are meaningful since (1) different criteria were employed to establish the presence of these infections during the studies and (2) some trials required genital examinations to exclude preexisting infections before acceptance into the study. The differences in efficacious dose reflected a combination of *in vitro* potency and human PK profile. At a sufficiently high dose all these agents were efficacious, exhibiting dose-dependent glucosuria which reduced glycemic levels and if the study was of sufficient duration, ultimately improved HbA1c.

SGLT2 inhibitors provide an insulin-independent means to reduce glycemic levels. Due to the resultant glucosuria, these glucose molecules are not available for conversion to fat or for metabolism by the glycolytic pathway. Hepatic and renal gluconeogenesis reconverts around two-thirds of the lactate and pyruvate generated by glycolysis back to glucose. Since this recycling of carbons from glucose is disrupted by SGLT2 inhibitor mediated glucosuria, consequently the impact on glucose availability as an energy source for tissues such as muscle, fat, etc. could be disproportionately greater than that which might be expected from the loss of 60–90 g of glucose eliminated in urine. The resultant greater utilization of fatty acids by these tissues reduces circulating and hepatic fatty acid levels, improving hepatic insulin resistance and leading to diminished hepatic glucose output. In addition, the lower glycemic levels reduce glucotoxicity which should be manifested by enhanced β-cell preservation.

The loss of ∼75 g of glucose or 300 calories per day due to SGLT2 inhibition appears to be only partially compensated by increased caloric intake.[135] Results at present indicate that patients may expect to achieve modest weight loss or at least no weight gain as well as reduction in glycemic levels. This reduction in weight not only may confer physiological benefits but also psychological rewards increasing the probability of patient compliance. An additional benefit, consistently noted across this inhibitor class, was the trend for modest reduction in blood pressure, particularly systolic pressure. Although this effect has been attributed to volume depletion arising from glucose-induced diuresis, dehydration has not become an issue in the clinical trials to date.

3.4.1   Clinical Evaluation of *O*-Glucosides

Clinical evaluation of SGLT2 inhibitors as anti-diabetic medicants first utilized *O*-glucosides. T-1095 was the first to enter clinical trials; however, no disclosures regarding clinical findings have been released after the studies were halted.[23] Subsequently, clinical trials were initiated with at least five other *O*-glucosides – sergliflozin, remogliflozin, BI 44847, TS-033, and AVE 2268.[30] None progressed beyond phase 2A/2B. Sanofi-Aventis conducted a phase 2B dose-ranging study with AVE 2268 in 300 patients for which the doses were 300, 600, and 1200 mg bid and 1200 mg qd. Results from this study have not been disclosed but AVE 2268 has been discontinued. Clinical progression of BI

44847, an *O*-glucoside licensed by Boehinger-Ingelheim from Ajinomoto, was halted after a phase 2 study in which the doses were 100, 400, and 800 mg bid.

Much more information is available pertaining to the GSK/Kissei phase 1 and 2 clinical studies to develop sergliflozin and remogliflozin etabonate. Sergliflozin was well tolerated in two phase 1 single dose studies when administered at doses of 5–500 mg to healthy males and diabetics. Dose-dependent urinary glucose excretion was observed to plateau at the higher doses in both studies.[101] The duration of glucose excretion paralleled plasma levels of **6a**. A 2-week study entailing tid administration of **6b** at 500 or 1000 mg/day to overweight subjects produced a dose-dependent increase in urinary glucose excretion accompanied by 1.5 kg reduction in body weight without any symptoms of hypoglycemia.[101] The most common AEs were headache, nausea, dizziness, and flatulence. When administered at 500 mg to healthy T2DM, **6b** decreased plasma glucose concentrations by 7 mmol/h/L over a 4 h period following an OGTT test.

Remogliflozin etabonate **6b** was well tolerated in a single ascending dose study at doses as high as 1000 mg when administered to healthy volunteers or type 2 diabetics.[28] A 12-day double-blind, placebo-controlled, multiple ascending dose study with **6b** was conducted with 35 type 2 diabetic patients who received either 100 mg bid, 1000 mg qd, or 1000 mg bid. Only the bid doses produced reductions in fasting plasma glucose levels. For many parameters such as glucose AUC there was a non-linear dose response. The most frequent adverse events were headache and flatulence.

In summary, the unfavorable PK profile of the *O*-glucoside based SGLT2 inhibitors required doses of 500–1000 mg/day to compensate. Moreover, the inherent susceptibility of these *O*-glucosides to α-glucosidase cleavage contributed to the clearance rate and necessitated administration as alkyl carbonate pro-drugs to achieve the plasma levels required for efficacy.

3.4.2 Clinical Evaluation of *C*-Glucosides

3.4.2.1 Dapagliflozin

Dapagliflozin, the most advanced of the nine *C*-glucosides undergoing clinical evaluation, is being co-developed by Bristol-Myers Squibb and Astra Zeneca. In addition to the reviews previously cited, a number exist focused solely on dapagliflozin.[102–104] Results from both a double-blind, placebo-controlled, single ascending dose study with healthy volunteers receiving doses of 2.5, 5, 10, 20, 50, 100, 250, and 500 mg qd and a 2-week multiple ascending dose study entailing administration of 2.5, 10, 20, 50, and 100 mg qd to healthy volunteers established the compound to be safe, well tolerated, and amenable to once daily administration.[105] Dose-dependent glucosuria was obtained with no evidence of hypoglycemia. Near maximum UGE of ~60 g per day was obtained with ≥20 mg as a result of sustained glucosuric output of ~ 3g/h of glucose. In a subsequent phase 2 double-blind 2-week multiple ascending dose study, 5, 25, and 100 mg or placebo was administered to 47 type 2 diabetics with unimpaired renal function who were either drug naïve or were on stable doses of metformin.[106] Dose-dependent UGE was obtained, plateauing at ~75 g with the 25 and 100 mg doses. The slight reduction in UGE over the 2-week study was attributed to an improvement in glycemic levels. Both fasting serum glucose levels and the AUC for glucose excursion following an oral glucose tolerance test (OGTT) exhibited significant improvement on day 14 thereby establishing dapagliflozin to be efficacious in diabetics.

A subsequent 12-week, double-blind placebo-controlled phase 2B study was conducted with 389 type 2 diabetics receiving 2.5, 5, 10, 20, or 50 mg qd.[107] Extended release metformin titrated up to 1500 mg from 750 mg by week 2 was included as an active comparator. Statistically significant reductions in HbA1c ranging from −0.55 to −0.9% were achieved for the dapagliflozin treated cohorts as compared to −0.18% for placebo and −0.73% for the metformin treated group. Mean weight reduction for the dapagliflozin treated groups ranged from −2.5% to −3.4% as compared to minor changes for the placebo and metformin groups. Improvements in fasting serum glucoses were observed after 1 week for the dapagliflozin treated cohorts. Relative to baseline, dapagliflozin treatment produced greater improvements for postprandial glucose than for fasting glucose levels – suggesting that renal glucose excretion is an effective mechanism to blunt postprandial glucose excursions.

When dapagliflozin was administered at 2.5, 5, and 10 mg/day in a double-blind, placebo-controlled study to 485 type 2 diabetics controlled by diet and exercise, dose-dependent HbA1c reductions of −0.58, −0.77, and −0.83% respectively were observed after 24 weeks versus −0.23% for the placebo group.[108] The respective body weight reductions were −3.3, −2.8, and −3.2 kg versus −2.2 kg for the placebo arm. A similar outcome was obtained following administration of dapagliflozin at 2.5, 5, and 10 mg/day to 546 poorly controlled metformin-treated type 2 diabetics in a double-blind, placebo-controlled study.[109] After 24 weeks, the respective HbA1c reductions were −0.67, −0.70, and −0.84% versus −0.3% for placebo; the corresponding decrease in body weight was 2.26, 3.1, and 2.96 kg versus −0.87 kg for the placebo arm treated with metformin alone. Rates of hypoglycemia and hypotension were similar for the dapagliflozin treated groups and placebo.

Results from several studies have been published in which dapagliflozin was administered to patients receiving insulin. For example, 71 type 2 diabetics sub-optimally controlled by insulin with either metformin or a thiazolidinedione were administered dapagliflozin at 10 or 20 mg/day.[110] After 12 weeks the placebo subtracted reduction in HbA1c was −0.7 and −0.78% respectively; the corresponding reductions in body weight were −2.6 and −2.4 kg versus placebo. In another study the response induced by dapagliflozin was compared following treatment of early versus late stage type 2 diabetes.[111] 58 late stage diabetics, poorly controlled (HbA1c = 8.4) by combination therapy comprising insulin, metformin, and a thiazolidinedione, were administered 10 or 20 mg/day of dapagliflozin. The dapagliflozin treated late stage cohorts became markedly glucosuric, spilling 87 g/day after 12 weeks; the respective reduction in HbA1c was −0.60% and −0.8% versus 0.0% for placebo; body weight decreases were −4.3 and −5.05 kg for the 10 and 20 mg treated cohorts versus −1.55 kg for the placebo. The corresponding results for the 151 early stage type 2 diabetics (starting HbA1c of 7.6%) were urinary glucose output of 55 and 71 g/day versus none for placebo; HbA1c reductions of −0.7% and −0.5% versus −0.2% for placebo; and body weight decreases of −2.0 and −2.5 kg versus −0.95 kg for placebo.

Encouraging results were obtained from a double-blind non-inferiority combination study in which either dapagliflozin or glipizide was administered to 814 type 2 diabetics who were maintained on metformin.[112] After 52 weeks of combination therapy, HbA1c had decreased by −0.52% for both groups; however, body weight had decreased on average 3.2 kg for the dapagliflozin cohorts but increased 1.42 kg for the glipizide group. In addition, 40.8% of the glipizide group had experienced at least one episode of hypoglycemia versus 3.5% for those receiving dapagliflozin. Greater reductions in systolic and diastolic blood pressures were also obtained with dapagliflozin.

An increased incidence of signs, symptoms and other reports of genital tract infections has been a consistent pattern for type 2 diabetics treated with dapagliflozin which can approach doubling that observed for the placebo treated group. These cases were self reported by patients; no clinical confirmation occurred. Virtually all were readily resolved by topical treatment necessitating few if any withdrawals. In contrast, urinary tract infections exhibited a trend to be slightly increased for the dapagliflozin treated patients over placebo.

3.4.2.2 Canagliflozin

Clinical trials have progressed to phase 3 for canagliflozin which was licensed from Mitsubishi Tanabe by Johnson & Johnson.[39] A phase 1 double-blind, placebo-controlled single ascending dose study established the compound to be well tolerated when administered to healthy volunteers at 10, 30, 100, 200, 400, 600, and 800 mg qd and 400 mg bid.[113] Glucosuria increased in a dose-dependent manner; doses >200 mg produced a maximum 24 h glucose output of ∼70 g. A 14-day double-blind, placebo-controlled multiple ascending dose study comprising 97 type 2 diabetic patients maintained on an isocaloric diet with a mean HbA1c of 8% was conducted with doses of 30, 100, 200, and 400 mg qd and 300 mg bid in addition to a placebo arm.[114] Dose-related decreases in body weight ranged from ∼0.7 to 1.5 kg after placebo subtraction for doses >30 mg. Urinary glucose output over 24 h increased in a dose-dependent manner from 69 g to plateauing at 88 g for the 400 mg qd and 300 mg bid cohorts. A dose-dependent decrease in renal threshold was observed that decreased from ∼240–250 mg/dL at baseline on day −1 to 152 mg/dL for the 30 mg dose on day 16, to 101 mg/dL for the 200 mg dose group and plateaued at 90 mg/dL for the 400 mg and 300 mg bid doses. Meaningful reductions in fasting plasma glucose and plasma glucose levels were obtained with doses ≥100 mg without evidence of hypoglycemia. Canagliflozin was generally well tolerated throughout the studies.

In a 12-week double-blind, placebo-controlled phase 2B study, 451 type 2 diabetic patients inadequately controlled on metformin received 50, 100, 200, or 300 mg qd or 300 mg bid or 100 mg sitagliptin qd.[115] Reductions in fasting plasma glucose values ranged from −16.2 to −32.4 mg/dL versus a placebo increase of 3 mg/dL and a −12.5 mg/dL decrease for the sitagliptin cohort. Mean body weight reductions ranged from −1.3% to −2.3% for all canagliflozin treated cohorts whereas the sitagliptin group experienced an increase of +0.4%. After placebo subtraction of −0.22%, dose related reductions in HbA1c for the canagliflozin treated cohorts ranged from −0.48% to −0.73% versus −0.52% for the sitagliptin group. An increase in genital tract infections was observed for the canagliflozin cohorts as well as a trend for an increase in urinary tract infections. The results supported dose selection of 100 and 300 mg qd for phase 3 studies. In a 4-week double-blind, placebo-controlled study, 29 type 2 diabetic patients inadequately controlled by insulin received 100 mg qd or 300 mg of canagliflozin bid.[116] Dose-dependent decreases in fasting plasma glucose were −38.1 and −42.4 mg/dL for the 100 and 300 mg cohorts; the corresponding reduction in HbA1c were −0.73% and −0.92% versus −0.19% for placebo. Although the 300 mg bid dose reduced the renal threshold from 222 mg/dL at baseline to 77 mg/dL on day 28, the incidence in non-severe hypoglycemia was not greater than placebo. A 2-week placebo-controlled multiple ascending dose study with 80 obese non-diabetic subjects (BMI range 30–39) receiving doses of 30, 100, 300, or 600 mg qd or 300 mg bid produced weight reductions ranging from −2.1 kg to −3.5 kg.[117]

3.4.2.3   *Empagliflozin*

Empagliflozin (BI 10773), being co-developed by Boehringer Ingelheim and Eli Lilly, has progressed to phase 3.[52] In a 12-week double-blind, placebo-controlled phase 2B study with 495 type 2 diabetic patients when administered at 1–50 mg/day, empagliflozin reduced HbA1c −0.09% to −0.56% versus −0.45% for sitagliptin at 100 mg/day and a 0.15% increase for the placebo arm; fasting plasma glucose reductions ranged from −1.7 to −27.9 mg/dL; weight reductions ranged from −1.5 to −2.8 kg.[118] There was a dose-related increase in genital tract infections and a trend for increase in urinary tract infections. In general, empagliflozin was well tolerated. Phase 3 clinical trials are in progress.

3.4.2.4   *Ipragliflozin*

Ipragliflozin (ASP1941), being developed by Astellas and Kotobuki, has progressed to phase 3. Results from double-blind, placebo-controlled single and multiple ascending dose studies with healthy volunteers revealed ipragliflozin to be well tolerated for doses of 5, 30, 100, 300, and 600 mg qd.[119] The dose-dependent increase in 24-h urinary glucose output plateaued at ~72 g for doses ≥300 mg. Ipragliflozin was rapidly absorbed and eliminated with a half-life of ~12 h. A placebo-controlled, double-blind, multiple ascending dose 28-day study was conducted with 61 type 2 diabetic patients after a 2-week washout period for patients treated with other anti-diabetic medication.[120] Upon completion of the study, fasting plasma glucose levels for cohorts receiving doses of 50, 100, 200, and 300 mg of ipragliflozin were reduced respectively −60, −49, −71, and −65 mg/dL versus −10 mg/dL for the placebo group; 24-h UGE was 61.6, 80.3, 82.7, and 90.8 g respectively versus 9.4 g for the placebo group; body weight reduction ranged from 3.2 to 4.2 kg for the ipragliflozin treated cohorts versus 1.8 for the placebo group. No incidences of genital tract infections were noted; however, two urinary tract infections were reported for patients receiving ipragliflozin. A 2-week placebo-controlled, double-blind study with 36 patients stabilized on metformin revealed ipragliflozin could be safely administered at 300 mg qd, causing 24-h urinary glucose excretion of 74.9 g without producing signs of hypoglycemia.[121] The effect of renal impairment on pharmacokinetics and urinary glucose excretion was assessed in a 40-person study with healthy volunteers and diabetics with non-impaired, mildly impaired, moderately impaired, and severely impaired renal function.[122] Following administration of 100 mg of ipragliflozin, AUC for ipragliflozin was 112%, 143%, 166%, and 174% respectively for diabetics with non- impaired, mildly impaired, moderately impaired, and severely impaired renal function relative to that of healthy volunteers; in addition 24-h UGE for the diabetic patients with mildly impaired, moderately impaired, and severely impaired renal function was 133%, 46%, and 25% of that of diabetics with non-impaired renal function. A number of phase 3 combination trials are in progress.

3.4.2.5 Luseogliflozin

Luseogliflozin (TS-071), being developed by Taisho, has progressed to phase 3 clinical trials. Following evaluation in a placebo-controlled, double-blind, single ascending dose study with 57 healthy volunteers at 1, 3, 9, 15, and 25 mg qd, approximate values for the dose-dependent 24-h UGE were respectively 19, 36, 55, 63, and 70 g respectively; half-life ranged from 9 to 13 h.[123] In a subsequent 7-day placebo-controlled, double-blind, multiple ascending dose study with 24 healthy volunteers at 5 and 10 mg qd, urinary glucose output did not significantly change from day 1 to day 7 nor induce hypoglycemia.[123] No food effect on the PK parameters was discerned at the 5 mg dose. The compound was deemed safe and well tolerated. Dose-dependent 24-h UGE was observed in a placebo-controlled, double-blind, 7-day multiple ascending dose study upon administration of 0.5, 1, 2.5, and 5 mg qd to 40 type 2 diabetic Japanese patients (HbA1c 6.9–10.4%) who had not received anti-diabetic medication during the four prior weeks.[124] The drug-treated cohorts exhibited dose-dependent UGE of 48.2, 68.3, 86.6, and 103 g respectively versus 0.2 g for the placebo group; no significant difference in UGE for day 1 versus day 7 was observed. Both fasting and postprandial plasma glucose levels were reduced for all doses. PK half-life was 9–10 h. Nine mild recoverable adverse events occurred; there were no instances of hypoglycemia.

Luseogliflozin was evaluated as monotherapy in a 12-week placebo-controlled, double-blind, phase 2B study conducted with 236 type 2 diabetic patients (HbA1c 6.9–10.4%) receiving 0.5, 2.5, and 5 mg qd.[125] The respective placebo-subtracted reductions in HbA1c were −0.43%, −0.70%, and −0.82%; placebo-subtracted reductions in fasting plasma glucose were −14.6, −25.9, and −27.9 mg/dL; placebo-subtracted body weight reductions were −0.5, −1.8, and −1.8 kg. The dose-dependent placebo-subtracted reduction in postprandial glucose ranged from −36.1–57.3 mg/dL. There was a trend for reduction in systolic blood pressure. No SAEs were reported; luseogliflozin significantly improved HbA1c and other glycemic parameters and was well tolerated and safe.

3.4.2.6   *PF-04971729 (Ertugliflozin)*

A 2-week placebo-controlled, double-blind phase 1 study revealed that UGE was increased following administration of 0.5–300 mg of PF-04971729.[126] A subsequent double-blind placebo-controlled 12-week phase 2 study was conducted with 338 type 2 diabetics inadequately controlled by metformin with HbA1c of 6.5–11%.[127] The statistically significant reductions in HbA1c for the cohorts administered 1, 5, 10, and 25 mg qd of PF-04971729 were −0.56, −0.80, −0.73, and −0.83% respectively versus −0.87% for the group administered 100 mg of sitagliptin and −0.11% for the placebo group. Fasting plasma glucose and fasting insulin levels exhibited dose-dependent reductions which plateaued at around −30 mg/dL and −15% for both the 10 and 25 mg doses; the corresponding changes for the placebo and sitagliptin were +2.76 and −17.29 mg/dL and +12.6% and +14%. The decrease in mean body weight ranged from −1.9 to −2.9 kg versus −0.75 kg and −0.3kg for the placebo and sitagliptin groups. In addition there was a trend for improvement in systolic and diastolic blood pressure. A double-blind placebo-controlled 4-week phase 2 study with Type 2 diabetic hypertensive patients administered 1, 5, or 25 mg of PF-04971729 or 12.5 mg of hydrochlorothiazide resulted in improved glycemic control and clinically meaningful decreases in blood pressure that were at least comparable to that achieved with hydrochlorothiazide.[128] Although increases in urinary tract and genital tract infections were observed in both studies, PF-04971729 was safe and well tolerated.

3.4.2.7   *LX4211*

Lexicon reported the results of a 4-week double-blind, placebo-controlled phase 2a study with 36 type 2 diabetics administered either 150 or 300 mg of the non-selective SGLT1/SGLT2 inhibitor LX4211 as monotherapy.[129] Dose-dependent glucosuria was observed which diminished 10–20% over the study.

On day 1 GLP-1 levels were increased. After 28 days HbA1c, fasting plasma glucose, and OGTT were significantly improved. A second study, conducted for 2 weeks with 12 type 2 diabetics, revealed similar responses to p.o. administration of 300 mg of LX4211 qd as a tablet versus solution.[130] Administration of 300 mg of LX4211 was reported to increase circulating levels of active and total GLP-1 and PYY and decrease insulin and glycemic levels. The PK profile supported qd administration. LX4211 was reported to be well tolerated with a favorable safety profile despite the fact that the 20-fold selectivity of LX4211 versus SGLT1 was much less than that of the other glucosides currently in development. Active recruitment of 285 type 2 diabetic patients for a phase 2B trial to assess combination therapy of LX-4211 with metformin began in June 2011 with a projected completion in April 2012.[131]

3.4.2.8 Other C-Glucoside SGLT2 Inhibitors in Clinical Development

Development of tofogliflozin (CSG452) is being pursued by Chugai alone after the co-development with Roche was terminated in 2010 following the decision by Roche to return development rights.[131,132] Chugai recently reported the results of a 12-week double-blind, placebo-controlled phase 2B study with 398 type 2 diabetics administered 2.5, 5, 10, 20 and 40 mg qd. Placebo-subtracted dose-dependent reductions in HbAic ranged from -0.17 to -0.54% and from -0.9 to -2.1 kg for decrease in body weight.[137]

Although in late 2009 Theracos began clinical development of EGT 1474, the choline complex of EGT 1442, no results have been disclosed.[131] The dose range for the phase 1 single ascending study was 2.5 to 150 mg qd. Subsequently a 2-week multiple ascending dose study was conducted with 78 healthy volunteers with doses of 10, 50, and 150 plus a placebo arm. A phase I, randomized, placebo-controlled study to assess the safety, tolerability, and pharmacokinetics after a single oral 25 mg dose has been completed with 24 subjects with type 2 diabetes. A 4-week double-blind, placebo-controlled phase 2 study was completed with 129 type 2 diabetic patients receiving daily doses of 5, 10, 20, or 50 mg. A 24-week double-blind, placebo-controlled phase 2B study with 300 type 2 patients is scheduled to begin in December 2011.

3.4.3 Antisense Inhibitors of SGLT2

ISIS Pharmaceuticals is pursuing development of ISIS 388626 (ISIS-SGLT2$_{RX}$), an RNAase chimeric antisense oligonucleotide containing 12 nucleotides.[133] Weekly injections for as long as 13 weeks suppressed expression of the SGLT2 gene in rodents, dogs, and primates by 75–90%, accompanied by an expected increase in glucosuria and improved insulin sensitivity.[134] No evidence of hypoglycemia, hepatotoxicity, renal toxicity, or diminution of renal function has been observed. An early clinical safety study with ISIS 388626, which began in 2009, has yet to be completed.[131]

3.5   Conclusions

An increase in genital tract infections appears to be the hallmark of SGLT2 inhibitors as the incidence in the drug-treated cohorts was nearly double that of the placebo groups. Although most SGLT2 inhibitors significantly increased the risk of genital tract infections, patients appeared to readily resolve this issue. Despite constant glucosuria, urinary tract infections appear not to be significantly elevated relative to placebo. Moreover, clinical experience with these agents has ameliorated concerns regarding polyuria and renal impairment. The glucose diuretic effect did not impact patient acceptance since typically this effect caused an increase of one void volume over a 24-h period. No evidence of SGLT2 inhibitors contributing to loss of renal function has appeared.

The results disclosed to date from the clinical trials with SGLT2 inhibitors as a class have been consistently encouraging. The compounds appear to be safe and well tolerated yet provide a novel effective means to decrease fasting and postprandial blood glucose. Clinical results reveal that efficacy is not diminished upon progression of type 2 diabetes since this mechanism unlike most is not insulin dependent. The glycemic improvement following treatment appears to be predictable as the reduction in HbA1c appears unchanged whether these SGLT2 inhibitors are used as monotherapy or in combination with the currently prescribed agents including insulin. This class of anti-hyperglycemic agents poses a low risk of hypoglycemia when utilized as monotherapy since they impose only a partial blockade of renal recovery and do not impact the hepatic counter-regulatory mechanisms. Moreover, since SGLT2 inhibition is not insulin dependent, SGLT2 inhibitors in combination with any of the existing classes of anti-diabetic medications did not markedly increase the hypoglycemic risk over that presented by the other agent. To date the clinical results support the expectation that this class of inhibitors could be safely used at any stage of T2DM either alone or in combination with other marketed anti-diabetic medications. The clinical experience gained over the next few years will determine whether SGLT2 inhibitors represent a new exciting paradigm-breaking approach for treatment of diabetes.

References

1. IDF Diabetes Atlas 5th Edition, November 14, 2011 http://www.idf.org/diabetesatlas/news
2. Centers for Disease Control. National Diabetes Fact Sheet United States, 2011. Available at: http://www.cdc.gov/diabetes/pubs/pdf/ndfs_2011.pdf released January 26, 2011
3. V. Poitout and R. P. Robertson, *Endocrinology*, 2002, **143**, 339.
4. The Diabetes Control and Complications Trial Research Group, *N. Engl. J. Med.*, 1993, **329**, 977.
5. UK Prospective Diabetes Study (UKPDS) Group, *Lancet*, 1998, **352**, 837.

6. C. J. Bailey, *Trends Pharmacol. Sci.*, 2011,**32**, 63.

7. J. E. Gerich, *Diabet. Med.*, 2010, **27**, 136.

8. A. Mather and C. Pollock, *Kidney Int.*, 2011, **79**, Suppl. 120, S1.

9. E. M. Wright, D. D. F. Loo and B. A. Hirayama, *Physiol Rev.*, 2011, **91**, 733.

10. Y. Kanai, W. S. Lee, G. You, D. Brown and M. A. Hediger, *J. Clin. Invest.*, 1994, **93**, 397.

11. P. P. Fromter, B. Hormann, R. Zwiebel and K. Baumann, *Pflugers Arch.*, 1970, **315**, 66.

12. V. Gorboulev, A. Schürmann, V. Vallon, H. Kipp, A. Jaschke, D. Klessen, A. Friedrich, S. Scherneck, T. Rieg, R. Cunard, M. Veyhl-Wichmann, A. Srinivasan, D. Balen, D. Breljak, R. Rexhepaj, H. E. Parker, F. M. Gribble, F. Reimann, F. Lang, S. Wiese, I. Sabolic, M. Sendtner and H. Koepsell, *Diabetes*, 2012, **61**, 187.

13. M. A. Hediger and D. B. Rhoads, *Physiol. Rev.*, 1994, **74**, 993.

14. L. P. van de Heuvel, K. Assink, M. Willemsen and L. Monnens, *Hum. Genet.*, 2002,**11**, 544.

15. S. R. Calado, *Clin. J. Am. Soc. Nephrol.*, 2010, **5**, 133.

16. J. Chen, J. Feder, I. Neuhaus and J. M. Whaley, *Diabetes*, 2008, **57**, Suppl 1A682.

17. E. M. Wright, B. A. Hirayama and D. F. Loo, *J. Intern. Med.*, 2007, **261**, 32.

18. W. N. Washburn, *Expert Opin. Ther. Patents*, 2009, **19**, 1485.

19. J. R. L. Ehrenkranz, N. G. Lewis, C. R. Kahn and J. Roth, *Diabetes Metab. Res. Rev.*, 2005, **21**, 31.

20. T. Asano, M. Anai, H. Sakoda, M. Fujishiro, H. Ono, H. Kurihara and Y. Uchijima, *Drugs Future*, 2004, **29**, 461.

21. K. Nunoi, K. Yasuda, T. Adachi, Y. Okamoto, N. Shihara, M. Uno, A. Tamon, N. Suzuki, A. Oku and K. Tsuda, *Clin. Exp. Pharmacol. Physiol.*, 2002, **29**, 386.

22. K. Ueta, T. Ishihara, Y. Matsumoto, A. Oku, M. Nawano, T. Fujita, A. Saito and K. Arakawa, *Life Sci.*, 2005, **76**, 2655.

23. M. Isaji, *Curr. Opin. Investig Drugs*, 2007, **8**, 285.

24. A. L. Handlon, *Expert Opin. Ther. Patents*, 2005, **15**, 1532.

25. K. Katsuno, Y. Fujimori, Y. Ishikawa-Takemura, M. Hiratochi, F. Itoh, Y. Komatsu, H. Fujikura and M. Isaji, *J. Pharmacol. Exp. Ther.*, 2007, **320**, 323.

26. Y. Fujimori, K. Katsuno, I. Nakashima, Y. Ishikawa-Takemura, H. Fujikura and M. Isaji, *J. Pharmacol. Exp. Ther.*, 2008, **327**, 268.

27. R. L. Dobbins, R. O'Connor-Semmes, A. Kapur, C. Kapitza, G. Golor, I. Mikoshiba, W. Tao and E. K. Hussey, *Diabetes, Obesity & Metab.*, 2012, **14**, 15.

28. E. K. Hussey, R. V. Clark, D. M. Amin, M. S. Kipnes, R. L. O'Connor-Semmes, E. C. O'Driscoll, J. Leong, S. C. Murray, R. L. Dobbins, D. Layko and D. J. R. Nunez, *J. Clin Pharmacol.*, 2010 **50**, 623.

29. S-P. Han, D. Hagan, J. Taylor, L. Xin, W. Meng, S. Biller, J. Wetterau, W. N. Washburn and J. M. Whaley, *Diabetes*, 2008, **57**, 1723.
30. W. N. Washburn, *J. Med. Chem.*, 2009, **52**, 1785.
31. J. T. Link and B. K. Sorensen, *Tetrahedron Lett.*, 2000, **41**, 9213.
32. B. E. Ellsworth, W. Meng, M. Patel, R. N. Girotra, G. Wu, P. Sher, D. Hagan, M. Obermeier, W. G. Humphreys, J. G. Robertson, A. Wang, S. Han, T. Waldron, N. Morgan, J. M. Whaley and W. N. Washburn, *Bioorg. Med. Chem. Lett.*, 2008, **18**, 4770.
33. W. Meng, B. A. Ellsworth, A. A. Nirschl, P. J. McCann, M. Patel, R. N. Girotra, G. Wu, P. M. Sher, E. P. Morrison, S. A. Biller, R. Zahler, P. P. Deshpande, A. Pullockaran, D. L. Hagan, N. Morgan, J. R. Taylor, M. T. Obermeier, W. G. Humphreys, A. Khanna, L. Discenza, J. G. Robertson, A. Wang, S. Han, J. R. Wetterau, E. B. Janovitz, O. P. Flint, J. M. Whaley and W. N. Washburn, *J. Med. Chem.*, 2008, **51**, 1145.
34. Bristol-Myers Squibb, *US Patent*, 6,414,126, 2002.
35. Bristol-Myers Squibb, *US Patent*, 6,515,117, 2003.
36. Bristol-Myers Squibb, *US Patent*, 7,589,193, 2005.
37. S. Nomura, S. Sakamaki, M. Hongu, E. Kawanishi, Y. Koga, T. Sakamoto, Y. Yamamoto, K. Ueta, H. Kimata, K. Nakayama and M. Tsuda-Tsukimoto, *J. Med. Chem.*, 2010, **53**, 6355.
38. Tanabe Seiyaku, *US Patent Application*, 2005/0233988.
39. E.C. Chao, *Drugs of the Future*, 2011, **36**, 351.
40. K.-S. Song, S. H. Lee, M. J. Kim, H. J. Seo, J. Lee, S.-H. Lee, M. E. Jung, E.-J. Son, M. Lee, J. Kim and J. Lee, *Med. Chem. Lett.*, 2011, **2**, 182.
41. J. Lee, S.-H. Lee, H. J. Seo, E.-J. Son, S. H. Lee, M. E. Jung, M. Lee, H.-K. Han, J. Kim, J. Kang and J. Lee, *Bioorg. Med. Chem.*, 2010, **18**, 2178.
42. M. J. Kim, J. Lee, S. Y. Kang, S.-H. Lee, E.-J. Son, M. E. Jung, S. H. Lee, K.-S. Song, M. Lee, H.-K. Han, J. Kim and J. Lee, *Bioorg. Med. Chem. Lett.*, 2010, **20**, 3420.
43. J. Lee, J. Y. Kim, J. Choi, S-H. Lee, J. Kim, and J. Lee, *Bioorg. Med. Chem. Lett.*, 2010, **20**, 7046.
44. Kotobuki Pharmaceutical Co Ltd, *US Patent*, 6,627,611, 2003.
45. Astellas/Kotobuki, *US Patent*, 7,169,761, 2007.
46. Astellas Pharmaceutical R&D Pipeline, May 2009, Astellas.com/en/ir/library/pdf/4q2009_rd_en.pdf/.
47. K. Ikegai, M. Imamura, T. Suzuki, K. Nakanishi, T. Murakami, T. Maruyama, E. Kurosaki, A. Tahara, M. Yokono, T. Ffurukawa, A. Noda, Y. Kobayashi, M. Yokota, T. Koide, K. Kosakai, Y. Ohkura, M. Takeuchi, H. Tomiyama, M. Ohta and S.-I. Tsukamaoto, Presented at the 11th AFMC International Medicinal Chemistry Symposium, November 30–December 2, 2011, Tokyo, Japan, Presentation 1P-162.
48. B. Xu, Y. Fang, H. Cheng, Y. Song, B. Lv, Y. Wu, C. Wang, S. Li, M. Xu, J. Du, K. Peng, J. Dong, W. Zhang, T. Zhang, L. Zhu, H. Ding, Z.

Sheng, A. Welihinda, J. Roberge, B. Seed and Y. Chen, *Bioorg. Med. Chem. Lett.*, 2011, **21**, 4465.

49. Astellas Pharmaceutical, *US Patent*, 7,202,350, 2007.

50. S. A. Veltkamp, T. Kadokura, W. J. J. Krauwinkel and R. A. Smulders, *Clin. Drug. Investig.*, 2011, **31**, 839.

51. E. Kurosaki, A. Tahara, M. Yokono, D. Yamajuku, T. Takasu, M. Imamura, T. Funatsu and Q. Li, Presented at the 70th Scientific Sessions of the American Diabetes Association, June 25–29, 2010, Orlando, FA, Presentation 570-P.

52. A. I, Calado, *Curr Opin Investig Drugs*, 2010, **11**, 1182.

53. R. Grempler, L. Thomas, M. Eckhardt, F. Himmelsbach, A. Sauer, D. E. Sharp, R. A. Bakker, M. Mark, T. Klein and P. Eickelmann, *Diabetes, Obes. Metab.*, 2012, **14**, 83.

54. W. Zhang, A. Weihinda, J. Mechanic, H. Ding, L. Zhu, Y. Lu, Z. Deng, Z. Sheng, B. Lv, Y. Chen, J. Y. Roberge, B. Seed and Y. X. Wang, *Pharmacol. Res.*, 2011, **63**, 284.

55. Taisho Pharmaceutical, WO2007136116.

56. H. Morita, J. Deguchi, Y. Motegi, S. Sato, C. Aoyama, J. Takeo, M. Shiro and Y. Hirasawa, *Bioorg. Med. Chem. Lett.*, 2010, **20**, 1070.

57. S. Y. Kang, M. J. Kim, J. S. Lee and J. Lee, *Bioorg. Med. Chem. Lett.*, 2011, **21**, 3759.

58. Chugai Pharmaceutical, *US Patent Application*, 2008/0319047.

59. Kissei Pharmaceutical, *US Patent Application*, 2007/0197449.

60. Kissei Pharmaceutical, *US Patent Application*, 2007/0197450.

61. Kissei Pharmaceutical, WO2006054629.

62. Tanabe Seiyaku, *US Patent Application*, 2008/0027122.

63. Kissei Pharmaceutical, *US Patent Application*, 2008/0139484.

64. S. Y. Kang, K.-S. Song, J. Lee, S.-H. Lee and J. Lee, *Bioorg. Med. Chem.*, 2010, **180**, 6069.

65. R. P. Robinson, V. Mascitti, C. M. Boustany-Kari, C. L. Carr, P. M. Foley, E. Kimoto, M. T. Leininger, A. Lowe, M. K. Klenotic, J. I. MacDonald, R. J. Maguire, V. M. Masterson, T. S. Maurer, Z. Miao, J. D. Patel, C. Preville, M. R. Reese, L. She, C. M. Steppan, B. A. Thuma and T. Zhu, *Bioorg. Med. Chem. Lett.*, 2010, **20**, 1569.

66. V. Mascitti, R. P. Robinson, C. Preville, B. A. Thuma, C. L. Carr, M. R. Reese, R. J. Maguire, M. T. Leininger, A. Lowe and C. M. Steppan, *Tetrahedron Lett.*, 2010, **51**, 1880.

67. V. Mascitti, T. S. Maurer, R. P. Robinson, J. Bian, C. M. Boustany-Kari, T. Brandt, B. M. Collman, A. S. Kalgutkar, M. K. Klenotic, M. T. Leininger, A. Lowe, R. J. Maguire, V. M. Masterson, Z. Miao, E. Mukaiyama, J. D. Patel, J. C. Pettersen, C. Preville, B. Samas, L. She, Z. Sobol, C. M. Steppan, B. D. Stevens, B. A. Thuma, M. Tugnait, D. Zeng and T. Zhu, *J. Med. Chem.*, 2011, **54**, 2952.

68. Pfizer Inc., *US Patent Application*, 2010/0056618.

69. N. C. Goodwin, R. Mabon, B. A. Harrison, M. K. Shadoan, Z. Y. Almstead, Y. Xie, J. Healy, L. M. Buhring, C. M. DaCosta, J. Bardenhagen, F. Mseeh, Q. Liu, A. Nouraldeen, A. G. E. Wilson, S. D. Kimball, D. R. Powell and D. B. Rawlins, *J. Med. Chem.*, 2009, **52**, 6201.

70. J. Freiman, D. A. Ruff, K. S. Frazier, K. Combs, A. Turnage, M. Shadoan, D. Powell, B. Zambrowicz and P. Brown, Presented at 70th Scientific Sessions of the American Diabetes Association, June 25–29, 2010, Orlando, FL, Presentation 17-LB.

71. Lexicon Pharmaceutical, *US Patent Application*, 2009/0030198.

72. E.-J. Park, Y. Kong, J. S. Lee, S.-H. Lee and J. Lee, *Bioorg. Med. Chem. Lett.*, 2011, **21**, 742.

73. C.-H. Yao, J.-S. Song, C.-T. Chen, T.-K. Yeh, M.-S. Hung, C.-C. Chang, Y.-W. Liu, M.-C. Yuan, C.-J. Hsieh, C.-Y. Huang, M.-H. Wang, C.-H. Chiu, T.-C. Hsieh, S.-H. Wu, W.-C. Hsiao, K.-F. Chu, C.-H. Tsai, Y.-S. Chao and J.-C. Lee, *J. Med Chem.*, 2011, **54**, 166.

74. H. Kakinuma, T. Oi, Y. Hashimoto-Tsuchiya, M. Arai, Y. Kawakita, Y. Fukasawa, I. Iida, N. Hagima, H. Takeuchi, Y. Chino, J. Asami, L. Okumura-Kitajima, F. Io, D. Yamamoto, N. Miyata, T. Takahashi, S. Uchida and K. Yamamoto, *J. Med. Chem.*, 2010, **53**, 3247.

75. Chugai Pharmaceutical, *US Patent Application*, 2008/0318874.

76. Y. Ohtake, T. Sato, H. Matsuoka, M. Nishimoto, N. Taka, K. Takano, K. Yamamoto, M. Ohmori, T. Higuchi, M. Murakata, T. Kobayashi, K. Morikawa, N. Shimma, M. Suzuki, H. Hagita, K. Ozawa, K. Yamaguchi, M. Kato and S. Ikeda, *Bioorg. Med. Chem.*, 2011, **19**, 5334.

77. Bristol-Myers Squibb, *US Patent*, 6,774,112, 2004.

78. B. Xu, B. Lv, Y. Feng, G. Xu, J. Du, A. Welihinda, Z. Sheng, B. Seed and Y. Chen, *Bioorg. Med. Chem. Lett.*, 2009, **19**, 5632.

79. Chugai Pharmaceutical, WO 2006/080421; US 2009/0030006.

80. Chugai Pharmaceutical, Oct 21, 2001, www.chugai-pharm.co.jp/hc/ss/english/ir/reports_downloads/pipeline.html

81. Chugai Pharmaceutical, EP 2048153 200.

82. B. Lv, B. Xu, Y. Feng, K. Peng, G. Xu, J. Du, L. Zhang, W. Zhang, T. Zhang, L. Zhu, H. Ding, Z. Sheng, A. Welihinda, B. Seed and Y. Chen, *Bioorg. Med. Chem. Lett.*, 2009, **19**, 6877.

83. B. Lv, Y. Feng, J. Dong, M. Xu, B. Xu, W. Zhang, Z. Sheng, A. Welihinda, B. Seed and Y. Chen, *ChemMedChem*, 2010, **5**, 827.

84. M. J. Kim, S. H. Lee, S. O. Park, H. Kang, J. S. Lee, K. N. Lee, M. E. Jung, J. Kim and J. Lee, *Bioorg. Med. Chem.*, 2011, **19**, 5468.

85. B. Xu, Y. Feng, B. Lv, G. Xe, L. Zhang, J. Du, K. Peng, M. Xu, J. Dong, W. Zhang, T. Zhang, L. Zhu, H. Ding, Z. Sheng, A. Welihinda, B. Seed and Y. Chen, *Bioorg. Med. Chem.*, 2010, **18**, 4422.

86. S. H. Lee, M. J. Kim, S.-H. Lee, J. Kim, H.-J. Park and J. Lee, *Eur. J. Med. Chem.*, 2011, **46**, 2662.

87. Merck GMBH, WO 2008071288.

88. Merck GMBH, WO 2008101586.

89. Merck GMBH, WO 2008046497.

90. Merck GMBH, WO2009049731.

91. A.-R. Li, J. Zhang, J. Greeenberg, T. W. Lee and J. Liu, *Bioorg. Med. Chem. Lett.*, 2011, **21**, 2472.

92. X. Du, M. Lizarzaburu, S. Turcotte, T. Lee, J. Greeenberg, B. Shan, P. Fan, Y. Ling, J. C. Medina and J. Houze, *Bioorg. Med. Chem. Lett.*, 2011, **21**, 3774.

93. J.-S. Wu, Y.-H. Peng, J.-M. Wu, C.-J. Hsieh, S.-H. Wu, M.-S. Coumar, J.-S. Song, J.-C. Lee, C.-H. Tsai, C.-T. Chen, Y.-W. Liu, Y.-S. Chao and S.-Y. Wu, *J. Med. Chem.*, 2010, **53**, 8770.

94. G. Musso, R. Gambino, M. Cassader and G. Pagano, *Ann. Med.*, 2011, DOI: 10.3109/07853890.2011.560181.

95. M. Isaji, *Kidney Int.*, 2011, **79**, Suppl. 120, S14.

96. S. Nair, F. Joseph, D. Ewins, J. Wilding and N. Goenka, *Pract. Diab. Int.*, 2010, **27**, 311.

97. E. C. Chao and R. R. Henry, *Nat. Rev. Drug Discov.*, 2010, **9**, 551.

98. S. Nair and J. P. H. Wilding, *J. Clin. Endocrinol. Metab.*, 2010, **95**, 34.

99. M. F. A. Borghese and M. P. Mojowicz, *Drugs of the Future*, 2009, **34**, 297.

100. I. Idris and R. Donnelly, *Diabetes, Obes. Metab.*, 2009, **11**, 79.

101. E. K. Hussey, R. L. Dobbins, R. R. Stoltz, N. L. Stockman, R. L. O'Connor-Semmes, A. Kapur, S. C. Murray, D. Layko and D. J. Nunez, *J. Clin. Pharmacol.*, 2010, **50**, 636.

102. N. Katsiki, N. Papanas and D. P. Mikhailidis, *Expert Opin. Investig. Drugs*, 2010, **19**, 1581.

103. J. Calado, *IDrugs*, 2009, **12**, 785.

104. V. C. Woo, *Expert Opin. Pharmacother.*, 2009, **10**, 2527.

105. B. Komoroski, N. Vachharajani, D. Boulton, D. Kornhauser, M. Geraldes, L. Li and M. Pfister, *Clin. Pharmacol. Ther.*, 2009, **85**, 520.

106. B. Komoroski, N. Vachharajani, Y. Feng, L. Li, D. Kornhauser and M. Pfister, *Clin. Pharmacol. Ther.*, 2009, **85**, 513.

107. J. F. List, V. Woo, E. Morales, W. Tang and F. T. Fiedorek, *Diabetes Care*, 2009, **32**, 650.

108. E. Ferrannini, S. J. Ramos, A. Salsali, W. Tang and J. F. List, *Diabetes Care*, 2010, **33**, 2217.

109. C. J Bailey, J. L Gross, A. Pieters, A. Bastien and J. F List, *Lancet*, 2010, **375**, 2223.

110. J. P. Wilding, P. Norwood, C. T'joen, A. Bastien, J. F. List and F. T. Fiedorek, *Diabetes Care*, 2009, **32**, 1656.

111. L. Zhang, Y. Feng, J. List, S. Kasichayanula and M. Pfister, *Diabetes, Obes. Metab.*, 2010, **12**, 510.

112. M. A. Nauck, D. P. Stefano Del Prato, J. J. Meier, S. Durán-García, K. Rohwedder, M. Elze and S. J. Parikh, *Diabetes Care*, 2011, **34**, 2015.

113. S. Sha, D. Devineni, A. Ghosh, D. Polidori, S. Chien, D. Wexler, K. Shalayda, K. Demarest and P. Rothenberg, *Diabetes, Obes. Metab.*, 2011, **13**, 669.

114. S. Sha, D. Devineni, A. Ghosh, D. Polidori, M. Hompesch, S. Arnolds, L. Morrow, H. Spitzer, J. Blake, D. Wexler, Y. Tan, K. Smulders, K. Demarest and P. Rothenberg, Presented at 70th Scientific Sessions of the American Diabetes Association, June 25–29, 2010, Orlando, FL, Presentation 568-P.

115. J. Rosenstock, D. Polidori, Y. Zhao, S. Sha, D. Arbit, K. Usiskin, G. Capuano and W. Canovatchel, Presented at 46th Annual Meeting of European Association for the Study of Diabetes, September 20–24, 2010, Stockholm, Sweden, Presentation 873.

116. S. Schwartz, L. Morrow, M. Hompesch, D. Devineni, D. Skee, A. Vandebosch, J. Murphy and M. Pfeifer, Presented at 70th Scientific Sessions of the American Diabetes Association, June 25–29, 2010, Orlando, FL, Presentation 564-P.

117. T. C. Sarich, D. Devineni, A. Ghosh, D. Wexler, K. Shalayda, J. Blake, M. Saraiva, M. J. Gutierrez, D. Polidori, K. Demarest and P. Rothenberg, Presented at 46th Annual Meeting of European Association for the Study of Diabetes, September 20–24, 2010, Stockholm, Sweden, Presentation 874.

118. J. Rosenstock, A. Jelaska, L. Seman, S. Pinnetti, S. Hantel and H. J. Woerle, Presented at 71th Scientific Sessions of the American Diabetes Association, June 24–28, 2011, San Diego, CA, Presentation 989-P.

119. S. A. Veltkamp, T. Kadokura, W. J. J. Krauwinkel and R. A. Smulders, Presented at 70th Scientific Sessions of the American Diabetes Association, June 25–29, 2010, Orlando, FL, Presentation 565-P.

120. S. Schwartz, B. Akinlade, S. Klasen, D. Kowalski, W. Zhang and W. Wilpshaar, *Diabetes Technol. Ther.*, 2011, **13**, 1219.

121. S. A. Veltkamp, J. V. Dijk, C. Collins and R. A. Smulders, Presented at 47th Annual Meeting of European Association for the Study of Diabetes, September 12–16, 2011, Lisbon, Portugal, Presentation 849.

122. S. A. Veltkamp, J. V. Dijk, W. J. J. Krauwinkel and R. A. Smulders, Presented at 71th Scientific Sessions of the American Diabetes Association, June 24–28, 2011, San Diego, CA, Presentation 1127-P.

123. T. Sasaki, Y. Seino, A. Fukatsu, Y. Samukawa, S. Sakai and T. Watanabe, Presented at 71th Scientific Sessions of the American Diabetes Association, June 24–28, 2011, San Diego, CA, Presentation 1140-P.

124. T. Sasaki, Y. Seino, A. Fukatsu, Y. Samukawa, S. Sakai and T. Watanabe, Presented at 47th Annual Meeting of European Association for the Study of Diabetes, September 12–16, 2011, Lisbon, Portugal, Presentation 846.

125. Y. Seino, T. Sasaki, A. Fukatsu, Y. Samukawa, S. Sakai and T. Watanabe, Presented at 71th Scientific Sessions of the American

Diabetes Association, June 24–28, 2011, San Diego, CA, Presentation 998-P.

126. A. S. Kalgutkar, M. Tugnait, T. Zhu, E. Kimoto, Z. Miao, V. Mascitti, X. Yang, B. Tan, R. L. Walsky, J. Chupka, B. Feng and R. P. Robinson, *Drug Metab. Dispos.*, 2011, **39**, 1609.

127. G. Nucci, N. B. Amin, X. Wang, D. S. Lee and J. M. Rusnak, Presented at 47th Annual Meeting of European Association for the Study of Diabetes, September 12–16, 2011, Lisbon, Portugal, Presentation 850.

128. N. B. Amin, X. Wang, G. Nucci and J. M. Rusnak, Presented at 47th Annual Meeting of European Association for the Study of Diabetes, September 12–16, 2011, Lisbon, Portugal, Presentation 844.

129. J. Freiman, D. A. Ruff, K. S. Frazier, K. Combs, A. Turnage, M. Shadoan, D. Powell, B. Zambrowicz and P. Brown, 70th Scientific Sessions of the American Diabetes Association, June 25–29, 2010, Orlando, FL, Abstract 17-LB.

130. D. Powell, J. Freiman, K. Frazier, A. Turnage, P. Banks, J. Bronner, K. A. Boehm, D. Ruff, A. Wilson, A. Sands and B. Zambrowicz, Presented at 71th Scientific Sessions of the American Diabetes Association, June 24–28, 2011, San Diego, CA, Presentation 982-P.

131. Clinical trials status obtained fromhttp://www.clinicaltrials.gov

132. Chugai Pharmaceutical Co, Oct 21, 2001, www.chugai-pharm.co.jp/hc/ss/english/ir/reports_downloads/pipeline.html

133. ISIS Pharmaceuticals, http://www.isip.com/Pipeline/index.htm

134. S. Bhanot, S. F. Murray, S. L. Booten, K. Chakravarty, T. Zanardi, S. Henry, L. M. Watts, E. V. Wancewicz and A. Siwkows, Presented at 69th Scientific Sessions of the American Diabetes Association, June 5–9, 2009, New Orleans, LA, Presentation 328-OR.

135. J. F. List and J. M. Whaley, *Kidney Int.*, 2011, **79** (Suppl. 120), S20.

136. K. Tsujihara, M. Hongu, K. Saito, H. Kawanishi, K. Kuriyama, M. Matsumoto, A. Oku, K. Ueta, M. Tsuda, A. Saito, *J. Med. Chem.*, 1999, **42**, 5311.

137. T. Takashi, S. Ikeda, Y. Takano, O. Cynshi, A. D. Christ, V Boerlin, U. Beyer and A. Beck, Presented at: 72th Scientific Sessions of the American Diabetes Association; June 8-12, 2012; Philadelphia, PA. Presentation 80-OR.

Glucokinase Activators in Development

KEVIN J. FILIPSKI, BENJAMIN D. STEVENS AND
JEFFREY A. PFEFFERKORN*

Cardiovascular, Metabolic & Endocrine Disease Research, Pfizer Worldwide
Research & Development, Cambridge, MA 02139, USA
*E-mail: jeffrey.a.pfefferkorn@pfizer.com

4.1 Introduction

Type 2 diabetes mellitus (T2DM) is a rapidly expanding public health problem
affecting over 220 million people worldwide.[1] The disease is characterized by
elevated fasting plasma glucose (FPG), insulin resistance, abnormally elevated
hepatic glucose production (HGP), and reduced glucose-stimulated insulin
secretion (GSIS).[2] The current standard of care for treating T2DM typically
begins with oral metformin as a first line therapy followed by sulfonylureas,
DPP-IV inhibitors, and thiazolidinones as second line therapies. As the
underlying disease progresses and patients are not adequately controlled on
these oral therapies, injectable agents such as GLP-1 analogs and, ultimately,
insulin are utilized to help maintain glycemic control. Despite current
treatment options, many patients are unable to achieve and maintain tight
glycemic control; hence, there remains a significant need for new therapies with
improved durability and safety to help patients achieve their treatment goals.[3]
Among the potential next generation therapies, small molecule glucokinase
activators offer a promising opportunity for the treatment of T2DM
patients.[4,5]

RSC Drug Discovery Series No. 27
New Therapeutic Strategies for Type 2 Diabetes: Small Molecule Approaches
Edited by Robert M. Jones

Published by the Royal Society of Chemistry, www.rsc.org

4.2 The Role of Glucokinase in the Regulation of Glucose Homeostasis

Glucokinase, also known as hexokinase IV, catalyzes the phosphorylation of glucose to glucose-6-phosphate (G-6-P).[6] Glucokinase is unique among the members of the hexokinase family given its low substrate binding affinity ($S_{0.5}$ ~ 8 mM), positive substrate cooperativity, and lack of product inhibition. As a monomeric enzyme, glucokinase achieves this cooperativity through equilibration between multiple conformations leading it to exhibit sigmoidal kinetics.[6] This intricate coupling of glucokinase activity to glucose concentrations, particularly over the physiologically relevant glucose range, enables the enzyme to effectively function as a glucose sensor regulating glucose homeostasis.

Glucokinase is expressed in the α- and β-cells of the pancreas, liver hepatocytes, glucose sensing neurons in the ventromedial hypothalamus, K and L cells of the gastrointestinal tract, and gonadotropes of the pituitary. Below, we examine the role of glucokinase in these individual tissues in both the normal and disease state.

In the β-cells of the pancreas, glucokinase acts as a glucostat, establishing the threshold for glucose-stimulated insulin secretion (GSIS),[6,7] whereas its role in the pancreatic α-cells is less well characterized but may be involved in regulating glucagon secretion.[8] During the progression of diabetes, this threshold for GSIS in β-cells becomes inappropriately right shifted requiring higher glucose concentrations to stimulate insulin secretion thus contributing to hyperglycemia. This loss of glucose responsiveness has been suggested to be due, in part, to reduced activity of the glucokinase enzyme,[9] and encouragingly, in *ex vivo* perifusion studies in human diabetic islets, pharmacological activation of glucokinase has been shown to partially restore the normal threshold for GSIS.[10,11]

In hepatocyctes, glucokinase represents the rate-determining step for hepatic glucose uptake and also plays important roles in glycogen synthesis and the regulation of hepatic glucose production.[12] Specifically in the liver, but not other tissues, the activity of glucokinase is regulated though an interaction with glucokinase regulatory protein (GKRP).[13] Physiologically, in a low glucose state, GKRP binds the inactive conformation of glucokinase and sequesters the enzyme to the nucleus. As intracellular glucose concentrations increase, glucokinase is released from GKRP, diffuses into the cytoplasm, and can be converted to its active form through binding to glucose. T2DM patients have been shown to have reduced capacity for hepatic glucose uptake and glycogen synthesis which may be attributable, in part, to reduced hepatic glucokinase activity.[14,15] In fact, one study found that T2DM patients lose up to 50% of their hepatic glucokinase activity, potentially contributing to reduced hepatic glycogen synthesis and improper regulation of hepatic glucose output (HGO); however, a separate study found that subjects with impaired fasting glucose, impaired glucose tolerance, or early diabetes tended to have normal or even elevated hepatic glucokinase activity.[16,17] Consistent with the former results, most diabetic animal models demonstrate loss of hepatic glucokinase activity with progression of diabetes and, encouragingly, normalization of hepatic enzyme activity through over-expression normalizes glycogen levels, restores HGO regulation, and reduces both fasting and postprandial glucose.[18]

Beyond the pancreas and liver, glucokinase is also found in glucose sensing neurons of the ventromedial hypothalamus where it regulates the counter regulatory response (CRR) to hypoglycemia.[19] Glucokinase is also expressed in the endocrine K and L cells of the gut[8] as well as certain pituitary cells[20] where its functions are less well characterized but may be involved in nutrient sensing.

4.3 Genetic Evidence for the Importance of Glucokinase in Diabetes

There is significant genetic evidence to support the importance of glucokinase in glucose control and diabetes.[21] Human loss of function mutations in glucokinase result in conditions associated with hyperglycemia. For example, heterozygous loss of function in the glucokinase gene is associated with maturity-onset diabetes of the young type 2 (MODY2) while homozygous loss of function is associated with the more severe condition of permanent neonatal diabetes.[22] By contrast, human gain of function mutations which activate glucokinase are associated with conditions of low blood glucose such as hyperinsulinemic hypoglycemia of infancy.[23,24]

4.4 Small Molecule Glucokinase Activation: Opportunities and Challenges

Given the central role of glucokinase in the glucose sensing network and the fact that its activity may be reduced in key tissues during the progression of diabetes, therapeutically, activation of glucokinase represents a promising opportunity for the treatment of diabetes. Specifically, activation of glucokinase is anticipated to up-regulate hepatic glucose utilization, down-regulate hepatic glucose output, and enhance glucose-stimulated insulin secretion. In 2003, researchers at Roche first reported that small molecule activators were capable of binding to glucokinase at an allosteric site 20 Å remote from the active site and modulating the enzyme's kinetic profile (i.e. $S_{0.5}$ and $v_{\max}$) by influencing the conformational distribution of the enzyme.[25] An endogenous agonist for this allosteric site has not yet been found. Since the Roche publication, a significant number of small molecule activators have been reported.[26–28] These glucokinase activators have been shown to effectively lower blood glucose in a variety of diabetic animal models. Furthermore, as described in Section 4.5, multiple candidates have advanced to clinical studies and were found to effectively lower fasting and postprandial glucose in healthy subjects as well as T2DM patients.[29–32]

While the clinical efficacy offered by this mechanism has proven promising in phase 1 and 2 studies, no candidates have progressed to phase 3 development to date and several risks remain to be resolved. First, during both preclinical and clinical studies (including normal volunteers and diabetic patients), hypoglycemia has been revealed as an important side effect.[5,29–33] Various strategies have been utilized to manage this hypoglycemia risk including dose titration and dosing activators with meals.[31,33] Beyond managing hypoglycemia risk, several strategies have emerged to design glucokinase activators with inherently reduced hypoglycemia risk. One strategy is the design of liver selective activators, as described in Sections 4.5.1, 4.5.6, and 4.5.9, which seek to minimize pancreatic enzyme activation, thereby reducing hypoglycemia from potentiation of GSIS albeit at the presumed expense of reduced overall efficacy. As described in more detail in Section 4.5.6, an alternative strategy for mitigating hypoglycemia risk is the design of systemically acting "partial activators" which reduce the glucokinase $S_{0.5}$ for glucose to a lesser degree, thereby retaining increased dependence of enzymatic activity on the prevailing glucose concentration. A second mechanistic risk is the potential for glucokinase activators to cause adverse circulating or hepatic lipid changes. Such concerns have generally arisen as a result of hepatic-specific over-expression studies;[34] however, to date, significant adverse lipid changes have not been reported with most small molecule activators with the notable exception of MK-0941 (Section 4.5.4) which was found to cause modest circulating triglyceride elevations in T2DM patients. A third potential concern is the possibility of loss of durability of this mechanism in long-term studies. In particular, recent clinical data reported for MK-0941 demonstrated a loss of efficacy after 14 weeks of treatment in diabetic patients on basal insulin background therapy.[31] The underlying reason for this loss of efficacy in this study is unclear, and longer term studies with other activators in various patient populations are awaited to further inform this issue.

4.5 Development Status of Glucokinase Activators

Based on the substantial genetic data supporting the role of glucokinase in regulating glucose homeostasis and the promising efficacy of small molecule activators in a variety of diabetic animal models, there has been significant investment in the medicinal chemistry design and clinical development of glucokinase activators for treating T2DM patients. The active and discontinued clinical glucokinase activator development candidates are summarized in Table 4.1. Descriptions of these candidates with structures, where available, are provided in Sections 4.5.1–4.5.10.

4.5.1 Advinus Glucokinase Activators

Advinus is a relatively recent entry in the field of glucokinase activation and has disclosed the discovery of GKM-001 as a hepatoselective activator candidate which has proceeded through a phase 1 single ascending dose study

Table 4.1   Glucokinase activator clinical development candidates.

Company	Clinical Candidates	Current Status	Section	References
Advinus	GKM-001	Phase 1	4.5.1	35–39
Array/Amgen	ARRY 403 (AMG-151)	Phase 1	4.5.2	40–50
AstraZeneca	AZD1656	Discontinued (phase 2)	4.5.3	51–67
	AZD5658	Discontinued (phase 1)		
	AZD6370	Discontinued (phase 2)		
	AZD6714	Discontinued (phase 1)		
Merck	MK-0599	Discontinued (phase 1)	4.5.4	31, 68–99
	MK-0941	Discontinued (phase 2)		
OSI Prosidion/Eli Lilly	PSN010	Discontinued (phase 2)	4.5.5	33, 100–111
	LY2599506			
	LY2608204	Phase 1		
Pfizer	PFE-GKA1	Phase 1	4.5.6	112–113
	PFE-GKA2	Phase 1		
Roche	R-1675 (RO0281675)	Discontinued (phase 1)	4.5.7	25, 114–144
	R-1440 (RO4389620, Piragliatin)	Discontinued (phase 2)		
	R-1511	Discontinued (phase 1)		
Takeda	TAK-329	Phase 1	4.5.8	145–162
TransTech Pharma/ Forest Laboratories/Novo Nordisk	TTP399	Phase 2	4.5.9	163–175
	TTP547	Phase 1		
	TTP355	Discontinued		
	NN9101	(phase 1)		
Zydus Cadila	ZYDK1	Phase 1	4.5.10	176–178

in healthy volunteers and a phase 1 multi-dose study in T2DM patients.[35,36] The structure of GKM-001 has not yet been reported; however, several patent applications detail the synthesis and activity of novel α-alkoxy amides and pyrrole carboxamides.[37–39] From these applications, **1** and **2** (Figure 4.1) are representative activators reported to have mid- to low-nanomolar potencies and to afford significant increases in *de novo* glycogen synthesis in rat hepatocytes at 10 μM concentration.[37,38]

4.5.2   Array Biopharma / Amgen Glucokinase Activators

Array Biopharma has filed multiple glucokinase activator patent applications, disclosing structures derived from an aminopyridine core bearing two pendent

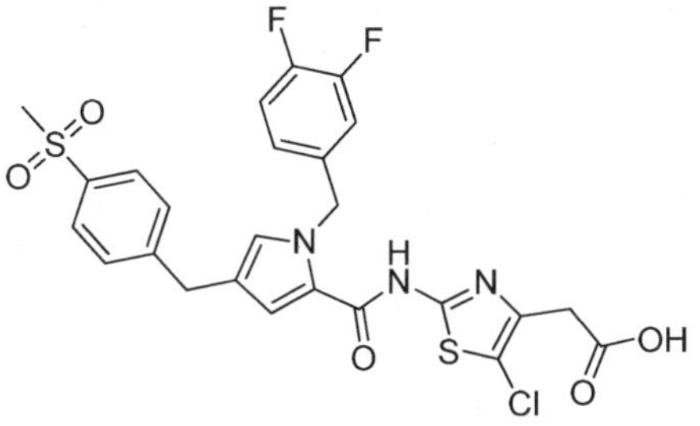

1: EC$_{50}$ = 300 nM

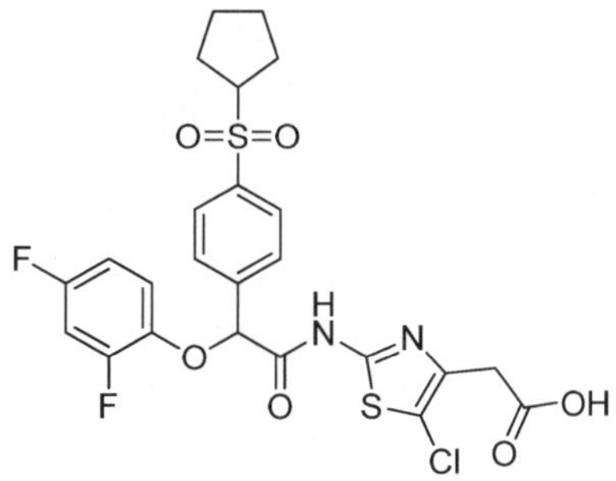

2: EC$_{50}$ = 370 nM

Figure 4.1   Representative Advinus glucokinase activators.

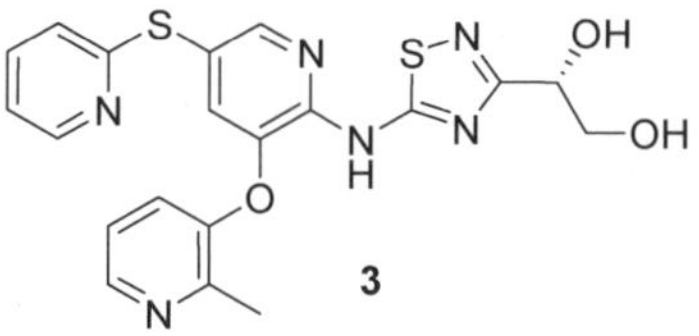

3

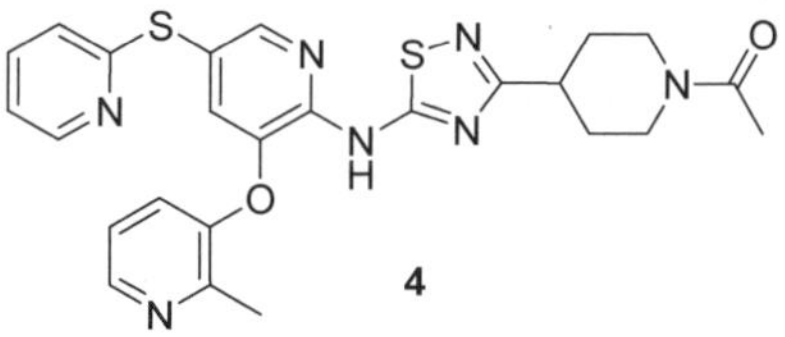

4

Figure 4.2   Representative Array/Amgen glucokinase activators.

aryl or heteroaryl rings linked through an ether or thioether; an optional third substituent is also exemplified in some claimed structures.[40–45] Representative structures **3** and **4** from the Array patent estate are provided in Figure 4.2. Array selected ARRY-403 (structure not disclosed) as a clinical development candidate that entered phase 1 in March 2009. This single ascending dose (25–400 mg qd) study of ARRY-403 was conducted in 41 T2DM patients.[46–48] In this study, ARRY-403 exhibited dose linear exposure and reduced fasting as well as postprandial glucose following a mixed meal tolerance test (MMTT). ARRY-403 was well tolerated although one subject in the 400 mg cohort experienced moderate symptomatic hypoglycemia. This candidate was subsequently advanced to a 10-day multiple ascending dose (10–100 mg qd and bid dosing) study in T2DM patients, the results of which have not yet been reported. In late 2009, Array entered a partnership with Amgen for further development of ARRY-403 (AMG-151) as well as other glucokinase activators in their portfolio.[49,50]

4.5.3   AstraZeneca Glucokinase Activators

AstraZeneca has conducted extensive research in the field of glucokinase activation, culminating in the identification of four clinical candidates. Among these candidates, the most advanced, AZD1656, reached phase 2 development and has been the subject of more than 20 clinical studies.[51,52] Phase 1 trials of

AZD1656 have included single ascending dose, multiple ascending dose, comparison of qd versus bid, drug–drug interaction, and human pharmacokinetics studies. Two 4-week studies were also performed with AZD1656 in T2DM patients on either insulin or metformin background theapies.[53,54] A large (530 subjects) phase 2 trial was then conducted to evaluate the efficacy, safety, and tolerability of AZD1656 administered for 4 months as add-on treatment to metformin.[55] A separate phase 2 trial was conducted in Japan evaluating 4 months treatment with AZD1656 as monotherapy.[56]

Two other Astra Zeneca candidates, AZD6714 and AZD5658, also entered phase 1 studies. A single ascending dose study of AZD6714 in healthy volunteers was terminated early citing an inability to identify relevant doses for progression into T2DM patients.[57] A phase 1 study in type 2 diabetics was completed in 2011 for AZD5658 and additional studies have not been reported.[58] A final candidate, AZD6370, had progressed to phase 2 studies in 2008;[59] however, further development work on this compound has not been reported.[51] Previously a phase 1 trial with AZD6370 was conducted in T2DM subjects.[60] In addition to these four candidates, AstraZeneca has recently published the structure of an additional development candidate AZD1092 (**5**, Figure 4.3)[61] which was in preclinical testing.[62,63]

In July 2011, AstraZeneca provided a pipeline update indicating that their three active clinical candidates (AZD1656, AZD5658, and AZD6714) had all been discontinued.[51] No specific explanation was offered for the termination.

While the structures of AstraZeneca's four clinical candidates have not yet been disclosed, representative examples from recent process and crystal form patent applications are illustrated in Figure 4.3. A 2006 process patent application describes the synthesis of disubstituted aryl amide **6** on multi-

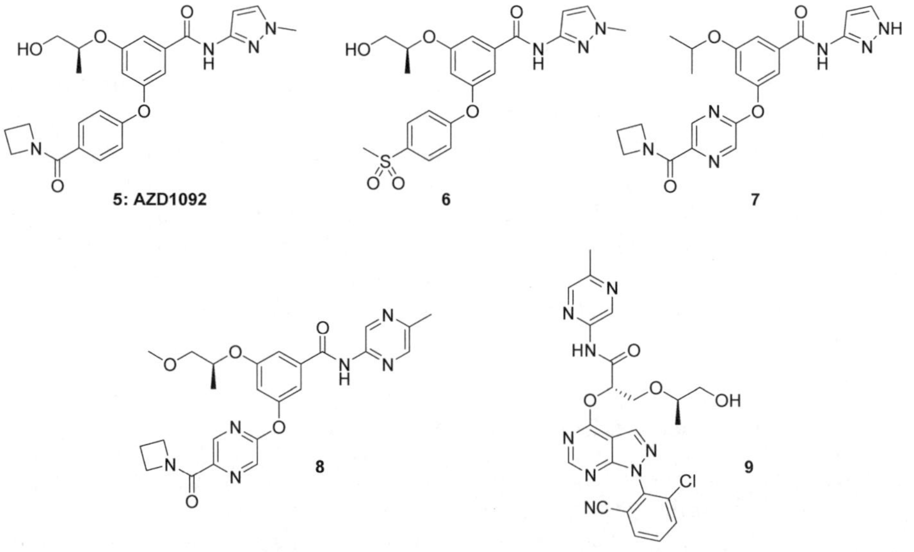

Figure 4.3 Representative AstraZeneca glucokinase activators.

kilogram scale.[64] Crystal form patent applications have also highlighted compounds **7** and **8**.[65,66] In 2010, a patent application with narrow scope disclosed the structurally unique compound **9** as a glucokinase activator with $EC_{50} = 0.442$ μM.[67]

4.5.4 Merck Glucokinase Activators

Merck has also had a significant interest in glucokinase activators developed principally through its acquisition of Banyu Pharmaceuticals. Banyu has numerous glucokinase activator patent applications published between 2004 and 2010. While the scope of these applications is substantial, the majority of the claimed compounds can be classified into three core groups: benzimidazoles,[68–73] quinazolines,[74–76] and benzamides.[77–86] From these efforts, Merck advanced two activators, MK-0941 and MK-0599, into clinical development; however, both candidates have now been discontinued.[87]

The structure of MK-0941 (**10**, Figure 4.4) was disclosed in a recent process chemistry publication.[88] It has also been claimed as a single compound in a recent patent application.[86] Merck conducted an extensive clinical development program for MK-0941, including seven phase 1 and two phase 2 trials. A phase 1 study of MK-0941, initiated in May 2007, examined safety and efficacy of multiple daily administration of MK-0941 before meals in T2DM patients.[89] A second study of similar design examined the effects of MK-0941 on postprandial glucose in T2DM patients on background insulin therapy.[90] Several other phase 1 studies examining MK-0941 ADME, its effects in Japanese patients, and its effects in combination with other oral anti-diabetic therapies have been conducted.[91–94] A planned study to evaluate MK-0941 in patients with renal impairment was terminated.[95] At least two phase 2 dose escalation (10 to 40 mg, tid) studies of MK-0941 were conducted in T2DM patients on metformin or insulin background therapy.[96,97] Most notably, in the phase 2 study of patients on insulin background therapy, significant reductions in HbA_{1c} were observed through 14 weeks of treatment, but subsequent up titration to the maximum tolerated dose over the subsequent 30 weeks resulted in a loss of durability of effect.[31] In addition to loss of efficacy, an increase in plasma triglycerides as well as increased incidence of hypoglycemia was observed in subjects treated with MK-0941. Increases in blood pressure were

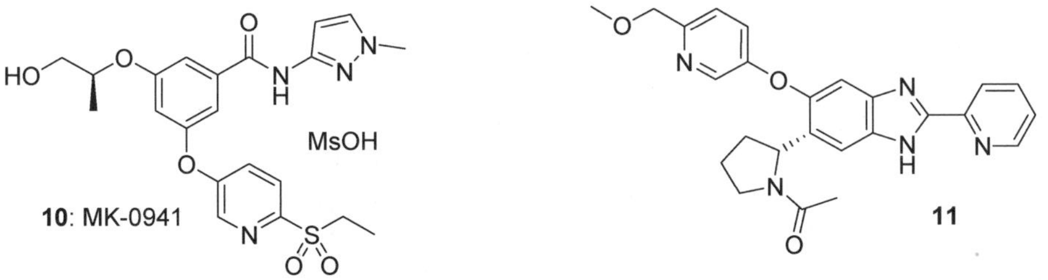

Figure 4.4 Merck glucokinase activators.

also noted for subjects treated with MK-0941. Based on these results, development of MK-0941 was terminated.[96]

The structure of Merck's second development candidate, MK-0599, is currently unknown, although a 2008 publication[98] highlights a process synthesis of 0.7 kg of a benzimidazole (11) originally claimed in a 2005 Banyu patent application.[73] Merck conducted a phase 1 evaluation of MK-0599 in healthy subjects evaluating 0.4 to 160 mg qd and 13 to 100 mg tid doses, observing glucose-lowering effects.[99] No subsequent development activities have been reported for MK-0599.

4.5.5   OSI Prosidion / Eli Lilly Glucokinase Activators

OSI Prosidion and Tanabe have been engaged in the discovery of glucokinase activators and have reported PSN-GK1 (12, Figure 4.5), as well as the clinical development candidate PSN010, whose structure has not been disclosed.[100] PSN010 activates GK *in vitro* by 2.1-fold at 5 mM glucose with an EC_{50} of 540 nM.[101] PSN010 entered phase 1 trials in 2006.[102] In 2007, Lilly acquired the exclusive rights to the OSI Prosidion glucokinase activator program.[103] Results of a phase 1 2-week multiple ascending dose study of PSN010 (also called LY2599506) in T2DM patients revealed that the candidate afforded significant reductions in both fasting and postprandial glucose; however, significant occurrences of hypoglycemia and changes in liver function tests were identified as key adverse events.[33] In late 2009, PSN010 (LY2599506) was progressed to a phase 2 12-week study in T2DM patients; however, this study was prematurely terminated citing nonclinical safety findings.[104,105]

Lilly also progressed a second candidate, LY2608204, into early development. The structure of this second candidate has not yet been reported; however, the patent literature indicates two compounds of interest (Figure 4.6). Specifically, a 2009 application covers a crystalline form of 13,[106] which was first disclosed in a 2004 OSI application.[107] A second activator, 14, was disclosed as a single compound in a 2010 Lilly patent application application.[108] This compound reportedly activated glucokinase with an EC_{50} of 42 nM at 10 mM glucose. In an *in vivo* rat oral glucose tolerance test (OGTT), 14 decreased plasma glucose in a

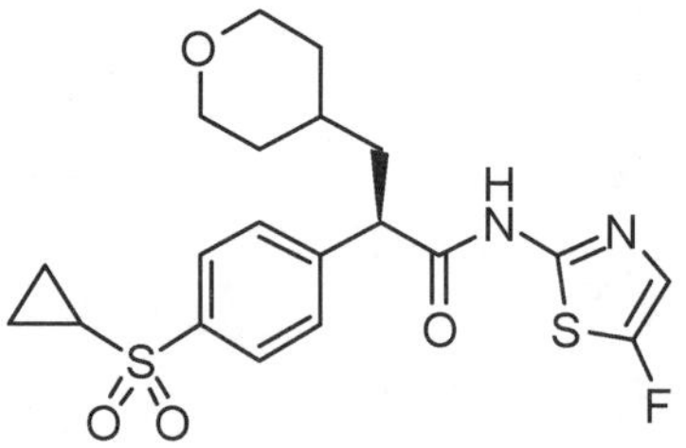

12

Figure 4.5   Structure of PSN-GK1.

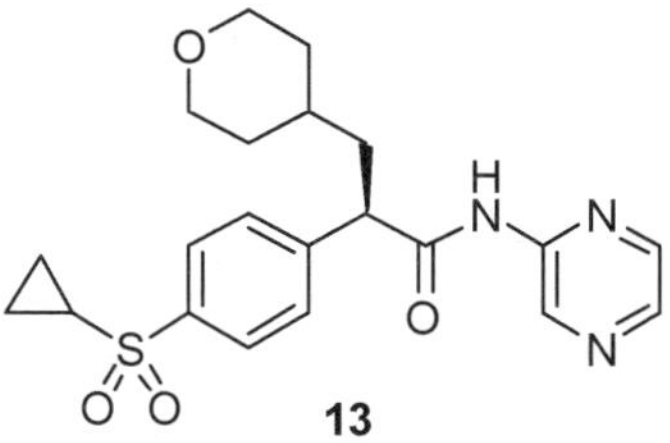

Figure 4.6 Representative Lilly glucokinase activators.

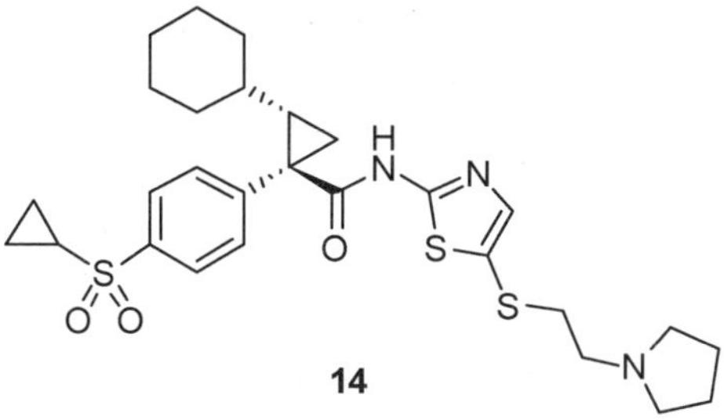

dose-dependent manner in both the fasted and postprandial states. This compound also was shown to have limited blood–brain barrier penetration.

A 2-week phase 1 study of LY2608204, dosed once daily, in type 2 diabetics was completed in 2010.[109] In 2011, a phase 1 study was completed whereby type 2 diabetics were treated for 4-weeks with ascending doses of LY2608204 starting at 160 mg qd and escalating weekly to doses of 240, 320, and 400 mg.[110] A subsequent phase 1 study was completed in 2011 testing a new formulation of LY2608204.[111]

4.5.6 Pfizer Glucokinase Activators

Pfizer have recently disclosed two glucokinase activators, **15** and **16** (Figure 4.7), as clinical development candidates. The discovery of **15** as a "partial activator" of glucokinase was recently reported by utilizing a screening cascade designed to establish a correlation between an activator's biochemical properties (i.e. changes in glucokinase K_m and v_{max}) and its *in vivo* efficacy and safety profile in order to identify activators with reduced hypoglycemia risk.[112] Activator **15** has a modest effect on K_m ($\alpha = 0.10$) and minimal effect on v_{max} ($\beta = 0.87$) with an EC_{50} of 188 nM. Consistent with its design for reduced hypoglycemia risk, when tested in dispersed rat islets, at a range of 1- to 3-fold its biochemical potency, activator **15** had no effects at low glucose (3 mM and 5 mM) but potentiated glucose-stimulated insulin secretion at higher glucose concentrations. In a rat OGTT, **15** offered dose-dependent

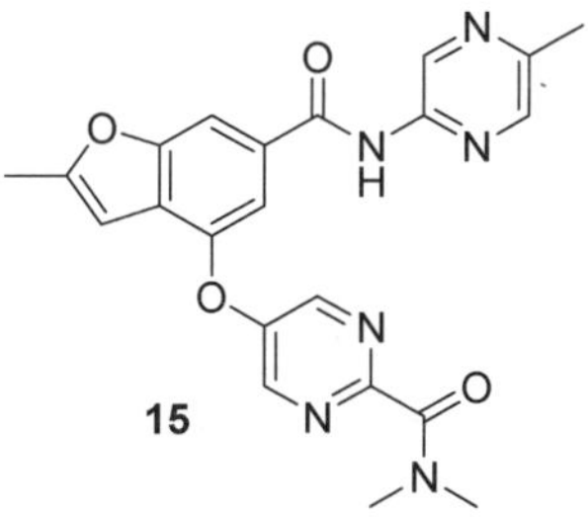

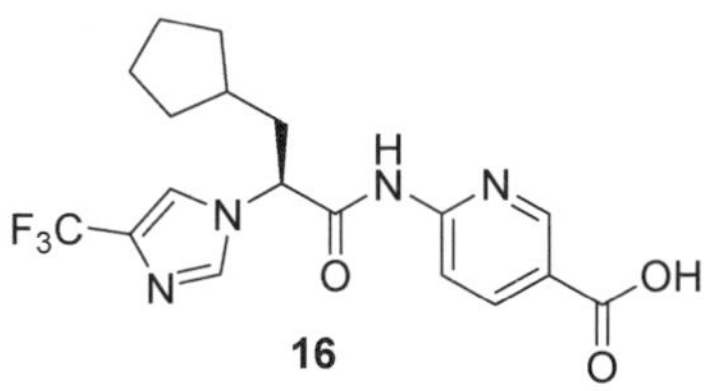

Figure 4.7 Pfizer glucokinase activator clinical candidates.

reductions in postprandial glucose with minimal risk of hypoglycemia. Activator **15** was reportedly advanced to a phase 1 single escalating dose study (10–640 mg, qd) in type 2 diabetic patients to evaluate safety, pharmacokinetics, and pharmacodynamics. The results of this study and further development activities have not yet been reported.

In parallel with the strategy described above, Pfizer has also reported the selection of hepatoselective activator **16** as a clinical development candidate.[113] Pfizer along with Advinus (see above) and TransTech Pharma (see below) are reportedly pursing liver-selective activators to mitigate the hypoglycemia risk previously reported with prototypical dual acting agents. Activator **16** (EC_{50} = 68 nM) was optimized as a substrate for active hepatic uptake via organic anion transporter polypeptides (OATP) which are selectively expressed on hepatocyctes. Concurrently, the passive permeability of **16** was reduced to minimize distribution into peripheral tissues such as pancreas. The repeat dose liver:pancreas distribution of **16** was reported to be 75-fold in rat and 94-fold in dog, and the compound was effective at reducing both fasting and postprandial glucose in the Goto-Kakizaki diabetic rat model. A single escalating dose phase 1 study of **16** in healthy volunteers was conducted, and the results of this study as well as future development activities for **16** have not yet been reported.

4.5.7   Roche Glucokinase Activators

Researchers at Roche have pioneered the discovery of glucokinase activators, which they first reported in 2003,[25] and an excellent account of these efforts has recently been published.[114] Roche has an extensive patent estate in the field, focused largely on an aryl propionamide template originally identified through HTS screening.[115–137] Roche has advanced three clinical candidates into development: piragliatin also known as RO4389620 (R-1440); RO0281675 (R-1675); and R-1511. The structures for RO0281675 and piragliatin have been reported as **17** and **18**, respectively (Figure 4.8).[25,122,126,127,137,138] The structure of R-1511 has not yet been reported although a recent process patent identified **19** as a compound of interest.[124,139]

RO-0281675 (**17**) was the first glucokinase activator advanced to clinical studies by Roche. In a phase 1 study conducted in healthy volunteers, RO-0281675 (25–

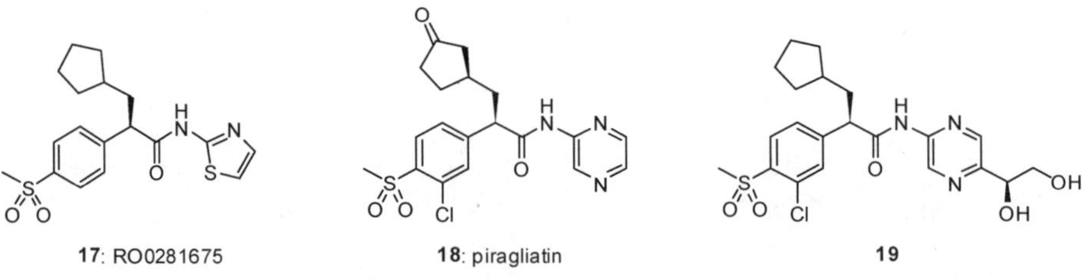

Figure 4.8   Roche glucokinase activator candidates.

400 mg, qd) was found to dose-dependently lower the glucose excursion associated with an oral glucose tolerance test. Hypoglycemia was the only adverse event observed in this study; however, observations of hepatic lipidosis in preclinical safety studies, attributed to metabolism of the aminothiazole motif of RO-0281675, resulted in termination of this candidate.[114,140]

Piragliatin (**18**), Roche's most advanced candidate, was advanced through phase 2 studies of up to 52 weeks in duration.[141] The discovery and clinical development of piragliatin (**18**) has recently been reviewed by Sarabu and co-workers.[114] Piragliatin was profiled in multiple phase 1 and phase 2 studies. Early clinical studies of piragliatin demonstrated reductions in fasting and postprandial glucose during a 6-day multiple ascending dose study in T2DM patients with hypoglycemia as the main adverse event. Further development of this candidate was terminated for reasons not yet reported. After the discontinuation of piragliatin, Roche advanced a third candidate, R-1511, into clinical development, including a phase 1 study in T2DM patients.[142,143] R-1511 was subsequently discontinued.[144]

4.5.8 Takeda Glucokinase Activators

Takeda have been an active participant in the glucokinase activator field, having published at least 16 patent applications since 2006.[145–160] Their medicinal chemistry strategy has focused on two major themes: exploitation of novel fused ring systems (e.g. indole, benzimidazole, etc.) to establish the glucokinase activator donor–acceptor pharmacophore and the development of α-amino amides, particularly those possessing N-linked pyridones. From this chemical estate, Takeda has reportedly advanced TAK-329 (structure not disclosed) into phase 1 development.[161,162] Interestingly, the currently reported phase 1 study of TAK-329 is a glucose clamp study conducted in type 1 diabetes mellitus (T1DM) patients. This is one of the few reported studies wherein a glucokinase activator has been evaluated in T1DM patients.

4.5.9 TransTech Pharma / Novo Nordisk / Forest Laboratories Glucokinase Activators

TransTech Pharma, in partnership with Novo Nordisk, has been extensively engaged in the development of hepatoselective glucokinase activators as a strategy to mitigate the hypoglycemia risk observed with previous dual acting activators. This work began in 2001 as a collaboration between TransTech Pharma and Novo Nordisk, and in 2007 Novo Nordisk licensed this portfolio to TransTech Pharma.[163,164] These efforts have resulted in the identification of three clinical candidates: TTP355, TTP399, and TTP547. The structures of these clinical candidates have not yet been reported; nevertheless, representative structures from the Novo Nordisk patent estate are shown in Figure 4.9.[165–167] The first candidate, TTP355 (NN9101), a liver-selective glucokinase activator, advanced to phase 1 in 2006 and demonstrated reductions in fasting and

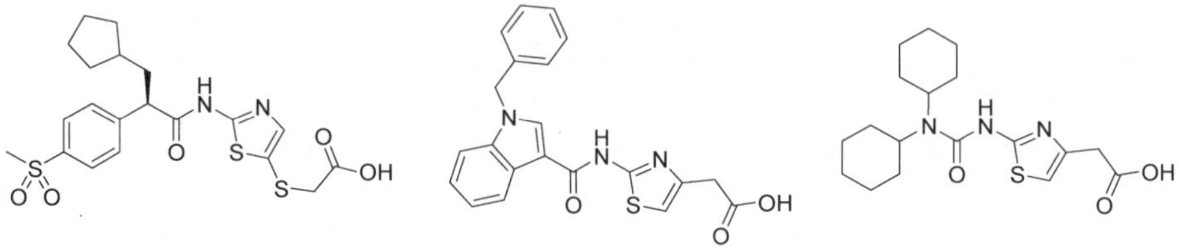

Figure 4.9 Representative Novo Nordisk / TransTech Pharma glucokinase activators.

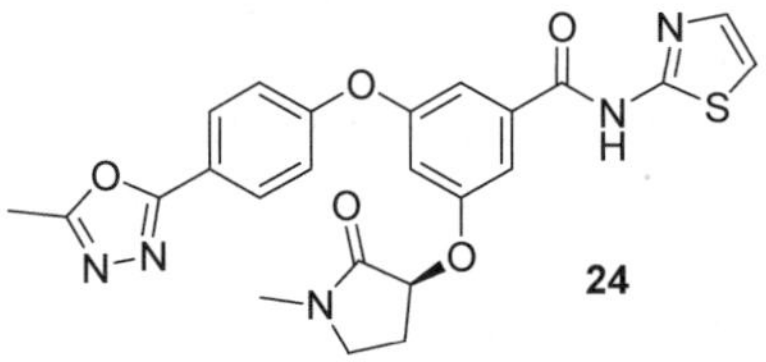

Figure 4.10 Representative structure of Zydus Cadila chemical matter.

postprandial glucose.[168–170] Preclinically, TTP355 was found to reduce HbA_{1C} levels in mildly diabetic rats and increase β-cell mass. No hypoglycemia was observed in fed or fasted, normal or diabetic animal models.[169] The clinical development of TTP355 was terminated in 2009.[170]

TTP399, also liver selective, represents the current lead candidate from this family of activators and is currently entering phase 2 development.[171,172] While its structure has not been disclosed, TTP399 reportedly originates from a different chemical series relative to TTP355.[173] A third candidate, TTP547, was also identified and completed a single ascending dose phase 1 study in healthy volunteers in 2009.[170] In 2010, Forest Laboratories entered into a licensing agreement with TransTech for the further development of TTP399 and other glucokinase activators in their portfolio.[172]

4.5.10 Zydus Cadila Glucokinase Activators

In 2011 Zydus Cadila reported its first glucokinase activator ZYDK1 which is currently in phase 1 development.[174,175] ZYDK1 was reportedly effective in controlling both fasting and non-fasting glucose levels in preclinical models. The structure of ZYDK1 has not yet been disclosed; however, a recent patient application has been published from which **24** (Figure 4.10) was selected as a representative example.[176]

References

1. *World Health Organization Diabetes Statistics*, http://www.who.int/ mediacentre/factsheets/fs312/en/, Accessed 02-17, 2011.

2. P. H. Bennett and W. C. Knowler, in *Definitions, Diagnosis and Classifications of Diabetes Mellitus and Glucose Homeostasis*, in Joslin's Diabetes Mellitus, 14th edn., 2004, pp. 331–339.

3. J. Xu, *IDrugs*, 2004, **7**, 249–256.

4. M. C. T. Fyfe and M. J. Procter, *Drugs Future*, 2009, **34**, 641–653.

5. F. M. Matschinsky, *Nat. Rev. Drug Discovery*, 2009, **8**, 399–416.

6. F. M. Matschinsky and M. A. Magnuson (eds.), *Glucokinase and Glycemic Disease: From Basics to Novel Therapeutics*, Basel, 2004.

7. F. M. Matschinsky, B. Glaser and M. A. Magnuson, *Diabetes*, 1998, **47**, 307–315.

8. H. Heimberg, A. De Vos, K. Moens, E. Quartier, L. Bouwens, D. Pipeleers, E. Van Schaftingen, O. Madsen and F. Schuit, *Proc. Natl. Acad. Sci. USA*, 1996, **93**, 7036–7041.

9. S. Del Guerra, R. Lupi, L. Marselli, M. Masini, M. Bugliani, S. Sbrana, S. Torri, M. Pollera, U. Boggi, F. Mosca, S. Del Prato and P. Marchetti, *Diabetes*, 2005, **54**, 727–735.

10. N. Doliba, W. Qin, H. Najafi, H. Collins, D. Wilson, J. Grimsby, R. Sarabu, C. Liu, A. Naji and F. M. Matschinsky, 71st American Diabetes Association Meeting, 0473-PP, San Diego, CA, 2011.

11. M. L. MacDougall, S. J. Orena, L. J. Sweet, C. M. Boustany-Kari, J. A. Pfefferkorn and O. Cabrera, 71st American Diabetes Association Meeting, 0126-LB, San Diego, CA, 2011.

12. P. B. Iynedjian, *Cell. Mol. Life Sci.*, 2009, **66**, 27–42.

13. E. Van Schaftingen, A. Vandercammen, M. Detheux and D. R. Davies, *Adv. Enzyme Regul.*, 1992, **32**, 133–148.

14. A. Basu, R. Basu, P. Shah, A. Vella, C. M. Johnson, M. Jensen, K. S. Nair, W. F. Schwenk and R. A. Rizza, *Diabetes*, 2001, **50**, 1351–1362.

15. G. Velho, K. F. Petersen, G. Perseghin, J.-H. Hwang, D. L. Rothman, M. E. Pueyo, G. W. Cline, P. Froguel and G. I. Shulman, *J. Clin. Invest.*, 1996, **98**, 1755–1761.

16. J. F. Caro, S. Triester, V. K. Patel, E. B. Tapscott, N. L. Frazier and G. L. Dohm, *Horm. Metab. Res.*, 1995, **27**, 19–22.

17. B. Willms, P. Ben-Ami and H. D. Soeling, *Horm. Metab. Res.*, 1970, **2**, 135–141.

18. T. P. Torres, R. L. Catlin, R. Chan, Y. Fujimoto, N. Sasaki, R. L. Printz, C. B. Newgard and M. Shiota, *Diabetes*, 2009, **58**, 78–86.

19. A. A. Dunn-Meynell, V. H. Routh, L. Kang, L. Gaspers and B. E. Levin, *Diabetes*, 2002, **51**, 2056–2065.

20. D. Zelent, M. L. Golson, B. Koeberlein, R. Quintens, L. van Lommel, C. Buettger, H. Weik-Collins, R. Taub, J. Grimsby, F. Schuit, K. H. Kaestner and F. M. Matschinsky, *Diabetes*, 2006, **55**, 1923–1929.

21. A. L. Gloyn, *Hum. Mutat.*, 2003, **22**, 353–362.

22. P. R. Njolstad, O. Sovik, A. Cuesta-Munoz, L. Bjorkhaug, O. Massa, F. Barbetti, D. E. Undlien, C. Shiota, M. A. Magnuson, A. Molven, F. M. Matschinsky and G. I. Bell, *N. Engl. J. Med.*, 2001, 1588–1592.

23. B. Glaser, P. Kesavan, M. Heyman, E. Davis, A. Cuesta, A. Buchs, C. A. Stanley, P. S. Thornton, M. A. Permutt, F. M. Matschinsky and K. C. Herold, *N. Engl. J. Med.*, 1998, **338**, 226–230.

24. A. L. Gloyn, K. Noordam, M. A. A. P. Willemsen, S. Ellard, W. W. K. Lam, I. W. Campbell, P. Midgley, C. Shiota, C. Buettger, M. A. Magnuson, F. M. Matschinsky and A. T. Hattersley, *Diabetes*, 2003, **52**, 2433–2440.

25. J. Grimsby, R. Sarabu, W. L. Corbett, N.-E. Haynes, F. T. Bizzarro, J. W. Coffey, K. R. Guertin, D. W. Hilliard, R. F. Kester, P. E. Mahaney, L. Marcus, L. Qi, C. L. Spence, J. Tengi, M. A. Magnuson, C. A. Chu, M. T. Dvorozniak, F. M. Matschinsky and J. F. Grippo, *Science*, 2003, **301**, 370–373.

26. M. Coghlan and B. Leighton, *Expert Opin. Invest. Drugs*, 2008, **17**, 145–167.

27. R. Sarabu, S. J. Berthel, R. F. Kester and J. W. Tilley, *Expert Opin. Ther. Patents*, 2008, **18**, 759–768.

28. R. Sarabu, S. J. Berthel, R. F. Kester and J. W. Tilley, *Expert Opin. Ther. Patents*, 2011, **21**, 13–33.

29. J. Zhi, S. Zhai, M.-E. Mulligan, J. Grimsby, C. Arbet-Engels, M. Boldrin and R. Balena, Presentation 42, EASD 44th Annual Meeting, Rome, Italy, 2008.

30. C. Bonadonna Riccardo, T. Heise, C. Arbet-Engels, C. Kapitza, A. Avogaro, J. Grimsby, J. Zhi, F. Grippo Joseph and R. Balena, *J. Clin. Endocrinol. Metab.*, 2010, **95**, 5028–5036.

31. G. E. Meininger, R. Scott, M. Alba, Y. Shentu, E. Luo, H. Amin, M. J. Davies, K. D. Kaufman and B. J. Goldstein, *Diabetes Care*, 2011, **34**, 2560–2566.

32. Y. Shentu, Y. Chen, Q. Yu, D. Sheng, B. J. Musser, M. Alba, B. B. Zhang, E. M. Migoya, J. Ehrhart, K. D. Kaufman and B. J. Goldstein, EASD 46th Annual Meeting, Stockholm, Sweden, 2010.

33. J. M. Bue-Valleskey, K. B. Schneck, V. P. Sinha, E. T. Wonddmagegnehu, C. Kapitza and J. W. Miller, 71st American Diabetes Association Meeting, 0993-P, San Diego, CA, 2011.

34. R. M. O'Doherty, D. L. Lehman, S. Telemaque-Potts and C. B. Newgard, *Diabetes*, 1999, **48**, 2022–2027.

35. *Advinus Company News GKM-001*, http://www.advinus.com/view_news. asp?id=151, Accessed 06-29, 2011.

36. *Advinus Phase 1 Clinical Trial*, http://ctri.nic.in/Clinicaltrials/pmaindet2. php?trialid=2869, Accessed 06-29, 2011.

37. *WO Patent*, 047798, 2009.

38. *WO Patent*, 149382, 2008.

39. *WO Patent*, 104994, 2008.

40. *WO Patent*, 042435, 2009.

41. *WO Patent*, 091770, 2008.

42. *WO Patent*, 117381, 2007.

43. *WO Patent*, 118718, 2008.

44. *WO Patent*, 089512, 2007.

45. *WO Patent*, 053345, 2007.

46. *Array Phase 1 ARRY-403 #1*, http://clinicaltrials.gov/ct2/show/ NCT00859755?term=NCT00859755&rank=1, Accessed 06-27, 2011.

47. R. J. Hinklin, T. D. Aicher, S. A. Boyd, K. R. Condroski, J. DeWolf, E. Walter, J. B. Fell, J. Fischer, M. L. Frank, M. Hompesch, J. Imm, A. James, S. Karan, P. A. Lee, K. Litwiler, M. McVean, N. A. Neitzel, S. Rojas-Caro, E. M. Wallace, L. M. Williams and J. M. Trevillyan, *ARRY-403, A Glucokinase Activator with Potent Glucose-dependent Anti-hyperglycemic Activity In Animal Models of Type 2 Diabetes Mellitus: First-in-Patient Clinical Results*, http://www.arraybiopharma.com/_documents/ Publication/PubAttachment361.pdf, Accessed 06-27, 2011.

48. *Results of ARRY-403 Phase 1*, http://www.arraybiopharma.com/ Documents/PDF/Slides.pdf, Accessed 06-27, 2011.

49. *Array Collaboration with Amgen*, http://www.arraybiopharma.com/ Collaboration/Amgen.asp, Accessed 06-27, 2011.

50. T. D. Aicher, D. Anderson, S. A. Boyd, M. Chicarelli, K. R. Condroski, J. DeWolf, E. Walter, J. B. Fell, J. Fischer, M. L. Frank, S. E. Galbraith, R. Garrey, I. Gunawardana, R. J. Hinklin, J. Hirsch, P. A. Lee, K. Litwiler, N. A. Neitzel, S. A. Pratt, A. Singh, F. X. Sullivan, L. K. Thomas, T. M. Turner, W. C. Voegtli, E. M. Wallace and L. M. Williams, *ARRY-403, A Novel Gluco-kinase Activator with Potent Glucose-Dependent Anti-Hyperglycemic Activity in Animal Models of Type 2 Diabetes Mellitus*, http://www.arraybiopharma. com/_documents/Publication/PubAttachment318.pdf, Accessed 06-27,2011.

51. *AstraZeneca Pipeline 28 July 2011*, http://www.astrazeneca.com/Research/ Our-pipeline-summary, Accessed 04-19, 2011.

52. *Clinical Trial List for AZD1656*, http://www.clinicaltrials.gov/ct2/ results?term=AZD1656&pg=1, Accessed 08-04, 2011.

53. *AstraZeneca AZD1656 4-Week Insulin Phase 2 Trial*, http://www.clinical-trials.gov/ct2/show/NCT00856908?term=AZD1656&rank=22, Accessed 08-04, 2011.

54. *AstraZeneca AZD1656 4-Week Metformin Phase 2 Trial*, http://www. clinicaltrials.gov/ct2/show/NCT00817778?term=AZD1656&rank=23, Accessed 08-04, 2011.

55. *AstraZeneca AZD1656 12-Week Phase 2 Trial*, http://www.clinicaltrials. gov/ct2/show/NCT01020123?term=AZD1656&rank=8, Accessed 08-04, 2011.

56. *AstraZeneca AZD1656 Japan Phase 2 Trial*, http://www.clinicaltrials.gov/ ct2/show/NCT01152385?term=AZD1656&rank=12, Accessed 08-04, 2011.

57. *AstraZeneca 6714*, http://www.clinicaltrials.gov/ct2/show/NCT00886366?term= AZD6714&rank=1.

58. *AstraZeneca 5658*, http://www.clinicaltrials.gov/ct2/show/NCT01176097?term= AZD5658&rank=1.

59. *AstraZeneca Directors Report*, http://www.astrazeneca-annualreports.
    com/2008/directors_report/therapy_area_review/index.html Accessed
    04-19, 2011.
60. *AstraZeneca AZD6370 Phase 1 DRF Study*, http://www.clinicaltrials.gov/
    ct2/show/NCT00690287?term=AZD6370&rank=1, Accessed 08-04, 2011.
61. M. J. Waring, C. Johnstone, D. McKerrecher, K. G. Pike and G. Robb,
    Med. Chem. Commun., 2011, **2**, 775–779.
62. *AstraZeneca Development Pipeline 2 Feb 2006*, http://moneyam.uk-wire.
    com/cgi-bin/articles/200602021100078116X.html?print=1;landscape=1,
    Accessed 07/05/2011, 2011.
63. *AstraZeneca Development Pipeline 8 June 2006*, http://en.astrazeneca.
    com.cn/_mshost379733/content/legacy-site-content/resources/media/
    1274415/1274417, Accessed 07/05/2011, 2011.
64. *WO Patent*, 040527, 2006.
65. *WO Patent*, 075073, 2008.
66. *WO Patent*, 092386, 2010.
67. *WO Patent*, 116176, 2010.
68. *WO Patent*, 024110, 2010.
69. *JP Patent*, 091324, 2009.
70. *JP Patent*, 022937, 2007.
71. *WO Patent*, 007910, 2007.
72. *WO Patent*, 049304, 2006.
73. *WO Patent*, 063738, 2005.
74. *WO Patent*, 063821, 2009.
75. *WO Patent*, 090332, 2005.
76. T. Iino, Y. Sasaki, M. Bamba, M. Mitsuya, A. Ohno, K. Kamata, H.
    Hosaka, H. Maruki, M. Futamura, R. Yoshimoto, S. Ohyama, K.
    Sasaki, M. Chiba, N. Ohtake, Y. Nagata, J.-I. Eiki and T. Nishimura,
    Bioorg. Med. Chem. Lett., 2009, **19**, 5531–5538.
77. *WO Patent*, 076420, 2004.
78. *WO Patent*, 081782, 2009.
79. *WO Patent*, 041475, 2009.
80. *WO Patent*, 044777, 2008.
81. *WO Patent*, 081001, 2004.
82. M. Mitsuya, K. Kamata, M. Bamba, H. Watanabe, Y. Sasaki, K. Sasaki,
    S. Ohyama, H. Hosaka, Y. Nagata, J.-i. Eiki and T. Nishimura, *Bioorg.
    Med. Chem. Lett.*, 2009, **19**, 2718–2721.
83. T. Iino, N. Hashimoto, T. Hasegawa, M. Chiba, J.-I. Eiki and T.
    Nishimura, *Bioorg. Med. Chem. Lett.*, 2010, **20**, 1619–1622.
84. T. Iino, N. Hashimoto, K. Sasaki, S. Ohyama, R. Yoshimoto, H. Hosaka,
    T. Hasegawa, M. Chiba, Y. Nagata, J.-i. Eiki and T. Nishimura, *Bioorg.
    Med. Chem.*, 2009, **17**, 3800–3809.
85. T. Iino, D. Tsukahara, K. Kamata, K. Sasaki, S. Ohyama, H. Hosaka, T.
    Hasegawa, M. Chiba, Y. Nagata, J.-i. Eiki and T. Nishimura, *Bioorg.
    Med. Chem.*, 2009, **17**, 2733–2743.

86. *WO Patent*, 107610, 2010.

87. *Merck Pipeline*, http://www.merck.com/research/pipeline/home.html, Accessed 06-28, 2011.

88. N. Yoshikawa, F. Xu, J. D. Arredondo and T. Itoh, *Org. Process Res. Dev.*, 2011, **15**, 824–830.

89. *Merck Phase 1 MK0941 #1*, http://clinicaltrials.gov/show/NCT00511667, Accessed 06-27, 2011.

90. *Merck Phase 1 MK0941 #2*, http://clinicaltrials.gov/show/NCT00511472, Accessed 06-27, 2011.

91. *Merck Phase 1 MK0941 #4*, http://clinicaltrials.gov/ct2/show/ NCT00754130, Accessed 06-27, 2011.

92. *Merck Phase 1 MK0941 #5*, http://clinicaltrials.gov/ct2/show/ NCT00912002, Accessed 06-27, 2011.

93. *Merck Phase 1 MK0941 #6*, http://clinicaltrials.gov/ct2/show/ NCT00873821, Accessed 06-27, 2011.

94. *Merck Phase 1 MK0941 #7*, http://clinicaltrials.gov/show/NCT01106287, Accessed 06-27, 2011.

95. *Merck Phase 1 MK0941 #8*, http://clinicaltrials.gov/show/NCT01106287, Accessed 06-27, 2011.

96. *Merck Phase 2 Trial #1*, http://clinicaltrials.gov/ct2/show/NCT00792935, Accessed 06-29, 2011.

97. *Merck Phase 2 Trial #2*, http://www.clinicaltrials.gov/ct/show/ NCT00824616, Accessed 06-29, 2011.

98. A. Klapars, K. R. Campos, J. H. Waldman, D. Zewge, P. G. Dormer and C. Chen, *J. Org. Chem.*, 2008, **73**, 4986–4993.

99. E. M. Migoya, J. Miller, T. Laetham, L. Maganti, K. Gottesdeiner and J. A. Wagner, Presentation Number 876, EASD 45th Annual Meeting, Vienna, Austria, 2009.

100. L. S. Bertram, D. Black, P. H. Briner, R. Chatfield, A. Cooke, M. C. T. Fyfe, P. J. Murray, F. Naud, M. Nawano, M. J. Procter, G. Rakipovski, C. M. Rasamison, C. Reynet, K. L. Schofield, V. K. Shah, F. Spindler, A. Taylor, R. Turton, G. M. Williams, P. Wong-Kai-In and K. Yasuda, *J. Med. Chem.*, 2008, **51**, 4340–4345.

101. M. C. T. Fyfe, J. R. White, A. Taylor, R. Chatfield, E. Wargent, R. L. Printz, T. Sulpice, J. G. McCormack, M. J. Procter, C. Reynet, P. S. Widdowson and P. Wong-Kai-In, *Diabetologia*, 2007, **50**, 1277–1287.

102. *OSI Pharmaceuticals PSN010 Phase 1 Announcement*, http://investor. osip.com/releasedetail.cfm?ReleaseID=372499 Accessed 05-22, 2011.

103. *OSI Pharmaceuticals Lilly Collaboration*, http://osip.client.shareholder. com/releasedetail.cfm?ReleaseID=372069 Accessed 05-22, 2011.

104. *OSI Pharmaceuticals PSN010 Phase 2 Trials*, http://www.osip.com/ PSN010, Accessed 05-22, 2011.

105. *LY2599506 Phase 2 Trial Termination*, http://www.clinicaltrials.gov/ct2/ show/NCT01029795?term=LY2599506&rank=1 Accessed 05-22, 2011.

106. *US Patent*, 0181981, 2009.

107. *WO Patent*, 072031, 2004.

108. *WO Patent*, 080333, 2010.

109. *Lilly Phase 1 Study of LY2608204* http://www.lillytrials.com/initiated/files/ 13000.pdf Accessed 05-22, 2011.

110. *Phase 1 Study of LY2608204 #1*, http://www.clinicaltrials.gov/ct2/show/ NCT01247363?term=glucokinase&rank=2 Accessed 05-22, 2011.

111. *Phase 1 Study of LY2608204 #2*, http://www.clinicaltrials.gov/ct2/show/ NCT01313286?term=glucokinase&rank=3 Accessed 05-22, 2011.

112. J. A. Pfefferkorn, A. Guzman-Perez, P. J. Oates, J. Litchfield, G. Aspnes, A. Basak, J. Benbow, M. A. Berliner, J. Bian, C. Choi, K. Freeman-Cook, J. W. Corbett, M. Didiuk, J. R. Dunetz, K. J. Filipski, W. M. Hungerford, C. S. Jones, K. K. Karaki, A. Ling, J.-C. Li, L. Patel, H. Risley, J. Saenz, W. Song, M. Tu, R. Aiello, K. Atkinson, N. Barucci, D. Beebe, P. Bourassa, F. Bourbounais, A. M. Brodeur, R. Burbey, J. Chen, T. D'Aquila, D. R. Derksen, N. Haddish-Berhane, C. Huang, J. Landro, A. L. Lapworth, M. MacDougall, D. Perregaux, J. Pettersen, A. Robertson, B. Tan, J. L. Treadway, S. Liu, X. Qiu, J. Knafels, X. Song, P. DaSilva-Jardine, S. Liras, L. Sweet and T. P. Rolph, *MedChemComm*, 2011, **2**, 828–839.

113. J. A. Pfefferkorn, A. Guzman-Perez, J. Litchfield, R. Aiello, J. L. Treadway, J. Petterson, M. L. Minich, K. J. Filipski, C. S. Jones, M. Tu, G. Aspnes, H. Risley, S. W. Wright, J.-C. Li, J. Bian, J. Benbow, R. L. Dow, P. Bourassa, T. D'Aquila, L. Baker, N. Barucci, A. Roberson, F. Bourbonais, D. R. Derksen, M. MacDougall, M. Van Volkenburg, O. Cabrera, J. Chen, A. L. Lapworth, J. A. Landro, K. Atkinson, B. Tan, L. Tao, R. E. Kosa, N. Haddish-Berhane, B. Feng, D. B. Duignan, A. El-Kattan, M. A. Berliner, J. R. Dunetz, S. Murdande, S. Liu, M. Ammirati and J. Knafels, Abstracts of Papers, 241st ACS National Meeting & Exposition, MEDI-164, Anaheim, CA, 2011.

114. R. Sarabu, J. W. Tilley and J. Grimsby, in *RSC Drug Discovery Series*, Editon 1, 2011, vol. 4, pp. 51–70.

115. *WO Patent*, 127544, 2009.

116. *US Patent*, 0248537, 2008.

117. *WO Patent*, 074694, 2008.

118. *US Patent*, 0146625,2008 .

119. *WO Patent*, 043701, 2008.

120. *US Patent*, 0021032, 2008.

121. *WO Patent*, 115967, 2007.

122. *WO Patent*, 115968, 2007.

123. *WO Patent*, 048717, 2007.

124. *WO Patent*, 052869, 2004.

125. *US Patent*, 0067939, 2004.

126. *WO Patent*, 095438, 2003.

127. *US Patent*, 6610846, 2003.

128. *WO Patent*, 048106, 2002.

129. *WO Patent*, 046173, 2002.
130. *WO Patent*, 014312, 2002.
131. *WO Patent*, 008209, 2002.
132. *WO Patent*, 085707, 2001.
133. *WO Patent*, 085706, 2001.
134. *WO Patent*, 083478, 2001.
135. *WO Patent*, 083465, 2001.
136. *WO Patent*, 044216, 2001.
137. *WO Patent*, 058293, 2000.
138. N.-E. Haynes, W. L. Corbett, F. T. Bizzarro, K. R. Guertin, D. W. Hilliard, G. W. Holland, R. F. Kester, P. E. Mahaney, L. Qi, C. L. Spence, J. Tengi, M. T. Dvorozniak, A. Railkar, F. M. Matschinsky, J. F. Grippo, J. Grimsby and R. Sarabu, *J. Med. Chem.*, 2010, **53**, 3618–3625.
139. *WO Patent*, 023706, 2011.
140. J. Grimsby, J. Zhi, M.-E. Mulligan, C. Arbet-Engels, R. Taub and R. Balena, Keystone Symposia: Diabetes Mellitus, Insulin Action and Resistance, poster 151, Breckenridge, CO, 2008.
141. *Roche Phase 2 Trial of Piragliatin*, http://clinicaltrials.gov/ct2/show/ NCT00266240, Accessed 06-30, 2011.
142. *Roche Phase 1 R-1511 #1*, http://clinicaltrials.gov/show/NCT00517465, Accessed 06-30, 2011.
143. *Roche Phase 1 R-1511 #2*, http://www.roche.com/irp080721.pdf, Accessed 06-30, 2011.
144. *Roche Pipeline*, http://www.roche.com/research_and_development/pipeline/ roche_pharma_pipeline.htm, Accessed 06-30, 2011.
145. *WO Patent*, 076884, 2010.
146. *WO Patent*, 140624, 2009.
147. *WO Patent*, 128481, 2009.
148. *WO Patent*, 125873, 2009.
149. *WO Patent*, 156757, 2008.
150. *WO Patent*, 136428, 2008.
151. *WO Patent*, 116107, 2008.
152. *WO Patent*, 079787, 2008.
153. *US Patent*, 0096877, 2008.
154. *WO Patent*, 047821, 2008.
155. *US Patent*, 0281942, 2007.
156. *WO Patent*, 104034, 2007.
157. *WO Patent*, 075847, 2007.
158. *WO Patent*, 061923, 2007.
159. *WO Patent*, 028135, 2007.
160. *WO Patent*, 112549, 2006.
161. H. Takahara, *Takeda Consolidated Results Second Quarter of Fiscal 2009*, http://www.takeda.com/pdf/usr/default/00_35286_4.pdf, Accessed 06-27, 2011.

162. *Takeda TAK-329 Glucose Clamp Study*, http://clinicaltrials.gov/ct2/show/ NCT01311076?term=TAK-329&rank=1, Accessed 06-27, 2011.
163. *TransTech Pharma and Novo Nordisk Collaboration*, http://www.ttpharma. com/PressReleases/2007/20070222BioWorldToday/tabid/194/Default.aspx, Accessed 05-22, 2011.
164. *Novo Nordisk GK Portfolio Licensed to TransTech Pharma*, http:// www.ttpharma.com/PressReleases/2007/tabid/152/Default.aspx, Accessed 05-22, 2011.
165. *WO Patent*, 058923, 2006.
166. *EP Patent*, 1532980, 2005.
167. *WO Patent*, 066145, 2005.
168. *Novo Nordisk Diabetes Portfolio*, http://www.novonordisk.com/images/investors/ capital-markets-day/2006/Diabetes%20care%20research%20pipeline.pdf, Accessed 05-22, 2011.
169. C. Valcarce, T. Bödvarsdóttir, P. Wahl, M. Larsen, K. Fosgerau, N. Blume, J. Selmer and A. Mjalli, *Diabetologia*, 2008, **51** (Suppl. 1), Abst 929.
170. *Discontinuation of TTP355*, http://www.bio.org/businessforum/pdfs/00002818.pdf, Accessed 05-20, 2011.
171. *TransTech Pharma TTP339*, http://www.ttpharma.com/TherapeuticAreas/ MetabolicDisorders/Diabetes/TTP399/tabid/108/Default.aspx, Accessed 05-19, 2011.
172. *TransTech Pharma TTP339 Phase 1 Trial*, http://www.ttpharma.com/ PressReleases/2010/20100608TransTechPharmaInc/tabid/213/Default.aspx.
173. J. W. Tilley, *Expert Opin. Ther. Patents*, 2009, **19**, 549–553.
174. *Zydus Cadila Press Note*, http://www.zyduscadila.com/press/PressNote07-04-11.pdf, Accessed 05/23/2011, 2011.
175. *Zydus Cadila Pipeline*, http://www.zyduscadila.com/discovery.html, Accessed 05/23/2011, 2011.
176. *WO Patent*, 013141, 2011.

11β-Hydroxysteroid Dehydrogenase Type 1 (11β-HSD1) Inhibitors in Development

JAMES S. SCOTT* AND JASEN CHOORAMUN

Cardiovascular & Gastrointestinal Innovative Medicines Unit, AstraZeneca Mereside, Alderley Park, Macclesfield, Cheshire, SK10 4TG, UK
*E-mail: jamie.scott@astrazeneca.com

5.1 Introduction to 11β-Hydroxysteroid Dehydrogenase Type 1 (11β-HSD1)

5.1.1 Glucocorticoids and the Metabolic Syndrome

The metabolic syndrome is a collection of abnormalities, including resistance to insulin, obesity, dyslipidemia, hyperglycemia, and hypertension, that represents a major risk factor for cardiovascular disease and type II diabetes.[1] The underlying pathogenesis of the metabolic syndrome is complex and influenced by a number of mechanisms. Amongst these, glucocorticoid synthesis and metabolism has been proposed to play a key role.[2,3]

Evidence to support this in humans is derived from patients with Cushing's syndrome, who exhibit elevated circulating glucocorticoid levels resulting in a phenotype similar to the metabolic syndrome.[4] Although patients with the metabolic syndrome do not exhibit elevated plasma glucocorticoid levels,[5] it has been hypothesised that elevated intracellular concentrations may play a crucial role. Therefore, the ability to modulate intracellular glucocorticoid

RSC Drug Discovery Series No. 27
New Therapeutic Strategies for Type 2 Diabetes: Small Molecule Approaches
Edited by Robert M. Jones

Published by the Royal Society of Chemistry, www.rsc.org

concentrations has been proposed as an attractive therapeutic paradigm for the metabolic syndrome.[6–9]

5.1.2 Role, Function, and Structure of 11β-HSD1

In humans, local glucocorticoid availability is primarily controlled by two enzymes; 11β-hydroxysteroid dehydrogenase type 1 (11β-HSD1) and type 2 (11β-HSD2). 11β-HSD1 is an enzyme expressed primarily in liver, adipose and brain that has the role of reducing the physiologically inactive cortisone to the active glucocorticoid cortisol.[10,11] In contrast, 11β-HSD2 is primarily expressed in the kidney where it catalyses the reverse reaction, converting cortisol to the physiologically inactive cortisone. Its function is to protect the mineralocorticoid receptor from cortisol activation which can lead to hypokalemia and hypertension.[12] In rodents, the enzymatic substrates are subtly different, lacking the 17-hydroxyl functionality, with 11β-HSD1 converting the inactive glutocorticoid 11-dihydrocorticosterone (11-DHC) to the active corticosterone. This is summarised in Figure 5.1.

11β-HSD1 is a member of the short chain alcohol dehydrogenases. The enzyme exists as a homodimer with the *N*-terminus of one subunit interacting with the active site of the other. It is tethered to the membrane of the endoplasmic reticulum with the catalytic domain located within the lumen. Crystal structures of human[13] and mouse[14] variants have been solved and a number of groups have published structures with a variety of inhibitor chemotypes. The key structural elements include a serine (S170) and tyrosine (Y183) that position and activate the 11-keto group of cortisone towards reduction by NADPH, with a second tyrosine residue (Y177) forming a key part of the active site. Significant differences between the amino acid sequence of human and mouse 11β-HSD1 have led to differences in potencies across species with various chemotypes. A crystal structure of human 11β-HSD1 showing the enzyme together with the topology of the active site is shown in Figure 5.2.

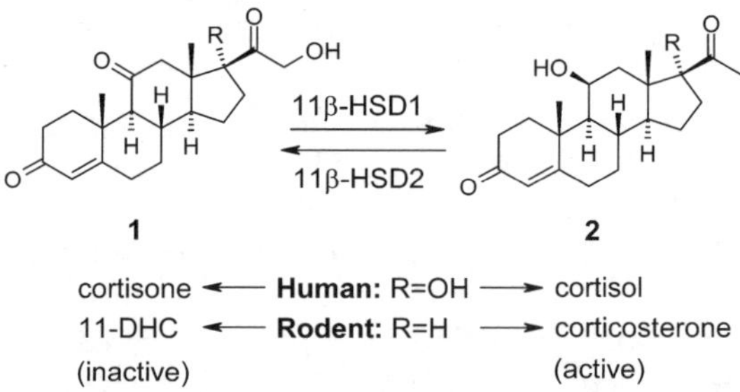

Figure 5.1 Interconversion by 11β-HSD1 and 11β-HSD2 of cortisone and cortisol in humans and 11-dihydrocorticosterone (11-DHC) and corticosterone in rodents.

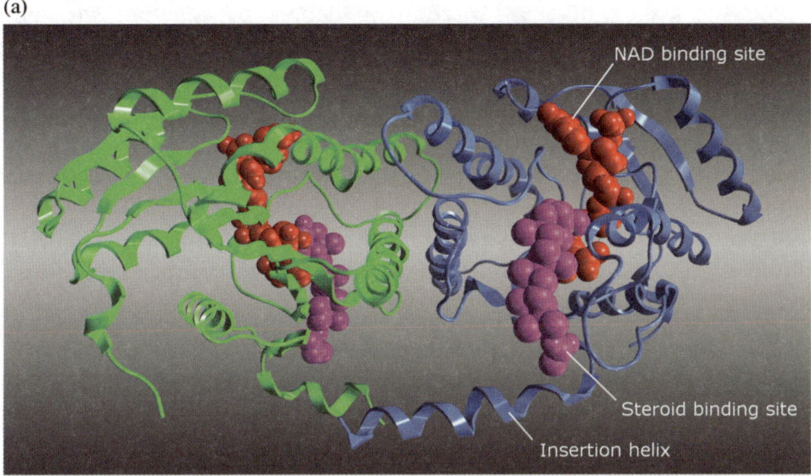

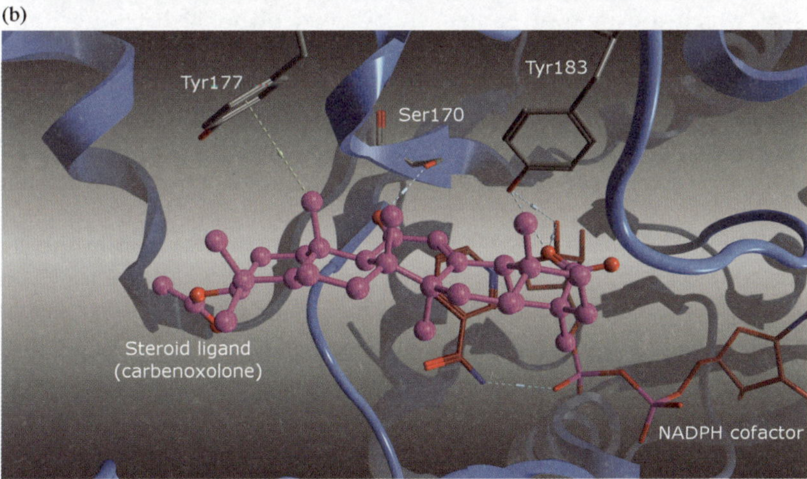

Figure 5.2   (a) Structure of 11β-HSD1 dimer (PDB entry 2BEL) with the bound NADPH in red and the bound steroid carbenoxolone in pink. (b) Detail of the active site highlighting the key residues tyrosine177, tyrosine183 and serine170 in relation to NADPH and carbenoxolone.

5.1.3 Preclinical Evidence for 11β-HSD1 in Treatment of the Metabolic Syndrome

Evidence to support the central role of 11β-HSD1 in the metabolic syndrome has been augmented by rodent studies in which the enzyme has been either deleted, in 11β-HSD1 knock-out animals, or over-expressed. 11β-HSD1 global knock-out mice have a metabolically improved phenotype being sensitised to insulin and resistant to stress-induced hyperglycemia.[15] Furthermore, they resist development of both visceral fat and the metabolic syndrome when subjected to high fat feeding.[16] In contrast, over-expression of 11β-HSD1 in adipose tissue led to increased visceral fat deposition together with glucose intolerance and insulin resistance which worsened upon high fat feeding.[17] Over-expression of 11β-HSD1 in liver caused hypertension and dyslipidemia in addition to insulin resistance, although no effects on glucose tolerance or body weight were observed.[18] Additionally, over-expression of 11β-HSD2 in adipose tissue led to resistance to weight gain in mice on a high fat diet as well as improved glucose tolerance and increased insulin senstitivity.[19] An increasing body of evidence is now emerging from clinical trials in humans of 11β-HSD1 inhibitors and this will be discussed in the relevant company sections that follow.

5.1.4 The Hypothalamic-Pituitary-Adrenal (HPA) Axis

Cortisol synthesis in the adrenal gland[20] is stringently regulated by adrenocorticotropic hormone (ACTH) secreted from the pituitary gland, the levels of which are in turn controlled by corticotrophin-releasing hormone (CRH) produced in the hypothalamus.[21] Elevated cortisol concentrations result in negative feedback signals from the adrenal gland, inhibiting the release of ACTH and CRH.[22] This neuroendocrine system is referred to as the hypothalamic–pituitary–adrenal (HPA) axis and together with the 11β-HSD1 and 11β-HSD2 enzymes is responsible for the tight control of circulating cortisol levels.[23] A key concern in the strategy of using 11β-HSD1 inhibition is that intracellular lowering of cortisol concentrations may result in the risk of compensatory cortisol production via up-regulation of the HPA axis. Evidence for this preclinically comes from observations of increased adrenal weights in rodent studies in 11β-HSD1 knock-out mice[24] and also that those humans deficient in 11β-HSD1 exhibit evidence of HPA axis activation.[25] This area has recently been reviewed[26] and will be addressed in the conclusion of this chapter with regard to emerging clinical data.

5.1.5 Carbenoxolone (CBX)

Carbenoxolone (CBX) (**3**) is a hemisuccinate ester of glycyrrhetinic acid, a naturally occurring compound found in the liquorice root (Figure 5.3).[27] It is a moderately potent inhibitor of 11β-HSD1 and has been evaluated in humans. It is unselective over 11β-HSD2 and this has limited its utility as either a research tool or as a clinical compound.[28]

Carbenoxolone (**3**) was shown to improve insulin sensitivity when dosed to healthy human males, providing the first evidence in humans that 11β-HSD1 inhibition could elicit this effect.[29] When carbenoxolone was dosed to lean type II diabetics, improved hepatic insulin sensitivity and reduced glucose production and glycogenolysis were reported. These encouraging results were tempered by the fact that no effects on gluconeogenesis, peripheral glucose uptake or insulin-mediated suppression of plasma free fatty acids were observed.[30]

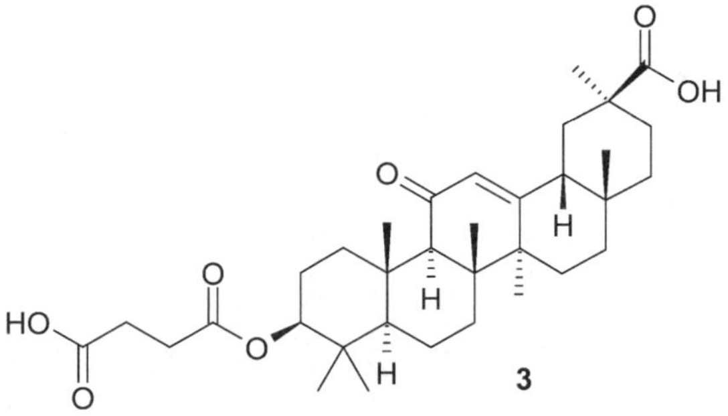

Carbenoxolone (CBX)

Figure 5.3   Structure of the unselective 11β-HSD1 and 11β-HSD2 inhibitor carbenoxolone (**3**).

In contrast to the encouraging effects seen in lean humans, when carbenoxolone was administered to obese individuals, no beneficial effects on insulin sensitivity were observed.[31] It has been postulated that the observed loss of efficacy in obese patients may be due to either the inability of carbenoxolone to effectively penetrate adipose tissue, or alternatively that 11β-HSD1 is down-regulated in the liver of obese individuals.

The lack of selectivity over 11β-HSD2 can lead to generation of excess mineralocorticoid in the kidneys and resultant hypertensive effects. This is a significant concern, particularly among patients with type II diabetes and other metabolic abnormalities.[32] Therefore, the majority of efforts in this field have been to develop an inhibitor that is intrinsically selective for 11β-HSD1 over 11β-HSD2.

5.2   Overview of 11β-HSD1 Inhibitors in Development

The extensive body of literature around 11β-HSD1 as a target has been reviewed and the reader is referred to the following references for a comprehensive account.[3,33–36]

Reflecting the keen interest in the therapeutic potential of this target, the intellectual property landscape for 11β-HSD1 has become increasingly crowded, with over 250 patent applications from over 25 pharmaceutical companies and academic groups now published. A graph of the number of patent applications from the fifteen largest institutions by patent volume is presented in Figure 5.4. Several of the top companies in terms of volume of patents published such as Biovitrum and Incyte have worked against 11β-HSD1 for over a decade. Other companies, such as Vitae, have joined the field relatively late but through prolific patenting now have a large numerical share of the published patent applications.

This patent landscape and chemical equity contained within has also been the subject of comprehensive reviews and will not be covered here.[37–39] The

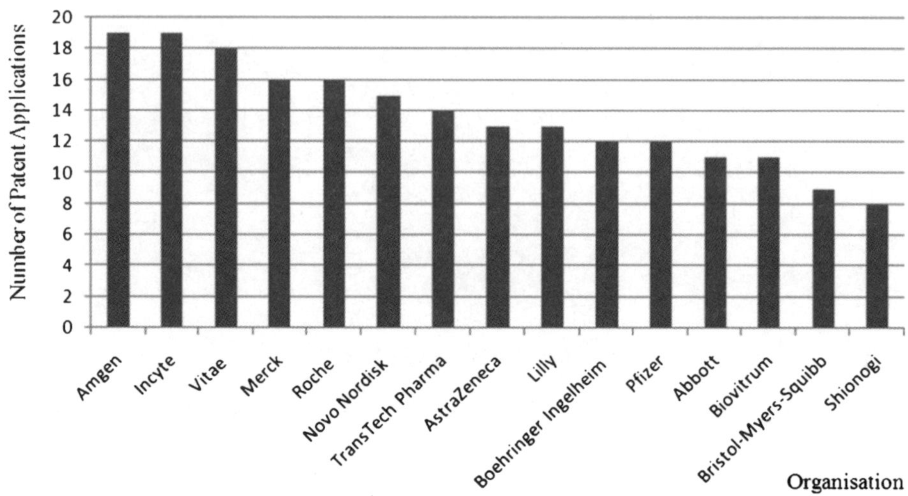

Figure 5.4 11β-HSD1 patent landscape showing the top fifteen organisations by patent volume.

focus of this chapter is rather to review the 11β-HSD1 selective inhibitors that have been taken to the clinic to date.

As a consequence of the significant research efforts in this area, a number of 11β-HSD1 inhibitors have now progressed into the clinic and this information is summarised in Table 5.1. A more detailed assessment of the published

Table 5.1 Summary of 11β-HSD1 inhibitors taken into development by company.

Company	Drug	Phase	Status
Amgen/Biovitrum	AMG-331 (BVT-3498)	II	Discontinued (2005)
Amgen/Biovitrum	AMG-221 (BVT-83370)	I	Discontinued (2011)
Merck & Co	MK-0916	II	No longer in company pipeline
Merck & Co	MK-0736	II	No longer in company pipeline
Pfizer	PF-915275	II	Discontinued (2007)
Incyte	INCB-13739	II	No longer in company pipeline
Incyte	INCB-20817	I	No longer in company pipeline
AstraZeneca	AZD-4017	I	Indication changed
AstraZeneca	AZD-8329	I	Active
Vitae/Boehringer Ingelheim	BI-135585	I	Active
Wyeth (now Pfizer)	HSD-016	I	Discontinued (2008)
Bristol-Myers-Squibb	BMS-816336	I	Active
Bristol-Myers-Squibb	BMS-770767	II	Active
Roche	RG-7234	I	Discontinued (2010)
Roche	RG-4929	II	Active
Japan Tobacco	JTT-654	II	Discontinued (2010)
Lilly	LY-2523199	II	Active

medicinal chemistry, clinical data and structures (where disclosed) are contained in the sections on 11β-HSD1 inhibitors by company below.

5.3 11β-HSD1 Inhibitors by Company

5.3.1 Amgen / Biovitrum

Biovitrum were the first company to disclose inhibitors BVT-2733 (**4**) and BVT-14225 (**5**) that were selective for 11β-HSD1 over 11β-HSD2 (Figure 5.5).[40] They entered into collaboration with Amgen and this led to the development of the first compound to enter clinical trials, BVT-3498 (AMG-311) (**6**), the structure of which was identified in a Biovitrum patent application.[41] This compound **6** completed phase I trials in late 2002 and entered phase II trials the following year but was subsequently stopped in 2005 and replaced by BVT-83370 (AMG-221) (**16**). This compound was progressed into phase I trials but development was halted in April 2011.[42]

The first reported selective 11β-HSD1 inhibitor, BVT-2733 (**4**), is potent against the mouse enzyme (IC_{50} = 96 nM), has reduced potency against the human enzyme (IC_{50} = 3341 nM) but significantly, shows no activity against human 11β-HSD2 (IC_{50} > 10 μM). Pharmacokinetic profiling of BVT-2733 (**4**) revealed that the compound had modest bioavailability (F = 21%) and an acceptable half life (2.5–3.5 h) in mice and was therefore suitable for use as a tool compound for *in vivo* profiling.[40]

Oral dosing of compound **4** (25, 50, 100 mg/kg) twice daily for 11 days in the hyperglycemic KKA^y mouse model, lowered blood glucose levels in a dose-dependent manner. The maximal reduction in glucose was 53% of the control after 11 days of treatment at 100 mg/kg twice daily. No effects on liver or heart marker enzymes were observed, suggesting that the pharmacological effect observed was not related to liver impairment.[40] In a separate experiment KKA^y mice were treated with compound for 7 days at a higher dose (167 mg/kg/day) using osmotic minipumps. This resulted in a reduction in mRNA levels encoding for two key enzymes in hepatic glucose production: phosphoenol-

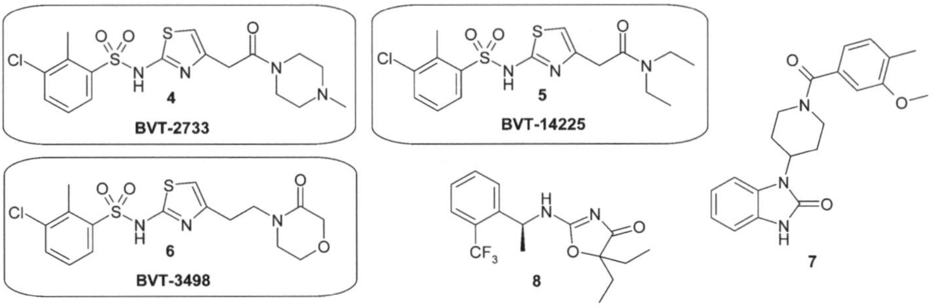

Figure 5.5 Structures of arylsulfonamidothiazole inhibitors including BVT-2733, BVT-14225, BVT-3498 (AMG-311) and alternative chemotypes disclosed by Biovitrum.

pyruvate carboxykinase (75% of control) and glucose-6-phosphatase (55% of control).[40] These data generated with BVT-2733 (**4**) were supportive of the key role of 11β-HSD1 as a mediator of gluconeogenesis and strengthened the case for inhibition of 11β-HSD1 as a treatment strategy for type II diabetes.

The structurally related compound BVT-14225 (**5**), which has a more flexible di-ethyl amide and lacks the basic group, is potent against mouse (IC_{50} = 284 nM) and in contrast to BVT-2733 (**4**), is also potent against the human form of the enzyme (IC_{50} = 52 nM). In common with BVT-2733 (**4**) it is also inactive against 11β-HSD2 (IC_{50} > 10 μM) and therefore had potential to offer therapeutic benefit without the risk of hypertension.

Biovitrum have also reported on two further structural series (Figure 5.5), piperidine amides[43] such as **7**, although these are only reported to have weak potency (IC_{50} = 110 nM in an enzyme assay for the example shown), and oxazolones,[44] as exemplified by **8** closely related to the thiazolones co-developed with Amgen.

A series of publications from Amgen have disclosed the optimisation of a series of aminothiazolones (Figure 5.6). Initial high throughput screening hits such as **9** were optimised in terms of the aryl substitution pattern, leading to

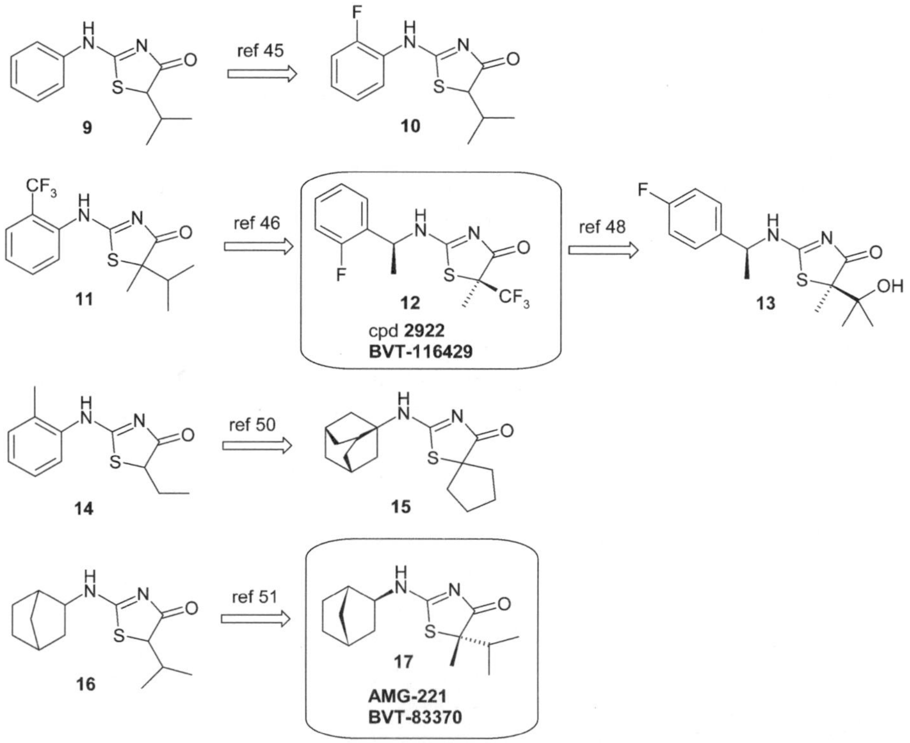

Figure 5.6 Optimisation of a series of aminothiazolones leading to the identification of research tool BVT-116429 (compound 2922) and BVT-83370 (AMG-221) by Amgen.

compounds such as **10**.[45] The phenyl ring of compounds such as **11** was replaced with a benzyl substituent leading to an important tool compound **12**, also referred to as 'compound 2922' or BVT-116429.[46]

Compound **12**, is reported as a potent inhibitor against human 11β-HSD1 in an enzyme assay (IC_{50} = 14 nM) with no drop off observed in a cellular assay (IC_{50} = 14 nM).[46,47] It was less potent against the mouse isoform of the enzyme (IC_{50} = 161 nM) but had low plasma protein binding (21% free in mouse) making it suitable for *in vivo* studies. Compound **12** was selective against 11β-HSD2, 17β-HSD1 and glucocorticoid receptor as well as a panel of 60 other human targets (less than 30% inhibition at 10 μM).[47]

Compound **12** was shown to dose dependently inhibit 11β-HSD1 in an acute mouse adipose tissue *ex vivo* pharmacodynamic model at doses of 0.1 to 10 mg/kg with >90% inhibition at 2 h post-dose at the top dose. Analysis indicated that compound levels in adipose tissue were approximately 2.5 times higher than in plasma suggesting an effective distribution into target tissue.[47] The compound was also examined for its ability to inhibit the conversion of prednisone to prednisolone (a surrogate biomarker for 11β-HSD1 activity in the liver). Compound **12** was shown to dose-dependently (1–10 mg/kg) inhibit conversion of prednisone to prednisolone, with a maximum of 70% reduction in the prednisolone:prednisone ratio at a dose of 10 mg/kg. Inhibition of 11β-HSD1 in adipose tissue was also observed during this experiment.[46]

Importantly, the potential of compound **12** to activate the HPA axis was examined and it was demonstrated that at levels of compound (1–10 mg/kg) that delivered systemic and adipose efficacy, no significant changes were observed in the levels of plasma corticosterone or ACTH, suggesting that activation of the HPA axis had not occurred.[46]

Further profiling in drug metabolism studies revealed that compound **12** activated the human pregnane X receptor (PXR), a nuclear receptor that up-regulates genes involved in drug metabolism particularly cytochrome P450 (CYP) 3A4. Activation of PXR is a concern in drug development as increased metabolic clearance may translate to reduced compound levels and in turn reduced efficacy. At concentrations of 20 μM, compound **12** showed induction of PXR at levels approximately 45% of control (effect of rifampicin at 12.5 μM = 100%) and this may have been a factor that prevented this compound being developed further. A strategy of removal of the PXR liability through incorporation of polar functionality has been reported, leading to compounds such as **13** that were devoid of this liability.[48]

The effects of compound **12** in diabetic KKA^y mice were examined at doses of 3–30 mg/kg for 10 days and found to increase adiponectin levels, a potentially useful biomarker of efficacy for 11β-HSD1 inhibitors. After 7 days decreased basal insulin levels but no changes in glucose tolerance were seen. After 10 days of treatment, fasting blood glucose level was decreased comparable to the effects of rosiglitazone at 5 mg/kg.[49]

Further optimisation of this series (Figure 5.6) involved replacement of the aryl ring of compounds such as **14** and replacing it with bridged cycloalkyl groups as exemplified by compound **15**.[50] A focus on norbornyl groups led, through the optimisation of compounds such as **16**, to the identification of a clinical candidate AMG-221 (**17**).[51]

AMG-221 (**17**), also known as BVT-83370,[52] is reported as a potent (K_i = 12.8 nM) inhibitor against human 11β-HSD1 in an enzyme assay and in a cellular assay (IC_{50} = 10.1 nM).[51] The compound was selective against 11β-HSD2 (IC_{50} > 10 μM), 17β-HSD1 (IC_{50} >10 μM) and glucocorticoid receptor (IC_{50} >10 μM) as well as a panel of 61 other human targets (less than 25% binding at 10 μM). The compound did not inhibit any of the five main cytochrome P450 enzymes, 3A4, 2D6, 1A2, 2C9 and 2C19 (IC_{50} > 15 μM) and there was no evidence for transport by P-glycoprotein. In terms of pharmacokinetic profile, the compound had moderate to high clearance in mouse (Cl = 3.2 L/h/kg), rat (Cl = 1.4 L/h/kg), dog (Cl = 0.9 L/h/kg) and monkey (Cl = 1.9 L/h/kg) and exhibited a biphasic profile. Bioavailability was reasonable in mouse (F = 54%), rat (F = 34%) and dog (F = 50%) but low in monkey (F = 7%), consistent with high first-pass metabolism as evidenced by low microsomal stability in this species.[51]

AMG-221 (**17**) was shown to inhibit 11β-HSD1 in an acute mouse adipose tissue pharmacodynamic model at doses of 5–50 mg/kg with significant inhibition remaining after 8 h at the 15 and 50 mg/kg doses of 36% and 39% respectively. The compound was also evaluated in a 14 day study in diet-induced obese (DIO) mice at doses of 25 and 50 mg/kg twice daily in order to ensure total inhibition of 11β-HSD1 for more than 16 h per day. At the end of the study, fed blood glucose showed a statistically significant reduction in comparison to the vehicle group at both doses and on day 14, after a 12 h fast, glucose tolerance was slightly improved at both doses compared with the vehicle group. On day 13, there were statistically significant decreases in insulin levels at both doses when compared with the vehicle control group. Body weight was also decreased in a dose-dependent fashion compared to the control group.[51]

AMG-221 (**17**) has been dosed to healthy, obese humans at doses of 3, 30 and 100 mg in an effort to establish a pharmacokinetic/pharmacodynamic relationship.[53] Adipose concentrations needed to achieve 50% inhibition (IC_{50}) or maximal inhibition (I_{max}) of enzyme activity were established as 0.975 ng/ mL and 1.19 ng/mL. The compound produced sustained inhibition over the 24 hour study as measured by *ex vivo* adipose samples. A delay between plasma and adipose concentrations was observed that was attributed to perfusion limited distribution to adipose tissue.

A structurally distinct class of inhibitors (Figure 5.7a) has also been disclosed by Amgen. A piperazine sulfonamide hit **18** was optimised to increase potency leading to compound **19**[54] and subsequently to improve water solubility and rat bioavailability as exemplified by compound **20**.[55,56] Further work led to the development of di-aryl sulfones[57] such as compound **21** that showed improved metabolic stability and oral bioavailability. Compound **21** displayed excellent potency in an enzyme assay (IC_{50} = 0.9 nM) with an

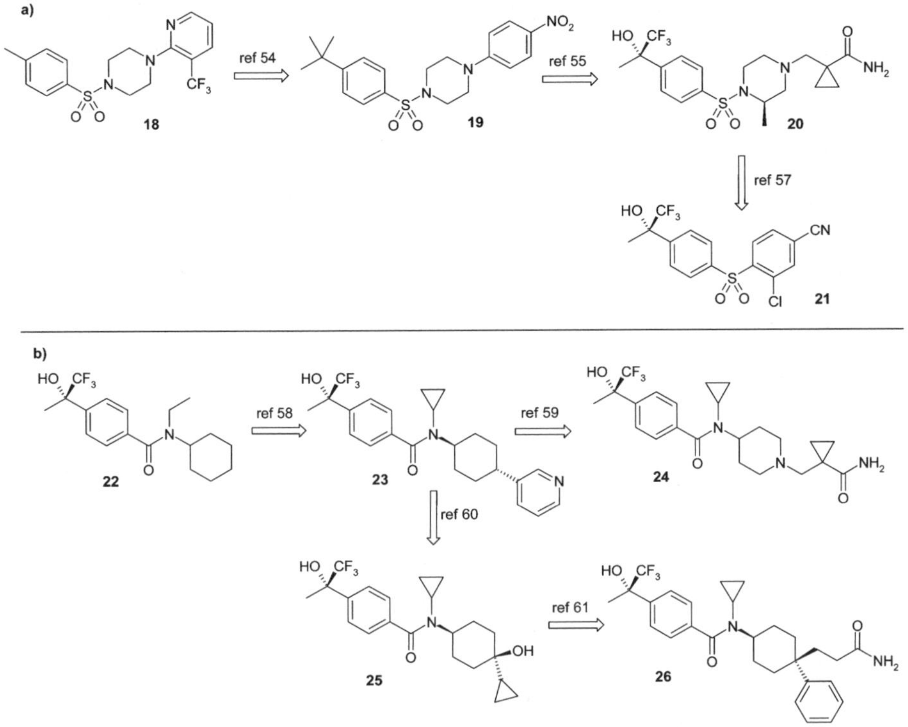

Figure 5.7 (a) Optimisation of a series of piperazine sulfonamides and di-aryl sulfones by Amgen. (b) Optimisation of a series of piperazine amides by Amgen.

approximate 10-fold drop off in a cell assay (IC_{50} = 11 nM). The compound was bioavailable in rat (F = 55%), dog (F = 73%) and to a lesser extent cynomolgus monkey (F = 13%). The compound retained potency against the cynomolgus monkey enzyme (IC_{50} = 3 nM) and demonstrated efficacy at 2 and 10 mg/kg in an acute *ex vivo* cynomolgus monkey model involving the conversion of [^{3}H]-cortisone in mesenteric fat samples collected 2 hours post dose.[57]

The sulfonamide series was also elaborated into a benzamide series typified by compound **22** and then further optimised to improve pharmacokinetics (Figure 5.7b).[58] The *trans* 3-pyridyl substituent present in compound **23** was found to be associated with *in vitro* cytotoxicity in HeLa cells (IC_{50} = 2.5 µM) and efforts to remove this whilst retaining favourable properties have been described, leading to amides such as compound **24**.[59] An alternative approach that also sought to remove PXR activation[60] evolved through hydroxy piperidine **25** and ultimately to amide **26**.[61]

Compound **26** displayed high potency in an enzyme assay (IC_{50} = 0.8 nM) with a small decrease in activity in a cell assay (IC_{50} = 3 nM). In the presence of 3% human serum albumin, the compound was less potent in both the cell assay (IC_{50} = 37 nM) and an adipocyte assay (IC_{50} = 17 nM) indicating a potential for high plasma protein binding. The compound was potent against the cynomolgus monkey enzyme (IC_{50} = 22 nM) and had good pharmaco-kinetics in this species in terms of clearance (Cl = 0.06 L/h/kg) and bioavailability (F = 81%). It was evaluated at doses of 4, 20 and 100 mg/kg in an acute *ex vivo* cynomolgus monkey model and was shown to inhibit the production of [^{3}H]-cortisol from [^{3}H]-cortisol in samples of mesenteric fat samples collected 2 hours post dose at all doses.[61] No reports of compounds from either of the sub-series described in Figure 5.7 entering development have been made to date.

5.3.2 Merck

Merck have described the medicinal chemistry of a series of adamantyl triazoles including an important tool compound **27**, also known as 'compound 544', which was identified from high throughput screening efforts. Merck have progressed two compounds into development, MK-0916, which entered phase II in 2004, and MK-0736 which was originally developed as a treatment for hypertension and entered phase II trials in 2005. The structures of these have not been disclosed although it is known that MK-0736 is the preferred candidate as MK-0916 is known to induce CYP3A4 at therapeutic doses.[62]

Compound **27** is a potent inhibitor against human 11β-HSD1 in an enzyme assay (IC_{50} = 7.8 nM) and has respectable potency against mouse (IC_{50} = 98 nM) (Figure 5.8a). It is highly selective against both human 11β-HSD2 (IC_{50} > 3 μM) and mouse 11β-HSD2 (IC_{50} > 10 μM).[63] The compound also exhibited *in vivo* activity in a mouse model at a dose of 10 mg/kg inhibiting the conversion of [^{3}H]cortisone to [^{3}H]cortisol at a level of 59% at 1 hour post dosing and 17% at 4 hours post dosing relative to control. The modest duration of action with this compound is reported to be due to sub-optimal pharmacokinetic properties. Accordingly, enhanced *in vivo* effects (85% inhibition of [^{3}H]cortisone conversion at 1 hour post dosing and 47% at 4 hours post dosing) were observed with less potent inhibitors such as **28** (human IC_{50} = 37 nM; mouse IC_{50} = 109 nM) and attributed to improved pharmacokinetic exposure.[63]

In a DIO mouse model, dosing compound **27** twice daily at 20 mg/kg for 11 days led to a reduction in body weight gain (7%) and cumulative food intake (12%) relative to control animals with preferential loss of central fat pad mass (retroperitoneal fat mass reduced but epididymal fat mass unchanged). Evidence of lowered fasting glucose levels (15%) and lower insulin levels were also observed relative to control. Reduction in triglycerides (18%) and cholesterol (24%) were also observed leading to improved lipid profiles.[64] In contrast, the compound had no significant effects on body weight in a high fat feeding, low dose streptozotocin (HF STZ) mouse model of Type II diabetes. However, lower fasting and postprandial glucose levels (37% lowering), lower glucagon levels and improved insulin sensitivity were observed after dosing

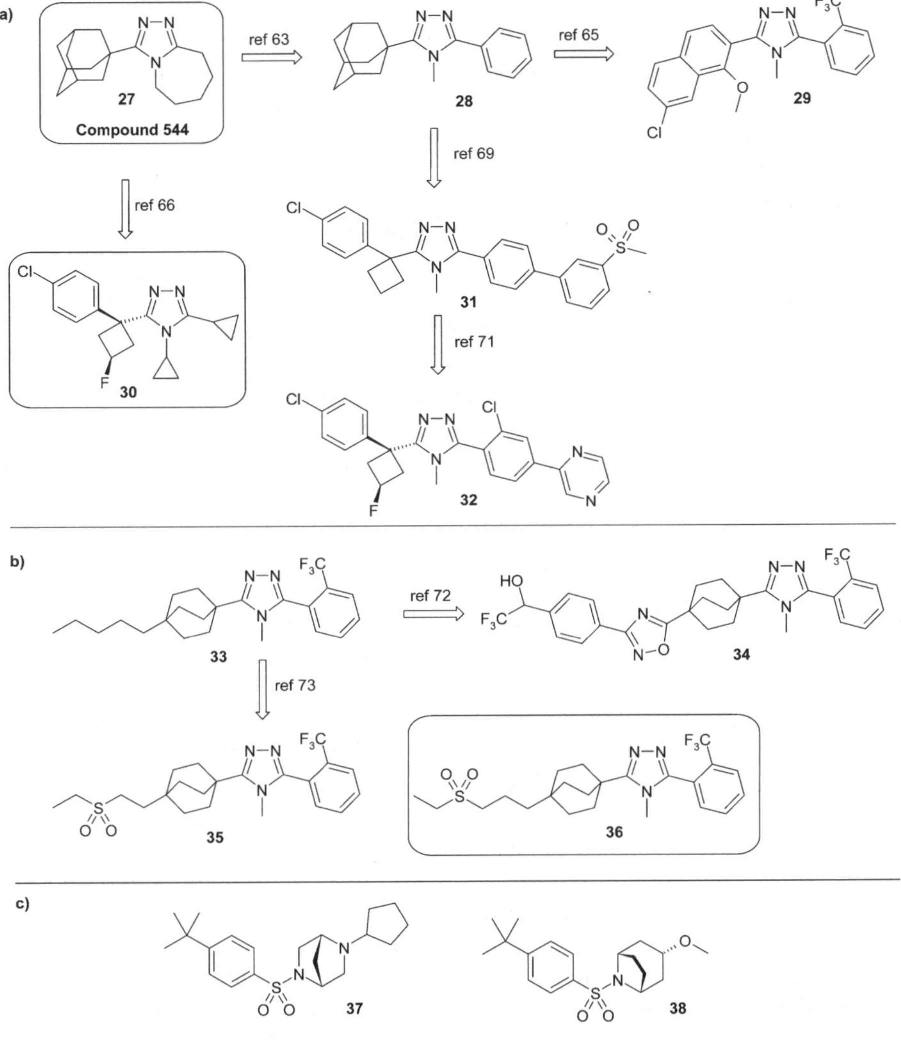

Figure 5.8   (a) Optimisation of a series of triazoles and structure of 'compound 544' disclosed by Merck. (b) Optimisation of a series of bicyclo[2.2.2]octyl-triazoles by Merck. (c) Alternative sulfonamide chemotypes disclosed by Merck.

twice daily at 30 mg/kg for 9 days. Normalisation of triglycerides and a reduction in free fatty acids were also observed.[64]

In an atherosclerosis (apolipoprotein E knock-out) mouse model, animals given compound **27** at 10 mg/kg/day in a high fat diet for 8 weeks exhibited lower accumulation of aortic total cholesterol (84%), as well as lower serum cholesterol and triglycerides (although no effects on body weight were observed). These significant findings highlighted the potential of 11β-HSD1 inhibition as an effective treatment for atherosclerosis, the key clinical consequence of metabolic syndrome on premature mortality.[64]

Compound **27** was not progressed to the clinic but paved the way for the development of analogues with improved pharmacokinetics. Replacement of the adamantly group with a bi-aryl system gave rise to compounds such as **29**.[65] Further optimisation led to the identification of compound **30**, a potent inhibitor against both human 11β-HSD1 (IC_{50} = 5 nM) and mouse 11β-HSD1 (IC_{50} = 16 nM) in an enzyme assay (Figure 5.8a).[66] It retained excellent selectivity against both human and mouse 11β-HSD2 (both IC_{50} > 4 μM). Compound **30** also exhibited *in vivo* activity in a mouse model at a dose of 10 mg/kg inhibiting the conversion of [^{3}H]cortisone to [^{3}H]cortisol at a level of 92% at 1 hour post dosing and 91% at 4 hours post dosing relative to control, indicating an extended duration of action relative to compound **27**. This was attributed to a superior pharmacokinetic profile in mouse with low clearance (11 mL/min/kg) leading to an extended half-life (1.5 h) and excellent oral bioavailability (F ∼ 100%). Data on analogous compounds indicated that the *anti* arrangement of the fluorine atom on the cyclobutyl ring relative to the triazole was instrumental in achieving low clearance and high oral exposures. Compound **30** was the subject of patent applications that identified novel crystalline forms of the compound[67] together with a synthesis described on kg scale[68] suggesting that this was a compound of significant interest to Merck.

Further optimisation studies have been published and detail the challenges that were overcome, notably the introduction of polar functionality in compounds such as **31** to reduce activity at the PXR activation associated with cytochrome P450 induction (Figure 5.8a).[69,70] Compound **32** is a potent inhibitor against human 11β-HSD1 in an enzyme assay (IC_{50} = 5.5 nM) and with similar potency against mouse (IC_{50} = 2.3 nM). The compound showed robust *in vivo* activity in a mouse model at a dose of 10 mg/kg inhibiting the conversion of [^{3}H]cortisone to [^{3}H]cortisol at a level of 95% at 4 hour post dosing and 80% at 16 hours post dosing relative to control, consistent with good mouse pharmacokinetics (Cl = 5 mL/min/kg) a long half-life (6.8 h) and excellent bioavailability (F = 96%). Importantly, the compound had low PXR activation (EC_{50} > 30 μM) with 23% activation at 10 μM and low levels of inhibition of the CYP450 enzymes 3A4 (IC_{50} = 36 μM), 2C9 (IC_{50} = 26 μM) and 2D6 (IC_{50} = 25 μM).[71]

Other work in this area reported by Merck has focused on bicyclo[2.2.2]octyltriazoles such as compound **33** and **34** (Figure 5.8b).[72] A development candidate has not been confirmed from this avenue of research, although compound **35** is reported to be a potent inhibitor of 11β-HSD1 (human IC_{50} = 7 nM; mouse IC_{50} = 9.7 nM) with good selectivity over 11β-HSD2 (both species IC_{50} > 4 μM) and good *in vivo* activity in a mouse model at a dose of 10 mg/kg inhibiting the conversion of [^{3}H]cortisone to [^{3}H]cortisol at a level of 89% at 1 hour post dosing and 76% at 4 hours post dosing relative to control.[73] Compound **35** was reported to have good solubility (>1 mg/mL), low lipophilicity (log D = 2.2) and a good pharmacokinetic profile with a low clearance in mouse (Cl = 2.1 mL/min/kg), rat (Cl = 2.9 mL/min/kg), dog (Cl = 1.9 mL/min/kg) and monkey (Cl = 4.0 mL/min/kg). This resulted in moderate bioavailability in mouse (F = 24%) but high values in rat (F = 87%), dog (F = 59%) and monkey (F = 89%). Importantly, the compound was shown not to induce CYP induction in human microsomes (human PXR activation at 10 μM < 10%). Efficacy was demonstrated in a DIO, apo-E knockout mouse model of atherosclerosis and an oral glucose tolerance test (OGTT) in B6-Ay mouse.[73] The homologated analogue **36** is a more potent inhibitor of 11β-HSD1 (human IC_{50} = 3.2 nM; mouse IC_{50} = 2.4 nM) with good selectivity over 11β-HSD2 (both species IC_{50} > 4 μM) and good *in vivo* activity in a mouse model at a dose of 10 mg/kg inhibiting the conversion of [^{3}H]cortisone to [^{3}H]cortisol at a level of 92% at 1 hour post dosing and 63% at 4 hours post dosing relative to control and was the subject of a recent patent application that identified novel crystalline forms of this compound.[74]

A structurally distinct series of azabicyclic sulfonamides (Figure 5.8c) have also been reported by Merck.[75] These compounds were notable by the fact that they were more active against the mouse than the human form of the enzyme. Reported potencies were human IC_{50} = 40 nM; mouse IC_{50} = 1 nM for compound **37** and human IC_{50} = 37 nM; mouse IC_{50} = 5 nM for compound **38**.

The structures of MK-0916 & MK-0736 have not yet been disclosed although they may be represented by the structures highlighted here. Some initial data has been reported from a 12 week, Phase IIa study of MK-0916 in patients with type II diabetes and metabolic syndrome at doses ranging from 0.5 to 6 mg/day.[76] The compound was well tolerated but showed no significant improvement in fasting plasma glucose at week 12 relative to placebo. Modest dose-dependent decreases in blood pressure and body weight were observed over the course of the study, together with a small but significant reduction of 0.3% in haemoglobin A1C (hBA1c) at week 12. At the top dose, an increase in low-density lipoprotein cholesterol (LDL-C) of 10.4% relative to placebo was reported which was unexpected and may be a compound rather than mechanism related effect, potentially related to the known CYP3A4 induction effect of this compound. Elevations of 20 – 30% from baseline in circulating adrenal androgens were observed, indicating modest activation of the HPA axis, but these remained within normal physiological levels.[76]

A 12 week study of both MK-0916 & MK-0736 in overweight and obese patients with hypertension has also been reported.[62] MK-0736 was the primary focus of this study as MK-0916 is known to be potent inducer of CYP3A4. At the top dose of 7 mg, MK-0736 was well tolerated but did not achieve its primary efficacy endpoint with no statistically significant reduction in trough sitting diastolic blood pressure. Modest effects on the metabolic syndrome were observed relative to placebo including a decrease in LDL-C of 12.3%, high-density lipoprotein cholesterol (HDL-C) by 6.3%, and body weight by 1.4 kg. An elevation of adrenal androgens indicative of HPA axis activation was also reported although the levels observed were not thought to be clinically meaningful being within two-fold of normal physiological levels.

5.3.3 Pfizer

Pfizer took an 11β-HSD1 inhibitor PF-915275 (**40**) into phase I trials in 2006. In 2007 it was announced that this compound had progressed to phase II; however, it was subsequently stopped due to tablet formulation issues.[80] The medicinal chemistry discovery story has been published and details how the initial lead molecule **39** was optimised in terms of ligand efficiency and physiochemical properties to compound PF-915275 (**40**) (Figure 5.9).[77]

PF-915275 (**40**) is reported to be an extremely potent inhibitor against human 11β-HSD1 in an enzyme assay ($K_i < 1$ nM) that maintains potency in a HEK393 cellular assay ($IC_{50} = 5$ nM) as well as being highly selective against human 11β-HSD2 (<1.5% inhibition at 10 μM). Against mouse and rat 11β-HSD1 the compound was significantly less active (mouse enzyme $K_i = 750$ nM; rat hepatoma cell $IC_{50} = 14,500$ nM) and this precluded demonstration of biomarker inhibition or efficacy in rodent models. Further species investigation in primary hepatocytes revealed that monkey ($IC_{50} = 100$ nM) and dog ($IC_{50} = 120$ nM) were closer in activity to human ($IC_{50} = 20$ nM). Selectivity was investigated using the Cerep Bioprint screening panel and displayed only weak affinity for the rodent choline transporter ($K_i = 9.6$ μM) and the hamster melatonin MT3 receptor ($K_i = 9.6$ μM). Compound **39** showed low clearance *in vitro* (liver microsome assays) with high permeability (CACO2 assay) and has an excellent pharmacokinetic profile in rat characterised by low clearance (0.87 mL/min/kg), long half-life (6.6 h) and good oral bioavailability (74%).[77]

With acceptable 11β-HSD1 potency in cynomolgus monkey, primates were used to demonstrate *in vivo* target inhibition using prednisone to prednisolone conversion as a surrogate biomarker for 11β-HSD1 cortisone to cortisol conversion.[78] PF-915275 (**40**) was shown to dose-dependently (0.1–3 mg/kg) inhibit 11β-HSD1 mediated conversion of prednisone to prednisolone, with a maximum of 87% inhibition at the top dose and a good correlation between circulating plasma exposures, suggesting no lag time or hysteresis between

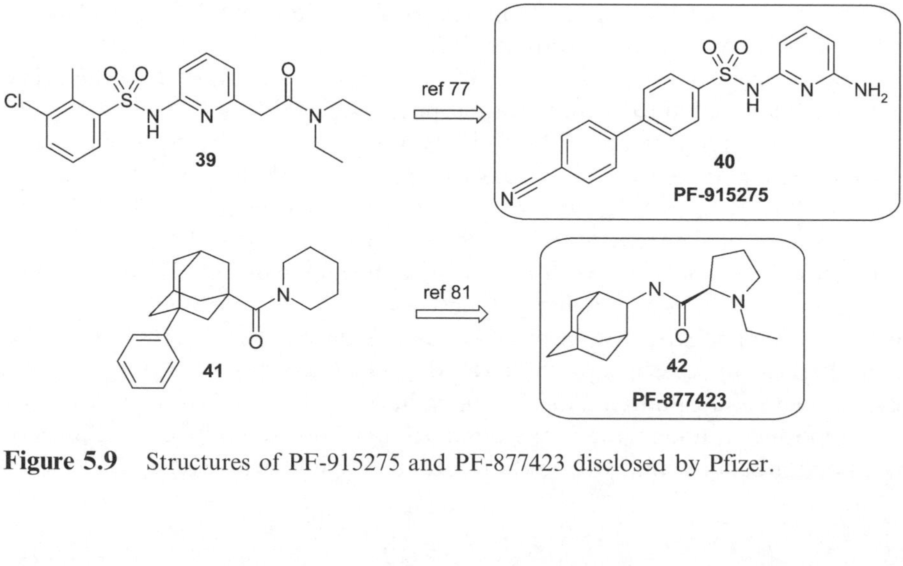

Figure 5.9 Structures of PF-915275 and PF-877423 disclosed by Pfizer.

drug levels and response. The exposure/response relationship allowed an IC_{50} of 391 nM (total) and 17 nM (free) to be determined. Observable dose-dependent reduction in plasma insulin levels were also reported with a maximum of 60% reduction at the top dose of 3 mg/kg although no effect was observed on plasma glucose or lipid levels.

PF-915275 (**40**) was progressed into phase I trials[79] with 60 healthy adult volunteers given oral doses of 0.3–15 mg over 14 days. The compound was well tolerated over this dose range. PF-915275 (**40**) was rapidly absorbed (median T_{max} = 1 h) and slowly eliminated (mean $t_{1/2}$ = 30 h) with clearance values in the range of 1.0–1.4 L/h and no significant renal elimination (<1% of the administered dose was excreted unchanged in the urine). The compound generally displayed dose-proportional increases in exposure and attained steady state by day 7 with a volume of distribution approximately the same as total body water (42 litres), indicating no significant penetration into peripheral tissues.[79]

PF-915275 (**40**) was shown to dose-dependently inhibit conversion of prednisone to prednisolone in humans, with a maximum of 37% inhibition at the top 15 mg dose. It also dose-dependently reduced the (5α- + 5β-tetrahydrocortisol) to tetrahydrocortisone ratio with a maximum reduction of 26% after 14 days. The urinary free cortisol to urinary free cortisone ratio, an indicator of 11β-HSD2 inhibition, did not change and no clinical signs of hypotension or hypokalemia were observed consistent with this being a selective 11β-HSD1 inhibitor. In order to address concerns over HPA axis activation, levels of ACTH, adrenal androgens and the urinary corticosteroid profile were examined. These were not found to be significantly altered, thus indicating that over-activation of the HPA axis was not occurring at these doses.[79]

Pfizer have subsequently reported a series of pyrrolidine carboxamide inhibitors developed from the adamantyl amide **41** (Figure 5.9).[81] A lead compound PF-877423 (**42**) is potent against both human (K_i = 1.4 nM) and mouse (K_i = 0.63 nM) and retains good activity in a cell assay (IC_{50} = 4.2 nM). The compound has moderate metabolism in human liver microsomes (Cl_{int} = 44 mL/min/kg) and showed dose dependent *in vivo* activity in a cortisone to cortisol mouse model with a maximal effect of 82% at 100 mg/kg and a calculated dose for 50% inhibition of 5 mg/kg. This compound has been investigated in human preadipocytes for the potential to treat obesity via adipogenesis blockade.[82]

5.3.4 Incyte

Incyte have taken a lead compound INCB-13739, the structure of which has not yet been disclosed, as far as phase II trials and data in patients with type II diabetes has recently been reported.[83] The study showed that after 12 weeks as an addition to metformin monotherapy, INCB-13739 at doses of 200 mg resulted in significant reductions in HbA1C (−0.6%), fasting plasma glucose

(−24 mg/dL) and homeostasis model assessment–insulin resistance (HOMA-IR) (−24%) compared with placebo. Modest body weight reduction (−1 kg at 200 mg dose) was observed as were reductions in cholesterol (total and low density lipoprotein) and triglycerides in hyperlipidemic patients. A reversible, dose-dependent elevation in ACTH levels was observed, potentially indicating HPA axis activation, but basal cortisol homeostasis, testosterone in men and free androgen index in women were unchanged. No increase in adverse events was reported, indicating that INCB-13739 was well tolerated at efficacious doses.[83]

A subsequent review[84] has highlighted the fact that despite the high potency of INCB-13739 (IC_{50} = 1.1 nM), efficacy was only observed at doses of 100 mg and 200 mg. It is postulated that these may be better understood in terms of the concentration required to achieve 90% maximal inhibition of the enzyme (IC_{90} rather than IC_{50}). Comments were also made on the degree of ACTH elevation in terms of the reversibility and magnitude of elevation (two-fold), suggesting that this was likely to represent the maximal degree of HPA axis activation. The observation that therapy appeared to be most effective in obese patients (BMI > 30 kg/m^2) was also discussed in relation to adipose tissue distribution, increased splanchnic cortisol exposure and the potential implications for 11β-HSD1 therapy in treatment of obesity and its cardiometabolic consequences.

A structurally distinct follow-on compound INCB-20817 was also being developed by Incyte and was well tolerated in phase I trials. However, INCB-20817 no longer appears on the company pipeline and its current status is unclear.[85]

5.3.5 AstraZeneca

AstraZeneca's company website[86] reports two compounds currently in clinical development that inhibit 11β-HSD1. AZD8329 is listed as being in phase I trials for the treatment of diabetes and obesity whilst AZD4017 went into phase I trials for diabetes and obesity and has since switched indications and is currently in phase II for treatment of raised intraocular pressure. Neither of the structures has been disclosed publically at the time of writing. The structure of another clinical candidate, AZD6925 (**44**) and its genesis from compound **43** were disclosed at the European Federation for Medicinal Chemistry 21st International Symposium on Medicinal Chemistry in 2010 (Figure 5.10).[87]

AZD6925 (**44**) is a potent inhibitor against human 11β-HSD1 enzyme (IC_{50} = 6 nM), retains efficacy in a human adipocyte assay (IC_{50} = 4 nM) and has good selectivity against 11β-HSD2 (IC_{50} > 30 μM). The compound shows no inhibition against a panel of CYP450 enzymes (CYP3A4, CYP2D6, CYP2C9, CYP2C19, CYP1A2, all IC_{50} > 30 μM) and displays no activity against the hERG ion channel (IC_{50} > 79 μM). The compound has good physical properties in terms of solubility (>3000 μM), lipophilicity (log D = 2.1) and permeability in a Madin Darby Canine Kidney (MDCK) assay, with no

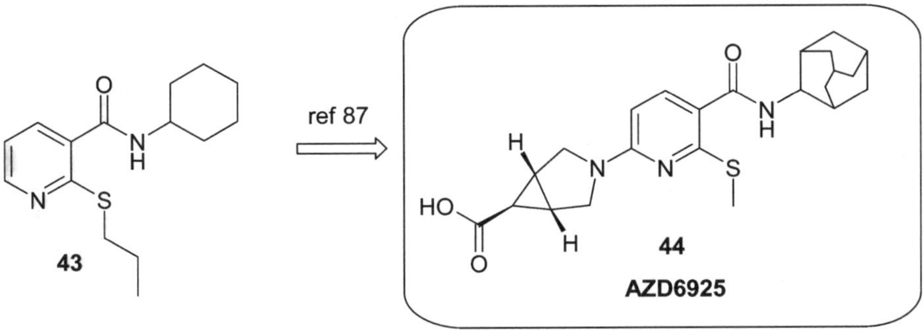

Figure 5.10   Structure of AZD6925 disclosed by AstraZeneca.

evidence of efflux. AZD6925 (**44**) has reasonable plasma protein binding free levels across species (1.7%, 1.5% and 1.1% free in rat, dog and human respectively). Pharmacokinetic profiles in both rat and dog were good with low clearance in rat (Cl = 6 mL/min/kg) and dog (Cl = 0.8 mL/min/kg) leading to good bioavailability in rat (F = 86%) and dog (F = 59%).

5.3.6   Vitae Pharmaceuticals / Boehringer Ingelheim

According to the company website, Vitae Pharmaceticals[88] have been in collaboration with Boehringer Ingelheim[89] since 2007 to develop a compound that inhibits 11β-HSD1 for the treatment of diabetes and associated metabolic diseases. A compound (BI-135585) entered phase I trials in 2010. Although the structure of this compound has not been confirmed, the medicinal chemistry leading to novel series of ureas and carbamates has been reported (Figure 5.11).

The urea **46**, developed from initial lead compound **45**, is a very potent inhibitor against human 11β-HSD1 enzyme (IC$_{50}$ = 1.1 nM), retained efficacy in a human adipocyte assay (IC$_{50}$ = 2.5 nM) and has good selectivity against

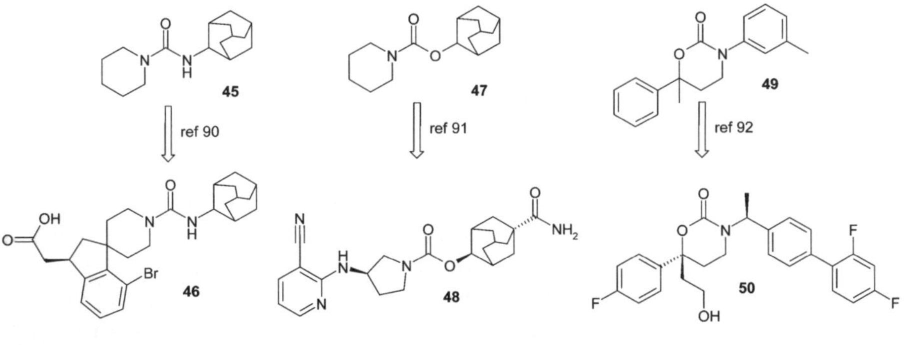

Figure 5.11   Optimisation of ureas and carbamates by Vitae.

11β-HSD2, 3β-HSD2 and 17β-HSD1 (all $IC_{50} > 10$ μM).[90] The compound showed no inhibition against CYP3A4, CYP2C9 and CYP2D6 ($IC_{50} > 15$ μM) and displayed no significant activity against a panel of 68 receptors and ion channels. The sodium salt of the compound had good pharmacokinetic properties in the rat, with moderate clearance (27 mL/min/kg) and half-life (3.5 h) and high bioavailability (90%). In the mouse, urea **46** had lower clearance relative to hepatic blood flow (28 mL/min/kg) and longer half-life (6.2 h) but more modest bioavailability (29%). At a dose of 30 mg/kg, the compound was shown to distribute predominantly into liver (approximately 18 times higher than plasma levels) but with significant concentrations being observed in adipose (0.3 times plasma levels). In monkey the compound had higher clearance relative to hepatic blood flow (14 mL/min/kg), similar half-life (6.0 h) and modest bioavailability (31%). At a dose of 10 mg/kg the plasma concentration of cortisol following a challenge with cortisone 21-acetate was shown to be reduced by approximately 80%, consistent with significant inhibition of 11β-HSD1 *in vivo*.[90]

The carbamate **48**, developed from initial lead compound **47**, is also a very potent inhibitor against human 11β-HSD1 enzyme ($IC_{50} = 0.9$ nM) and has good selectivity against 11β-HSD2 ($IC_{50} = 7350$ nM) as well as 3β-HSD2 and 17β-HSD1 (both $IC_{50} > 10$ μM).[91] The compound retained efficacy in a human adipocyte assay ($IC_{50} = 2.5$ nM) and in an enzyme assay in the presence of 50% human plasma ($IC_{50} = 2.4$ nM). Carbamate **48** exhibits weak inhibition of CYP3A4 ($IC_{50} = 17.1$ μM) but has no effects on CYP2C9 and CYP2D6 (both $IC_{50} > 30$ μM). The compound is stable in rat liver microsomes ($t_{1/2} > 60$ min) and orally available in rat with high clearance (42 mL/min/kg), low volume (0.70 L/kg) and moderate bioavailability (42%).[91]

The cyclic carbamate **50** was derived from structure-based design from lead **49** and is a very potent inhibitor against human 11β-HSD1 enzyme ($IC_{50} = 0.75$ nM) with good selectivity against 11β-HSD2 and 17β-HSD1 (both $IC_{50} > 10$ μM) and only weak activity against 3β-HSD2 ($IC_{50} = 4.7$ μM).[92] Compound **50** retained efficacy in a human adipocyte assay ($IC_{50} = 2$ nM) but is highly protein bound with less than 1% free drug available in dog, monkey and human plasma. This is consistent with a pronounced (18-fold) reduction in enzyme potency when assayed in the presence of 50% human plasma. The compound exhibits inhibition of CYP3A4 ($IC_{50} = 12$ μM) and CYP2C9 ($IC_{50} = 4.9$ μM) but showed no activity against the hERG ion channel ($IC_{50} > 10$ μM). Compound **50** is stable in rat liver microsomes ($t_{1/2} = 119$ min) leading to high bioavailability in rat ($F = 42\%$), and to a lesser extent in dog ($F = 16\%$) and monkey ($F = 26\%$). Compound **50** was much less potent against mouse ($IC_{50} = 670$ nM) which precluded the use of this species for *in vivo* efficacy studies, although an oral dose of 30 mg/kg confirmed that the compound distributed into adipose and liver tissue. The compound retained potency against cynomolgus monkey enzyme ($IC_{50} = 1.0$ nM) and subsequently showed efficacy in an *in vivo* model at a dose of 10 mg/kg, which reduced cortisol production by approximately 85%.[92]

5.3.7   Wyeth

Wyeth had a compound HSD-016 in phase I trials in 2008; however, following the acquisition of Wyeth by Pfizer, this program does not appear in the Pfizer development pipeline and is thought to be discontinued.[93]

Wyeth initially reported the serendipitous discovery that β-keto sulfones[94] such as **51** are inhibitors of 11β-HSD1, but subsequently went on to demonstrate that these are substrates for the enzyme and are turned over by the enzyme to products which are not inhibitors (Figure 5.12).[95] Work on an alternative sulfonamide series has also been reported describing optimisation of a high throughput screening hit **52** to improve pharmacokinetics in the rat leading to compound **53**.[96] This was followed by further efforts to improve *in vivo* efficacy leading to compound **54**.[97]

Compound **54** is reported to be an equipotent inhibitor against both human and mouse 11β-HSD1 in a cell assay (IC_{50} = 10 nM) and showed no shift in the presence of 10% human serum.[97] The compound was completely selective against 11β-HSD2 (IC_{50} > 100 μM) and 11β-HSD1 activity was demonstrated in an *ex vivo* mouse model with 80% inhibition in epididymal fat at 10 mg/kg. Compound **53** was stable in both human and mouse microsomes ($t_{1/2}$ > 30 min) and the pharmacokinetics in mouse showed low clearance (5 mL/min/kg), a lengthy half-life (7.3 h) and good oral bioavailability (F = 50%). *In vivo* studies with the compound in diet showed a 95% *ex vivo* inhibition in liver after 4 days and analysis of tissue samples showed that appreciable compound levels were present in both liver and fat. A 28-day study in DIO mice was carried out which showed a significant reduction in fed glucose and fasting insulin levels at a dose of 0.5 mg compound/g diet. No effect was observed on fasted glucose or fed insulin, nor on food intake or body weight, though compound **53** appeared to be well tolerated with no adverse events reported.

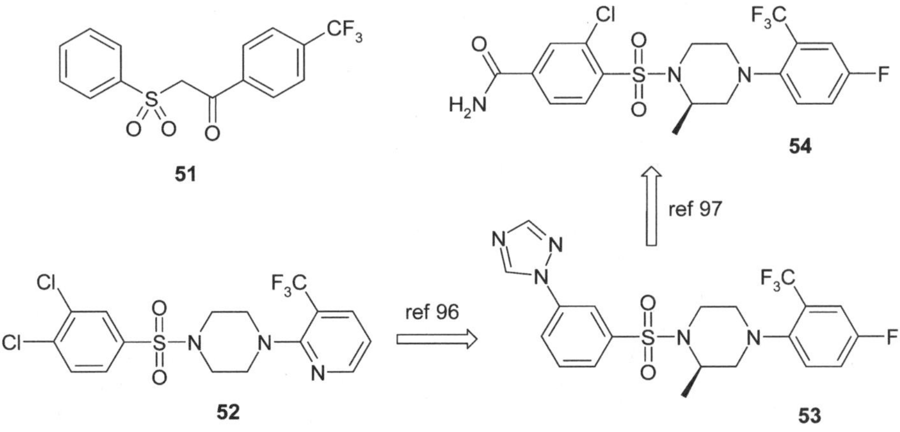

Figure 5.12   β-Keto sulfones and optimisation of a series of piperazine sulfonamides by Wyeth.

Wyeth has published pharmacokinetic data comparing the exposure of 11β-HSD1 inhibitors in DIO mice relative to normal lean controls. Increases of up to three fold in oral bioavailability of the compounds in the DIO mice were observed. This was attributed to the lower first-pass clearance effects reflecting the compromised metabolic capacity of these animals.[98]

5.3.8   Bristol-Myers-Squibb

Bristol-Myers-Squibb have progressed two compounds into development (BMS-816336 and BMS-770767) with BMS-770767 believed to be currently progressing in phase II trials and BMS-816336 in phase I.[99] The structures of these compounds have not yet been disclosed. Bristol-Myers-Squibb have reported the optimisation of a high throughput screening hit **55** which showed moderate potency (IC_{50} = 235 nM) to a series of chemotypes that are extremely potent against the enzyme (e.g. IC_{50} = 0.1 nM for compound **56**) as shown in Figure 5.13.[100]

Compound **55** has also been developed into an alternative 1,2,4-triazolo-pyridine series as exemplified by compound **57** which showed good potency against human 11β-HSD1 (IC_{50} = 12 nM) but with lower activity against mouse 11β-HSD1 (IC_{50} = 1026 nM), precluding use in murine *in vivo* models.[101] Compound **57** showed some activity against human 11β-HSD2 (IC_{50} = 2.52 μM) but with reasonable selectivity (>200 fold). Solubility was measured at 11 mg/mL and other *in vitro* properties such as CYP inhibition, ion channel activity and cytotoxicity were described as favourable, although the compound was noted to be a weak activator of the PXR receptor (EC_{50} > 50 μM). These compounds are described as 'leads' and at present it is unclear as to the structural relationship between them and BMS-770767 or BMS-816336.

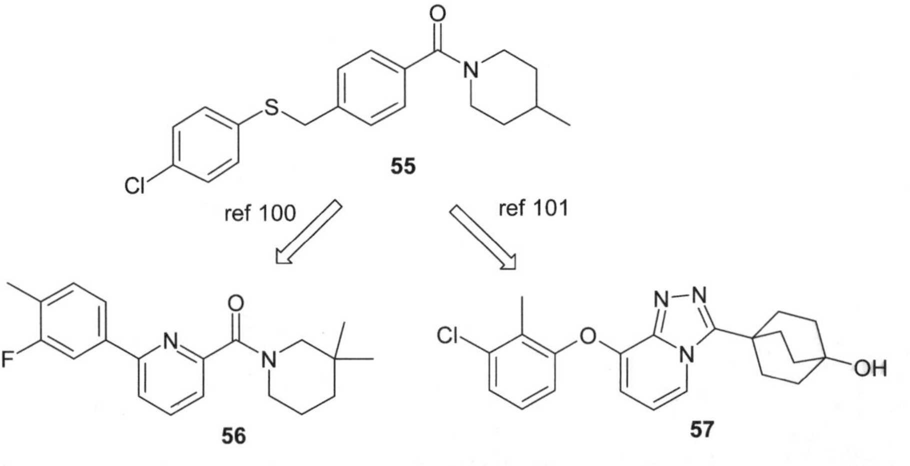

Figure 5.13   Structures of alternative chemotypes disclosed by Bristol-Myers-Squibb.

5.3.9 Roche

Roche took two 11β-HSD1 inhibitors (RG-7234 and RG-4929) into the clinic in a head to head, proof-of-concept study in 2009. It was subsequently announced that RG-7234 would be discontinued at phase I (Oct. 2010) and, according to the company website,[102] RG-4929 is continuing into phase II trials for metabolic diseases. The structures of these compounds have not yet been disclosed.

The optimisation of the pyrazolone **58** to RO-506 (**59**) was described at the 2011 Frontiers in Medicinal Chemistry meeting in Saarbrüken (Figure 5.14).[103] RO-506 (**59**) is reported as having good potency against human 11β-HSD1 enzyme ($IC_{50} = 5$ nM) that translates into cell ($IC_{50} = 6$ nM). RO-506 (**59**) has low lipophilicity ($\log D = 2.2$) leading to good solubility (183 μg/mL) and high free levels across species (5%, 35% and 13% free in mouse, cyno and human respectively). The compound is active against the cynomolgus monkey enzyme ($IC_{50} = 30$ nM) and had good pharmacokinetics in this species with moderate clearance ($Cl = 28$ mL/min/kg), volume of distribution ($V_{ss} = 1.1$ L/kg) and high bioavailability ($F = 84\%$). It was dosed in a 12-week study in cynomolgus monkey at doses of 2, 5 and 10 mg/kg and showed weight loss of 7% at the top dose. It is not clear at present how this structure is related to RG-4929.

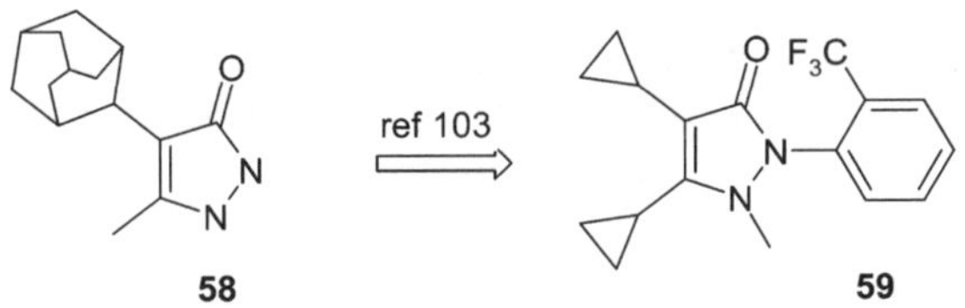

Figure 5.14   Structure of RO-506 disclosed by Roche.

5.3.10 Japan Tobacco / Akros Pharma

Akros Pharma, a subsidiary of Japan Tobacco, was investigating the development of JTT-654, an 11β-HSD1 inhibitor. This progressed into phase II trials in 2009 but the trial was terminated in May 2010 and no longer appears in the Japan Tobacco pipeline.[104] The structure of this compound has not been disclosed.

5.3.11 Lilly

According to the company website, Lilly[105] currently has an 11β-HSD1 inhibitor LY-2523199 in phase II development for the treatment of diabetes. The structure of this compound has not been disclosed, nor have there been any public disclosures outside the patent literature.

5.3.12   Other Companies and Institutions

Other institutions, including the University of Edinburgh,[106,107] the University of Innsbruck,[108] the Korea Research Institute of Chemical Technology,[109–111] the Chinese Academy of Sciences,[112–114] and the Universidad Nacional Autónoma de México,[115] have made contributions to the field in terms of alternative chemotypes and pharmacophore modelling. Other companies, including Novartis,[116] Merck-Serono,[117,118] Sanofi-Aventis,[119] Schering-Plough[120] and in particular Abbott,[121–128] have also published work in this area but have not, to the best of our knowledge, progressed candidates to the clinic.

5.4   Conclusions

11β-HSD1 is a long standing and attractive target in the treatment of the metabolic syndrome. Medicinal chemistry optimisation efforts, often facilitated by crystallographic protein structural information, have delivered a range of structurally diverse chemotypes that are potent inhibitors of the enzyme. Issues such as achieving selectivity over 11β-HSD2, which hampered the development of earlier 11β-HSD1 inhibitors such as carbenoxolone, have now been largely overcome. Considerable progress has also been made in delivering compounds that are potent against both the human and rodent forms of the enzyme which have enabled *in vivo* evaluation of preclinical candidates. Challenges such as pregnane X receptor (PXR) activation associated with cytochrome P450 induction have emerged in a number of diverse structural series. This has led to some elegant medicinal chemistry involving the incorporation of polar substituents in positions accommodated by the enzyme in order to overcome this issue.

As compounds have progressed in development, clinical data on the effects of 11β-HSD1 inhibition in humans has recently been published and is leading to a greater understanding of the pharmacological effects of this mechanism. Up-regulation of the HPA axis in response to 11β-HSD1 inhibition has been a long standing concern for this mechanism; however, an initial report from Pfizer indicated this was not observed and subsequent studies from Incyte and Merck have observed elevations in adrenal androgens, but within the normal range, potentially ameliorating this risk.

The failure to achieve primary efficacy endpoints in studies with MK-0916 and MK-0736 at low doses (6 mg and 7 mg respectively) is disappointing, but is counterbalanced by some positive effects of parameters relevant to the metabolic syndrome. Data from INCB-13739 at high dose (200 mg) is more encouraging with effects on a range of metabolic syndrome parameters, albeit lower doses did not produce the desired effects. This has led to the hypothesis that a high drive on the target may be required (i.e. coverage of IC_{90} rather than IC_{50}).

The challenge to establish a pharmacokinetic/pharmacodynamic understanding that will allow clinically meaningful efficacy at an acceptable dose will face subsequent compounds that are progressing in development. The authors remain optimistic that with the considerable number of companies with compounds in active development, an 11β-HSD1 inhibitor may be progressed through trials and subsequently prove to be a novel, safe and effective therapy for patients with the metabolic syndrome.

Acknowledgements

Thanks to Brendan Leighton for useful discussions on the manuscript, Graeme Robb for generation of the images in Figure 5.2 and Darren McKerrecher for proofreading the manuscript.

References

1. S. M. Grundy, H. B. Brewer, J. I. Cleeman, S. C. Smith and C. Lenfant, *Circulation*, 2004, **109**, 433.
2. P. Björntorp and R. Rosmond, *Nutrition*, 2000, **16**, 924–936.
3. M. Wamil and J. R. Seckl, *Drug Discov. Today*, 2007, **12**, 504–520.
4. G. Arnaldi, A. Angeli, A. B. Atkinson, X. Bertagna, F. Cavagnini, G. P. Chrousos, G. A. Fava, J. W. Findling, R. C. Gaillard, A. B. Grossman, B. Kola, A. Lacroix, T. Mancini, F. Mantero, J. Newell-Price, L. K. Nieman, N. Sonino, M. L. Vance, A. Giustina and M. Boscaro, *J. Clin. Endocrinol. Metab.*, 2003, **88**, 5593.
5. G. W. Strain, B. Zumoff, J. J. Strain, J. Levin and D. K. Fukushima, *Metab. Clin. Exp.*, 1980, **29**, 980–985.
6. R. Thieringer and A. Hermanowski-Vosatka, *Expert Rev. Cardiovasc. Ther.*, 2005, **3**, 911–924.
7. P. Björntorp, G. Holm and R. Rosmond, *Diabetic Med.*, 1999, **16**, 373–383.
8. B. R. Walker, *Eur. J. Endocrinol.*, 2007, **157**, 545–559.
9. J. W. Tomlinson, E. A. Walker, I. J. Bujalska, N. Draper, G. G. Lavery, M. S. Cooper, M. Hewison and P. M. Stewart, *Endocr. Rev.*, 2004, **25**, 831–866.
10. C. R. W. Edwards, R. Benediktsson, R. S. Lindsay and J. R. Seckl, *Steroids*, 1996, **61**, 263–269.
11. E. A. Walker and P. M. Stewart, *Trends Endocrinol. Metab.*, 2003, **14**, 334.
12. R. C. Wilson, S. Dave-Sharma, J. Wei, V. R. Obeyesekere, K. Li, P. Ferrari, Z. S. Krozowski, C. H. L. Shackleton, L. Bradlow, T. Wiens and M. I. New, *Proc. Natl. Acad. Sci. USA*, 1998, **95**, 10200–10205.
13. D. J. Hosfield, Y. Wu, R. J. Skene, M. Hilgers, A. Jennings, G. P. Snell and K. Aertgeerts, *J. Biol. Chem.*, 2005, **280**, 4639.
14. J. Zhang, T. D. Osslund, M. H. Plant, C. L. Clogston, R. E. Nybo, F. Xiong, J. M. Delaney and S. R. Jordan, *Biochemistry (NY)*, 2005, **44**, 6948.

15. N. M. Morton, M. C. Holmes, C. Fievet, B. Staels, A. Tailleux, J. J. Mullins and J. R. Seckl, *J. Biol. Chem.*, 2001, **276**, 41293.
16. N. M. Morton, J. M. Paterson, H. Masuzaki, M. C. Holmes, B. Staels, C. Fievet, B. R. Walker, J. S. Flier, J. J. Mullins and J. R. Seckl, *Diabetes*, 2004, **53**, 931–938.
17. H. Masuzaki, J. Paterson, H. Shinyama, N. M. Morton, J. J. Mullins, J. R. Seckl and J. S. Flier, *Science*, 2001, **294**, 2166.
18. J. M. Paterson, N. M. Morton, C. Fievet, C. J. Kenyon, M. C. Holmes, B. Staels, J. R. Seckl and J. J. Mullins, *Proc. Natl. Acad. Sci. USA*, 2004, **101**, 7088–7093.
19. E. E. Kershaw, N. M. Morton, H. Dhillon, L. Ramage, J. R. Seckl and J. S. Flier, *Diabetes*, 2005, **54**, 1023.
20. W. Arlt and P. M. Stewart, *Endocrinol. Metab. Clin. North Am.*, 2005, **34**, 293.
21. L. Jacobson, *Endocrinol. Metab. Clin. North Am.*, 2005, **34**, 271.
22. J. P. Warne, *Mol. Cell. Endocrinol.*, 2009, **300**, 137–146.
23. M. S. Cooper and P. M. Stewart, *J. Clin. Endocrinol. Metab.*, 2009, **94**, 4645–4654.
24. Y. Kotelevtsev, M. C. Holmes, A. Burchell, P. M. Houston, D. Schmoll, P. Jamieson, R. Best, R. Brown, C. R. W. Edwards, J. R. Seckl and J. J. Mullins, *Proc. Natl. Acad. Sci. USA*, 1997, **94**, 14924–14929.
25. A. Jamieson, A. M. Wallace, R. Andrew, B. S. Nunez, B. R. Walker, R. Fraser, P. C. White and J. M. C. Connell, *J. Clin. Endocrinol. Metab.*, 1999, **84**, 3570–3574.
26. E. Harno and A. White, *Trends Endocrinol. Metab.*, 2010, **21**, 619–627.
27. C. Monder, P. M. Stewart, V. Lakshmi, R. Valentino, D. Burt and C. R. W. Edwards, *Endocrinology*, 1989, **125**, 1046–1053.
28. S. Diederich, C. Grossmann, B. Hanke, M. Quinkler, M. Herrmann, V. Bahr and W. Oelkers, *Eur. J. Endocrinol.*, 2000, **142**, 200–207.
29. B. R. Walker, A. A. Connacher, R. M. Lindsay, D. J. Webb and C. R. W. Edwards, *J. Clin. Endocrinol. Metab.*, 1995, **80**, 3155.
30. R. C. Andrews, O. Rooyackers and B. R. Walker, *J. Clin. Endocrinol. Metab.*, 2003, **88**, 285.
31. T. C. Sandeep, R. Andrew, N. Z. M. Homer, R. C. Andrews, K. Smith and B. R. Walker, *Diabetes*, 2005, **54**, 872.
32. Y. Kotelevtsev, R. W. Brown, S. Fleming, C. Kenyon, C. R. W. Edwards, J. R. Seckl and J. J. Mullins, *J. Clin. Invest.*, 1999, **103**, 683.
33. K. A. Hughes, S. P. Webster and B. R. Walker, *Expert Opin. Investig. Drugs*, 2008, **17**, 481–496.
34. C. Fotsch and M. Wang, *J. Med. Chem.*, 2008, **51**, 4851–4857.
35. C. Hale and M. Wang, *Mini Rev. Med. Chem.*, 2008, **8**, 702–710.
36. N. M. Morton, *Mol. Cell. Endocrinol.*, 2010, **316**, 154–164.
37. S. P. Webster and T. D. Pallin, *Expert Opin. Ther. Patents*, 2007, **17**, 1407–1422.

38. C. D. Boyle and T. J. Kowalski, *Expert Opin. Ther. Patents*, 2009, **19**, 801–825.
39. S. A. Morgan and J. W. Tomlinson, *Expert Opin. Investig. Drugs*, 2010, **19**, 1067–1076.
40. T. Barf, J. Vallgårda, R. Emond, C. Häggström, G. Kurz, A. Nygren, V. Larwood, E. Mosialou, K. Axelsson, R. Olsson, L. Engblom, N. Edling, Y. Rönquist-Nii, B. Öhman, P. Alberts and L. Abrahmsén, *J. Med. Chem.*, 2002, **45**, 3813–3815.
41. L. Abrahmsen, J. Nilsson, U. Opperman and S. Svensson, *WO Patent Application*, WO2005068646, 2005.
42. http://www.amgen.com/media/media_pr_detail.jsp?year=2011&releaseID=1553298, April 2011.
43. K. Flyrén, L. O. Bergquist, V. M. Castro, C. Fotsch, L. Johansson, D. J. St. Jean Jr., L. Sutin and M. Williams, *Bioorg. Med. Chem. Lett.*, 2007, **17**, 3421–3425.
44. L. Sutin, S. Andersson, L. Bergquist, V. M. Castro, E. Danielsson, S. James, M. Henriksson, L. Johansson, C. Kaiser, K. Flyrén and M. Williams, *Bioorg. Med. Chem. Lett.*, 2007, **17**, 4837–4840.
45. C. Yuan, D. J. St. Jean Jr., Q. Liu, L. Cai, A. Li, N. Han, G. Moniz, B. Askew, R. W. Hungate, L. Johansson, L. Tedenborg, D. Pyring, M. Williams, C. Hale, M. Chen, R. Cupples, J. Zhang, S. Jordan, M. D. Bartberger, Y. Sun, M. Emery, M. Wang and C. Fotsch, *Bioorg. Med. Chem. Lett.*, 2007, **17**, 6056–6061.
46. C. Hale, M. Véniant, Z. Wang, M. Chen, J. McCormick, R. Cupples, D. Hickman, X. Min, A. Sudom, H. Xu, G. Matsumoto, C. Fotsch, D. J. St. Jean and M. Wang, *Chem. Biol. Drug Design*, 2008, **71**, 36–44.
47. D. J. St. Jean, C. Yuan, E. A. Bercot, R. Cupples, M. Chen, J. Fretland, C. Hale, R. W. Hungate, R. Komorowski, M. Veniant, M. Wang, X. Zhang and C. Fotsch, *J. Med. Chem.*, 2007, **50**, 429–432.
48. C. Fotsch, M. D. Bartberger, E. A. Bercot, M. Chen, R. Cupples, M. Emery, J. Fretland, A. Guram, C. Hale, N. Han, D. Hickman, R. W. Hungate, M. Hayashi, R. Komorowski, Q. Liu, G. Matsumoto, D. J. St. Jean, S. Ursu, M. Véniant, G. Xu, Q. Ye, C. Yuan, J. Zhang, X. Zhang, H. Tu and M. Wang, *J. Med. Chem.*, 2008, **51**, 7953–7967.
49. M. Sundbom, C. Kaiser, E. Bjorkstrand, V. Castro, C. Larsson, G. Selen, C. Nyhem and S. James, *BMC Pharmacol.*, 2008, **8**, 3.
50. L. Johansson, C. Fotsch, M. D. Bartberger, V. M. Castro, M. Chen, M. Emery, S. Gustafsson, C. Hale, D. Hickman, E. Homan, S. R. Jordan, R. Komorowski, A. Li, K. McRae, G. Moniz, G. Matsumoto, C. Orihuela, G. Palm, M. Veniant, M. Wang, M. Williams and J. Zhang, *J. Med. Chem.*, 2008, **51**, 2933–2943.
51. M. M. Veniant, C. Hale, R. W. Hungate, K. Gahm, M. G. Emery, J. Jona, S. Joseph, J. Adams, A. Hague, G. Moniz, J. Zhang, M. D. Bartberger, V. Li, R. Syed, S. Jordan, R. Komorowski, M. M. Chen, R. Cupples, K. W. Kim, D. J. St. Jean Jr., L. Johansson, M. A. Henriksson,

M. Williams, J. Vallgarda, C. Fotsch and M. Wang, *J. Med. Chem.*, 2010, **53**, 4481–4487.

52. S. Caille, S. Cui, T. L. Hwang, X. Wang and M. M. Faul, *J. Org. Chem.*, 2009, **74**, 3833–3842.

53. J. P. Gibbs, M. G. Emery, I. McCaffery, B. Smith, M. A. Gibbs, A. Akrami, J. Rossi, K. Paweletz, M. R. Gastonguay, E. Bautista, M. Wang, R. Perfetti and O. Daniels, *J. Clin. Pharmacol.*, 2011, **51**, 830–841.

54. D. Sun, Z. Wang, Y. Di, J. C. Jaen, M. Labelle, J. Ma, S. Miao, A. Sudom, L. Tang, C. S. Tomooka, H. Tu, S. Ursu, N. Walker, X. Yan, Q. Ye and J. P. Powers, *Bioorg. Med. Chem. Lett.*, 2008, **18**, 3513–3516.

55. D. Sun, Z. Wang, M. Cardozo, R. Choi, M. DeGraffenreid, Y. Di, X. He, J. C. Jaen, M. Labelle, J. Liu, J. Ma, S. Miao, A. Sudom, L. Tang, H. Tu, S. Ursu, N. Walker, X. Yan, Q. Ye and J. P. Powers, *Bioorg. Med. Chem. Lett.*, 2009, **19**, 1522–1527.

56. H. Tu, J. P. Powers, J. Liu, S. Ursu, A. Sudom, X. Yan, H. Xu, D. Meininger, M. DeGraffenreid, X. He, J. C. Jaen, D. Sun, M. Labelle, H. Yamamoto, B. Shan, N. P. C. Walker and Z. Wang, *Bioorg. Med. Chem.*, 2008, **16**, 8922–8931.

57. X. Yan, Z. Wang, A. Sudom, M. Cardozo, M. DeGraffenreid, Y. Di, P. Fan, X. He, J. C. Jaen, M. Labelle, J. Liu, J. Ma, D. McMinn, S. Miao, D. Sun, L. Tang, H. Tu, S. Ursu, N. Walker, Q. Ye and J. P. Powers, *Bioorg. Med. Chem. Lett.*, 2010, **20**, 7071–7075.

58. L. D. Julian, Z. Wang, T. Bostick, S. Caille, R. Choi, M. DeGraffenreid, Y. Di, X. He, R. W. Hungate, J. C. Jaen, J. Liu, M. Monshouwer, D. McMinn, Y. Rew, A. Sudom, D. Sun, H. Tu, S. Ursu, N. Walker, X. Yan, Q. Ye and J. P. Powers, *J. Med. Chem.*, 2008, **51**, 3953–3960.

59. Y. Rew, D. L. McMinn, Z. Wang, X. He, R. W. Hungate, J. C. Jaen, A. Sudom, D. Sun, H. Tu, S. Ursu, E. Villemure, N. P. C. Walker, X. Yan, Q. Ye and J. P. Powers, *Bioorg. Med. Chem. Lett.*, 2009, **19**, 1797–1801.

60. D. L. McMinn, Y. Rew, A. Sudom, S. Caille, M. DeGraffenreid, X. He, R. Hungate, B. Jiang, J. Jaen, L. D. Julian, J. Kaizerman, P. Novak, D. Sun, H. Tu, S. Ursu, N. P. C. Walker, X. Yan, Q. Ye, Z. Wang and J. P. Powers, *Bioorg. Med. Chem. Lett.*, 2009, **19**, 1446–1450.

61. D. Sun, Z. Wang, S. Caille, M. DeGraffenreid, F. Gonzalez-Lopez de Turiso, R. Hungate, J. C. Jaen, B. Jiang, L. D. Julian, R. Kelly, D. L. McMinn, J. Kaizerman, Y. Rew, A. Sudom, H. Tu, S. Ursu, N. Walker, M. Willcockson, X. Yan, Q. Ye and J. P. Powers, *Bioorg. Med. Chem. Lett.*, 2011, **21**, 405–410.

62. S. Shah, A. Hermanowski-Vosatka, K. Gibson, R. A. Ruck, G. Jia, J. Zhang, P. M. T. Hwang, N. W. Ryan, R. B. Langdon and P. U. Feig, *J. Am. Soc. Hypertension*, 2011, **5**, 166–176.

63. S. Olson, S. D. Aster, K. Brown, L. Carbin, D. W. Graham, A. Hermanowski-Vosatka, C. B. LeGrand, S. S. Mundt, M. A. Robbins, J. M. Schaeffer, L. H. Slossberg, M. J. Szymonifka, R. Thieringer, S. D.

Wright and J. M. Balkovec, *Bioorg. Med. Chem. Lett.*, 2005, **15**, 4359–4362.

64. A. Hermanowski-Vosatka, J. M. Balkovec, K. Cheng, H. Y. Chen, M. Hernandez, G. C. Koo, C. B. Le Grand, Z. Li, J. M. Metzger, S. S. Mundt, H. Noonan, C. N. Nunes, S. H. Olson, B. Pikounis, N. Ren, N. Robertson, J. M. Schaeffer, K. Shah, M. S. Springer, A. M. Strack, M. Strowski, K. Wu, T. Wu, J. Xiao, B. B. Zhang, S. D. Wright and R. Thieringer, *J. Exp. Med.*, 2005, **202**, 517–527.

65. S. D. Aster, D. W. Graham, D. Kharbanda, G. Patel, M. Ponpipom, G. M. Santorelli, M. J. Szymonifka, S. S. Mundt, K. Shah, M. S. Springer, R. Thieringer, A. Hermanowski-Vosatka, S. D. Wright, J. Xiao, H. Zokian and J. M. Balkovec, *Bioorg. Med. Chem. Lett.*, 2008, **18**, 2799–2804.

66. Y. Zhu, S. H. Olson, D. Graham, G. Patel, A. Hermanowski-Vosatka, S. Mundt, K. Shah, M. Springer, R. Thieringer, S. Wright, J. Xiao, H. Zokian, J. Dragovic and J. M. Balkovec, *Bioorg. Med. Chem. Lett.*, 2008, **18**, 3412–3416.

67. Y. Bereznitski, M. A. Huffman, J. E. Lynch and M. Zhao, *WO Patent Application*, WO2005073200, (2005).

68. M. M. Zhao, *WO Patent Application*, WO2007038452, (2007).

69. Y. Zhu, S. H. Olson, A. Hermanowski-Vosatka, S. Mundt, K. Shah, M. Springer, R. Thieringer, S. Wright, J. Xiao, H. Zokian and J. M. Balkovec, *Bioorg. Med. Chem. Lett.*, 2008, **18**, 3405–3411.

70. Y. D. Gao, S. H. Olson, J. M. Balkovec, Y. Zhu, I. Royo, J. Yabut, R. Evers, E. Y. Tan, W. Tang, D. P. Hartley and R. T. Mosley, *Xenobiotica*, 2007, **37**, 124–138.

71. W. Sun, M. Maletic, S. S. Mundt, K. Shah, H. Zokian, K. Lyons, S. T. Waddell and J. Balkovec, *Bioorg. Med. Chem. Lett.*, 2011, **21**, 2141–2145.

72. X. Gu, J. Dragovic, G. C. Koo, S. L. Koprak, C. Legrand, S. S. Mundt, K. Shah, M. S. Springer, E. Y. Tan, R. Thieringer, A. Hermanowski-Vosatka, H. J. Zokian, J. M. Balkovec and S. T. Waddell, *Bioorg. Med. Chem. Lett.*, 2005, **15**, 5266–5269.

73. M. Maletic, A. Leeman, M. Szymonifka, S. S. Mundt, H. J. Zokian, K. Shah, J. Dragovic, K. Lyons, R. Thieringer, A. H. Vosatka, J. Balkovec and S. T. Waddell, *Bioorg. Med. Chem. Lett.*, 2011, **21**, 2568–2572.

74. J. E. Lynch, F. T. Mattrey, Y. V. Bereznitski, J. L. Leazer Jr., R. R. Ferlita, J. Liu, J. Yin and R. M. Wenslow Jr., *WO Patent Application*, WO2010068580, (2010).

75. U. Shah, C. D. Boyle, S. Chackalamannil, H. Baker, T. Kowalski, J. Lee, G. Terracina and L. Zhang, *Bioorg. Med. Chem. Lett.*, 2010, **20**, 1551–1554.

76. P. U. Feig, S. Shah, A. Hermanowski-Vosatka, D. Plotkin, M. S. Springer, S. Donahue, C. Thach, E. J. Klein, E. Lai and K. D. Kaufman, *Diabetes, Obesity Metab.*, 2011, **13**, 498–504.

77. M. Siu, T. O. Johnson, Y. Wang, S. K. Nair, W. D. Taylor, S. J. Cripps, J. J. Matthews, M. P. Edwards, T. A. Pauly, J. Ermolieff, A. Castro, N. A. Hosea, A. LaPaglia, A. N. Fanjul and J. E. Vogel, *Bioorg. Med. Chem. Lett.*, 2009, **19**, 3493–3497.

78. B. G. Bhat, N. Hosea, A. Fanjul, J. Herrera, J. Chapman, F. Thalacker, P. M. Stewart and P. A. Rejto, *J. Pharmacol. Exp. Ther.*, 2008, **324**, 299–305.

79. R. Courtney, P. M. Stewart, M. Toh, M. N. Ndongo, R. A. Calle and B. Hirshberg, *J. Clin. Endocrinol. Metab.*, 2008, **93**, 550–556.

80. http://clinicaltrials.gov/ct2/show/NCT00427401, September 2008.

81. H. Cheng, J. Hoffman, P. Le, S. K. Nair, S. Cripps, J. Matthews, C. Smith, M. Yang, S. Kupchinsky, K. Dress, M. Edwards, B. Cole, E. Walters, C. Loh, J. Ermolieff, A. Fanjul, G. B. Bhat, J. Herrera, T. Pauly, N. Hosea, G. Paderes and P. Rejto, *Bioorg. Med. Chem. Lett.*, 2010, **20**, 2897–2902.

82. I. J. Bujalska, L. L. Gathercole, J. W. Tomlinson, C. Darimont, J. Ermolieff, A. N. Fanjul, P. A. Rejto and P. M. Stewart, *J. Endocrinol.*, 2008, **197**, 297–307.

83. J. Rosenstock, S. Banarer, V. A. Fonseca, S. E. Inzucchi, W. Sun, W. Yao, G. Hollis, R. Flores, R. Levy, W. V. Williams, J. R. Seckl, R. Huber and for the INCB13739-202 Principal Investigators, *Diabetes Care*, 2010, **33**, 1516–1522.

84. G. Hollis and R. Huber, *Diabetes, Obesity Metab.*, 2011, **13**, 1–6.

85. http://www.incyte.com/drugs_product_pipeline.html, Last accessed August 2011.

86. http://www.astrazeneca.com/Research/Our-pipeline-summary, Last accessed August 2011.

87. P. R. O. Whittamore, EFMC-ISMC 2010, Brussels, Belgium, 2010.

88. http://www.vitaepharma.com/view.cfm/1/Homepage, Last accessed August 2011.

89. http://www.boehringer-ingelheim.com/, Last accessed August 2011.

90. C. M. Tice, W. Zhao, Z. Xu, S. T. Cacatian, R. D. Simpson, Y. Ye, S. B. Singh, B. M. McKeever, P. Lindblom, J. Guo, P. M. Krosky, B. A. Kruk, J. Berbaum, R. K. Harrison, J. J. Johnson, Y. Bukhtiyarov, R. Panemangalore, B. B. Scott, Y. Zhao, J. G. Bruno, L. Zhuang, G. M. McGeehan, W. He and D. A. Claremon, *Bioorg. Med. Chem. Lett.*, 2010, **20**, 881–886.

91. C. M. Tice, W. Zhao, P. M. Krosky, B. A. Kruk, J. Berbaum, J. A. Johnson, Y. Bukhtiyarov, R. Panemangalore, B. B. Scott, Y. Zhao, J. G. Bruno, L. Howard, J. Togias, Y. Ye, S. B. Singh, B. M. McKeever, P. R. Lindblom, J. Guo, R. Guo, H. Nar, A. Schuler-Metz, R. E. Gregg, K. Leftheris, R. K. Harrison, G. M. McGeehan, L. Zhuang and D. A. Claremon, *Bioorg. Med. Chem. Lett.*, 2010, **20**, 6725–6729.

92. Z. Xu, C. M. Tice, W. Zhao, S. Cacatian, Y. Ye, S. B. Singh, P. Lindblom, B. M. McKeever, P. M. Krosky, B. A. Kruk, J. Berbaum, R. K. Harrison,

J. A. Johnson, Y. Bukhtiyarov, R. Panemangalore, B. B. Scott, Y. Zhao, J. G. Bruno, J. Togias, J. Guo, R. Guo, P. J. Carroll, G. M. McGeehan, L. Zhuang, W. He and D. A. Claremon, *J. Med. Chem.*, 2011, **54**, 6050–6062.

93. http://www.pfizer.com/research/product_pipeline/product_pipeline.jsp, Last accessed August 2011.

94. J. Xiang, M. Ipek, V. Suri, W. Massefski, N. Pan, Y. Ge, M. Tam, Y. Xing, J. F. Tobin, X. Xu and S. Tam, *Bioorg. Med. Chem. Lett.*, 2005, **15**, 2865–2869.

95. J. Xiang, M. Ipek, V. Suri, M. Tam, Y. Xing, N. Huang, Y. Zhang, J. Tobin, T. S. Mansour and J. McKew, *Bioorg. Med. Chem.*, 2007, **15**, 4396–4405.

96. J. Xiang, Z. K. Wan, H. Q. Li, M. Ipek, E. Binnun, J. Nunez, L. Chen, J. C. McKew, T. S. Mansour, X. Xu, V. Suri, M. Tam, Y. Xing, X. Li, S. Hahm, J. Tobin and E. Saiah, *J. Med. Chem.*, 2008, **51**, 4068–4071.

97. Z. Wan, E. Chenail, J. Xiang, H. Li, M. Ipek, J. Bard, K. Svenson, T. S. Mansour, X. Xu, X. Tian, V. Suri, S. Hahm, Y. Xing, C. E. Johnson, X. Li, A. Qadri, D. Panza, M. Perreault, J. F. Tobin and E. Saiah, *J. Med. Chem.*, 2009, **52**, 5449–5461.

98. M. Wang, X. Tian, L. Leung, J. Wang, N. Houvig, J. Xiang, E. Saiah, S. Hahm, V. Suri and X. Xu, *Drug Metab. Lett.*, 2011, **5**, 55–63.

99. http://www.bms.com/research/pipeline/Pages/default.aspx, Last accessed August 2011.

100. H. Wang, Z. Ruan, J. J. Li, L. M. Simpkins, R. A. Smirk, S. C. Wu, R. D. Hutchins, D. S. Nirschl, K. Van Kirk, C. B. Cooper, J. C. Sutton, Z. Ma, R. Golla, R. Seethala, M. E. K. Salyan, A. Nayeem, S. R. Krystek Jr., S. Sheriff, D. M. Camac, P. E. Morin, B. Carpenter, J. A. Robl, R. Zahler, D. A. Gordon and L. G. Hamann, *Bioorg. Med. Chem. Lett.*, 2008, **18**, 3168–3172.

101. H. Wang, J. A. Robl, L. G. Hamann, L. Simpkins, R. Golla, Y. Li, R. Seethala, T. Zvyaga, D. A. Gordon and J. J. Li, *Bioorg. Med. Chem. Lett.*, 2011, **21**, 4146–4149.

102. http://www.roche.com/research_and_development/research/pipeline/roche_pharma_pipeline.htm, Last accessed August 2011.

103. W. Neidnart, *Frontiers in Medicinal Chemistry 2011*, Saarbruken, Germany, 2011.

104. http://www.jt.com/investors/results/forecast/pdf/B.S.20101028_E.pdf, Last accessed August 2011.

105. http://www.lilly.com/research/Pages/pipeline.aspx, Last accessed August 2011.

106. S. P. Webster, M. Binnie, K. M. M. McConnell, K. Sooy, P. Ward, M. F. Greaney, A. Vinter, T. D. Pallin, H. J. Dyke, M. I. A. Gill, I. Warner, J. R. Seckl and B. R. Walker, *Bioorg. Med. Chem. Lett.*, 2010, **20**, 3265–3271.

107. S. P. Webster, P. Ward, M. Binnie, E. Craigie, K. M. M. McConnell, K. Sooy, A. Vinter, J. R. Seckl and B. R. Walker, *Bioorg. Med. Chem. Lett.*, 2007, **17**, 2838–2843.
108. D. Schuster, E. M. Maurer, C. Laggner, L. G. Nashev, T. Wilckens, T. Langer and A. Odermatt, *J. Med. Chem.*, 2006, **49**, 3454–3466.
109. S. H. Kim, R. Ramu, S. W. Kwon, S. Lee, C. H. Kim, S. K. Kang, S. D. Rhee, M. A. Bae, S. H. Ahn, D. C. Ha, H. G. Cheon, K. Y. Kim and J. H. Ahn, *Bioorg. Med. Chem. Lett.*, 2010, **20**, 1065–1069.
110. S. W. Kwon, S. K. Kang, J. H. Lee, J. H. Bok, C. H. Kim, S. D. Rhee, W. H. Jung, H. Y. Kim, M. A. Bae, J. S. Song, D. C. Ha, H. G. Cheon, K. Y. Kim and J. H. Ahn, *Bioorg. Med. Chem. Lett.*, 2011, **21**, 435–439.
111. J. H. Lee, N. S. Kang and S. Yoo, *Bioorg. Med. Chem. Lett.*, 2008, **18**, 2479–2490.
112. Y. Ye, Z. Zhou, H. Zou, Y. Shen, T. Xu, J. Tang, H. Yin, M. Chen, Y. Leng and J. Shen, *Bioorg. Med. Chem.*, 2009, **17**, 5722–5732.
113. X. Zhang, Z. Zhou, H. Yang, J. Chen, Y. Feng, L. Du, Y. Leng and J. Shen, *Bioorg. Med. Chem. Lett.*, 2009, **19**, 4455–4458.
114. H. Yang, W. Dou, J. Lou, Y. Leng and J. Shen, *Bioorg. Med. Chem. Lett.*, 2008, **18**, 1340–1345.
115. H. Moreno-Díaz, R. Villalobos-Molina, R. Ortiz-Andrade, D. Díaz-Coutiño, J. L. Medina-Franco, S. P. Webster, M. Binnie, S. Estrada-Soto, M. Ibarra-Barajas, I. León-Rivera and G. Navarrete-Vázquez, *Bioorg. Med. Chem. Lett.*, 2008, **18**, 2871–2877.
116. G. M. Coppola, P. J. Kukkola, J. L. Stanton, A. D. Neubert, N. Marcopulos, N. A. Bilci, H. Wang, H. C. Tomaselli, J. Tan, T. D. Aicher, D. C. Knorr, A. Y. Jeng, B. Dardik and R. E. Chatelain, *J. Med. Chem.*, 2005, **48**, 6696–6712.
117. F. Lepifre, S. Christmann-Franck, D. Roche, C. Leriche, D. Carniato, C. Charon, S. Bozec, L. Doare, F. Schmidlin, M. Lecomte and E. Valeur, *Bioorg. Med. Chem. Lett.*, 2009, **19**, 3682–3685.
118. D. Roche, D. Carniato, C. Leriche, F. Lepifre, S. Christmann-Franck, U. Graedler, C. Charon, S. Bozec, L. Doare, F. Schmidlin, M. Lecomte and E. Valeur, *Bioorg. Med. Chem. Lett.*, 2009, **19**, 2674–2678.
119. O. Venier, C. Pascal, A. Braun, C. Namane, P. Mougenot, O. Crespin, F. Pacquet, C. Mougenot, C. Monseau, B. Onofri, R. Dadji-Faihun, C. Leger, M. Ben-Hassine, T. Van-Pham, J. Ragot, C. Philippo, S. Guessregen, C. Engel, G. Farjot, L. Noah, K. Maniani and E. Nicolai, *Bioorg. Med. Chem. Lett.*, 2011, **21**, 2244–2251.
120. S. F. Neelamkavil, C. D. Boyle, S. Chackalamannil, W. J. Greenlee, L. Zhang and G. Terracina, *Bioorg. Med. Chem. Lett.*, 2009, **19**, 4563–4565.
121. V. S. C. Yeh, J. R. Patel, H. Yong, R. Kurukulasuriya, S. Fung, K. Monzon, W. Chiou, J. Wang, D. Stolarik, H. Imade, D. Beno, M. Brune, P. Jacobson, H. Sham and J. T. Link, *Bioorg. Med. Chem. Lett.*, 2006, **16**, 5414–5419.

122. V. S. C. Yeh, R. Kurukulasuriya, D. Madar, J. R. Patel, S. Fung, K. Monzon, W. Chiou, J. Wang, P. Jacobson, H. L. Sham and J. T. Link, *Bioorg. Med. Chem. Lett.*, 2006, **16**, 5408–5413.

123. V. S. C. Yeh, R. Kurukulasuriya, S. Fung, K. Monzon, W. Chiou, J. Wang, D. Stolarik, H. Imade, R. Shapiro, V. Knourek-Segel, E. Bush, D. Wilcox, P. T. Nguyen, M. Brune, P. Jacobson and J. T. Link, *Bioorg. Med. Chem. Lett.*, 2006, **16**, 5555–5560.

124. S. Richards, B. Sorensen, H. Jae, M. Winn, Y. Chen, J. Wang, S. Fung, K. Monzon, E. U. Frevert, P. Jacobson, H. Sham and J. T. Link, *Bioorg. Med. Chem. Lett.*, 2006, **16**, 6241–6245.

125. B. Sorensen, J. Rohde, J. Wang, S. Fung, K. Monzon, W. Chiou, L. Pan, X. Deng, D. Stolarik, E. U. Frevert, P. Jacobson and J. T. Link, *Bioorg. Med. Chem. Lett.*, 2006, **16**, 5958–5962.

126. B. Sorensen, M. Winn, J. Rohde, Q. Shuai, J. Wang, S. Fung, K. Monzon, W. Chiou, D. Stolarik, H. Imade, L. Pan, X. Deng, L. Chovan, K. Longenecker, R. Judge, W. Qin, M. Brune, H. Camp, E. U. Frevert, P. Jacobson and J. T. Link, *Bioorg. Med. Chem. Lett.*, 2007, **17**, 527–532.

127. J. J. Rohde, M. A. Pliushchev, B. K. Sorensen, D. Wodka, Q. Shuai, J. Wang, S. Fung, K. M. Monzon, W. J. Chiou, L. Pan, X. Deng, L. E. Chovan, A. Ramaiya, M. Mullally, R. F. Henry, D. F. Stolarik, H. M. Imade, K. C. Marsh, D. W. A. Beno, T. A. Fey, B. A. Droz, M. E. Brune, H. S. Camp, H. L. Sham, E. U. Frevert, P. B. Jacobson and J. T. Link, *J. Med. Chem.*, 2007, **50**, 149–164.

128. J. R. Patel, Q. Shuai, J. Dinges, M. Winn, M. Pliushchev, S. Fung, K. Monzon, W. Chiou, J. Wang, L. Pan, S. Wagaw, K. Engstrom, F. A. Kerdesky, K. Longenecker, R. Judge, W. Qin, H. M. Imade, D. Stolarik, D. W. A. Beno, M. Brune, L. E. Chovan, H. L. Sham, P. Jacobson and J. T. Link, *Bioorg. Med. Chem. Lett.*, 2007, **17**, 750–755.

Recent Advances in PTP1B Inhibitor Development for the Treatment of Type 2 Diabetes and Obesity

RONGJUN HE, LI-FAN ZENG, YANTAO HE AND ZHONG-YIN ZHANG*

Department of Biochemistry and Molecular Biology, Indiana University School of Medicine, 635 Barnhill Drive, Indianapolis, IN 46202, USA
*E-mail: zyzhang@iupui.edu

6.1 Introduction

Reversible phosphorylation of protein residues is a key strategy for cells to convey and conduct cellular signals, such as growth, differentiation, migration, and apoptosis. Aberrant phosphorylations either initiate inappropriate or block functional signal pathways, which results in the pathogenesis of various human diseases, including cancer, diabetes/obesity, inflammation, and infectious diseases.[1] The reactions of phosphorylation and dephosphorylation are catalyzed by protein kinases and protein phosphatases respectively. Kinases as promoters of protein phosphorylation are among the most successful therapeutic targets for human cancer therapy with eight small molecule kinase inhibitor drugs approved by the US FDA from 2001 to 2008 and a large number of compounds are presently in clinical trials.[2] Due to the opposite role to kinases in controlling the level of protein phosphorylation,

RSC Drug Discovery Series No. 27
New Therapeutic Strategies for Type 2 Diabetes: Small Molecule Approaches
Edited by Robert M. Jones

Published by the Royal Society of Chemistry, www.rsc.org

protein phosphatases, including protein tyrosine phosphatases (PTPs), are highly anticipated as the next generation drug targets.[3–6]

PTPs are a large family of enzymes composed by 107 members, of which 99 are class I cysteine-based phosphatases sharing a conserved signature motif sequence $(H/V)C(X)_5R(S/T)$. These are divided into two major groups: classic tyrosine-specific PTPs with 38 members, and tyrosine and serine/threonine dual-specific phosphatases (DSP) with 61 members. The classic group contains 17 intracellular or non-receptor PTPs and 21 trans-membrane or receptor tyrosine phosphatases.[5,6]

PTP1B belongs to the classic non-receptor PTP group, which was the first mammalian PTP isolated from human placenta.[7] PTP1B is a ubiquitously expressed protein with 435 amino acids. The catalytic domain is composed by residues 30–278, where the signature motif (PTP loop conserved in all PTPs for binding the phosphate group) lies between residues 214 and 221, and Cys215 is the key nucleophilic residue responsible for the catalytic activity.[8] The sequence at the C-terminal localizes PTP1B to the cytosolic side of endoplasmic reticulum in the cell.[9]

The significance of PTP1B is that it has been implicated to play critical roles in human diseases such as cancer, diabetes, and obesity. For example, PTP1B is able to delay lymphomagenesis in p53-deficient mice by promoting B-cell differentiation, suggesting its tumor suppressing role.[10] In contrast, PTP1B could also act as a tumor promoter: breast cancer mice model induced by ErbB-2 activation shows delayed tumor progression and a protective effect in lung metastasis when PTP1B was deficient or inhibited.[11,12] These studies indicate that PTP1B has two roles in the development of cancer, depending on the cellular substrates and localizations.[13] Perhaps more importantly, PTP1B has been demonstrated to be a key negative regulator in both insulin and leptin signaling pathways, which has triggered tremendous research interest in establishing PTP1B as a therapeutic target for type 2 diabetes and obesity, and the development of small molecule inhibitors of PTP1B for treatments.[14] In this chapter, we will briefly introduce the biochemistry of PTP1B and its association to type 2 diabetes and obesity. Then we will focus on the recent advances in PTP1B inhibitor development, especially during the past five years.

6.2 Biochemistry of PTP1B

A number of crystal structures of PTP1B have been reported. This wealth in structure information coupled with detailed mechanistic studies has contributed to our understanding of PTP catalysis. For example, the crystal structure of PTP1B C215S mutant in complex with a p-Tyr substrate was determined (Figure 6.1A, PDB bank code: 1PTV),[15] which shows that PTP loop Ser215 and Arg221 are in close proximity to the p-Tyr phosphate group, and Asp181 from the closed conformation WPD loop (composed by residues tryptophan (W), proline (P) and aspartic acid (D)) and Gln262 from Q loop

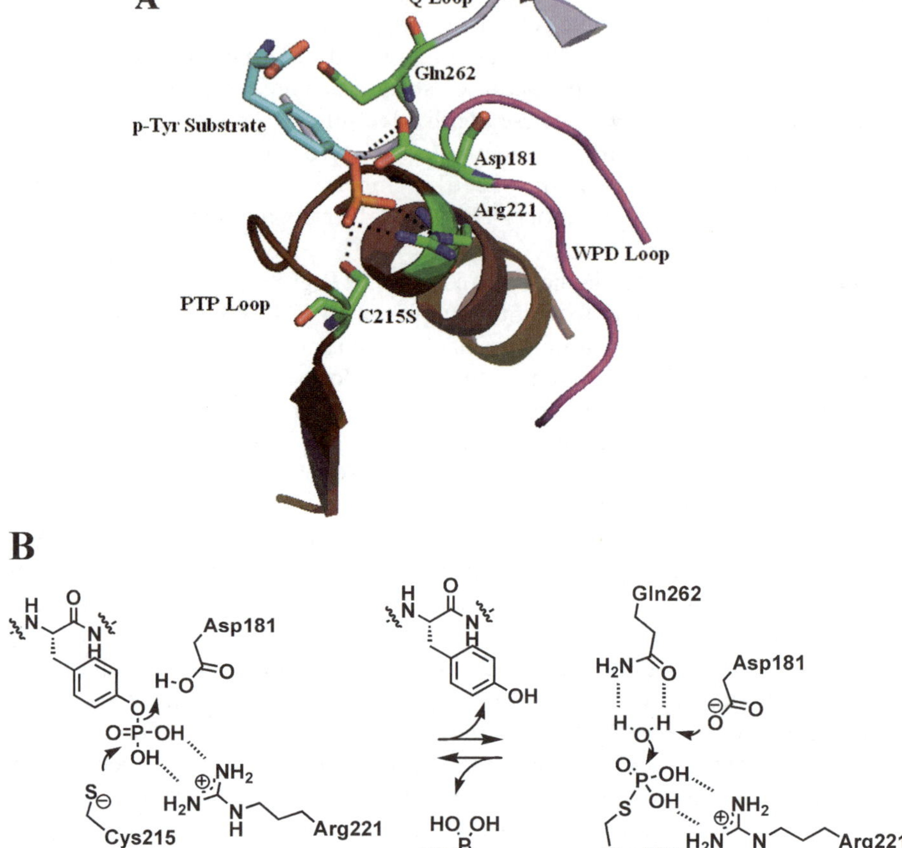

Figure 6.1   Structure of substrate bound PTP1B C215S mutant (PDB code: 1PTV) and PTP1B catalytic mechanism. (A) X-ray crystallographic structure of catalytic inactive C215S mutant of PTP1B in complex with p-Tyr substrate. PTP loop C215S and Arg221 are in close proximity to the p-Tyr phosphate group, and Asp181 from the closed conformation WPD loop and Gln262 from Q loop are adjacent to the oxygen atom of Tyr residue. (B) In a catalytic cycle, residue Cys215 removes the phosphate group from substrate with the assistance of Arg221 and forms an S–P bonded intermediate in the aid of general acid catalyst Asp181 from WPD loop; this intermediate is hydrolyzed by water mediated by Asp181 with the help of Q loop residue Gln262 to restore Cys215.

are adjacent to the phenolic oxygen atom of the Tyr residue. Hence during catalysis, when a p-Tyr substrate binds to the active site, the WPD loop closes down to align the phosphate to Cys215 and Arg221, and Asp181 to the tyrosine oxygen atom (Figure 6.1B). Cys215 attacks the phosphate group to form an S–P bond with the concurrent departure of the Tyr leaving group. This S–P bonded intermediate is subsequently hydrolyzed by water, which is mediated by Asp181 with the assistance of the Q loop residue Gln262, to release a naked phosphate ion and to complete a catalytic cycle.[16,17] In this process, Cys215, Arg221, Asp181, and Gln262 are the key residues for catalysis. Nevertheless, other residues which form hydrophobic, electrostatic, and hydrogen bonding interactions with substrates are also important.

6.3 Association of PTP1B with Type 2 Diabetes and Obesity

Diabetes, a chronic disease due to the inadequate production of, or decreased response to, insulin, is a major public health threat affecting more than 180 million people worldwide.[18,19] Diabetes-related conditions kill up to 3 million people annually. More than 90% incidence belongs to type 2 diabetes, which results from insulin resistance, a condition that the body uses insulin improperly or secretes an inadequate amount of insulin.[20] Obesity is characterized by increased body weight due to the excessive fat accumulation in the body. Globally, more than 400 million people are obese and this number is projected to reach 700 million by 2015. The consequence of obesity is an increased risk of developing diabetes, cancers, cardiovascular diseases, and musculoskeletal disorders.[21] This prevailing crisis has triggered research efforts from academia, biotech, and pharmaceutical industry to fight with type 2 diabetes and obesity. To this end, PTP1B has been found to be associated with type 2 diabetes and obesity and is a new and exciting drug target for their treatments.

6.3.1 PTP1B in Insulin Signaling

Insulin signaling plays a key role in the regulation of glucose homeostasis and metabolism which begins with insulin, a hormone secreted by pancreatic β-cells, binding to insulin receptor (IR) in response to increased nutrient level in the blood.[22] IR is a trans-membrane kinase with two extracellular α subunits and two trans-membrane β subunits (Figure 6.2). When insulin binds to IR, the kinase domain located at the β subunits' cytoplasmic region is activated, resulting in the autophosphorylation of multiple tyrosine residues in the β subunits.[23] The consequence of this autophosphorylation is the increase of IR's kinase activity, which carries out tyrosine phosphorylation on its substrates, such as insulin receptor substrate (IRS) 1–4, and several adaptor proteins like Grbs and Shc. The phosphorylation of IRS proteins activates the phosphatidylinositol 3-kinase (PI3K)-AKT pathway, which leads to the promotion of glucose storage as glycogen by phosphorylation and inactivation of glycogen synthase kinase 3 (GSK-3), and stimulation of glucose uptake by translocation of glucose transporter 4 (GLUT4) from intracellular storage to plasma membrane. The phosphorylation of Grbs and Shc proteins activates the Ras-mitogen-activated protein kinase (MAPK) pathway and promotes mitogenic responses which control cell growth and division.[24–27]

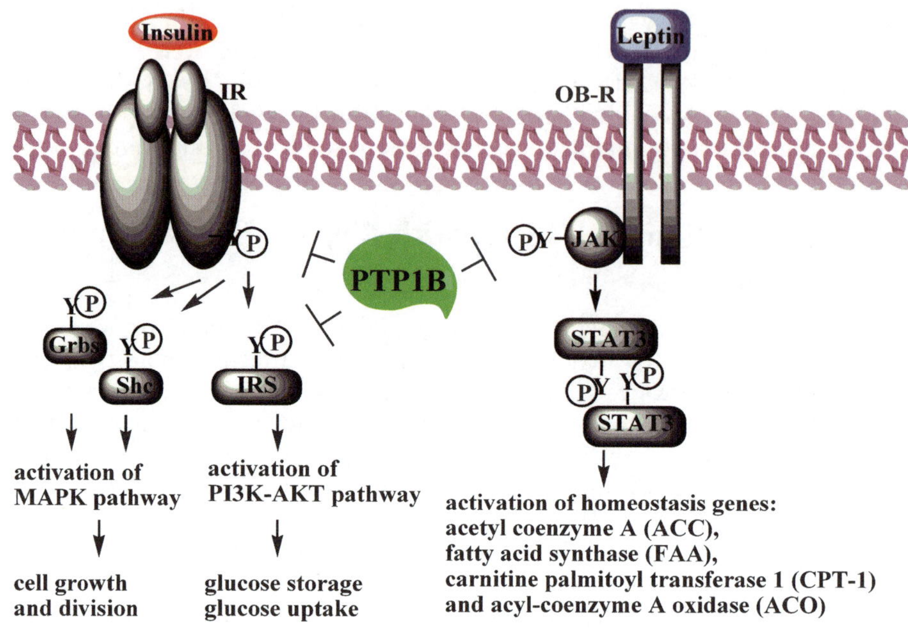

Figure 6.2 PTP1B in insulin and leptin signaling. In insulin signaling, the binding of insulin to insulin receptor (IR) results the autophosphorylation on its β subunits and the activation of its kinase domain, which in turn, phosphorylates insulin receptor substrates (IRS), and adaptor proteins Grbs and Shc. As a result, PI3K-AKT and MAPK pathways are activated and lead to glucose uptake and storage, cell growth, and cell division. In leptin signaling, the binding of leptin to its receptor OB-R causes the phosphorylation and activation of Janus kinase (JAK), which phosphorylates signal transducers and activators of transcription 3 (STAT3) protein. Phosphorylated STAT3 dimerizes, and travels into the nucleus to activate many genes regulating homeostasis, such as ACC, FAA, CPT-1, and ACO. PTP1B dephosphorylates phosphorylated IR, IRS, as well as JAK, thus interrupting both insulin signaling and leptin signaling.

Various studies showed that both phosphorylated IR and IRS are substrates of PTP1B.[14,28,29] PTP1B over-expression in different cells resulted in reduced phosphorylation levels of IR, IRS1, AKT, and MAPK, decreased activity of PI3K and MAPK, GLUT4 translocation, and impaired glycogen synthesis and glucose uptake.[30–32] In contrast, inhibition of PTP1B activity showed opposite effects: PTP1B antibodies were able to increase insulin-stimulated IR and IRS-1 phosphorylation, hence promoted PI3K activation and DNA synthesis.[33] Also, PTPs are known to be vulnerable to oxidation of its catalytic cysteine residue in the active site. Generation of H_2O_2 by Nox4 over-expression significantly reversed PTP1B-inhibited IR phosphorylation.[34] Vanadium compounds, which are potent inhibitors of PTPs including PTP1B, also displayed insulin mimetic or enhancing effects.[35]

Perhaps the most convincing data in clarifying the negative role of PTP1B in insulin signaling are PTP1B gene knockout mouse studies. It was shown in 1999 that PTP1B gene disruption gave normal and healthy mice with lower blood glucose concentration and 50% reduced circulating insulin level in fed mice.[36] After insulin stimulation, PTP1B−/− mice showed increased IRS-1 phosphorylation in muscle and IR phosphorylation in muscle and liver. These mice also exhibited enhanced insulin sensitivity in oral glucose and intraperitoneal insulin tolerance tests. In addition, PTP1B−/− and PTP1B−/+ mice were protected from weight gain and remained insulin sensitive when subjected to high fat diet, in contrast to PTP1B+/+ mice that gained weight rapidly and then were resistant to insulin. Similar results were obtained from a second, independent study in 2000.[37] It showed that mice lacking the PTP1B gene had increased tissue-specific insulin sensitivity, for example, skeletal muscle, and significantly less adiposity due to reduced fat cell mass with an unchanged adipocyte count. These mice were also resistant to diet-induced weight gain with elevated basal metabolic rate and total energy expenditure, but without appreciable uncoupling of mRNA expression level variations. Subsequently, it was reported that diabetic mice had increased insulin sensitivity and normalized blood glucose level after treatment with PTP1B neutralizing antisense oligonucleotides.[38] A more recent study suggested that liver specific PTP1B re-expression in PTP1B−/− mice showed attenuation of enhanced insulin sensitivity.[39] Taken together, these studies implicated that PTP1B is a major negative regulator in insulin signaling pathway by dephosphorylating IR and IRS proteins.

6.3.2　PTP1B in Leptin Signaling

Since leptin resistance usually occurs in type 2 diabetes and obesity patients, the observation that PTP1B-deficient mice were resistant to diabetes and diet-induced obesity has promoted research in defining the role of PTP1B in leptin signaling. Leptin is an important protein hormone (16 kD) that regulates metabolism and body weight.[40] When leptin binds to its receptor OB-R1, Janus kinase 2 (JAK2) is activated, and carries out phosphorylation on signal transducers and activators of transcription 3 (STAT3) protein. Phosphorylated STAT3 dimerizes and travels into the nucleus to activate genes involved in the regulation of homeostasis, including acetyl coenzyme A (ACC), fatty acid synthase (FAA), carnitine palmitoyl transferase 1 (CPT-1), and acyl-coenzyme A oxidase (ACO).[41]

Studies indicated that PTP1B plays a negative role in the leptin signaling cascade. Leptin stimulation of GT1-7 cells resulted in phosphorylation of STAT3 and activation of a STAT-dependent luciferase reporter gene. However, PTP1B over-expression in these cells reduced the phosphorylation level of JAK2 and STAT3, impaired the activation of leptin-stimulated, STAT-dependent luciferase reporter gene, and decreased expression of other genes induced by leptin, such as suppressor-of-cytokine-signaling-3 (SOCS3) and cfos.[42] In another study, PTP1B, not other PTPs such as CD45, PTPα, and LAR, dephosphorylated activated JAK2 and STAT3 *in vitro*.[43] The activation of STAT3 by leptin can be blocked by co-transfection with PTP1B in a dose-dependent manner, while a selective PTP1B inhibitor was able to reverse this process and re-activate STAT3.

In addition, treatment of hyperglycemic obese (ob/ob) mice with PTP1B antisense oligonucleotide ISIS-113715 reduced adiposity with down-regulation of lipogenesis genes, for example, sterol regulatory element-binding protein 1 and its downstream proteins spot14 and fatty acid synthase.[44] Mice with leptin and PTP1B deficiency had attenuation in weight gain, a decrease in adipose tissue, and an increase in resting metabolic rate.[45] PTP1B knockout mice exhibited reduced leptin/body fat ratio, leptin hypersensitivity, and enhanced leptin-induced hypothalamic STAT3 phosphorylation.[46] Resistance to obesity in these mice can be partially solved by gold thioglucose treatment, which ablated leptin-responsive hypothalamic neurons. On the other hand, diet-induced resistance to insulin and leptin are linked to elevated hepatic PTP1B levels. Over-expression of hepatic PTP1B in ob/ob mice blocked exogenous leptin function in reducing food intake and plasma glucose levels.[47] Also, hypothalamic PTP1B level increased with age and contributed to leptin resistance. Taken together, *in vitro* and *in vivo* studies indicated a negative role of PTP1B in leptin signaling. In conjunction with the role of PTP1B in insulin signaling, it is clear that PTP1B represents an exciting target for both type 2 diabetes and obesity mellitus.

6.4   Development of PTP1B Inhibitors

PTP1B plays major negative roles in both insulin and leptin signaling and is associated with type 2 diabetes and obesity. Small molecule inhibitors of PTP1B therefore possess potential therapeutic values for the treatment of these diseases. Given that PTP hydrolyzes phosphorylated tyrosine residue (p-Tyr), design of nonhydrolyzable p-Tyr surrogates as PTP inhibitors targeting the catalytic site has been a common strategy. Targeting a second allosteric pocket or both catalytic and allosteric sites is also a frequently used strategy. In this field, selectivity and bioavailability are the two major challenges attributed to the conserved active site configuration among all PTPs and the highly positively charged catalytic pocket favoring negatively charged inhibitors. The selectivity issue for PTP1B is critically important, as its closest relative, Tc-PTP, shares high structure and sequence similarities with PTP1B, and is required for proper hematopoiesis and immune functions as Tc-PTP-deficient mice died within 3–5 weeks after birth.[48] In the following section, reported PTP1B inhibitors are classified by their chemical nature that mimic the p-Tyr substrate. Their selectivities over other PTPs especially Tc-PTP, as well as bioavailability, are described. Although several reviews on PTP1B inhibitors have been published,[49–51] we will cover the progress made since 2005 in a comprehensive manner.

6.4.1 Phosphonic Acid and F2PMP Derivatives

The phosphonodifluoromethyl phenylalanine (F2PMP) moiety is the earliest successful p-Tyr surrogate that exists in numerous PTP1B inhibitors.[52] Earlier research focused on the inhibitors that targeted active site with unavoidable poor selectivity to other PTPs. Since the discovery of a second p-Tyr binding site adjacent to the active site in PTP1B, it was proposed to design molecules that could target both active and adjacent peripheral sites for achieving both potency and specificity.[53] This led to the discovery of compound **1** as a most potent PTP1B inhibitor with K_i at 2.4 nM and selectivity of 10-fold over Tc-PTP, which shares >70% similarity with PTP1B.[54] Given its exceptional potency and specificity, a crystal structure of PTP1B in complex with its derivative **2** was determined,[55] which shows that compound **2** simultaneously binds to the active site and a unique proximal noncatalytic site formed by Lys-41, Arg-47, and Asp-48. This work not only furnishes a structural basis for potent and selective PTP1B inhibition, but also suggests that potent, yet highly selective, PTP1B inhibitory agents can be acquired by targeting the area defined by residues Lys-41, Arg-47, and Asp-48, in addition to the previously identified second aryl phosphate-binding pocket.[53] However, the highly negatively charged nature of compound **1** gave it poor cell permeability under physiological conditions, hence strategies that included adding a fatty acid chain (Figure 6.3, compound **3**),[56] utilizing AB protein delivery (Figure 6.3,

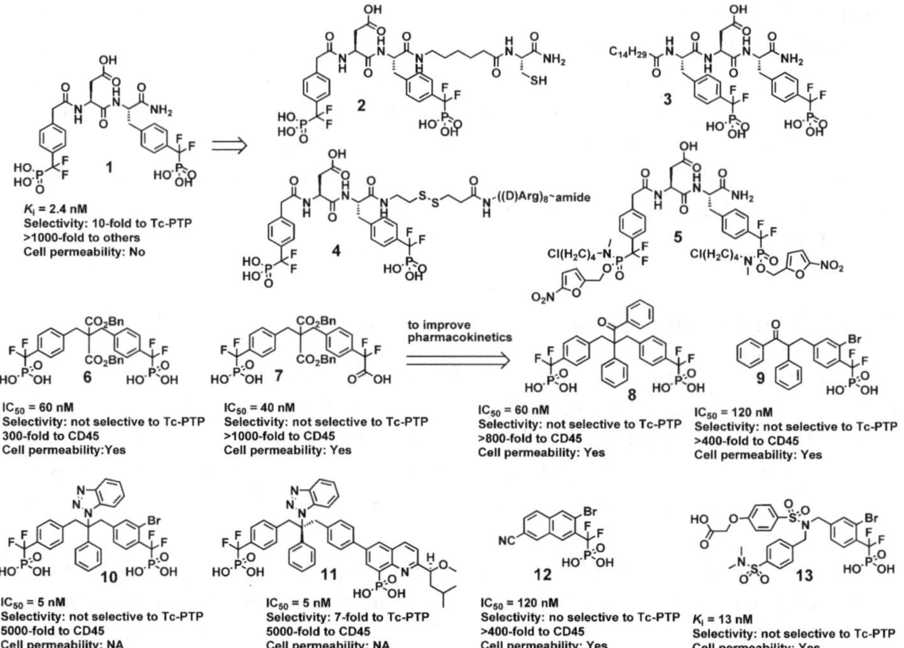

Figure 6.3 Structures of phosphonic acid and F2PMP-based PTP1B inhibitors.

compound **4**),[57] or shielding the negatively charged phosphonic acids with a nitrofuryl moiety have been developed in order to enable cell membrane penetration by the compound (Figure 6.3, compound **5**).[58]

Given the high potency of the F2PMP-based compounds, recent progress in developing the F2PMP mimetics were largely from the pharmaceutical industry. For example, Merck reported malonate ester **6** containing two F2PMP groups with an IC_{50} of 60 nM.[59] Switching one F2PMP group into a carboxylic acid resulted in a slightly more potent compound **7** which had an IC_{50} of 40 nM. Unfortunately, compounds **6** and **7** displayed poor pharmacokinetics in rats presumably due to their susceptibility to esterase in the blood. To improve the physicochemical properties, the malonate scaffold was replaced with deoxybenzoin, which provided compound **8** with similar potency. Deleting one F2PMP from **8** gave a structurally simplified compound **9** but with impaired potency. Although these compounds showed no selectivity over Tc-PTP, compounds **8** and **9** were orally available and were active in animal models of NIDDM. At the same time, tri- and tetra-substituted benzotriazoles were explored and compound **10** was the most potent one among them, exhibiting an IC_{50} of 5 nM with no selectivity over Tc-PTP.[60] The X-ray structure of one of these compounds bound with PTP1B was solved to provide the binding information, and a subsequent molecular modeling effort led to the discovery of a moderately selective inhibitor (7-fold over Tc-PTP) without a sacrifice in potency (compound **11**). These studies illustrated a successful example of utilizing structural information in achieving potent and selective PTP1B inhibitors. Recently, this team also found that quinoline/naphthalene-based F2PMP molecules are potent PTP1B inhibitors, and compound **12** possessed the best overall potency (IC_{50} = 120 nM) and pharmacokinetic profile, and it showed dose-dependent anti-diabetic and anti-cancer effects in animal models.[61] In 2005, scientists from Affymax reported a series of sulfonamide F2PMP inhibitors of PTP1B, with the most potent compound being **13**, which had a K_i of 13 nM but again without selectivity against Tc-PTP.[62]

6.4.2 Carboxylic Acids

Although early efforts in PTP1B inhibitor development focused on using F2PMP as a p-Tyr surrogate, the trend has shifted to carboxylic acids recently, due to the large abundance and easy access and manipulation of carboxylic acids. Specifically, salicylic acid was first discovered by Zhang as a novel p-Tyr mimetic by structure-based docking methodology (Figure 6.4, compound **14**).[63] The importance of the salicylic acid p-Tyr mimetic is that inhibitors based on it are cell permeable, as the adjacent hydroxyl group is able to form internal hydrogen bonds with the carboxylic acid and thus stabilize and reduce its negative charge.[64–66]

Recently, Choi synthesized a series of cyclopenta[*d*][1,2]-oxazine derivatives and evaluated their inhibitory activity against PTP1B.[67] Compound **15** showed

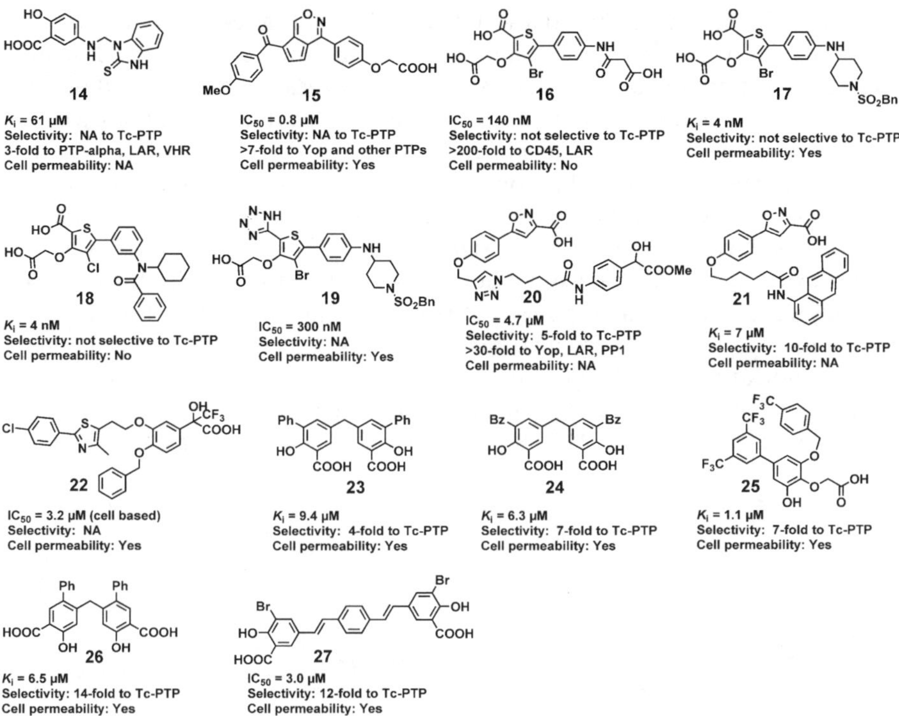

Figure 6.4 Structures of carboxylic acid-based PTP1B inhibitors, part 1.

an IC$_{50}$ of 0.8 µM and was able to normalize the plasma glucose level in ob/ob mice via i.v. administration. However, it was only weakly active when orally administrated, suggesting its poor bioavailability. With the aid of X-ray crystallography and computational modeling, a team at Wyeth has developed several reversible and competitive nanomolar PTP1B inhibitors with a monocyclic thiophene scaffold.[68–70] Compounds **16**, **17**, and **18** have K_i values at 140 nM, 4 nM, and 4 nM towards PTP1B, respectively. However, these three compounds showed no selectivity to Tc-PTP and only compound **17** appeared to be cell permeable: it could be actively transported into hepatocytes. Subsequent work on replacement for one carboxylic acid group afforded compound **19** with improved cell permeability but a loss in potency in compensation.[71] Esterification of the carboxylic acid groups in these compounds afforded bioavailable prodrugs with *in vivo* activities in inhibiting PTP1B and enhancing insulin sensitivity.[72] In 2006, Yao assembled a library of 66 compounds from isoxazole carboxylic acid using click chemistry and found **20** as a PTP1B inhibitor with an IC$_{50}$ of 4.7 µM and specificity of 5-fold over Tc-PTP.[73] With the same p-Tyr mimetic group, they also obtained compound **21** using solid-phase chemistry, which has a K_i at 7 µM with a specificity of 10-fold over Tc-PTP.[74] Adams discovered 2-aryl-3,3,3-trifluoro-2-hydroxypropionic acids as a new class of PTP1B inhibitors using a structure-based approach.[75] These compounds exhibited PTP1B inhibitory activity in a Sf9 cell-based assay, and the best compound (**22**) had an IC_{50} of 3.2 μM against PTP1B, which provided a starting point for a new series of cell permeable, non-peptidic, mono-acid PTP1B inhibitors. As salicylic acid has been reported as a novel and cell permeable p-Tyr mimetic,[63] Cho synthesized a series of salicylic acid derivatives as PTP1B inhibitors, and found compounds **23**, **24**, **25**, **26**, and **27** had IC_{50} or K_i values at sub μM range against PTP1B with 4- to 14-fold specificities over Tc-PTP.[76–79] The significance of these compounds was that they showed good oral bioavailability and were able to suppress high-fat-diet induced weight gain and adipocyte fat storage in mouse models. Interestingly, compounds **23** and **26** were also reported to be potent inhibitors of kinase IKK-β, indicating their *in vivo* targets may not be limited to PTP1B.[80]

Maccari synthesized 5-arylidene-2,4-thiazolidinediones and evaluated their inhibition activity against PTP1B (Figure 6.5).[81] Compound **28** was the most

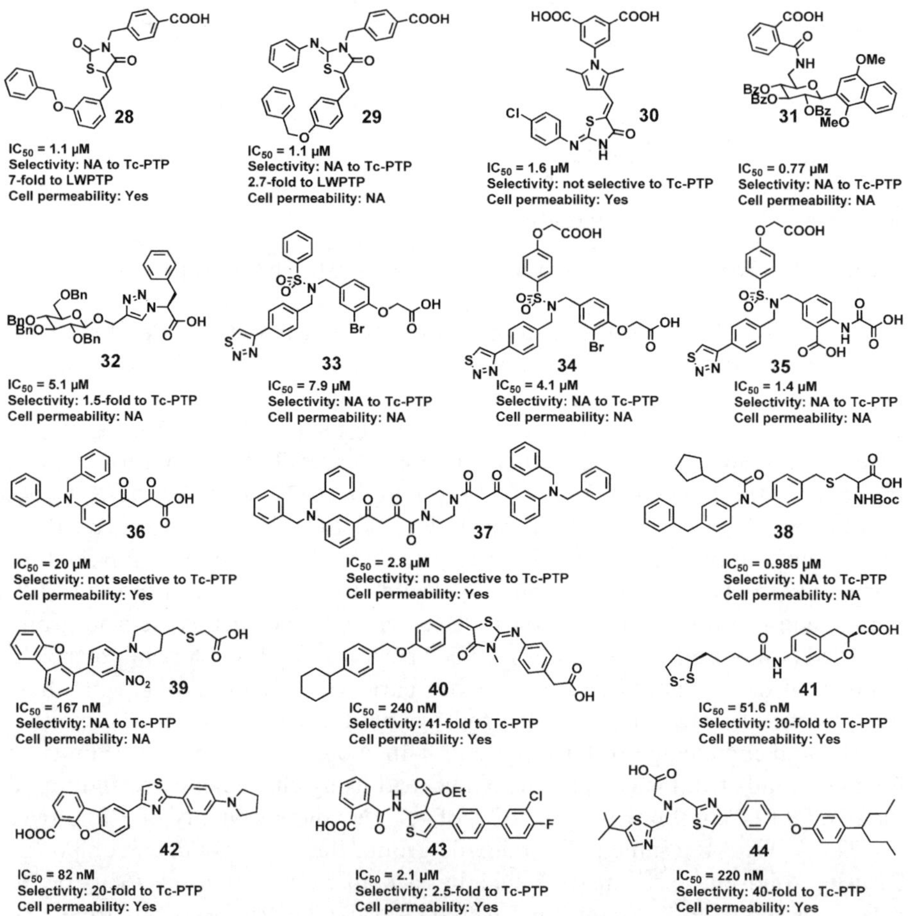

Figure 6.5   Structures of carboxylic acid based PTP1B inhibitors, part 2.

potent, with an IC_{50} of 1.1 μM. Although it was 7-fold selective over the low molecular weight PTP, its specificity over Tc-PTP was not reported. Based on this result, Ottana reported compound **29** as a new generation PTP1B inhibitor.[82] Unfortunately, it showed a similar potency compared to **28** with even decreased selectivity, indicating that the phenyl aniline group disrupted the original bind mode. In a high throughput screening campaign, Li identified compound **30** from a set of 48 000 compounds as a novel and competitive inhibitor against PTP1B.[83] It exhibited an IC_{50} of 1.6 μM but without selectivity to Tc-PTP. This compound was able to increase the insulin-induced tyrosine phosphorylation level of IRβ in a dose-dependent manner in CHO/ *h*IR cells. Xie synthesized a class of sugar-based PTP1B inhibitors, the most potent one presented an IC_{50} at 0.77 μM (compound **31**).[84] This study demonstrated the potential of *C*-glycosyl compounds as small molecular inhibitors of PTP1B. Subsequently, they employed a microwave-assisted click reaction for the assembly of sugar-based compound library, and found compound **32** to be the most potent inhibitor against PTP1B with IC_{50} value of 5 μM and poor selectivity against Tc-PTP.[85] In probing F2PMP bioisosteres on a sulfonamide scaffold, a team from Affymax discovered compounds **33**, **34**, **35** as PTP1B inhibitors at low μM range, demonstrating the feasibility of this strategy in identifying non-phosphonate p-Tyr mimetics in a small molecule scaffold.[86] The selectivity and cell permeability data of these compounds were not reported. Recently, Zhang identified aryl diketoacids as novel p-Tyr surrogates and showed that neutral amide-linked aryl diketoacid dimers have excellent PTP inhibitory activity (compounds **36** and **37**).[87] Kinetic studies established that they act as noncompetitive inhibitors of PTP1B. Crystal structures of ligand-bound PTP1B revealed that both the aryl diketoacid and its dimeric derivative bind PTP1B at the active site, with distinct modes of interaction, in the catalytically inactive, WPD loop open conformation. Importantly, these compounds were cell permeable and were able to enhance insulin signaling in hepatoma cells, suggesting that targeting the inactive conformation may provide a unique opportunity for creating active site-directed PTP1B inhibitors with improved pharmacological properties. Cysteine derivatives and piperidin-methylthio acetic acids were reported by the Institute of Pharmaceutical Discovery to be potent PTP1B inhibitors: compounds **38** and **39** exhibited IC_{50} values of 0.98 μM and 167 nM, respectively.[88] Unfortunately, their specificity to Tc-PTP and cell permeability were not disclosed. Lupin Limited also patented their results of phenyl acetic acids on heterocyclic thiazolidine scaffold as PTP1B inhibitors.[89] The most potent inhibitor displayed an IC_{50} at 240 nM with 41-fold preference over Tc-PTP (compound **40**). Given the similarity between PTP1B and Tc-PTP, this level of selectivity is extraordinarily high, suggesting that it may target some pockets that are unique to PTP1B. In addition, this class of compounds was able to improve oral glucose tolerance in diet-induced obese mice, and decrease plasma glucose and triglyceride levels, indicating its good *in vivo* efficacy. Prasad and co-workers reported isochroman carboxylic acids as a novel class of PTP1B inhibitors.[90,91] Analysis of the structure–activity relationships (SAR) showed that a dithiolane ring with a spacer of 5 carbons to the isochroman ring was essential for the high potency of this class of compounds, and this led to the identification of compound **41** as a inhibitor PTP1B with an IC_{50} of 51.6 nM and selectivity of 30-fold over Tc-PTP. This group also synthesized a series of dibenzo[*b,d*]furan carboxylic acids as inhibitors of PTP1B and identified compound **42** with an IC_{50} of 82 nM and 20-fold specificity over Tc-PTP. *In vivo* studies showed that both compounds normalized the hyperglycemic condition in ob/ob mice and significantly reduced the fasting plasma glucose but not triglyceride levels. Liu designed and synthesized a series of novel thiophene derivatives, and evaluated their activities towards PTP1B.[92] They found that compound **43** was the most potent inhibitor among them, with an IC_{50} of 2.1 µM and 2.5-fold specificity over Tc-PTP. In addition, the molecule was a competitive and reversible PTP1B inhibitor, and was able to increase the AKT phosphorylation level in CHO-K1 cells at 10 µM. A team in Japan Tobacco reported compound **44** as a novel PTP1B inhibitor with a mixed binding mode.[93] This compound had a K_i of 0.2 µM with great preference over Tc-PTP (K_i = 9.3 µM). Moreover, it increased the insulin-stimulated glucose uptake when treated in L6 cells. In ob/ob mice, a single dose administration of compound **44** enhanced the IR phosphorylation in liver and reduced the glucose level. Interestingly, in db/db mice, chronic administration exhibited a hypoglycemic effect without an acceleration of body weight gain. These *in vivo* studies indicated that further work is needed to improve its pharmacological properties.

6.4.3 Sulfonic Acids

Sulfonic acids represent another class of p-Tyr mimetic in addition to phosphonic acids and carboxylic acids (Figure 6.6). Recently, scientists from P&G Pharmaceuticals identified the sulfamic acid moiety as a potential p-Tyr mimetic during a HTS campaign.[94] By incorporating this moiety into a tetrahydroisoquinoline scaffold, they discovered compound **45** as a PTP1B inhibitor with IC_{50} of 4.8 µM and specificity of 2-fold over Tc-PTP. The X-ray crystal structure of the ligand-bound PTP1B showed that compound **45** lies in the active site of the protein with the second sulfamic acid in the P+1 pocket. Inspired by a previously reported F2PMP-based PTP1B inhibitor (compound **46**, IC_{50} = 8 nM),[95] Taylor designed and synthesized similar molecules using difluoromethylenesulfonic acid (DFMS) as a p-Tyr mimetic.[96] Unfortunately,

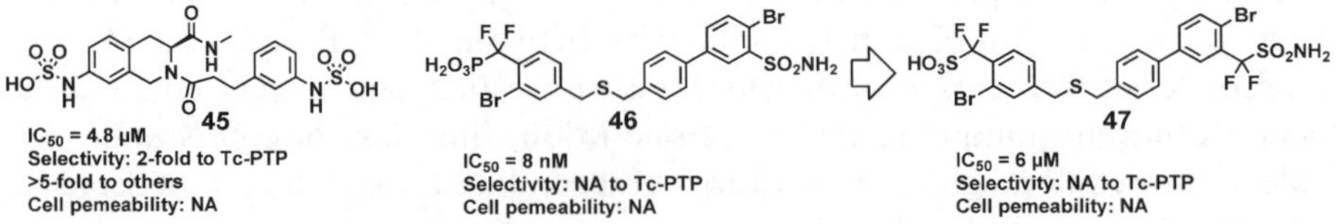

Figure 6.6 Structures of sulfonic acid-based PTP1B inhibitors.

they found that this class of compounds was about 1000-fold less potent against PTP1B: compound **47** had an IC_{50} at 6 μM. They also demonstrated that the methylene fluorines had only a modest effect on inhibitor potency. Despite the poor potency, this study showed that DFMS is a new type of readily accessible p-Tyr mimetic.

6.4.4 Imides

Nonhydrolyzable phosphonic, carboxylic and sulfonic acids are good p-Tyr mimetics; however, molecules with these anionic moieties often exhibit poor cell permeability. Hence *in vitro* activity of these molecules is hardly converted into the corresponding *in vivo* efficacy. In this regard, the search for less acidic (higher pK_a value) or delocalized anionic heterocyclic p-Tyr mimetics has been one of the new directions in developing cell permeable PTP1B inhibitors in recent years. In 2005, scientists from AstraZeneca identified 1,2,5-thiadiazolidin-3-one-1,1-dioxide (TDZ, compound **48**) as a novel PTP1B inhibitor with an IC_{50} at 1.6 mM.[97] With the aid of protein NMR and crystallography technologies, they further developed compound **49** as a sub-micromolar PTP1B inhibitor (IC_{50} = 2.5 μM), which is 1000-fold more potent than the parent compound **48**. SAR studies showed that the methoxy and phenyl groups were essential for the potency, as deleting either of them resulted in a 100-fold loss of activity. A ligand-bound PTP1B crystal structure revealed that the TDZ part of compound **49** heads into the active site, confirming its role in mimicking a p-Tyr substrate. However, the specificity and biological data of this compound are not available. In probing the replacement for carboxylic acid, Wyeth scientists also developed a series of TDZ derivatives as PTP1B inhibitors on the thiophene scaffold.[98–104] Compound **50** appeared to be the best among them with an IC_{50} at 1.7 μM. Cell permeability tests of this class of compounds on PAMPA suggested that TDZ has better permeability than the mono- and di-acid parent compounds and than tetrazole, an isostere of carboxylic acid.[64] Given the potential of TDZ as a cell-permeable p-Tyr mimetic, Novatis has explored various TDZ derivatives on different scaffolds, and their representative compounds are shown in Figure 6.7 (**51**–**55**). These compounds are generally very potent against PTP1B with IC_{50} values between 5 and 300 nM. Interestingly, this set of compounds showed much better potency and ligand efficiency than those similar compounds (**48** and **49**) from AstraZeneca, indicating that the *O*-hydroxyl group is crucial for the improved activity. Significantly, *in vivo* tests showed that these molecules were cell permeable: they were able to increase the phosphorylation level of AKT, which resulted in a physiological increase of myotube diameter in a mouse model for the treatment of musculoskeletal diseases. The Institutes for Pharmaceutical Discovery also reported several TDZ derivatives that were slightly different from the aforementioned compounds. For example, compound **56** incorporates a methylene bridge between the TDZ head and the aryl group, and compound **57** contains a fused tricyclic TDZ moiety.[105] In addition, TransTech also disclosed their research results in developing the TDZ-based PTP1B inhibitors without

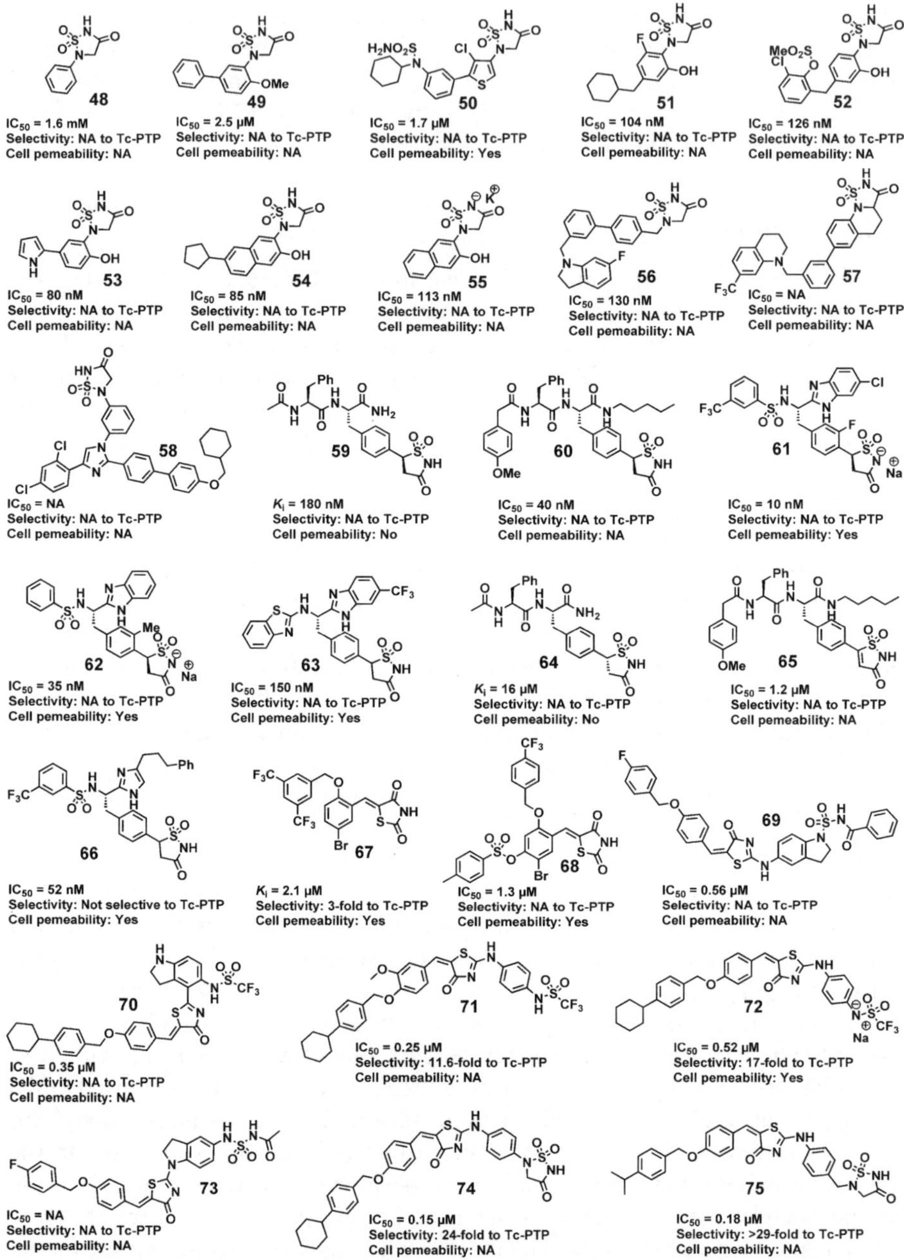

Figure 6.7 Structures of imide based PTP1B inhibitors.

reporting the activity (compound **58**).[106] These molecules with new scaffolds revealed that TDZ is a versatile p-Tyr mimetic and that it provides opportunities in developing various potent PTP1B inhibitors with better cell permeability and oral bioavailability profile.

Isothiazolidinone (IZD) is another type of heterocyclic p-Tyr mimetic that differs from TDZ only in one atom, where the nitrogen is replaced by a carbon. Incyte has published a series of molecules with the IZD group incorporated on various scaffolds, such as peptides, sulfonamides, and heterocycles.[107–112] Representative compounds from each class are summarized in Figure 6.7 (compounds **59–63**). Their IC_{50} values for PTP1B range from 10 to 180 nM, suggesting that IZD is an effective p-Tyr mimetic in inhibiting PTP1B. It is interesting to note that the stereochemistry on the IZD carbon plays a vital role in retaining activities, as switching from *S*-IZD to *R*-IZD on the same molecule abolished activity (**59** versus **64**).[107] In addition, the IZD ring has to be saturated, because the unsaturated molecule displayed a 30-fold decrease in potency (**60** versus **65**).[109] With regard to cell permeability, only a few compounds, such as **61** and **62**, were able to increase the phosphorylation level of IR in a dose-dependent manner.[110] And this class of compounds displayed poor selectivity over Tc-PTP, for example, with compound **66** having an identical IC_{50} value against both PTP1B and Tc-PTP.[113]

Very recently, thiazolidinedione (TZD) has been reported as the third type of heterocyclic p-Tyr mimetic. Cho synthesized a series of molecules based on TZD and evaluated their inhibition activities against PTP1B.[114,115] Compounds **67** and **68** were the best among them and had IC_{50} values of 2.1 and 1.3 µM, respectively, for PTP1B, with moderate selectivity over Tc-PTP. They significantly suppressed weight gain and improved blood parameters such as TG, total cholesterol, and NEFA, indicating good cell permeability and bioavailability.

Additionally, scientists from Lupin Limited have explored various imides as p-Tyr mimetics, such as linear trifluromethanesulfonamide and cyclic TDZ (compounds **69–75**).[116] These molecules were based on a common thiazole scaffold and generally displayed IC_{50} values at less than 1 µM with great than 10-fold selectivity over Tc-PTP. Compound **72**, a sodium salt of trifluro-methanesulfonamide, showed good *in vivo* efficacy in reducing glucose, triglyceride, and insulin levels, and in improving oral glucose tolerances in mouse models. Although the biological data was not available for the rest of compounds, they were very likely to be cell permeable.

Taken together, TDZ, IZD, TZD, and other imides represent new and less acidic p-Tyr mimetics which can be used for the development of potent PTP1B inhibitors. They exhibit improved cell permeability and oral availability when tested *in vivo*. Nevertheless, challenges still exist, as many of molecules within these groups failed to show *in vivo* efficacy, suggesting that additional work will be required to turn these less acidic p-Tyr mimetics into highly bioavailable molecules.

6.4.5 Neutral Molecules

Since molecules based on acidic p-Tyr mimetics are usually not cell permeable, additional strategies such as prodrugs and delivery vehicles are required to translate their *in vitro* activity into *in vivo* efficacy. Imides represent a class of less acidic p-Tyr mimetics with improved cellular activity, but many of these molecules still lack *in vivo* efficacy. Recent efforts in addressing the bioavailability problem include the development of essentially neutral molecules as PTP1B inhibitors.

In 2005, Goel found that functionalized acetophenones could function as a class of PTP1B inhibitors.[117] The best one among them was compound **76**, which exhibited 54% inhibition against PTP1B at 100 μM concentration (Figure 6.8). However, any additional data, such as IC_{50} value, specificity, and cell permeability, were not provided. Maurya synthesized uncharged benzo-furan isoxazolines via 1,3-dipolar cycloaddition reactions.[118] The activities of these compounds against PTP1B were evaluated and compound **77** displayed a K_i at 30 μM in a competitive mode. Coincidently, compounds **76** and **77** shared a common benzofuran scaffold, although there was no correlated origin between the two studies. Roche filed two patents that uncovered their findings of aminoquinazoline and pyridopyrimidinediamine derivatives as PTP1B

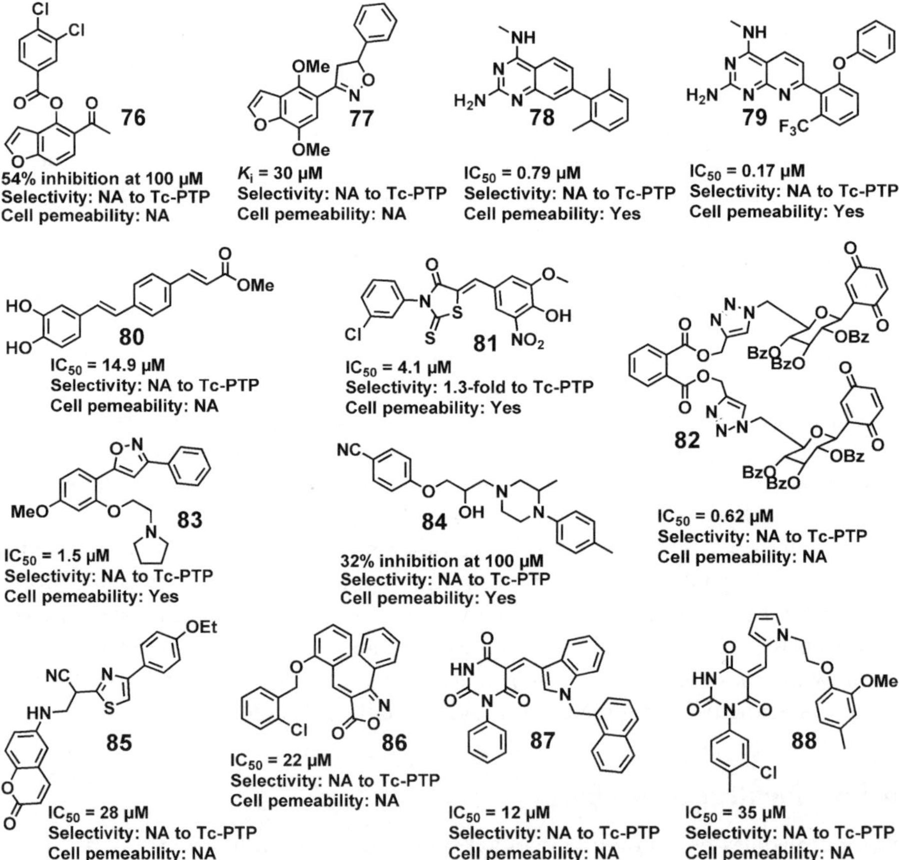

Figure 6.8   Structures of neutral molecules as PTP1B inhibitors.

inhibitors.[119,120] Compounds **78** and **79** were the most potent representatives with IC_{50} at 0.79 μM and 0.17 μM, respectively. Due to the existence of multiple amine groups, these classes of compounds are positively charged, which may offer them excellent cell permeability and bioavailability. Indeed, *in vivo* tests in mouse models showed their anti-diabetic effects, such as lower glucose, insulin, and triglyceride levels, improved oral glucose tolerance, etc. Inspired by lithospermic acid B, Jung designed and synthesized dihydroxyl stilbene derivatives from 4-(chloromethyl) benzoic acid, and tested their effects in inhibiting PTP1B activity.[121] Among these stilbenes, compound **80** had the best activity with IC_{50} at 14.9 μM. The two hydroxy groups were crucial for activity, as protecting them with methyl groups afforded a compound without any activity towards PTP1B. Tremblay designed and synthesized a series of uncharged thioxothiazolidione derivatives, and identified compound **81** as a competitive inhibitor against many PTPs, with IC_{50} at 4.1 μM towards PTP1B.[122] This neutral PTP1B inhibitor successfully sensitized wild-type but not PTP1B-null fibroblasts to insulin stimulation, and prevented PTP1B-dependent FLT3-ITD receptor tyrosine kinase dephosphorylation. Xie synthesized dimeric benzoylated β-C-D-glycosyl 1,4-dimethoxybenzenes and 1,4-benzoquinones and found that they were PTP1B inhibitors with compound **82** exhibiting the lowest IC_{50} values.[123] Since quinone compounds have strong oxidation potential and have been reported as irreversible cdc25 phosphatase inhibitors,[124,125] these compounds could likely inhibit PTP1B in the same way. Kumar designed and synthesized a class of 3,5-diarylisoxazole derivatives and screened them directly for *in vivo* anti-hyperglycemic activity in sucrose-loaded model (SLM), sucrose-challenged streptozotocin-induced diabetic rat model (STZ-S) as well as db/db mice model.[126] These molecules showed promising *in vivo* anti-hyperglycemic as well as moderate lipid lowering activity. In searching for the mechanism of their actions, they found that the compounds did inhibit PTP1B activity (compound **83**), hence they proposed that these molecules may exhibit their anti-diabetic effect through PTP1B pathway. Saxena have synthesized a series of substituted phenoxy-3-piperazin-1-yl-propan-2-ols and evaluated their *in vitro* PTP1B inhibitory activity and *in vivo* anti-diabetic activity.[127] Compound **84** showed a 32% inhibition against PTP1B at 100 μM concentration and 40.3% normalization of plasma glucose levels at 100 mg/kg in sucrose-loaded model (SLM) and 32% activity in streptozotocin model (STZ). The accurate IC_{50} data and specificity over Tc-PTP for these compounds were not reported. Cho identified several neutral PPP1B inhibitors via a structure-based virtual screening approach.[128] These compounds were structurally diverse and generally had IC_{50} values less than 50 μM (compounds **85–88**). Docking studies showed that these inhibitors could be stabilized by multiple hydrogen bonds and van der Waals contacts in the active site. Again, specificity and cell permeability data for these compounds were not reported. Finally, diketo compound **37**, listed in Figure 6.5, is a neutral and cell-permeable PTP1B inhibitor which is described in Section 6.4.2.

As discussed above, neutral molecules have emerged as a class of new PTP1B inhibitors. All compounds that selected for *in vivo* testing exhibited excellent activity, suggesting that neutral molecules with PTP inhibitory activity may provide an effective strategy in addressing the long-standing bioavailability issue. However, it is important to point out that additional work will be required to establish that the observed *in vivo* activity is in fact due to direct inhibition of PTP1B activity. It is also clear that many of these molecules display only moderate PTP1B inhibitory activity. Thus future work should be focused on improving both the potency and selectivity of these neutral molecules as PTP1B inhibitors.

6.4.6   Natural Products and Derivatives

Natural products are a rich source of biologically active molecules and are indispensable throughout drug discovery history. Interestingly, several natural products have been found to possess PTP1B inhibitory activity. The reason that they are described here separately is because in many situations they interact with PTP1B via their unique scaffolds, rather than relying on functionalities that mimic p-Tyr. Since 2005, several groups have reported that ursolic acid (**89**),[129,130] oleanolic acid (**90**),[131] maslinic acid (**91**),[132] and other triterpenoids (**92**, **93**),[131,133] which exist in many plants, are capable of inhibiting PTP1B with IC_{50} values at the low micromolar range. These molecules may serve as good starting points for developing natural product-derived potent PTP1B inhibitors. For example, extension on the carboxylic acid group of ursolic acid afforded compound **94** with a 10-fold increase in potency.[129] Although not selective over Tc-PTP, this molecule was able to enhance IR phosphorylation in CHO/hIR cell and stimulate glucose uptake in L6 myotubes. Derivatization of oleanolic acid provided compound **95** (NPLC441) with improved selectivity over Tc-PTP although with decreased potency.[134] It enhanced insulin-stimulated phosphorylation of IR and AKT in HepG2 cell, indicating good cell permeability. Compound **96** was also originated from oleanolic acid but without any increase in potency.[135] Several derivatives of maslinic acid have been synthesized and evaluated for PTP1B inhibition.[134] The best compound, **97**, showed an IC_{50} of 0.64 µM with selectivity of 6.9-fold over Tc-PTP. In addition, compound **98** was developed from another triterpenoid saponin. It has an IC_{50} as low as 270 nM towards PTP1B with 3-fold selectivity over Tc-PTP. Recently, ilekudinol A and B were isolated from the leaves and stems of *Weigela subsessilis* (Caprifoliaceae) and found to be PTP1B inhibitors. Ilekudinol B (compound **99**, IC_{50} = 5.3 µM) possesses better activity than ilekudinol A and is a non-competitive inhibitor of PTP1B.[136] Simpler diterpenoids from roots of *Acanthopanax koreanum* (Araliaceae) were also found to inhibit PTP1B, and the best compound, **100**, displayed an IC_{50} at 7.1 µM in a non-competitive manner. Introduction of a hydroxyl group at C-16 or reduction of the carboxylate at C-19 abolished the inhibition, indicating the importance of these residues.[137] Jiang discovered that Hyrtiosal (**101**), a natural product from marine sponge *Hyrtios erectus*, is a non-competitive PTP1B inhibitor with an IC_{50} value of 42 µM.[138] This compound exhibited excellent *in vivo* activities, such as abolishing the retardation of AKT membrane translocation and enhancing the membrane translocation of the key glucose transporter Glut4 in PTP1B over-expressed CHO cells. Genaera Corporation has published a patent that covered more than 100 steroid compounds for the treatment of diabetes by targeting PTP1B.[139] Representative compounds with various functional substituents are summarized in Figure 6.9. As shown, the anionic group such as sulfonic acid is not necessary for the activity as compound **102**, without the sulfonic acid head group, and compound **103**, with it, exhibited a similar level of inhibition against PTP1B. In some cases, the existence of an anionic group even reduced the potency of a compound (**105** versus **106**), suggesting that the steroid scaffold may play a critical role in maintaining the potency of this class of compounds. Moreover, these molecules can reduce the glucose level and enhance the IR phosphorylation level in animal models. Indeed, one of these molecules, Trodusquemine (**104**), has progressed into clinical trial for the treatment of type 2 diabetes, which will be discussed further in a later section.

Besides steroids, many other structurally different natural products have been identified to be PTP1B inhibitors. Cho isolated sanggenon C, sanggenon G, mulberrofuran C, kuwanon L, moracin O, and moracin P from Morus root

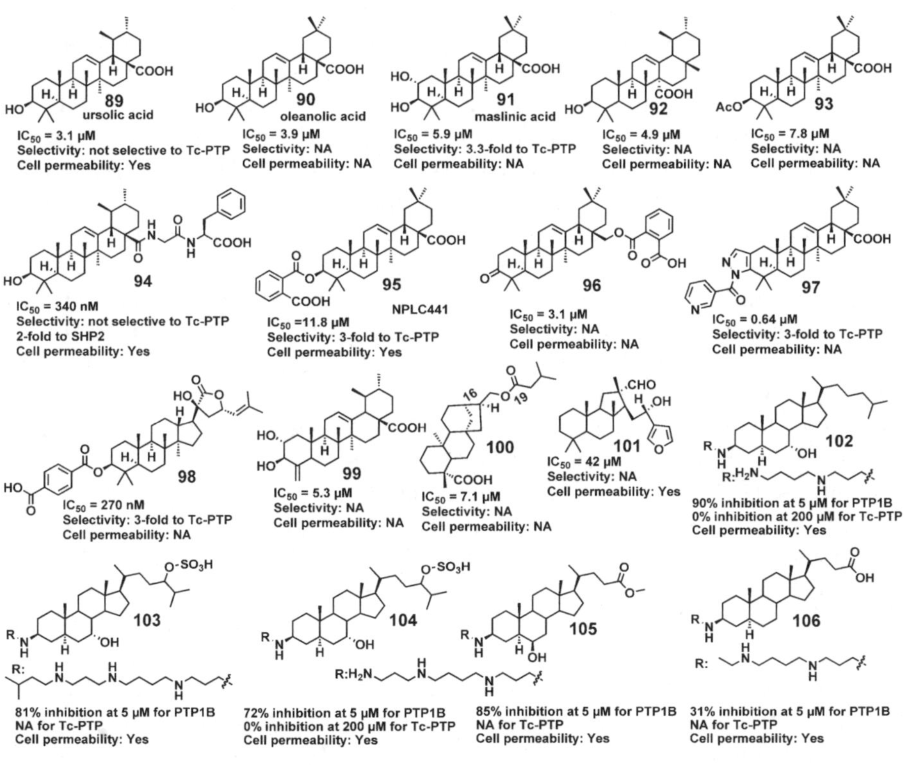

Figure 6.9 Structures of steroid natural products and derivatives as PTP1B inhibitors.

bark and studied their inhibitory activity against PTP1B.[140] Moracin O and moracin P showed no activity at 80 μM concentration, while the remaining four compounds inhibited PTP1B at the μM range with the most potent, sanggenon G (**107**), having a IC$_{50}$ of 1.6 μM. From dried semen of *M. fragrans*, *meso*-dihydroguaiaretic acid (**108**) and otobaphenol were identified as non-competitive inhibitors of PTP1B with IC$_{50}$ at 19.6 and 48.9 μM, respectively. Treatment of 32D cells over-expressing the insulin receptor (IR) with compound **108** resulted in a dose-dependent increase in the tyrosine phosphorylation of IR, indicating its cellular efficacy in blocking PTP1B activity.[141] They also isolated several 2-arylbenzofurans from the extract of stem bark of *Erythrina addisoniae* that exhibited inhibition towards PTP1B.[142] The representative compound **109** has an IC$_{50}$ of 13.6 μM; however, little is known about its specificity to other PTPs and cellular activity. In addition, they have obtained several pterocarpan derivatives from the stem bark of *Erythrina abyssinica* (Leguminosae).[143] Several of them showed good inhibitory effect on PTP1B, for example, Erybreadin B (**110**) has an IC$_{50}$ at 4.2 μM. Since these compounds also exhibited strong inhibitory activity against several breast cancer lines, they proposed that it could result from the inhibition of PTP1B.

Guo isolated two novel valerenane-type sesquiterpenes along with one known metabolite, caulerpin (**111**), from the Chinese green alga *Caulerpa taxifolia* (Vahl) C.[144] These molecules were evaluated for inhibitory activity against PTP1B and only caulerpin was found to exhibit a strong PTP1B inhibitory activity with an IC$_{50}$ value at 3.77 μM. They also separated several sesquiterpene quinones from Hainan sponge *Dysidea villosa* and found that dysidine (**112**) was a PTP1B inhibitor with an IC$_{50}$ value of 6.7 μM.[145] Subsequent studies showed that it effectively activated the insulin signaling pathway and greatly promoted glucose uptake in 3T3-L1 cells.[146] From the roots of *Suussurea lappa*, Hu obtained one new lignan glycoside along with 20 known compounds, which were tested against PTP1B.[147] Rhein-8-*O*-β-D-glucopyrariosidc (**113**) showed the best activity with an IC$_{50}$ of 11.5 μM. And from the whole plant of *Ardisia japonica*, four ¹,⁴-benzoquinones were isolated by means of bioassay-directed fractionation.[148,149] These compounds exhibited moderate inhibition against PTP1B with **114** having an IC$_{50}$ at 3 μM. Mangiferin (**115**) is a xanthone glucoside that exists in many plants and exhibits anti-diabetic effects. When it was tested against PTP1B, Wu found that it has only 24% inhibition against PTP1B at 500 μM concentration.[150] However, protection with aliphatic carbon chains afforded compound **116**, exhibiting 100% inhibition towards PTP1B at 50 μM concentration, and so indicating a more than 50-fold increase in potency.

Oh isolated four benzonaphthoxanthenone natural products from the MeOH extract of Antarctic moss *Polytrichastrum alpinum* by various chromatographic methods.[151] Ohioensins F (**117**) was found to be a new compound that inhibits PTP1B in a competitive manner with a K_i at 1.5 μM. They also tested seven phenolic lichen metabolites from the extract of the

Antarctic lichen *Stereocaulon alpinum* that exhibited inhibitory activity toward PTP1B.[152] Lobaric acid (**118**) appeared to be the most potent among them with an IC_{50} at 0.87 μM. Protecting the carboxylic or phenoxyl group with a methyl group led to only 2- to 3-fold decrease in potency, suggesting that the naturally formed scaffold plays a more important role in inhibiting PTP1B. Several chalcones have been isolated by Cheon from the CH_2Cl_2 extract of *Glycyrrhiza inflata* and were evaluated for their PTP1B inhibitory activity.[153] Licochalcone A (**119**) inhibited PTP1B with an IC_{50} at 19.1 μM, and methylation of one hydroxyl group provided a derivative (**120**) with 2-fold increased potency. Bae identified three 2-arylbenzofurans and three chalcone-derived Diels-Alder products from the chloroform-soluble fraction of *Morus bombycis*.[154] These compounds showed good inhibitory activity against PTP1B, with IC_{50} values ranging from 2.7 to 13.8 μM. Kinetics analysis suggested that they inhibited PTP1B in a mixed-type manner. The structures of two representative compounds (**121**, **122**) are shown in Figure 6.10.

In the course of a bioassay-guided study, Ahn isolated aquastatin A (**123**) from the EtOAc extract of a culture broth of the marine-derived fungus *Cosmospora* sp. SF-5060.[155] This compound exhibited potent inhibitory activity against PTP1B with IC_{50} value of 0.19 μM, and kinetic analyses revealed that it was a competitive inhibitor of PTP1B. Aquastatin A showed modest specificity over other PTPs, such as Tc-PTP, SHP2, LAR, and CD45. In addition, the benzoic acid moiety of the molecule is responsible for the inhibitory activity, as methyl esterification caused a 100-fold decrease in IC_{50} value, while deleting the glucose group did not affect the activity. Li has isolated two new 2-arylbenzofurans and ten known flavonoids from the roots of *Glycyrrhiza uralensis*.[156] Among them, glycybenzofuran (**124**) and glisoflavone (**125**) showed PTP1B inhibitory activity with IC_{50} values at 25.5 and 27.9 μM, respectively. The structure–activity relationship indicated that the presence of the prenyl group and *ortho*-hydroxy group is important for the inhibitory activity. Kinetic studies indicated that **124** inhibits PTP1B in a competitive mode, whereas **125** does it in a mixed mode. 1,2,3,4,6-Penta-*O*-galloyl-D-glucopyranose (**126**) was isolated by Heiss and co-workers from the roots of *Paeonia lactiflora* as an inhibitor of PTP1B, with an IC_{50} value of 4.8 μM.[157] **126** was shown to act as an insulin sensitizer in human hepatoma cells (HCC-1.2) at a concentration of 10 μM, which may explain the anti-diabetic effects of *P. lactiflora*.

A novel vanillic acid derivative (**127**) and its sulfate adduct (**128**) were isolated from a green algae, *Cladophora socialis*, by Quinn.[158] Their structures were elucidated from NMR and HRESIMS experiments and are depicted in Figure 6.10. Both compounds showed potent inhibitory activity against PTP1B, with IC_{50} values of 3.7 and 1.7 μM, respectively. Novel lipidyl pseudopteranoids, lipidyl pseudopteranes A–F, have been isolated from the soft coral *Pseudopterogorgia acerosa* collected from the Bahamas by Kerr.[159] They were subjected to a primary screen (at a concentration of 100 μg/mL) for inhibition of a panel of PTPs. The most active compound against PTP1B was

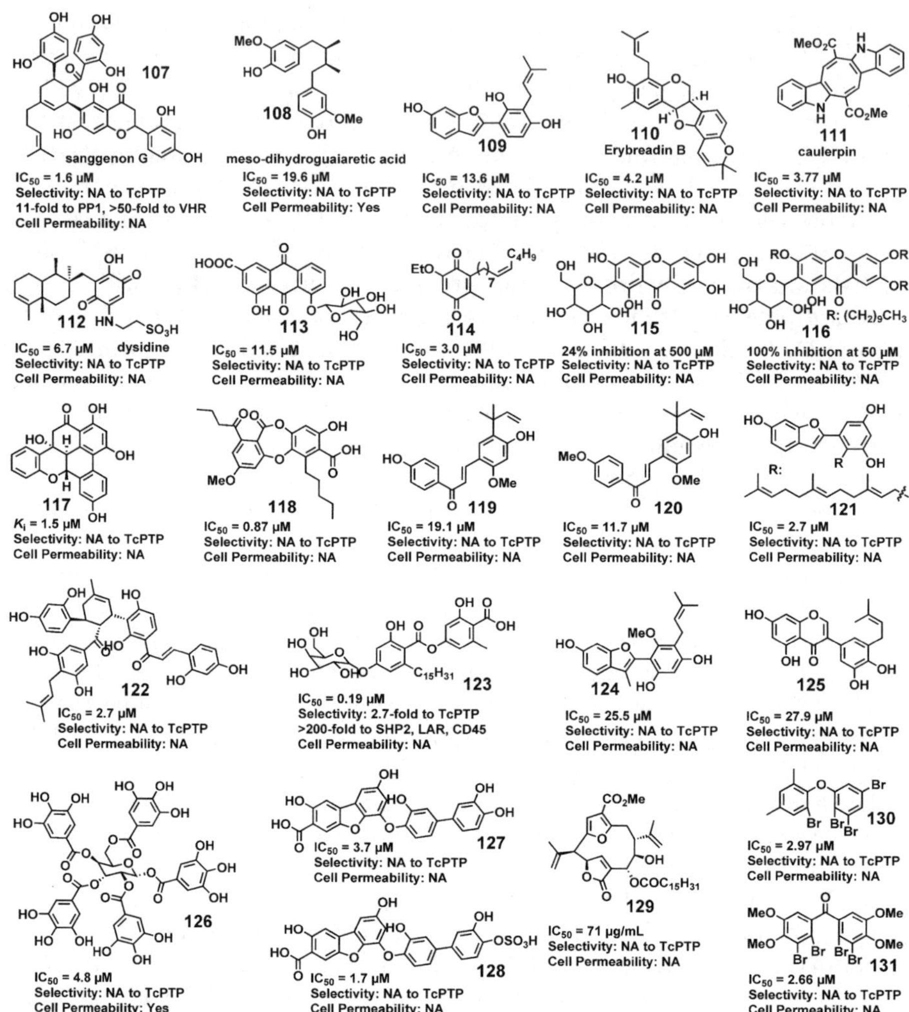

Figure 6.10 Structures of miscellaneous natural products and derivatives as PTP1B inhibitors.

lipidyl pseudopterane A (**129**), with an IC$_{50}$ of 71 µg/mL. It was also inactive against other members of the PTP family (including LAR, SHP1, and MKPX), indicating excellent preference to PTP1B. Li has isolated five new highly brominated metabolites from the red alga *Laurencia similis*.[160] Their structures were elucidated by NMR as well as HREIMS analysis. Two of them, **130** and **131**, showed good inhibitory activities against PTP1B with IC$_{50}$ values of 2.97 and 2.66 µM, respectively. They represent a class of novel PTP inhibitors that are different from aforementioned carboxylic acid, poly phenol, and quinone-based natural products.

6.5 Clinical Development of PTP1B Inhibitors

A number of *in vitro* and *in vivo* studies indicated that PTP1B plays a prominent role in type 2 diabetes and obesity mellitus. The pharmaceutical industry has been eager to test clinically the anti-diabetic and anti-obesity efficacy of PTP1B inhibitors. Ertiprotafib (**132**, Figure 6.11) was the first PTP1B inhibitor to be evaluated by Wyeth for the treatment of type 2 diabetes.[161–163] It inhibited PTP1B and normalized the plasma glucose and insulin levels in diabetic animal models, and progressed to phase 2 clinical trials. Unfortunately, it failed at this stage in 2002 due to the poor clinical efficacy and undesirable side effects. Recently, the Wyeth team reported that Ertiprotafib did not behave as a simple, competitive inhibitor of PTP1B.[164] Instead, inhibition followed a time-dependent, irreversible mechanism that displayed non-classical kinetics, with greater inhibition observed at lower enzyme levels and later time points. The IC_{50} value of Ertiprotafib hence was from 1.6 to 29 μM depending on the assay conditions. Ertiprotafib was also found to activate peroxisome proliferator-activated receptor PPARα and PPARγ and drive adipocyte differentiation of C3II10T$_{1/2}$ cells, a hallmark of PPARγ activation. In addition, Ertiprotafib was shown to be a potent inhibitor of IκB kinase β, with an IC_{50} of 400 nM.[165] Hence it is easy to interpret the failed clinical trials in light of these complexities.

ISIS Pharmaceuticals are developing ISIS 113715 as an antisense drug targeting PTP1B. It is a 20-mer antisense oligonucleotide (ASO) that is complementary to the mRNA of PTP1B and hence is able to reduce the translation of this protein. ISIS 113715 has a sequence of "GCTC-CTTCCACTGATCCTGC" where a ten-deoxynucleotide gap region is flanked on its 3′ and 5′ ends with five 2′-*O*-(2-methoxyethyl) nucleotides, all cytosines nucleotides are 5-methylcytosines, and all linkages are phosphorothioates.[44,166,167] It showed excellent *in vivo* anti-diabetic effects in animals models such as increased insulin sensitivity, and reduced body glucose level

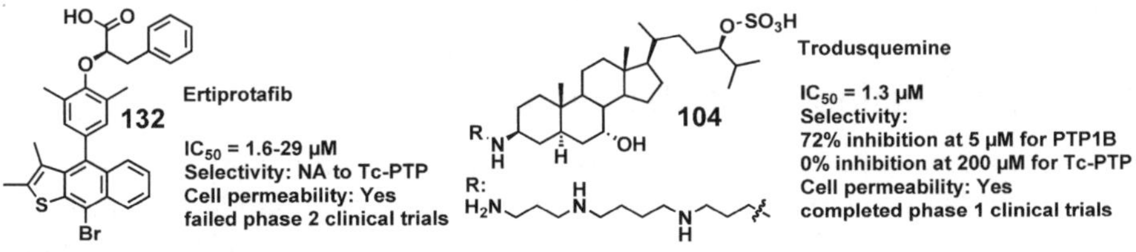

Figure 6.11 PTP1B inhibitors in clinic trials. Ertiprotafib (**132**) is a carboxylic acid based PTP1B inhibitor developed by Wyeth which has been tested clinically for the treatment of type 2 diabetes. However, it failed at phase 2 clinical trials due to the poor clinical efficacy and undesired side effects. Trodusquemine (MSI-1436, **104**) is a natural product derived sulfonic acid that was discovered by Genaera for the treatment of type 2 diabetes and obesity. It showed excellent efficacy in preclinical studies and good pharmacokinetics profile in a phase 1 clinical trial.

and weight gain, and hence was selected for clinical development. In 2009, ISIS Pharmaceuticals announced positive results from a phase 2 study evaluating the safety and efficacy of ISIS 113715 in patients with type 2 diabetes. As a single agent in newly diagnosed type 2 diabetic patients, ISIS 113715 led to statistically significant reductions in multiple measures of glucose control. A tendency toward weight loss was observed, even in this short-term study without strict dietary control, and it was preceded by a statistically significant increase in circulating adiponectin, a hormone that increases with weight loss. In addition, treatment of ISIS 113715 resulted in the reduction of LDL-C, suggesting it may reduce the cardiovascular disease risks for patients with type 2 diabetes. Importantly, the studies showed a favorable safety profile of ISIS 113715, in which no exacerbation of sulfonylurea-induced hypoglycemia or other clinically significant adverse effects was recorded. Since ISIS-PTP1B$_{RX}$ was discovered as a newer antisense drug with significantly better potency than ISIS 113715, ISIS Pharmaceuticals initiated a phase 1 clinical study with this new molecule in July 2011.[168]

Another molecule is Trodusquemine (MSI-1436, **104**), a drug candidate discovered by Genaera for the treatment of type 2 diabetes and obesity.[139,165,169] It is a reversible, allosteric, noncompetitive, and highly selective inhibitor of PTP1B that significantly enhanced insulin-stimulated tyrosine phosphorylation of IR and STAT3, substrates of PTP1B, in HepG2 cells and/or hypothalamic tissue. By inhibiting PTP1B, Trodusquemine decreased appetite, caused weight loss without metabolic rebound, and normalized both fasting blood glucose, blood cholesterol, and triglyceride levels in obese animal models.[170] From the phase 1 clinical trial data, Trodusquemine was well tolerated with no serious adverse effects and no dose-limiting toxicities, and its pharmacokinetic profile was consistent with the results obtained in preclinical studies.[171] Unfortunately, Genaera is unable to advance Trodusquemine into the next phase of clinical development due to limited financial resources. Nevertheless, Trodusquemine showed at least proof-of-concept that PTP1B small molecule inhibitors offer potential treatment for type 2 diabetes and obesity.

6.6 Conclusions and Perspectives

Type 2 diabetes and obesity are worldwide public health threats that affect nearly one billion people. The existing treatments, for example Meglitinides, have severe side effects such as hypoglycemia, weight gain, and other complications. Therapeutic approaches that target PTP1B have emerged as novel solutions to the treatment of these diseases with minimum side effects, based on a number of *in vitro* and *in vivo* studies that PTP1B negatively regulates both insulin and leptin signaling, and PTP1B knockout mice were healthy with decreased blood glucose level, increased insulin sensitivity, and higher tolerance to a high-fat diet. In this regard, the efforts in developing PTP1B inhibitors have shifted from early nonhydrolyzable phosphonic acid derivatives to various molecules with diverse chemical and structural properties. For instance, carboxylic acids are a class of equally efficient p-Tyr mimetics offering the benefits of abundance, low cost, and easy manipulation albeit with the same poor cell permeability in many cases. Less acidic or even neutral molecules provide alternatives in developing less potent PTP1B inhibitors with improved cell permeability and bioavailability. Natural products and derivatives, although moderately active, are serving as the foundation for the development of compounds with increased potency and specificity. Considering the multiple challenges, including potency, specificity, bioavailability, and druggability in the field, a combination of multiple approaches such as targeting both active and allosteric sites, designing less anionic or neutral molecules, and developing natural product based inhibitors may be a productive direction in which to go. A good example is Trodusquemine (MSI-1436, **101**), a steroid-based sulfonic acid that showed promising efficacy in preclinical and phase 1 clinical studies for the treatment of type 2 diabetes and obesity. This molecule, together with the antisense compound ISIS 113715, are paving the way for the eventual successful application of PTP1B inhibitors as novel anti-diabetic and anti-obesity agents.

Acknowledgment

This work was supported by National Institutes of Health Grants CA69202, CA126937, and CA152194.

References

1. T. Hunter, *Cell*, 1995, **80**, 225.
2. R. Li and J. A. Stafford (eds.), *Kinase Inhibitor Drugs*, Wiley, Hoboken, 2009.
3. N. K. Tonks and B. G. Neel, *Curr. Opin. Cell Biol.*, 2001, **13**, 182–195.
4. Z.-Y. Zhang, *Curr. Opin. Chem. Biol.*, 2001, **5**, 416–423.
5. A. Alonso, J. Sasin, N. Bottini, I. Friedberg, I. Friedberg, A. Osterman, A. Godzik, T. Hunter, J. Dixon and T. Mustelin, *Cell*, 2004, **117**, 699–711.
6. N. K. Tonks, *Nat. Rev. Mol. Cell Biol.*, 2006, **7**, 833–846.
7. N. K. Tonks, C. D. Diltz and E. H. Fisher, *J. Biol. Chem.*, 1988, **263**, 6722–6730.
8. K. L. Guan and J. E. Dixon, *J. Biol. Chem.*, 1991, **266**, 17026–17030.
9. J. V. Frangioni, P. H. Beahm, V. Shifrin, C. A. Jost and B. G. Neel, *Cell*, 1992, **68**, 545–560.
10. N. Dube, A. Bourdeau, K. M. Heinonen, A. Cheng, A. L. Loy and M. L. Tremblay, *Cancer Res.*, 2005, **65**, 10088–10095.
11. S.G. Julien, N. Dube, M. Read, J. Penney, M. Paquet, Y. Han, B. P. Kennedy, W. J. Muller and M. L. Tremblay, *Nat. Genet.*, 2007, **39**, 338–346.

12. M. Bentires-Alj and B. G. Neel, *Cancer Res.*, 2007, **67**, 2420–2424.
13. L. Lessard, M. Stuible and M. L. Tremblay, *Biochim. Biophys. Acta*, 2010, **1804**, 613–619.
14. Z.-Y. Zhang and S.-Y. Lee, *Expert Opin. Investig. Drugs*, 2003, **12**, 223–233.
15. Z. Jia, D. Barford, A. J. Flint and N. K. Tonks, *Science*, 1995, **268**, 1754–1758.
16. D. Barford, A. J. Flint and N. K. Tonks, *Science*, 1994, **263**, 1397–1404.
17. A. D. B. Pannifer, A. J. Flint, N. K. Tonks and D. Barford, *J. Biol. Chem.*, 1998, **273**, 10454–10462.
18. S. Wild, G. Roglic, A. Green, R. Sicree and H. King, *Diabetes Care*, 2004, **27**, 1047–1053.
19. G. Roglic, N. Unwin, P. H. Bennett, C. Mathers, J. Tuomilehto, S. Nag, V. Connolly and H. King, *Diabetes Care*, 2005, **28**, 2130–2135.
20. S. I. Taylor, *Cell*, 1999, **92**, 9–12.
21. D. W. Haslam, W. Philip and T. James, *Lancet*, 2005, **366**, 1197–1209.
22. C. M. Taniguchi, B. Emanuelli and C. R. Kahn, *Nat. Rev. Mol. Cell Biol.*, 2006, **7**, 85–96.
23. M. F. White and C. R. Kahn, *J. Biol. Chem.*, 1994, **269**, 1–4.
24. L. T. Brozinick, B. A. Berkemeier and J. S. Elmendorf, *Curr. Diabetes Rev.*, 2007, **3**, 111–122.
25. L. Fritsche, C. Weigert, H. U. Häring and R. Lehmann, *Curr. Med. Chem.*, 2008, **15**, 1316–1329.
26. H. Zaid, C. N. Antonescu, V. K. Randhawa and A. Klip, *Biochem. J.*, 2008, **413**, 201–215.
27. Z. Cheng, Y. Tseng and M. F. White, *Trends Endocrinol. Metab.*, 2010, **21**, 589–598.
28. D. Bandyopadhyay, A. Kusari, K. A. Kenner, F. Liu, J. Chernoff, T. A. Gustafson and J. Kusari, *J. Biol. Chem.*, 1997, **272**, 1639–1645.
29. B. J. Goldstein, A. Bitner-Kowalczyk, M. F. White and M. Harbeck, *J. Biol. Chem.*, 2000, **275**, 4283–4289.
30. H. Chen, S. J. Wertheimer, C. H. Lin, S. L. Katz, K. E. Amrein, P. Burn and M. J. Quon, *J. Biol. Chem.*, 1997, **272**, 8026–8031.
31. C. L. Venable, E. U. Frevert, Y.-B. Kim, B. M. Fischer, S. Kamatkar, B. G. Neel and B. B. Kahn, *J. Biol. Chem.*, 2000, **275**, 18318–18326.
32. K. Egawa, H. Maegawa, S. Shimizu, K. Morino, Y. Nishio, M. Bryer-Ash, A. T. Cheung, J. K. Kolls, R. Kikkawa and A. Kashiwagi, *J. Biol. Chem.*, 2001, **276**, 10207–10211.
33. F. Ahmad, P.-M. Li, J. Meyerovitch and B. J. Goldstein, *J. Biol. Chem.*, 1995, **270**, 20503–20508.
34. K. Mahadev, H. Motoshima, X. Wu, J. M. Ruddy, R. S. Arnold, G. Cheng, J. D. Lambeth and B. J. Boldstein, *Mol. Cell Biol.*, 2004, **24**, 1844–1854.
35. D. A. Barrio and S. B. Etcheverry, *Curr. Med. Chem.*, 2010, **17**, 3632–3642.

36. M. Elchebly, P. Payette, E. Michaliszyn, W. Cromlish, S. Collins, A. L. Loy, D. Normandin, A. Cheng, J. Himms-Hagen, C.-C. Chan, C. Ramachandran, M. J. Gresser, M. L. Tremblay and B. P. Kennedy, *Science*, 1999, **283**, 1544–1548.
37. L. D. Klaman, O. Boss, O. D. Peroni, J. K. Kim, J. L. Martino, J. M. Zabolothy, N. Moghal, M. Lubkin, Y. B. Kim, A. H. Sharpe, A. Stricker-Krongrad, G. I. Shulman, B. G. Neel and B. B. Kahn, *Mol. Cell Biol.*, 2000, **20**, 5479–5489.
38. B. A. Zinker, C. M. Rondinone, J. M. Trevillyan, R. J. Gum, J. E. Clampit, J. F. Waring, N. Xie, D, Wilcox, P. Jacobsen, L. Frost, P. E. Kroeger, R. M. Reilly, S. Koterski, T. J. Opgenorth, R. G. Ulrich, S. Crosby, M. Butler, S. F. Murray, R. A. Mckay, S. Bhanot, B. P. Monia and M. R. Jirousek, *Proc. Natl. Acad. Sci. USA*, 2002, **99**, 11357–11362.
39. F. G. Haj, J. M. Zabolotny, Y. M. Kim, B. B. Kahn and B. G. Neel, *J. Biol. Chem.*, 2005, **280**, 15038–15046.
40. J. M. Friedman, *Nature*, 2000, **404**, 632–634.
41. R. Yang and L. A. Barouch, *Circ. Res.*, 2007, **101**, 545–559.
42. W. Kaszubska, H. D. Falls, V. G. Schaefer, D. Haasch, L. Frost, P. Hessler, P. E. Kroeger, D. W. White, M. R. Jirousek and J. M. Trevillyan, *Mol. Cell Endocrinol.*, 2002, **195**, 109–118.
43. I. K. Lund, J. A. Hansen, H. S. Andersen, N. P. H. Moller and N. Billestrup, *J. Mol. Endocrinol.*, 2005, 34, 339–351.
44. C. M. Rondinone, J. M. Trevillyan, J. Clampit, R. J. Gum, C. Berg, P. Kroeger, L. Frost, B. A. Zinker, R. Reilly, R. Ulrich, M. Butler, B. P. Monia, M. R. Jirousek and J. F. Waring, *Diabetes*, 2002, **51**, 2405–2411.
45. A. Cheng, N. Uetani, P. D. Simoncic, V. P. Chaubey, A. Lee-Loy, C. J. McGlade, B. P. Kennedy and M. L. Tremblay, *Dev. Cell*, 2002, **2**, 497–503.
46. J. M. Zabolotny, K. K. Bence-Hanulec, A. Stricker-Krongrad, F. Haj, Y. Wang, Y. Minokoshi, Y.-B. Kim, J. K. Elmquist, L. A. Tartaglia, B. B. Kahn and B. G. Beel, *Dev. Cell*, 2002, **2**, 489–495.
47. C. D. Morrison, C. L. White, Z. Wang, S.-Y. Lee, D. S. Lawrence, W. T. Cefalu, Z.-Y. Zhang and T. W. Gettys, *Endocrinology*, 2007, **148**, 433–440.
48. M. Stuible, K. M. Doody and M. L. Tremblay, *Cancer Metastasis Rev.*, 2008, **27**, 215–230.
49. S. Lee and Q. Wang, *Med. Res. Rev.*, 2007, **27**, 553–573.
50. S. Zhang and Z.-Y. Zhang, *Drug Discov. Today*, 2007, **12**, 373–381.
51. A. P. Combs, *J. Med. Chem.*, 2010, **53**, 2333–2344.
52. T. R. Burke, Jr., H. K. Kole and P. P. Roller, *Biochem. Biophys. Res. Commun.*, 1994, **204**, 129–134.
53. Y. A. Puius, Y. Zhao, M. Sullivan, D. S. Lawrence, S. C. Almo and Z.-Y. Zhang, *Proc. Natl. Acad. Sci. USA*, 1997, **94**, 13420–13425.
54. K. Shen, Y.-F. Keng, L. Wu, X.-L. Guo, D. S. Lawrence and Z.-Y. Zhang, *J. Biol. Chem.*, 2001, **276**, 47311–47319.

55. J.-P. Sun, A. A. Fedorov, S.-Y. Lee, X.-L. Guo, K. Shen, D. S. Lawrence, S. C. Almo and Z.-Y. Zhang, *J. Biol. Chem.*, 2003, **278**, 12406–12414.

56. L. Xie, S.-Y. Lee, J. N. Andersen, S. Waters, K. Shen, X.-L. Guo, N. P. H. Moller, J. M. Olefsky, D. S. Lawrence and Z.-Y. Zhang, *Biochemistry*, 2003, **42**, 12792–12804.

57. S.-Y. Lee, F. Liang, X.-L. Guo, L. Xie, S. M. Cahill, M. Blumenstein, H. Yang, D. S. Lawrence and Z.-Y. Zhang, *Angew. Chem. Int. Ed.*, 2005, **44**, 4242–4244.

58. T. G. Boutselis, X. Yu, Z.-Y. Zhang and R. F. Borch, *J. Med. Chem.*, 2007, **50**, 856–864.

59. C. Dufresne, P. Roy, Z. Wang, E. Asante-Appiah, W. Cromlish, Y. Boie, F. Forghani, S. Desmarais, Q. Wang, K. Skorey, D. Waddleton, C. Ramachandran, B. P. Kennedy, L. Xu, R. Gordon, C. C. Chan and Y. Leblanc, *Bioorg. Med. Chem. Lett.*, 2004, **14**, 1039–1042.

60. C. K. Lau, C. I. Bayly, J. Y. Gauthier, C. S. Li, M. Therien, E. Asante-Appiah, W. Cromlish, Y. Boie, F. Forghani, S. Desmarais, Q. Wang, K. Skorey, D. Waddleton, P. Payette, C. Ramachandran, B. P. Kennedy and G. Scapin, *Bioorg. Med. Chem. Lett.*, 2004, **14**, 1043–1048.

61. Y. Han, M. Belley, C. I. Bayly, J. Colucci, C. Dufresne, A. Giroux, C. K. Lau, Y. Leblanc, D. McKay, M. Therien, M.-C. Wilson, K. Skorey, C.-C. Chan, G. Scapin and B. P. Kennedy, *Bioorg. Med. Chem. Lett.*, 2008, **18**, 3200–3205.

62. C. P. Holmes, X. Li, Y. Pan, C. Xu, A. Bhandari, C. M. Moody, J. A. Miguel, S. W. Ferla, M. N. De Francisco, B. T. Frederick, S. Zhou, N. Macher, L. Jang, J. D. Irvine and J. R. Grove, *Bioorg. Med. Chem. Lett.*, 2005, **15**, 4336–4341.

63. M. Sarmiento, L. Wu, Y.-F. Keng, L. Song, Z. Luo, Z. Huang, G.-Z. Wu, A. K. Yuan and Z.-Y. Zhang, *J. Med. Chem.*, 2000, **43**, 146–155.

64. X. Yu, J.-P. Sun, Y. He, X.-L. Guo, S. Liu, B. Zhou, A. Hudmon and Z.-Y. Zhang, *Proc. Natl. Acad. Sci. USA*, 2007, **104**, 19767–19772.

65. B. Zhou, Y. He, X. Zhang, J. Xu, Y. Luo, Y. Wang, S. G. Franzblau, Z. Yang, R. J. Chan, Y. Liu, J. Zheng and Z.-Y. Zhang, *Proc. Natl. Acad. Sci. USA*, 2010, **107**, 4573–4578.

66. X. Zhang, Y. He, S. Liu, Z. Yu, Z.-X. Jiang, Z. Yang, Y. Dong, S. C. Nabinger, L. Wu, A. M. Gunawan, L. Wang, R. J. Chan and Z.-Y. Zhang, *J. Med. Chem.*, 2010, **53**, 2482–2493.

67. S. Y. Cho, J. Y. Baek, S. S. Han, S. K. Kang, J. D. Ha, J. H. Ahn, D. Lee, K. R. Kim, H. G. Cheon, S. D. Rhee, S. D. Yang, G. H. Yon, C. S. Pak and J.-K. Choi, *Bioorg. Med. Chem. Lett.*, 2006, **16**, 499–502.

68. Z.-K. Wan, J. Lee, W. Xu, D. V. Erbe, D. Joseph-McCarthy, B. C. Follows and Y.-L. Zhang, *Bioorg. Med. Chem. Lett.*, 2006, **16**, 4941–4945.

69. D. P. Wilson, Z.-K. Wan, W.-X. Xu, S. J. Kirincich, B. C. Follows, D. Joseph-McCarthhy, K. Foreman, A. Moretto, J. Wu, M. Zhu, E.

Binnun, Y.-L. Zhang, M. Tam, D. V. Erbe, J. Tobin, X. Xu, L. Leung, A. Shinlling, S. Y. Tam, T. S. Mansour and J. Lee, *J. Med. Chem.*, 2007, **50**, 4681–4698.

70. Z. K. Wan, J. Lee, R. Hotchandani, A. Moretto, E. Binnun, D. P. Wilson, S. J, Kirncich, B. C. Follows, M. Ipek, W. Xu, D. Joseph-McCarthy, Y.-L. Zhang, M. Tam, D. V. Erbe, J. F. Tobin, W. Li, S. Y. Tam, T. S. Mansour and J. Wu, *ChemMedChem*, 2008, **3**, 1525–1529.

71. Z.-K. Wan, B. Follows, S. Kirincich, D. Wilson, E. Binnun, W. Xu, D. Joseph-McCarthy, J. Wu, M. Smith, Y.-L. Zhang, M. Tam, D. Erbe, S. Tam, E. Saiah and J. Lee, *Bioorg. Med. Chem. Lett.*, 2007, **17**, 2913–2920.

72. D. V. Erbe, L. D. Klaman, D. P. Wilson, Z.-K. Wan, S. J. Kirincich, S. Will, X. Xu, L. Kung, S. Wang, S. Tam, J. Lee and J. F. Tobin, *Diabetes Obes. Metab.*, 2009, **11**, 579–588.

73. R. Srinivasan, M. Uttamchandani and S. Q. Yao, *Org. Lett.*, 2006, **8**, 713–716.

74. R. Srinivasan, L. P. Tan, H. Wu and S. Q. Yao, *Org. Lett.*, 2008, **10**, 2295–2298.

75. D. R. Adams, A. Abraham, J. Asano, C. Breslin, C. A. J. Dick, U. Ixkes, B. F. Johnston, D. Johnston, J. Kewnay, S. P. Mackay, S. J. MacKenzie, M. McFarlane, L. Mitchell, D. Spinks and Y. Takano, *Bioorg. Med. Chem. Lett.*, 2007, **17**, 6579–6583.

76. S. Shrestha, B. R. Bhattarai, K. J. Chang, K.-H. Lee and H. Cho, *Bioorg. Med. Chem. Lett.*, 2007, **17**, 2760–2764.

77. B. R. Bhattarai, S. Shrestha, S. W. Ham, K. R. Kim, H. G. Cheon, K.-H. Lee and H. Cho, *Bioorg. Med. Chem. Lett.*, 2007, **17**, 5357–5360.

78. S. Shrestha, B. R. Bhattarai, K.-H. Lee and H. Cho, *Bioorg. Med. Chem.*, 2007, **15**, 6535–6548.

79. S. Shrestha, B. R. Bhattarai, B. Kafle, K.-H. Lee and H. Cho, *Bioorg. Med. Chem.*, 2008, **16**, 8643–8652.

80. B. R. Bhattarai, J.-H. Ko, S. Shrestha, B. Kafle, H. Cho, J.-H. Kang and H. Cho, *Bioorg. Med. Chem. Lett.*, 2010, **20**, 1075–1077.

81. R. Maccari, P. Paoli, R. Ottana, M. Jacomelli, R. Ciurleo, G. Manao, T. Steindl, T. Langer, M. G. Vigorita and G. Camici, *Bioorg. Med. Chem. Lett.*, 2007, **15**, 5137–5149.

82. R. Ottana, R. Maccari, R. Ciurleo, P. Paoli, M. Jacomelli, G. Manao, G. Camici, C. Laggner and T. Langer, *Bioorg. Med. Chem.*, 2009, **17**, 1928–1937.

83. L. Shi, H.-P. Yu, Y.-Y. Zhou, J.-Q. Du, Q. Shen, J.-Y. Li and J. Li, *Acta Pharmacol. Sin.*, 2008, **29**, 278–284.

84. L. Lin, Q. Shen, G. R. Chen and J. Xie, *Bioorg. Med. Chem. Lett.*, 2008, **18**, 6348–6351.

85. J. W. Yang, X.-P. He, C. Li, L.-X. Gao, L. Sheng, J. Xie, X.-X. Shi, Y. Tang, J. Li and G.-R. Chen, *Bioorg. Med. Chem. Lett.*, 2011, **21**, 1092–1096.

86. C. P. Holmes, X. Li, Y. Pan, C. Xu, A. Bhandari, C. M. Moody, J. A. Miguel, S. W. Ferla, M. N. De Francisco, B. T. Frederick, S. Zhou, N. Macher, L. Jang, J. D. Irvine and J. R. Grove, *Bioorg. Med. Chem. Lett.*, 2008, **18**, 2719–2724.

87. S. Liu, L.-F. Zeng, L. Wu, X. Yu, T. Xue, A. M. Gunawan, Y.-Q. Long and Z.-Y. Zhong, *J. Am. Chem. Soc.*, 2008, **130**, 17075–17084.

88. M. C. Van Zandt, *PCT Int. Appl.*, WO 2008/033931, 2008.

89. S. K. Arora, R. Banerjee, R. K. Kamboj, R. Loriya, B. Suthar, R. Dixit, A. Waghchoure and R. Goel, *PCT Int. Appl.*, WO 2009/109998, 2009.

90. N. Lakshminarayana, Y. R. Prasad, L. Gharat, A. Thomas, P. Ravikumar, S. Narayanan, C. V. Srinivasan and B. Gopalan, *Eur. J. Med. Chem.*, 2009, **44**, 3147–3157.

91. N. Lakshminarayana, Y. R. Prasad, L. Gharat, A. Thomas, S. Narayanan, A. Raghuram, C. V. Srinivasan and B. Gopalan, *Eur. J. Med. Chem.*, 2010, **45**, 3709–3718.

92. D. Ye, Y. Zhang, F. Wang, M. Zheng, X. Zhang, X. Luo, X. Shen, H. Jiang and H. Liu, *Bioorg. Med. Chem.*, 2010, **18**, 1773–1782.

93. S. Fukuda, T. Ohta, S. Sakata, H. Morinaga, M. Ito, Y. Nakagawa, M. Tanaka and M. Matsushita, *Diabetes Obes. Metab.*, 2010, **12**, 299–306.

94. S. R. Klopfenstein, A. G. Evdokimov, A.-O. Colson, N. T. Fairweather, J. J. Neuman, M. B. Maier, J. L. Gray, G. S. Gerwe, G. E. Stake, B. W. Howard, J. A. Farmer, M. E. Pokross, T. R. Downs, B. Kasibhatla and K. G. Peters, *Bioorg. Med. Chem. Lett.*, 2006, **16**, 1574–1578.

95. Y. Romsicki, M. Reece, J.-Y. Gauthier, E. Asante-Appiah and B. P. Kennedy, *J. Biol. Chem.*, 2004, **279**, 12868.

96. M. Hussain, V. Ahmed, B. Hill, Z. Ahmed and S. D. Taylor, *Bioorg. Med. Chem.*, 2008, **16**, 6764–6777.

97. E. Black, J. Breed, A. L. Breeze, K. Embrey, R. Garcia, T. W. Gero, L. Godfrey, P. W. Kenny, A. D. Morley, C. A. Minshull, A. D. Pannifer, J. Read, A. Rees, D. J. Russell, D. Toader and J. Tucker, *Bioorg. Med. Chem.*, 2005, **15**, 2503–2507.

98. D. Barnes, G. M. Coppola, T. Stams and S. W. Topiol, *PCT Int. Appl.*, WO 2007/067612, 2007.

99. D. Barnes, G. R. Bebernitz, G. M. Coppola, T. Stams, S. W. Topiol, T. R. Vedananda and J. R. Wareing, *PCT Int. Appl.*, WO 2007/067613, 2007.

100. D. Barnes, G. M. Coppola, R. E. Damon, K. Nakajima, B. C. Raudenbush, T. Stams, S. W. Topiol and T. R. Vedananda, *PCT Int. Appl.*, WO 2007/067614, 2007.

101. D. Barnes, G. R. Bebernitz, G. M. Coppola, K. Nakajima, T. Stams, S. W. Topiol, T. R. Vedananda and J. R. Wareing, *PCT Int. Appl.*, WO 2007/067615, 2007.

102. A. Neubert, D. Barnes, Y.-S. Kwak, K. Nakajima, G. R. Bebernitz, G. M. Coppola, L. Kirman, M. H. Serrano-Wu, T. Stams, S. W. Topiol,

T. R. Vedananda and J. R. Wareing, *PCT Int. Appl.*, WO 2007/115058, 2007.

103. A. C. Paul, R. Halse, A. Duttaroy, L. E. Holton, D. Kemp, D. J. Glass and D. W. Barnes, *PCT Int. Appl.*, WO 2008/067527, 2008.

104. D. Barnes, G. M. Coppola, T. Stams, S. W. Topiol and J. R. Wareing, *PCT Int. Appl.*, WO 2008/148744, 2008.

105. M. C. Van Zandt, *PCT Int. Appl.*, WO 2008/033455, 2008.

106. A. M. M. Mjalli, D. R. Polisetti, J. C. Quada, R. R. Yarragunta, R. C. Andrews, R. Xie and G. Subramanian, *PCT Int. Appl.*, WO 2007/089857, 2007.

107. A. P. Combs, E. W. Yue, M. Bower, P. J. Ala, B. Wayland, B. Douty, A. Takvorian, P. Polam, Z. Wasserman, W. Y. Zhu, M. L. Crawley, J. Pruitt, R. Sparks, B. Glass, D. Modi, E. McLaughlin, L. Bostrom, M. Li, L. Galya, K. Blom, M. Hillman, L. Gonneville, B. G. Reid, M. Wei, M. Becker-Pasha, R. Klabe, R. Huber, Y. L. Li, G. Hollis, T. C. Burn, R. Wynn, P. Liu and B. Metcalf, *J. Med. Chem.*, 2005, **48**, 6544–6548.

108. P. J. Ala, L. Gonneville, M. C. Hillman, M. Becker-Pasha, M. Wei, B. G. Reid, R. Klabe, E. W. Yue, B. Wayland, B. Douty, P. Polam, Z. Wasserman, M. Bower, A. P. Combs, T. C. Burn, G. F. Hollis and R. Wynn, *J. Biol. Chem.*, 2006, **281**, 32784–32795.

109. E. W. Yue, B. Wayland, B. Douty, M. L. Crawley, E. McLaughlin, A. Takvorian, Z. Wasserman, M. J. Bower, M. Wei, Y. L. Li, P. J. Ala, L. Gonneville, R. Wynn, T. C. Burn, P. C. C. Liu and A. P. Combs, *Bioorg. Med. Chem.*, 2006, **14**, 5833–5849.

110. A. P. Combs, W. Y. Zhu, M. L. Crawley, B. Glass, P. Polam, R. B. Sparks, D. Modi, A. Takvorian, E. McLaughlin, E. W. Yue, Z. Wasserman, M. Bower, M. Wei, M. Rupar, P. J. Ala, B. M. Reid, D. Ellis, L. Gonneville, T. Emm, N. Taylor, S. Yeleswaram, Y. L. Li, R. Wynn, T. C. Burn, G. Hollis, P. C. C. Liu and B. Metcalf, *J. Med. Chem.*, 2006, **49**, 3774–3789.

111. P. J. Ala, L. Gonneville, M. Hillman, M. Becker-Pasha, E. W. Yue, B. Douty, B. Wayland, P. Polam, M. L. Crawley, E. McLaughlin, R. B. Sparks, B. Glass, A. Takvorian, A. P. Combs, T. C. Burn, G. F. Hollis and R. Wynn, *J. Biol. Chem.*, 2006, **281**, 38013–38021.

112. R. B. Sparks, P. Polam, W. Zhu, M. L. Crawley, A. Takvorian, E. McLaughlin, M. Wei, P. J. Ala, L. Gonneville, N. Taylor, Y. Li, R. Wynn, T. C. Burn, P. C. C. Liu and A. P. Combs, *Bioorg. Med. Chem. Lett.*, 2007, **17**, 736–740.

113. B. Douty, B. Wayland, P. J. Ala, M. J. Bower, J. Pruitt, L. Bostrom, M. Wei, R. Klabe, L. Gonneville, R. Wynn, T. C. Burn, P. C. C. Liu, A. P. Combs and E. W. Yue, *Bioorg. Med. Chem. Lett.*, 2008, **18**, 66–71.

114. B. R. Bhattarai, B. Kafle, J. S. Hwang, D. Khadka, S. M. Lee, J. S. Kang, S. W. Ham, I. O. Han, H. Park and H. Cho, *Bioorg. Med. Chem. Lett.*, 2009, **19**, 6161–6165.

115. B. R. Bhattarai, B. Kafle, J. S. Hwang, S. W. Ham, K. H. Lee, H. Park, I. O. Han and H. Cho, *Bioorg. Med. Chem. Lett.*, 2010, **20**, 6758–6763.
116. S. K. Arora, R. Banerjee, R. K. Kamboj, R. Loriya, S. Mathai, M. Joshi, B. Suthar, R. Cheerlavancha, G. Gote, R. Bagul, R. Wetal, S. Patel, R. Dixit, A. Waghchoure, R. Goel and K. H. Sreedhara Swamy, *PCT Int. Appl.*, WO 2009/109999, 2009.
117. M. Dixit, B. K. Tripathi, A. K. Srivastava and A. Goel, *Bioorg. Med. Chem. Lett.*, 2005, **15**, 3394–3397.
118. G. Ahmad, P. K. Mishra, P. Gupta, P. P. Yadav, P. Tiwari, A. K. Tamrakar, A. K. Srivastava and R. Maurya, *Bioorg. Med. Chem. Lett.*, 2006, **16**, 2139–2143.
119. S. J. Berthel, A. W.-H. Cheung, K. C. Thakkar and W. Yun, *PCT Int. Appl.*, WO 2006/050843, 2006.
120. S. J. Berthel, W.-H. Cheung, K. Kim, S. Li, K. C. Thakkar and W. Yun, *PCT Int. Appl.*, WO 2007/009911, 2007.
121. M. Jung, Y. Lee, M. Park, H. Kim, H. Kim, E. Lim, J. Tak, M. Sim, D. Lee, N. Park, W. K. Oh, K. Y. Hur, E. S. Kang and H. C. Lee, *Bioorg. Med. Chem. Lett.*, 2007, **17**, 4481–4486.
122. M. Stuible, L. Zhao, I. Aubry, D. Schmidt-Arras, F. D. Bohmer, C. J. Li and M. L. Tremblay, *ChemBioChem*, 2007, **8**, 179–186.
123. L. Lin, Q. Shen, G. R. Chen and J. Xie, *Bioorg. Med. Chem.*, 2008, **16**, 9757–9763.
124. A. Vogt, P. R. McDonald, A. Tamewitz, R. P. Sikorski, P. Wipf, J. J. Skoko and J. S. Lazo, *Mol. Cancer Ther.*, 2008, **7**, 330–340.
125. Y. Han, H. Shen, B. I. Carr, P. Wipf, J. S. Lazo and S. S. Pan, *J. Pharmacol. Exp. Ther.*, 2004, **309**, 64–70.
126. A. Kumar, R. A. Maurya, S. Sharma, P. Ahmad, A. B. Singh, A. K. Tamrakar and A. K. Srivastava, *Bioorg. Med. Chem.*, 2009, **17**, 5285–5292.
127. S. Gupta, G. Pandey, N. Rahuja, A. K. Srivastava and A. K. Saxena, *Bioorg. Med. Chem. Lett.*, 2010, **20**, 5732–5734.
128. H. Park, B. R. Bhattarai, S. W. Ham and H. Cho, *Eur. J. Med. Chem.*, 2009, **44**, 3280–3284.
129. W. Zhang, D. Hong, Y. Zhou, Y. Zhang, Q. Shen, J. Y. Li, L. H. Hu and J. Li, *Bba-Gen Subjects*, 2006, **1760**, 1505–1512.
130. P. T. Thuong, C. H. Lee, T. T. Dao, P. H. Nguyen, W. G. Kim, S. J. Lee and W. K. Oh, *J. Nat. Prod.*, 2008, **71**, 1775–1778.
131. M. Na, L. Cui, B. S. Min, K. Bae, J. K. Yoo, B. Y. Kim, W. K. Oh and J. S. Ahn, *Bioorg. Med. Chem. Lett.*, 2006, **16**, 3273–3276.
132. W. W. Qiu, Q. Shen, F. Yang, B. Wang, H. Zou, J. Y. Li, J. Li and J. Tang, *Bioorg. Med. Chem. Lett.*, 2009, **19**, 6618–6622.
133. J. H. Kwon, M. J. Chang, H. W. Seo, J. H. Lee, B. S. Min, M. Na, J. C. Kim, M. H. Woo, J. S. Choi, H. K. Lee and K. Bae, *Phytother. Res.*, 2008, **22**, 1303–1306.

134. Z. Lin, Y. Zhang, Y. Zhang, H. Shen, L. Hu, H. Jiang and X. Shen, *Biochem. Pharm.*, 2008, **76**, 1251–1262.
135. S. Qian, H. Li, Y. Chen, W. Zhang, S. Yang and Y. Wu, *J. Nat. Prod.*, 2010, **73**, 1743–1750.
136. M. Na, P. T. Thuong, I. H. Hwang, K. Bae, B. Y. Kim, H. Osada and J. S. Ahn, *Phytother. Res.*, 2010, **24**, 1716–1719.
137. M. Na, W. K. Oh, Y. H. Kim, X. F. Cai, S. Kim, B. Y. Kim and J. S. Ahn, *Bioorg. Med. Chem. Lett.*, 2006, **16**, 3061–3064.
138. T. Sun, Q. Wang, Z. G. Yu, Y. Zhang, Y. W. Guo, K. X. Chen, X. Shen and H. L. Jiang, *ChemBioChem*, 2007, **8**, 187–193.
139. M. Mclane, *PCT Int. Appl.*, WO 2009/032321, 2009.
140. L. Cui, M. Na, H. Oh, E. Y. Bae, D. G. Jeong, S. E. Ryu, S. Kim, B. Y. Kim, W. K. Oh and J. S. Ahn, *Bioorg. Med. Chem. Lett.*, 2006, **16**, 1426–1429.
141. S. Yang, M. K. Na, J. P. Jang, K. A. Kim, B. Y. Kim, N. J. Sung, W. K. Oh and J. S. Ahn, *Phytother. Res.*, 2006, **20**, 680–682.
142. M. Na, D. M. Hoang, D. Njamen, J. T. Mbafor, Z. T. Fomum, P. T. Thuong, J. S. Ahn and W. K. Oh, *Bioorg. Med. Chem. Lett.*, 2007, **17**, 3868–3871.
143. P. H. Nguyen, T. V. Le, P. T. Thuong, T. T. Dao, D. T. Ndinteh, J. T. Mbafor, K. W. Kang and W. K. Oh, *Bioorg. Med. Chem. Lett.*, 2009, **19**, 6745–6749.
144. S. C. Mao, Y. W. Guo and X. Shen, *Bioorg. Med. Chem. Lett.*, 2006, **16**, 2947–2950.
145. Y. Li, Y. Zhang, X. Shen and Y. W. Guo, *Bioorg. Med. Chem. Lett.*, 2009, **19**, 390–392.
146. Y. Zhang, Y. Li, Y. W. Guo, H. L. Jiang and X. Shen, *Acta Pharmacol. Sin.*, 2009, **30**, 333–345.
147. S. Li, T.-Y. An, J. Li, Q. Shen, F.-C. Lou and L.-H. Hu, *J. Asian Nat. Prod. Res.*, 2006, **8**, 281–286.
148. T. Miura, H. Ichiki, N. Iwamoto, M. Kato, M. Kubo, H. Sasaki, M. Okada, T. Ishida, Y. Seino and K. Tanigawa, *Biol. Pharm. Bull.*, 2001, **24**, 1009–1011.
149. T. Miura, H. Ichiki, I. Hashimoto, N. Iwamoto, M. Kato, M. Kubo, E. Ishihara, Y. Komatsu, M. Okada, T. Ishida and K. Tanigawa, *Phytomedicine*, 2001, **8**, 85–87.
150. H.-G. Hu, M.-J. Wang, Q.-J. Zhao, H.-L. Liao and L.-Z. Cai, *Chem. Nat. Compd.*, 2007, **43**, 663–666.
151. C. Seo, Y. H. Choi, J. H. Sohn, J. S. Ahn, J. H. Yim, H. K. Lee and H. Oh, *Bioorg. Med. Chem. Lett.*, 2008, **18**, 772–775.
152. C. Seo, J. H. Sohn, J. S. Ahn, J. H. Yim, H. K. Lee and H. Oh, *Bioorg. Med. Chem. Lett.*, 2009, **19**, 2801–2803.
153. G. Yoon, W. Lee, S. N. Kim and S. H. Cheon, *Bioorg. Med. Chem. Lett.*, 2009, **19**, 5155–5157.

154. D. M. Hoang, T. M. Ngoc, N. T. Dat, T. Ha do, Y. H. Kim, H. V. Luong, J. S. Ahn and K. Bae, *Bioorg. Med. Chem. Lett.*, 2009, **19**, 6759–6761.
155. C. Seo, J. H. Sohn, H. Oh, B. Y. Kim and J. S. Ahn, *Bioorg. Med. Chem. Lett.*, 2009, **19**, 6095–6097.
156. S. Li, W. Li, Y. Wang, Y. Asada and K. Koike, *Bioorg. Med. Chem. Lett.*, 2010, **20**, 5398–5401.
157. R. R. Baumgartner, D. Steinmann, E. H. Heiss, A. G. Atanasov, M. Ganzera, H. Stuppner and V. M. Dirsch, *J. Nat. Prod.*, 2010, **73**, 1578–1581.
158. Y. Feng, A. R. Carroll, R. Addepalli, G. A. Fechner, V. M. Avery and R. J. Quinn, *J. Nat. Prod.*, 2007, **70**, 1790–1792.
159. A. S. Kate, I. Aubry, M. L. Tremblay and R. G. Kerr, *J. Nat. Prod.*, 2008, **71**, 1977–1982.
160. J. Qin, H. Su, Y. Zhang, J. Gao, L. Zhu, X. Wu, H. Pan and X. Li, *Bioorg. Med. Chem. Lett.*, 2010, **20**, 7152–7154.
161. R. H. V. Huijsduijnen, A. Bombrun and B. Swinnen, *Drug Discov. Today*, 2002, **7**, 1013–1019.
162. L. Bialy and H. Waldmann, *Angew. Chem. Int. Ed.*, 2005, **44**, 3814–3839.
163. V. V. Vintonyak, A. P. Antonchick, D. Rauh and H. Waldmann, *Curr. Opin. Chem. Biol.*, 2009, **13**, 272–283.
164. D. V. Erbe, S. Wang, Y. L. Zhang, K. Harding, L. Kung, M. Tam, L. Stolz, Y. Xing, S. Furey, A. Qadri, L. D. Klaman and J. F. Tobin, *Mol. Pharmacol.*, 2005, **67**, 69–77.
165. S. Shrestha, B. R. Bhattarai, H. Cho and J. K. Choi, *Bioorg. Med. Chem. Lett.*, 2007, **17**, 2728–2730.
166. S. Bhanot, B. P. Monia, R. S. Geary and L. L. Kjems, *PCT Int. Appl.*, WO 2006/044531, 2006.
167. R. S. Geary, J. D. Bradley, T. Watanabe, Y. Kwon, M. Wedel, J. J. van Lier and A. A. VanVliet, *Clin. Pharmacokinet.*, 2006, **45**, 789–801.
168. ISIS Pharmaceuticals, Inc., http://www.isip.com.
169. K. A. Lantz, S. G. Hart, S. L. Planey, M. F. Roitman, I. A. Ruiz-White, H. R. Wolfe and M. P. McLane, *Obesity*, 2010, **18**, 1516–1523.
170. M. P. Mclane, I. Ruiz-White and H. R. Wolfe, *PCT Int. Appl.*, WO 2009/091609, 2009.
171. Genaera, Co, http://www.genaera.com.

Recent Advances in the Discovery of GPR119 Agonists

UNMESH SHAH*, SCOTT EDMONDSON AND
JASON W. SZEWCZYK

Merck Research Laboratories, 2015 Galloping Hill Road, Kenilworth,
NJ 07033, USA
*E-mail: unmesh.shah@merck.com

7.1   Introduction

The increased worldwide prevalence of diabetes mellitus (DM) in recent decades has been accompanied by an increase in the number of available pharmacologic therapies used to treat the disease. Differentiation of the available drugs is therefore an important factor for managing the disease effectively in different patient populations. A measure of the relative clinical efficacies of anti-hyperglycemic agents may be done by comparing effects of these drugs on the reduction of hemoglobin A1c levels. Though insulin can reduce hemoglobin A1c levels by more than 3%, most other monotherapies only reduce hemoglobin A1c levels in a range of 0.5 to 2%.[1] Combination therapy is therefore commonly used in diabetic patients with high hemoglobin A1c levels (e.g. >8.5%) in order to achieve their desired glycemic targets. Combination therapies offer additive and sometimes synergistic glycemic efficacy relative to monotherapies. In addition to glycemic efficacy, different pharmacotherapies exhibit variable effects on body weight. More than 80% of adults with DM are obese/overweight and the prevalence of DM is greater in people who possess high body mass indexes.[2] Drugs that lower body weight

RSC Drug Discovery Series No. 27
New Therapeutic Strategies for Type 2 Diabetes: Small Molecule Approaches
Edited by Robert M. Jones

Published by the Royal Society of Chemistry, www.rsc.org

therefore provide an additional benefit to patients with DM. With respect to the various adverse effects caused by anti-diabetes medications, hypoglycemia is a major concern because it limits the effectiveness of many current medications. Drugs that can achieve effective glycemic control without risk of hypoglycemia are therefore preferred therapies for treating DM.

Incretin-based therapies such as glucagon-like peptide 1 (GLP-1) mimetics and DPP-4 inhibitors are effective type 2 diabetes (T2D) drugs that not only improve glycemic control but also have beneficial effects on body weight. In this regard, GLP-1 mimetics have demonstrated weight loss while DPP-4 inhibitors are weight neutral in patients with T2D.[3] Furthermore, incretin-based therapies exhibit glucose-dependent insulin secretion (GDIS) which confers a low risk of hypoglycemia thus offering an improved overall safety profile compared to other diabetes medications.[1–3] On the other hand, the overall efficacy of DPP-4 inhibitors still remains modest compared to other available oral pharmacotherapies, and although the efficacy of GLP-1 mimetics is more robust, they must be delivered by injection. Orally delivered therapies with robust glycemic efficacy, beneficial effects on body weight loss, and a low risk of hypoglycemia are therefore currently an unmet medical need for patients with DM.

A series of lipid/fatty acid G-protein coupled receptors (GPCRs) have emerged as drug targets that exhibit GDIS action and have demonstrated beneficial effects on glycemic control and body weight/food intake in preclinical species.[4,5] G-Protein-coupled receptor 119 (GPR119) is one of these exciting new targets and this report summarizes the biology, pharmacology, chemical matter, and current clinical status of GPR119 agonists.

7.2   GPR119: Receptor Expression, Signaling, and Deorphanization

GPR119 is a class A (rhodopsin-like) GPCR that has been assigned to the subfamily of biogenic amine and MECA (**m**elanocortin, **e**ndothelial, **c**annabinoid, and **a**denosine) receptors.[6,7] There is little overall sequence homology to other receptors and the closest homologues of GPR119 are the adenosine (A1 and A3) and cannabinoid receptors.[7,8] The human receptor is comprised of 335 amino acids and is highly conserved across several vertebrate species including the mouse, rat, zebrafish, monkey, and dog.[9]

The expression of GPR119 is restricted mainly to the pancreas, fetal liver, and GI tract, with some expression also seen in various regions of the brain in rodents.[10–12] Within the pancreas, GPR119 mRNA is localized to the islets of Langerhans;[13–15] *in situ* hybridization[14,16] and immunohistochemical studies[14] have demonstrated that the receptor is expressed in insulin producing β-cells. This is further supported by the evidence of GPR119 mRNA expression in insulinoma derived β-cell lines (MIN6, HIT-T15, NIT-1, RIN5).[13–15] While there is evidence supporting GPR119 expression in glucagon-secreting α-cells,[17] there are limited data that implicate a direct action of GPR119 signaling on glucagon release.[17,18] There has also been a report showing predominant localization of GPR119 in pancreatic polypeptide-secreting rat and mouse islet cells, with no definitive immunoreactivity observed in β-cells;[19] however, these findings have not been replicated. Significant expression of GPR119 has also been detected in the duodenal, jejunal, ileal, and colonic regions of the human and rodent GI tract.[10,11,20] Colocalization of GPR119 with preproglucagon-expressing L-cells in rat ileum and in mouse colon has demonstrated that a significant fraction of GLP-1 producing cells contain GPR119; northern blot analysis showed expression of the receptor in GLP-1 producing GLUTag and STC-1 cell lines.[20] In addition, GPR119 expression has also been observed in the GIP secreting murine K-cells.[21]

GPR119 couples efficiently to $G\alpha_s$ as indicated by the increase in intracellular cAMP levels via activation of adenylate cyclase in transfected HEK293 cells.[10,11,14,22] Using this constitutively active GPR119-expressing cell line, Chu *et al.*[14] failed to observe increases in inositol phosphate with or without coexpression of a $G\alpha_q/G\alpha_i$ chimera, suggesting that GPR119 exhibits poor coupling efficiency to $G\alpha_i$ and $G\alpha_q$.

The endogenous ligands of GPR119, identified through deorphanization efforts from different research groups, can be classified mainly into phospholipids and fatty acid amides. Among the various phospholipids that have been shown to increase cAMP accumulation in human GPR119 transfected cells (EC_{50} range = 1.5–5.7 μM), Soga *et al.* demonstrated that oleoyl-lysophospatidylcholine (LPC; Figure 7.1) mediates a glucose-dependent increase in insulin release.[15] Although the significant decrease in LPC-induced insulin release in the presence of a GPR119-specific siRNA suggests this phenomenon is promoted via GPR119 signaling, other groups have failed to demonstrate the GPR119-dependent nature of LPC-induced insulin release.[16,23] In the fatty acid amide class, oleoylethanolamide (OEA; Figure 7.1) has been indicated to be a major endogenous ligand of GPR119.[22] OEA has been shown to produce a concentration-dependent increase in cAMP levels in stably transfected and endogenous GPR119-expressing cell lines with potency greater than LPC, and to increase GDIS in MIN6 cells. OEA-induced increases in GLP-1 secretion have also been observed in intestinal L-cells and this was shown to be a GPR119-dependent phenomenon since the effect was significantly reduced in the presence of a GPR119-specific siRNA.[24] Interestingly, the effects of OEA on GLP-1 release were diminished at higher concentrations, suggesting possible GPR119 desensitization as $G\alpha_s$-coupled receptors are known to undergo homologous desensitization.[24,25] Finally, the ability of OEA to reduce food intake and body weight gain has been hypothesized to be associated, at least in part, with OEA-induced GPR119 activation.[22] The importance of GPR119 signaling in the feeding effects of OEA is questionable, however, since the hypophagic effect of OEA is maintained in GPR119 null mice[16] but is absent in PPARα null animals.[26] *N*-Oleoyldopamine (OLDA; Figure 7.1) and other hydroxybenzyl lipid amides have also been identified as potential activators of GPR119.[23]

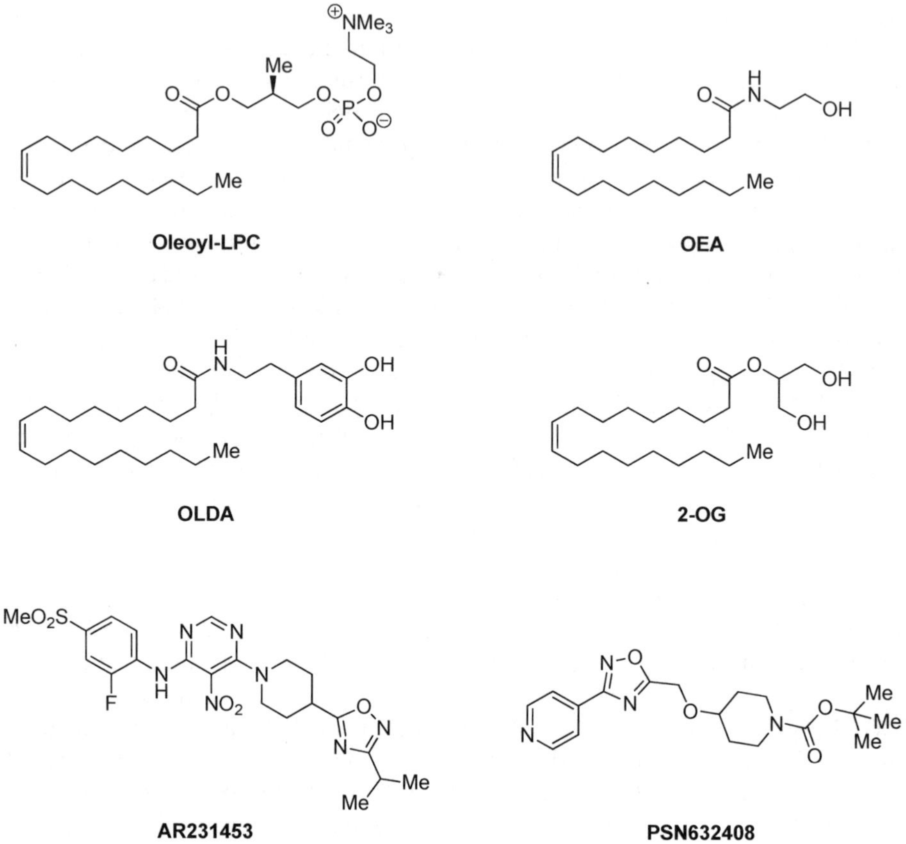

Oleoyl-LPC

OEA

OLDA

2-OG

AR231453

PSN632408

Figure 7.1 Potential endogenous ligands and prototype synthetic agonists of GPR119.

OLDA was demonstrated to increase insulin release *in vitro*, to improve oral glucose tolerance, and to enhance the release of glucose-dependent insulinotropic peptide (GIP) upon oral administration in mice (100 mg/kg). These effects were absent or attenuated in GPR119 null mice. This finding raises the possibility that other lipid amides may play a physiological role in GPR119 signaling. More recently, the naturally occurring lipid 2-oleoylglycerol (2-OG; Figure 7.1) was shown to activate transiently expressed GPR119 in COS-7 cells.[27] In fasting humans, administration of 2-OG significantly increased GLP-1 and GIP secretion.

7.3 Prototype Small-Molecule GPR119 Agonists and Glucose Homeostasis

The pioneering research on two prototype synthetic GPR119 agonists discovered at Arena Pharmaceuticals Inc. and Prosidion Ltd. has established the role of GPR119 on metabolic homeostasis and the use of agonists for therapeutic benefit in T2D and possibly for obesity. The Arena-discovered small molecule agonist AR231453 (Figure 7.1) is a highly potent and selective GPR119 agonist that showed no off-target activity when screened against a panel of receptors and enzymes, including pancreatic islet receptors, known and orphan GPCRs, and enzymes such as DPP-4.[14,28] AR231453 significantly increased cAMP levels in transfected HEK293 cells (EC_{50} = 5.7 nM) and in HIT-T15 cells (EC_{50} = 4.7 nM), and enhanced insulin release in transfected RIN-5F insulinoma cells and in HIT-T15 cells (EC_{50} = 3.5 nM).[14] The effects on cAMP and insulin release were similar to those elicited by forskolin, and AR231453 was essentially inactive in the presence of a GPR119-selective siRNA. In addition, using isolated rat and mouse islets it was demonstrated that AR231453 stimulates insulin secretion in a glucose-dependent manner and with a magnitude similar to that of GLP-1. Upon oral administration in mice, AR231453 enhanced plasma insulin levels and improved oral glucose tolerance with an efficacy (20 mg/kg) that was comparable to the sulfonylurea glyburide (30 mg/kg). However, unlike glyburide the observed effects with AR231453 were glucose dependent and no hypoglycemia risk was observed even at a high dose of 100 mg/kg in fasted mice. Furthermore, the lack of activity in GPR119-deficient mice confirmed that the observed *in vivo* effects were indeed mediated by GPR119 and confirmed the role of pancreatic cell-expressed GPR119 in glucose homeostasis.[14]

The role of intestinal endocrine cell-expressed GPR119 in glycemic control was demonstrated with the ability of AR231453 to stimulate incretin hormone release.[20] In GLUTag cells, AR231543 enhanced GLP-1 release via increased cAMP accumulation, and upon oral administration to lean mice (10 mg/kg, p.o.) it stimulated a glucose-dependent increase in active GLP-1 levels. Co-administration of AR231453 with the DPP-4 inhibitor sitagliptin enhanced both plasma active GLP-1 levels and oral glucose tolerance significantly more than with either agent alone, and conversely, the ability of AR231453 to improve glucose tolerance was reduced in the presence of the GLP-1 antagonist exendin-(9–39).[20,29,30] AR231453 also increased plasma total GIP levels prior to an oral glucose load and augmented the glucose-induced elevation of plasma GIP levels upon oral administration to mice (10 mg/kg). The lack of the above observed effects on plasma GLP-1 and GIP levels in GPR119$^{-/-}$ mice again confirmed that AR231453 acts via GPR119 in improving glucose homeostasis. AR231453 has also been shown to have a potential in protecting pancreatic β-cells through increased cAMP levels in MIN6 pancreatic β-cells expressing GPR119.[31] Separately, researchers at Prosidion Ltd. have demonstrated beneficial effects of a representative GPR119 agonist in delaying or preventing a diabetic condition following chronic oral administration in pre-diabetic db/db mice and ZDF rats.[32]

The research group at Prosidion has demonstrated that synthetic GPR119 agonists can modulate food intake and body weight in addition to improving glucose tolerance.[22,33,34] In a HEK-293 cell line expressing hGPR119,

Prosidion's prototype GPR119 agonist PSN632408 (Figure 7.1) produced a concentration-dependent increase in cAMP levels (EC_{50} = 1.9 µM) similar to the endogenous ligand OEA (EC_{50} = 2.9 µM).[22] PSN632408 reduced 24 h cumulative food intake at an acute dose of 100 mg/kg in rats (i.p. and p.o.), and upon subchronic administration (100 mg/kg, p.o.) to both diet-induced obese (DIO) and high fat-fed rats, the compound reduced food intake, body weight gain, and white adipose tissue deposition. Although PSN632408 was shown to be a selective GPR119 agonist, data from GPR119-knockout mice confirming the specificity of the PSN632408-induced hypophagia and weight loss were not reported.[22] Nonetheless, these initial results from Prosidion's prototype compound suggest further opportunities for investigation of the effects of GPR119 agonists in controlling obesity and related metabolic disorders.

7.4   GPR119 Agonists: Medicinal Chemistry

Several pharmaceutical research organizations have been actively investigating the therapeutic potential of orally active GPR119 agonists for T2D and related metabolic disorders.[35–38] Many of the known GPR119 agonists have evolved from the original prototype compounds described by Arena and Prosidion (e.g. AR231453 and PSN632408; Figure 7.1). The two important pharmacophores consist of: (a) an aryl or heteroaryl moiety substituted with a hydrogen bond accepting group on one part of the molecule, and (b) a piperidine moiety N-capped with a carbamate or an isosteric heterocycle on the opposite side of the molecule. These two structural motifs are connected through an appropriate central spacer containing a heterocyclic ring or an acyclic chain. Multiple classes of GPR119 agonists have been developed based on this general pharmacophore model and the available SAR information is summarized in the following sections according to the research entity.

7.4.1   Arena Pharmaceuticals

The medicinal chemistry efforts leading to the discovery of Arena's tool compound AR231453 (Figure 7.1)[28] have been described in earlier published review articles.[35–38] Although AR231453 demonstrated robust activity in acute studies in mice, it was associated with significant toxicity upon multiple dosing. Furthermore, the compound suffered from poor solubility and relatively poor exposure in rats (10 mg/kg, p.o.: t_{max} = 1 h, C_{max} = 0.25 µM, AUC = 263 h·ng/mL, $t_{1/2}$ = 1.1 h), which resulted in lack of activity in a rat oral glucose tolerance test (oGTT). AR231453 was determined to be not suitable for further evaluation in chronic models of diabetes.[39]

A new series of GPR119 agonists was developed through removal of both the undesirable aromatic nitro group and the aniline functionality from AR231453, and by introducing a 6,5-fused pyrimidine ring system that would maintain some of the critical hydrogen bond acceptors and key functional groups (e.g. sulfone) in a similar position as in AR231453.[39] The pyrazolopyrimidine based analog **1** (Figure 7.2) retained significant agonist activity at human GPR119 but was around 50-fold less potent (hEC_{50} = 42 nM) than AR231453 (hEC_{50} = 0.65 nM) in a melanophore dispersion assay.[39] Further exploration involved replacement of the oxadiazole piperidine moiety in **1** with oxygen-linked reverse piperidine carbamates, which resulted in substantial improvements in potency. For example, in contrast to **1** the potency of compound **2** (hEC_{50} = 1.1 nM; Figure 7.2) was comparable to that of AR231453. Other analogs based on isomeric imidazopyrimidine, triazolopyrimidine, isoxazolopyrimidine, quinoline, and pyridopyrimidine central cores were also investigated, but the pyrazolopyrimidine-fused system was determined to be optimal. Compound **3** (Figure 7.2) was discovered as the lead compound (hEC_{50} = 2.7 nM; rEC_{50} = 33 nM) and was advanced for further evaluation.[39,40] Compound **3** increased cAMP levels in a concentration-dependent manner in HEK cells transfected with human GPR119 but not in non-transfected cells; it also demonstrated enhanced GDIS in rat and human isolated pancreatic islets. Oral administration of single doses of **3** showed that the bioavailability was good in mice, rats, and monkeys (F = 44–79%), but somewhat lower in dogs (F = 22%; t_{max} = 6 h); the iv half-life in mice, rats, dogs, and monkeys was short to moderate (0.8–3.9 h). Interestingly, compound **3** exhibited longer $t_{1/2}$ and t_{max}, and also possessed higher AUC and oral bioavailability in Zucker fa/fa rats than in Sprague Dawley (SD) rats. The dose-escalation pharmacokinetics data (10–1000 mg/kg, p.o.) generated in SD rats has also been described for this compound.[41,42]

Compound **3** was effective in reducing blood glucose levels in a dose-dependent manner during an oGTT in mice (MED = 3 mg/kg) and SD rats (MED = 10 mg/kg); an acute effect was also observed in an oGTT in cynomolgus monkeys (10 mg/kg, p.o.).[39] In a hyperglycemic clamp model in SD rats, **3** showed no effect on plasma insulin and blood glucose levels under euglycemic conditions, but it significantly stimulated insulin release upon raising blood glucose levels up to 300 mg/dL. Finally, chronic treatment with **3**

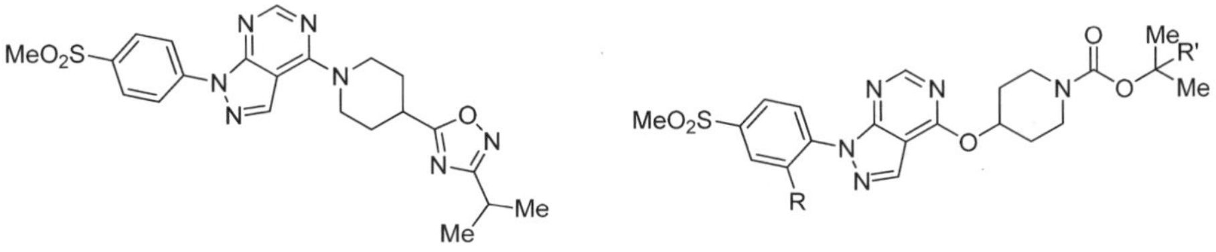

1
hEC_{50} = 42 nM

2 R = H, R' = Me
hEC_{50} = 1.1 nM

3 R = F, R' = H
(APD668, JNJ28630368)
hEC_{50} = 2.7 nM

Figure 7.2　First generation GPR119 agonists from Arena Pharmaceuticals.

once daily to Zucker diabetic fatty (ZDF) rats over 8 weeks (30 mg/kg, p.o.) produced significant reduction in blood glucose and glycosylated hemoglobin (HbA1c) levels without any apparent desensitization of the acute drug response. Compound **3** exhibited a moderate inhibition of the hERG channel in the patch clamp assay (IC_{50} = 3 μM) and possessed a CYP2C9 enzyme inhibition K_i value of 100 nM; it was clean against other major CYP isoforms. Compound **3** (APD668, JNJ28630368), in collaboration with Johnson and Johnson, was progressed into clinical development for the treatment of T2D.

The second generation GPR119 agonists from Arena were derived from a non-fused (piperidin-4-yloxy)pyrimidine structural class (Figure 7.3).[43–47] Several small functional groups were shown to be tolerated at the C5 position of the central pyrimidine ring, and the C5 unsubstituted analog **4** and the C5 cyano analog **5** reduced glucose excursion by 39% (10 mg/kg, p.o.) in rat oGTTs.[43,45,46] The left-hand side structural motif in the non-fused (piperidin-4-yloxy)pyrimidine series consisted of an aryl moiety attached to the central ring either via a nitrogen or an oxygen linker (e.g. **6**, **7**). Compound **7** (hEC_{50} = 0.47 nM), which has been reported as a prototype compound in this series,

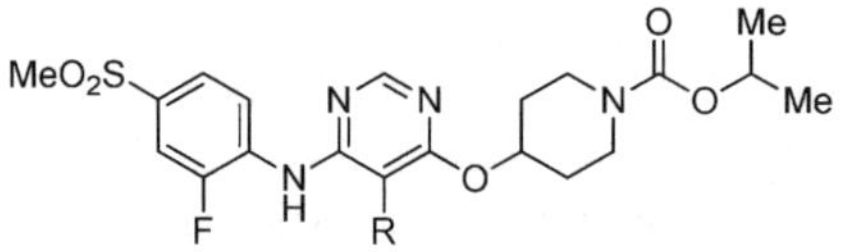

4 R = H
hEC_{50} = 6.9 nM

5 R = CN
hEC_{50} = 4.5 nM

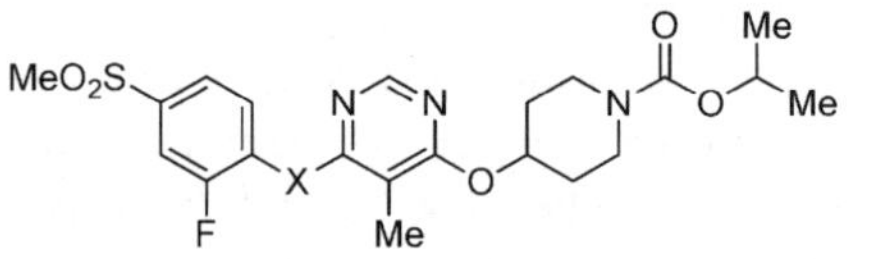

6 X = NH
hEC_{50} = 0.81 nM

7 X = O
hEC_{50} = 0.47 nM

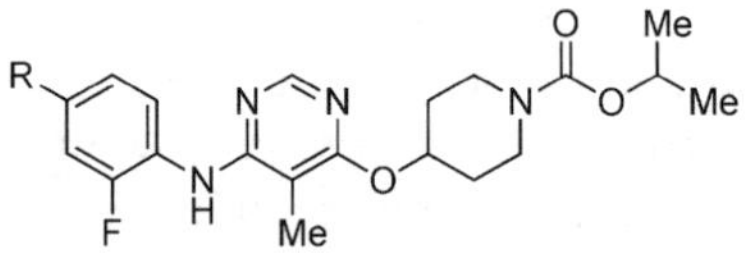

8 R = CN
hEC_{50} = 0.79 nM

9 R = CH$_2$CH$_2$OH
hEC_{50} = 0.11 nM

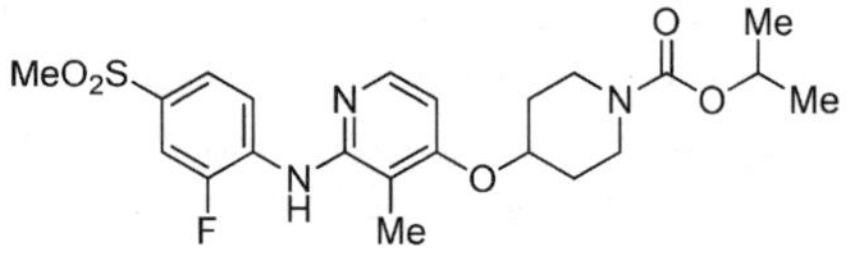

10
hEC_{50} = 9.84 nM

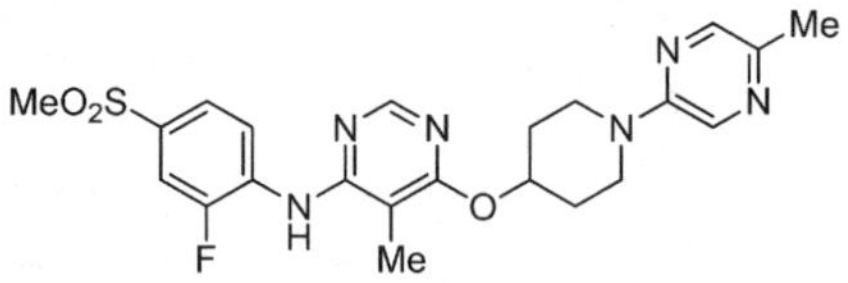

11
hEC_{50} = 6.0 nM

Figure 7.3 Representative non-fused (piperidin-4-yloxy)pyrimidine-based GPR119 agonists from Arena.

demonstrated an oral bioavailability of 72% with an iv half-life of 4.3 h (V_{dss} = 3.1 L/kg, Cl = 0.578 L/h/kg) in SD rats (2 mg/kg, iv/po).[43] This compound was a partial agonist at the human receptor but a full agonist at the rat ortholog, and it effectively lowered blood glucose levels in rat and mouse oGTT studies (30 mg/kg and 10 mg/kg, p.o.). Upon chronic administration in pre-diabetic ZDF rats (10, 30 mg/kg, qd, 8 weeks) compound **7** facilitated significant improvements in baseline glucose, plasma HbA1c, triglyceride, and free fatty acid levels.

In addition to the more preferred methylsulfone moiety, other substituents at the *para* position of the aryl ring also provided highly potent compounds (e.g. **8**, **9**). Replacement of the central pyrimidine with a pyridine, as exemplified by compound **10**, was also tolerated (compound **10**: 43% inhibition of glucose AUC, rat oGTT, 30 mg/kg, p.o.).[43,46] Although the isopropyl carbamate has been highlighted as a preferred nitrogen-capping group, isosteric *N*-linked heterocycles have been also shown to be tolerated. The methylpyrazine analog **11** was described as a full agonist that increased cAMP levels in CHO cells expressing human GPR119 with an EC_{50} of 6 nM (rat EC_{50} = 65 nM) and stimulated GLP1 release from mouse GLUTag cells with an EC_{50} of 54 nM.[44] Compound **11** significantly reduced blood glucose levels in

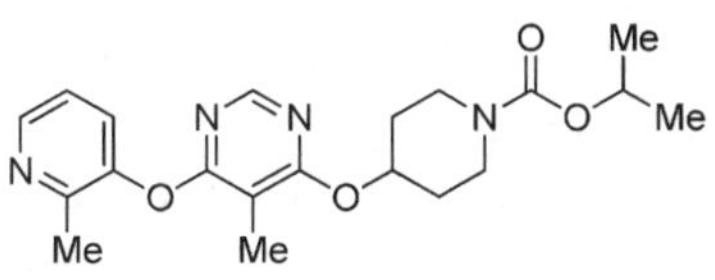

12

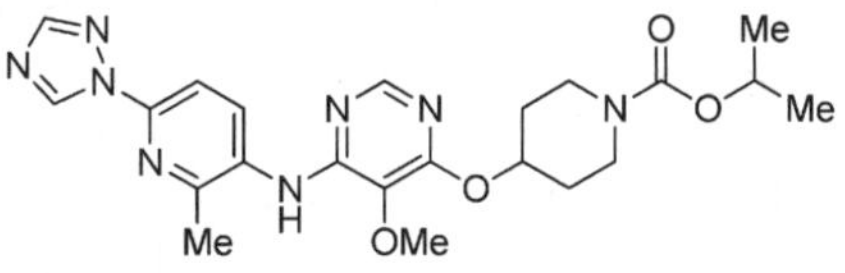

13

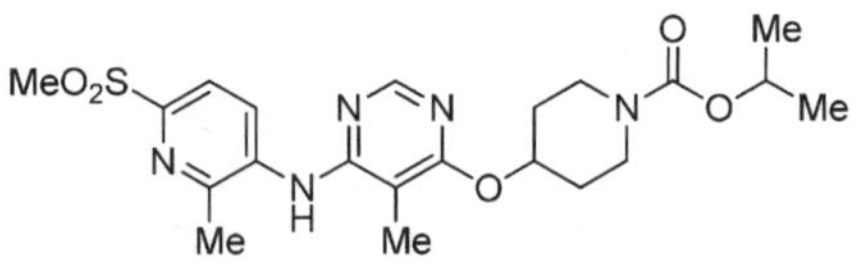

14

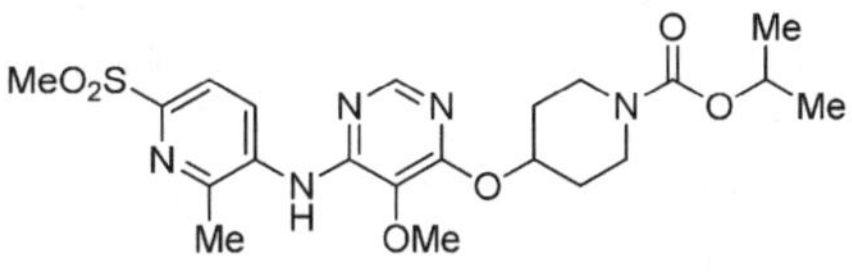

15

Compound	EC$_{50}$ (nM)				Rat oGTT, % I of glucose excursion
	Human	Dog	Mouse	Rat	
12	2	1	41	44	70, 24, 22 % (30, 3, 0.3 mg/kg)
13	2	4	57	81	60%
14	2	8	43	42	38% (10 mg/kg)

Figure 7.4 SAR of methylpyridine analogs discovered at Arena.

oGTT studies in male SD rats (0.3 and 3 mg/kg, p.o.) and cynomolgus monkeys (1.5 mg/kg, p.o.).

A notable improvement in solubility was achieved by replacing the left-hand side phenyl ring with a pyridyl ring (e.g. **12**: aq. solubility = 40 mg/mL). The pyridyl analogs shown in Figure 7.4 exhibited promising activity in rat oGTT studies despite the relatively higher rat EC_{50} values. These analogs were reported to be much more potent at the human and dog receptors.[41,42,47] Furthermore, compounds **13–15** exhibited a linear rising-dose pharmacokinetic profile in fasted male SD rats – AUC (3, 30, 300 mg/kg) = **13**: 3.59, 79.82, 285.99 μg·h/mL; **14**: 4.73, 55.9, 515.32 μg·h/mL; **15**: 14.91, 65.91, 418.53 μg·h/mL.[41,42]

7.4.2   Prosidion Ltd.

Researchers at Prosidion developed a series of GPR119 agonists based on the central oxadiazole core present in their initial prototype compound PSN632408, and much of this effort has been summarized in detail.[35,36] More recent work at Prosidion involved replacing the oxadiazole core in PSN632408 with an acyclic spacer, which resulted in identification of compound **16** (Figure 7.5) as an initial hit with modest GPR119 potency (EC_{50} = 0.5 μM) and a poor agonist response (E_{max} = 33% compared to OEA E_{max} = 100%). Replacement of the left-hand side pyridine ring with the more established GPR119 aryl motifs provided analogs with improved potency and efficacy. PSN119-1 (Figure 7.5) has been described as a lead compound from this work[33,48,49] and it possessed an EC_{50} value of 0.5 μM and exhibited a full agonist response (E_{max} = 407%). In high fat diet fed rats (20 mg/kg, p.o.), PSN119-1 demonstrated a positive glucose-lowering effect after administration of an oGTT with no evidence of hypoglycemia. Similar antihyperglycemic effects were noted during an oGTT in diabetic ZDF rats. In a 21-day chronic study in db/db mice PSN119-1 (100 mg/kg, qd, p.o.) effectively improved glucose tolerance at the end of the treatment period, suggesting its potential in slowing the progression of diabetes. Furthermore, PSN119-1 enhanced GLP-1 release in murine GLUTag cells (EC_{50} = 153 ± 70 nM) and also inhibited

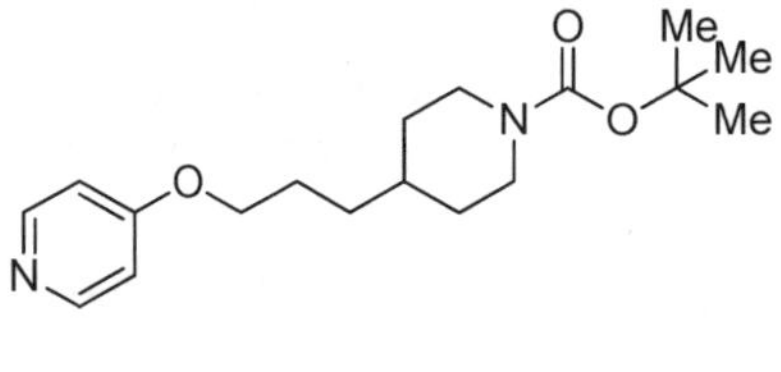

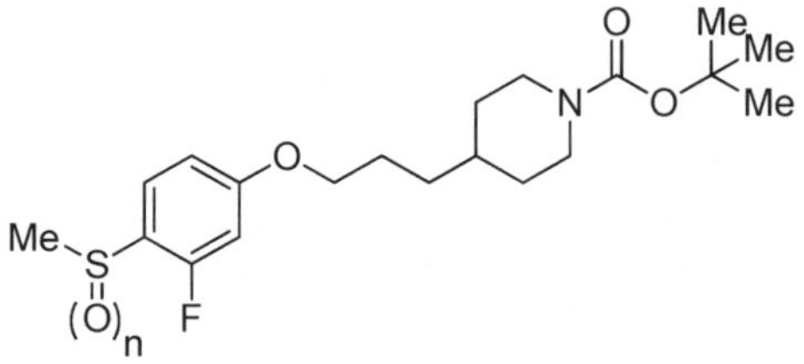

Figure 7.5   Early GPR119 agonists from Prosidion containing an alkoxylene spacer.

gastric emptying at an oral dose of 30 mg/kg in SD rats. In a rat study evaluating its hypophagic activity, PSN119-1 demonstrated a statistically significant reduction in 24-h food intake (50 mg/kg, p.o.) that was comparable to the anti-obesity agent sibutramine (5 mg/kg, p.o.). *In vivo*, PSN119-1 was

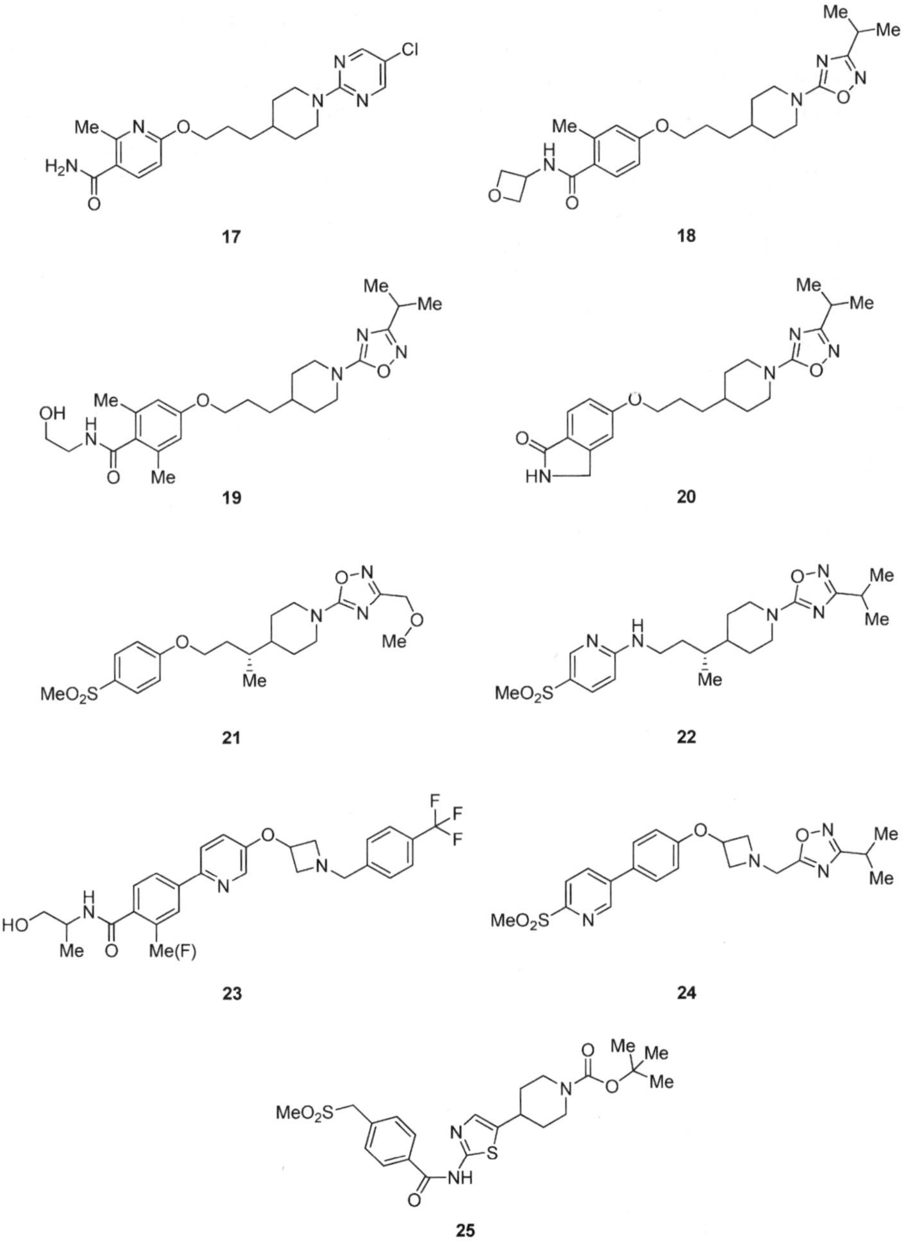

Figure 7.6   GPR119 agonists from Prosidion.

metabolized to the corresponding sulfone PSN119-1M, which demonstrated an equipotent GPR119 agonist activity (EC_{50} = 0.2 μM, E_{max} = 392%; Figure 7.5).

Additional work from Prosidion involved extensive exploration at both the piperidine side and the aryl side of molecules containing a central acyclic spacer (Figure 7.6). The carbamates on the piperidine nitrogen were replaced with aromatic heterocycles, and a variety of substituents, especially amides, were explored on the left-hand side aryl group (e.g. **17–20**).[50–53] Introduction of a methyl group on the spacer and exchanging the ether with an aniline functionality has also been reported (e.g. **21**, **22**).[54,55] Novel azetidine-based GPR119 agonists featuring appropriately substituted *O*-linked biaryl groups on the left-hand side also have been developed by Prosidion (Figure 7.6); the azetidine nitrogen was typically capped with a benzyl group (e.g. **23**) or with a five-membered heterocycle (e.g. **24**).[56,57] Finally, a small series of piperidinyl thiazole-based analogs containing an amide functionality at the 2-position of the thiazole was recently claimed as agonists of GPR119 (e.g. **25**; Figure 7.6).[58] Biological data on representative compounds shown in Figure 7.6 have not been disclosed by Prosidion; preferred analogs have been claimed to be agonists of GPR119 with an EC_{50} value less than 1 μM and to statistically reduce glucose excursion in rat oGTT studies.

The discovery team at Prosidion has explored molecules having dual activity as GPR119 agonists and DPP-4 inhibitors based on the observation that co-administration of a GPR119 agonist with a DPP-4 inhibitor improves glucose

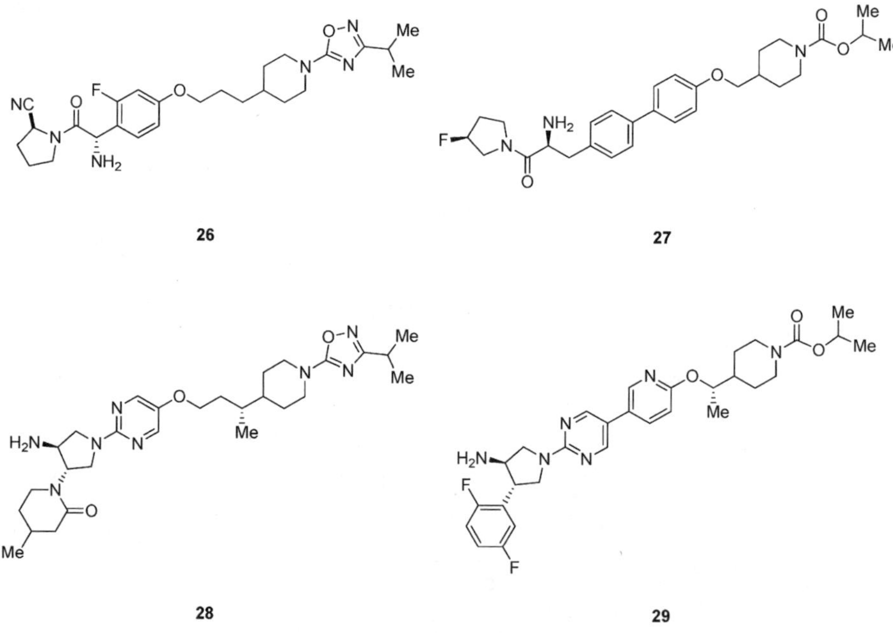

26

27

28

29

Figure 7.7   Combined GPR119 agonists/DPP-4 inhibitors from Prosidion.

tolerance and enhances plasma active GLP-1 levels more than with either agent alone. Figure 7.7 shows representative analogs (e.g. **26–29**) that maintain, on the right-hand side, an appropriately *N*-capped piperidine motif required for GPR119 agonist activity. A variety of substituted aryl ethers were linked to the piperidine group with an aim of combining GPR119 activation with DPP-4 inhibition. Biological data is yet to be published from this effort.[59–61]

7.4.3 Metabolex

Metabolex discovered a series of GPR119 agonists with a piperidine ring linked at the 4-position to a five- or six-membered heteroaromatic ring which in turn is tethered to an aryl substituent through a methoxymethylene linker (Figure 7.8). Thiazoles (e.g. **30** and **31**) and triazoles (e.g. **32**) comprised the majority of the five-membered heteroaromatic ring cores[62–64] while pyridine (e.g. **33**) was the dominant six-membered heteroaromatic core.[65] Other core aromatics included oxazoles, triazoles, pyrazoles, oxadiazoles, pyrimidines, and a phenyl ring. The piperidine nitrogen was capped with either a carbamate or a bioisosteric heterocycle, most frequently a 4-substituted pyrimidine. On the opposite side of the molecule, the tethered aromatic group was typically a phenyl or pyridine ring substituted with a hydrogen bond acceptor at the 4-position. Most frequently cited hydrogen bond acceptors include: methyl sulfones (e.g. **31**) and *N*-linked tetrazoles (e.g. **30** and **33**). Using a FRET assay, Metabolex demonstrated that examples in Figure 7.8 stimulated increases in intracellular cAMP levels in cells that stably express GPR119. Furthermore, selected GPR119 agonists demonstrated enhanced insulin secretion in isolated rat islets in response to glucose and they induced reductions in glucose excursion in mice in response to an oGTT.

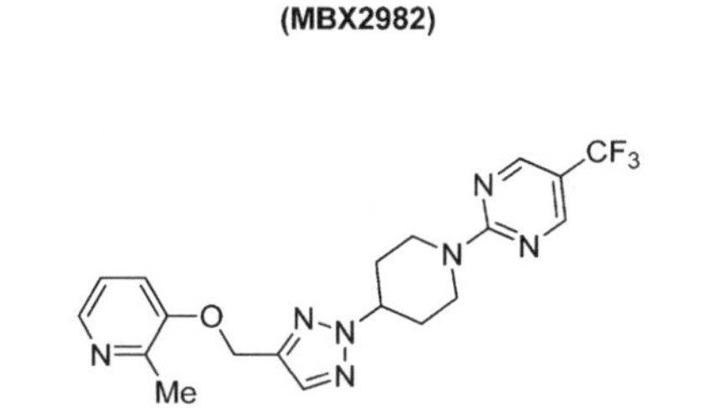

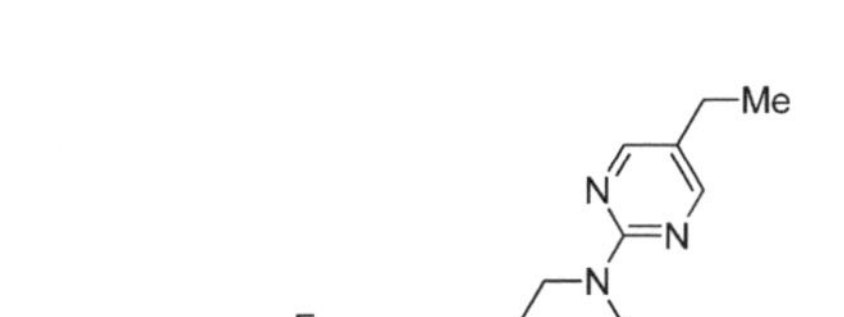

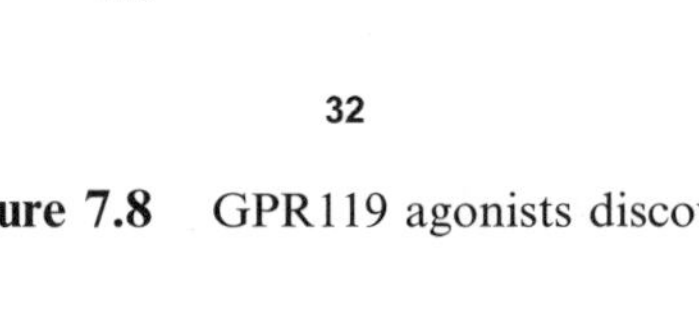

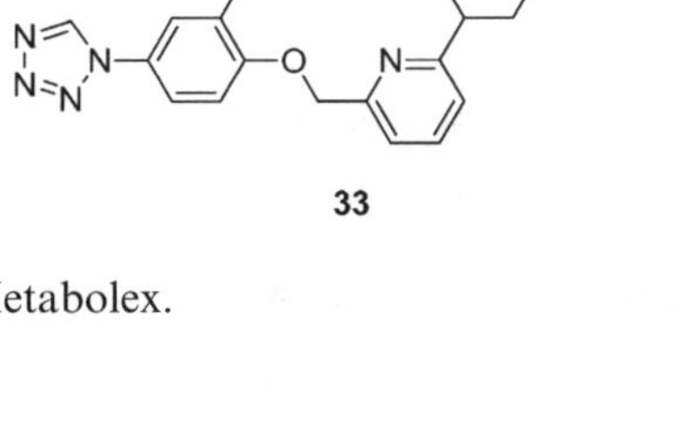

Figure 7.8 GPR119 agonists discovered at Metabolex.

Compound **30** potently stimulated increases in intracellular cAMP with EC_{50} values of 3.9 nM, 66 nM, and 40 nM in CHO cells expressing human, mouse, and rat GPR119 receptors, respectively.[66] In addition to stimulating GDIS in isolated islets, this compound enhanced insulin secretion in normal rats during a hyperglycemic clamp. Administration of an oGTT to rats and mice treated with compound **30** resulted in a reduction of glucose excursion and an increase in plasma levels of the incretins GLP-1 and GIP. Incretin levels were further augmented when **30** was co-administered with the DPP-4 inhibitor sitagliptin. To demonstrate efficacy in a multiple dose setting, **30** was orally administered once daily for 32 days to female ZDF rats placed on a high fat diet. Vehicle-administered rats on this diet developed symptoms consistent with T2D while rats treated with **30** (30 and 100 mg/kg, p.o.) exhibited a dose-dependent reduction in these symptoms. Compound **30** (MBX2982) entered clinical trials for the treatment of T2D and is currently in phase 2.[66–68]

7.4.4　GlaxoSmithKline

GlaxoSmithKline (GSK) reported a series of dihydropyrrolopyrimidines that are structurally similar to Arena's pyrimidine-derived GPR119 agonists.[69–71] Representative compound GSK252A (Figure 7.9) increased cAMP in CHO cells expressing human GPR119 at an $EC_{50} = 40$ nM and was studied extensively in preclinical assays.[69] In isolated rat islets, GSK252A stimulated GDIS and in GLUTag cells it stimulated the release of GLP-1. In rats, GSK252A augmented insulin secretion at a dose of 10 mg/kg during a hyperglycemic clamp and at a dose of 1 mg/kg during an iv GTT. A reduction

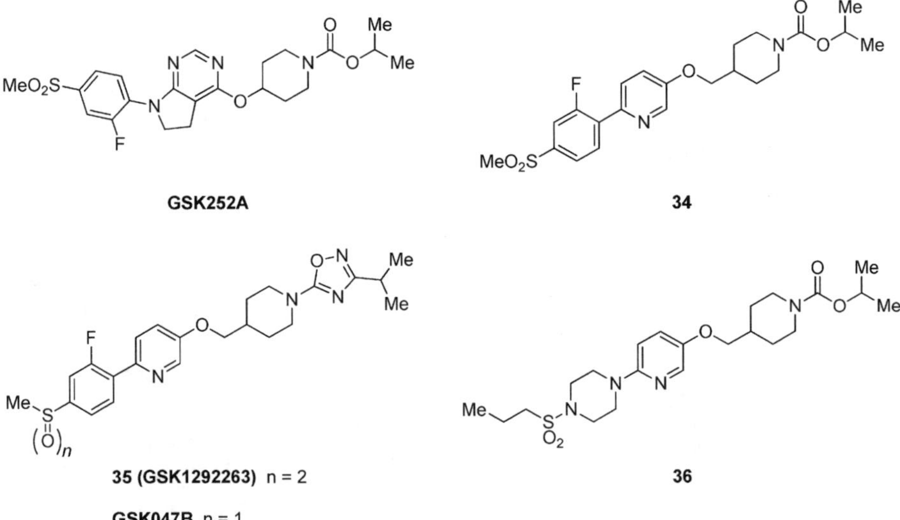

Figure 7.9　GPR119 agonists discovered at GlaxoSmithKline.

in glucose AUC by 30% after an oGTT was observed in rats that were dosed with this compound at 10 mg/kg. Similarly, in mice a reduction in glucose AUC by 28% after an oGTT was observed after dosing at 30 mg/kg. Furthermore, dose-dependent increases in GLP-1 and GIP were observed with GSK252A in rodents. In GPR119 deficient mice, the observed increases in GLP-1/GIP and oGTT effects were ablated, thereby demonstrating that these effects were GPR119 mediated. Other biological effects noted with GSK252A in rodents included a reduction in gastric emptying in rats, and a reduction in food intake in rats and mice (60 mg/kg and 30 mg/kg, respectively). The effects on food intake induced by GSK252A occurred at doses higher than those required to elicit other beneficial effects on glucose, insulin, and incretins. In mice, rats, and dogs, GSK252A exhibited oral bioavailabilities of 40%, 66%, and 37% and half-lives of 3.9 h, 8.7 h, and 6.8 h, respectively. The solubility of GSK252A in fasted state simulated intestinal fluid (FaSSIF) was only 4 μg/mL, however, and rapid oxidation of the parent compound to a pyrrole metabolite with similar potency at GPR119 was observed *in vivo* in rodents.

GSK has also explored a series of GPR119 agonists in which a typical GPR119 piperidine motif is connected to biaryl moieties through an oxomethylene linker.[72,73] The piperidine nitrogen was generally capped with a carbamate or a heteroaromatic bioisostere such as an isopropyl oxadiazole. Typical compounds from this series included the isopropyl carbamate **34** and the isopropyl oxadiazole **35**, which has been profiled extensively in preclinical assays (Figure 7.9). Compound **35** exhibited a pEC_{50} = 6.9 at increasing cAMP in GPR119-expressing cells and was shown to increase insulin secretion and decrease glucose AUCs in response to GTTs in mice and rats. Compound **35** enhanced insulin secretion in rodents after a GTT and augmented incretin secretion in rodents in both a nutrient-independent and nutrient-dependent manner. In contrast to GSK252A, however, this compound did not inhibit gastric or gallbladder emptying in rodents nor does it inhibit food intake or decrease body weight. Furthermore, compound **35** decreased HbA_{1c} levels in ZDF rats after 7 weeks of dosing and was reported to show sustained effects on an oGTT after 8 weeks of dosing. In rats, the oral bioavailability of **35** was 39%, but was much lower in dogs (3%) and monkeys (7%). A micronized formulation of this compound increased oral bioavailability in dogs by 5–10-fold, but had no effect on oral bioavailability in rats; the solubility of **35** in FaSSIF was ∼1.4 μg/mL. Compound **35** (GSK1292263) progressed into clinical trials and is currently in phase 2.[74]

In comparison to the sulfone **35**, the corresponding sulfoxide GSK047B exhibited increased solubility and increased oral drug exposures in rats. At equivalent doses of compound **35**, GSK047B showed improved efficacy by augmentation of insulin and reduction of glucose AUCs after oGTTs in rodents. The team noted that there was a strong correlation between increased exposures in rodents and the increased insulin response and glucose disposal rate observed with GSK047B. GSK has also reported analogs in which the distal phenyl group is replaced with a saturated piperazine ring capped with an alkyl sulfonamide. These analogs offered no significant improvement in FaSSIF solubility (3.4 µg/mL for **36**), however, and the exposure of **36** after oral administration was low in preclinical species.[75,76]

7.4.5 Merck and Co.

Merck and Co. has described bispiperidine-based GPR119 agonists that contain an aryl or heteroaryl moiety on one piperidine ring and a carbamate or bioisosteric heterocycle on the other piperidine ring of the molecule (Figure 7.10).[77] Compound **37** was an early lead which demonstrated an EC_{50} value of 22 nM in a GPR119 HTRF cAMP assay but possessed a poor pharmacokinetic profile (Cl = 120 mL/min/kg; AUC = 0.09 µM·h·kg). Replacement of the Boc group with metabolically stable heteroaromatics was shown to be tolerated and further modifications on the left-hand side piperidine provided compounds that were active *in vivo*. Compound **38** (hEC_{50} = 19 nM) was an optimized analog that demonstrated efficacy in a mouse oGTT study at an oral dose ranging from 3 to 30 mg/kg. The bispiperidine

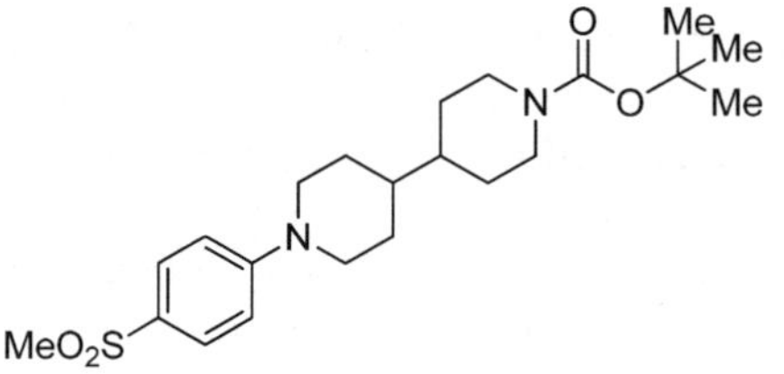

37
EC_{50} = 22 nM

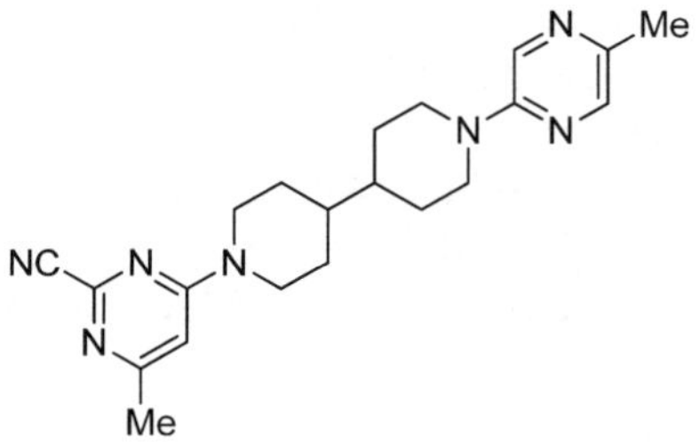

38
EC_{50} = 19 nM

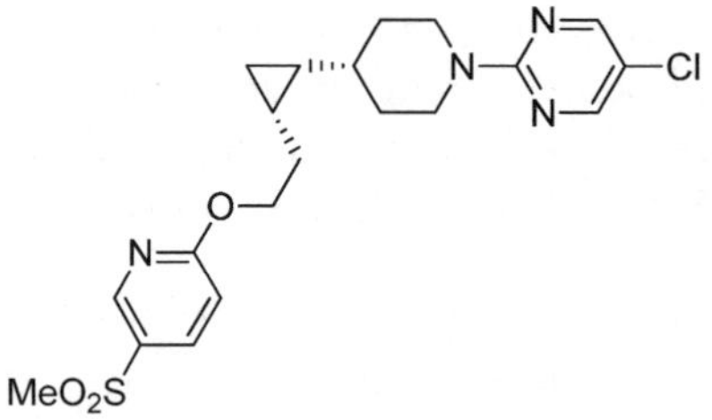

39
hEC_{50} = 4 nM

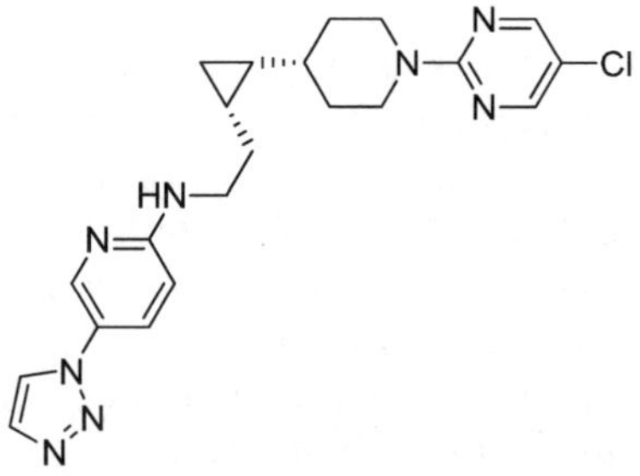

40

Figure 7.10 GPR119 agonists derived from bispiperidine and cyclopropane cores at Merck.

analogs, however, possessed poor physical properties which hindered their further development; several compounds also exhibited a potent hERG signal.

Researchers at Merck have also described GPR119 agonists based on a linear core similar to Prosidion's analogs (in Figure 7.6) and which featured a cyclopropyl restricted ether chain connecting the aryl and piperidine motifs (e.g. **39**; Figure 7.10).[77] Compounds from this series typically possessed improved physical properties and were highly efficacious in oGTT studies (MED = 0.01–10 mg/kg). Compound **39** (hEC$_{50}$ = 4 nM) is a representative analog which demonstrated an MED ranging from 1 to 3 mg/kg in mouse oGTT studies. Lastly, analogs from this series in which the aryl ether linker is modified to an amine linker (e.g. **40**; Figure 7.10) have also been reported.[78]

Potent GPR119 agonists derived from fused pyrimidinone ring systems (Figure 7.11) have been described by Schering-Plough (now Merck). These included 7-substituted 5,6,7,8-tetrahydropyrido[3,4-*d*]pyrimidin-4(3*H*)-one and 6-substituted 5,6,7,8-tetrahydropyrido[4,3-*d*]pyrimidin-4(3*H*)-one analogs.[79–81] In both series an optionally substituted benzhydryl group was shown to be an optimal pyridone-nitrogen substituent (e.g. **41**, **42**); replacement of one of the phenyl groups with a small cycloalkyl group was also shown to be tolerated (e.g. **43**: hEC$_{50}$ = 55 nM). While *N*-benzyl analogs were highlighted in the pyrido[4,3-*d*]pyrimidinone series, aryl carbamates have been described to be the desired nitrogen capping groups on the tetrahydropyridine moiety in the pyrido[3,4-*d*]pyrimidinone series; oxadiazoles were

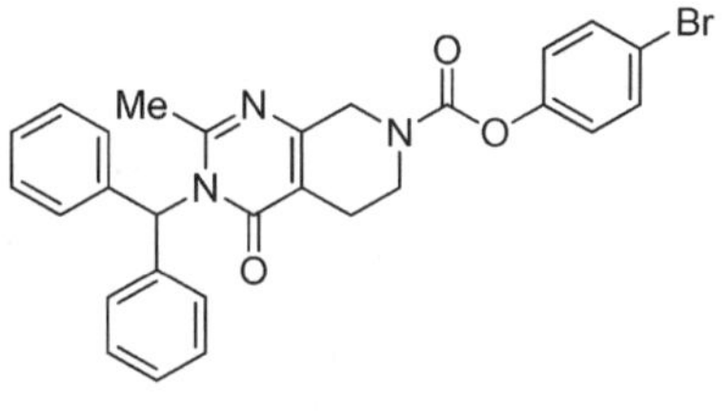

41

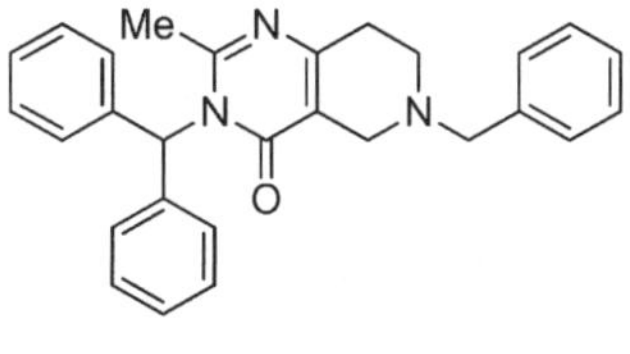

42

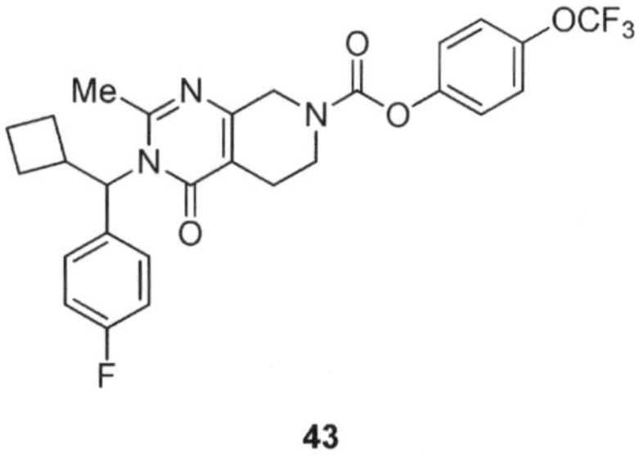

43

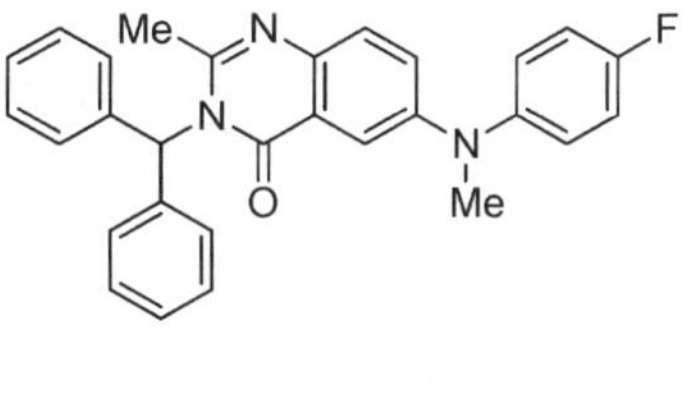

44

Figure 7.11 Fused pyrimidinone-based GPR119 agonists from Schering-Plough (now Merck).

also shown to be good isosteres of carbamates. SAR optimization in the pyrido[3,4-*d*]pyrimidinone series provided compound **41** (hEC_{50} = 86 nM; mEC_{50} = 1.1 µM), which demonstrated acute oGTT activity in lean mice at a dose of 3 mg/kg.[81] Combination of **41** (10 mg/kg) with a GLP-1 analog (Ex-4; 0.3 mg/kg) or a DPP-4 inhibitor (KR-62436; 10 mg/kg) produced an additive effect in lowering blood glucose in oGTT studies in lean mice. Furthermore, compound **41** effectively lowered blood glucose in STZ-DIO mice in a 14-day study (10–30 mg/kg) and exhibited good oral pharmacokinetics in rats and

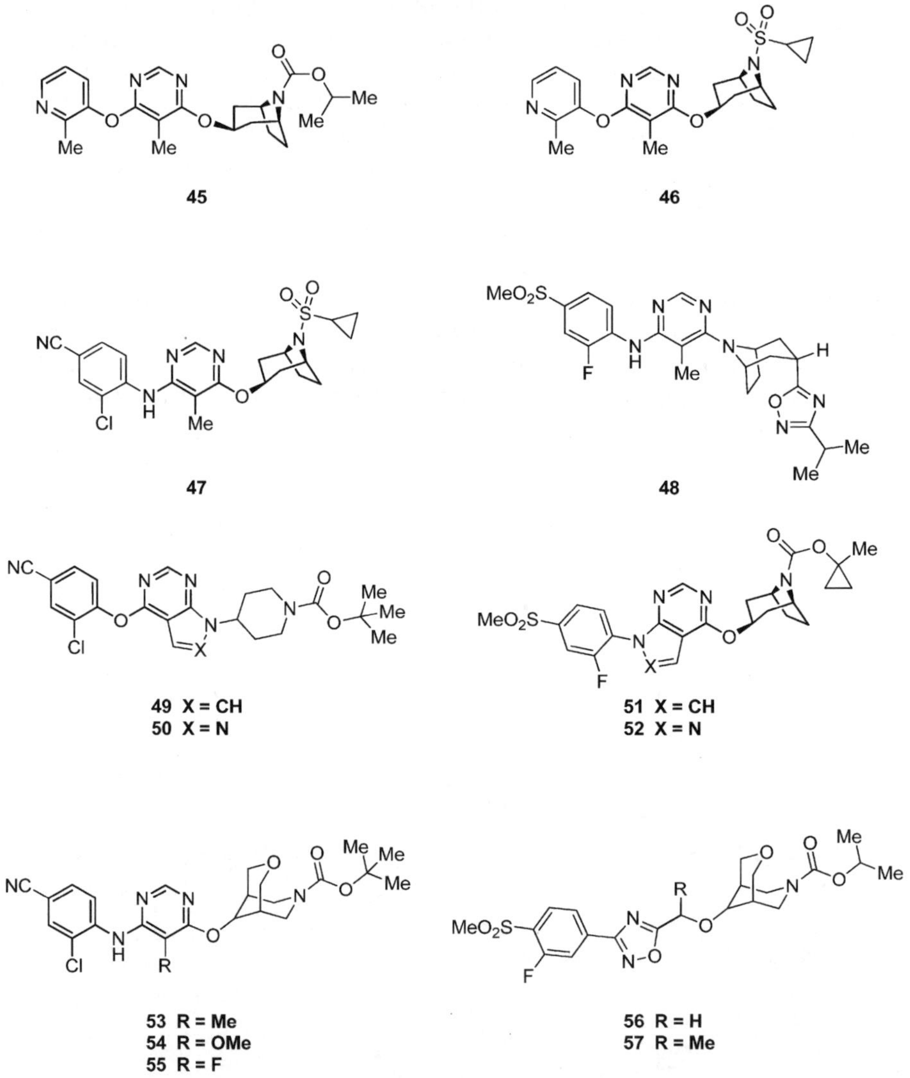

Figure 7.12   Fused pyrimidine- and bicyclic piperidine-based GPR119 agonists from Schering-Plough (now Merck).

monkeys (rat AUC_{0-6h} = 2654 nM·h, 10 mg/kg; monkey AUC_{0-6h} = 1000 nM·h, 3 mg/kg). Fused quinazolinones represented by compound **44** (Figure 7.11) have also been claimed as agonists of GPR119, although limited data has been disclosed from this series.[82]

A nortropanol-based series of GPR119 agonists (Figure 7.12) derived from Arena's compound **12** was also developed by Merck.[83] Library evaluation of various *N*-capping groups on the nortropane moiety showed carbamates and sulfonamides to be more potent over amides, ureas, and amines in this series. The carbamate **45** exhibited good *in vitro* potency (hEC_{50} = 24 nM, mEC_{50} = 55 nM), but was inactive in the mouse oGTT assay; this was consistent with the low agonist activity of this compound at the mouse GPR119 (mE_{max} = 20%). On the other hand, the sulfonamide **46** (hEC_{50} = 24 nM, mEC_{50} = 63 nM), which demonstrated a full agonist response against the mouse receptor (mE_{max} = 76%), was orally active in lowering blood glucose in the mouse oGTT study (3 mg/kg, p.o.). Further optimization of the methylpyridinol portion of **46** furnished the more potent lead **47** (hEC_{50} = 3 nM) which showed a dose-dependent lowering of blood glucose excursion with statistically significant lowering of glucose AUC by 30% and 39% at 10 and 30 mg/kg, respectively; this compound was also shown to increase plasma insulin levels (1 mg/kg, iv) in a rat hyperglycemic model. Furthermore, compound **47** demonstrated a good oral pharmacokinetic profile in rat (AUC_{0-6h} = 2170 nM·h) and monkey (AUC_{0-6h} = 16,548 nM·h), and was clean in the CYP450 enzyme inhibition, PXR, hERG, and AMES assays. Figure 7.12 shows additional examples of bicyclic piperidine- and fused pyrimidine-based GPR119 agonists claimed within recent patent applications.[84–88] Although biological data on specific analogs has not been reported, several undisclosed compounds were claimed to be efficacious in mouse oGTT studies.

7.4.6 Novartis

The Genomics Institute of the Novartis Research Foundation (GNF) has developed a series of GPR119 agonists structurally similar to some of Prosidion's GPR119 agonists but with piperidine or piperazine alkyl sulfonamides replacing the alkyl phenyl sulfone pharmacophore.[89–92] In CHO cells expressing GPR119, representative tetrahydropyrido[4,3-*d*]pyrimidine ethers such as **58** and amines such as **59** were potent at increasing intracellular cAMP levels (Figure 7.13).[89] Analogs in which the carbamate capping group is replaced by typical bioisosteres such as an oxadiazole or a pyrimidine also showed potent EC_{50} values in this cAMP GPR119 assay. A series in which the fused pyrimidine ring is replaced by a phenyl group (i.e. tetrahydroisoquinoline) also furnished potent GPR119 agonists.[90] In a separate report, piperazinyl pyrazines such as **60** (Figure 7.13) demonstrated that the sulfonamide need not be fused to a central aromatic ring;[91] a structurally similar series of GPR119 agonists has also been reported by GlaxoSmithKline.[75,76] The GNF team further demonstrated that the spacer

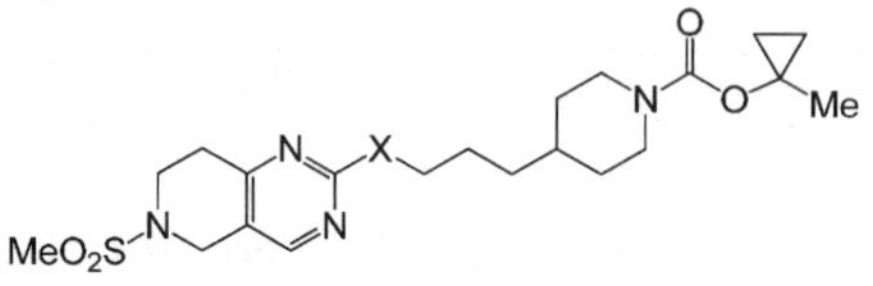

58 X = O
EC_{50} = 2 nM

59 X = NH
EC_{50} = 8 nM

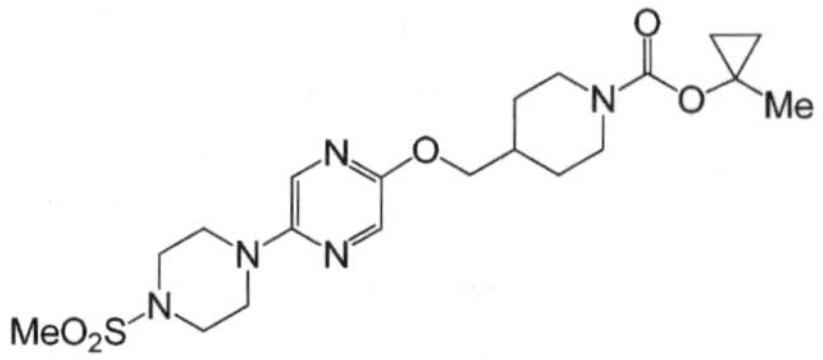

60

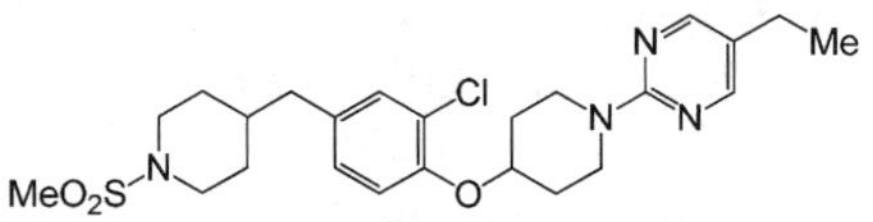

61
EC_{50} = 3 nM

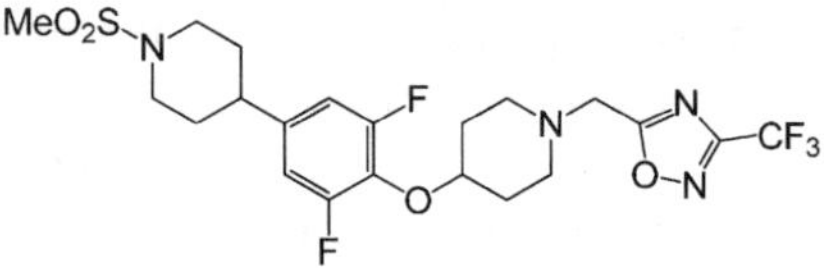

62
EC_{50} = 7 nM

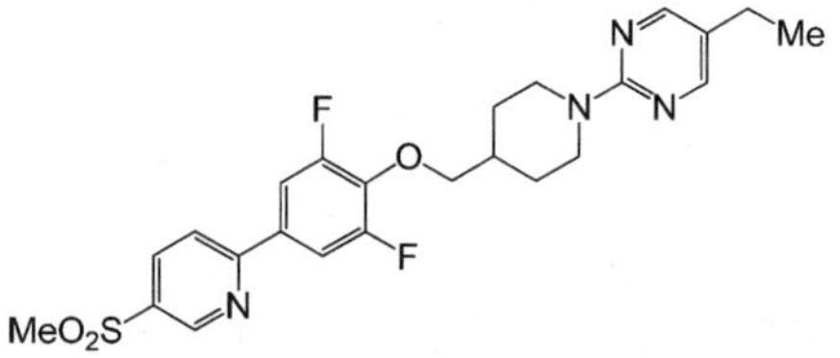

63
EC_{50} = 1 nM

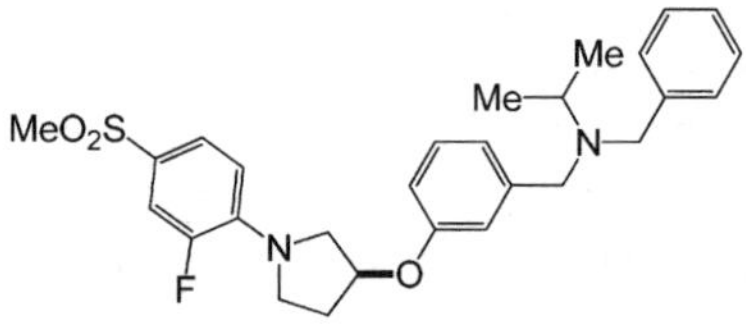

64
EC_{50} = 33 nM

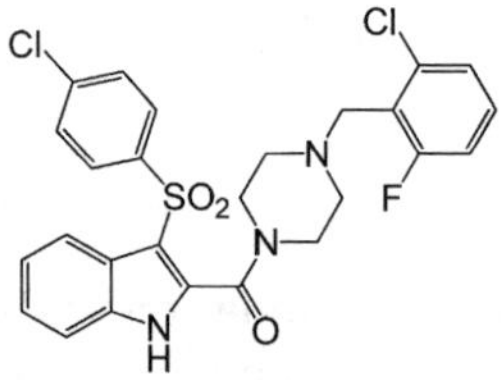

65
EC_{50} = 7 nM

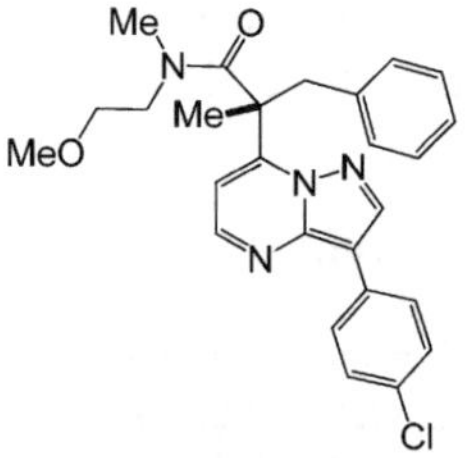

66
EC_{50} = 7 nM

Figure 7.13   GPR119 agonists discovered at Novartis.

length between the piperidine capping group on the right-hand side and the piperidine/piperazine sulfonamide on the left-hand side of **60** (Figure 7.13) can be varied while still retaining potent activity at GPR119. For example, when the distance between the central piperidine and the halophenyl group is reduced by one methylene unit, potent GPR119 agonists such as **61** (Figure 7.13) were identified in which an extra methylene group is placed between the halophenyl ring and the piperidine sulfonamide ring.[92] Alternatively, the distance between the central piperidine ring and the heterocycle cap can also be lengthened by one methylene group to afford potent GPR119 agonists such as oxadiazole **62** (Figure 7.13).[93]

GNF has also claimed GPR119 agonists that are structurally distinct from previously reported chemical classes. In one series, the common pharmacophore contained the familiar phenyl methylsulfone moiety linked at the *para* position to a pyrrolidine nitrogen.[94] The agonists were then linked at the 3-position of the pyrrolidine ring to a phenol oxygen which in turn was linked to a hydrophobic acyclic tertiary methylamine, as in the case of representative analog **64** (Figure 7.13). The SAR surrounding the phenol ring was somewhat permissive as the acyclic tertiary amine was successfully replaced by non-basic carbamates and carbonates substituted at either the *meta* or *para* position relative to the phenol oxygen. A second structurally distinct series of GPR119 agonists contained an indole bearing an amide at the 2-position and a sulfone at the 3-position of the indole ring.[95] The most potent examples disclosed from this effort contained amides derived from either a benzylpiperazine (e.g. **65**) or a 4-substituted piperidine. At the 3-position of the indole, a *para* chlorophenyl sulfone appeared frequently in the most potent analogs and the remainder of the indole ring was unsubstituted in the majority of the examples. Finally, a series of pyrazolopyrimidine linked amides such as **66** were identified as potent GPR119 agonists.[96] The amide SAR was somewhat permissive in this class, but the disposition of the chiral center was quite important as the antipodes of the more potent stereoisomers were often >100-fold less potent at increasing cAMP in GPR119 expressing cells.

7.4.7 Bristol-Myers Squibb

Bristol-Myers Squibb (BMS) reported a series of GPR119 agonists based on [6,5]-, [6,6]-, and [6,7]-bicyclic central cores.[97,98] Compounds **67–69** (Figure 7.14) are illustrative examples that were derived from pyrimidine-fused pyrazole, triazole, and morpholine ring systems. Analogs from this class exhibited EC_{50} values ranging from 19 nM to 3546 nM in a luciferase assay utilizing transfected HEK293 cells. BMS has also reported GPR119 agonists featuring a pyridone central core that is *N*-substituted with the left-hand side aryl motif and linked to the right-hand side piperidine motif at the 4-position through an oxygen linker (**70–73**; Figure 7.14); pyridazone analogs have also been claimed as GPR119 modulators.[99–101] For the pyridone-based analogs, the EC_{50} values ranged from 1 nM to 7 μM in a human tetracycline inducible cAMP assay. In both the BMS series, the general substitution pattern on the aryl and the piperidine groups remained consistent with competitors' GPR119 agonists.

7.4.8 Boehringer Ingelheim

The first series of GPR119 agonists from Boehringer Ingelheim has structural features reminiscent of the earlier Prosidion pharmacophore in which the central core has been modified to incorporate twin piperidines linked via a central oxime.[102] In a cAMP assay using transfected HEK293 cells,

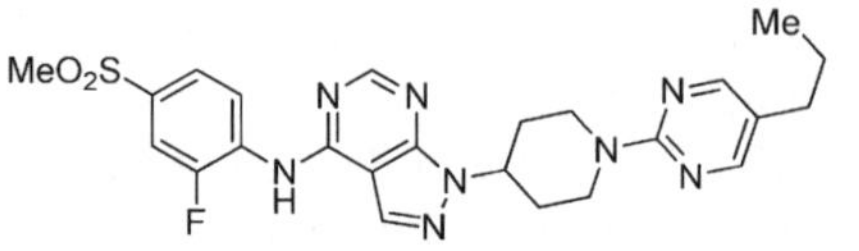

67

EC$_{50}$ = 82 nM

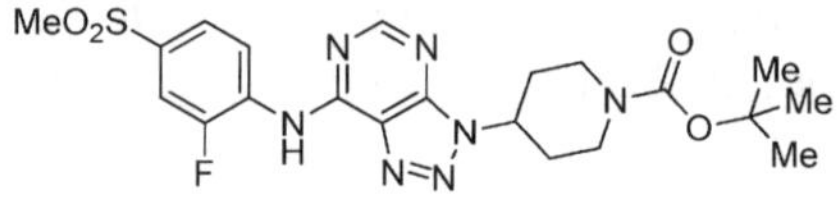

68

EC$_{50}$ = 19 nM

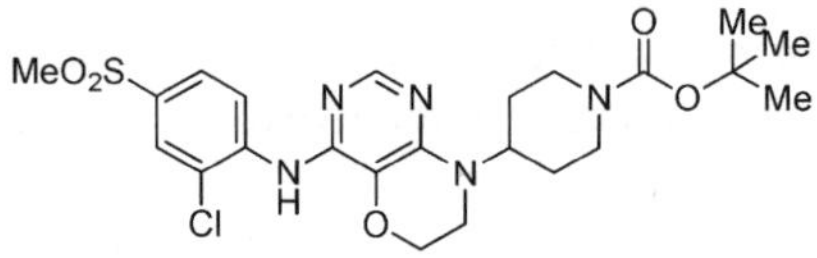

69

EC$_{50}$ = 3 nM

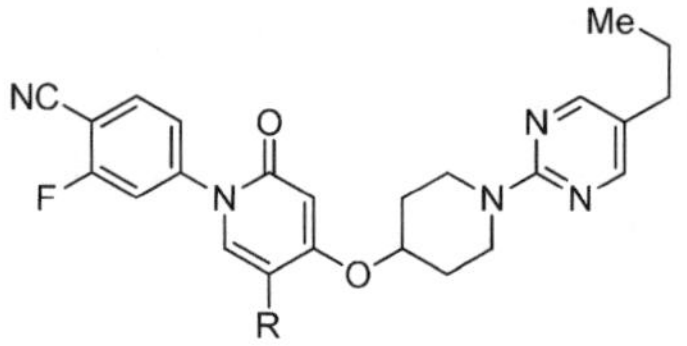

70 R = Cl

EC$_{50}$ = 1 nM

71 R = H

EC$_{50}$ = 7 nM

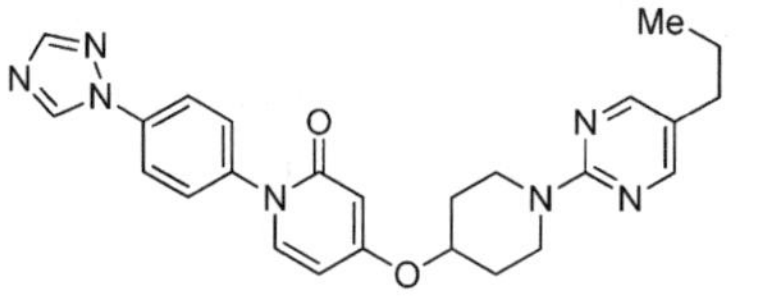

72

EC$_{50}$ = 87 nM

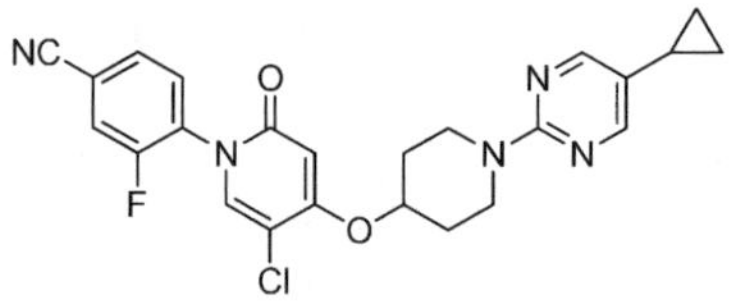

73

EC$_{50}$ = 4 nM

Figure 7.14 Representative GPR119 agonists from Bristol-Myers Squibb based on fused-pyrimidine and pyridone scaffolds.

compounds from this series were reported to have EC$_{50}$ values ranging from 5 nM to >10 µM. Compounds **74** and **75** (Figure 7.15) are representative examples that exhibited EC$_{50}$ values of 538 nM and 34 nM, respectively, and reduced glucose excursion by 70% and 75%, respectively, in oGTT studies conducted in male SD rats (30 mg/kg, p.o.). The second series from Boehringer Ingelheim[103] featured a diamino-pyrimidine central core (similar to agonists from Arena) that was *N*-substituted with an arene and linked to a piperidine motif at the 4-position through an amino linker (e.g. **76**; Figure 7.15). The general substitution pattern on the aryl and the piperidine groups was consistent with the known SAR of GPR119 agonists, and EC$_{50}$ values ranging from 23 nM to 906 nM has been reported for compounds from this series.

7.4.9 Pfizer

Researchers at Pfizer have also explored conformationally restricted bicyclic analogs of Arena's compound **12** (Figure 7.4). The *syn* oxazabicyclic analog **77**

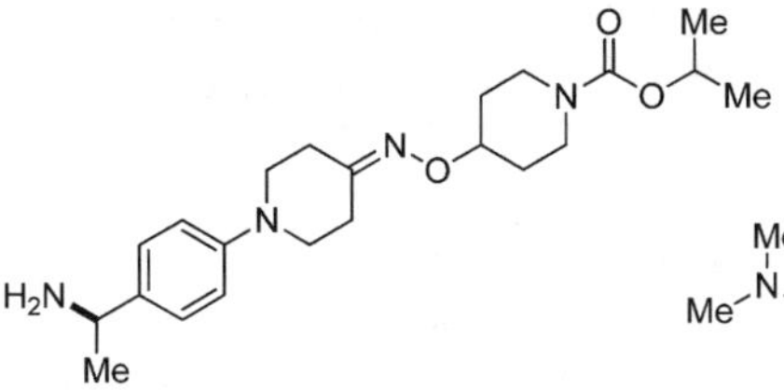

74
EC$_{50}$ = 538 nM

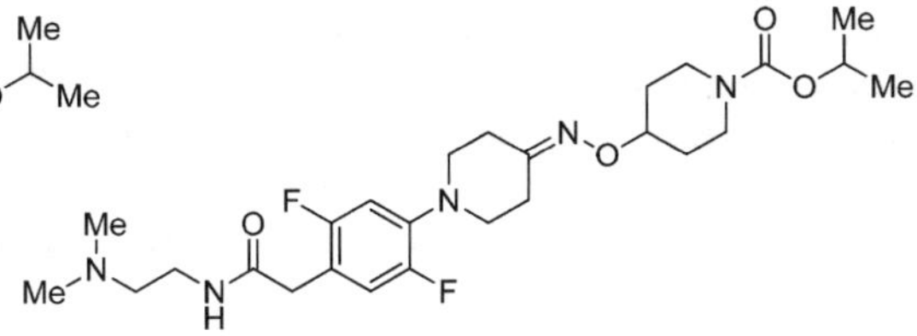

75
EC$_{50}$ = 34 nM

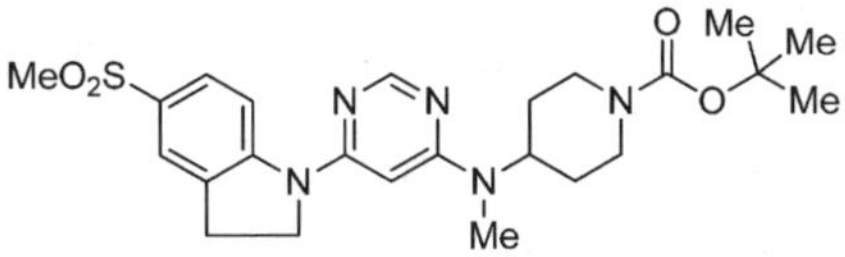

76
EC$_{50}$ = 34 nM

Figure 7.15   SAR of selected GPR119 agonists from Boehringer Ingelheim.

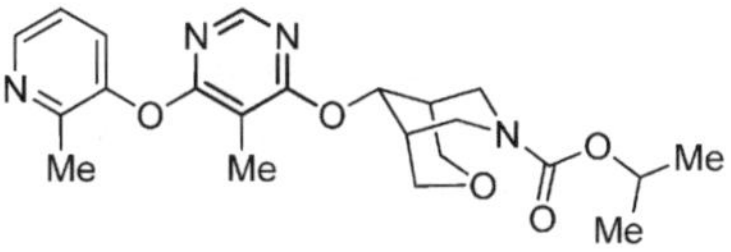

77
hEC$_{50}$ > 10 μM
antagonist

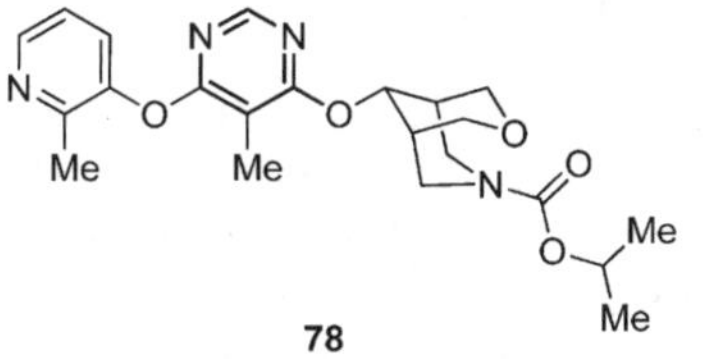

78
hEC$_{50}$ = 65 ± 31 nM
IA = 78%

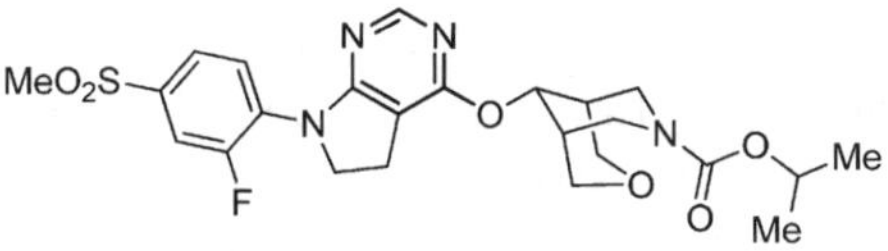

79
hEC$_{50}$ = 16 nM
E$_{max}$ = 88%

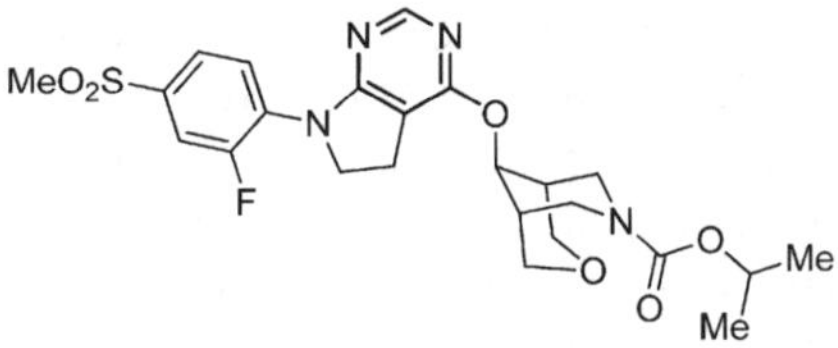

80
hEC$_{50}$ = 25 nM
E$_{max}$ = 93%

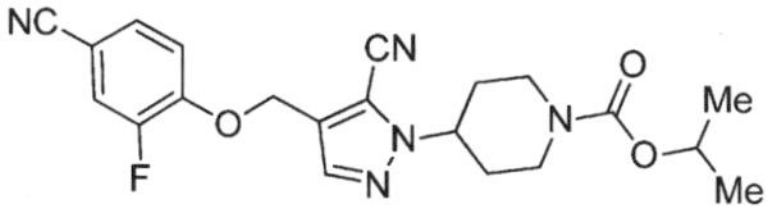

81
hEC$_{50}$ = 7 nM
E$_{max}$ = 79%

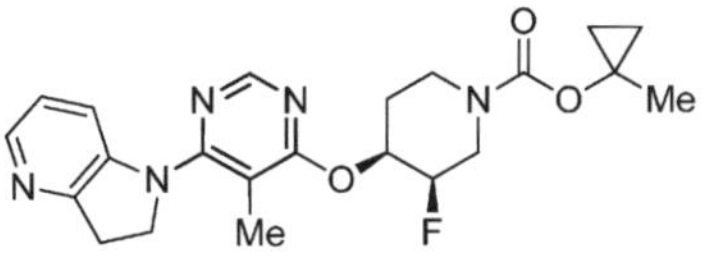

82
hEC$_{50}$ = 22 nM
E$_{max}$ = 68%

Figure 7.16   Representative GPR119 agonists from Pfizer.

was found to be an antagonist at human GPR119 and the corresponding *anti* isomer **78** exhibited an agonist response; both **77** and **78** demonstrated similar binding affinities at the human receptor (**77**: $K_i = 33 \pm 38$ nM; **78**: $K_i = 20 \pm 17$ nM).[104] Compound **12** (Figure 7.4), which can adopt both axial and equatorial conformations, showed partial agonist activity in the same assay. Since the agonist (compound **78**) and antagonist (compound **77**) conformations place the carbamates in different spatial regions, it was hypothesized that the carbamates in each may have different hydrogen bond interactions with the receptor thereby resulting in different functional profiles. Alternatively, the conformational changes in the bicyclic system may be causing changes on the other side of the molecule in a way that agonist activity is observed for **78** and not for **77**. At the rat receptor, however, compound **77** exhibited an agonist response, which translated into lowering of blood glucose and increases in GLP-1 levels in rat oGTT studies. This data suggests that the results in rodent models may not be predictive of human efficacy. Figure 7.16 shows additional examples (**79–82**) of piperidine-containing analogs from Pfizer's recent GPR119 patent applications.[105–107]

7.4.10 Astellas

Astellas has reported the discovery of several 2,4,6-trisubstituted pyrimidine-based GPR119 agonists (Figure 7.17). AS1269574 was identified through a library screening effort and this agonist exhibited a modest EC_{50} value of 2.5 μM in HEK293 cells transiently expressing human GPR119.[108] In a mouse MIN-6 pancreatic β-cell line AS1269574 enhanced insulin secretion under high-glucose conditions (16.8 mM) but not under low-glucose conditions (2.8 mM), suggesting a low risk of hypoglycemia with such a compound. At a dose of 100 mg/kg AS1269574 reduced blood glucose levels in normal mice after oral glucose challenge; importantly, AS1269574 did not affect fed and fasting plasma glucose levels in normal mice. AS1535907 is another structurally related agonist which increased intracellular cAMP levels in human GPR119 transfected HEK293 cells ($EC_{50} = 1.5$ μM) and enhanced insulin secretion in NIT-1 cells and in perfused rat pancreas.[109,110] A statistically significant improvement in oral glucose tolerance was observed upon administration of a single dose of 10 mg/kg in both normal and db/db mice. Furthermore, upon multiple dosing for 2 weeks in

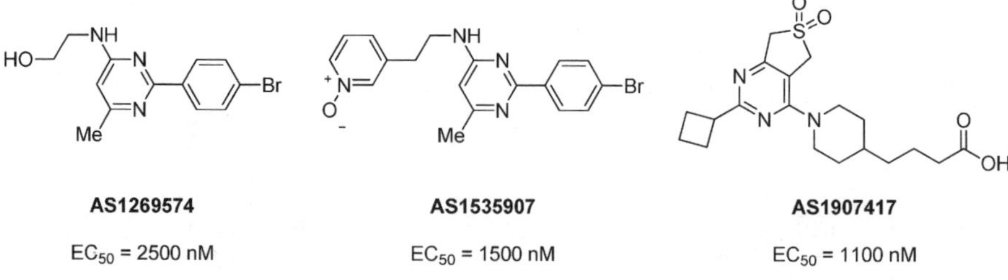

Figure 7.17   GPR119 agonists from Astellas.

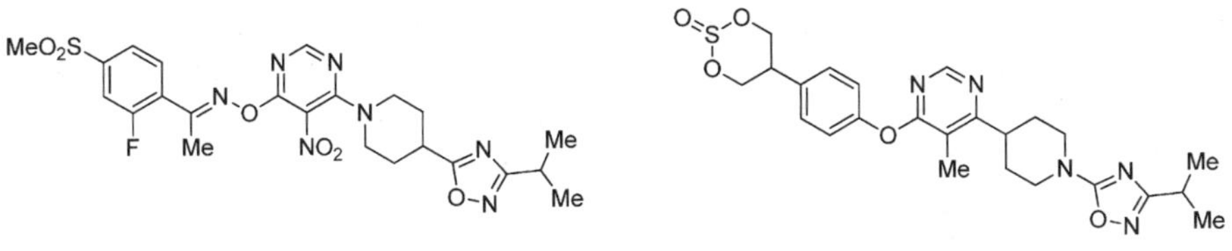

Figure 7.18   GPR119 agonists discovered at Cadila Healthcare.

db/db mice, AS1535907 significantly increased plasma insulin levels and decreased blood glucose levels. AS1535907 was also shown to exert a potential protective effect on pancreatic β-cell function via regulation of transcription factors. Finally, the fused pyrimidine analog AS1907417 has also been reported to possess a profile of continued interest.[111] This compound enhanced intracellular cAMP (hEC$_{50}$ = 1.1 μM), GDIS, and human insulin promoter activity *in vitro* and improved glucose tolerance in fasted normal mice; there was no effect on plasma glucose or insulin levels without a glucose challenge. In rodent models of diabetes and obesity, twice-daily doses of AS1907417 for 4 weeks resulted in reduced hemoglobin A1c levels, and furthermore, in db/db mice it was shown to possibly preserve the pancreatic β-cell function.

7.4.11   Cadila Healthcare

Cadila Healthcare has claimed two series of GPR119 agonists with structures reminiscent of Arena's pyrimidines.[112,113] In the first series, the aniline nitrogen linker of AR231453 (Figure 7.1) was replaced with an oxime linker to afford compound **83** (Figure 7.18), which increased cAMP in cells expressing human GPR119.[112] When administered to SD rats intraperitoneally at a dose of 25 mg/kg, **83** reduced cumulative food intake by 37% over an 8-hour period. Furthermore, after an oral dose of 30 mg/kg, this compound reduced glucose excursion in mice by 20% during the 2 hours following an oGTT.

The second series of GPR119 agonists from Cadila included compounds in which the piperidine ring was reversed and the phenyl sulfone substituent was replaced by a dioxathiane 2-oxide.[113] Compound **84** (Figure 7.18) is a representative example which increased cAMP in human GPR119 expressing cells at an EC$_{50}$ = 11 nM. When **84** was administered intraperitoneally at a dose of 25 mg/kg to SD rats, a reduction of cumulative food intake by 21% over a 6-hour period was observed. Additionally, at an oral dose of 30 mg/kg, **84** reduced glucose excursion in mice by 31% during the 2 hours after an oGTT.

7.5   Clinical Status of GPR119 Agonists

Several GPR119 agonists have progressed into clinical trials, with many showing encouraging results from single and multiple dose studies. Much of the information regarding these studies has been released during presentations at conferences or in press releases, and thus the available data for these compounds vary substantially depending on the source. The next sections summarize the available data from known GPR119 agonists that have progressed into human clinical studies.

7.5.1   APD668 and APD597

Although phase 1 single dose data for APD668 (JNJ28630368, compound **3**; Figure 7.2) was reported to be encouraging,[114] this compound was put on hold in order to advance a more potent analog APD597 (JNJ38431055; structure not yet disclosed).[115] Available clinical data for APD597 is scarce, but it was disclosed that the compound exhibited evidence for incretin (GLP-1, GIP, and PYY) increases and reductions in post-meal glucose increases both alone and in combination with sitagliptin.[116] After completion of phase 1 clinical trials with multiple day dosing, Ortho-McNeil-Janssen elected to return the rights of APD597 to Arena.[116]

7.5.2   PSN821

The structure of PSN821 has not yet been disclosed, but preclinical and clinical data describing this compound have been reported by OSI/Prosidion. PSN821 was shown to stimulate insulin and GLP-1 secretion from HIT-T15 cells and GLUTag cells, respectively, at EC_{50} values of 153 and 135 nM. In male diabetic ZDF rats, PSN821 dose-dependently reduced glucose excursion after an oGTT at single doses of 3, 10, and 30 mg/kg, with the lowest dose affording reductions in glucose AUCs similar in magnitude to a DPP-4 inhibitor (P32/98). Additionally, this compound retained its oGTT efficacy after 10 days of treatment, indicating that tachyphylaxis had not occurred. In an 8-week study in pre-diabetic ZDF rats, PSN821 significantly lowered non-fasting blood glucose levels and attenuated their HbA_{1c} gain. Furthermore, treatment of this compound to weight-neutral female DIO Wistar rats at a dose of 30 mg/kg/day resulted in a body weight loss comparable to sibutramine administered at 5 mg/kg/day.[117]

In phase 1 single-dose studies, PSN821 was well tolerated at doses up to 3000 mg in healthy subjects and up to 1000 mg in diabetic subjects. In a 14-day study in T2D patients, PSN821 alone (250 mg BID) and in combination with metformin (250 mg BID, 500 mg qd) was evaluated for pharmacodynamic effects in comparison to placebo-treated control patients and baseline.[118] In all treated groups, plasma glucose was reduced at day 14 in the fasting state compared to placebo and after a liquid meal challenge compared to baseline and placebo. Additionally, energy intake was lowered substantially (-40%) in the 500 mg PSN821 + metformin group after administration of a test meal. On day 14, GLP-1 and GIP levels were highly variable with no changes observed while an apparent increase in PYY levels was observed relative to both baseline and placebo. In all treatment groups, there appeared to be reductions in total cholesterol, LDL, and triglycerides on day 14 relative to baseline and no differences were noted in HDL in any group. Moreover, a reduction of serum leptin and elevation of adiponectin were reported in all active treatment groups. At all doses in the study, PSN821 was safe and well tolerated with only a low incidence of gastrointestinal upset compared to placebo. Non-statistically significant reductions in body weight were noted in all active treatment groups compared to placebo.[118,119]

7.5.3　MBX2982

In single dose studies in healthy male and female volunteers, MBX2982 (compound **30**; Figure 7.8) exhibited a half-life suitable for once-daily oral dosing ($\sim$14 h) and was safe and well tolerated through a dose range of 10 to 1000 mg.[66] A dose-dependent reduction in glucose excursion was noted during mixed meal tolerance tests (MMTTs) with the most robust glucose reduction occurring at the highest dose. Additionally, volunteers receiving MBX2982 exhibited an increase in total GLP-1 after a MMTT, consistent with the effects of this compound in rodents. In a phase 1 multiple-dose study with MBX2982, 100 mg and 300 mg qd were administered for 4 days in pre-diabetic volunteers and decreased glucose excursion was observed after a MMTT and an oGTT with both doses.[66] In a separate phase 1 study, a second generation formulation of MBX2982 also demonstrated PK consistent with once-daily oral dosing and repeat dose exposures were significantly higher (>8-fold) than the original formulation.[67] After 4 days of qd dosing in subjects with impaired glucose tolerance, the new formulation demonstrated a reduction in glucose excursion after an MMTT at all doses studied (25–600 mg). Further analysis of the data revealed that greater efficacy was observed with the most glucose-intolerant subjects. In all three phase 1 studies to date, no safety or tolerability concerns were noted, including hypoglycemia.[67] A 4-week phase 2 study of MBX2982 in T2D patients was performed to evaluate its efficacy, safety, tolerability, and pharmacokinetic profile at doses of 25 mg, 100 mg, and 300 mg. Sitagliptin (100 mg qd) was used as an active comparator in this study and no data has been reported from this study to date.[68]

7.5.4　GSK1292263

In a rising single-dose phase 1 study, GSK1292263 (compound **35**; Figure 7.9) was reported to be safe and well tolerated at oral doses ranging from 10 to 400 mg and exhibited a PK profile consistent with once-daily oral dosing with a mean half-life of approximately 12–18 hours. GSK1292263 caused a dose-dependent reduction in glucose AUCs and an increase in circulating gut hormone levels after administration of an oGTT. When GSK1292263 was co-administered with sitagliptin, active GLP-1 plasma concentrations were augmented while total GLP-1, GIP, and PYY levels were lowered and no effects on gastric emptying were observed.[120]

In a multiple-day study in T2D subjects, GSK1292263 (50 to 300 mg BID and 600 mg qd) or placebo was dosed for 14 days and then co-dosed on day 14 with sitagliptin (100 mg).[121] In another study, GSK1292263 (75 or 300 mg BID or 600 mg qd) was co-administered for 14 days to T2D patients on metformin monotherapy. As an active comparator, sitagliptin (100 mg qd) was also administered for 14 days to T2D patients. Twenty-four-hour profiles of glucose, insulin, C-peptide, glucagon, GLP-1, GIP, and PYY were measured in these studies. Lastly, a single-dose crossover design study assessing the pharmacodynamic effects of GSK1292263 (25, 150, or 800 mg) on an oGTT was performed. All doses of GSK1292263 were well tolerated and no significant adverse effects were noted among the studies. Drug exposures were less than dose proportional in the ranges studied, and administration with food increased drug exposures (positive food effect). Steady state PK was achieved after 4 days of dosing and co-administration with sitagliptin did not alter the PK of either drug.

A dose-dependent decrease in glucose AUC (0–3 h) was observed during an oGTT after a single dose of GSK1292263. In the multi-dose studies, however, there was no effect of GSK1292263 treatment on glucose AUC (0–24 h) after dosing for 13 days either alone or with metformin, nor was there an additive effect when co-dosed with sitagliptin on day 14. When combined with a single dose of sitagliptin, GSK1292263 caused an increase in active plasma GLP-1 concentrations but a decrease in total GLP-1, GIP, and PYY levels, which is consistent with previous results from a single dose study. In comparison, sitagliptin alone reduced glucose AUC (0–24 h).[121]

7.6　Summary

Preclinically, GPR119 agonists show promise as anti-hyperglycemic agents. Investigators report a direct effect on the β-cell that results in GDIS. Additionally, GLP-1 and GIP release induced by GPR119 agonists from intestinal enteroendocrine cells are also anticipated to have a beneficial effect on the maintenance of blood glucose homeostasis via the incretin axis. Taken together, these two independent biological pathways suggest that GPR119 agonists will carry a low risk for hypoglycemia and therefore offer an advantage over sulfonylureas and insulins with respect to patient safety.

Hypophagic effects of GPR119 agonists have been reported by some groups and these effects are often linked to delayed gastric emptying. Reduced food intake and delayed nutrient absorption are expected to augment the anti-hyperglycemic effects of these compounds. The precise mechanism of these effects is unclear, however, and more work is needed to better understand whether these effects are mediated by GPR119 activation. Furthermore, additional work is needed to understand the PK/PD relationships of hypophagia and delayed gastric emptying relative to the GDIS effects of GPR119 agonists. Finally, the magnitude of these effects varies widely with different GPR119 agonists, and it is uncertain whether this variability is due to a difference between compounds, doses tested, or simply different assays used by the various investigators. Nevertheless, the promise of a mechanism that can control blood glucose levels while maintaining body weight neutrality or possibly reducing body weight is an attractive possibility for T2D patients and healthcare providers.

Although it has been reported that GPR119 can become desensitized *in vitro*, many investigators report promising durability of GPR119 agonists after multiple-day doses in ZDF rat models or db/db mouse models. Specifically, beneficial effects were noted on blood glucose and blood insulin levels after dosing for 14 days or longer in these rodent models. Significant caveats exist, however, in the ability to translate the observed pharmacological effects in these dynamic rodent models to the treatment of humans with T2D.[122] Moreover, it is difficult to discern the relative contributions of GDIS and hypophagia to the observed effects on blood glucose/insulin that have been reported in these models.

A distinct pharmacophore appears to be common to potent GPR119 agonists reported to date, and many agonists possess structural features reminiscent of either Arena or Prosidion's early compounds. Nevertheless, novel structure classes are beginning to emerge as even more companies enter into this crowded competitive field. One common issue that is frequently cited for GPR119 agonists is poor physicochemical properties (PPs), most notably with respect to poor aqueous solubility. Poor PPs not only contribute to difficulty in evaluating the preclinical pharmacology of these agonists, but also present substantial challenges in the development space. Companies that have progressed GPR119 agonists into the clinic have reported challenges with PPs of GPR119 agonists and frequently they have investigated the use of enabled formulations in order to achieve sufficient pharmacokinetic profiles of their compounds in humans. It is currently not clear whether these poor PPs have compromised the ability to effectively study the pharmacology of these compounds in humans, but it is clear that GPR119 agonists with improved PPs are needed in order to offer sufficient flexibility to probe mechanistic questions in humans even more thoroughly.

In phase 1 clinical studies, GPR119 agonists appear to be safe and well tolerated, with no reports of hypoglycemia. With respect to efficacy, GPR119 agonists have demonstrated an ability to reduce blood glucose excursion after single dose oGTTs or MMTs in diabetic and pre-diabetic patients. On the other hand, available clinical efficacy data is mixed from studies in which GPR119 agonists were dosed for 14 days or longer. Although negative data was reported by GSK with GSK1292263, positive data was reported by Prosidion for PSN821. The study designs and endpoints were different, however, and therefore it is difficult to compare these studies with each other. The development of APD597 and MBX2982 are currently stalled as Arena and Metabolex continue to seek partners for co-development of these compounds in phase 2 clinical trials.

Perhaps the greatest promise of GPR119 agonists lies in combination therapy with DPP-4 inhibitors, where increases in active GLP-1 may be achieved due to the GLP-1 secretion effects of GPR119 agonists and the prevention of the degradation of active GLP-1 by the action of the DPP-4 inhibitor. Quantification of the magnitude of increase in total and active GLP-1 in preclinical species and humans is nevertheless a challenging goal with combination therapy as many factors can affect levels of active and total GLP-1 in humans. Notably, GSK reported that GSK1292263 does not show any further beneficial effect on MMT efficacy when co-administered with a single dose of sitagliptin. Unfortunately, detailed data sets from this study are not yet available, and thus questions remain regarding whether a combination therapy of a GPR119 agonist with a DPP-4 inhibitor can offer additive or synergistic efficacy in type 2 diabetic patients. If substantial augmentation of active GLP-1 can be achieved in humans, efficacy comparable to injectable GLP-1 mimetics both on glucose and on body weight would be expected from an orally administered combination of a GPR119 agonist with a DPP-4 inhibitor. Such a combination would offer the safe and robust efficacy profile of a GLP-1 mimetic but with the increased convenience and compliance of a once-daily oral dosing regime, and therefore this mechanism continues to be of interest as an attractive anti-diabetes therapy.

Much is now known about the biochemistry and pharmacology of GPR119 agonists; however, many key questions remain to be elucidated. Some work has appeared that describes the promising effects of GPR119 agonists on the pancreatic β-cell, and additional work is needed to support the hypothesis that these compounds might improve β-cell function over time. If a GPR119 agonist is capable of treating β-cell degeneration, it may hold promise for delaying the progression of the pre-diabetic state or T2D in patients. Additionally, further studies are necessary to better understand GPR119 distribution and function within the pancreatic islets and between species. Finally, further studies on the biology of GPR119 agonists on the enteroendocrine system are also warranted in order to gain a better understanding of the gut effects of GPR119 agonists. An improved overall understanding of GPR119 biology and species differences will lead to an improved ability to design studies in preclinical species and in humans that can ultimately reveal the promise of this mechanism for the treatment of T2D.

Notes

Since the completion of this manuscript, valuable information regarding the structure of APD597 (JNJ38431055) and its clinical data have been published.[123–125]

Acknowledgment

The authors thank Dr Timothy Kowalski for his valuable input in the preparation of this chapter.

References

1. E. A. Nyenwe, T. W. Jerkins, G. E. Umpierrez and A. E. Kitabchi, *Metab. Clin. Exp.*, 2011, **60**, 1.
2. S. S. Schwartz and B. A. Kohl, *Mayo Clin. Proc.*, 2010, **85**(Suppl.), S15.
3. W. Kim and J. M. Egan, *Pharmacol. Rev.*, 2008, **60**, 460.
4. V. Vangaveti, V. Shashidhar, G. Jarrod, B. T. Baune and R. L. Kennedy, *Ther. Adv. Endocrinol. Metab.*, 2010, **1**, 165.
5. S. Dhayal and N. G. Morgan, *Drug News Perspectives*, 2010, **23**, 418.
6. R. Fredriksson, P. J. Höglund, D. E. Gloriam, M. C. Lagerström and H. B. Schiöth, *FEBS Lett.*, 2003, **554**, 381.
7. S. Costanzi, S. Neumann and M. C. Gershengorn, *J. Biol. Chem.*, 2008, **283**, 16269.
8. J. Brown, *Br. J. Pharmacol.*, 2007, **152**, 567.
9. H. A. Overton, M. C. Fyfe and C. Reynet, *Br. J. Pharmacol.*, 2008, **153** (Suppl. 1), S76.
10. J. A. Bonini, B. E. Borowsky, N. Adham, N. Boyle and T. O. Thompson (Synaptic Pharmaceutical Corp.), *US Pat.*, 06221660, 2001.
11. J. A. Bonini, B. E. Borowsky, N. Adham, N. Boyle and T, O. Thompson (Synaptic Pharmaceutical Corp.), *US Pat. Appl. Publ.*, 20030125539, 2003.
12. G. Griffin (Prosidion), *US Pat.*, 07083933, 2006.
13. R. M. Jones, G. Semple, B. Fioravanti, G. Pereira, I. Calderon, J. Uy, K. Duvvuri, J. S. K. Choi, Y. Xiong. and V. Dave (Arena Pharmaceuticals Inc.), *WO Pat. Appl.*, 2004065380, 2004.
14. Z.-L. Chu, R. M. Jones, H. He, C. Carroll, V. Gutierrez, A. Lucman, M. Moloney, H. Gao, H. Mondala, D. Bagnol, D. Unett, Y. Liang, K. Demarest, G. Semple, D. P. Behan and J. Leonard, *Endocrinology*, 2007, **148**, 2601.
15. T. Soga, T. Ohishi, T. Matsui, T. Saito, M. Matsumoto, J. Takasaki, S. Matsumoto, M. Kamohara, H. Hiyama, S. Yoshida, K. Momose, Y. Ueda, H. Matsushime, M. Kobori and K. Furuichi, *Biochem. Biophys. Res. Commun.*, 2005, **326**, 744.
16. H. Lan, G. Vassileva, A. Corona, L. Liu, H. Baker, A. Golovko, S. J. Abbondanzo, W. Hu, S. Yang, Y. Ning, R. A. Del Vecchio, F. Poulet, M. Laverty, E. Gustafson, J. Hedrick and T. Kowalski, *J. Endocrinol.*, 2009, **201**, 219.
17. N. Whalley, L. Pritchard, D. M. Smith and A. White, *J. Endocrinol.*, 2011, **211**, 99.
18. J. E. Ayala, D. K. Croom, A. J. Carpenter, D. J. N. Nunez, A. A. Young, P. L. Feldman, F. D. James, D. P. Bracy, W. H. Mayes, D. H. Wasserman and K. K. Brown, *American Diabetes Association 70th Annual Scientific Sessions*, Orlando, FL, USA, 25–29 June 2010, Abstract 1391-P.
19. Y. Sakamoto, H. Inoue, S. Kawakami, K. Miyawaki, T. Miyamoto, K. Mizuta and M. Itakura, *Biochem. Biophys. Res. Commun.*, 2006, **351**, 474.

20. Z.-L. Chu, C. Carroll, J. Alfonso, V. Gutierrez, H. He, A. Lucman, M. Pedraza, H. Mondala, H. Gao, D. Bagnol, R. Chen, R. M. Jones, D. P. Behan and J. Leonard, *Endocrinology*, 2008, **149**, 2038.

21. H. E. Parker, A. M. Habib, G. J. Rogers, F. M. Gribble and F. Reimann, *Diabetologia*, 2009, **52**, 289.

22. H. A. Overton, A. J. Babbs, S. M. Doel, M. C. Fyfe, L. S. Gardner, G. Griffin, H. C. Jackson, M. J. Procter, C. M. Rasamison, M. Tang-Christensen, P. S. Widdowson, G. M. Williams and C. Reynet, *Cell Metab.*, 2006, **3**, 167.

23. Z.-L. Chu, C. Carroll, R. Chen, J. Alfonso, V. Gutierrez, H. He, A. Lucman, C. Xing, K. Sebring, J. Zhou, B. Wagner, D. Unett, R. M. Jones, D. P. Behan and J. Leonard, *Mol. Endocrinol.*, 2010, **24**, 161.

24. L. M. Lauffer, R. Iakoubov and P. L. Brubaker, *Diabetes*, 2009, **58**, 1058.

25. E. Kelly, C. P. Bailey and G. Henderson, *Br. J. Pharmacol.*, 2008, **153** (Suppl. 1), S379.

26. J. Fu, S. Gaetani, S. F. Oveisi, J. Lo Verme, A. Serrano, F. Rodríguez De Fonseca, A. Rosengarth, A. H. Luecke, H. B. Di Giacomo, G. Tarzia and D. Piomelli, *Nature*, 2003, **425**, 90.

27. K. B. Hansen, M. M. Rosenkilde, F. K. Knop, N. Wellner, T. A. Diep, J. F. Rehfeld, U. B. Andersen, J. J. Holst and H. S. Hansen, *J. Clin. Endocrinol. Metab.*, 2011, **96**, E1409.

28. G. Semple, B. Fioravanti, G. Pereira, I. Calderon, J. Uy, K. Choi, Y. Xiong, A. Ren, M. Morgan, V. Dave, W. Thomsen, D. J. Unett, C. Xing, S. Bossie, C. Carroll, Z. L. Chu, A. J. Grottick, E. K. Hauser, J. Leonard and R. M. Jones, *J. Med. Chem.*, 2008, **51**, 5172.

29. R. Göke, H. C. Fehmann, T. Linn, H. Schmidt, M. Krause, J. Eng and B. Göke, *J. Biol. Chem.*, 1993, **268**, 19650.

30. F. Kolligs, H. C. Fehmann, R. Göke and B. Göke, *Diabetes*, 1995, **44**, 16.

31. R. M. Jones, *232nd American Chemical Society National Meeting*, San Francisco, CA, USA, 10–14 September 2006, MEDI 275.

32. M. C. T. Fyfe and P. Widdowson (Prosidion), *WO Pat. Appl.*, 2007138362, 2007.

33. M. C. T. Fyfe, A. J. Babbs, L. S. Bertram, S. E. Bradley, S. M. Doel, S. Gadher, W. T. Gattrell, R. P. Jeevaratnam, J. F. Keily, J. G. McCormack, H. A. Overton, C. M. Rasamison, C. Reynet, P. J. Rushworth, C. P. Sambrook Smith, V. K. Shah, D. F. Stonehouse, S. A. Swain, J. R. White, P. S. Widdowson, G. M. Williams and M. J. Procter, *234th American Chemical Society National Meeting*, Boston, MA, USA, 19–23 August 2007, MEDI 062.

34. M. C. T. Fyfe, A. J. Babbs, L. S. Bertram, S. E. Bradley, S. M. Doel, S. Gadher, W. T. Gattrell, R. P. Jeevaratnam, J. F. Keily, J. G. McCormack, H. A. Overton, C. M. Rasamison, C. Reynet, P. J. Rushworth, C. P. Sambrook Smith, V. K. Shah, D. F. Stonehouse, S. A. Swain, J. R. White, P. S. Widdowson, G. M. Williams and M. J. Procter,

    236th American Chemical Society National Meeting, Philadelphia, PA, USA, 17–21 August 2008, MEDI 197.

35. U. Shah and T. J. Kowalski, *Vitam. Horm.*, 2010, **84**, 415.

36. U. Shah, *Curr. Opin. Drug Discov. Devel.*, 2009, **12**, 519.

37. R. M. Jones, J. N. Leonard, D. J. Buzard and J. Lehmann, *Expert Opin. Ther. Pat.*, 2009, **19**, 1339.

38. R. M. Jones and J. N. Leonard, *Annu. Rep. Med. Chem.*, 2009, **44**, 149.

39. G. Semple, A. Ren, B. Fioravanti, G. Pereira, I. Calderon, K. Choi, Y. Xiong, Y. J. Shin, T. Gharbaoui, C. R. Sage, M. Morgan, C. Xing, Z.-L. Chu, J. N. Leonard, A. J. Grottick, H. Al-Shamma, Y. Liang, K. T. Demarest and R. M. Jones, *Bioorg. Med. Chem. Lett.*, 2011, **21**, 3134.

40. T. Gharbaoui, D. Sengupta, E. A. Lally, N. S. Kato, M. Carlos and N. Rodriguez (Arena Pharmaceuticals Inc.), *US Pat. Appl.*, 2006154940, 2006.

41. R. M. Jones and J. Lehmann (Arena Pharmaceuticals Inc.), *WO Pat. Appl.*, 2008005576, 2008.

42. R. M. Jones, J. Lehmann and A. Siu-Ting Wong (Arena Pharmaceuticals Inc.), *WO Pat. Appl.*, 2008005569, 2008.

43. A. S. Wong, J. Lehmann, I. Calderon, G. Semple, Y. Xiong, Y.-J. Shin, A. S. Ren, K. J. S. Choi, T. Gharbaoui, M. Morgan, A. J. Grottick, D. J. Unett, Z.-L. Chu, H. Al-Shamma, J. N. Leonard and R. M. Jones, *236th American Chemical Society National Meeting*, Philadelphia, PA, USA, 17–21 August 2008, MEDI 116.

44. R. M. Jones and J. Lehmann (Arena Pharmaceuticals Inc.), *WO Pat. Appl.*, 2011005929, 2011.

45. R. M. Jones, G. Semple, Y. Xiong, Y.-J, Shin, A. S. Ren, J. Lehmann, B. Fioravanti, M. A. Bruce and J. S. K. Choi (Arena Pharmaceuticals Inc.), *WO Pat. Appl.*, 2005121121, 2005.

46. R. M. Jones, G. Semple, Y. Xiong, Y.-J. Shin, A. S. Ren, I. Calderon, J. S. K. Choi, B. Fioravanti, J. Lehmann and M. A. Bruce (Arena Pharmaceuticals Inc.), *WO Pat. Appl.*, 2005007647, 2005.

47. R. M. Jones and J. Lehmann (Arena Pharmaceuticals Inc.), *WO Pat. Appl.*, 2007035355, 2007.

48. M. C. T. Fyfe, J. White, P. Widdowson, H. A. Overton and C. Reynet, *American Diabetes Association 67th Annual Scientific Sessions*, Chicago, IL, USA, 22–26 June 2007, Abstract 0532-P.

49. M. C. T. Fyfe, J. White, P. Widdowson, H. A. Overton and C. Reynet, *Diabetes*, 2007, **56** (Suppl. 1), 532.

50. L. S. Bertram, M. C. T. Fyfe, R. P. Jeevaratnam, J. Keily, T. M. Krulle, C. M. Rasamison, C. P. Sambrook-Smith and S. A. Swain (Prosidion), *WO Pat. Appl.*, 2010004343, 2010.

51. L. S. Bertram, M. C. T. Fyfe, R. P. Jeevaratnam and J. Keily (Prosidion), *WO Pat. Appl.*, 2010004344, 2010.

52. L. S. Bertram and M. C. T. Fyfe (Prosidion), *WO Pat. Appl.*, 2010004346, 2010.

53. L. S. Bertram, P. G. Clarke, G. J. Dawson, P. T. Fry, M. C. T. Fyfe, W. Gattrell, R. P. Jeevaratnam, J. Keily, T. M. Krulle, M. J. Procter, C. M. Rasamison, C. P. Sambrook-Smith and S. A. Swain (Prosidion), *WO Pat. Appl.*, 2010004347, 2010.
54. L. S. Bertram, M. C. T. Fyfe, W. Gattrell, R. P. Jeevaratnam, J. Keily and M. J. Procter (Prosidion), *WO Pat. Appl.*, 2010004348, 2010.
55. L. S. Bertram and M. C. T. Fyfe (Prosidion), *WO Pat. Appl.*, 2010004345, 2010.
56. M. C. T. Fyfe, W. Gattrell, C. P. Sambrook-Smith and S. A. Swain (Prosidion), *WO Pat. Appl.*, 2009050522, 2009.
57. M. C. T. Fyfe, W. Gattrell and C. P. Sambrook-Smith (Prosidion), *WO Pat. Appl.*, 2009050523, 2009.
58. L. Elster, T. Hoegberg, A. Murray and J. M. Receveur (Prosidion), *WO Pat. Appl.*, 2010001166, 2010.
59. O. Barba, P. T. Fry, M. C. T. Fyfe, W. Gattrell, R. P. Jeevaratnam, T. M. Krulle, M. J. Procter, C. P. Sambrook-Smith, K. L. Schofield, D. Smyth, A. J. W. Stewart, D. F. Stonehouse and S. A. Swain (Prosidion), *WO Pat. Appl.*, 2010103333, 2010.
60. O. Barba, T. B. Dupree, P. T. Fry, M. C. T. Fyfe, R. P. Jeevaratnam, T. M. Krulle, K. L. Schofield, D. Smyth, T. Staroske, A. J. W. Stewart, D. F. Stonehouse, S. A. Swain and D. M. Withall (Prosidion), *WO Pat. Appl.*, 2010103334, 2010.
61. O. Barba, S. H. Davis, M. C. T. Fyfe, R. P. Jeevaratnam, K. L. Schofield, T. Staroske, A. J. W. Stewart, S. A. Swain and D. M. Withall (Prosidion), *WO Pat. Appl.*, 2010103335, 2010.
62. X. Chen, P. Cheng, L. E. Clemens, J. D. Johnson, J. Ma, A. Murphy, I. Nashashibi, C. J. Rabbat, J. Song, M. E. Wilson, Y. Zhu and Z. Zhao (Metabolex Inc.), *WO Pat. Appl.*, 2008083238, 2008.
63. X. Chen, P. Cheng, L. E. Clemens, J. D. Johnson, J. Ma, A. Murphy, I. Nashashibi, C. J. Rabbat, J. Song, M. E. Wilson, Y. Zhu and Z. Zhao (Metabolex Inc.), *US Pat. Appl.*, 2009054475, 2009.
64. M. E. Wilson, J. Johnson, L. E. Clemens, Z. Zhao and X. Chen (Metabolex Inc.), *WO Pat. Appl.*, 2009123992, 2009.
65. J. Song, J. Ma, C. J. Rabbat, I. Nashashibi, X. Chen and Z. Zhao (Metabolex Inc.), *WO Pat. Appl.*, 2009008739, 2009.
66. B. Roberts, F. M. Gregoire, D. B. Karpf, E. Clemens, B. Lavan, J. Johnson, C. A. Mcwherter, R. Martin and M. Wilson, *American Diabetes Association 69th Annual Scientific Sessions*, New Orleans, LA, USA, 5–9 June 2009, Abstract 164-OR.
67. C. McWherter, *32nd Annual National Medicinal Chemistry Symposium*, Minneapolis, MN, USA, 6–9 June 2010.
68. Metabolex Inc. Press Release, June 25, 2010, Hayward, CA, http://www.metabolex.com/news/jun252010.html; last accessed August 18, 2011.

69. C. Ammala and C. Briscoe (Smithkline Beecham Corp.), *WO Pat. Appl.*, 2008008895, 2008.

70. S. R. Katamreddy, R. D. Caldwell, D. Heyer, V. Samano, J. B. Thompson, A. J. Carpenter, C. R. Conlee, E. E. Boros and B. D. Thompson (SmithKline Beecham Corp.), *WO Pat. Appl.*, 2008008887, 2008.

71. Y. Wu, J. D. Kuntz and A. J. Carpenter, *240th American Chemical Society National Meeting*, Boston, MA, USA, August 22–26, 2010, MEDI-203.

72. J. Fang, J. Tang, A. J. Carpenter, G. Peckham, C. R. Conlee, K. S. Du and S. R. Katamreddy (SmithKline Beecham Corp.), *WO Pat. Appl.*, 2008070692, 2008.

73. A. J. Carpenter, J. Fang and G. Peckham (SmithKline Beecham Corp.), *WO Pat. Appl.*, 2010014593, 2010.

74. A. J. Carpenter, *32nd Annual National Medicinal Chemistry Symposium*, Minneapolis, MN, USA, June 6–9th, 2010.

75. G. Peckham, C. Ammala, S. Bullard, A. Carpenter, B. Chauder, K. Du, J. Fang, S. Hart, S. Katamreddy, D. Minick, E. Pullen, S. Roller, H. Sauls and S. Tadepalli, *32nd Annual National Medicinal Chemistry Symposium*, Minneapolis, MN, USA, June 6–9th, 2010.

76. Y. Wu, J. D. Kuntz, A. J. Carpenter, J. Fang, H. R. Sauls, D. J. Gomez, C. Ammala, Y. Xu, S. Hart and S. Tadepalli, *Bioorg. Med. Chem. Lett.*, 2010, **20**, 2577.

77. J. W. Szewczyk, J. Acton, A. D. Adams, G. Chicchi, S. Freeman, A. D. Howard, Y. Huang, C. Li, P. T. Meinke, R. Mosely, E. Murphy, R. Samuel, C. Santini, M. Yang, Y. Zhang, K. Zhao and H. B. Wood, *Bioorg. Med. Chem. Lett.*, 2011, **21**, 2665.

78. H. B. Wood, A. D. Adams, J. W. Szewczyk, Y. Zhang and M. Yang (Merck), *WO Pat. Appl.*, 2011019538, 2011.

79. C. D. Boyle, S. F. Neelamkavil, S. Chackalamannil, B. R. Neustadt, J. Hao, U. Shah, J. Harris, H. Liu and A. W. Stamford (Schering Corp.), *WO Pat. Appl.*, 2008130581, 2008.

80. C. D. Boyle, S. Chackalamannil, C. M. Lankin, U. G. Shah, B. R. Neustadt, H. Liu and A. W. Stamford (Schering Corp.), *WO Pat. Appl.*, 2008130584, 2008.

81. S. F. Neelamkavil, C. D. Boyle, S. Chackalamannil, A. W. Stamford, W. J. Greenlee, B. R. Neustadt, J. Hao, B. Hawes, K. O'Neill, H. Baker and T. J. Kowalski, *239th American Chemical Society National Meeting*, San Francisco, CA, USA, 21–25 March 2010, MEDI-182.

82. J. M. Harris, S. F. Neelamkavil, B. R. Neustadt, C. D. Boyle, H. Liu, J. Hao, A. Stamford, S. Chackalamannil and W. J. Greenlee (Schering Corp.), *WO Pat. Appl.*, 2009143049, 2009.

83. Y. Xia, S. Chackalamannil, W. J. Greenlee, C. Jayne, B. Neustadt, A. Stamford, H. Vaccaro, X. K. Xu, H. Baker, K. O'Neill, M. Woods, B. Hawes and T. Kowalski, *Bioorg. Med. Chem. Lett.*, 2011, **21**, 3290.

84. S. F. Neelamkavil, C. D. Boyle, J. M. Harris, A. W. Stamford, J. Hao, B. R. Neustadt, S. Chackalamannil, Y. Xia and W. J. Greenlee (Schering Corp.), *WO Pat. Appl.*, 2010009207, 2010.
85. S. F. Neelamkavil, C. D. Boyle, S. Chackalamannil and W. J. Greenlee (Schering Corp.), *WO Pat. Appl.*, 2010009195, 2010.
86. Y. Xia, C. D. Boyle, W. J. Greenlee, S. Chackalamannil, C. L. Jayne, A. W. Stamford, X. Dai, J. M. Harris, B. R. Neustadt, S. F. Neelamkavil, U. G. Shah, C. M. Lankin and H. Liu (Schering Corp.), *WO Pat. Appl.*, 2009055331, 2009.
87. J. M. Harris, A. Stamford, W. J. Greenlee and S. F. Neelamkavil (Schering Corp.), *WO Pat. Appl.*, 2011053688, 2011.
88. U. Shah, C. D. Boyle, S. Chackalamannil, B. R. Neustadt, J. M. Harris, A. Stamford and W. J. Greenlee (Schering Corp.), *WO Pat. Appl.*, 2010075273, 2010.
89. M. Azimioara, C. Cow, R. Epple, S. Jiang, G. Lelais, D. Mutnick and B. Wu (IRM LLC), *WO Pat. Appl.*, 2009105717, 2009.
90. P. Alper, M. Azimioara, C. Cow, R. Epple, S. Jiang, G. Lelais, P.-Y. Michellys, T. N. Nguyen, L. Westscott-Baker and B. Wu (IRM LLC), *WO Pat. Appl.*, 2008097428, 2008.
91. P. Alper, M. Azimioara, C. Cow, R. Epple, S. Jiang, G. Lelais, P.-Y. Michellys, D. Mutnick, V. Nikulin and L. Westcott-Baker (IRM LLC), *WO Pat. Appl.*, 2009038974, 2009.
92. C. Cow and R. Epple (IRM LLC), *WO Pat. Appl.*, 2009126535, 2009.
93. M. Azimioara, C. Cow, R. Epple, G. Lelais, J. Mecom and V. Nikulin (IRM LLC), *WO Pat. Appl.*, 2011044001, 2011.
94. P. B. Alper, C. Cow, R. Epple, P.-Y. Michellys and D. Mutnick (IRM LLC), *WO Pat. Appl.*, 2009105715, 2009.
95. P. B. Alper, R. Epple, P.-Y. Michellys, D. Mutnick, V. Nikulin, H. Petrassi and J. Michael (IRM LLC), *WO Pat. Appl.*, 2009105722, 2009.
96. P. Alper, M. Azimioara, C. Cow, R. Epple, G. Lelais, D. Mutnick and V. Nikulin (IRM LLC), *WO Pat. Appl.*, 2011014520, 2011.
97. J. M. Fevig and D. A. Wacker (Bristol Myers Squibb), *WO Pat. Appl.*, 2008137435, 2008.
98. J. M. Fevig and D. A. Wacker (Bristol Myers Squibb), *WO Pat. Appl.*, 2008137436, 2008.
99. D. A. Wacker, K. A. Rossi and Y. Wang (Bristol Myers Squibb), *WO Pat. Appl.*, 2009012275, 2009.
100. D. A. Wacker, K. A. Rossi and Y. Wang (Bristol Myers Squibb), *WO Pat. Appl.*, 2009012277, 2009.
101. D. A. Wacker, K. A. Rossi, Y. Wang and G. Wu (Bristol Myers Squibb), *WO Pat. Appl.*, 2010009183, 2010.
102. B. Dyck, J. A. Tran, J. Tamiya, F. Jovic, T. Vickers, C. Chen, N. Harriott, T. Coon and N. J. Ashweek (Boehringer Ingelheim), *WO Pat. Appl.*, 2010149684, 2010.

103. J. A. Tran and C. Chen (Boehringer Ingelheim), *WO Pat. Appl.*, 2010149685, 2010.
104. K. F. McClure, E. Darout, C. R. Guimarães, M. P. DeNinno, V. Mascitti, M. J. Munchhof, R. P. Robinson, J. Kohrt, A. R. Harris, D. E. Moore, B. Li, L. Samp, B. A. Lefker, K. Futatsugi, D. Kung, P. D. Bonin, P. Cornelius, R. Wang, E. Salter, S. Hornby, A. S. Kalgutkar and Y. Chen, *J. Med. Chem.*, 2011, **54**, 1948.
105. M. P. Deninno, K. Futatsugi, B. A. Lefker, V. Mascitti, K. F. McClure and M. J. Munchhof (Pfizer Inc.), *WO Pat. Appl.*, 2010106457, 2010.
106. V. Mascitti, K. F. McClure, M. JMunchhof and R. P. Robinson (Pfizer Inc.), *WO Pat. Appl.*, 2010140092, 2010.
107. E. Darout, M. P. Deninno, K. Futatsugi, C. R. W. Guimaraes, B. A. Lefker, V. Mascitti, K. F. Mcclure, M. J. Munchhof and R. P. Robinson (Pfizer Inc.), *WO Pat. Appl.*, 2010128425, 2010.
108. S. Yoshida, T. Ohishi, T. Matsui and M. Shibasaki, *Biochem. Biophys. Res. Commun.*, 2010, **400**, 437.
109. S. Yoshida, T. Ohishi, T. Matsui, H. Tanaka, H. Oshima, Y. Yonetoku and M. Shibasaki, *Biochem. Biophys. Res. Commun.*, 2010, **402**, 280.
110. S. Yoshida, T. Ohishi, T. Matsui, H. Tanaka, H. Oshima, Y. Yonetoku and M. Shibasaki, *Diab. Obesity Metab.*, 2011, **13**, 34.
111. S. Yoshida, H. Tanaka, H. Oshima, T. Yamazaki, Y. Yonetoku, T. Ohishi, T. Matsui and M. Shibasaki, *Biochem. Biophys. Res. Commun.*, 2010, **400**, 745.
112. H. Pingali, M. R. Jain and P. Zaware (Cadila Healthcare Ltd.), *WO Pat. Appl.*, 2010084512, 2010.
113. H. Pingali, P. Zaware and M. R. Jain (Cadila Healthcare Ltd.), *WO Pat. Appl.*, 2010146605, 2010.
114. Arena Pharmaceuticals Inc., Press Release, January 07, 2008, San Diego, CA, http://invest.arenapharm.com/releases.cfm; last accessed August 18, 2011.
115. Arena Pharmaceuticals Inc., Press Release, December 15, 2008, San Diego, CA, http://invest.arenapharm.com/releases.cfm; last accessed August 18, 2011.
116. Arena Pharmaceuticals Inc., Press Release, November 04, 2010, San Diego, CA, http://invest.arenapharm.com/releases.cfm; last accessed August 18, 2011.
117. M. Fyfe, J. Mccormack, H. Overton, M. Procter and C. Reynet, *American Diabetes Association 68th Annual Scientific Sessions*, San Francisco, CA, USA, 6–10 June 2008, Abstract 297-OR.
118. M. L. Goodman, J. Dow, A. A. van Vliet, S. Hadi, D. Karbiche and J. A. Lockton, *American Diabetes Association 71st Annual Scientific Sessions*, San Diego, CA, USA., 24–28th June 2011, Abstract 0306-OR.
119. M. L. Goodman, J. Dow, A. A. van Vliet, A. Pleszko and J. A. Lockton, *47th European Association for the Study of Diabetes Meeting*, Lisbon, Portugal, 12–16 Sept. 2011, Presentation 188.

120. D. J. Nunez, E. W. Lewis, S. Swan, M. A. Bush, C. Cannon, S. L. McMullen, D. A. Collins and P. L. Feldman, *American Diabetes Association 70th Annual Scientific Sessions*, Orlando, FL, USA, 25–29 June 2010, Abstract 80-OR.

121. D. J. Nunez, M. A. Bush, C. Cannon, D. A. Collins, S. L. McMullen, P. L. Feldman, T. D. Jackson and S. A. Ross, *American Diabetes Association 71st Annual Scientific Sessions*, San Diego, CA, USA, 24–28 June 2011, Abstract 0996-P.

122. W. T. Cefalu, *ILAR J.*, 2006, **47**, 186.

123. G. Semple, J. Lehmann, A. Wong, A. Ren, M. Bruce, Y. J. Shin, C. R. Sage, M. Morgan, W. C. Chen, K. Sebring, Z.-L. Chu, J. N. Leonard, H. Al-Shamma H, A. J. Grottick, F. Du, Y. Liang, K. Demarest and R. M. Jones, *Bioorg. Med. Chem. Lett.*, 2012, **22**, 1750.

124. L. B. Katz, J. J. Gambale, P. L. Rothenberg, S. R. Vanapalli, N. Vaccaro, L. Xi, D. C. Polidori, E. Vets, T. C. Sarich and P. P. Stein, *Clin. Pharmacol. Ther.*, 2011, **90**, 685.

125. L. B. Katz, J. J. Gambale, P. L. Rothenberg, S. R. Vanapalli, N. Vaccaro, L. Xi, T. C. Sarich and P. P. Stein, *Diabetes Obes. Metab.*, 2012, Epub ahead of print: http://www.ncbi.nlm.nih.gov/pubmed/22340428

Acyl-CoA: Diacylglycerol Acyltransferase-1 Inhibition as an Approach to the Treatment of Type 2 Diabetes

ROBERT L. DOW

Cardiovascular and Metabolic Diseases Chemistry, Pfizer Inc., 620 Memorial Drive, Cambridge, MA, USA, 02139
E-mail: robert.l.dow@pfizer.com

8.1 Introduction

Obesity and type 2 diabetes have been on a dramatic rise in Westernized countries in recent years, resulting in significant societal burdens. The increased prevalence of these conditions is largely driven by a net imbalance between energy intake and expenditure. This imbalance leads to increased fat disposition resulting from storage of triglycerides (TG). Triglyceride stores are ultimately converted to fatty acids and glycerol via a series of enzymatically driven deacylation reactions. These components are then available for transport to other tissues where they are utilized for energy production and the synthesis of other lipid types for storage.[1] The resulting increases in TG stores in peripheral tissues, especially skeletal muscle, have been implicated in the development of insulin resistance[2] and the constellation of lipotoxic disorders associated with metabolic syndrome X.[3] Intracellular concentrations of lipids in muscle have been shown to be tightly negatively correlated with

RSC Drug Discovery Series No. 27
New Therapeutic Strategies for Type 2 Diabetes: Small Molecule Approaches
Edited by Robert M. Jones
© The Royal Society of Chemistry 2012
Published by the Royal Society of Chemistry, www.rsc.org

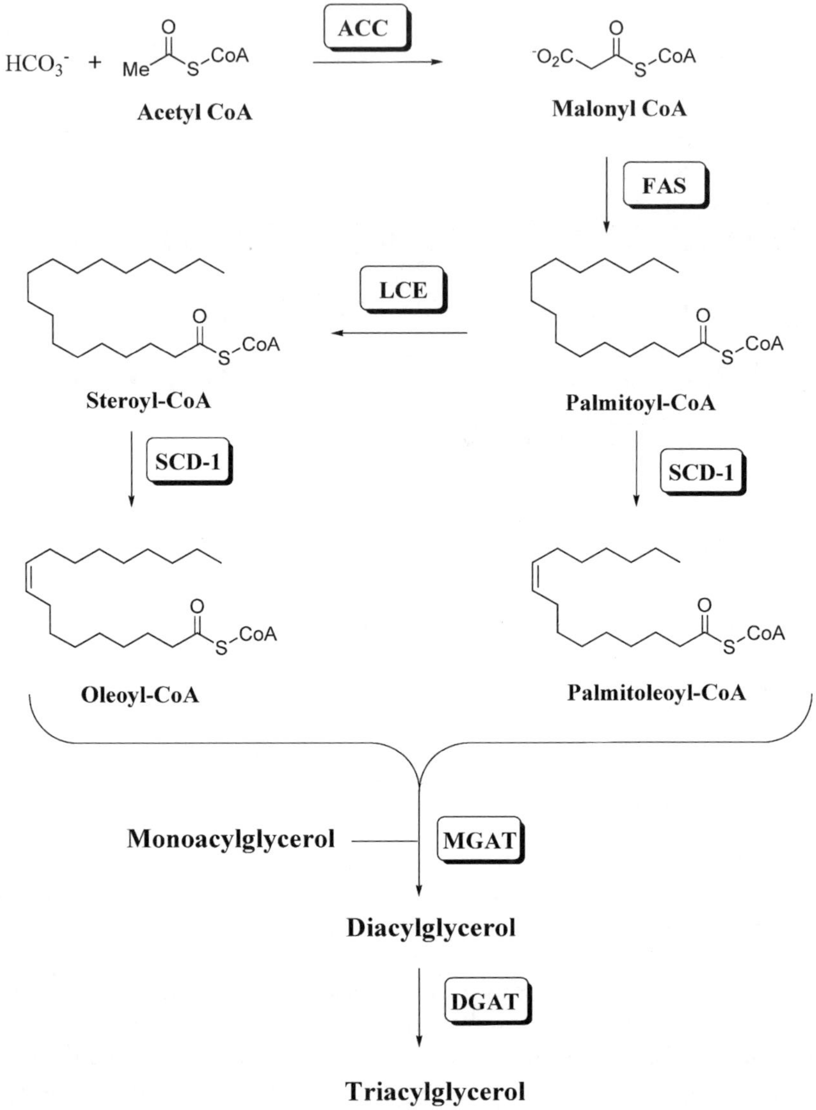

Figure 8.1 Monoacylglyceride-based triglyceride synthesis pathway. ACC = acetyl-CoA carboxylase, FAS = fatty acid synthase, LCE = long chain elongase, SCD-1 = steroyl-CoA desaturase, MGAT = acyl-CoA:monoacylglycerol acyltransferase-1, DGAT = acyl-CoA:diacylglycerol acyltransferase.

insulin-stimulated glucose disposal in healthy males.[4] This relationship is also found in humans with impaired glucose tolerance and type 2 diabetes.[5] Intramyocellular concentrations of lipid are more highly correlated with these disease states relative to a variety of other parameters such as body mass index, waist-to-hip ratio, or total body adiposity.

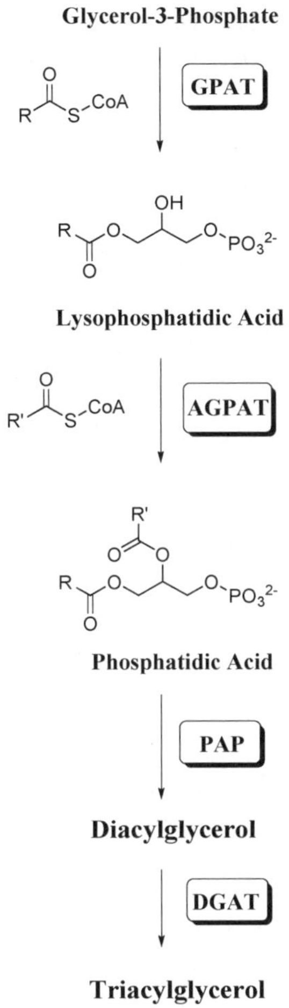

Glycerol-3-Phosphate

GPAT

Lysophosphatidic Acid

AGPAT

Phosphatidic Acid

PAP

Diacylglycerol

DGAT

Triacylglycerol

Figure 8.2 Glycerol phosphate-based triglyceride synthesis pathway. GPAT = glycerol 3-phosphate acyltransferase, AGPAT = 1-acyl-*sn*-glycerol-3-phosphate, PAP = phosphatidate phosphatase, DGAT = acyl-CoA:diacylglycerol acyltransferase.

Another line of evidence supporting the lipotoxicity hypothesis is the metabolic state of people suffering from the genetic disorders leading to lipodystrophy.[6] Lipodystrophy is a medical condition associated with the absence of fat tissue; however, it is also characterized by excessive accumulation of lipids in tissues such as skeletal muscle and the liver. This aberrant distribution of lipids leads to an extreme insulin-resistant state. In animal models of lipodystrophy, treatment with the adipose regulatory hormone leptin reduces intracellular lipid levels in skeletal muscle and other tissues, leading to improved insulin sensitivity.[7,8]

Triglyceride biosynthesis and the resulting TG burden on tissues are largely controlled by two major pathways in humans. The monoacylglyceride pathway

is typically operative in tissues where dietary monoacylglycerides are re-
esterified, such as small intestine, liver, and adipose.[9] Fatty acids that feed into
this pathway can come from dietary absorption or via *de novo* fatty acid
synthesis from acetyl CoA, via a series of enzyme-catalyzed homologation
reactions (Figure 8.1). The second pathway, found in most cell types, is the
glycerol phosphate pathway (Figure 8.2), which sequentially adds two fatty
acyl chains to glycerol-3-phosphate generating a phosphatidic acid intermediate that is subsequently converted to TG.[10] Both of these pathways
converge at the intermediate diacylglycerols (DAG) which are then converted
to triglycerides through acylation by a fatty acid acyl-CoA catalyzed by the
acyl-CoA:diacylglycerol acyltransferases (DGAT) and to a lesser extent by
acyl-CoA:monoacylglycerol transferases (MGAT). The former set of enzymes
is composed of DGAT-1 and DGAT-2, which while carrying out the same
chemical transformation, arise from distinct gene families.

Elucidation of the lipid biosynthetic pathways and the characterization of
the enzymes involved presents multiple entry points for inhibiting lipid
biosynthesis.[11] The enzymes involved in these pathways have attracted
substantial interest as potential targets for the treatment of obesity and
diabetes.[12,13] Of these approaches, inhibition of acyl-CoA:diacylglycerol
acyltransferase-1 (DGAT-1) appears to be the most advanced with compounds
in phase I/II human clinical trials.

This chapter begins with a discussion of the molecular and pharmacological
characterization of the DGAT family of enzymes. This leads into a discussion
of the role DGAT-1 plays in the etiology of obesity and diabetes. The author
then summarizes the various structural classes of small molecule inhibitors of
DGAT-1, with a focus on both *in vitro* and *in vivo* preclinical pharmacology.
The chapter concludes with a summary of the clinical status of this class of
lipid synthesis inhibitors.

8.2   Characterization of DGAT Enzymes

8.2.1   DGAT Enzymatic Activity

Acyl-CoA:diacylglycerol acyltransferase catalyzes the final step in triglyceride
biosynthesis, the acylation of a DAG by an acyl-CoA.[14] DGAT is the only
enzyme in the glycerolipid pathway that is unique to TG synthesis and is the
rate-limiting step in the synthesis of neutral lipids.[15] The first demonstration of
DGAT activity was reported around 50 years ago, utilizing a chicken liver
membrane preparation.[16] Because DGAT activity is associated with microsomal membranes, purification of enzymatically active protein has been
challenging. Purification to near homogeneity has been achieved utilizing rat
liver microsomal preparations.[17,18] These efforts resulted in the isolation of a
protein with high specific DGAT activity. DGAT enzymatic activity is
associated with the endoplasmic recticulum (ER) and is localized on the
cytoplasmic side of the ER membrane.[19]

DGAT can accept a wide range of acyl-CoA and diacylglycerol substrates in the synthesis of TG.[20] Both 1,2- and 1,3-DAG can serve as substrates for DGAT. Studies looking at the stereospecificity of DGAT-catalyzed acylation have revealed little discrimination between DAG enantiomers.[21]

8.2.2 Cloning of DGAT Enzymes

As part of an effort to identify additional members of the acyl-CoA:cholesterol acyltransferase (ACAT) enzyme family, two research groups independently carried out cDNA homology searches against a carboxyl terminus sequence of ACAT.[22,23] One of the hits from this exercise was a cDNA encoding a protein with 22% sequence identity to ACAT. In the course of examining the cholesterol esterase activity of this new protein it was found that it induced substantial increases in triglyceride content but had no effect on cholesterol ester levels.[22] Consistent with this substrate profile was the finding that the protein possesses a diacylglycerol binding site.[23] Based on this set of data it was concluded that this protein was DGAT (ultimately renamed DGAT-1).

Following this discovery, a DGAT-1 deficient (DGAT-1$^{-/-}$) strain of mice was generated and these animals were found to have reduced body fat and were resistant to diet-induced obesity (DIO), but paradoxically had normal plasma and adipose tissue levels of TG.[24] This suggested that an alternative mechanism of TG synthesis was operative in these mice. Around the same time investigators had identified a DGAT enzyme from a fungus species which shared no sequence homology with previously reported DGATs from a range of plant and mammalian sources.[25] Searches of the expressed sequence tag database within GenBank® identified a new candidate gene (DGAT-2), which shares no sequence homology with the previously identified DGAT-1.[26] DGAT-2, like DGAT-1, possesses a putative DAG binding site and exhibits high substrate selectively against a range of fatty acyl acceptors. DGAT-1 differs from DGAT-2 in that it has been shown to process a range of acyl acceptors, such as retinol, long chain alcohols and monoacylglycerols, in addition to DAG.[27]

8.2.3 Structural Characterization of DGAT-1

The human DGAT-1 gene encodes for a 488-amino acid protein with a predicted mass of 55 kDa.[23] Based on hydropathy plots, this ER bound protein is predicted to have 6–12 trans-membrane domains.[22,23] A recent study utilizing a series of protease protection experiments found that the murine protein has three trans-membrane domains.[28] The vast majority of DGAT-1 is located on the ER lumen face with a single loop extending into/through the membrane near the middle of the protein and approximately 110 amino acids of the N-terminus extending through the ER and into the cytosol.

Employing a range of mapping techniques, the functional domains of DGAT-1 have been defined. For the other members of the ACAT family it has been demonstrated that the catalytic active site resides in the C-terminus.[29] For hACAT-1, a histidine residue (His-460) in the C-terminus has been implicated as a participant in the catalytic cycle. Consistent with this being a key domain for catalytic activity in DGAT-1 is the finding that the C-terminal 250 amino acids share the highest sequence identity with ACAT-1.[23] Mutation of the corresponding histidine residue in DGAT-1 leads to a loss of enzymatic activity.[28] Further support for the catalytic machinery of DGAT-1 residing in the C-terminal region comes from a C-truncated variant of DGAT-1 isolated from human fat cells.[30] Cloning and expression of this mutant lacking the first 101 amino acids of the C-terminal domain revealed this variant to be catalytically inactive.

Members of the ACAT family, including DGAT-1, have been shown to exist in oligomeric form within the ER lumen.[30,31] While well removed from the catalytic site, the N-terminus of these proteins does appear to play an important role in the expression of enzymatic competency via control of the self-association. DGAT-1 appears to reside in a tetrameric form based on multiple analyses.[28,30] When amino acids 2–84 of the N-terminus of DGAT-1 are deleted, the enzyme fails to establish itself in the homotetrameric state. An increase in the catalytic activity of the dimer, relative to the tetramer, suggests that DGAT-1 may be regulated via switching between the dimeric (active) and tetrameric (inactive) forms.[28]

8.3 Role of DGAT-1 in Tissue Physiology and Disease States

8.3.1 Tissue Distribution of DGAT Enzymes

DGAT-1 mRNA is expressed across a wide range of human tissues, with the highest levels in organs responsible for producing large amounts of TG, such as the small intestine, adipose tissue, liver, and mammary gland.[22] DGAT-2 is also expressed in most tissues in a pattern consistent with the output of TG.[26] Highest levels are found in the liver, adipose tissue, and mammary gland. DGAT-2 expression is important in neonatal development based on the non-viability of DGAT-2 deficient (DGAT-2$^{-/-}$) mice after a few hours of life. The difference in viability between DGAT-1$^{-/-}$ and DGAT-2$^{-/-}$ mice is due to a combination of nearly complete elimination of the TG fuel source in the liver and erosion of the skin barrier leading to dehydration in the latter strain.[32] This dramatic impact on key lipid stores observed in DGAT-2$^{-/-}$ animals has in part been responsible for a majority of subsequent drug discovery efforts being focused on DGAT-1, rather than DGAT-2, as a target for inhibition of TG synthesis.

8.3.2 Role of DGAT-1 in Intestine

The absorption of energy-dense triglycerides in the small intestine is a very efficient process, with a very low portion ($<5\%$) being excreted in feces.

Dietary fats are converted to monoacylglycerol (MAG) and free fatty acids (FFA) by the action of pancreatic lipase in the intestinal lumen. These products then pass into the absorptive cells of the small intestine, enterocytes, where they are repackaged into TG which are then distributed throughout the body via the lymphatic system. The repackaging of TG requires the stepwise esterification of a MAG with two FFA molecules.

Despite a substantial reduction in DGAT activity in intestinal tissues of DGAT-1$^{-/-}$ mice, there were no overt signs of lipid malabsorption.[24] This genetic ablation of DGAT-1 activity does not completely eliminate TG synthesis in these animals. It has subsequently been shown that a tetrad of enzymes consisting of DGAT-1, DGAT-2, MGAT-2[33] and MGAT-3[34] drive the reconversion of MAG to TG in the intestine. All of these enzymes are capable of independently converting both MAG to DAG and DAG to TG, suggesting a fair degree of substrate promiscuity at the active sites of these enzymes.[27] Based on kinetic analyses it has been shown that DGAT-1 affinities for both MAG and DAG are roughly comparable.[35] This is not the case for DGAT-2 where the relative affinity for DAG is much higher, resulting in preferential conversion of this intermediate to TG irrespective of MAG concentration. Utilizing the selective DGAT-1 inhibitor XP620 (Figure 8.3), the components of MAG versus DAG acylation have been dissected *in vitro*.[35,36] XP620 selectively inhibits the acylation of MAG by hDGAT-1 (IC$_{50}$ = 17 nM) relative to acylation of DAG (IC$_{50}$ = 1499 nM) and shows no inhibitory activity against DGAT-2, MGAT-2, or MGAT-3. In rat intestinal mucosa membranes and Caco-2 cells it was determined that a majority (76–89%) of TG synthesized from MAG was mediated by DGAT-1.

DGAT-1$^{-/-}$ mice on a high-fat diet accumulate neutral lipid droplets in their enterocytes, which are composed of both TG and DAG.[37] When intestinal-specific expression of DGAT-1 is re-established in these mice, the enterocyte lipid content is normalized.[38] These results suggest there is a transient pool of lipids in the enterocytes as TG are transported from the gastrointestinal tract to the plasma compartment. The buildup of intracellular triglyceride concentrations in DGAT-1$^{-/-}$ mice suggests that DGAT-1 plays a role in the repackaging of these lipids prior to incorporation into chylomicron particles and subsequent systemic distribution.

Bariatric surgery techniques currently in practice achieve their benefits through restriction and/or diversion of nutrients to more distal parts of the

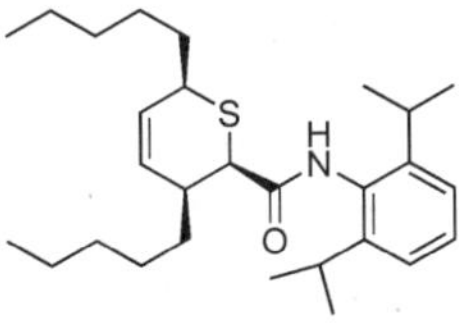

XP620

Figure 8.3   Structure of XP620.

intestinal tract.[39] These benefits not only include weight loss, but also remission of diabetes in a majority of patients. The dramatic and rapid reduction in the hyperglycemic state of these patients means these effects are not driven by weight loss, which occurs in a longer timeframe. It is believed that the improvement in the diabetic state is mainly driven by the release of gut hormones involved in carbohydrate metabolism.[40] In certain bariatric surgery patients there are increased levels of the gut hormone glucagon-like peptide (GLP-1), which has been demonstrated to have a positive effect on the glycemic state.[41] Concomitantly, many of these bariatric surgery patients also have reduced levels of another hormone, gastric inhibitory peptide (GIP). These lowered levels of GIP are consistent with the favorable weight and carbohydrate metabolic effects of synthetic GIP antagonists in preclinical animal models.[42] The altered TG partitioning observed in the intestinal tract of DGAT-1$^{-/-}$ mice and data showing that fatty acid sensors likely exist in the intestinal wall[43] raises the possibility that inhibition of DGAT-1 might alter gut hormone secretion. Interestingly, researchers at Janssen first noted that small molecule inhibitors of DGAT-1 could elevate levels of GLP-1 in dogs, though no specific results were provided.[44] Subsequently, a detailed evaluation of various gut hormone levels in the DGAT-1$^{-/-}$ mice has been carried out.[45] Following an oral triglyceride challenge in these mice, GLP-1 and peptide YY levels initially increased to levels comparable to those observed in the wild-type (WT) mice. However, levels of these two hormones in the DGAT-1$^{-/-}$ mice remained elevated 2 h post challenge, unlike the WT where concentrations of these peptides had returned to baseline. While an oral TG load in DGAT-1$^{-/-}$ mice increased GIP levels, the levels at 30 min and 2 h were significantly blunted relative to the WT controls. Taken together these effects on gut hormone levels in the DGAT-1$^{-/-}$ animals are strikingly similar to those observed in humans who have undergone bariatric surgery and support the potential of small molecule DGAT-1 inhibitors in the treatment of obesity and diabetes.

8.3.3　Role of DGAT-1 in Liver

A hallmark of nonalcoholic fatty liver disease (NAFLD) is excessive accumulation of lipids in the hepatocytes of people who consume little or no alcohol.[46] Chronic accumulation of lipids in the liver can lead to nonalcoholic steatosis (NASH), one of the more extreme forms of NAFLD, and is regarded as the major cause of cirrhosis of the liver.[47] DGAT-1 may play a role in driving NAFLD based on the findings that DGAT-1 mRNA levels are increased two-fold in these patients.[48] This potential link has been investigated in the preclinical setting utilizing both DGAT-1 knockout animals and the use of DGAT-1 antisense. Hepatic triglyceride levels in DGAT-1$^{-/-}$ mice were 80% lower than the wild-type littermates when fed a high-fat diet.[49] Fatty acid oxidation was also substantially increased in the KO animals, suggesting that inhibition of DGAT-1 in the liver diverts FA away from triglyceride synthesis in a high-fat diet setting. To separate the liver from the global effects of DGAT-1 knockdown, an adenovirus-mediated knockdown of DGAT-1 in the liver was carried out. Following 3 weeks on a high-fat diet these animals had 50% lower hepatic TG levels and visibly healthier livers relative to the parental strain. These results are in contradiction to earlier work utilizing liver-targeted over-expression or antisense oligonucleotides (ASO) knockdown of DGAT-1.[50–52] Results from the independent liver-targeted over-expression of DGAT-1 and DGAT-2 in mice suggests that DGAT-2 drives initial cytosolic lipid accumulation and then DGAT-1 plays a role in very low density lipoprotein assembly/secretion from the hepatocyte.[50] This functional role for DGAT-1 in hepatocytes is consistent with what has been observed in enterocytes of DGAT-1$^{-/-}$ mice.[37,38] Based on the results from hepatic over-expression of DGAT-1 and DGAT-2, it was concluded that the latter enzyme is the main driver of hepatic steatosis. In a high-fat fed rat model of NAFLD, liver-specific reduction of DGAT-1 with ASO did not impact the disease state in these animals; however, DGAT-2 ASO treatment did improve hepatic steatosis.[51] In an independent study of reduction of hepatic DGAT-1 activity via ASO in a mouse model of NASH revealed improvements in hepatic fibrosis but, however, no improvement in steatosis or inflammation.[52] In a third experiment employing DGAT-1 ASO the investigators found that treatment in mice did produce improvements in hepatic steatosis.[49] These studies differed from earlier work in that the mice had a 4-week pretreatment with the ASO prior to commencing the high-fat diet and the diet utilized also varied between the studies. Given the conflicting results and caveats associated with the above studies, investigations with small molecule inhibitors of DGAT-1 will be of high importance in sorting out its potential role in the treatment or prevention of NALFD. Towards this end, preliminary insights from studies with small molecule inhibitors are discussed in a later section of this chapter.

While global tissue knockout of DGAT-1 in normal and insulin-resistant strains of mice restores insulin and leptin sensitivity, it is unclear which tissue(s) are responsible for driving these improvements.[53] The changes in hepatic and extra-hepatic lipids driven by DGAT-1 suggest the liver may be an organ involved in driving insulin resistance in the diabetic state. One can draw a connection between the role of DGAT-1 in hepatic steatosis and the well-established relationship between this disease state and insulin resistance in humans.[54] Whether hepatic steatosis drives insulin resistance or the reverse is operative has yet to be determined. To address this question in the preclinical setting, mice over-expressing the DGAT-1 enzyme, specifically in the liver, were generated.[55] While hepatic DGAT-1 mRNA was increased 90-fold relative to the WT animals, there were no changes in their glucose levels or insulin tolerance. These results support the conclusion that DGAT-1 plays a role in hepatic steatosis, but that this organ disease state is not a major driver of the insulin resistance in this model.

8.3.4 Role of DGAT-1 in Adipose Tissue

The finding that excessive visceral fat is associated with an increased risk of cardiovascular disease, diabetes, dyslipidemia, and metabolic syndrome demonstrates that it is not simply an energy depot tissue.[56,57] On a mechanistic level it has been demonstrated that adipose tissue serves an important role in regulating energy and glucose homeostasis through secretion of a number of regulatory factors such as tumor necrosis factor-α (TNF-α), leptin, and adiponectin.[58] These relationships, combined with the high expression levels of DGAT-1 in white adipose tissue (WAT), suggest that this tissue may be an important driver of the enhanced leptin and insulin sensitivity in DGAT-1$^{-/-}$ mice.[53]

Tissue-specific over-expression of DGAT-1 in WAT of mice produces greater fat pad size and increased weight gain compared to the WT animals.[59] However, these animals do not have impaired glucose disposal, suggesting that deposition of TG in adipocytes is not responsible for insulin resistance. Consistent with this lack of correlation between obesity and insulin-resistant state via WAT over-expression of DGAT-1 was the absence of changes in the circulating levels of adiponectin, resistin, or TNF-α. WAT-specific up-regulation of DGAT-1 in a different strain of mice produces an insulin-resistant phenotype, which contrasts with the earlier study.[60] It was concluded that the differences in the backgrounds of the mice strains were responsible for these contradictory results. As in the earlier study there were no changes in adiponectin, resistin, or TNF-α, suggesting that an adipose-secreted factor is not responsible for the improved insulin sensitivity.

An alternative approach to defining the role of adipose-based DGAT-1 in the insulin-resistant state has been to transplant DGAT-1-deficient WAT into wild-type animals.[61] On a high-fat diet DGAT-1$^{+/+}$ mice implanted with fat from DGAT-1$^{-/-}$ mice have reduced body weight gain relative to WT mice implanted with fat from littermates. Increased energy expenditure appears to account for this differential weight gain. While increases in adiponectin were implicated in driving the increased energy expenditure, subsequent studies revealed that these two are not linked.[62] This suggests there are other, yet to be identified, biomolecules secreted by DGAT-1-deficient adipocytes that are responsible for driving enhanced insulin and glucose tolerance.

8.3.5   Role of DGAT-1 in Muscle

Muscle accounts for ∼40% of total body mass and is a major tissue of energy expenditure in humans.[63] It is responsible for approximately 80% of insulin-stimulated glucose disposal[64] and a major proportion of fatty acid (FA) uptake and oxidation.[65] Excess lipid accumulation in muscle is associated insulin resistance,[66] with acute elevations of FA and diacylglycerol being associated with this outcome.[67] Since DGAT-1 is an important regulatory element in establishing both the absolute and relative intracellular concentrations of TG, FA, and DAG in myocytes there has been considerable effort directed towards understanding the relationship between DGAT-1 activity and the insulin-resistant state in muscle tissue. Genetic models of DGAT-1 over-expression[68–72] or knockout[24,73,74] and pharmacologic inhibition of DGAT-1[74] have been utilized to gain insight as to whether inhibition of DGAT-1 will have a beneficial impact on insulin sensitization in muscle.

In contrast to the positive association between excess lipid accumulation and insulin resistance in muscle tissue, which is associated with obesity and type 2 diabetes, the opposite relationship occurs in athletes[75] or when individuals engage in a short-term exercise regime.[76] While exercise increases muscle fat stores there is an apparent contradictory improvement in insulin resistance post training, which is referred to as "the athlete's paradox".[75,77] Consistent with the increases in muscle TG levels induced by acute sessions of exercise are observations that DGAT-1 mRNA[78] and protein[79] expression are significantly increased. While TG levels are high in the myocytes of these individuals, the overall improvement in insulin sensitivity is driven by increased flux of fatty acids through the TG pathway. Transgenic over-expression of DGAT-1 in rodent models support these results observed in humans.[68] Selective over-expression of DGAT-1 in skeletal muscle of mice produced a 3-fold increase in DGAT activity, resulting in increased TG levels and improved insulin sensitivity. There is also increased energy expenditure in these animals, which supports the findings in humans of an increased flux of fatty acids.[69] Increased flux through this pathway is also consistent with the paradoxical finding that over-expression of DGAT-1 in rat muscle leads to high levels of DAG which should be lipotoxic in this tissue rather than having the observed insulin-sensitizing effect.[70] Also, heart-specific over-expression of DGAT-1, in a mouse model of lipotoxic cardiomyopathy, provides a protective effect against reduction of cardiac output.[71] In this study heart levels of TG were increased while DAG concentrations were decreased. However, a recent analysis of myocyte-specific over-expression of DGAT-1 in mice led to very different conclusions.[72] As these mice reached one year of age they developed cardiac wall thickening and chamber dilation. During weeks 52 through 62 of age a subset of these animals developed both clinical (lethargy, piloerection, and tachypnea) and histological (increase heart weight and wall thinning and chamber dilation) signs of heart failure. One potential explanation for the discrepancy between these results and the earlier study that showed a cardioprotective response could reside in their corresponding durations. The study showing cardiac benefit was carried out in mice of three to four months of age.[71] The mice in the second study began to show modest cardiac changes at three months of age, but required a year to show severe symptoms. From these results one can infer that acute over-expression of DGAT-1 is cardioprotective through conversion of FA loads to triglycerides, which is in line with the acute exercise-induced cardioprotective effects observed in humans. In turn, chronic cardiac steatosis leads to cellular toxicity via the continuous release of harmful metabolites (FA, DAG, etc.) from intracellular TG stores.

Contrary to the acute DGAT-1 over-expression studies discussed above, DGAT-1$^{-/-}$ mice have improved insulin sensitivity.[53] Utilizing this genetic model and pharmacological inhibition of DGAT-1, follow-on investigations have provided insights into the relationship between DGAT-1 and insulin sensitivity.[73,74] Paralleling the whole animal response, insulin-stimulated glucose transport is increased in the soleus muscle of chow-fed DGAT-1$^{-/-}$

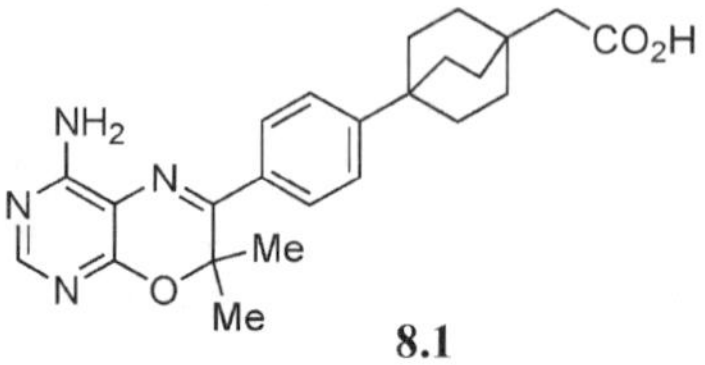

8.1

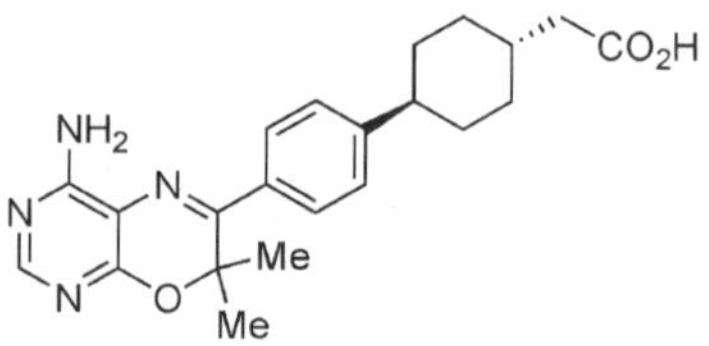

8.2

Figure 8.4 Structures of aminopyrimidine-based DGAT-1 inhibitors.

mice relative to DGAT-1$^{+/+}$. There was also a 50% reduction in phosphorylation of insulin receptor substrate-1 (IRS-1), a molecule implicated in insulin resistance, in the DGAT-1$^{-/-}$ animals.[73] A similar impact on IRS-1 was observed in DGAT-1$^{+/+}$ mice transplanted with fat from the knockout animals. A potential explanation for why high levels of the potentially lipotoxic intermediates such as DAG and FA do not materialize in the DGAT-1$^{-/-}$ model or in acute inhibition in DGAT-1$^{+/+}$ mice has been proposed to occur through counter-regulatory mechanisms.[74] Suppression of members of the peroxisome proliferator-activated receptor (PPAR) gene family and key lipases are observed in both of these models, supporting the findings of low levels of the TG precursors DAG and FA. In concert, up-regulation of glucose transporter mRNAs is observed, suggesting a shift in energy utilization from FA consumption to glucose utilization. A similar profile is observed when DGAT-1$^{+/+}$ mice are treated with the DGAT-1 selective inhibitor **8.1** (Figure 8.4).[74] The reductions in lipotoxicity mediators in skeletal and heart muscle of these animals supports the hypothesis that inhibition of DGAT-1 will have a beneficial impact on insulin sensitivity.

8.3.6 Role of DGAT-1 in Hepatitis C Infectivity

Hepatitis C virus (HCV) is a member of a family of viruses that infect the liver and is found in approximately 3% of the world's population. Half of the individuals infected with HCV go on to develop hepatic steatosis. Though hepatic lipid accumulation was originally thought to be a byproduct of the infection, more recent studies suggest a dependency between HCV lifecycle and lipid droplet formation.[80] To determine whether DGAT-1 and/or DGAT-2 play a role in the lipid-dependent viral infectivity of the HCV, each was specifically knocked down with silencing RNA in liver cells infected with HCV.[81] The DGAT-1 silencing RNA treated cells saw infection spread reduced by up to 90%, whereas suppression of DGAT-2 had no significant impact. The selective DGAT-1 inhibitor **8.2** (Figure 8.4) was also shown to suppress infectivity to a similar degree. Inhibition of DGAT-1 does not alter the overall content of the lipid droplets, rather DGAT-1 associates with the HCV core protein and directs incorporation of the core proteins into DGAT-1 generated lipid particles.[81] These results suggest that DGAT-1 inhibition may be an effective antiviral approach for the treatment of HCV.

8.4 DGAT-1 Inhibitors

Pharmacologic intervention via inhibition of DGAT-1 holds potential for the treatment of obesity, type 2 diabetes, and other diseases associated with metabolic syndrome. Identification of selective, orally active inhibitors of this key lipid synthesis enzyme has been the focus of intensive efforts. More than 20 research groups have filed patents on small molecule DGAT-1 inhibitors, with sustained levels of patent application filings from 2004 to the present.[82] These inhibitors cover a range of chemotypes encompassing neutral, acidic, and basic molecules. While many of these series possess similar overall structural topology, the paucity of detailed enzyme kinetic data makes it difficult to determine whether they share similar binding sites/modes or pharmacophore features. For the purposes of this discussion, inhibitor series have been grouped into broad chemotype classes based on common structural elements.

8.4.1 Biarylamines

One of the first disclosures of selective DGAT-1 inhibitors was a series of biarylamines reported by Bayer Pharmaceuticals and is exemplified by BAY 74-4113 (Figure 8.5).[83] This series was derived from optimization of ketoacid lead **8.3**,[84] which was identified in a DGAT-1 high-throughput screen (HTS).[85] Gem-dimethyl substitution alpha to the acid (**8.4** in Table 8.1) provided comparable potency to the more lipophilic phenethyl side chain present in **8.3**. Phenacetamide- or benzamide-based replacements (**8.5** and **8.6**) of the aliphatic amide moiety provide an ∼10-fold improvement in DGAT-1 inhibitory potency. While conversion of the benzamide functionality to an *N*-arylurea (**8.7**) led to a further improvement in potency, this modification led to reduced solubilities and oral exposures. Masking of the urea functionality via heterobicyclic isosteres (**8.8** and **8.9**) improved solubility properties.[85] Optimization of the ketoacid and biaryl moieties of **8.9** ultimately led to the discovery of BAY 74-4113. Following the decision by Bayer Pharmaceuticals

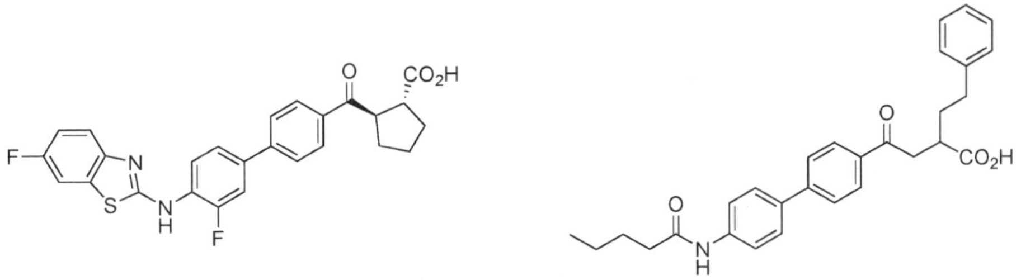

BAY 74-4113
DGAT-1 IC$_{50}$ = 83 nM
DGAT-2 IC$_{50}$ > 10 μM

8.3
DGAT-1 IC$_{50}$ = 3200 nM

Figure 8.5 Structure of BAY 74-4113 and Bayer HTS lead.

Table 8.1   Bayer DGAT-1 inhibitor lead development.

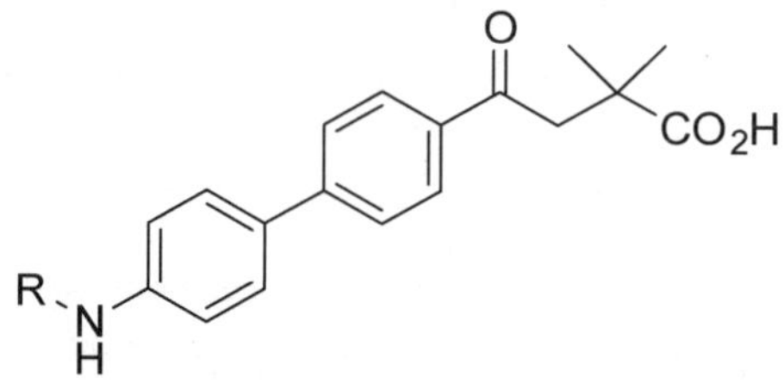

	R	*DGAT-1 IC$_{50}$ (nM)*
8.4	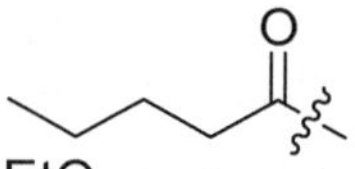 (butyl ketone)	3500
8.5	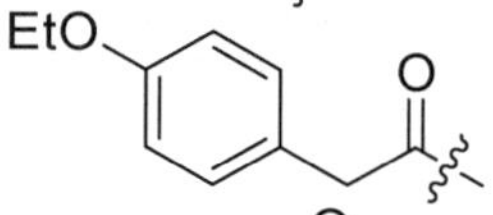 EtO–(phenyl)–CH$_2$–C(=O)–	250
8.6	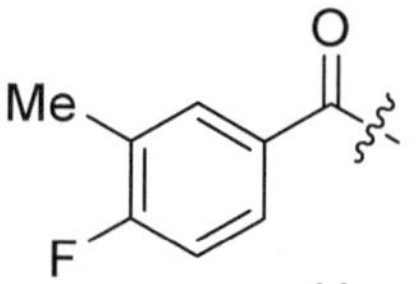 Me,F-substituted benzoyl	430
8.7	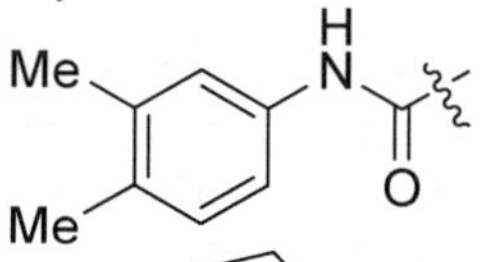 Me,Me-substituted phenyl–NH–C(=O)–	82
8.8	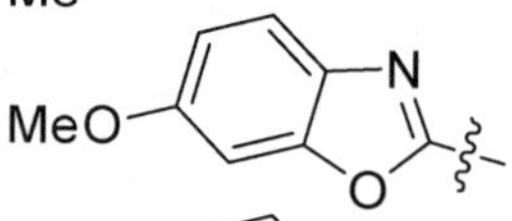 MeO-benzoxazol-2-yl	380
8.9	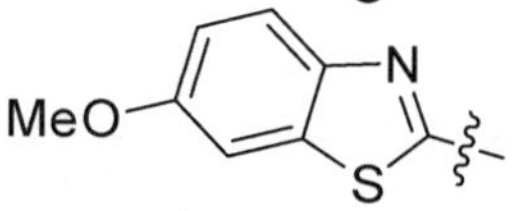 MeO-benzothiazol-2-yl	410

to shift their research focus away from metabolic diseases, BAY74-4113 was licensed to Pfizer Inc. (Pfizer code – PF-04415060).[86]

Evaluation of BAY 74-4113 in a number of acute and chronic preclinical models has demonstrated robust effects on obesity and diabetes endpoints. Administration of a 30 mg/kg oral dose of BAY 74-4113 30 minutes prior to treatment with a bolus of corn oil (5 mL/kg) in male C57BL/6J mice produced near complete suppression of the plasma triglyceride excursion observed in control animals.[87] Treatment of diet-induced obese (DIO) mice on a high-fat (45% calories from fat) diet with BAY 74-4113 (5 mg/kg, p.o., bid) for 28 days produced a 10% reduction in weight gain (Figure 8.6).[88] The magnitude of this response was significantly ($p < 0.05$) greater than that produced by sibutramine/Meridia® (6% at 5 mg/kg, p.o., qd).

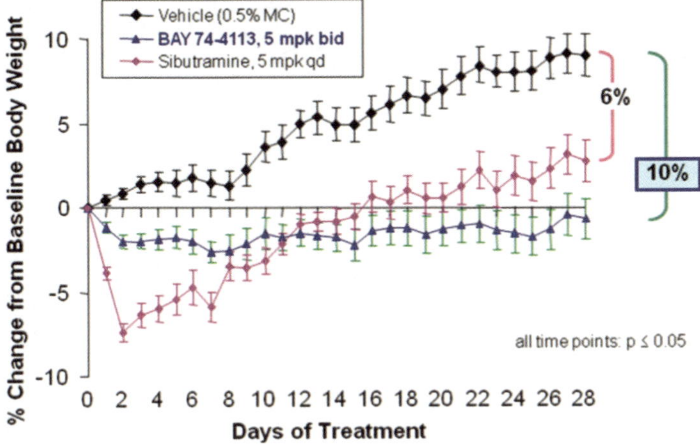

Figure 8.6 Weight gain inhibition of BAY 74-4113 in DIO mice.

The potential impact of treatment with BAY 74-4113 on diabetic endpoints was evaluated in ob/ob mice on a Western diet (Figure 8.7).[87] Daily administration of 25 or 50 mg/kg of BAY 74-4113 (po, qd) for 11 days reduced plasma glucose concentrations to those observed in lean controls. The glycemic improvement observed in these animals is likely driven in part through beneficial changes in gut incretin secretion and is supported by a recent study with BAY 74-4113.[89] Two consecutive daily doses of BAY 74-4113 (30 mg/kg, p.o.) to Sprague Dawley rats produced significant and sustained reductions in GIP levels (-62% and -52% at 6 and 8 h post 2nd dose, respectively). This result is consistent with the changes in GIP observed in DGAT-1$^{-/-}$ mice relative to their parental strain and that seen in humans following bariatric surgery. In addition, plasma ghrelin levels in the treatment group were 10-fold higher in the BAY 74-4113 treatment group relative to control animals 1 h post dose and were 3-fold increased after 24 h.

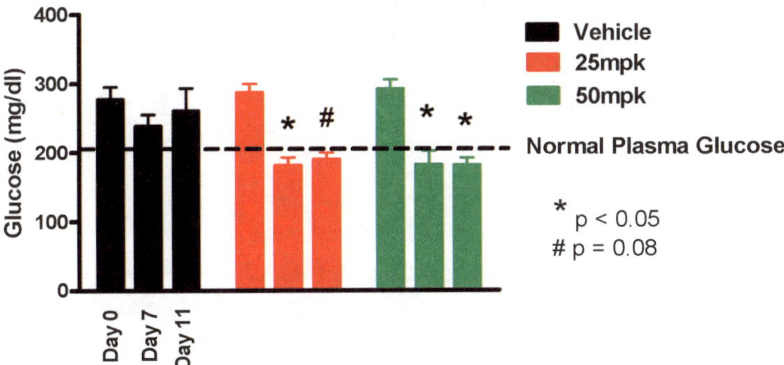

Figure 8.7 Glucose lowering effect of BAY 74-4113 in ob/ob mice on western diet.

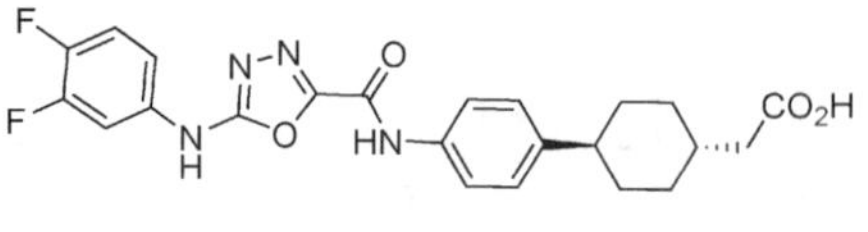

8.10
DGAT-1 IC_{50} = 2.5 nM

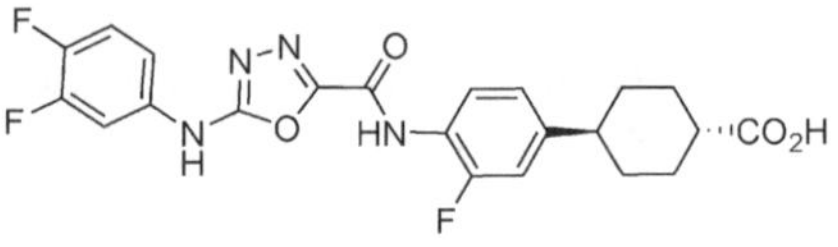

8.11
DGAT-1 IC_{50} = 5.5 nM

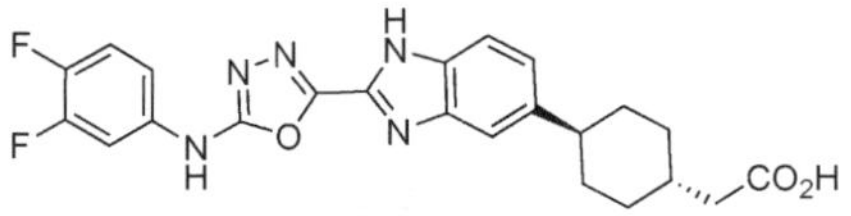

8.12
DGAT-1 IC_{50} = 10 nM

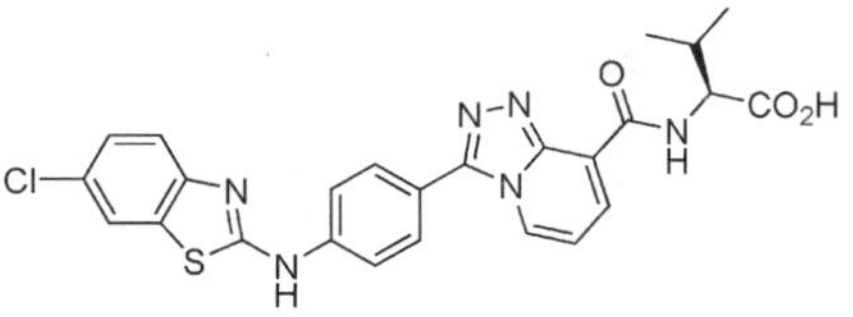

8.13
DGAT-1 IC_{50} = 7 nM

Figure 8.8   Structures of biarylamine-based DGAT-1 inhibitors.

An alternative biarylamine motif containing an oxadiazole carboxamide-based replacement for one of the *N*-aryl substituents of the Bayer series has been reported to afford potent DGAT-1 inhibitors.[90] A patent covering several crystalline forms of **8.10** (DGAT-1 IC_{50} = 2.5 nM) has published.[91] More recently, a selection invention from AstraZeneca disclosing analogs with improved properties has been filed.[92] Compound **8.11** (DGAT-1 IC_{50} = 5.5 nM) is the primary focus of this patent, with a 1 kg synthesis, solid form characterizations, formulations, and salt forms being detailed. While this compound appears to be of high import to the AstraZeneca team, they have yet to disclose the structure of their lead DGAT-1 inhibitor, AZD7687, which is currently in phase I human clinical trials.[93] Additional patent filings have followed which disclose variants of the cyclohexylphenyl region of these biarylamine-based inhibitors.[94–97] This team has also prepared a series of DGAT-1 inhibitors in which the amide functionality of compounds such as **8.10** has been conformationally restrained in a benzimidazole ring (e.g. **8.12**) (Figure 8.8).[98] Disclosure[92] of efforts to find amides with improved physical properties suggests that compounds such as **8.12** were likely pursued as an approach to optimizing oral bioavailability. Bristol-Myers Squibb has disclosed a series of potent inhibitors, exemplified by **8.13**, in which one of the phenyl groups of BAY 74-4113 is replaced with a triazolopyridine.[99]

8.4.2   Ureas

Since the initial disclosure[84] of urea-based DGAT-1 inhibitors such as **8.7**, there have been a number of reports on structurally related series.[100–106] Modifications include heterocyclic replacements for the aryl rings of the biphenyl core[101–104] and conformational restriction of the urea moiety as in dihydrobenzo-1,4-oxazine **8.14** (Figure 8.9).[105,106] Presumably, many of these

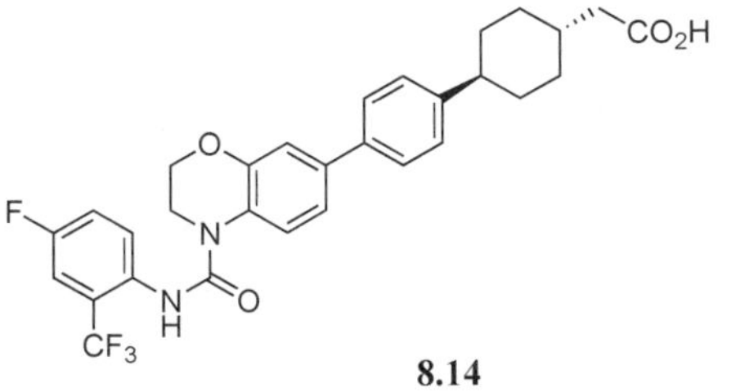

8.14

DGAT-1 IC_{50} = 20 nM

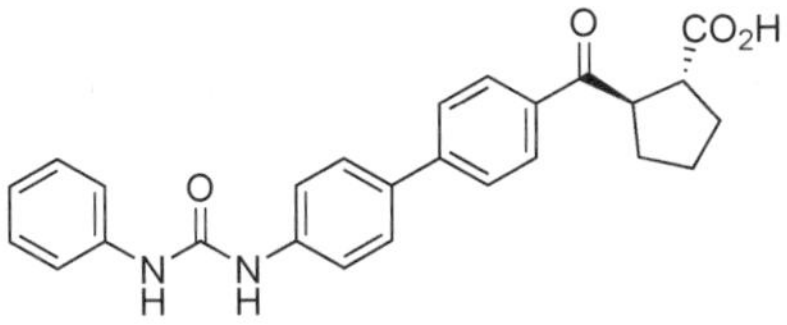

A-922500

DGAT-1 IC_{50} = 7 nM

Figure 8.9   Structures of urea-based DGAT-1 inhibitors.

modifications were directed towards improving the physical properties within this urea series, since many of the analogs related to **8.7** suffer from poor solubility and oral bioavailability.[85] However, it has been found that the parent urea A-922500 (Figure 8.9) is moderately orally bioavailable in rodent models.[107] Based on this profile, A-922500 has been utilized to define the pharmacodynamic impact of small molecule DGAT-1 inhibitors on a range of metabolic syndrome endpoints in preclinical animal models.[107-109] In diet-induced obese mice treated with A-922500 (3 mg/kg, bid, p.o.) there was an 8.5% reduction in body weight relative to the vehicle controls after 4 weeks.[107] The lack of a significant effect on food intake in the drug-treated group suggests either enhanced energy expenditure or reduced intestinal fat absorption driving weight loss. As observed in DGAT-1$^{-/-}$ mice, hepatic triglycerides were significantly reduced in DIO mice treated with A-922500. In a separate report, administration of A-922500 for 2 weeks significantly reduced serum triglycerides in a diet-induced dyslipidemic hamster model and in Zucker fatty rats.[107] Inhibition of DGAT-1 in Zucker fatty rats suppressed (32%) the increase in free fatty acid levels observed in control animals. The increases in free fatty acids prior to the onset of diabetes in these animals suggest the potential for DGAT-1 inhibition to improve insulin sensitivity in this model. In the leptin-deficient db/db mouse model, 5 weeks of treatment with A-922500 (10 mg/kg, qd, p.o.) suppressed weight gain by 50% and food intake by 6.8%.[109] A number of endpoints associated with the hepatic steatosis observed in the control animals were favorably impacted, including a 25% reduction in liver triglyceride levels, histological improvements, and reductions in liver enzyme levels (ALT/AST). Expression levels of a number of genes involved in hepatic lipid metabolism were evaluated at the end of this study as a means of defining the mechanism by which DGAT-1 inhibition produces beneficial effects on obesity and hepatic steatosis in this model. Genes associated with fatty acid oxidation, such as carnitine palmitoyltransferase 1a and PPARα, were significantly up-regulated in the drug treatment group.[109] Genes associated with fatty acid synthesis were not altered by DGAT-1 inhibition. These changes are consistent with the reduction in FFA observed in the treatment group.

8.4.3   Amides

Screening of a diverse chemical library identified ketoamide **8.15** as a modestly potent ($\sim$5 μM) inhibitor of DGAT-1.[110] A synthetic library designed around modifications of the arylamino substituent of **8.15** led to the identification of **8.16**. From related analogs it was shown that the 4,5-diaryl substitution on the thiazole ring was required for activity. Further structure–activity relationship (SAR) analysis around the thiazole of **8.16** revealed that incorporation of a 5-benzyl substituent (**8.17**) led to $\sim$20-fold improvement in DGAT-1 inhibitory potency (Figure 8.10). Compound **8.17** has been evaluated in the genetically diabetic and obese KKA^y mouse model.[111] Given the high lipophilicity (c log P = 7.3) associated with this DGAT-1 inhibitor, solubility/intestinal permeability enhancing formulations or admixture with food were required for *in vivo* assessments. Mice on either a high-fat or low-fat/high-carbohydrate diet experienced significant reductions in both subcutaneous and visceral fat pads following 4 weeks of treatment with **8.17** (30 mg/kg, qd, p.o.). Hepatic TG trended lower in these animals, while hepatic cholesterol was significantly reduced with most of the change arising from the cholesterol ester fraction as compared to free cholesterol.

While **8.17**, with the aid of solubilizing vehicles, had sufficient pharmacokinetic properties to serve as a proof of concept tool, it did not have sufficient solubility and metabolic stability to support further development. The

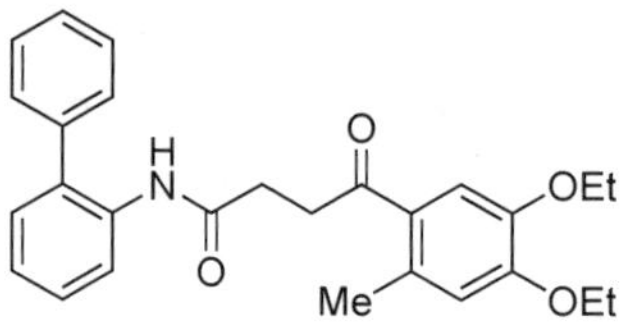

8.15
DGAT-1 IC$_{50}$ = 5200 nM

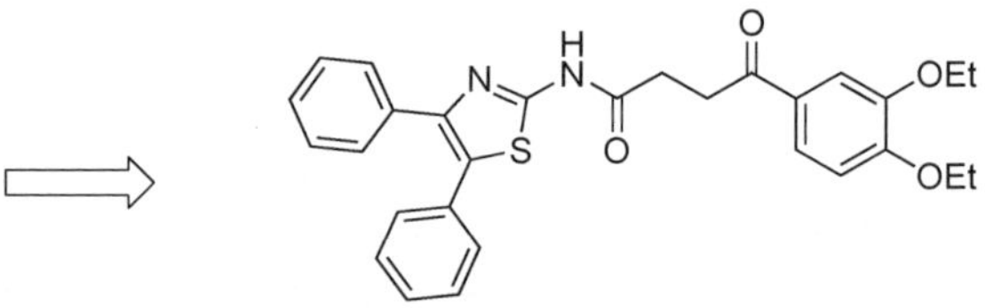

8.16
DGAT-1 IC$_{50}$ = 150 nM

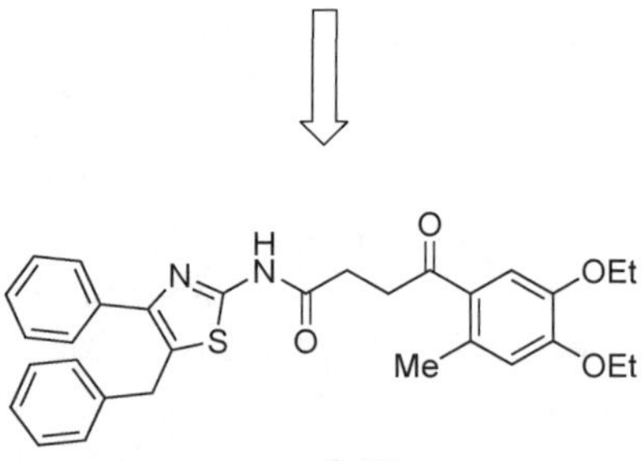

8.17
DGAT-1 IC$_{50}$ = 8.5 nM

Figure 8.10   Structures of acylaminothiazole-based DGAT-1 inhibitors.

aminothiazole present in this lead raises an additional concern in that biooxidative ring opening of this functionality can lead to the formation of potentially toxic thiourea metabolites.[112] Also, **8.17** contains an aromatic ketone functionality that may have contributed to the high clearance via reduction of the ketone mediated by carbonyl reductase enzymes.[113] Research efforts focused on addressing the structural and physical property issues associated with this lead compound have resulted in the identification of an analog with vastly superior properties.[114] Conversion of the ketone functionality to an amide and reversing the amide linkage on the thiazole provided **8.18**, which retains significant DGAT-1 inhibitory potency. These modifications, however, did not improve the poor solubility or high clearance associated with the lead compound. Optimization of the thiazole and benzamide moieties afforded the potent inhibitor **8.19**, though this analog still suffered from high *in vivo* clearance. Replacement of the ethoxy substituent of **8.19** with trifluoroethoxy reduced rat clearance by 10-fold while retaining excellent DGAT-1 inhibitory potency.[114] In summary, significant reductions in overall lipophilicity in going from **8.17** ($c \log P = 7.3$) to **8.20** ($c \log P = 3.1$) afforded a significant improvement in oral bioavailability (**8.20**, $F = 50\%$), while retaining comparable DGAT-1 potency (Figure 8.11).

In an *ex vivo* setting, 4 days of exposure to **8.20** reduced triglyceride content in a dose-dependent manner in both adipocytes and myotubes.[115] Treatment of myotubes with 1 μM concentrations of **8.20** for 4 days produced mRNA increases in glucose transporter-1 and -4 expression of 60% and 100%, respectively. The up-regulation of these transporters is consistent with the 34% increase in 2-deoxyglucose levels observed in these cells. In addition, a 40% reduction in TG was also seen, but this change was not associated with a

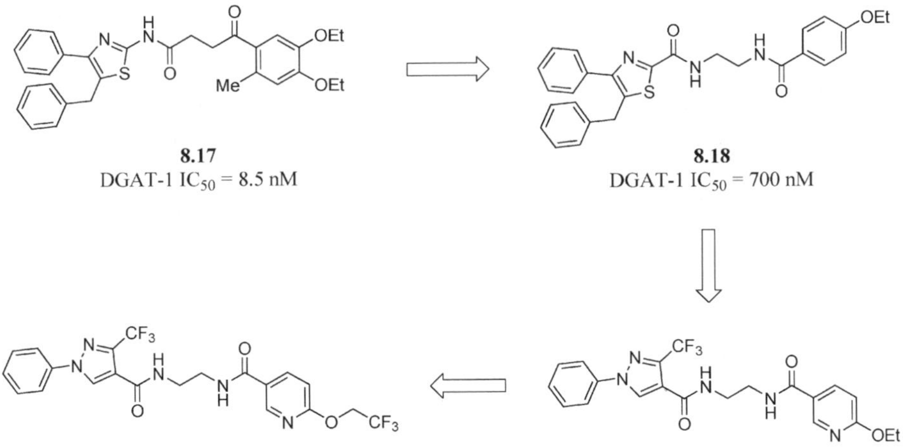

8.17
DGAT-1 IC$_{50}$ = 8.5 nM

8.18
DGAT-1 IC$_{50}$ = 700 nM

8.20
DGAT-1 IC$_{50}$ = 22 nM

8.19
DGAT-1 IC$_{50}$ = 25 nM

Figure 8.11 Structures of trifluoromethylpyrazole-based DGAT-1 inhibitors.

compensatory increase in free fatty acids. This could be explained by the increase (50%) in uncoupling protein-3 mRNA , which codes a skeletal muscle protein involved in regulation of energy metabolism, suggesting that the FFA shunted away from TG synthesis are being oxidized. This conclusion is consistent with the finding that myotube FFA levels are initially elevated after treatment with **8.20**, but return to normal levels after 4 days. A similar effect was observed in diet-induced (C57BL/6J) and genetic (KKA^y) mouse models of obesity following treatment with **8.20** (30 mg/kg for 4 weeks), which produced soleus muscle FFA oxidation increases of 37% and 60%, respectively. In both animal models, compound **8.20** was able to dose dependently reduce weight gain over the 4 weeks of treatment. These changes were associated with reductions in fat pad weights and TG levels in the liver. In summary, the pharmacodynamic profile of **8.20** supports the hypothesis that inhibition of DGAT-1 holds potential for amelioration of a number of endpoints associated with metabolic syndrome. Takeda Pharmaceuticals and Array BioPharma have not disclosed whether **8.20** or a related analog has been advanced to human clinical trials.

Researchers at Hoffmann-La Roche have reported on a structurally distinct series of amide-based DGAT-1 inhibitors (Figure 8.12).[116] The acylhydrazide lead **8.21** was identified as part of a corporate library screening exercise. Concerns around the high *in vivo* clearance of **8.21** and potential safety concerns associated with the acylhydrazide functionality[112] led the team to design molecules containing isosteric five-membered heterocycles. Incorporation of a 2-phenyl-5-trifluoromethyl-3-carboxamide replacement provided the potent DGAT-1 inhibitor **8.22**. The close structural similarity of this heterocycle with the pyrazole present in **8.20** suggests that these two series share a common mode of binding to DGAT-1. Further optimization of the amine portion of **8.22** afforded RO-6036, which exhibits good pharmacokinetic and pharmacodynamic profiles in preclinical models. Further evaluation of RO-6036 revealed that it blocks the hERG potassium channel and causes QT prolongation in the guinea pig. This cardiovascular safety concern led the team to design analogs with neutral, polar modifications of the aminopyridyl unit of RO-6036. Phenylpiperidine amides and carbamates such as **8.23** retain DGAT-1 inhibitory potency and have significantly reduced hERG inhibitory activity. However, these neutral compounds suffer from poor exposure in preclinical models. It is not disclosed whether this poor *in vivo* performance is due to high clearance, low solubility, or a combination of both. As a means of improving the pharmacokinetic profile of this series, carboxylic acid substituents were incorporated. From this effort, compound **8.24** was identified as a potent inhibitor with good pharmacokinetic profiles in preclinical species. In DIO rats, **8.24** produced a dose-responsive decrease in body weight gain at 0.3, 1, and 3 mg/kg (po, qd) for 3 weeks of 5%, 6%, and 8%, respectively.[116] Unlike studies with other DGAT-1 inhibitors, part of this effect on weight is likely driven by reductions in food intake (17%, 19%, and 27%, respectively) in these animals. The authors speculate that this food effect

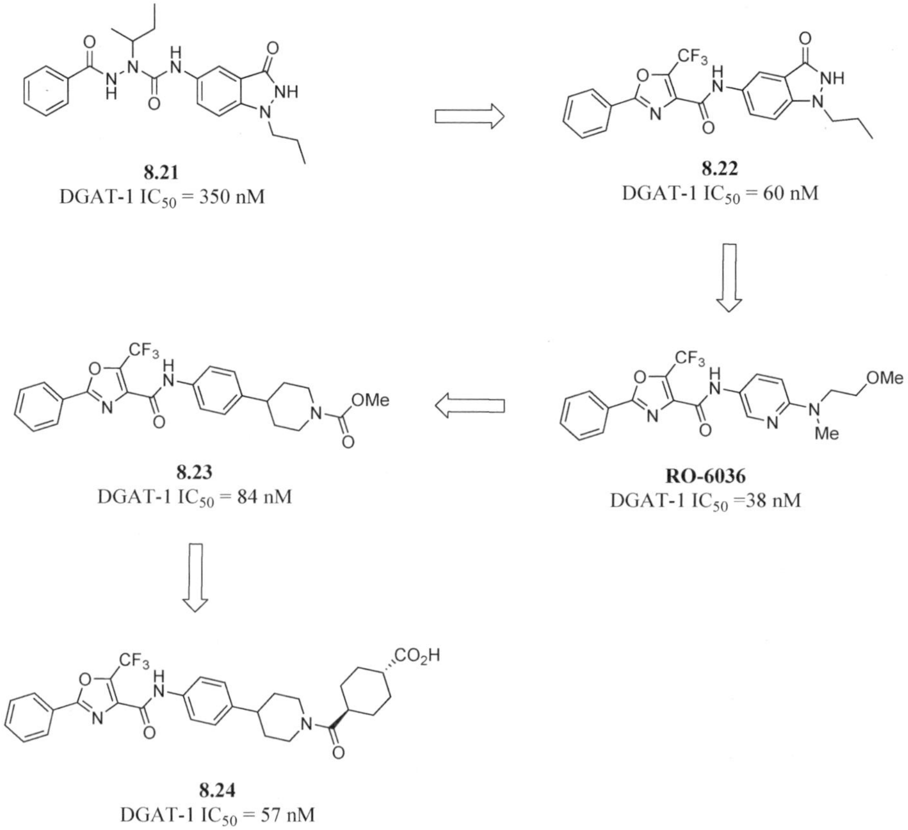

8.21
DGAT-1 IC_{50} = 350 nM

8.22
DGAT-1 IC_{50} = 60 nM

8.23
DGAT-1 IC_{50} = 84 nM

RO-6036
DGAT-1 IC_{50} = 38 nM

8.24
DGAT-1 IC_{50} = 57 nM

Figure 8.12   Structures of trifluoromethyloxazole-based DGAT-1 inhibitors.

may be driven by the release of gut hormones GLP-1 and PYY, though incretin levels were not measured in this study. To gauge the potential effect of DGAT-1 inhibition on glucose homeostasis, an oral glucose tolerance test was administered at the completion of the 3-week study. At the 3 mg/kg dose of **8.24**, the glucose excursion was blunted by 15% ($p < 0.05$) relative to the control animals. While day 1 and day 21 basal glucose levels in the DIO rats treated with 3 mg/kg **8.24** were not significantly different, basal insulin levels were significantly lower following the 3 weeks of treatment. These data suggest a significant insulin-sensitizing effect of DGAT-1 inhibition in this model and is consistent with the result seen in DGAT-1 knockout models.[24]

8.4.4   Aminopyrimidines

Researchers at Japan Tobacco and Amgen (Tularik) have disclosed a series of aminopyrimidooxazine-based DGAT-1 inhibitors (Figure 8.13).[117] Compound **8.2**, a prototypical member of this series, is a potent and selective inhibitor of

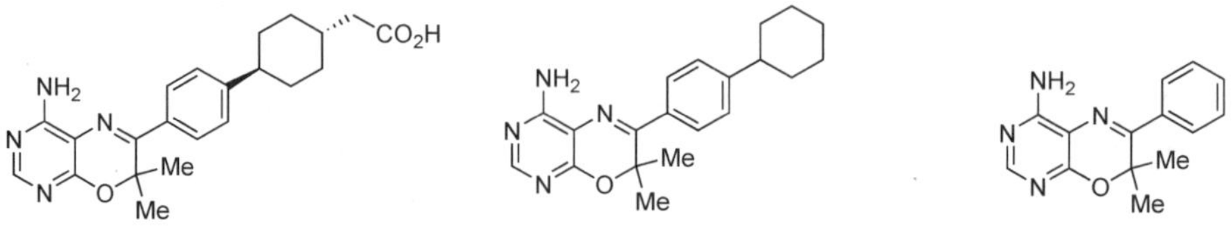

8.2
DGAT1 IC$_{50}$ = 72 nM

8.25
DGAT1 IC$_{50}$ = 240 nM

8.26
DGAT1 IC$_{50}$ = 12 M

Figure 8.13 SAR of lead aminopyrimidine-based DGAT-1 inhibitor.

DGAT-1. The synthetic route utilized to prepare these analogs (condensation of 4,5-diamino-6-hydroxypyrimidine with an α-bromoacetophenone) has limited the development of SAR around the aminopyrimidooxazine core. However, it has been shown that reduction of the carbon–nitrogen double bond of the oxazine led to 10-fold reduction in potency.[117] Substitution of the nitrogen of the oxazine ring with carbon has little or no effect on potency. These SAR findings suggest that the in-plane display of the phenyl cyclohexyl side chain relative to the pyrimidooxazine core is a critical feature of this series. The carboxylic acid functionality is not critical to potency in this series with the des-acetic acid analog **8.25** being only 3-fold less active than **8.2**.[87] However, further truncation of the side chain to the phenyl substituent (**8.26**) leads to a substantial reduction (50-fold) in DGAT-1 inhibitory potency. The good physical property space occupied by this series (**8.2**, log $D_{7.4}$ = 1.4, PSA = 111 and moderate passive permeability) support the low human microsomal clearance (<8 mL/min/kg) and excellent DGAT-1 cellular inhibitory potency (EC$_{50}$ = 16 nM, HT-29 cell line) observed.[118] Compound **8.2** has low clearance and excellent oral bioavailability in both the rat and dog.[119] Though this series has outstanding pharmacologic/pharmacokinetic profiles, concerns around both photoinstability and the potential for idiosyncratic toxicity, induced by reactive acyl glucuronide-derived intermediates, has led to alternative designs based on **8.2**.[119,120]

To address the concern around acyl glucuronide-based reactive intermediates of **8.2** being formed *in vivo*, a steric block approach was pursued.[119] Steric encumbrance of a carboxylic acid has been shown to be an effective strategy for suppressing acyl glucuronide formation.[121] In addition, hindered acyl glucuronides and their associated rearrangement products, if formed, are less reactive as electrophilic species. Increased steric hindrance in either the α- or β-positions of the carboxylic moiety of **8.2** did not negatively impact DGAT-1 potency, with bicyclo[2.2.2]octane-based analog **8.1** (Figure 8.14) being equipotent to **8.2**.[119] The acyl glucuronide of **8.1** was shown to undergo hydrolytic cleavage or rearrangement to acyl migrated products very slowly under physiological conditions. These results suggest that this carboxylic acid would have low probability of eliciting idiosyncratic toxicology via acyl glucuronide metabolites. Compound **8.1** demonstrated an improved pharmacokinetic profile relative to **8.2**, with reduced plasma clearances observed in both rats

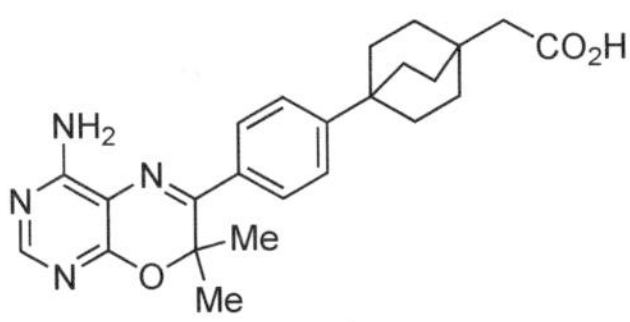

8.1
DGAT1 IC_{50} = 15 nM

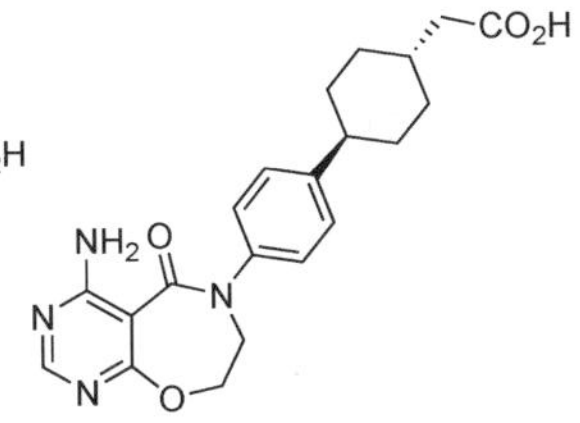

PF-04620110
DGAT1 IC_{50} = 19 nM

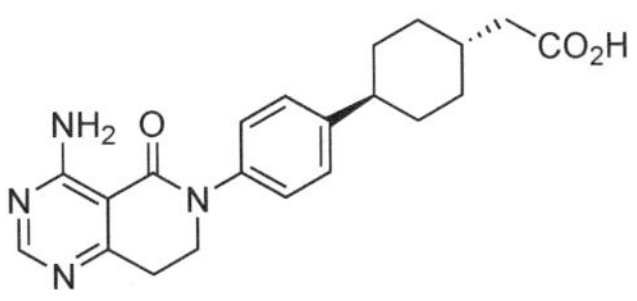

8.27
DGAT1 IC_{50} = 71 nM

Figure 8.14 Structures of aminopyrimidine-based DGAT-1 inhibitors.

and dogs.[119] Based on this profile **8.1** was advanced to models of intestinal lipid absorption and diet-induced obesity. In an oral lipid challenge in fasted rats, **8.1** was able to suppress plasma TG excursion dose-dependently with 3 mg/kg producing a 57% reduction. Treatment of diet-induced obese mice with 20 mg/kg (po, qd) for 3 days produced a statistically significant (3%) reduction in weight compared to the vehicle treatment group.

A second concern with this lead series is the potential for phototoxicity associated with **8.2** since it suffers from both solution and solid photoinstability.[120] Based on the assumption that the aminopyrimidooxazine core of these compounds is responsible for this issue, alternative bicyclic designs were pursued. Pyrimidooxazepinone-based analog PF-04620110 retains the key pharmacophore features and their respective spatial arrangements present in **8.2**.[120] PF-04620110 (Figure 8.14) is a potent inhibitor of DGAT-1 in isolated membrane preparations (IC_{50} = 16 nM) and HT-29 intestinal cells (EC_{50} = 8 nM). Oral doses as low as 0.1 mg/kg of PF-04620110 are effective in completely suppressing the plasma TG excursion in rats following an oral lipid challenge. PF-04620110 has been shown in preclinical models to modulate gut hormones in a manner consistent with that observed in patients showing improved glycemic control following bariatric surgery.[89] A single oral dose of PF-04620110 to fasted rats significantly increases plasma concentrations of GLP-1, PYY, and GIP (60%, 60%, and 23%, respectively) and produces a 23% decrease in GIP.

The six-membered lactam analog (**8.27**) of PF-04620110 has also been prepared and is also a potent inhibitor of DGAT-1. However, **8.27** suffers from high turnover in human liver microsomes and poorer solubility relative to PF-04620110. Human clearance projections for PF-04620110 based on liver microsomes are consistent with rat and dog single species scaling results which predict low clearance and a moderate half-life in humans.[118]

PF-04620110, in contrast to **8.2**, has minimal absorbance in the near visible light region and exhibits excellent solution and solid state stabilities.[120] To assess the potential risk of reactive metabolites arising from the acetic acid moiety of PF-04620110, the acyl glucuronide was prepared. Under physiological conditions the acyl glucuronide was shown to be very stable towards both direct electrophilic reactions and to intramolecular rearrangement reactions.

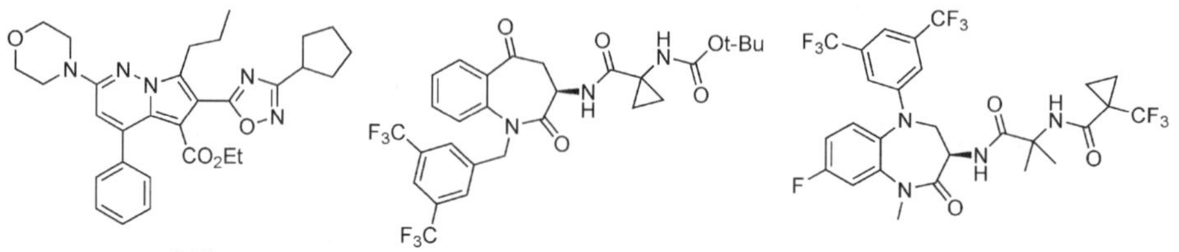

8.28
DGAT1 IC$_{50}$ = 48 nM
ACAT1 IC$_{50}$ = 230 nM

8.29
DGAT1 IC$_{50}$ = 37 nM

8.30
DGAT1 IC$_{50}$ =16 nM

Figure 8.15 Structures of pyrrolopyridazine-, benzazepinone-, and benzodiazepinone-based DGAT-1 inhibitors.

These results combined with the lack of formation of the acyl glucuronide in human liver microsomes supports the conclusion of a low risk of idiosyncratic toxicology driven by the acetic acid group of PF-04620110. Based on its preclinical efficacy and safety profile PF-04620110 has been advanced to human clinical trials for the treatment of type 2 diabetes.

8.4.5 Additional Lead Series

Optimization of a pyrrolotriazine-based HTS hit by researchers at Japan Tobacco and Amgen led to the identification of a series of pyrrolopyridazines exemplified by **8.28** (Figure 8.15).[122] While these efforts led to significantly improved DGAT-1 inhibitory potency relative to the lead structure, all of the compounds prepared are relatively non-selective with regards to inhibition of ACAT-1 enzymatic activity. Pharmacokinetic or pharmacodynamic profiles for members of this structural class have not been disclosed.

Collaborators at Banyu and Merck Sharp & Dohme have disclosed a series of DGAT-1 inhibitors containing a benzazepinone core.[123] Representative analog **8.29** (Figure 8.15) is a potent inhibitor of DGAT-1. All of the examples contain an N-1 benzyl substituent and the (*R*)-configuration of the acylamino substituent appears to be preferred (5–10×) over the (*S*)-enantiomer. This team has also disclosed two series of benzodiazepinone-based inhibitors, encompassing structures such as **8.30** (Figure 8.15).[124] Given the broad range of pharmacologies expressed by benzazepinone-/benzazepinone-based structures it would be of interest to determine whether compounds such as **8.29** and **8.30** selectively inhibit DGAT-1.[125]

8.5 Human Clinical Trials with DGAT-1 Inhibitors

Table 8.2 summarizes the current status of DGAT-1 inhibitors that have been advanced to human clinical trials. Clinical indications being pursued include obesity, type 2 diabetes, and hypertriglyceridemia. With a majority of these programs still in phase I clinical trials or having been discontinued, there have

Table 8.2 Clinical trials of DGAT-1 inhibitors.

Company (Drug)	Current Status	Clinical Endpoints	References
Bayer/Pfizer (BAY 74-4113)	Discontinued	Obesity	126
Japan Tobacco (JTT-553)	Discontinued	Obesity	127,128
AstraZeneca (AZD7687)	Phase I	Type 2 diabetes mellitus Obesity	93,129–131
Pfizer (PF-04620110)	Phase I	Type 2 diabetes mellitus	132–136
Novartis (LCQ-908)	Phase II	Type 2 diabetes mellitus Severe hypertriglyceridemia	137–139

not been any published reports on the clinical efficacy against any of these endpoints yet.

The first DGAT-1 inhibitor to be advanced to human studies is BAY 74-4113, which was discovered at Bayer Pharmaceuticals and later co-developed with Pfizer Inc.[126] In a dose-escalation safety study (2 mg to 20 mg, p.o.) in healthy male volunteers, BAY 74-4113 exhibited no adverse effects on vital signs and clinical endpoints. At the 20 mg dose C_{max}, AUC and $T_{1/2}$ were 761 mg/L, 6.9 mg·h/L and 25 h, respectively. As seen in preclinical models there was significant conversion to the alcohol metabolite ((S)-stereoconfiguration) corresponding to reduction of the ketone functionality of BAY 74-4113. The AUC of this alcohol metabolite (DGAT-1 IC_{50} = 276 nM) in humans was approximately 3-fold higher than that of the parent drug and had a $T_{1/2}$ of 39 h. Based on the excellent single-dose pharmacokinetic and safety profiles of BAY 74-4113, multi-dose regulatory safety studies were initiated. Safety findings from these studies led to the projection of a relatively low therapeutic index in humans and therefore further development of BAY 74-4113 was halted.

As part of a collaborative DGAT-1 inhibitor discovery program Japan Tobacco and Amgen advanced JTT-553 to phase I clinical studies in mid-2007.[127] Though the structure of JTT-553 has never been disclosed, it is likely to be **8.2** or a closely related analog based on the patenting activity of these research groups.[117,122] Development of JTT-553 was halted[128] approximately 1 year after the initiation of phase I trials; no reason for the discontinuation has been reported to date.

AstraZeneca's lead compound (AZD7687) has completed both phase I single-[129] and multi-dose trials.[130] The multiple ascending dose study was carried out in overweight patients starting at 1 mg/kg (qd, 8 days) with up to four dose escalations. Pharmacodynamic biomarkers (TG, DAG, FFA, and insulin in plasma, TG and DAG in adipose tissue) were measured on days 2–8. The results from this multi-dose study were scheduled to read out in March 2011.[130] Parallel single- and multi-dose studies assessing pharmacokinetic and pharmacodynamic effects of AZD7687 are being carried out in Japanese males.[131] A secondary outcome measure of this study is the determination of plasma concentrations of the acyl glucuronide metabolite (AZ13128940) of

AZD7687. While the structure of AZD7687 has not been disclosed, both **8.10** and **8.11** contain carboxylic acid functionalities and as mentioned earlier in this chapter have been the focus of significant intellectual property protection.[91,92]

Phase I clinical trials with Pfizer's DGAT-1 inhibitor PF-04620110 are currently ongoing.[132] Both single- and multiple-dose studies in overweight/obese healthy volunteers have been completed.[133–135] Dose-proportional exposures and a $T_{1/2}$ of ~ 8 h have been observed in humans, which is consistent with predictions from single species scaling in rat and dog.[118] A 4-week phase 1B, placebo-controlled study in type 2 diabetics with poor glycemic control previously treated with metformin is in progress.[136] Doses of PF-04620110 in this study are 5mg/day as either a single dose or in a split dose format. In addition to determining pharmacokinetic parameters in the target disease population, a number of pharmacodynamic endpoints are under evaluation.

The most advanced DGAT-1 inhibitor clinical candidate is Novartis' LCQ-908, which is currently in phase II efficacy trials.[137,138] The first of these studies has a primary goal of measuring the impact on glycemic control in type 2 diabetics treated with LCQ-908. Secondary endpoints that will be evaluated include weight loss effects and changes in insulin sensitivity. Up to five LCQ-908 dose groups have been built into this study, as well as a sitagliptin treatment arm.[137] This is a sizeable study with an estimated enrolment of 720 patients. A phase II multiple-dose study evaluating DGAT-1 inhibition in patients with severe hypertriglyceridemia (familial hyperchlomicronemia phenotypes I and V) has also been initiated.[138] The primary outcome measurement following 3 weeks dosing of LCQ-908 is fasting and postprandial plasma triglycerides. Additional endpoint measurements include blood lipid endpoints such as phospholipids, FFA, and apolipoproteins. The structure and preclinical profile of LCQ-908 have yet to be disclosed. Novartis is projecting a NDA filing in 2014 for LCQ-908, if the desired clinical endpoints are achieved.[139]

8.6   Conclusions

The findings discussed above suggest that DGAT-1 inhibition may represent an effective intervention point for the treatment of type 2 diabetes, obesity, and hypertriglyceridemia. In the decade since DGAT-1 was first characterized, a number of potent and selective inhibitors have been identified. These agents are serving as valuable tools for dissecting the key target tissues for these various disease endpoints and evaluating the potential safety implications of DGAT-1 inhibition. Because of the conflicting data obtained in DGAT-1 knockout and over-expression models, human clinical trials will be key in determining the value of DGAT-1 inhibition in metabolic disease states. It is also of import to note that the full potential of DGAT-1 inhibition in the treatment of human diseases is still evolving, evidenced by the recent report that this enzyme is a key host factor for hepatitis C infectivity.[81]

Acknowledgment

I would like thank David Edmonds for proofreading this manuscript.

References

1. P. Arner, *Best Pract. Res. Clin. Endocrinol. Metab.*, 2005, **19**, 471.
2. J. D. McGarry, *Diabetes*, 2002, **51**, 7.
3. K. G. Alberti, P. Zimmet and J. Shaw, *Lancet*, 2005, **366**, 1059.
4. D. T. Stein, R. Dobbins, L. Szczepaniak, C. Malloy and J. D. McGarry, *Diabetes*, 1997, **46** (Suppl. 1), 23A.
5. L. S. Szsczepaniak, E. E. Babcock, F. Schick, R. L. Dobbins, A. Garg, D. K. Burns and J. D. McGarry, *Am. J. Physiol.*, 1999, **276**, E977.
6. I. Huang-Doran, A. Sleigh, J. J. Rochford, S. O'Rahilly and D. B. Savage, *J. Endocrinol.*, 2010, **207**, 245.
7. R. H. Unger, Y.-T. Zhou and L. Orci, *Proc. Natl. Acad. Sci. USA*, 1999, **96**, 2327.
8. I. Shimomura, R. Hammer, S. Ikemoto, M. Brown and J. Goldstein, *Nature*, 1999, **401**, 73.
9. H. J. Kayden, J. R. Senior and F. H. Mattson, *J. Clin. Invest.*, 1967, **46**, 1695.
10. E. Kennedy, *Annu. Rev. Biochem.*, 1957, **49**, 119.
11. R. A. Coleman and D. P. Lee, *Prog. Lipid Res.*, 2004, **43**, 134.
12. D. O. Koltun and J. Zablocki, *Annu. Rep. Med. Chem.*, 2010, **45**, 109.
13. The reader is referred to discussions of SCD-1 and ACC inhibitors which are the subject of other chapters in this book.
14. R. Lehner and A. Kuksis, *Prog. Lipid Res.*, 1996, **35**, 169.
15. N. Mayorek, I. Grinstein and J. Bar-Tana, *Eur. J. Biochem.*, 1989, **182**, 395.
16. S. B. Weiss, E. P. Kennedy and J. Y. Kiyasu, *J. Biol. Chem.*, 1960, **235**, 40.
17. M. A. Polokoff and R. M. Bell, *Biochim. Biopyhys. Acta*, 1980, **618**, 129.
18. M. Andersson, M. Wettesten, J. Boren, A. Magnusson, A. Sjaberg, S. Rustaeus and S.-O. Olofsson, *J. Lipid Res.*, 1994, **35**, 535.
19. L. M. Ballas and R. M. Bell, *Biochim. Biophys. Acta*, 1980, **602**, 578.
20. R. A. Coleman, *Biochim. Biophys. Acta*, 1988, **963**, 367.
21. R. Lehner, A. Kuksis and Y. Itabashi, *Lipids*, 1993, **28**, 29.
22. S. Cases, S. J. Smith, Y. Zheng, H. M. Myers, S. R. Lear, E. Sande, S. Novak, C. Collins, C. B. Welch, A. J. Lusis, S. K. Erickson and R. V. Farese, Jr., *Proc. Natl. Acad. Sci. USA*, 1998, **95**, 13018.
23. P. Oelkers, A. Behari, D. Cromley, J. T. Billheimer and S. L. Sturley, *J. Biol. Chem.*, 1998, **273**, 26765.
24. S. J. Smith, S. Cases, D. R. Jensen, H. C. Chen, E. Sande, B. Tow, D. A. Sanan, J. Raber, R. H. Eckel and R. V. Farese, Jr., *Nature Genet.*, 2000, **25**, 87.

25. K. D. Lardizabal, J. T. Mai, N. W. Wagner, A. Wyrick, T. Voelker and D. J. Hawkins, *J. Biol. Chem.*, 2001, **276**, 38862.

26. S. Cases, S. J. Stone, P. Zhou, E. Yen, B. Tow, K. D. Lardizabel, T. Voelker and R. V. Farese, Jr., *J. Biol. Chem.*, 2001, **276**, 38870.

27. C.-L. E. Yen, M. Monetti, B. J. Burri and R. V. Farese, Jr., *J. Lipid Res.*, 2005, **46**, 1502.

28. P. J. McFie, S. L. Stone, S. L. Banman and S. J. Stone, *J. Biol. Chem.*, 2010, **285**, 37377.

29. Z. Y. Guo, S. Lin, J. A. Heinen, C. C. Chang and T. Y. Chang, *J. Biol. Chem.*, 2005, **280**, 37814.

30. D. Cheng, R. L. Meegalla, B. He, D. A. Cromley, J. T. Billheimer and P. R. Young, *Biochem. J.*, 2001, **359**, 707.

31. C. Yu, Y. Zhang, X. Lu, J. Chen, C. C. Y. Chang and T.-Y. Chang, *Biochemistry*, 2002, **41**, 3762.

32. S. J. Stone, H. Myers, B. E. Brown, S. M. Watkins, K. R. Feingold, P. M. Elias and R. V. Farese, Jr., *J. Biol. Chem.*, 2004, **279**, 11767.

33. C. L. Yen and R. V. Farese, Jr., *J. Biol. Chem.*, 2003, **278**, 18532.

34. D. Cheng, T. C. Nelson, J. Chen, S. G. Walker, J. Wardwell-Swanson, R. Meegalla, R. Taub, J. T. Billheimer, M. Ramaker and J. N. Feder, *J. Biol. Chem.*, 2003, **278**, 13611.

35. D. Cheng, J. Iqbal, J. Devenny, C. Chu, L. Chen, J. Dong, R. Seethala, W. J. Keim, A. V. Azzara, R. M. Lawrence, M. A. Pelleymounter and M. M. Hussain, *J. Biol. Chem.*, 2008, **283**, 29802.

36. M. D. Orland, K. Anwar, D. Cromley, C. Chu, L. Chen, J. T. Billheimer, M. M. Hussain and D. Cheng, *Biochim. Biophys. Acta*, 2005, **1737**, 76.

37. K. K. Buhman, S. J. Smith, S. J. Stone, J. J. Repa, J. S. Wong, F. F. Knapp, B. J. Burri, R. L. Hamilton, N. A. Abumrad and R. V. Farese, Jr., *J. Biol. Chem.*, 2002, **277**, 25474.

38. B. Lee, A. M. Fast, J. Zhu, J. Cheng and K. K. Buhman, *J. Lipid Res.*, 2010, **51**, 1770.

39. H. Buchwald, Y. Avidor, E. Braunwald, M. D. Jensen, W. Pories, K. Fahrbach and K. Schoelles, *JAMA*, 2004, **292**, 1724.

40. J. P. Thaler and D. E. Cummings, *Endocrinology*, 2009, **150**, 2518.

41. L. L. Baggio and D. J. Drucker, *Gastroenterology*, 2007, **132**, 2131.

42. P. L. McClean, N. Irwin, R. S. Cassidy, J. J. Holst, V. A. Gault and P. R. Flatt, *Am. J. Physiol. Endocrinol. Metab.*, 2007, **293**, E1746.

43. J. McLaughlin, *Biochem. Soc. Trans.*, 2007, **35**, 1199.

44. J.-P. A. M. Bongartz, L. Meerpoel, G. M. Boeckx, G. R. E. Van Lommen, C. F. R. N. Buyck, D. Obrecht, P. Ermert and A. Luther, WO2008/148840.

45. M. Okawa, K. Fujii, K. Ohbuchi, M. Okumoto, K. Aragane, H. Sato, Y. Tamai, T. Seo, Y. Itoh and R. Yoshimoto, *Biochem. Biophys. Res. Commun.*, 2009, **390**, 377.

46. P. Angulo, *N. Engl. J. Med.*, 2002, **346**, 1221.

47. J. D. Browning and J. D. Horton, *J. Clin. Invest.*, 2004, **114**, 147.

48. M. Kohjima, M. Enjoji, N. Higuchi, M. Kato, K. Kotoh, T. Yoshimoto, T. Fujino, M. Yada, R. Yada, N. Harada, R. Takaynagi and M. Nakamuta, *Int. J. Mol. Med.*, 2007, **20**, 351.

49. C. J. Villanueva, M. Monetti, M. Shih, P. Zhou, S. M. Watkins, S. Bhanot and R. V. Farese, Jr., *Hepatology*, 2009, **50**, 434.

50. T. Yamazaki, E. Sasaki, C. Kakinuma, T. Yano, S. Miura and O. Ezaki, *J. Biol. Chem.*, 2005, **280**, 21506.

51. C. S. Choi, D. B. Savage, A. Kulkarni, X. X. Yu, Z. Liu, K. Morino, S. Kim, A. Distefano, V. T. Samuel, S. Neschen, D. Zhang, A. Wang, X. Zhang, M. Kahn, G. W. Cline, S. K. Pandey, J. G. Geisler, S. Bhanot, B. P. Monia and G. I. Shulman, *J. Biol. Chem.*, 2007, **282**, 22678.

52. K. Yamaguchi, L. Yang, S. McCall, J. Huang, X. X. Yu, S. K. Pandey, S. Bhanot, B. P. Monia, Y. Li and A. M. Diehl, *Hepatology*, 2008, **47**, 625.

53. H. C. Chen, S. J. Smith, Z. Ladha, D. R. Jensen, L. D. Ferreira, L. K. Pulawa, J. G. McGuire, R. E. Pitas, R. H. Eckel and R. V. Farese, Jr., *J. Clin. Invest.*, 2002, **109**, 1049.

54. H. Tilg and A. R. Moschen, *Trends Endocrinol. Metabol.*, 2008, **19**, 371.

55. M. Monetti, M. C. Levin, M. J. Watt, M. P. Sajan, S. Marmor, B. K. Hubbard, R. D. Stevens, J. R. Bain, C. B. Newgard, R. V. Farese. Sr. and R. V. Farese, Jr., *Cell Metabolism*, 2007, **6**, 69.

56. C. S. Fox, J. M. Massaro, U. Hoffmann, K. M. Pou, P. Maurovich-Horvat, C.-Y. Liu, R. S. Vasan, J. M. Murabito, J. B. Meigs, L. A. Cupples, R. B. D'Agostino and C. J. O'Donnell, *Circulation*, 2007, **116**, 39.

57. A. H. Kissebah and G. R. Krakower, *Physiol. Rev.*, 1994, **74**, 761.

58. G. Fruhbeck, J. Gomez-Ambrosi, F. J. Muruzabal and M. A. Burrell, *Am. J. Physiol. Endocrinol. Metab.*, 2001, **280**, E827.

59. H. C. Chen, S. J. Stone, P. Zhou, K. K. Buhman and R. V. Farese, Jr., *Diabetes*, 2002, **51**, 3189.

60. N. Chen, L. Liu, Y. Zhang, H. N. Ginsberg and Y. Yu, *Diabetes*, 2005, **54**, 3379.

61. H. C. Chen, D. R. Jensen, H. M. Myers, R. H. Eckel and R. V. Farese, Jr., *J. Clin. Invest.*, 2003, **111**, 1715.

62. R. S. Streeper, S. K. Koliwad, C. J. Villanueva and R. V. Farese, Jr., *Am. J. Physiol. Endocrinol. Metab.*, 2006, **291**, E388.

63. O. E. Owen, G. A. Reichard, G. Boden, M. S. Patel and V. E. Trapp, *Adv. Mod. Nutr.*, 1978, **2**, 517.

64. A. D. Baron, G. Brechtel, P. Wallace and S. V. Edelman, *Am. J. Physiol.*, 1988, **255**, E769.

65. B. B. Rasmussen and R. R. Wolfe, *Annu. Rev. Nutr.*, 1999, **19**, 463.

66. M. Krssak, K. Falk Petersen, A. Dresner, L. DiPietro, S. M. Vogel, D. L. Rothman, G. I. Schulman and M. Roden, *Diabetologia*, 1999, **42**, 113.

67. C. Yu, Y. Chen, G. W. Cline, D. Zhang, H. Zong, Y. Wang, R. Bergeron, J. K. Kim, S. W. Cushman, G. J. Cooney, B. Atcheson, M. F. White, E. W. Kraegen and G. I. Schulman, *J. Biol. Chem.*, 2002, **277**, 50230.

68. L. Liu, Y. Zhang, N. Chen, X. Shi, B. Tsang and Y. Yu, *J. Clin. Invest.*, 2007, **117**, 1679.
69. L. Liu, X. Shi, C. S. Choi, G. I. Shulman, K. Klaus, K. S. Nair, G. Schwartz, Y. Zhang, I. J. Goldberg and Y. Yu, *Diabetes*, 2009, **58**, 2516.
70. S. Timmers, J. de Vogel-van den Bosch, M. K. C. Hesselink, D. van Beurden, G. Schaart, M. J. Ferraz, M. Losen, P. Martinez-Martinez, M. H. De Baets, J. M. F. G. Aerts and P. Schrauwen, *PLoS One*, 2011, **6**, doi: 10.1371/journal.pone.0014503.
71. L. Liu, X. Shi, K. G. Bharadwaj, S. Ikeda, H. Yamashita, H. Yagyu, J. E. Schaffer, Y. Yu and I. J. Goldberg, *J. Biol. Chem.*, 2009, **284**, 36312.
72. D. J. Glenn, F. Wang, M. Nishimoto, M. C. Cruz, Y. Uchida, W. M. Holleran, Y. Zhang, Y. Yeghiazarians and D. G. Gardner, *Hypertension*, 2011, **57**, 216.
73. H. C. Chen, M. Rao, M. P. Sajan, M. Standaert, Y. Kanoh, A. Miura, R. V. Farese, Jr. and R. V. Farese, *Diabetes*, 2004, **53**, 1445.
74. L. Liu, S. Yu, R. S. Khan, G. P. Ables, K. G. Bharadwaj, Y. Hu, L. A. Huggins, J. W. Eriksson, L. K. Buckett, A. V. Turnbull, H. N. Ginsberg, W. S. Blaner, L.-S. Huang and I. J. Goldberg, *J. Lipid Res.*, 2011, **52**, 732.
75. B. H. Goodpaster, J. He, S. Watkins and D. E. Kelley, *J. Clin. Endocrinol. Metab.*, 2001, **86**, 5755.
76. S. Schenk, J. N. Cook, A. E. Kaufman and J. F. Horowitz, *Am. J. Physiol. Endocrinol. Metab.*, 2005, **288**, E519.
77. A. P. Russell, *Int. J. Obes. Relat. Metab. Disord.*, 2004, **28**, S66.
78. S. Ikeda, H. Miyazaki, T. Nakatani, Y. Kai, Y. Kamei, S. Miura, N. Tsuboyama-Kasaoka and O. Ezaki, *Biochem. Biophys. Res. Commun.*, 2002, **296**, 395.
79. S. Schenk and J. F. Horowitz, *J. Clin. Invest.*, 2007, **117**, 1690.
80. E. Herker and M. Ott, *Trends Endocrinol. Metab.*, 2011, **22**, doi:10.1016/j.tem.2011.03.004.
81. E. Herker, C. Harris, C. Hernandez, A. Carpentier, K. Kaehlcke, A. R. Rosenberg, R. V. Farese, Jr. and M. Ott., *Nature Med.*, 2010, **16**, 1295.
82. P. A. Carpino and B. Goodwin, *Expert. Opin. Ther. Patents*, 2010, **20**, 1627.
83. R. Smith, A.-M. Campbell, P. Coish, M. Dai, S. Jenkins, D. Lowe, S. O'Connor, N. Su, G. Wang, M. Zhang and L. Zhu, *W.O. Patent Application*, 100881, 2004.
84. R. Smith, D. Lowe, P. Coish, A.-M. Wang, M. Patel and G. Bondar, *W.O. Patent Application*, 113919, 2006.
85. Personal communication: Roger A. Smith, Bayer HealthCare, Pharmaceutical Division.
86. Pfizer signs agreement with Bayer to license compounds for potential treatment of obesity, diabetes and related disorders. June 14, 2006. http://www.pharma.bayer.com/scripts/pages/en/news_room/news_room/archive/newsroom_archive26.php.

87. Unpublished results, Cardiovascular & Metabolic Diseases Research Unit, Pfizer Inc.

88. Unpublished results, Bayer HealthCare, Pharmaceutical Division.

89. D. G. Perregaux, J. L. Treadway, C. Steppan, R. Dow, E. M. Gibbs, W. Zavadoski, D. Mather, J. Wisiniewski, S. Boyer, J. Joshi, J. LaPerle and T. B. Manion, Keystone Scientific Symposia, Triglycerides and Triglyceride Rich Particles in Health and Disease, Big Sky, Montana, January 9–14, 2010.

90. A. M. Birch, S. S. Bowker, R. J. Butlin, C. S. Donald, W. McCoull, T. Nowak and A. Plowright, *W.O. Patent Application*, 064189, 2006.

91. A. H. Dobson and W. Grundy, *W.O. Patent Application*, 144571, 2007.

92. A. M. Birch, S. Pal, A. Pettersen and G. P. Tomkinson, *W.O. Patent Application*, 070343, 2010.

93. AstraZeneca Development Pipeline, 27 January 2011: http://www.astrazeneca. com/cs/Satellite?blobcol=urldata&blobheader=application%2Fpdf& blobheadername1=Content-Disposition&blobheadername2=MDT-Type& blobheadervalue1=inline%3B+filename%3DDownload-pipeline- summary.pdf&blobheadervalue2=abinary%3B+charset%3DUTF-8& blobkey=id&blobtable=MungoBlobs&blobwhere=1285619440126& ssbinary=true

94. A. M. Birch, S. S. Bowker, R. J. Butlin, C. S. Donald, W. McCoull, T. Nowak and A. Plowright, *W.O. Patent Application*, 064189, 2006.

95. C. Johnstone and A. Plowright, *W.O. Patent Application*, 138304, 2007.

96. R. J. Butlin and A. Plowright, *W.O. Patent Application*, 138311, 2007.

97. R. J. Butlin, R. Davies and W. McCoull, *W.O. Patent Application*, 141502, 2007.

98. A. M. Birch, R. J. Butlin and A. Plowright, *W.O. Patent Application*, 141517, 2007.

99. Y. Huang, C. Sun, R. M. Lawrence, W. R. Ewing and H. Turdi, *W.O. Patent Application*, 126624, 2009.

100. R. R. Iyengar, G. Zhao, J. C. Freeman, J. Gao, A. S. Judd, P. R. Kym, J. K. Lynch, M. M. Mulhern and A. J. Souers, *W.O. Patent Application*, 137107, 2007.

101. A. S. Judd, M. M. Mulhern, R. R. Iyengar, P. R. Kym and A. J. Souers, *W.O. Patent Application*, 134693, 2008.

102. I. Hayakawa, M. Yoshida, Y. Sugano, H. Kurata, H. Karasawa, Y. Uto and T. Yaguchi, *W.O. Patent Application*, 019020, 2006.

103. Y. Uto, H. Karasawa and K. Takaishi, *W.O. Patent Application*, 119534, 2009.

104. H. Umemiya, M. Takahashi, M. Bohno, K. Kawabe, S. Shirokawa, T. Nagatsuka, S. Sasako, R. Satou, S. Itoh, T. Shimizu and K. Asamura, *W.O. Patent Application*, 011285, 2009.

105. G. Zhou, G. Wishart, P. C. Ting, R. G. Aslanian, N. Zorn and D. W.-S. Kim, *W.O. Patent Application*, 107765, 2010.

106. G. Zhou, G. Wishart, P. C. Ting, R. G. Aslanian, N. Zorn and J. Cao, *W.O. Patent Application*, 107768, 2010.
107. G. Zhao, A. J. Souers, M. Voorbach, H. D. Falls, B. Droz, S. Brodjian, Y. Y. Lau, R. R. Iyengar, J. Gao, A. S. Judd, S. H. Wagaw, M. M. Ravn, K. M. Engstrom, J. K. Lynch, M. M. Mulhern, J. Freeman, B. D. Dayton, X. Wang, N. Grihalde, D. Fry, D. W. A. Beno, K. C. Marsh, Z. Su, G. J. Diaz, C. A. Collins, H. Sham, R. M. Reilly, M. E. Brune and P. R. Kym, *J. Med. Chem.*, 2008, **51**, 380.
108. A. J. King, J. A. Segreti, K. J. Larson, A. J. Souers, P. R. Kym, R. M. Reilly, G. Zhao, S. W. Mittelstadt and B. F. Cox, *J. Pharmacol. Exp. Ther.*, 2009, **330**, 526.
109. X. Zhang, J. Yan, G. Yan, X. Sun, J. Ji, Y. Li, Y. Hu and H. Wang, *Acta Pharmacol. Sin.*, 2010, **31**, 1470.
110. (a) Y. Nakada, M. Ogino, K. Asano, K. Aoki, H. Miki, Y. Yamamoto, K. Kato, M. Masago, N. Tamura and M. Shimada, *Chem. Pharm. Bull.*, 2010, **58**, 673; (b) M. Ogino, Y. Nakada, M. Shimada, K. Asano, N. Tamura and M. Masago, *W.O. Patent Application*, 082952, 2006.
111. T. Yamamoto, H. Yamaguchi, H. Miki, M. Shimada, Y. Nakada, M. Ogino, K. Asano, K. Aoki, N. Tamura, M. Masago and K. Kato, *Eur. J. Pharmacol.*, 2010, **640**, 243.
112. A. S. Kalgutkar, I. Gardner, R. S. Obach, C. L. Shaffer, E. Callegari, K. R. Henne, A. E. Mutlib, D. D. Dalvie, J. S. Lee, Y. Nakai, J. P. O'Donnell, J. Boer and S. P. Harriman, *Curr. Drug Metab.*, 2005, **6**, 161.
113. P. Malatkova, E. Maser and V. Wsol, *Curr. Drug. Metab.*, 2010, **11**, 639.
114. (a) Y. Nakada, T. D. Aicher, Y. L. Huerou, T. Turner, S. A. Pratt, S. S. Gonzales, S. A. Boyd, H. Miki, T. Yamamoto and H. Yamaguchi, *Bioorg. Med. Chem. Lett.*, 2010, **18**, 2785; (b) S. Kitamura, T. D. Aicher, S. Gonzales, Y. Le Heurou, S. A. Pratt, T. Turner and Y. Nakada, *W.O. Patent Application*, 011131, 2008.
115. T. Yamamoto, H. Yamaguchi, H. Miki, S. Kitamura, Y. Nakada, T. D. Aicher, S. A. Pratt and K. Kato, *Eur. J. Pharmacol.*, 2011, **650**, 663.
116. (a) W. Yun, D. R. Bolin, S. Li, M. Ahmand, J. Cai, R. Goodnow, Jr., S. J. Wertheimer, K.Conde-Knape, Y. Chen, S. Kazmer, T. Whitard, C. Rondinone and R. Taub, presented at *237th American Chemical Society National Meeting*, Salt Lake City, UT, March 22–26, 2009, MEDI-256; (b) Y. Qian, S. J. Wertheimer, M. Ahmad, A. W.-H. Cheung, F. Firooznia, M. M. Hamilton, S. Hayden, S. Li, N. Marcopulos, L. McDermott, J. Tan, W. Yun, L. Guo, A. Pamidimukkala, Y. Chen, K.-S. Huang, G. B. Ramsey, T. Whittard, K. Conde-Knape, R. Taub, C. M. Rondinone, J. Tilley and D. Bolin, *J. Med. Chem.*, 2011, **54**, 2433.
117. B. M. Fox, N. Furukawa, X. Hao, K. Iio, T. Inaba, S. M. Jackson, F. Kayser, M. Labelle, K. Li, T. Matsui, D. L. McMinn, N. Ogawa, S. M. Rubenstein, S. Sagawa, K. Sugimoto, M. Suzuki, M. Tanaka, G. Ye, A. Toshida and J. Zhang, *W.O. Patent Application*, 047755, 2004.

118. R. L. Dow, J.-C. Li, L. Patel, C. Perreault, M. J. Munchhof, D. W. Piotrowski, E. M. Gibbs, W. J. Zavadoski, T. B. Manion, J. L. Treadway and J. L. LaPerle, *239th American Chemical Society Meeting*, San Francisco, CA, March 21–25, MEDI-315.

119. A. M. Birch, S. Birtles, L. K. Buckett, P. D. Kemmitt, G. J. Smith, T. J. D. Smith, A. V. Turnbull and S. J. Y. Wang, *J. Med. Chem.*, 2009, **52**, 1558.

120. R. L. Dow, J.-C. Li, M. P. Pence, E. M. Gibbs, J. L. LaPerle, J. Litchfield, D. W. Piotrowski, M. J. Munchhof, T. B. Manion, W. J. Zavadoski, G. S. Walker, R. K. McPherson, S. Tapley, E. Sugarman, A. Guzman-Perez and P. DaSilva-Jardine, *ACS Med. Chem. Lett.*, 2011, **2**, 407.

121. A. V. Stachulski, J. R. Harding, J. C. Lindon, J. L. Maggs, B. K. Park and I. D. Wilson, *J. Med. Chem.*, 2006, **49**, 6931.

122. (a) B. M. Fox, K. Iio, K. Li, R. Choi, T. Inaba, S. Jackson, S. Sagawa, B. Shan, M. Tanaka, A. Yoshida and F. Kayser, *Bioorg. Med. Chem. Lett.*, 2010, **20**, 6030; (b) B. M. Fox, K. Iio, T. Inaba, F. Kayser, K. Li, S. Sagawa, M. Tanaka and A. Yosida, *W.O. Patent Application*, 013907, 2005.

123. O. Okamoto, Y. Sasaki, H. Watanabe, H. Jona and K. D. Dykstra, *W.O. Patent Application*, 056496, 2010.

124. (a) M. Asai, T. Haketa, S. Inamura, M. Ishikawa, H. Jona, H. Kawamoto, H. Kurihara, J. Shibata, T. Shimamura, T. Suga and H. Watanabe, *W.O. Patent Application*, 084979, 2010; (b) M. Asai, T. Haketa, S. Inamura, H. Kurihara, T. Nishimura and T. Shimamura, *W.O. Patent Application*, 095766, 2010.

125. (a) D. A. Claremon, N. Liverton, H. G. Selnick and G. R. Smith, *W.O. Patent Application*, 40653, 1996; (b) K. Shinozaki, M. Murata, H. Taguchi and N. Miura, *W.O. Patent Application*, 40197, 2001; (c) R. E. Olson, H. Liu and L. A. Thompson, *W.O. Patent Application*, 19797, 2001.

126. Bayer HealthCare Pharmaceuticals News Release, June 14, 2006: http://www.pharma.bayer.com/scripts/pages/en/news_room/news_room/archive/newsroom_archive26.php.

127. Japan Tobacco Inc. Clinical Development, August 9, 2007: http://www.jt.com/investors/results/pharmaceuticals/pdf/P.L.20070809_E.pdf.

128. Japan Tobacco Inc. Clinical Development, July 31, 2008: http://www.jt.com/investors/results/pharmaceuticals/pdf/P.L.20080731_E.pdf.

129. A Study in Healthy Volunteers to Assess the Tolerability and Blood Levels of a Single Dose of AZD7687. ClinicalTrials.gov Identifier: NCT01046357: http://clinicaltrials.gov/ct2/show/NCT01046357?term=AZD7687&rank=2

130. AZD7687 Multiple Ascending Dose Study. ClinicalTrials.gov Identifier: NCT01119352: http://clinicaltrials.gov/ct2/show/NCT01119352?term=AZD7687&rank=3.

131. Japanese Single and Multiple Ascending Dose, Safety, Tolerability, Pharmacokinetic (PK) & Pharmacodynamic (PD) Study of AZD7687. ClinicalTrials.gov Identifier: NCT01217905: http://clinicaltrials.gov/ct2/show/NCT01217905?term=AZD7687&rank=1.

132. Pfizer development pipeline, May 12, 2011: http://www.pfizer.com/research/product_pipeline/product_pipeline.jsp.

133. A Single Dose Study of PF-04620110 in Overweight and Obese, Otherwise Healthy Volunteers. ClinicalTrials.gov Identifier: NCT00799006: http://clinicaltrials.gov/ct2/show/NCT00799006?term=pf-04620110&rank=4.

134. A Multiple Dose Study of PF-04620110 in Overweight and Obese, Otherwise Healthy Volunteers. ClinicalTrials.gov Identifier: NCT00959426: http://clinicaltrials.gov/ct2/show/NCT00959426?term=pf-04620110&rank=3.

135. A Multiple Dose Study of PF-04620110 in Healthy Overweight or Obese Subjects. ClinicalTrials.gov Identifier: NCT011246327: http://clinicaltrials.gov/ct2/show/NCT01146327?term=PF-04620110&rank=3.

136. A Multiple Dose Study of PF-04620110 in Type 2 Diabetes Patients. ClinicalTrials.gov Identifier: NCT01298518: http://clinicaltrials.gov/ct2/show/NCT01298518?term=pf-04620110&rank=2.

137. A 12 Week Study in Patients with Type 2 Diabetes Mellitus (T2DM). ClinicalTrials.gov Identifier: NCT00901979: http://clinicaltrials.gov/ct2/show/NCT00901979?term=LCQ908&rank=2.

138. Safety, Tolerability, Pharmacokinetics (PK) and Pharmacodynamics (PD) Assessment of LCQ908 in Patients with Severe Hypertriglyceridemia. ClinicalTrials.gov Identifier: NCT01146522: http://clinicaltrials.gov/ct2/show/NCT01146522?term=LCQ908&rank=1.

139. Novartis Clinical Pipeline, April 19, 2011: http://www.novartis.com/downloads/innovation/planned-filings.pdf.

Stearoyl-CoA Desaturase 1 (SCD1) Inhibitors: Bench to Bedside Must Only Go Through Liver

GANG LIU

Ambit Biosciences, San Diego, CA 92130, USA
E-mail: gliu@ambitbio.com

9.1 Introduction

9.1.1 Type 2 Diabetes, Obesity, and Dyslipidemia Epidemics

Type 2 diabetes is a metabolic disorder characterized by high blood glucose in the context of insulin resistance and relative insulin deficiency.[1] With readily available energy-dense, high-fat fast food, more sedentary lifestyle, and increased life expectancy, the number of people developing type 2 diabetes globally has been growing rapidly. The latest study estimated 347 million people currently afflicted, compared to 153 million in 1980.[2] The disease exerts a heavy toll on human lives and healthcare systems. In 2004, an estimated 3.4 million people died from consequences of high blood glucose globally and that number will double between 2005 and 2030.[3] In the US, this incurable, chronic, and often debilitating illness costs the healthcare system a staggering $174 billion a year.[4]

RSC Drug Discovery Series No. 27
New Therapeutic Strategies for Type 2 Diabetes: Small Molecule Approaches
Edited by Robert M. Jones
© The Royal Society of Chemistry 2012
Published by the Royal Society of Chemistry, www.rsc.org

There is a strong risk factor for developing type 2 diabetes in obese individuals. Of the people diagnosed with type 2 diabetes, about 80–90% are also diagnosed as obese. Obesity has been found to contribute to approximately 55% of cases of type 2 diabetes.[5] Therefore, effective anti-obesity agents can potentially address the underlining problems associated with insulin resistance.

Dyslipidemia is a disorder of lipoprotein metabolism, including lipoprotein overproduction or deficiency in the blood. The symptoms can include elevation of "bad" low-density lipoprotein (LDL), triglycerides (TG), and total cholesterol and decrease of "good" high-density lipoprotein (HDL).[6] Dyslipidemia can lead to symptomatic vascular disease, including athero-sclerosis, coronary artery disease (CAD), and peripheral arterial disease. A large body of evidence now supports the concept that insulin resistance contributes to the dysregulation of lipid metabolism.[7]

Over the past decade, the list of therapeutics treating type 2 diabetes has grown substantially − helping people gain better blood glucose control. Incidentally, all the newly approved drugs target the glucagon-like peptide-1 (GLP-1) axis. Injectable GLP-1 analogs, such as exenatide and liraglutide, have been real advances for type 2 diabetes treatment. Dipeptidyl peptidase IV (DPPIV) inhibitors, such as sitagliptin and saxagliptin, are providing treatment breakthroughs with the easiness of oral drugs. However, additional treatments with novel mechanism of action, particularly the ones addressing simultaneously the triad of insulin resistance, obesity, and dyslipidemia, are absolutely crucial for effectively combating the epidemic trend of these highly related disorders. Targeting stearoyl-CoA desaturase 1 (SCD 1) appears to be one of the promising approaches for fulfilling such critical needs.

9.1.2 Stearoyl-CoA Desaturases

SCD (Figure 9.1) is a lipogenic enzyme that catalyzes the critical committed step in the biosynthesis of monounsaturated fatty acids (MUFA). The function of SCD is to introduce a *cis*-double bond between carbons 9 and 10 of long-chain saturated fatty acyl-CoAs either derived from the diet or synthesized *de novo*.[8] The major desaturation substrates are palmitoyl(16:0)-CoA and stearoyl(18:0)-CoA, which are converted to palmitoleoyl(16:1)-CoA and oleoyl(18:1)-CoA, respectively.[9] The products, palmitoleic acid and oleic acid, are the major components of various lipids, including circulating TG, cholesterol esters (CE), skin lipid, wax esters, and cell membrane phospholipids (PL). Recently, the functions of MUFA have expanded beyond the realm of lipid building blocks. It is demonstrated that adipose tissue uses lipokines such as C16:1-palmitoleate, but not C18:1-oleate, to communicate with distant organs and regulate systemic metabolic homeostasis.[10] MUFA is also implicated as mediators in signal transduction and cellular differentiation.[11]

Two human SCD genes have been identified, SCD1 and SCD2 (aka SCD5). SCD1 is primarily found in liver, adipose tissue, eyelid, and skin, and is well-

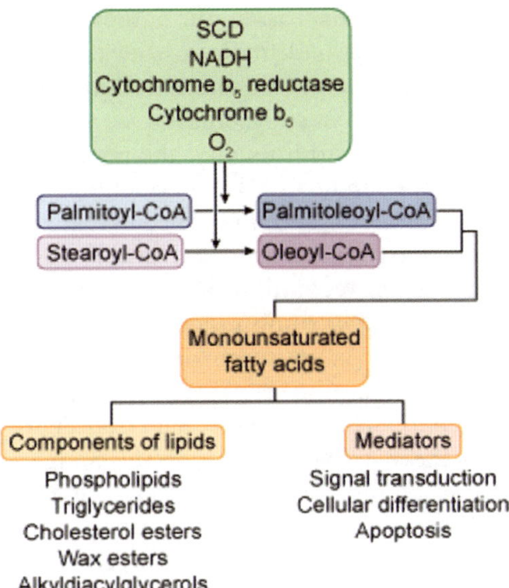

Figure 9.1 Functions of SCD. Adapted from http://mutagenetix.scripps.edu/phenotypic/phenotypic_rec.cfm?pk=169.

characterized compared to SCD2, which is found primarily in the brain. The SCD gene products are iron-containing, integral endoplasmic reticulum membrane proteins with four trans-membrane domains. The SCD enzymes have a 30-aa N-terminal sequence responsible for their rapid degradation and short half-life of 3–4 hours. SCD genes are tightly regulated by signals such as insulin, leptin, carbohydrate, fatty acids, nutritional status, and temperature.

Four SCD genes (SCD 1–4) are expressed in mouse and two SCD genes in rat. In adult mice, the SCD1 is prominently expressed in lipogenic tissues (liver and adipose) and sebaceous glands, SCD2 in most tissues except liver,[12] SCD3 in skin, preputial, and harderian gland,[13] and SCD4 in the heart.[14] The two human SCD genes show 85% homology to the murine SCD1 gene.[4,15,16]

9.2 Target Validation

9.2.1 Target Validation in Rodents

Through genetic deletion in rodents, SCD1 has been shown to play a crucial role in lipid synthesis and energy metabolism regulation. Asebia (ab^J/ab^J) mice with a natural knockout in the SCD1 gene[17,18] have deficiencies in hepatic CE and TG and very low levels of TG in the VLDL and LDL lipoprotein fractions compared with the normal animal.[19] Mice with targeted global deletion of SCD1 isoform have been shown to have reduced body adiposity, increased insulin sensitivity, increase energy expenditure, and are resistant to diet-induced weight gain.[20]

When ab^J/ab^J mice were backcrossed onto the diabetic C57/BL6 ob/ob background, the double mutant had the following phenotypes in comparison to the obese and leptin-deficient ob/ob mice: reduced weight and fat mass, increased percentage of lean mass, increased consumption of food, and increased energy expenditure.[21] Additionally, the hepatomegaly, steatosis, and elevated liver triglycerides and VLDL production of ob/ob mice were normalized in the double knock animals. The increased energy expenditure is due to increased fatty acid oxidation via activation of AMP-activated kinase (AMPK) in liver,[22] and down-regulation of sterol regulatory binding protein 1c (SREBP-1c), a lipogenic transcription factor.[23]

Inhibition of SCD1 activity via antisense oligonucleotide (ASO) knockdown of gene expression in diet-induced obese (DIO) mice also resulted in reduced adiposity, improved hepatic steatosis, and increased energy expenditure.[24,25] ASO treatment led to the re-programming of the lipogenic genes and FAO genes.[24] In Sprague Dawley rats fed a high-fat diet, SCD activity knockdown with ASOs improved hepatic insulin sensitivity, as indicated by increased glucose infusion rate and decreased hepatic glucose output during a hyperinsulinemic clamp study.[25] Application of RNA interference (RNAi) knocking down of SCD1 expression in the livers of ob/ob mice did not affect body weight or food intake, but significantly changed hepatic and systemic lipid profiles, including robustly decreased hepatic neutral lipid content in TAGs, diacylglycerols (DAGs), and CEs.[26]

In addition to lipid metabolism, SCD1 deficiency also improved glucose tolerance and insulin signaling. Increased basal phosphorylation of insulin receptor, insulin receptor substrate-1, and Akt were observed in muscle and brown adipose tissue of SCD1-deficient mice, accompanied by decreased message, protein and activity of protein tyrosine phosphatase 1B (PTP-1B).[27]

Recently, it was discovered that mice with a liver-specific knockout of SCD1 were protected from high-carbohydrate, but not high-fat diet-induced adiposity and hepatic steatosis.[28] Furthermore, inhibition of SCD1 via ASOs or RNAi recapitulated only a subset of the phenotypes observed in the globally SCD1-deficient mouse, suggesting the involvement of multiple tissues in the regulation of metabolism by SCD1.

9.2.2   Adverse Events Associated with SCD1 Gene Deletion

Although SCD1 inhibition has profoundly positive effects on many aspects of the metabolic syndrome, the accumulation of SCD1 substrates, i.e. saturated fatty acids (SFAs), appears to have three problematic consequences.

First of all, SCD1 plays a major role in the *de novo* synthesis of TGs, CEs, and wax esters required for normal skin and eyelid function. Like the Asebia (ab^J/ab^J) mice, the SCD1$^{-/-}$ mice exhibited cutaneous abnormalities with atrophic sebaceous glands and narrow eye fissure with atrophic meibomian glands. The SCD1$^{-/-}$ mice were deficient in eyelid TGs, CEs, and wax esters, and had significant elevation of free cholesterol in the skin and eyelid.[29]

Dietary supplement with oleate could not reverse the skin and eyelid phenotypes in SCD1-deleted animals, suggesting compartmentalization in the production and utilization of the MUFA by the body.

SCD1 is intimately involved in the regulation of cell membrane fluidity, an important feature of the cell structure. In 2007, Binczek *et al.* published a detailed study linking the obesity resistance of the SCD1$^{-/-}$ mouse with disruption of the epidermal lipid barrier and adaptive thermoregulation.[30] It was demonstrated that SCD1 deficiency disrupted the epidermal lipid barrier and led to uncontrolled trans-epidermal water loss, breakdown of adaptive thermoregulation and cold resistance, as well as a metabolic wasting syndrome. A confirmatory study using skin-specific deletion of SCD1 resulted in increased energy expenditure and resistance to high-fat diet-induced obesity and glucose intolerance, illustrating an underappreciated cross-talk between skin SCD1 and peripheral tissue in maintenance of energy homeostasis.[31]

Secondly, when SCD1 ASOs were tested in a mouse model of hyperlipidemia and atherosclerosis (LDLr(−/−)Apob(100/100)), increased aortic atherosclerosis was unexpectedly found and could not be reversed by dietary oleate.[32] Further analyses revealed that SCD1 inhibition promoted accumulation of saturated fatty acids in plasma and tissues and reduced plasma triglyceride. This was corroborated by another study, in which SCD1-deficient mice on the hyperlipidemic low-density lipoprotein receptor (LDLR)-deficient mice background have increased inflammation and atherosclerosis.[33] Conversely, upregulation of human SCD1 led to a desaturation of saturated fatty acids and facilitated their esterification and storage, thereby preventing downstream effects of lipotoxicity in primary human arterial endothelial cells.[34]

Thirdly, SCD1 inhibitors may also promote pancreatic β-cell death. It was shown that SCD deficiency in BTBR (black and tan, brachyuric; BTBR T+ *tf/tf*) ob/ob mice accelerated the progression to severe diabetes due to insufficient glucose-stimulated insulin secretion in the islet caused by a marked inhibition of hepatic lipogenesis that diverts lipid to the pancreas,[35] despite the fact that these animals have improved insulin sensitivity and reduced adiposity.

All these concerns are derived from the extreme cases of complete knockout of SCD1 gene. On the other hand, SCD1 knockout heterozygotes appear normal, which alleviates the concern associated with SCD1 inhibitors, since complete knockout of the SCD1 activity is not expected with SCD1 inhibitors. The pivotal 10-week SCD1 knockdown experiment with ASO did not induce the hair, skin, and eye abnormalities, unlike the global SCD$^{-/-}$ mice, since the distribution of ASO is restricted mostly to liver.[24] Additionally, the liver-specific knockout (LKO) of SCD1 mice using Cre-lox technology were indistinguishable from both the wild-type and Lox mice.[28] Such observations do suggest the need for viable SCD1 inhibitors to have a very narrow tissue distribution profile (liver, muscle, and adipose tissues) to avoid the aforementioned adverse effects manifested in the SCD1-deficient animals, even though such restriction may translate into attenuated efficacy for metabolic diseases.

9.2.3   Human Correlation

In contrast to mouse SCD1, information on the involvement of human SCD1 in type 2 diabetes, obesity, and lipid disorder is very preliminary; so far the relationship appears to be more associative than causative. A recent review summarized the role of SCD-1 in human metabolic disease,[36] where there appears to be an inverse relationship between the ratio of 18:1/18:0 or 16:1/16:0 and insulin sensitivity.[37,38] Positive correlations between SCD1 activity and plasma triglycerides in human hypertriglyceridemia and dyslipidemia in familial combined hyperlipidemia (FCHL) were established.[39,40] Recently a link between high muscle SCD1 activity and severe obesity in humans has been revealed.[41] Paradoxically, low hepatic SCD1 activity has been associated with fatty liver and insulin resistance in obese humans.[42] Even through the connection between SCD1 activity and lipid metabolism/energy expenditure appears solid, the true test will come when SCD1 inhibitors can be studied in well-designed human clinical trials.

9.3   First-Generation Systemically Distributed SCD Inhibitors

Even though SCD1 has been known since 1970s, the development of SCD1 inhibitors is a recent phenomenon following the improved understanding of SCD1 biological function in rodents and humans. Since SCD is an integral membrane protein, no crystal structure of the enzyme is available for guiding structure-based drug design. The discovery of novel leads for inhibiting SCD1 relied on the traditional high-throughput screening and subsequent scaffold hopping. The evolution of small molecule SCD1 inhibitors has been through a couple of reiterative cycles to date. This progression was propelled by the characterization of the pharmacological effects of the existing inhibitors. The first-generation SCD1 inhibitors roughly included the ones discovered before the year of 2009. A very comprehensive review of these compounds has appeared recently in the literature.[43] The new-generation SCD1 inhibitors were characterized by a concerted effort to target liver specifically for achieving adequate therapeutic window. This chapter will highlight mostly the compounds described in the primary literature.

9.3.1   Novel SCD Inhibitors

9.3.1.1   Pyridazine Carboxamides and Related Analogs

The first bona fide SCD1 inhibitor was disclosed by Xenon Pharmaceuticals in 2005 and started the subsequent explosive growth of small molecule SCD1 inhibitors.[44,45] The basic molecular scaffold of this series of compounds was a central pyridazine (**1**)/pyridine (**2**) substituted by functionalized piperazine benzamide on the one end, and a carboxamide on the other end (Table 9.1).

Table 9.1 Pyridazine carboxamides and related analogs.

1

2

3

4

5

6

7

8 R = CF$_3$
9 R = F

10

Compound **2** was highlighted in one of the patent applications as a moderately potent mouse SCD1 inhibitor (IC$_{50}$ = 100 nM).[45]

Based on **1**, an SAR campaign at Abbott Laboratories directed at addressing the potency and metabolic stability issue led to the discovery of highly potent, selective, orally bioavailable SCD1 inhibitors with robust cellular activity in HepG2 cells, as represented by compound **3**.[46] Compound **3** was not only a potent SCD1 inhibitor against mouse and human SCD1s with IC_{50} values of 4.5 and 26 nM, respectively, it also potently inhibited the long chain fatty acid (LCFA)-CoA desaturation in HepG2 cell with an IC_{50} value of 6.8 nM. Compound **3** also exhibited favorable rat PK profiles suitable for *in vivo* studies (AUC = 10.66 µg·h/mL, and F = 59%) when dosed orally at 10 mg/kg.

A similar compound to **3** was published by Merck Frosst, using a thiadiazole as the carboxamide bioisostere. Compound **4** (MF-438) had an IC_{50} value of 2.3 nM against rat SCD1, good PK in rodents with bioavailability of 73% and 38% in mice and rats, respectively.[47] It exhibited an ED_{50} between 1 and 3 mg/kg in inhibiting mouse liver SCD1 activity, as measured by inhibiting the conversion of ^{14}C-stearic acid to ^{14}C-oleic acid.

Another compound **5** (Table 9.1) was identified from the optimization of an initial lead thiazole compound, MF-152,[48] by the same lab.[49] Compound **5** was a potent SCD inhibitor with IC_{50} values of 1 nM against both rat SCD enzyme and HepG2 cells, and selective over both Δ5 and Δ6 desaturases (IC_{50} > 2 µM). It had improved bioavailability and whole blood exposure (F = 48%, AUC0–24 h = 16.8 µM·h) when dosed orally at 10 mg/kg. Compound **5** dose-dependently decreased the liver SCD activity index (ratio of ^{14}C-oleic acid/^{14}C-stearic acid) with an ED_{50} of ∼0.3 mg/kg.

Similarly to compound **3**, compound **6**, a complementary azetidine variation of the piperidine, was also reported by the same group from Merck.[50] It was based on the optimization of an earlier analog, MF-438.[51] The *in vitro* profile of **6** was somewhat worse than that of **5** (IC_{50} values of 4.9 and 50 nM against rSCD1 and HepG2 cells, respectively). Compound **6** appeared to be slightly more active *in vivo* – it achieved greater than 90% suppression of oleic acid production at a dose of 1 mg/kg in a mouse liver pharmacodynamic model.

Daiichi Sankyo also explored the SAR of pyridazine carboxamide extensively. The optimization process led to compound **7** with IC_{50} values of 28 nM (enzymatic) and 43 nM (293A cell) versus hSCD1.[52] In a PK-PD study, **7** inhibited the conversion of ^{14}C-stearate to ^{14}C-oleate in the liver of db/db mice with an ED_{50} of 3 mg/kg. The combination of hydroxy and 3-pyridyl groups improved the microsomal stability and oral absorption of **7** in mice (F = 68%, AUC of 7.8 µM·h at 20 mg/kg dosing).

Further potency improvement led to the discovery of compounds **8** and **9**. Compound **8** has IC_{50} values of 0.03 nM (hSCD) and 0.8 nM (293A cells).[53] Oral exposure of **8** in mice was AUC of 7.8 µM·h after a single 20 mg/kg dosing. Analog **9** had similar *in vitro* potency (0.3 nM versus hSCD1, and 0.7 nM in 293A cells).[54] Oral bioavailability of the compound was improved to greater than 95%, although the oral AUC remained the same (6.8 µM·h) as **8**. Spiro[1,5-benzoxazepine-2,4-piperidine] derivatives were also discovered as SCD1 inhibitors by Daiichi Sankyo, albeit somewhat less potent than compound **9**. For example, compound **10** had IC_{50} values of 10 nM against both mSCD1 and hSCD1 in 293A cells.[55]

9.3.1.2 Thiazole Carboxamides and Related Analogs

Xenon and Novartis have been collaborating on discovering novel SCD1 inhibitors since 2004 (Table 9.2). It appeared that thiazolyl-5-carboxamide scaffolds attracted most of the attention in the collaboration so far, based on the number of patent applications filed. One of the initial carboxamides **11**[56] was modified to give an interesting 2-oxopyridin-1(2*H*)-yl thiazole carboxamide derivative **12**, with IC_{50} value of 30 nM against mouse SCD1.[57]

A number of other motifs have been introduced as carboxamide replacements and yielded potent mouse SCD1 inhibitors. 2-(Pyrazin-2-yl)-thiazole derivative **13**[58] and triazolyl thiazole derivative **14**[59] had IC_{50} values of 42 nM and 10 nM, respectively. More recently, an imidazolinone derivative **15** was claimed to be a potent (IC_{50} = 8 nM) mouse SCD1 inhibitor.[60] The latest Novartis patent application depicted compound **16**, by switching the thiazole to pyrazole to give a potent mouse SCD1 inhibitor (IC_{50} = 3 nM).[61]

Starting from a screening hit, Daiichi Sankyo's lead optimization led to the discovery of potent SCD1 inhibitor **17** with IC_{50} values of 4 nM versus hSCD and 9 nM in the human 293A cell assay.[62] Mouse PK of **17** (dosed at 20 mg/kg) was pretty reasonable with AUC of 10 μg·h/mL, even though the bioavailability was somewhat low (12%). The ED_{50} in lowering the desaturation index (DI) in the liver of db/db mice was 1 mg/kg.

Table 9.2 Thiazole carboxamides and related analogs.

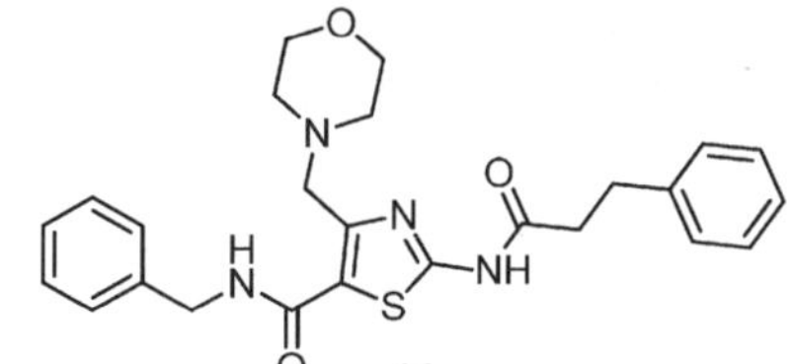

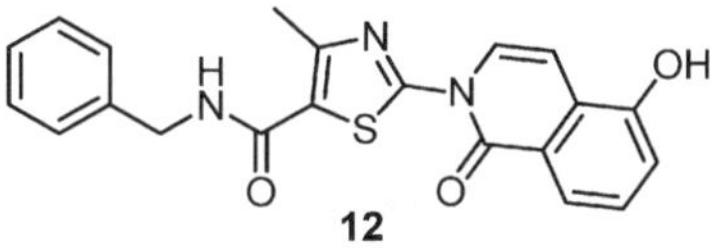

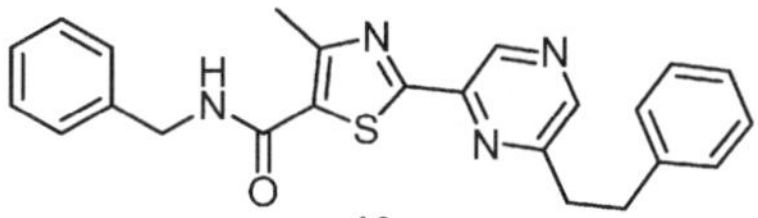

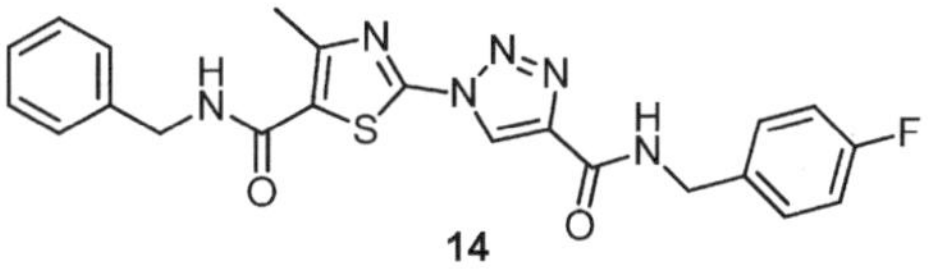

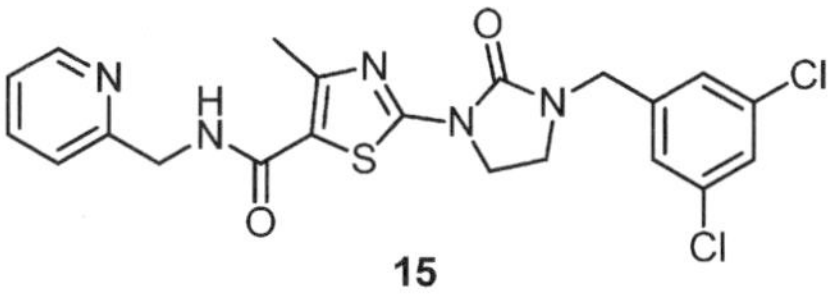

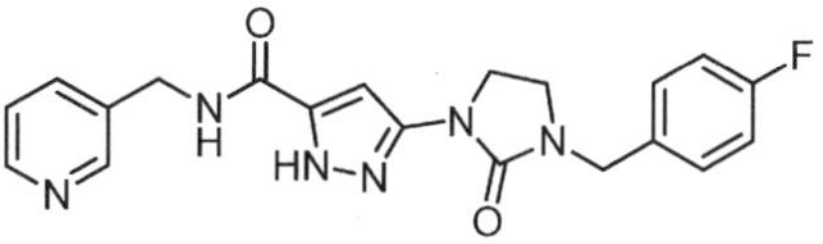

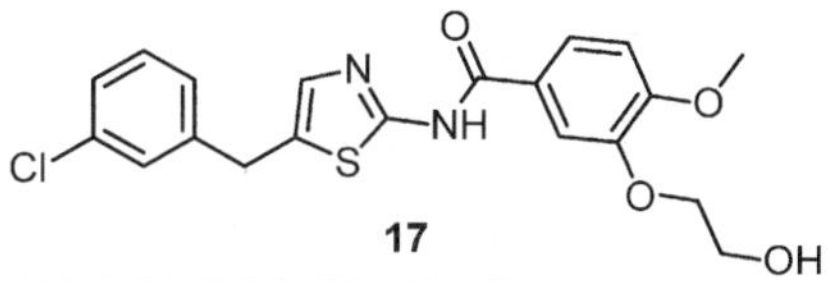

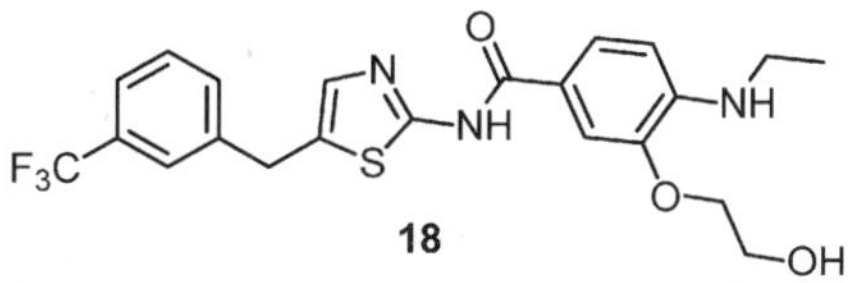

Further effort in improving oral bioavailability led to the identification of compound **18**.[63] Not only was the compound more potent *in vitro* (0.04 nM versus human SCD, 2 nM in a cell assay using A293A cells), it was also shown to have somewhat better oral bioavailability (27%), in comparison to **17**. However, the oral exposure (AUC) was actually less than half of that of **17**.

9.3.1.3 *Pteridinones and Related Analogs*

Recently, a number of publications featuring different pteridinone scaffold-based SCD1 inhibitors have been reported by CV Therapeutics (Table 9.3). Starting from a pteridinone derivative as the initial lead, systematic modification of the core template led to improvement in both potency and *in vitro* ADME profiles. One of the most potent pteridinone analogs is **19** (IC_{50} = 0.6 nM versus rSCD, IC_{50} = 0.05 nM in HepG2 cells) with good liver microsome stability.[64] Compound **20** (CVT-11,563) was discovered by scaffold-hopping from a HTS hit using a quinazolin-4-one core template. It has IC_{50} values of 267 nM (rSCD) and 79 nM (HepG2), and reasonable exposure (AUC = 935 ng·h/mL) and high oral bioavailability (90%).[65]

Merging select structural features of the two series led to the discovery of compound **21** (CVT-12,012), which is highly potent in a human cell-based (HEPG2) SCD assay (IC_{50} = 6 nM).[66] This compound has 78% oral bioavailability in rats and is preferentially distributed into liver (76 times versus plasma). In a 5-day study with sucrose-fed rats, compound **21** significantly reduced SCD activity, as measured by the DI lowering of the total lipid in a dose-dependent manner in plasma and liver (5, 10, 20 mg/kg),

Table 9.3   Thiazole carboxamides and related analogs.

and significantly reduced liver triglycerides versus the control group ($\sim$50%) at 20 mg/kg.

9.3.1.4 Urea and Amide-based Inhibitors

A series of piperidine urea-based SCD1 inhibitors was developed using a scaffold hopping strategy from the nicotinamide template of **2** (Table 9.4). The series was exemplified by lead compound **22**, which exhibited excellent *in vitro* profiles (IC_{50} < 4 nM versus mSCD1, 37 nM versus hSCD1) and PK properties in mice (CL = 0.4 L/h·kg, V_{ss} = 0.4 L/kg, oral AUC = 13.3 µg·h/mL, oral F = 92%).[67] Japan Tobacco claimed a similar series of urea compounds as SCD1 inhibitors. One representative compound **23**, for example, was claimed to have an IC_{50} value below 0.1 µM against rat SCD1.[68]

Scaffold hopping from the pyridazine template by Abbott resulted in a different series of potent SCD1 inhibitors. The pyridine glycine amide **24** potently inhibited human SCD1 with an IC_{50} value of 90 nM, while a pyrazine analog **25** demonstrated an IC_{50} value of 37 nM against human SCD1.[69]

A tool compound **26** (GSK993), with IC_{50} values of 0.18 µM versus rSCD1, and 0.17 µM in HepG2 cells, was published by GSK.[70] Compound **26** at 2 µM plasma exposure strongly inhibited SCD1 activity in primary human hepatocytes, differentiated human adipocytes, and human skeletal muscle cells. Good plasmatic exposure following oral administration of the compound was demonstrated.

Table 9.4   Urea and amide-based inhibitors.

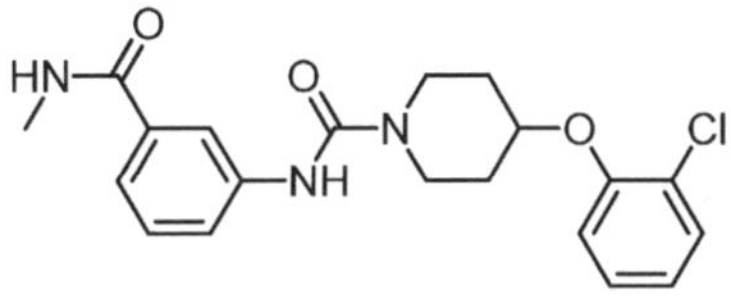

22

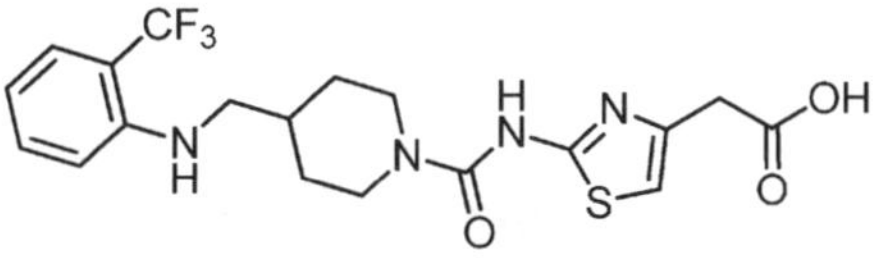

23

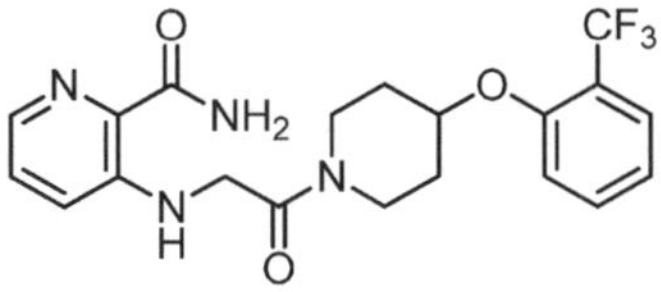

24

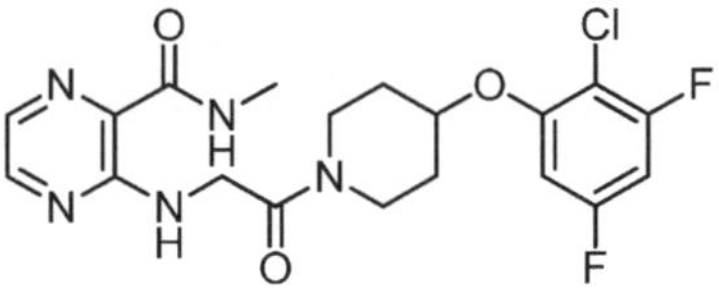

25

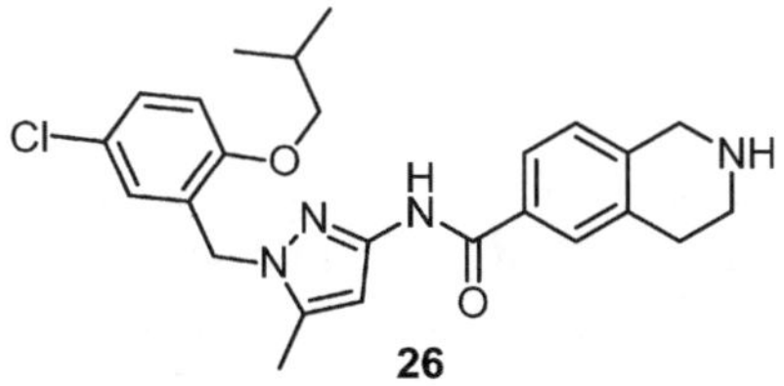

26

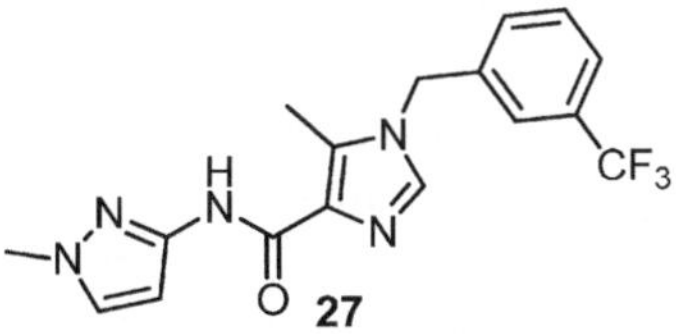

27

Table 9.5   SCD1 selective inhibitor.

28

A series of imidazole-4-carboxamides has been optimized from a HTS hit at Pfizer to provide a tool compound for discerning mechanism-based adverse effects from potential chemotype derived toxicity. A single dose of compound **27** (IC_{50} = 11.7 nM for rSCD1, 52% oral bioavailability in rat) showed a dose-dependent decrease in DI of total plasma lipid with a 55% reduction at 1 mg/kg and 85% reduction at the 10 and 30 mg/kg doses.[71]

9.3.1.5   SCD1 Selective Inhibitors

The only series of inhibitors selective for hSCD1 over hSCD2 was reported recently by Merck Frosst.[72] Benzimidazoles were identified from a HTS campaign and later confirmed in a selectivity assay against hSCD1 and hSCD2. The compound **28** appeared to be the more interesting one with IC_{50} values of 65 nM versus rSCD1, 27 nM versus hSCD1, and >20 μM versus hSCD2. Compound **28** (Table 9.5) possessed suitable cellular activity (1.3 μM in HepG2 cell) and mice pharmacokinetic properties (AUC = 8 μM·h and F = 39% at 10 mg/kg dose), and demonstrated liver SCD activity inhibition (36% at 30 mg/kg and 78% at 100 mg/kg) in a mouse pharmacodynamic assay.

9.3.2   Pharmacology and Adverse Events

A desaturation index (DI), defined as the ratio of 16:0/16:1n7 or 18:0/18:1n9 in any lipid class, has been used as an *in vivo* biomarker for SCD1 activity. A dose-dependent liver TG DI lowering effect was observed in ob/ob mice with the treatment of SCD1 inhibitor **22** dosed at 0.3, 3, and 10 mg/kg BID for 4.5 days, with the high dose demonstrating normalization of liver TG DI.[68] Similar degrees of DI decreases were observed with other liver lipid classes, including PL, CE, and DAG. These observations demonstrated a clear and general correlation between *in vitro* SCD1 inhibition and *in vivo* lipid DI lowering.

The profiles of compound **22** in an efficacy study in diet-induced obese (DIO) mice were reported.[73] When animals were treated with compound **22** (10 mg/kg, BID, p.o.) for 3 weeks, approximately 10% body weight loss over the vehicle group and reduced food intake were observed. The insulin sensitivity was restored to the level of lean animals, as measured by an insulin sensitivity test (ITT) on day 17. Unfortunately, after about 2 weeks treatment, the animals developed alopecia on the neck and around the eye, as well as eye ptosis (squinting). Skin histology confirmed that atrophy of the sebaceous glands and thickening of the epidermal layer in the drug-treated group. Apparently, SCD1 inhibition by compound **22** has disrupted the epidermal and follicular homeostasis. The TG DIs in the same skin samples were significantly lowered, and the total TG level was reduced to the lean level. More importantly, the total CE in the skin was reduced beyond lean level, while the free cholesterol was increased. All these observations recapitulated the SCD1 gene depletion phenotypes, suggesting the interconnection between metabolic phenotypes and adverse skin/hair/ocular effects upon SCD1 inhibition.

The anti-diabetic and anti-dyslipidemic efficacy and adverse effects observed with compound **22** were confirmed by other inhibitors. Mice treated with compound **5** at 0.2 mg/kg/day showed a robust 24% reduction in body weight gain with no significant change in food consumption as compared to the control mice on high-fat diet (HFD).[49] In addition, the mice treated with **5** had body weight gain comparable to that of mice fed on a normal chow diet during the study. Furthermore, the resistance to HFD-induced body weight gain with **5** was associated with an improved metabolic profile as exemplified by a $\sim 24\%$ reduction in plasma glucose and a $\sim 4\%$ reduction in insulin levels. The mice treated with **5** also developed partial eye closure and progressive alopecia after 7 days of drug treatment.

Compound **6** was evaluated in a 1-month growing diet-induced obesity model (gDIO) in mice in an attempt to separate efficacy from adverse effects. Unfortunately, even at a very low dose of 0.1 mg/kg/day, both statistically significant improvement in body weight over the control group and skin adverse effects were observed after 3 weeks of treatment.[50]

Interestingly, for both **5** and **6**, relative tissue distribution after oral dosing was reported. For **5**, 4.5-fold and 6-fold higher exposures in the liver than those in plasma were observed at 2 and 6 hours post dosing, respectively. For **6**, 4.5-fold higher drug concentration was observed in the liver compared to that of plasma at 2 hours post dosing. Apparently, such partition is not enough to separate the pharmacological effects in the liver from adverse effect in the skin and eyelids.

Compound **7** was tested in a 7-day efficacy study using Zucker fatty rats. Once-daily administration of **7** dose dependently reduced the plasma triglyceride levels, with a 56% reduction at 30 mg/kg.[52] The duration of the study is too short to claim this compound behaved differently from other systemically distributed compounds in term of avoiding any skin or eye adverse events.

Compound **9** was profiled in C57BL/6 mice on a high carbohydrate diet. Significant plasma DI lowering effect of **9** after 7 days (qd) was noted, with 32% and 66% reduction at 0.3 and 10 mg/kg, respectively. Mild eye fissure was also noted at 3 and 10 mg/kg dose groups.[54]

Compound **20** was studied in an *in vivo* efficacy study in Sprague Dawley rats that were kept on a high-carbohydrate diet for 4 weeks. A highly significant reduction in SCD product fatty acids in both plasma and liver was observed, based on DI change following bid dosing of **20** for 5 days.[65] Even though compound **20** was claimed to be liver preferred, the 3–5-fold separation from plasma concentration was not expected to translate into a significant therapeutic window.

In Zucker fa/fa rats, compound **26** (50 mg/kg for 14 days qd) exerted a marked reduction in hepatic lipid DIs and lipid contents (CE and TG) as well as a significant improvement of glucose tolerance as measured in a glucose tolerance test.[70] Furthermore, in a diet-induced insulin resistant rat model, **26** (50 mg/kg for 4 weeks qd) induced a very strong reduction in Triton-induced hepatic VLDL-TG production. In addition, the whole body insulin sensitivity was improved in a hyperinsulinemic–euglycemic clamp study, consistent with a selective and significant reduction in hepatic DI. Interestingly, no skin or eyelid liabilities were observed after 4 weeks of treatment with GSK993, although relative tissue distribution information of the compound was not disclosed.

The mechanism-based toxicity observed on the skin, hair, and eyelid with compounds **22**, **5**, and **6** recapitulated the SCD1 KO phenotypes without complete inhibition of SCD1 inhibition. Therefore, it remains a challenge for a globally distributed small molecule inhibitor to achieve adequate therapeutic efficacy without affecting normal human skin and eye functions. Given the non-life threatening nature of metabolic diseases and the increasingly tight regulatory environment, it is paramount to establish a sufficient therapeutic window for SCD1 inhibitors in metabolic indications before embarking on expensive human clinical trials.

9.4 Second-Generation Liver-Targeted SCD Inhibitors

9.4.1 Passively Liver Selective Inhibitors

The disclosure of skin and ocular adverse effects of systemically distributed SCD1 inhibitors, such as **22**, ushered in a new requirement for small molecule SCD1 inhibitors: they must be selective for insulin-responsive tissues (liver, adipocyte, and muscle) to be viable therapeutic agents.

Many of the first-generation compounds were profiled for their relative tissue distributions. Some compounds are found to have higher exposure in the liver than that in the plasma. In one extreme case, the liver concentration is 76-fold higher than that of plasma.[66] Even in those situations, no clear separation of efficacy and adverse effects has been documented. In some cases, much shorter duration of treatment has obscured the interpretation of the pharmacological results.[52] Ultimately, achieving an adequate therapeutic index must be demonstrated in longer term studies.

9.4.2 SCD Inhibitors Actively Transported into Liver

The 10-week SCD1 ASO treatment resulted in 60% reduction of desaturase activity, prevention of diet-induced obesity, and reduction in adiposity without the observed skin and eyelid issues and so encouraged the field to carry on despite the setbacks suffered by the first-generation SCD1 inhibitors. That study suggests that it is possible to harness the benefit of complete loss of SCD1 activity without adverse effects, *as long as the tissue distribution of the inhibitors is restricted to the liver.*[24]

A recent landmark publication from Merck Frosst illustrated the possibility of selectively targeting liver with small molecule SCD1 inhibitors via rational design to establish a therapeutic window. The key strategy is to take advantage of molecular recognition by liver-specific organic anion-transporting polypeptides (OATPs) so that a poorly permeable SCD1 inhibitor can be delivered to liver while minimizing its exposure in the skin and eyes associated with mechanism-based adverse effects. This strategy entails two important aspects: first one is to precisely incorporate key transporting elements, i.e. acidic moieties, into the designed molecule to enable recognition by the OATP transport proteins. Meanwhile, the presence of acidic functional groups usually reduces the passive cell penetration to off-target tissues (skin and eye).

Starting with MF-438, compound **29** (MK-8245, Table 9.6) was identified as a clinical candidate compound by placing acidic moieties to the thiadiazole side of the molecule while maintaining SCD1 potency.[74] It has IC_{50} values of 3 nM against rat SCD1 and 60 and 5 nM in rat and human primary hepatocytes expressing OATP1, respectively. The activity is much weaker in HepG2 cells (IC_{50} = 1066 nM) due to the lack of OATP expression and poor passive cell permeability. When MK-8245 was dosed in mice, rat, dog, and rhesus monkeys, moderate oral bioavailability (12%, 28%, and 40% in mice, rats, and dogs, respectively) was demonstrated. Compound **29** distributes mainly to the liver, with low exposures in skin and harderian gland. The liver-to-skin ratios

Table 9.6   Second-generation liver-targeted SCD1 inhibitors.

were >30:1 in all four species. Compound **29** was found to be a substrate for both OATP1B1 and OATP1B3.

Other similar compounds were recently claimed in the patent literature by the same group. Some of the best structural features of the first-generation SCD1 inhibitors were combined with the liver-targeting acid moieties to generate potent SCD1 inhibitors, such as compound **30** (IC_{50} = 10 nM),[75] and spiro analog **31** (IC_{50} = 0.92 nM) (Table 9.6).[76]

9.4.3 Pharmacological Characterization of Liver-Targeted SCD Inhibitors

The liver-targeted SCD1 inhibitor demonstrated excellent preclinical *in vivo* efficacy. Compound **29** improved glucose clearance dose dependently with an ED_{50} value of 7 mg/kg in an acute oral glucose tolerance test in eDIO mice.[74] When **29** dosed at 20 or 60 mg/kg bid p.o. in a 4-week oral dosing study in eDIO mice, a modest prevention of body weight gain (~5% BW gain reduction) was observed while hepatic steatosis was significantly reduced, which was confirmed by a reduction of liver triglyceride levels. There was no effect on food consumption with compound **29**. More importantly, the liver-targeted **29** did not induce any meaningful skin or eye adverse events during the 4-week treatment period at 20 mg/kg dose. Consistently, there was very significant reduction in liver DI (63%) while no significant reduction in harderian gland DI was detected.

To demonstrate improvement in whole body insulin sensitivity, a hyperinsulinemic-euglycemic glucose clamp experiment in overfed rats was carried out. Compound **29** dosed at 30 mg/kg significantly increased the glucose infusion rate (GIR) required to maintain euglycemia, indicative of improved insulin sensititivity.[74]

At a 3-fold higher dose of **29** (60 mg/kg bid) during a 4-week study, a clear skin and eyelid adverse event profile was observed.[74] Therefore, a therapeutic window, albeit a somewhat narrow one, could be obtained by limiting the SCD1 inhibitor to the liver. Compound **29** is currently in phase 2 clinical trials, and future reports on the safety and efficacy of the compound in human clinical studies are being anxiously anticipated.

9.5 Conclusions

Over the past decade, major advances have been made in establishing SCD1 inhibition as a novel approach for treating type 2 diabetes, obesity, and dyslipidemia. Many potent and selective SCD1 inhibitors have now been discovered and evaluated preclinically. The first-generation compounds were plagued with lack of adequate therapeutic window in treating metabolic disorders. The more recent advancement of liver-targeted SCD1 inhibitors, e.g. MK-8245, into human clinical trials, has rejuvenated the field. Even though such compounds will only partially fulfill the promise of originally envisioned

SCD1 inhibition based on genetic deletion, the potential payoff with this novel mechanism of action is worthy the investment. The coming years will be fascinating with the unveiling of the clinical data from MK-8245 and related compounds, and continued advancement of additional liver-targeted SCD1 inhibitors.

References

1. V. Kumar, N. Fausto, A. K. Abbas, R. S. Cotran and S. L. Robbins, in *Robbins and Cotran Pathologic Basis of Disease*, 7th edn, Saunders, Philadelphia, PA, 2005, p. 1194.
2. G. Danaei, M. M. Finucane, Y. Lu, G. M. Singh, M. J. Cowan, C. J. Paciorek, J. K. Lin, F. Farzadfar, Y.-H. Khang, G. A. Stevens, M. Rao, M. K. Ali, L. M. Riley, C. A. Robinson and M. Ezzati, *The Lancet*, 2011, **378**, 31; http://www.thelancet.com/popup?fileName=cite-using-doi
3. Diabetes Fact sheet No. 312, January 2011, World Health Organization.
4. http://www.diabetes.org/diabetes-basics/diabetes-statistics/
5. Centers for Disease Control and Prevention (CDC), MMWR. Morbidity and Mortality Weekly Report, 2004, **53**, 1066; http://en.wikipedia.org/wiki/PubMed_Identifier
6. C.F. Semenkovich in *Cecil Medicine*, 23rd edn, Chapter 217, ed. L. Goldman and D. Ausiello, Saunders Elsevier, Philadelphia, Pa, 2007.
7. R. K. Semple, A. Sleigh, P. R. Murgatroyd, C. A. Adams, L. Bluck, S. Jackson, A. Vottero, D. Kanabar, V. Charlton-Menys, P. Durrington, M. A. Soos, T. A. Carpenter, D. J. Lomas, E. K. Cochran, P. Gorden, S. O'Rahilly and D. B. Savage, *J. Clin. Invest.*, 2009, **119**, 315.
8. M. Miyazaki and J. M. Ntambi, *Prostaglandins Leukot. Essent. Fatty Acids*, 2003, **68**, 113.
9. H. G. Enoch, A. Catala and P. Strittmatter, *J. Biol. Chem.*, 1976, **251**, 5095.
10. H. Cao, K. Gerhold, J. R. Mayers, M. M. Wiest, S. M. Watkins and G. S. Hotamisligil, *Cell*, 2008, **134**, 933.
11. T. Yonezawa, S. Haga, Y. Kobayashi, K. Katoh and Y. Obara. *Biochem. Biophys. Res. Commun.*, 2008, **367**, 729.
12. K. H. Kaestner, J. M. Ntambi, T. J. Kelly, Jr. and M. D. Lane, *J. Biol. Chem.*, 1989, **264**, 14755.
13. Y. Zheng, S. M. Prouty, A. Harmon, J. P. Sundberg, K. S. Stenn and S. Parimoo, *Genomics*, 2001, **71**, 182.
14. M. Miyazaki, M. J. Jacobson, W. C. Man, P. Cohen, E. Asilmaz, J. M. Friedman and J. M. Ntambi, *J. Biol. Chem.*, 2003, **278**, 33904.
15. L. Zhang, L. Ge, S. Parimoo, K. Stenn and S. M. Prouty, *Biochem. J.*, 1999, **340**, 255.
16. S. Beiraghi, M. Zhou, C. B. Talmadge, N. Went-Sumegi, J. R. Davis, D. Huang, H. Saal, T. A. Seemayer and J. Sumegi, *Gene*, 2003, **309**, 11.
17. Y. Zheng, K. J. Eilertsen, L. Ge, L. Zhang, J. P. Sundberg, S. M. Prouty, K. S. Stenn and S. Parimoo, *Nat. Genet.*, 1999, **23**, 268.

18. J. P. Sundberg, D. Boggess, B. A. Sundberg, K. Eilertsen, S. Parimoo, M. Filippi and K. Stenn, *Am. J. Pathol.*, 2000, **156**, 2067.
19. M. Miyazaki, Y. C. Kim, M. P. Gray-Keller, A. D. Attie and J. M. Ntambi, *J. Biol. Chem.*, 2000, **275**, 30132.
20. J. M. Ntambi, M. Miyazaki, J. P. Stoehr, H. Lan, C. M. Kendziorski, B. S. Yandell, Y. Song, P. Cohen, J. M. Friedman and A. D. Attie, *Proc. Natl. Acad. Sci. USA*, 2002, **99**, 11482.
21. P. Cohen, M. Miyazaki, N. D. Socci, A. Hagge-Greenberg, W. Liedtke, A. A. Soukas, R. Sharma, L. C. Hudgins, J. M. Ntambi and J. M. Friedman, *Science*, 2002, **297**, 240.
22. P. Dobrzyn, A. Dobrzyn, M. Miyazaki, P. Cohen, E. Asilmaz, D. G. Hardie, J. M. Friedman and J. M. Ntambi, *Proc. Natl. Acad. Sci. USA*, 2004, **101**, 6409.
23. M. Miyazaki, Y. C. Kim and J. M. Ntambi, *J. Lipid Res.*, 2001, **42**, 1018.
24. G. Jiang, Z. Li, F. Liu, K. Ellsworth, Q. Dallas-Yang, M. Wu, J. Ronan, C. Esau, C. Murphy, D. Szalkowski, R. Bergeron, T. Doebber and B. B. Zhang, *J. Clin. Invest.*, 2005, **115**, 1030.
25. R. Gutiérrez-Juárez, A. Pocai, C. Mulas, H. Ono, S. Bhanot, B. P. Monia and L. Rossetti, *J. Clin. Invest.*, 2006, **116**, 1686.
26. H. Xu, D. Wilcox, P. Nguyen, M. Voorbach, H. Smith, S. Brodjian, T. Suhar, R. M. Reilly, P. B. Jacobson, C. A. Collins, K. Landschulz and T. K. Surowy, *Front. Biosci.*, 2007, **12**, 3781.
27. S. M. Rahman, A. Dobrzyn, S. H. Lee, P. Dobrzyn, M. Miyazaki and J. M. Ntambi, *Am. J. Physiol. Endocrinol. Metab.*, 2005, **288**, E381.
28. M. Miyazaki, M. T. Flowers, H. Sampath, K. Chu, C. Otzelberger, X. Liu and J. M. Ntambi, *Cell Metab.*, 2007, **6**, 484.
29. M. Miyazaki, W. C. Man and J. M. Ntambi, *J. Nutr.*, 2001, **31**, 2260.
30. E. Binczek, B. Jenke, B. Holz, R. H. Günter, M. Thevis and W. Stoffel, *Biol. Chem.*, 2007, **388**, 405.
31. H. Sampath, M. T. Flowers, X. Liu, C. M. Paton, R. Sullivan, K. Chu, M. Zhao and J. M. Ntambi, *J. Biol. Chem.*, 2009, **284**, 19961.
32. J. M. Brown, S. Chung, J. K. Sawyer, C. Degirolamo, H. M. Alger, T. Nguyen, X. Zhu, M.-N. Duong, A. L. Wibley, R. Shah, M. A. Davis, K. Kelley, M. D. Wilson, C. Kent, J. S. Parks and L. L. Rudel, *Circulation*, 2008, **118**, 1467.
33. M. L. MacDonald, M. van Eck, R. B. Hildebrand, B. W. Wong, N. Bissada, P. Ruddle, A. Kontush, H. Hussein, M. A. Pouladi, M. J. Chapman, C. Fievet, T. J. van Berkel, B. Staels, B. M. McManus and M. R. Hayden, *Arterioscler. Thromb. Vasc. Biol.*, 2009, **29**, 341.
34. A. Peter, C. Weigert, H. Staiger, K. Rittig, A. Cegan, P. Lutz, F. Machicao, H.-U. Häring and E. Schleicher, *Am. J. Physiol. Endocrinol. Metab.*, 2008, **295**, E339.
35. J. B. Flowers, M. E. Rabaglia, K. L. Schueler, M. T. Flowers, H. Lan, M. P. Keller, J. M. Ntambi and A. D. Attie, *Diabetes*, 2007, **56**, 1228.
36. H. Sampath and J. M. Ntambi, *Future Lipidology*, 2008, **3**, 163.

37. F. Paillard, D. Catheline, F. L. Duff, M. Bouriel, Y. Deugnier, M. Pouchard, J. C. Daubert and P. Legrand. *Nutr. Metab. Cardiovasc. Dis.*, 2008, **18**, 436.

38. E. Warensjö, E. Ingelsson, P. Lundmark, L. Lannfelt, A. C. Syvänen, B. Vessby and U. Risérus, *Obesity (Silver Spring)*, 2007, **15**, 1732.

39. A. D. Attie, R. M. Krauss, M. P. Gray-Keller, A. Brownlie, M. Miyazaki, J. J. Kastelein, A. J. Lusis, A. F. H. Stalenhoef, J. P. Stoehr, M. R. Hayden and J. M. Ntambi, *J. Lipid. Res.*, 2002, **43**, 1899.

40. R. Mar-Heyming, M. Miyazaki, D. Weissglas-Volkov, N. A. Kolaitis, N. Sadaat, C. Plaisier, P. Pajukanta, R. M. Cantor, T. W. A. de Bruin, J. M. Ntambi and A. J. Lusis, *Arterioscler. Thromb. Vasc. Biol.*, 2008, **28**, 1193.

41. M. W. Hulver, J. R. Berggren, M. J. Carper, M. Miyazaki, J. M. Ntambi, E. P. Hoffman, J. P. Thyfault, R. Stevens, G. L. Dohm, J. A. Houmard and D. M. Muoio, *Cell Metab.*, 2005, **2**, 251.

42. N. Stefan, A. Peter, A. Cegan, H. Staiger, J. Machann, F. Schick, C. D. Claussen, A. Fritsche, H.-U. Häring and E. Schleicher, *Diabetologia*, 2008, **51**, 648.

43. G. Liu, *Curr. Top. Med. Chem.*, 2010, **10**, 419.

44. M. Abreo, M. Chafeev, N. Chakka, S. Chowdhury, J.-M. Fu, H. W. Gschwend, M. W. Holladay, D. Hou, R. Kamboj, V. Kodumuru, W. Li, S. Liu, V. Raina, S. Sun, S. Sun, S. Sviridov, C. Tu, M. D. Winther and Z. Zhang, WO2005011655, 2005.

45. J.-M. Fu, V. Kodumuru, S. Sun, M. D. Winther, R. M. Fine, D. F. Harvey, B. Klebansky, M. P. Gray-keller, H. W. Gschwend and W. Li, US20050119251, 2005.

46. G. Liu, J. K. Lynch, J. Freeman, B. Liu, Z. Xin, H. Zhao, M. D. Serby, P. R. Kym, T. S. Suhar, H. T. Smith, N. Cao, R. Yang, R. S. Janis, J. A. Krauser, S. P. Cepa, D. W. Beno, H. L. Sham, C. A. Collins, T. K. Surowy and H. S. Camp, *J. Med. Chem.*, 2007, **50**, 3086.

47. S. Léger, W. C. Black, D. Deschenes, S. Dolman, J.-P. Falgueyret, M. Gagnon, S. Guiral, Z. Huang, J. Guay, Y. Leblanc, C.-S. Li, F. Massé, R. Oballa and L. Zhang, *Bioorg. Med. Chem. Lett.*, 2010, **20**, 499.

48. C.-S. Li, L. Belair, J. Guay, R. Murgasva, W. Sturkenboom, Y. K. Ramtohul, L. Zhang and Z. Huang, *Bioorg. Med. Chem. Lett.*, 2009, **19**, 5214.

49. Y. K. Ramtohul, C. Black, C.-C. Chan, S. Crane, J. Guay, S. Guiral, Z. Huang, R. Oballa, L.-J. Xu, L. Zhang and C.-S. Li, *Bioorg. Med. Chem. Lett.*, 2010, **20**, 1593.

50. E. Isabel, D. A. Powell, W. C. Black, C.-C. Chan, S. Crane, R. Gordon, J. Guay, S. Guiral, Z. Huang, J. Robichaud, K. Skorey, P. Tawa, L. Xu, L. Zhang and R. Oballa, *Bioorg. Med. Chem. Lett.*, 2011, **21**, 479.

51. S. Léger, W. C. Black, D. Deschênes, S. Dolman, J.-P. Falgueyret, M. Gagnon, S. Guiral, Z. Huang, J. Guay, Y. Leblanc, C.-S. Li, F. Masse, R. Oballa and L. Zhang, *Bioorg. Med. Chem. Lett.*, 2010, **20**, 499.

52. Y. Uto, T. Ogata, Y. Kiyotsuka, Y. Ueno, Y. Miyazawa, H. Kurata, T. Deguchi, N. Watanabe, M. Konishi, R. Okuyama, N. Kurikawa, T. Takagi, S. Wakimoto and J. Ohsumi, *Bioorg. Med. Chem. Lett.*, 2010, **20**, 341.

53. Y. Uto, Y. Kiyotsuka, Y. Ueno, Y. Miyazawa, H. Kurata, T. Ogata, T. Deguchi, M. Yamada, N. Watanabe, M. Konishi, N. Kurikawa, T. Takagi, S. Wakimoto, K. Kono and J. Ohsumi, *Bioorg. Med. Chem. Lett.*, 2010, **20**, 746.

54. Y. Uto, Y. Ueno, Y. Kiyotsuka, Y. Miyazawa, H. Kurata, T. Ogata, M. Yamada, T. Deguchi, M. Konishi, T. Takagi, S. Wakimoto and J. Ohsumi, *Eur. J. Med. Chem.*, 2010, **45**, 4788.

55. Y. Uto, Y. Ueno, Y. Kiyotsuka, Y. Miyazawa, H. Kurata, T. Ogata, T. Takagi, S. Wakimoto and J. Ohsumi, *Eur. J. Med. Chem.*, 2011, **46**, 1892.

56. J.-M. Fu, D. Hou, R. Kamboj, V. Kodumuru, N. Pokrovskaia, V. Raina, S. Sun, S. Sviridov and Z. Zhang, WO2007130075, 2007.

57. N. Dales, J. Fonarev, J.-M. Fu, D. Hou, R. Kamboj, V. Kodumuru, N. Pokrovskaia, V. Raina, S. Sun and Z. Zhang, WO2007143597, 2007.

58. N. Dales and Z. Zhang, WO2008024390, 2008.

59. N. Dales, Z. Zhang and J. Fonarev, WO2008074835, 2008.

60. S. Chowdhury, N. Dales, J. Fonarev, J.-M. Fu, D. Hou, Q. Jia, V. Kodumuru, N. Pokrovskaia, S. Sun and Z. Zhang, WO 2009103739, 2009.

61. N. Dales, J.-M. Fu, Q. Jia, N. Pokrovskaia, S. Sun and Z. Zhang, WO 2011039358, 2011.

62. Y. Uto, T. Ogata, J. Harada, Y. Kiyotsuka, Y. Ueno, Y. Miyazawa, H. Kurata, T. Deguchi, N. Watanabe, T. Takagi, S. Wakimoto, R. Okuyama, M. Abe, N. Kurikawa, S. Kawamura, M. Yamato and J. Osumi, *Bioorg. Med. Chem. Lett.*, 2009, **19**, 4151.

63. Y. Uto, T. Ogata, Y. Kiyotsuka, Y. Miyazawa, Y. Ueno, H. Kurata, T. Deguchi, M. Yamada, N. Watanabe, T. Takagi, S. Wakimoto, R. Okuyama, M. Konishi, N. Kurikawa, K. Kono and J. Osumi, *Bioorg. Med. Chem. Lett.*, 2009, **19**, 4159.

64. D. O. Koltun, E. Q. Parkhill, N. I. Vasilevich, A. I. Glushkov, T. M. Zilbershtein, A. V. Ivanov, A. G. Cole, I. Henderson, N. A. Zautke, S. A. Brunn, N. Mollova, K. Leung, J. W. Chisholm and J. Zablocki, *Bioorg. Med. Chem. Lett.*, 2009, **19**, 2048.

65. D. O. Koltun, N. I. Vasilevich, E. Q. Parkhill, A. I. Glushkov, T. M. Zilbershtein, E. I. Mayboroda, M. A. Boze, A. G. Cole, I. Henderson, N. A. Zautke, S. A. Brunn, N. Chu, J. Hao, N. Mollova, K. Leung, J. W. Chisholm and J. Zablocki, *Bioorg. Med. Chem. Lett.*, 2009, **19**, 3050.

66. D. O. Koltun, T. M. Zilbershtein, V. A. Migulin, N. I. Vasilevich, E. Q. Parkhill, A. I. Glushkov, M. J. McGregor, S. A. Brunn, N. Chu, J. Hao, N. Mollova, K. Leung, J. W. Chisholm and J. Zablocki, *Bioorg. Med. Chem. Lett.*, 2009, **19**, 4070.

67. Z. Xin, H. Zhao, M. D. Serby, B. Liu, M. Liu, B. G. Szczepankiewicz, L. T. Nelson, H. T. Smith, T. S. Suhar, R. S. Janis, N. Cao, H. S. Camp, C. A.

Collins, H. L. Sham, T. K. Surowy and G. Liu, *Bioorg. Med. Chem. Lett.*, 2008, **18**, 4298.

68. M. Ubukata, K. Maeda, T. Iida and I. Mitani, WO2008120759, 2008.

69. H. Zhao, M. D. Serby, H. T. Smith, N. Cao, T. S. Suhar, T. K. Surowy, H. S. Camp, C. A. Collins, H. L. Sham and G. Liu, *Bioorg. Med. Chem. Lett.*, 2007, **17**, 3388.

70. M. Issandou, A. Bouillot, J.-M. Brusq, M.-C. Forest, D. Grillot, R. Guillard, S. Martin, C. Michiels, T. Sulpice and A. Daugan, *Eur. J. Pharmacol.*, 2009, **618**, 28.

71. K. A. Atkinson, E. E. Beretta, J. A. Brown, M. Castrodad, Y. Chen, J. M. Cosgrove, P. Du, J. Litchfield, M. Makowski, K. Martin, T. J. McLellan, C. Neagu, D. A. Perry, D. W. Piotrowski, C. M. Steppan and R. Trilles, *Bioorg. Med. Chem. Lett.*, 2011, **21**, 1621.

72. D. A. Powell, Y. Ramtohul, M.-E. Lebrun, R. Oballa, S. Bhat, J.-P. Falgueyret, S. Guiral, Z. Huang, K. Skorey, P. Tawa and L. Zhang, *Bioorg. Med. Chem. Lett.*, 2010, **20**, 6366.

73. G. Liu, *The 233th ACS National Meeting*, Chicago, IL, March 2007, MEDI-382.

74. R. M. Oballa, L. Belair, W. C. Black, K. Bleasby, C. C. Chan, C. Desroches, X. Du, R. Gordon, J. Guay, S. Guiral, M. J. Hafey, E. Hamelin, Z. Huang, B. Kennedy, N. Lachance, F. Landry, C.-S. Li, J. Mancini, D. Normandin, A. Pocai, D. A. Powell, Y. K. Ramtohul, K. Skorey, D. Sørensen, W. Sturkenboom, A. Styhler, D. M. Waddleton, H. Wang, S. Wong, L. Xu and L. Zhang, *J. Med. Chem.*, 2011, **54**, 5082.

75. N. Lachance, S. Leger, R. M. Oballa, D. Powell, G. K. Tranmer, E. Martins and Y. Gareau, WO2010108268, 2010.

76. J.-P. Leclerc, C.-S. Li and O. M. Moradei, WO2011011872, 2011.

TGR5 Agonists in Development

ANTONIO MACCHIARULO, ANTIMO GIOIELLO AND
ROBERTO PELLICCIARI*

Dipartimento di Chimica e Tecnologia del Farmaco, University of Perugia, via
del liceo 1, 06123, Perugia, Italy
*E-mail: rp@unipg.it

10.1 Introduction

Lipid receptors are a growing family of druggable targets that include cell
surface receptors and ligand-dependent transcription factors.[1] They mediate a
plethora of signaling pathways involved in the fine tuning of important
physiological functions such as the control of metabolism, organ physiology,
cell differentiation, and homeostasis. As a consequence, small-molecule drug
development for lipid receptors is attracting a great deal of interest in academia
and pharmaceutical companies.

TGR5, also known as M-BAR, AXOR109, BG37, or GPR131, is a
membrane lipid receptor activated by bile acids (BAs) and G-protein coupled
to the production of cAMP. The activation of TGR5 bestows on BAs the
ability to modulate non-genomic signaling pathways that complement their
genomic actions mostly mediated by the interaction with the nuclear receptor
FXR.[2] Major non-genomic actions of BAs include immunosuppressive
properties and the regulation of glucose metabolism as well as energy
homeostasis.[3] While the therapeutic relevance of the immune properties of
TGR5 activation is pending further appraisals, the effect of the receptor on
glucose metabolism and energy homeostasis have thrust TGR5 into the
limelight as an attractive therapeutic target in the arena of metabolic disorders,
including type 2 diabetes (T2D) and obesity.

RSC Drug Discovery Series No. 27
New Therapeutic Strategies for Type 2 Diabetes: Small Molecule Approaches
Edited by Robert M. Jones
© The Royal Society of Chemistry 2012
Published by the Royal Society of Chemistry, www.rsc.org

Accordingly, recent years have witnessed an intense research activity on part of both academia and pharmaceutical companies towards the identification of potent and selective modulators of TGR5 for drug development. In this framework, early TGR5 ligands have been derived using the screening of natural and synthetic compound libraries. These attempts have resulted in the disclosure of several classes of TGR5 lead compounds that are being further modified and tested using iterative optimization cycles to allow the potency for the target to be increased, and the pharmacokinetic profile for the ligand to be tailored in order to produce efficacious drugs in humans. These efforts combine with attempts to understand how ligands interact within the binding cleft of the receptor in order to enable the prioritization of analogs for screening and the design of focused small molecule libraries around active TGR5 ligands.

Beginning with an overview on the early pharmacological characterization, structure, and physiological functions of TGR5, in this chapter we provide a discussion on the burgeoning evidence that settles the pharmacological modulation of TGR5 as a novel therapeutic opportunity in T2D. A thorough analysis of the emerging classes of TGR5 modulators is then reported, along with the knowledge gained on the structure–activity relationships of the different classes of ligands. Finally, we discuss some of the unsettled questions concerning TGR5 modulation that, once addressed, will certainly boost TGR5 ligand development.

10.2   TGR5 at a Glance

10.2.1   Early Pharmacological Characterization

At the beginning of this century, two Japanese research groups at Banyu Pharmaceuticals and Takeda Chemical Industries, independently reported the discovery of TGR5 as a membrane receptor activated by BAs and G-protein coupled to the production of cAMP.[4] As early pharmacological characterization of the receptor, it was found that lithocholic acid (LCA, **1**) and deoxycholic acid (DCA, **2**) markedly increased the production of cAMP in a dose-dependent manner, followed by chenodeoxycholic acid (CDCA, **3**), cholic acid (CA, **4**), and ursodeoxycholic acid (UDCA, **5**) (Table 10.1).

While this former study pinpointed that the hydroxylation pattern of BA steroid nucleus affected the activity at the receptor, later works evidenced that also the conjugation of the BA side chain with glycine or taurine influences the potency at TGR5. Accordingly, tauro-conjugates of endogenous BAs were found to slightly improve TGR5 potency over the glyco-conjugates and unconjugated BAs (Table 10.1).[5]

10.2.2   Sequence, Structure, and Gene Variants

The sequence of TGR5 is encoded by a gene that maps to position 2q35 of the human chromosome, and counts 330 amino acids that fold into the canonical

Table 10.1 TGR5 agonist potency of natural bile acids and their tauro- or glyco-conjugated forms.

			Acid-form		Tauro-form		Glyco-form	
Trivial Name	R_1	R_2	EC_{50} *(μM)*	*Efficacy (%)*	EC_{50} *(μM)*	*Efficacy (%)*	EC_{50} *(μM)*	*Efficacy (%)*
LCA (**1**)	-H	-H	0.58	101	0.29	106	0.54	92
DCA (**2**)	-H	-H	1.25	105	0.79	103	1.18	105
CDCA (**3**)	α-OH	-H	6.71	105	1.92	103	3.88	105
CA (**4**)	α-OH	-OH	13.6	101	4.95	104	13.6	103
UDCA (**5**)	β-OH	-H	36.4	74.9	30.0	97	33.9	91

Adapted from *J. Med. Chem.*, 2008, **51**, 1831.[5]

structure of family A GPCRs, being composed of a N-terminal extracellular region, seven trans-membrane helices and a C-terminal cytosolic tail region.

According to sequence similarity and phylogenetic studies, TGR5 belongs to the group of MECA (melanocortin, endothelial, cannabinoid, and adenosine) receptors within the family of lipid-activated GPCRs.[6] Although it is known that family A GPCRs holds the binding cleft for the endogenous ligand within the seven trans-membrane helices, the exact residues of TGR5 involved in the interaction with BAs are not yet disclosed.

At this regard, however, genetic and functional characterization of TGR5 in patients with primary sclerosing cholangitis (PSC) has led Johannes R. Hov and co-workers to identify five non-synonymous mutations of the receptor (Trp83Arg, Val178Met, Ala217Pro, Ser272Gly, and Gln296X) that diversely affect the targeting of TGR5 to the membrane, agonist binding, the propagation of conformational changes across the trans-membrane domains, and the intracellular G protein coupling (Figure 10.1).[7]

In particular, the construction of a homology model of TGR5 on the basis of the crystal structure of adenosine A2a receptor (PDB code: 3EML) was instrumental to suggest the effects of each of the mutations on the receptor. Accordingly, it was observed that the five non-synonymous variants are spread along the structure of TGR5, with Trp83 being located in the first extracellular loop, Val178 and Ser272 in the fifth and seventh trans-membrane helix, and Ala212 and Gln296 being localized on the third intracellular loop and C-terminal region, respectively.

Since the mutations of Trp83 and Val178 into arginine and methionine respectively reduced TGR5 response to taurolithocholic acid (TLCA), it was suggested

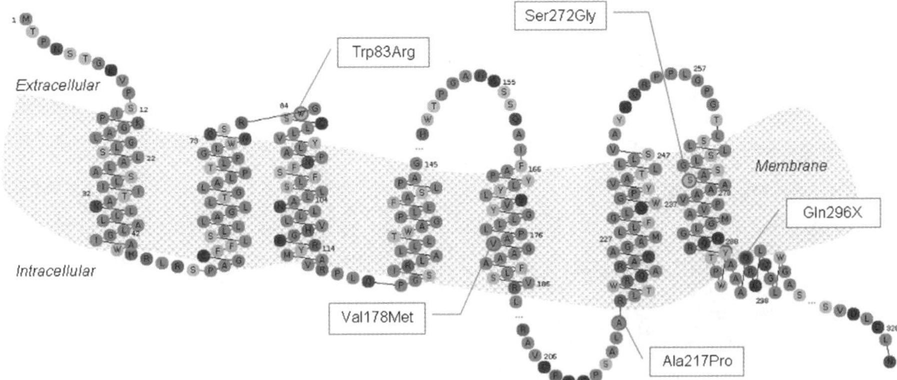

Figure 10.1 Snake representation of TGR5 receptor and gene variants. Non-synonymous mutations identified by Hov *et al.* (2010)[7] in patients with primary sclerosing cholangitis (PSC) as reducing or abolishing TGR5 function are labeled and highlighted in frames.

that these residues are involved in ligand binding (Trp98) and propagation of conformational changes across the trans-membrane domains (Val178).

Remarkably, the replacement of Ser272 with glycine was able to abolish TGR5 response to TLCA and diminish the cellular response to forskolin stimulation, suggesting that Ser272Gly is a constitutive active variant of the receptor with altered downstream signaling interactions. The mutation of Ala217 into proline abolished TGR5 activity. Being localized on the third intracellular loop, the effect of this variant was ascribed to the lack of coupling of the mutant receptor with the G protein. Finally, the premature stop codon Gln296X resulted in major TGR5 expression changes and plasma membrane localization, indicating the requirement of the C-terminal tail for the correct shuttling of the receptor to the cellular surface membrane.

10.3   Expression and Physiological Functions

TGR5 is ubiquitously expressed in human body, including the gallbladder, BAT, skeletal muscles, liver, adrenal cortex, intestine, and selected areas of the central nervous system.[8] Depending on the tissue where it is expressed, TGR5 mediates different physiological functions in response to circulating levels of BAs (Figure 10.2).

In the following paragraphs, it is reported a survey of the so far disclosed physiological functions of TGR5, according to the district of expression.

10.3.1   TGR5 in Brown Adipose Tissue (BAT) and Skeletal Muscle: Control of Energy Homeostasis and Body Weight

In 2006, the research group of Johan Auwerx firstly demonstrated that the activation of TGR5 by CA (**4**) was able to significantly induce a reduction of

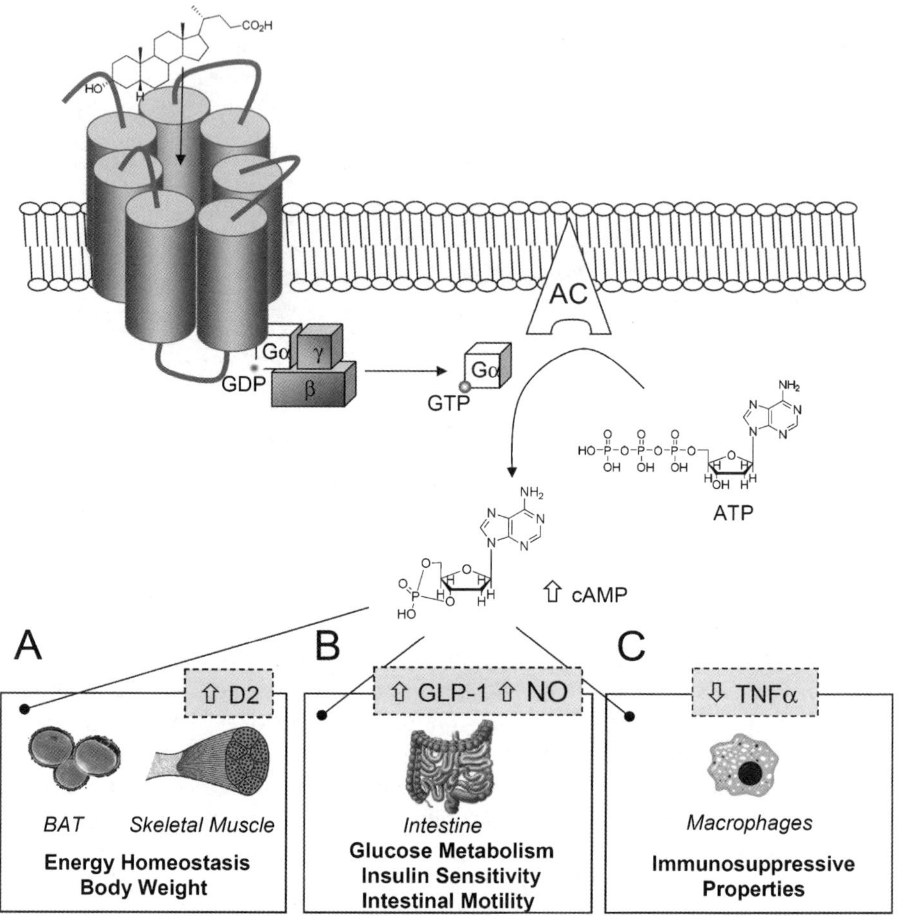

Figure 10.2 Schematic representation of TGR5 signaling pathway and functions in selected target tissues. (A) In BAT and skeletal muscle, activation of TGR5 increases cAMP production which triggers the protein kinase A (PKA), promoting the phosphorylation of the cAMP-response element binding protein (CREB). CREB is a ligand-independent transcription factor that induces the expression of D2, leading to the formation of the active 3,5,3′-triiodothyronine (T3). T3 binds thyroid hormone receptor (TR), modulating the expression of target genes involved in the regulation of energy homeostasis. (B) In the intestine, TGR5 activation stimulates GLP-1 secretion from L-enteroendocrine cells. GLP-1 is a member of the incretin family that exerts glucagonostatic actions, reduction of gastrointestinal motility, and appetite. Furthermore, activation of TGR5 in enteric neurons promotes the release of nitric oxide and suppression of intestinal motility. (C) TGR5 activation inhibits LPS-stimulated secretion of TNFα in rabbit alveolar macrophages. In liver resident macrophages (Kupffer cells), TGR5 activation elevates cytosolic cAMP, leading to the inhibition of LPS-induced cytokine expression.

the body weight of mice fed with a high fat diet. In particular, the effects of CA (**4**) were ascribed to the enhancement of energy expenditure rather than the reduction of caloric intake, and it was shown that they were independent from the activation of the nuclear BA receptor FXR.

Using iodothyronine deiodinase-2 (D2) deficient mice, it was then possible to ascertain the cellular mechanism at the basis of TGR5-mediated energy expenditure. Accordingly, it was suggested that the activation of TGR5 by BAs leads to the accumulation of cytosolic cAMP that triggers the protein kinase A (PKA), thereby promoting the phosphorylation of the cAMP-response element binding protein (CREB). CREB is a ligand-independent transcription factor that induces the expression of D2. D2 is one of the two enzymatic isoforms involved in the activation of thyroid hormones (T3 and T4). Accordingly, D2-mediated deiodination of thyroxine (T4) leads to the formation of the active 3,5,3′-triiodothyronine (T3) which, by binding to the thyroid hormone receptor (TR), eventually modulates the expression of target genes involved in the regulation of energy homeostasis, resulting in the increase of oxygen consumption and heat production, the stimulation of mitochondrial functions as well as the induction of catabolism of carbohydrates, fats, and proteins.[3,9] These findings were further corroborated by analogous results obtained at Banyu Pharmaceuticals, where researchers found that female TGR5 knockout mice showed significant fat accumulation and body weight gain compared with wild-type mice.[10]

Interestingly, these studies indicated murine BAT and human skeletal muscle as districts of action of the TGR5-cAMP-D2 mediated pathway. BAT, in particular, was formerly thought to be relevant only in animals, small mammals, and infants for the regulation of body weight and adaptive thermogenesis. The independent discoveries by three research groups that functional BAT is also present in adult humans, albeit reduced in obese subjects with its amount being inversely correlated with body mass index,[11] has contributed to increasing the attractiveness of TGR5 as drug target for metabolic disorders.

A further breakthrough in the relationships between TGR5 signaling pathway, energy metabolism, and body weight in humans came from two distinct studies that evidenced high levels of plasma BAs in patients after Roux-en-Y gastric bypass surgery (RYGB).[12] RYGB is a surgery providing body weight loss and improved metabolism in obese subjects through multifactorial and hitherto unclear mechanisms. Despite previous works that have investigated the effects of RYGB on energy expenditure of obese people with different results that may also vary depending on gender, the works of Nakatani and Patti provide interesting clues to the still patchy knowledge of factors mediating the metabolic effects of RYGB.[13] It should be mentioned, however, that at least two papers investigated and failed to observe a clear association between TGR5 signaling pathway, energy expenditure, and body weight loss.

In the first paper, investigating the role of TGR5 in diet-induced obese mice, Vassileva and co-workers found similar effects on body weight and food intake between TGR5-deficient mice and wild-type mice fed with chow or high-fat diet. Conversely, they evidenced a gender-dependent regulation of TGR5 metabolic functions, with female TGR5-null mice showing low plasma cholesterol and triglyceride levels as well as improved insulin sensitivity on high-fat diet, and male TGR5-null mice displaying increased hepatic steatosis and impaired insulin sensitivity on high-fat diet.[14]

The second work addressed the relationship between plasma BA levels and energy expenditure in humans.[15] The authors used a sequestrant resin to lower plasma BA concentrations in patients with type 2 diabetes and liver cirrhosis, monitoring the relative changes in energy metabolism. As a result, they found a lack of correlation between circulating levels of BAs and energy expenditure.

Although these results question a role of BAs-mediated activation of TGR5 in energy homeostasis, criticisms have arisen concerning the conception of the study.[16] In particular, the variation of circulating BA levels in diabetic patients following resin treatment has been considered too low to provide conclusive evidence with respect to correlations of BA levels and energy expenditure. Furthermore, it has been argued that the absence of control of additional symptoms of liver cirrhosis in the study may have negatively affected the neatness of the results.

Overall, these data suggest a subtle level of regulation of energy metabolism and body weight, where other factors, such as different diets, animal models, gender, and conditions, may affect the functional outcome of the experiment. In this scenario, it is clear that further work is required to better understand the influence of BAs mediated activation of TGR5 on energy metabolism and body weight in humans.

10.3.2 TGR5 in the Intestine: Control of Glucose Metabolism and Insulin Sensitivity

High levels of TGR5 have been detected in the intestine, mainly in the ileum and colon,[4b] where the receptor is expressed by enteroendocrine STC-1 cells and enteric nervous system. Two decades after the seminal observations that BAs were able to promote glucagon-like peptide-1 (GLP-1) release in dog, human, and rat,[17] the research group of Gozoh Tsujimoto at Kyoto University, Japan, showed that this effect was mediated by the activation of TGR5 in murine enteroendocrine STC-1 cell lines.[18]

GLP-1 is a member of the incretin family, a group of hormone peptides that includes the glucose-dependent insulinotropic polypeptide (GIP). It is secreted by L-enteroendocrine cells within minutes after meal ingestion, with plasma levels reaching approximately 25 to 40 pmol/L and returning to baseline within 3 hours. Despite both GLP-1 and GIP having additive insulinotropic effects,[19] it is currently thought that it is the former that is responsible for most of the incretin effects,[20] which include also glucagonostatic actions, reduction of gastrointestinal motility, and appetite.

In 2009, a study published in *Cell Metabolism* and conducted by a large industry/academic partnership involving CNRS / INSERM / ULP (France), University of Perugia (Italy), Ecole Polytechnique Fédérale de Lausanne (Switzerland), and Intercept Pharmaceuticals (New York, USA),[21] provided groundbreaking results for the understanding of the mechanism underlying BAs mediated stimulation of GLP-1.

As a first result, it was shown that INT-777 (**10**), a BA derivative endowed with remarkable metabolic stability as well as potency and selectivity at TGR5 (see Section 10.5.1.1),[22] was able to stimulate GLP-1 secretion in STC-1 cells and in human intestinal NCI-H716 cells. This effect was abolished by using RNA silencing of TGR5, therefore evidencing the implication of the receptor in the pharmacological effect of INT-777 (**10**). As for the cellular mechanism at the basis of TGR5-mediated stimulation of GLP-1 secretion, the activation of the receptor by **10** was linked to the increase of intracellular ATP/ADP ratio and a following rise in calcium mobilization, eventually leading to GLP-1 secretion. Secondly, it was found that diet-induced obese mice with a gain-of-function of TGR5 improved glucose tolerance, showing strong postprandial GLP-1 release and insulin secretion. In contrast, TGR5-deficient mice showed a delayed glucose clearance. Finally, the effect of INT-777 (**10**) on GLP-1 release was shown to be abolished when the compound was administered in TGR5-null mice, but not in TGR5-expressing mice.

Researchers at the University of California reported that TGR5 is also expressed in enteric neurons, prevailing in inhibitory motor neurons and descending interneurons which express nitric oxide synthase.[23] As a result, it was found that the activation of TGR5 by DCA (**2**) released nitric oxide and suppressed intestinal motility, suggesting that the receptor is involved in the "ileal brake", whereby BAs slow transit to allow a complete food digestion and absorption.[24] Considering that the inhibition of intestinal motility is also a known effect of GLP-1,[25] one could wonder whether the effects of TGR5 on GLP-1 secretion and intestinal motility are causally linked or if they occur independently of one another.

Nevertheless, these findings collectively suggest that, in consequence of meal ingestion, the secretion of BAs from gallbladder to the intestine boosts GLP-1 secretion through the activation of TGR5 in L-enteroendocrine cells. The ensuing high plasma levels of GLP-1 accounts for insulinotropic and glucagonostatic effects, satiety, and reduction of gastrointestinal motility, with the latter effect being synergistically (or casually) linked to the BAs mediated activation of TGR5 in the enteric nervous system.

10.3.3 TGR5 in Monocytes and Macrophages: Immunosuppressive Properties

The immunosuppressive effects of BAs on cell-mediated immunity and macrophage functions have been known since the middle of the 1970s.[26] Further studies demonstrated that BAs exert inhibitory actions on the production of cytokines that are induced by lipopolysaccharide (LPS) in macrophages.[27]

The discovery of TGR5 as membrane receptor of BAs, prompted the researchers at Takeda Chemical Industries to investigate the possibility that BAs could suppress macrophage functions via TGR5 activation.[4a] As a result, they found that the activation of TGR5 by BAs resulted in the increase of cytosolic cAMP and the inhibition of LPS-stimulated TNFα secretion in rabbit alveolar macrophages. In line with this observation, the researchers also showed a suppression of LPS-induced secretion of TNFα when BAs were applied to human monocytic leukemia THP-1 cells transfected with TGR5, but no effect was observed in parental THP-1 cells.

Unfortunately, they were not able to directly support these results with loss-of-function experiments, since the use of small interfering RNAs was ineffective to knock out TGR5 functions. Thus, it was not possible to assess at that time whether the immunosuppressive effects of BAs were directly linked to TGR5 activation or if other BA-regulated signaling pathways could contribute to the final outcome.

A further evidence for a direct link between TGR5 and cytokine production came from a subsequent study of a research group at the Heinrich-Heine-University (Düsseldorf, Germany) that investigated the role of TGR5 in Kupffer cells.[28] Kupffer cells are resident macrophages in the liver and play a central role in the production of inflammatory cytokines.[29] In agreement with previous observations reporting that BAs influence the immune functions of Kupffer cells,[30] the authors found that BA-mediated activation of TGR5 was able to elevate intracellular cAMP and inhibit LPS-induced cytokine expression in Kupffer cells.

Although they did not use loss-of-function experiments to determine whether the immunosuppressive effects of BAs were mediated by TGR5 activation, the researchers showed that the incubation of Kupffer cells with dibutyryl cAMP and forskolin, activating adenylate cyclase, increased cAMP levels and decreased LPS-induced cytokine expression, providing evidence that the effect was dependent on intracellular cAMP concentration.

10.3.4   TGR5 Functions in Other Tissues

A paper published in *Hepatology* provided evidence that TGR5 is expressed in sinusoidal endothelial cells (SECs).[31] SECs are cells that separate hepatocytes from the passing blood, playing an important role in the hepatic microcirculation.[32] Owing to their position, SECs are exposed to variable concentrations of nutrients and high levels of BAs from the enterohepatic circulation.

The research team of Ralf Kubitz at the Heinrich-Heine-University (Düsseldorf, Germany), showed that BAs were able to raise the levels of cAMP in SECs, by activating TGR5. The ensuing high intracellular concentration of cAMP enhanced the expression of the endothelial nitric oxide synthase (eNOS), while promoting eNOS activation by favoring the phosphorylation of the enzyme. The production and secretion of nitric oxide (NO) by SECs mediates vasodilatation, maintaining the vasomotor control and protecting against liver microcirculation dysfunction which is a determinant for subsequent hepatic injury.[33] Taken together, the results of Ralf Kubitz and co-workers indicate that the expression and activation of TGR5 in SECs connect BAs to liver blood flow regulation via NO production.

Remarkably, it has been shown that the administration of the potent and selective TGR5 agonist INT-777 (**10**) to high-fat diet-fed mice reduces liver steatosis and improves hepatocyte damage.[21] Although TGR5 is not expressed in hepatocytes, it is very likely that the activation of TGR5 in SECs combined with the activation of the receptor in Kupffer cells may contribute to the underlying mechanism of TGR5-mediated hepatoprotection.

The gallbladder is a small organ that stores bile produced by the liver during fasting and excretes fluid after food intake.[34] Dysfunctions in the gallbladder processes of absorption and secretion have been associated to the development of gallstone disease.[35] TGR5 expression was detected in human gallbladder epithelial cells as well as gallbladder smooth muscle. In the former, the activation of the receptor promotes chloride and fluid secretion in response to BAs.[36] Although it was reported that TGR5-deficient mice were not able to develop cholesterol gallstones when fed with a lithogenic diet,[37] no relationship was observed between protein levels of the receptor in human gallbladder and the presence of gallstones, making elusive the link between TGR5 signaling pathway in gallbladder and gallstone disease.

In a subsequent study, it was shown that the activation of TGR5 on gallbladder smooth muscle by hydrophobic BAs increased cytosolic cAMP, promoting the opening of K(ATP) channels via PKA pathway activation, and membrane hyperpolarization. The authors suggested that the ensuing reduction of gallbladder smooth muscle activity could be a contributing factor in the formation of gallstones, providing a causative link between TGR5 activation and the manifestation of gallstone disease.[38] A further study on the effect of TGR5 stimulation in the gallbladder provided evidence that gallbladder volume was increased in wild-type but not TGR5-deficient mice by administration of the selective TGR5 agonist INT-777 (**10**).[39] Taken together, these results suggests a role for TGR5 in the regulation of gallbladder contractility, with the activation of TGR5 promoting the filling of the gallbladder with bile.

Since the early characterization of the receptor, TGR5 expression was also found in the brain. Recently, the research group of Häussinger at the Heinrich-Heine-University (Düsseldorf, Germany) investigated the function of TGR5 in astrocytes and neurons.[40] As a result, they found that TGR5 may act as a neurosteroid receptor, being potently activated by 5β-pregnan-3α-ol-20-one, 5β-pregnan-3α-17α-21-triol-20-one and 5α-pregnan-3α-ol-20-one. The activation of TGR5 by the endogenous neurosteroids increased the intracellular levels of cAMP and calcium, enhancing the generation of reactive nitrogen and oxygen species (RNOS) in astrocytes. Combining these results with the observation that the levels of TGR5 expression were reduced in the brain of patients with hepatic encephalopathy, Häussinger and co-workers suggested a role for the receptor in the pathogenesis of such neuropsychiatric disorder, which can develop in the course of acute and chronic liver disease.[41]

10.4   TGR5 Modulation as Therapeutic Opportunity in Type 2 Diabetes (T2D)

In the past few decades, many research activities have struggled to unravel the regulatory network underlying glucose homeostasis, with the aim of identifying new potential targets for novel therapeutic treatments of T2D. Recent studies have focused on GLP-1 as key player in T2D, opening new avenues to the development of novel glucose-lowering drugs.[20] In particular, two therapeutic approaches have been so far devised to enhance GLP-1 insulinotropic effects: (i) the design of stable synthetic analogs of GLP-1; (ii) the development of small molecule inhibitors of DPP-IV, an enzyme that is involved in GLP-1 degradation.[42] Both approaches have proved to be safe and tolerable in humans, though differences have been observed in a range of glycemic and metabolic parameters from head-to-head clinical trials.[19b]

DPP-IV inhibitors, in particular, do not increase total levels of circulating GLP-1 and reduce GLP-1 secretion, suggesting the presence of a feedback mechanism of regulation wherein inhibition of degradation of GLP-1 is linked to suppression of substrate secretion by L-enteroendocrine cells.[43] Synthetic GLP-1 analogs, on the other hand, show superior glycemic control as well as weight loss and improvement of blood pressure in clinical trials, evidencing the presence of a greater stimulation of GLP-1 receptor at systemic level than that achieved by DPP-IV inhibitors.[44] Nevertheless, the two approaches do not provide a complete remission of T2D in clinical studies, and there is still uncertainty concerning the ability of these classes of anti-diabetic drugs to block the progression of the disease.[42]

Embracing the hypothesis that major effects of secreted GLP-1 may occur through a local stimulation of sensory afferent nerves that signal onward to the hypothalamus and in turn to the pancreas,[45] some researchers have arguably attributed the above weaknesses to the poor efficacy of current GLP-1 based therapies in supplying a sustained stimulation of GLP-1 receptors in the enteric parasympathetic nervous system and hepatoportal sensors.[46]

In this scenario, the pharmacological activation of TGR5 offers unprecedented opportunities to further validate the hypothesis, providing a direct strategy to the regulation of GLP-1 secretion through a local action on L-enteroendocrine cells. If the hypothesis proves correct, then TGR5 agonists may well represent a superior therapeutic alternative to synthetic GLP-1 analogs and DPP-IV inhibitors in T2D, provided that they are safe and tolerable in humans.

Further evidence sustaining a putative beneficial effect of BA-mediated activation of TGR5 in T2D stems from observations in diabetic obese patients undergoing bariatric surgery. The seminal observation is that two major types of bariatric surgery, namely malabsorptive surgical procedures as well as combined restrictive and malabsorptive procedures (Roux-en-Y gastric bypass surgery, RYGB), are able to induce a sustained remission of diabetes in more than 80% of obese patients suffering from T2D.[47] Notably, two studies in 2009 reported that plasma levels of GLP-1 and BAs were increased in obese patients after bariatric surgery.[12] Although these works did not provide a causality link, it is plausible to conceive that BA-mediated activation of TGR5, stimulating GLP-1 secretion, may be part of the underlying mechanisms leading to the improvement of glucose tolerance after bariatric surgery.

Other studies have found that BA sequestrants increase GLP-1 secretion from the intestine of diabetic rats,[48] accounting for the improvement of glycemic control in patients with T2D after treatment with colesevelam or colestimide.[49]

While it is known that BA sequestrants reduce BA levels in the liver and induce alterations in the composition of BA pool, further studies are required to ascertain whether they may also affect TGR5 activation, thus enhancing GLP-1 secretion or, alternatively, other convergent signaling pathways may take over in the regulation of GLP-1 secretion, including fatty acids activation of GPR119 and/or GPR120.

Finally, it should be mentioned that burgeoning data point out a role for alterations of BA homeostasis in T2D and control of glucose tolerance,[50] though some controversial results have been reported.[15,51] Again, further appraisals are required to clarify the extent of which alterations of BA pool can affect the modulation of TGR5 and/or other signaling pathways such as those mediated by FXR and/or LXRα.[52]

10.5 Ligands in Development

10.5.1 Classification of TGR5 Ligands

Depending on the chemical structure, TGR5 ligands can be either steroidal or non-steroidal compounds. Steroidal ligands share a common steroidal nucleus and include BAs, bile alcohols, some steroid hormones, and their derivatives; non-steroidal compounds have different chemical structures, including synthetic ligands, natural products, and semisynthetic compounds.

It is worth noting that literature and patent applications have so far reported TGR5 ligands behaving as agonists or partial agonists at the receptor, regardless of the structural class. However, in spite of the large variety of TGR5 ligands that has been disclosed and developed, no agonist has still advanced in clinical trials, with only a single study currently appraising the efficacy of a combination therapy of UDCA and a DPP-IV inhibitor, namely sitagliptin, in T2D patients (ClinicalTrials.gov identifier: NCT01337440).

10.5.1.1 Steroidal Ligands: Bile Acids, Bile Alcohols, and Derivatives

At the beginning of the last decade, two independent research groups opened the way to the modulation of the non-genomic pathways of BAs by identifying TGR5 as a novel BA receptor.[14,15] They found that the most potent endogenous ligands of TGR5 were LCA (**1**) and DCA (**2**), followed by CDCA (**3**) and CA (**4**), with these compounds showing activities in the low micromolar range of potency. While the first synthesis of the enantiomeric forms of CDCA and LCA allowed the assessment of the specificity of BA interaction at TGR5,[53] chemical intuition and screening of semisynthetic BA libraries proved successful to disclose the first selective modulators of this lipid GPCR. The breakthrough came with the finding that the incorporation of a methyl moiety at the C23 position of the CDCA side chain, with the preference of the *S* over the *R* chiral configuration, conferred selectivity, albeit not potency, towards TGR5 over FXR (Figure 10.3).[54] This finding was explained with docking experiments of 23(*S*)- and 23(*R*)-methyl-CDCA (**6**, **7**) into the crystal structure of the ligand binding domain of FXR. As a result, steric clashes were observed between the 23(*S*)-methyl moiety and the side chain of Met262 that impaired the binding of this ligand to the nuclear receptor. When the 23(*S*)-methyl group was next introduced in 6-ECDCA (obeticolic acid, INT747, **8**), a potent FXR agonist in phase 3 clinical studies for primary biliary cirrhosis (PBC) and non-alcoholic steatohepatitis (NASH),[55] a reversal of the activity profile was observed, with 6α-ethyl-23(*S*)-methyl-chenodeoxycholic acid (**9**) showing a remarkable TGR5 activity in the nanomolar range of potency (TGR5 EC_{50} = 0.095 μM) and a moderate FXR activity (FXR EC_{50}

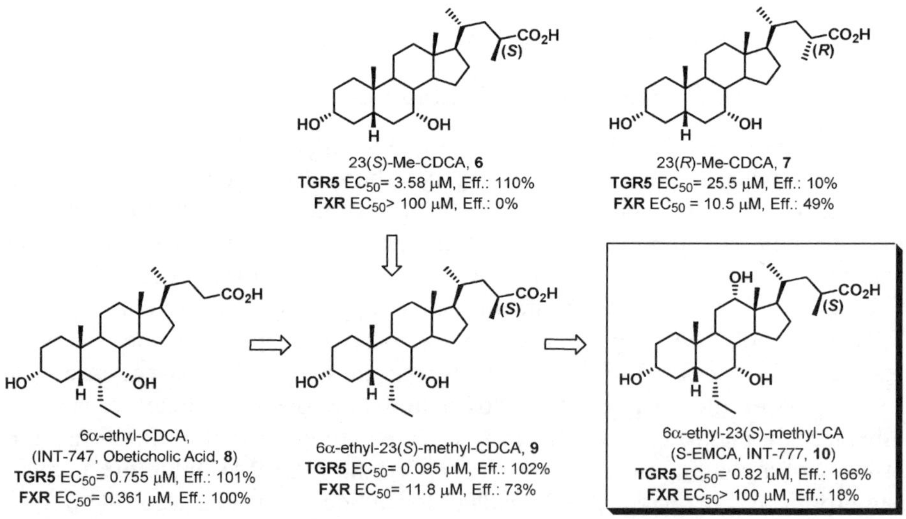

Figure 10.3 Structure and TGR5/FXR activity of C23-methyl bile acid derivatives.

= 11.80 μM) (Figure 10.3).[54] Next, the incorporation of the crucial 6α-ethyl- and 23(*S*)-methyl moieties into the chemical scaffold of CA (**4**) led to the synthesis of 6α-ethyl-23(*S*)-methyl-cholic acid (*S*-EMCA, INT-777, **10**).[22]

Unexpectedly, the biological appraisals of INT-777 (**10**) showed a complete abrogation of FXR activity, keeping a good potency towards TGR5. Preliminary pharmacokinetic studies demonstrated that **10** was endowed with a potent hypercholeretic effect, higher than the relative 23(*R*)-methyl isomer, CDCA (**3**) and CA (**4**).[22] At the same time, INT-777 (**10**) was found resistant to conjugation, with the compound being secreted into the bile as unconjugated form for the 90% rate. In view of the good pharmacodynamic and pharmacokinetic properties, INT-777 (**10**) was orally administered in high-fat fed mice at a dose of 30 mg/kg/day for the proof-of-concept that the pharmacological activation of TGR5 could be effective in controlling the intestinal GLP-1 secretion and maintaining glucose homeostasis.[21] Thus, after 10 weeks of treatment, a significant reduction of body weight as a consequence of the enhanced energy expenditure was found. Furthermore, the administration of **10** led to the improvement of liver functions and insulin sensitivity, with reduced levels of plasma triglycerides and free fatty acids. It is worth noting that the absence of the regulation of FXR target genes indicated the specificity of action of INT-777 (**10**) towards TGR5. The screening of focused libraries of semisynthetic BA derivatives was also instrumental to depict structure–activity relationships (SAR) of TGR5 agonists.[5] As a result of these studies, it was shown that the C3 position of the BA nucleus is important for TGR5 activity (Table 10.2). Firstly, the substitution of the 3α-hydroxyl group either with a sulfate (**11**) or an acetoxy (**12**) group was shown to be detrimental for the activity and efficacy of the compound. Likewise, UDCA derivatives bearing a methyl group at the C3 position (**23**, **24**) proved to be less active than the parent compound (Table 10.2).[56] Secondly, while the oxidation of the 3α-hydroxyl group led to the enhancement of the activity in CDCA (**3**) and LCA (**1**), its elimination or epimerization decreased the potency of the compound at the receptor (Table 10.2).[5] Interestingly, a different trend was observed in the case of UDCA derivatives, with either the epimerization or the oxidation proving detrimental for TGR5 agonist activity.[56] The authors ascribed this discrepancy to the presence of the 7β-hydroxyl group that might promote a different binding mode of the 3-dehydro-UDCA (**25**) to the receptor. The study of Sato *et al.* showed that the introduction of alkyl groups in C7-β position of CDCA improved TGR5 agonist activity, with the potency being directly correlated with the size of the substituent (Table 10.2). In agreement with the observation that the 7-methyl derivative of LCA (**15**) was a very potent TGR5 agonist, Iguchi and co-workers reported that the insertion of a methyl group in C7-α position of UDCA improved TGR5 agonism. Remarkably, the presence of a fluorine atom in the C7-α position but not in the C7-β position of LCA was able to double the potency of the compound.

Although the progressive shortening of the BA side chain reduced TGR5 activity, different bioisosteric replacements of the carboxylic group were found

Table 10.2   TGR5 agonist potency of body-modified bile acids derivatives.

Trivial Name	R_1	R_2	R_3	R_4	EC_{50} (μM)	Efficacy (%)
LCA (**1**)	-OH	-H	-H	-H	0.58	101
LCA-S (**11**)	-OSO$_3$H	-H	-H	-H	>100	0 (100μM)
LCA-Ac (**12**)	-OCOCH$_3$	-H	-H	-H	>102	0 (50μM)
Dehydro-LCA (**13**)	=O		-H	-H	0.27	106
Iso-LCA (**14**)	-H	-OH	-H	-H	1.25	99
7ξ-Me-LCA (**15**)	-OH	-H	-CH$_3$		0.076	106
7α-F-LCA (**16**)	-OH	-H	-F	-H	0.25	99
7β-F-LCA (**17**)	-OH	-H	$-H$	β-F	2.29	92
CDCA (**3**)	-OH	-H	-OH	-H	6.71	105
3-Dehydro-CDCA (**18**)	=O		-OH	-H	3.98	107
3-Deoxy-CDCA (**19**)	-H	-H	-OH	-H	14.5	81
7β-Me-CDCA (**20**)	-OH	-H	-OH	-CH$_3$	6.18	105
7β-Et-CDCA (**21**)	-OH	-H	-OH	-CH$_2$CH$_3$	2.63	99
7β-Pr-CDCA (**22**)	-OH	-H	-OH	-CH$_2$CH$_2$CH$_3$	0.78	108
UDCA (**5**)	-OH	-H	-H	-OH	14.9	81
3β-Me-UDCA (**23**)	-H	-CH$_3$	-H	-OH	59.4	72
3α-Me-UDCA (**24**)	-CH$_3$	-H	-H	-OH	26.6	60
3-Dehydro-UDCA (**25**)	=O		-H	-OH	13.4	49
Iso-UDCA (**26**)	-H	-OH	-H	-OH	16.3	57
7α-Me-UDCA (**27**)	-OH	-H	-CH$_3$	-OH	1.9	101

Biological data related to LCA and CDCA derivatives (**1**, **3**, **11**–**22**) were taken from *J. Med. Chem.*, 2008, **51**, 1831.[5] Biological data related to UDCA derivatives (**5**, **23**–**27**) were taken from *Biol. Pharm. Bull.*, 2011, **34**, 1.[56]

beneficial for TGR5 agonistic activity.[5] Of particular interest was the replacement of the carboxylic group with the hydroxyl and sulfate moieties (**31**, **32**), resulting in about one order of magnitude increase in potency than the methyl ester derivative **28** (Table 10.3).

On the basis of higher affinities of tauro- and glyco-conjugated BAs with the respect to the unconjugated forms, the design of UDCA analogs bearing side chain conjugations with various amino acids was pursued (Table 10.3).[56] As a result, it was found that conjugation with L-serine and *p*-aminobenzoic acid (**37**, **39**) enhanced TGR5 activity, though the former displayed lower efficacy. Conversely, conjugation of UDCA side chain with L-cysteic acid (**38**) abolished TGR5 activity.

The biological appraisals of the four C22-C23 cyclopropyl isomers of CDCA (**33**–**36**) provided useful information on the bioactive conformation adopted by

Table 10.3 TGR5 agonist potency of side chain-modified bile acids derivatives.

Trivial Name	R_1	R_2	EC_{50} (μM)	*Efficacy* (%)
CDCA (**3**)	α-OH	CO_2H	6.71	105
CDCA Me ester (**28**)	α-OH	CO_2CH_3	1.39	114
Nor-CDCA (**29**)	α-OH	CO_2H	10.4	102
Dinor-CDCA (**30**)	α-OH	CO_2H	>100	50 (316 μM)
CDC-OH (**31**)	α-OH	OH	0.12	103
CDC-Sul (**32**)	α-OH	SO_3H	0.44	103
22*S*,23*S*-CCDCA (**33**)	α-OH	CO_2H	1.33	110
22*S*,23*R*-CCDCA (**34**)	α-OH	CO_2H	2.91	102
22*R*,23*R*-CCDCA (**35**)	α-OH	CO_2H	75.7	5
22*R*,23*S*-CCDCA (**36**)	α-OH	CO_2H	>100	4
UDCA (**5**)	β-OH	CO_2H	14.9	81
S-UDCA (**37**)	β-OH	CONH … CO_2H, OH	5.8	71
C-UDCA (**38**)	β-OH	CONH … CO_2H, SH	>100	0
UDC-PABA (**39**)	β-OH	CONH … CO_2H	0.9	79

Biological data related to CDCA derivatives (**3**, **28**–**36**) were taken from *J. Med. Chem.*, 2008, **51**, 1831.[5] Biological data related to UDCA derivatives (**5**, **37**–**39**) were taken from *Biol. Pharm. Bull.*, 2011, **34**, 1.[56]

the χ-torsional angle of BA side chain into the binding site of TGR5 (Table 10.3).[5] On the basis of the better activity of 22*S*,23*S*-CCDCA (**33**) over the remaining stereoisomers, it was possible to infer a folded conformation of the χ-dihedral angle as slightly preferred over an extended conformation for TGR5 binding.

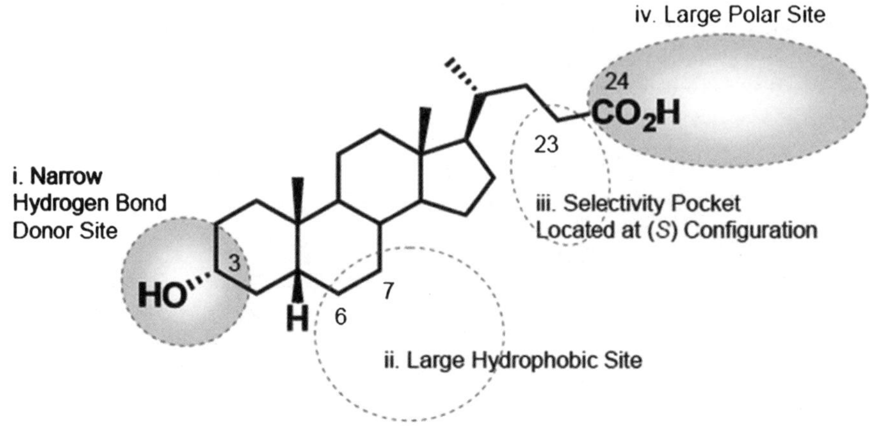

Figure 10.4 Schematic representation of TGR5 binding site according to structure–activity relationships of BAs and derivatives. BA binding site within TGR5 is composed of (i) a large polar site recognizing the acidic side chain and its conjugated forms, (ii) a hydrophobic pocket lining the C6 and C7 positions of the steroid nucleus, (iii) a narrow hydrogen bond donor site binding the BA 3-hydroxyl group, and (iv) a selectivity pocket located at the C23(*S*) position.

Collectively, these studies depict a map of BA recognition sites within TGR5 as composed of four pockets (Figure 10.4): (i) a large polar site recognizing the acidic BA side chain and its conjugated forms, (ii) a hydrophobic pocket lining the C6 and C7 positions of the steroid nucleus, (iii) a narrow hydrogen bond donor site binding the BA 3-hydroxyl group, and (iv) a selectivity pocket located at the C23(*S*) position.

In another study, Iguchi *et al.* reported the evaluation of bile alcohols as TGR5 ligands.[57] Bile alcohols are intermediate metabolites of BA biosynthesis as well as elimination products of cholesterol catabolism. Beside their previously reported FXR activity, bile alcohols appeared to possess TGR5 agonistic properties, with a selectivity over FXR that increased with the number of hydroxyl groups in the side chain. At odds with BAs, however, the authors found that the activity of bile alcohols towards TGR5 was not significantly affected by the shortening of the side chain, whereas it varied with the number, position, and configuration of the hydroxyl groups.

10.5.1.2 *Steroidal Ligands: Steroid Hormones*

Screening of libraries of natural products for TGR5 ligands pinpointed that some steroid hormones, including the most potent pregnandiol (**40**) and 5α-pregnandione (**41**), were either full or partial agonists at TGR5, with affinities in the low micromolar range of potency (Figure 10.5).[5] Noteworthy, this result was consistent with previously reported anti-obesity and anti-diabetic proper-

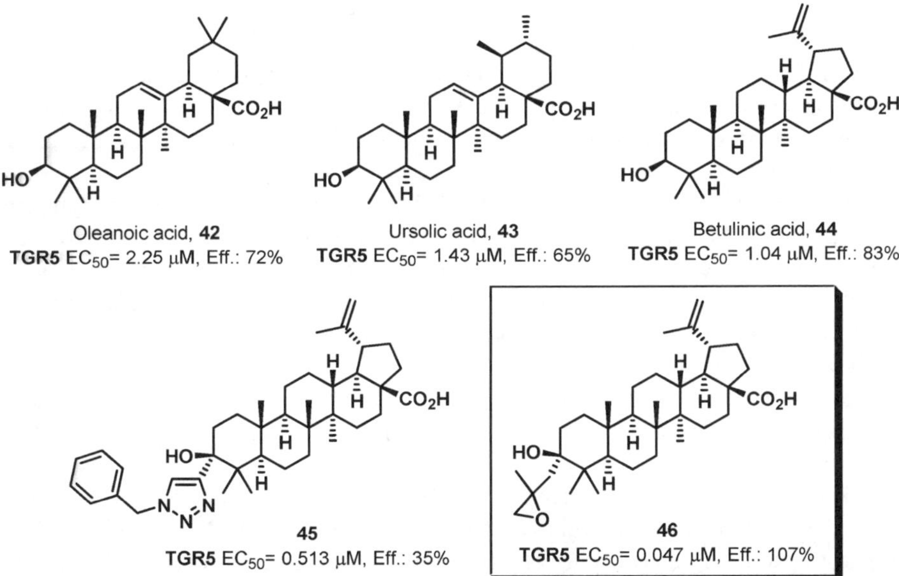

Pregnandiol, **40**
TGR5 EC$_{50}$= 0.85 µM, Eff.: 84%

5α-Pregnandione, **41**
TGR5 EC$_{50}$= 0.29 µM, Eff.: 72%

Figure 10.5   TGR5 activity of pregnan derivatives.

ties of some steroid hormones that were administered at supra-physiological concentrations in rodent models.

10.5.1.3   Non-Steroidal Ligands: Natural Products and Derivatives

In line with the reported anti-hyperglycemic and anti-diabetic properties of *Olea europaea*, an extract obtained from its leaves was found to have agonist properties towards TGR5.[58] The active ingredient was identified in a triterpenoid molecule, namely oleanoic acid (**42**) (Figure 10.6), that showed a selective activation of TGR5, with its EC$_{50}$ being comparable to that of LCA.

Oleanoic acid, **42**
TGR5 EC$_{50}$= 2.25 µM, Eff.: 72%

Ursolic acid, **43**
TGR5 EC$_{50}$= 1.43 µM, Eff.: 65%

Betulinic acid, **44**
TGR5 EC$_{50}$= 1.04 µM, Eff.: 83%

45
TGR5 EC$_{50}$= 0.513 µM, Eff.: 35%

46
TGR5 EC$_{50}$= 0.047 µM, Eff.: 107%

Figure 10.6   Structure and TGR5 activity of triterpenoid derivatives.

Further screening of a focused library containing naturally occurring triterpenoid molecules led to the identification of ursolic acid (**43**) and betulinic acid (**44**) as two additional selective TGR5 ligands.[58] Chemical manipulation around the scaffold of betulinic acid (**44**) allowed the same authors to disclose more potent TGR5 ligands and to develop SAR of this class of derivatives. Accordingly, modifications of the hydroxyl group at the C3(*S*) position, such as esterification and oxidation, proved detrimental for the activity, thereby suggesting the importance of a hydrogen bond donor group in that position. Remarkably, the same authors found that the addition of an allylic group at the C3 position of **44** resulted in more potent derivatives that showed an inversion of the enantiomeric specificity of the 3-hydroxyl group, with the equatorial configuration being preferred over the axial configuration for the activity.

Starting from the good reactivity of the allylic side chain, further explorations of the C3 position with aryl and heterocycle moieties led to the identification of a benzyl-triazole derivative **45** that was active in the low micromolar range of potency (Figure 10.6).[59] This finding suggested the presence of an accessory pocket located at that position that was still further probed with analogs containing hydrogen acceptor or donor groups. As a result, it was found that the epoxide derivative **46** was the most potent TGR5 agonist of the series, with an EC_{50} value in the nanomolar range of activity (TGR5 EC_{50} = 0.047 µM). Interestingly, either the replacement of the C3 epoxide group with an aldehyde or ketone slightly reduced the potency of the compound, whilst its substitution with a carboxylic group abolished TGR5 activity.

A number of chemical modifications were also designed for the carboxylic group and alkene moiety of betulinic acid (**44**).[58] Accordingly, bioisosteric replacements of the carboxylic group systematically dropped the activity of the resulting compounds, while modifications of the alkene group into ketone, alkane, epoxide, and alcohol were almost allowed, showing a marginal worsening of TGR5 activity.

Taken together, these data chart four distinct pockets that line the binding site harboring betulinic acid into TGR5 (Figure 10.7): (i) a narrow hydrogen bond acceptor site located at the equatorial C3 position, (ii) a larger hydrogen bond donor pocket at the axial C3 position, (iii) a solvent-exposed region containing the alkene moiety, and (iv) a narrow polar site harboring the carboxylic group.

Beside very few exceptions, it is worth noting that SAR of betulinic acid analogs are in remote agreement with similar studies arising from BA derivatives. The same authors found it difficult to obtain a reasonable superposition of the pharmacophoric elements of the most potent betulinic acid derivatives with the relative elements of LCA (**1**). Nevertheless, they suggested that these compounds could be still harbored into the BA binding site of TGR5, albeit adopting a different orientation.

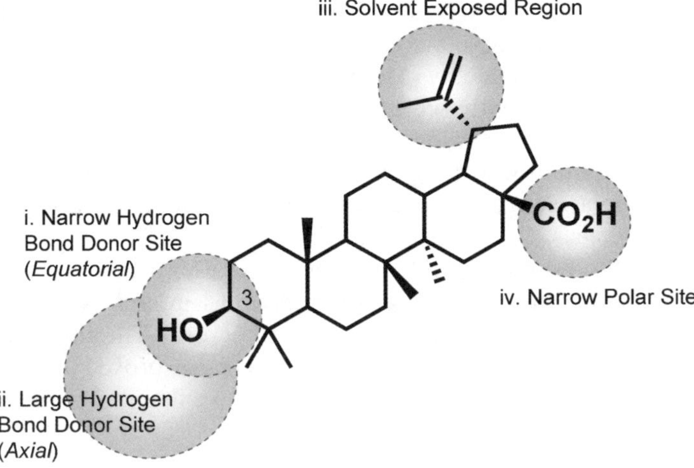

Figure 10.7   Schematic representation of TGR5 binding site according to structure–activity relationships of betulinic acid and derivatives. Betulinic acid binding site within TGR5 is composed of (i) a narrow hydrogen bond acceptor site located at the equatorial C3 position, (ii) a larger hydrogen bond donor pocket at the axial C3 position, (iii) a solvent-exposed region containing the alkene moiety, and (iv) a narrow polar site harboring the carboxylic group.

However, beside recognizing structural similarities between the C3 hydroxyl groups of betulinic acid and BAs, further studies are needed to determine the binding mode of these non-steroidal derivatives to TGR5, and thus provide conclusive evidence on the competitive or allosteric mechanism of action of

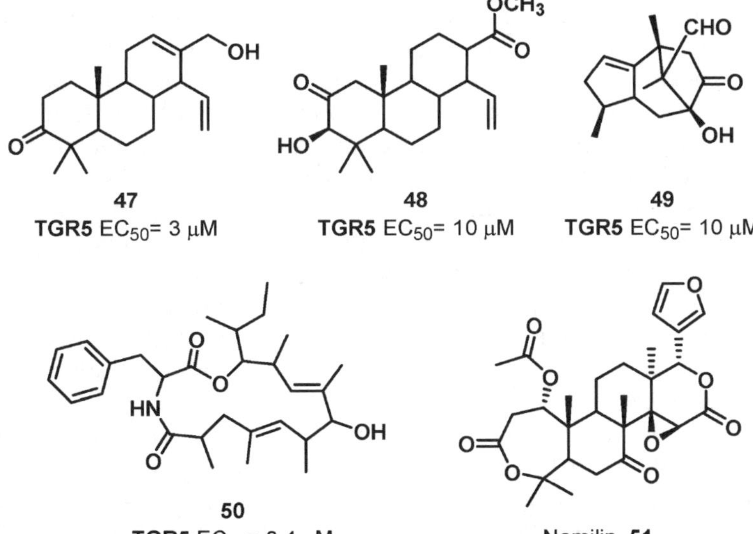

Figure 10.8   Structure and TGR5 activity of triterpenes and macrocyclic compounds.

these compounds at the receptor. Additional natural and semisynthetic terpenes as well as macrocycle compounds were disclosed as TGR5 ligands in two patent applications by Merck.[60] In the more recent filing, eight examples were depicted, with two of them resembling truncated forms of rings A–C of oleanoic acid and betulinic acid (**47**, **48**, Figure 10.8). Interestingly, the most potent compound was designated as a macrocycle exhibiting an EC_{50} in the sub-micromolar range of potency in cAMP production assay of TGR5 expressing BHK cells.

More recently, nomilin (**51**), another naturally occurring triterpenoid originally isolated from the seeds of oranges and lemons,[61] was disclosed as a TGR5 agonist, showing no activity at FXR receptor.[62] The administration of nomilin (**51**) (0.2% w/w) for 77 days to obese mice proved to lower body weight, serum levels of glucose, and insulin, and to enhance glucose tolerance.

10.5.1.4 Non-Steroidal Ligands: Compounds from Synthetic Chemical Libraries

One of the earliest screening of small molecule libraries for TGR5 ligands was carried out by researchers at Takeda Pharmaceuticals. Their work afforded the identification of 6-methyl-2-oxo-4-thiophen-2-yl-1,2,3,4-tetrahydropyrimidine-5-carboxylic acid benzyl ester (**52**) (Figure 10.9) as the first synthetic agonist towards the BA membrane receptor.[63] Interestingly, this compound was used as a chemical tool in the study of Watanabe *et al.* to demonstrate that the effects of BAs on cAMP levels and type 2 deiodinase activity were indeed mediated by TGR5 activation.[9] Further work allowed the same company to file patent applications describing bicyclic systems constituted by a substituted aromatic ring fused with five- to eight-membered rings as potent and selective TGR5 agonists.[63] About 200 compounds were exemplified in one of these patent applications, 11 of which were reported as active at human TGR5-expressing CHO cells with potencies in the same order of LCA. Among these, the *trans* mixture of **53** resulted in one of the most active compounds, showing a 128% rate of cAMP production over the control (1μM, LCA).

In 2005, Arena Pharmaceuticals filed another patent reporting agonists of TGR5 (termed RUP43 in the patent) that proved to increase glucose uptake in human adipocytes and skeletal muscle.[64] Two compounds (**54**, **55**) (Figure 10.9) are exemplified that led to a dose-dependent increase of intracellular cAMP in HEK293 cells transfected with TGR5. Compound **55**, in particular, stimulated glucose uptake in human primary adipocytes and skeletal muscle cells.

More recently, a number of patent applications have been reported describing additional heterocyclic molecules as TGR5 agonists. The first in a series of applications filed by Kalypsys disclosed pyrrolo- and pyrazole-diazepine compounds tethered to substituted benzoyl moieties (**56**, **57**).[65] Four out of 13 structures exemplified in the patent were reported with an EC_{50} below 10μM in a cAMP production assay using TGR5-expressing HEK293

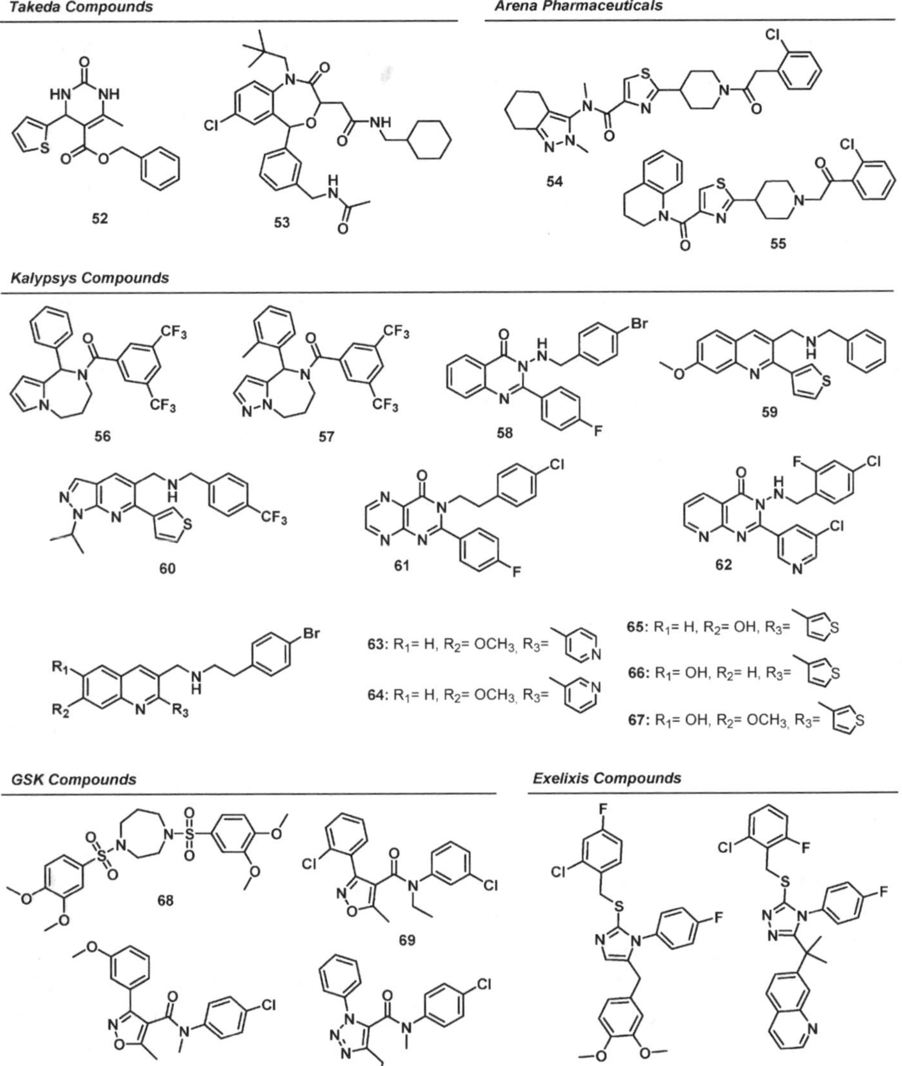

Figure 10.9 Non-steroidal TGR5 modulators.

cell lines. In a subsequent patent,[66] the same company reported a novel series of compounds containing a quinazolin-4-one scaffold (**58**). Remarkably, at odds with the previous application, a larger number of compounds was depicted. Of the 119 structures exemplified, more than 80% molecules were found as endowed with an EC_{50} below 10μM in a similar cAMP production assay. Additional patent applications published by Kalypsys include the disclosure of quinoline derivatives **59** and bicyclic systems containing a central pyridine or pyrimidin-4-one moiety (**60–62**) (Figure 10.9).[67] Common features in all of the disclosed compounds are an aromatic ring at the C2 position and a long benzyl chain at the C3 position that resemble similar substitutions in the quinazolin-4-one derivatives. Although almost all bicyclic pyridine and pyrimidin-4-one compounds were found with EC_{50} activity below 10μM in a cAMP production assay of TGR5-expressing HEK293 cells, it is worth noting that only 42% of 149 exemplified quinoline derivatives showed EC_{50} in that range of potency. As a consequence, in view also of the good activities shown by many quinazolin-4-ones, it is likely that having a hydrogen acceptor group at the C4 position of the heterocyclic system may favor TGR5 activity.

Starting from the observation that the quinoline derivative **59** was less potent on mouse TGR5 than human, researchers at Kalypsys reported in a separate communication the results of their synthetic efforts aimed at optimizing mouse and human TGR5 activities of **59**, as well as drawing structure–activity relationships.[68] As a first result, they succeed in developing a series of quinoline derivatives endowed with good potency at both human and mouse receptors (**63–67**). Secondly, the analysis of this series of compounds allowed the researchers to draw a scheme of structure–activity relationships. Accordingly, they found that *ortho-* and *meta*-substituted benzyl amines were generally inactive, whilst *para* substitutions improved the activity at both human TGR5 and mouse paralog. The extension of the amine linker to phenyl ethylamines was further beneficial for the activity at human receptor, but it was marginal for the potency at rat TGR5. Concerning the 3-thiophenyl group of **59**, different bioisosteric replacements were attempted. While 2-thiophenyl, 3-furanyl, 2-furanyl, and phenyl groups slightly reduced the activity, thiazole and oxazole moieties led to a significant drop in potency. Interestingly, the replacement of 3-thiophenyl group with 4-pyridyl or 3-pyridyl moieties provided the most active compounds of the series at human TGR5 (**63**, hTGR5 EC_{50} = 0.044 μM, mTGR5 EC_{50} = 3.7 μM; **64**, hTGR5 EC_{50} = 0.065 μM, mTGR5 EC_{50} = 5.0 μM). Further substitutions were explored around the quinoline ring, evidencing positions 6 and 7 as interesting for the activity. In particular, it was found that 7-hydroxyl and 6-hydroxyl derivatives showed an inverted profile of activity at human TGR5 and mouse paralog, with the former being more active at mouse receptor (**65**, hTGR5 EC_{50} = 5.1 μM, mTGR5 EC_{50} = 0.28 μM) while the latter being more portent at human TGR5 (**66**, hTGR5 EC_{50} = 0.082 μM, mTGR5 EC_{50} = 1.6 μM). Remarkably, the combination of 7-methoxy substitution with hydroxylation in position 6 succeeded in giving a compound that was almost equally active on both the human and rat receptor (**67**, hTGR5 EC_{50} = 0.43 μM, mTGR5 EC_{50} = 0.32 μM).

In another filing from GlaxoSmithKline, bis-aryl-sulfonyl derivatives linked with an amide bond to central diaminoalkyl or heterocyclic rings were exemplified as TGR5 ligands in transfected melanophore cells.[69] Only 8% of the reported compounds showed pEC_{50} values above 6.9. Among these, compound **68** proved to increase GLP-1 secretion in rodent models, preventing hyperglycemia and improving glucose tolerance. As part of their ongoing efforts in the field, GlaxoSmithKline recently published the results of a novel screening for TGR5 ligands, reporting 3-aryl-4-isoxazolecarboxamides as new agonists of the receptor.[70] The hit compound **69** was designated as having a pEC$_{50}$ of 5.3 in BacMam-transduced human osteosarcoma cell line (U2-OS). Further optimization work led the investigators to disclose more potent analogs (**70, 71**) and develop structure–activity relationships.[71] As a result, it was found that *para* substitution on the amide phenyl ring was favored over substitutions at *ortho* and *meta* positions. Moreover, while chlorine and small alkyl groups such as methyl and trifluoromethyl were preferred at the *para* position, the combination of *para*-methyl with further halo and methyl substitutions at *ortho* or *meta* positions increased the potency. Concerning the isoxazole phenyl ring, *ortho* and *meta* substitutions were preferred over *para* substitutions. Again, favored substituents at the *ortho* position were methyl, trifluoromethyl, and chlorine. Conversely, electron-withdrawing groups such as methoxyl and fluorine proved effective for the activity when inserted at the *meta* position, with the 3-fluoro derivative being the most potent compound of the series. While replacements of the isoxazole phenyl ring with pyridine and thiophene moieties reduced the activity, the substitution of the isoxazole ring with the triazole scaffold slightly increased the potency, in spite of the decrease of activity observed with benzene and furan replacements.

Interestingly, preferred compounds showed selectivity towards TGR5 when tested over more than 100 protein targets including GPCRs, ion channels, enzymes, transporters, and nuclear receptors. Although the lack of hERG binding contributed to the good profile of these compounds, the activity at some cytochrome P450 isoforms (2C19 and 3A4) and the high clearance pinpointed the presence of ADME issues that resulted in poor systemic exposure. However, pursuing the idea that systemic exposure was not necessary to achieve the desired effect on TGR5-mediated stimulation of GLP-1 *in vivo*, the authors tested these compounds in an acute conscious dog animal model via intracolonic administration. As a result, they found a significant improvement of GLP-1 secretion and reduction of glucose levels in the portal vein. Exelixis filed two patent applications on TGR5 ligands containing a central imidazole or triazole ring (**72, 73**). In the first of these,[72] they reported the activity of more than 700 compounds, with around 29% of them showing nanomolar potencies at human TGR5 in HEK 293 CRE-luciferase assay. In the second filing,[73] 160 compounds were exemplified with a higher rate of ligands (96%) being endowed with activities in the nanomolar range at human TGR5.

Overall, some of the non-steroidal synthetic TGR5 ligands seem to be compliant to a common pharmacophoric model constituted by two aromatic moieties linked to a central scaffold (Figure 10.10).

Although this may be a very preliminary model of interaction of these compounds to TGR5 and additional studies are required in the field, the model provides a map of the receptor binding site that appears different from those

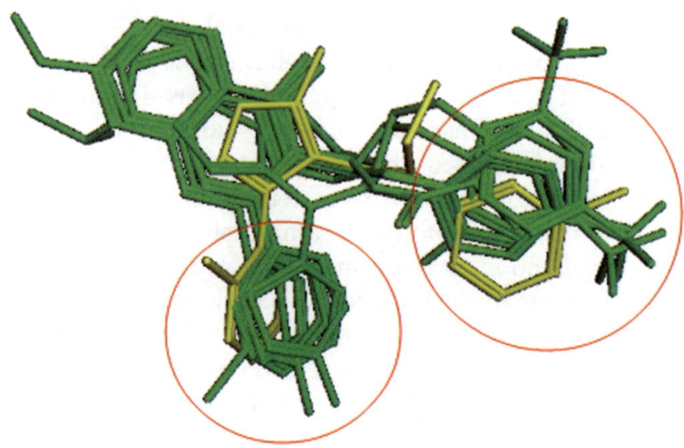

Figure 10.10 Pharmacophoric model of selected non-steroidal synthetic ligands. The pharmacophoric model was developed using MOE and compounds **56**, **58**, **60–62**, **65**, and **69**. The two conserved aromatic features are highlighted with red circles. Compound **69** is shown in yellow.

arising from structure–activity relationships of BAs and betulinic acid derivatives. Accordingly, it is plausible that the binding sites of steroidal and non-steroidal ligands locate on different regions of TGR5, or they are only in part overlapped involving diverse interacting residues.

10.5.2 Steroidal and Non-Steroidal TGR5 Ligands in the Drug-like Property Space

It is well acknowledged that the presence of specific physicochemical properties and structural features makes small molecules more or less drug-like.[74] The quest of drug-like compounds in the early stages of drug discovery is motivated by the need of decreasing the costly risk of late-stage preclinical and clinical failures. In this context, many prediction rules have been developed, suggesting optimal ranges and/or cut-off values for key molecular descriptors.[75] For instance, Lipinski's "rule of five" was introduced as one of the earliest rules of thumb to develop oral bioavailable compounds, pinpointing the relationship between poor oral bioavailability and high values of molecular mass (MW $>$ 500), $c \log P$ ($>$ 5) and number of hydrogen bond donors and acceptors (Don. $>$ 5; Acc. $>$ 10).[76]

However, optimal ranges and/or cut-off values of key molecular properties for better drug-likeness may not always account for favorable interactions with the drug target. On the basis of such consideration, researchers at Eli Lilly and Company investigated the profile of molecular properties of drugs and highly active compounds from literature according to the proteomic family of their main target.[77] As a result, they found that drugs and active compounds for specific proteomic families of targets are poorly compliant to standard rules of drug-likeness. Notably, the researchers reported that drugs and active

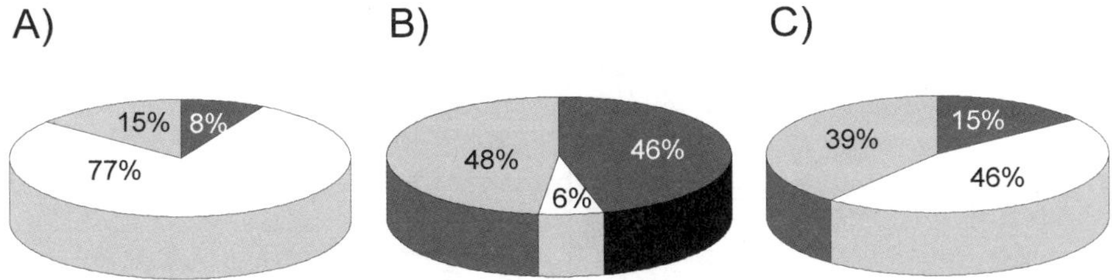

Figure 10.11 Compliance to Lipinski's "rule of five" for TGR5 ligands in development: (A) bile acids and derivatives; (B) non-steroidal natural products; (C) non-steroidal synthetic ligands. Colors used: white = no violation; light gray = one violation; dark gray = more than one violation.

compounds for lipid GPCRs have significantly higher $c \log P$, showing the lowest rule of five compliance.

This is also clearly evidenced in the case of BAs, non-steroidal natural products and non-steroidal synthetic TGR5 ligands. An inspection of Figure 10.11, indeed, reveals that the percentages of these compounds passing all four original Lipinski rules are low, with 77% and 46% for the BAs and non-steroidal ligands, respectively. These percentages are improved when compounds violating no more than one of the four rules are included in the computation.

Overall, it is surprising that non-steroidal compounds are less compliant than BAs to Lipinski rules as far as all the four criteria are considered. One explanation would be that the application of standard rules of drug-likeness would bias the screening for active TGR5 ligands away from the required molecular properties for achieving high affinity to the receptor. Conversely, the steroid scaffold of BAs would demand less effort to balance affinity and bioavailability, being endowed with intrinsic amphiphilic properties.

A careful analysis of the average values and percentiles of size, lipophilicity, and hydrogen bonding properties (Table 10.4) reveals that all the classes of compounds are characterized by relatively high molecular weight and lipophilicity. BAs, in particular, have higher numbers of hydrogen bond donors than non-steroidal ligands. Indeed, 90% of BAs (90th percentile) have values falling in a range from 1 to 4 hydrogen bond donor groups, whilst 90% of non-steroidal ligands fall in a shorter range from 0 to 1. Interestingly, some classes of non-steroidal ligands seem to compensate for the lack of these groups with a higher number of hydrogen bond acceptors. Accordingly, 90% of non-steroidal synthetic ligands lie in a range of hydrogen bond acceptors between 1 and 12, with an average value of 6.02 ± 2.80.

10.6 Predictive Models of TGR5 Affinity

The development of predictive models of target affinity is pivotal in drug discovery to enable, before synthesis, the prioritization of analogs resulting

Table 10.4   Average values, standard deviations, and 90th percentile of size (MW), lipophilicity (c log P) and hydrogen bonding properties (HB Acc., HB Don.) for TGR5 ligands according to the chemical class.

Class (Number of Compounds)	MW	HB Acc.	HB Don.	c log P	90% MW (Min. Value)	90% Acc. (Min. Value)	90% Don. (Min. Value)	90% c log P (Min. Value)
Bile acids (110)	432.84 $\pm$ 47.54	6.03 $\pm$ 1.49	3.15 $\pm$ 0.79	4.23 $\pm$ 0.89	505.54 (350.54)	7.40 (2)	4.00 (1)	5.47 (2.52)
Non-steroidal ligands (654)	448.82 $\pm$ 66.59	6.02 $\pm$ 2.80	0.54 $\pm$ 0.66	4.64 $\pm$ 1.70	546.13 (326.78)	12.00 (1.50)	1.00 (0)	6.71 (-0.18)
Triterpenes (48)	489.37 $\pm$ 62.03	3.78 $\pm$ 1.15	1.74 $\pm$ 0.58	6.85 $\pm$ 1.36	543.24 (232.32)	5.50 (2.75)	2.00 (0)	7.93 (1.94)

from the next round of virtual screening as well as the design of focused small molecule libraries around active ligands.[78] Furthermore, being mostly ligand-based approaches, their results provide invaluable clues on the pharmacophoric elements featuring the recognition of ligands at the molecular target that complement those arising from structure–activity relationships.

Depending on the definition of TGR5 activity with categorical or continuous values, classification analysis and quantitative structure–activity relationship (QSAR) studies have been carried out on BAs as TGR5 agonists. A linear discriminant analysis (LDA), in particular, was instrumental to generate a classification model of TGR5 activity on a dataset of 69 BA derivatives, using molecular shape descriptors as independent variables.[5] The resulting model was composed of two linear functions of discrimination and five selected variables that grouped BAs into "active" and "moderately active" ligands ($EC_{50} < 10$ μM, class 1), and "inactive" ($EC_{50} \geq 10$ μM, class 2) TGR5 agonists. Selected shape descriptors were the total hydrophobic surface area (TASA), the relative hydrophobic surface area of the compound (RASA, total hydrophobic surface area divided by the total molecular solvent-accessible surface area), the relative negative charged surface area (RNCS, solvent-accessible surface area of most negative atom divided by the relative charge of most negative atom), the common overlapping volume (COSV) calculated with the tauro-conjugated form of LCA, and the non-common overlapping volume (NCOSV) calculated with the tauro-conjugated form of LCA. After leave-one-out cross-validation, the LDA model was able to correctly classify 79.17% of active TGR5 agonists and 85.71% of inactive TGR5 agonists, thereby showing a good discriminating performance with a value of the area under the ROC curve (AUC) of 0.928. The inspection of contributions of the selected variables to the classification model pinpointed steric complementarities (COSV and NCOSV) and the lipophilic profile (RASA, TASA, RNCS) of BAs as relevant to TGR5 activation.

In another study, a 3D-QSAR approach was pursued for the generation of a predictive model of TGR5 affinity, based on the description of BAs in terms of molecular interaction fields (MIFs).[79] While being able to explain much of the variance of the 43 training set BAs and satisfactorily predicting the 70% rate of an external test set of 20 BAs, the 3D-QSAR model disclosed essential interactions implicated in the recognition of the TGR5-binding site that were in agreement with the structure–activity relationship scheme of BAs (Figure 10.12).

Accordingly, the presence of hydrophobic interactions near the C6 and C7 positions of the BA steroid nucleus was evidenced as important to gain potency. An additional hydrophobic interaction was also observed at C24, accounting for TGR5 activity of conjugated, C24-hydroxy and methyl ester derivatives. Polar and hydrogen bonding interactions were further highlighted near to positions C20–C22 of the side chain and C12-α of the steroid scaffold, respectively. More importantly, the predictive model emphasized the stereo-specificity in the recognition of a TGR5 accessory pocket by the methyl group

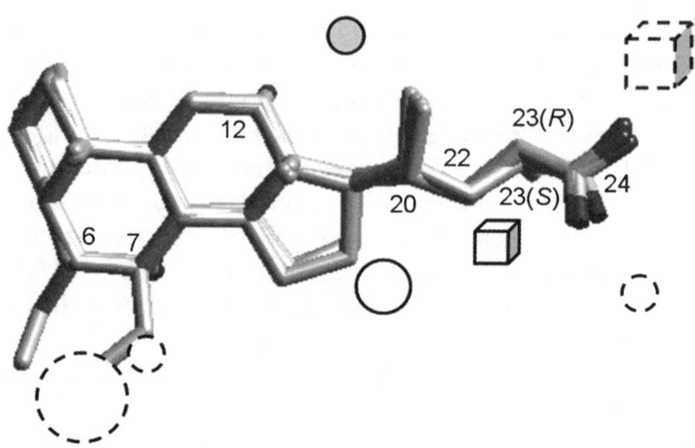

Figure 10.12   Predictive model of TGR5 affinity for bile acids and derivatives. Map of the interaction points according to the 3D-QSAR equation, see Macchiarulo *et al.* (2008).[79] White dashed circles depict unfavorable polar interactions (that is favorable hydrophobic interactions). White circles show favorable polar interactions. The gray circle indicates a hydrogen bonding interaction. The white dashed box depicts a sterically forbidden interaction. The white box shows a sterically allowed interaction. The size of the circles and boxes is in direct relation to the value of the coefficients of the independent variables that encode the relative interactions in the 3D-QSAR equation.

at position C23, showing an unfavorable steric interaction around the (*R*) configuration, while a favored steric interaction around the (*S*) configuration. Unfortunately, no predictive models have been published to date concerning the affinity of non-steroidal synthetic ligands at TGR5.

10.7   Conclusions

The structural diversity of TGR5 ligands so far disclosed in patents and the literature is remarkable. Combining it with the results of structure–activity relationship studies and predictive models of target affinity, we may firstly suggest that TGR5 features a large binding site composed of several interacting residues, akin to the BA binding pocket of FXR. Alternatively, more than one functionally active binding site may regulate G-protein coupling in TGR5. While both scenarios are in agreement with the emerging complexity of GPCR signaling,[80] they open unsettled questions concerning TGR5 modulation. First of all, it is currently accepted that GPCR modulation passes through a continuum of conformational states. Different ligands may stabilize diverse conformations of the GPCR, thereby leading to differential modulation of signaling across the cell membrane, such as in the case of cannabinoid receptors.[81] TGR5 belongs to a subfamily of GPCRs activated by intracellular lipid ligands, with cannabinoid receptors being the closest phylogenetically related GPCRs. Whether the concept of selective signaling applies to TGR5, with BAs and non-steroidal ligands modulating different downstream events, still remains to be ascertained.

A second question that is germane to TGR5 modulation, is the availability of pure antagonist and/or inverse agonist. No ligands with these pharmacological profiles have been disclosed for TGR5 yet. Pursuing these chemical entities will certainly provide invaluable chemical probes to explore the role of TGR5 in a broader context.

Additional questions concern the therapeutic opportunities that TGR5 activation may offer in metabolic-related disorders in humans, including T2D. The recent demonstration of the *in vivo* efficacy of INT-777 (**10**) in stimulating GLP-1 secretion via TGR5, thereby controlling energy metabolism and glucose homeostasis, is encouraging and provides the proof-of-concept that TGR5 may indeed be a novel drug target for treating diabesity. However, whether TGR5 stimulation may have other (un)desired effects in human is pending further appraisals. For instance, TGR5 is also expressed in acinar pancreatic cells, gallbladder, and liver macrophages in addition to cells of the GI tract. While the activation of TGR5 in the latter is responsible of the desired effect of stimulating GLP-1 secretion, the decrease of cytokine production ensuing TGR5 stimulation in liver macrophage and the activation of the receptor in pancreatic cells need additional work to determine the potential effects. Recently, it has been shown that the pharmacological activation of TGR5 with INT-777 (**10**) stimulates the increase of gallbladder volume in mice, regulating gallbladder contractility. It remains to be ascertained whether this effect applies also in humans, as TGR5 agonists are developed for treating diabesity.

Nevertheless, it should be mentioned that some researchers believe that the development of TGR5 agonists with poor systemic exposure may represent a viable strategy for a direct targeting of the receptor in the GI tract, avoiding potential bad systemic effects while keeping the good of TGR5 stimulation.[70] Accordingly, it is very likely that the future trends in medicinal chemistry of TGR5 modulators will be also directed to the development of novel ligands that are able to selectively target the receptor in distinct districts of human body.

In conclusion, while the outstanding results so far achieved show that the effort is worthwhile, several questions need to be addressed, making TGR5 a very active field of research in the next few years for both academia and companies. Activation of TGR5 may represent a promising novel strategy for the treatment of diabesity. This is supported by the recent demonstration of the *in vivo* efficacy of INT-777 (**10**), a BA derivative that via TGR5 stimulates GLP-1 secretion, thereby controlling energy metabolism and glucose homeostasis. However, while the reported efficacy of INT-777 (**10**) in mice is encouraging, further studies with this agent are needed to determine its potential in humans. Likewise, additional work is required to address some controversial questions regarding the potential adverse effects of TGR5 stimulation. In this context, the development of novel potent and selective modulators of this receptor will prove useful to further unravel the complexity of TGR5 signaling.

References

1. J. F. Evans and J. H. Hutchinson, *Nature Chem. Biol.*, 2010, **6**, 476.
2. C. Thomas, R. Pellicciari, M. Pruzanski, J. Auwerx and K. Schoonjans Nature Rev. *Drug Discov.*, 2008, **7**, 678.
3. C. Thomas, J. Auwerx and K. Schoonjans, *Thyroid*, 2008, **18**, 167.
4. (a) Y. Kawamata, R. Fujii, M. Hosoya, M. Harada, H. Yoshida, M. Miwa, S. Fukusumi, Y. Habata,T. Itoh, Y. Shintani, S. Hinuma, Y. Fujisawa, M. Fujino, *J. Biol. Chem.*, 2003, 278, 9435; (b) T. Maruyama, Y. Miyamoto, T. Nakamura, Y. Tamai, H. Okada, E. Sugiyama, H. Itadani and K. Tanaka, *Biochem. Biophys. Res. Commun.*, 2002, **298**, 714.
5. H. Sato, A. Macchiarulo, C. Thomas, A. Gioiello, M. Une, A. F. Hofmann, R. Saladin, K. Schoonjans, R. Pellicciari and J. Auwerx, *J. Med. Chem.*, 2008, **51**, 1831.
6. S. Costanzi, S. Neumann and M. C. Gershengorn, *J. Biol. Chem.*, 2008, **283**, 16269.
7. J. R. Hov, V. Keitel, J. K. Laerdahl, L. Spomer, E. Ellinghaus, A. ElSharawy, E. Melum, K. M. Boberg, T. Manke, T. Balschun, C. Schramm, A. Bergquist, T. Weismüller, D. Gotthardt, C. Rust, L. Henckaerts, C. M. Onnie, R. K. Weersma, M. Sterneck, A. Teufel, H. Runz, A. Stiehl, C. Y. Ponsioen, C. Wijmenga, M. H. Vatn; IBSEN Study Group, P. C. Stokkers, S. Vermeire, C. G. Mathew, B. A. Lie, U. Beuers, M. P. Manns, S. Schreiber, E. Schrumpf, D. Häussinger, A. Franke and T. H. Karlsen, *PLoS One*, 2010, **5**, e12403.
8. A. Tiwari and P. Maiti, *Drug Discov. Today*, 2009, **14**, 523.
9. M. Watanabe, S. M. Houten, C. Mataki, M. A. Christoffolete, B. W. Kim, H. Sato, N. Messadeq, J. W. Harney, O. Ezaki, T. Kodama, K. Schoonjans, A. C. Bianco and J. Auwerx, *Nature*, 2006, **439**, 484.
10. T. Maruyama, K. Tanaka, J. Suzuki, H. Miyoshi, N. Harada, T. Nakamura, Y. Miyamoto, A. Kanatani and Y. Tamai, *J. Endocrinol.*, 2006, **191**, 197.
11. (a) W. D. van Marken Lichtenbelt, J. W. Vanhommerig, N. M. Smulders, J. M. Drossaerts, G. J. Kemerink, N. D. Bouvy, P. Schrauwen and G. Teule, *N. Engl. J. Med.*, 2009, **360**, 1500; (b) A. M. Cypess, S. Lehman, G. Williams, I. Tal, D. Rodman, A. B. Goldfine, F. C. Kuo, E. L. Palmer, Y. H. Tseng, A. Doria, G. M. Kolodny and C. R. Kahn, *N. Engl. J. Med.*, 2009, **360**, 1509; (c) K. A. Virtanen, M. E. Lidell, J. Orava, M. Heglind, R. Westergren, T. Niemi, M. Taittonen, J. Laine, N. J. Savisto, S. Enerback and P. Nuutila, *N. Engl. J. Med.*, 2009, **360**, 1518.
12. (a) H. Nakatani, K. Kasama, T. Oshiro, M. Watanabe, H. Hirose and H. Itoh, *Metabolism*, 2009, **58**, 1400; (b) M. E. Patti, S. M. Houten, A. C. Bianco, R. Bernier, P. R. Larsen, J. J. Holst, M. K. Badman, E. Maratos-Flier, E. C. Mun, J. Pihlajamaki, J. Auwerx and A. B. Goldfine, *Obesity*, 2009, **17**, 1671.
13. (a) M. de Castro Cesar, M. I. de Lima Montebelo, I. Rasera, A. V. de Oliveira, P. R. Gomes Gonelli and C. G. Aparecida, *Obes. Surg.*, 2008, **18**,

1376; (b) S. Buscemi, G. Caimi and S. Verga, *Int. J. Obes. Relat. Metab. Disord.*, 1996, **20**, 41; (c) L. Flancbaum, P. S. Choban, L. R. Bradley and J. C. Burge, *Surgery*, 1997, **122**, 943; (d) D. G. Carey, G. Pliego and R. L. Raymond, *Obes. Surg.*, 2006, **16**, 1602; (e) F. Carrasco, K. Papapietro, A. Csendes, G. Salazar, C. Echenique, C. Lisboa, E. Díaz and J. Rojas, *Obes. Surg.*, 2007, **17**, 608; (f) G. Benedetti, *J. Am. Coll. Nutr.*, 2000, **19**, 270; (g) L. Mika Horie, M. C. González, M. Raslan, R. Torrinhas, N. L. Rodrigues, C. C. Verotti, I. Cecconello, S. B. Heymsfield and D. L. Waitzberg, *Nutr. Hosp.*, 2009, **24**, 676; (h) R. A. Tamboli, H. A. Hossain, P. A. Marks, A. W. Eckhauser, J. A. Rathmacher, S. E. Phillips, M. S. Buchowski, K. Y. Chen and N. N. Abumrad, *Obesity*, 2010, **18**, 1718.

14. G. Vassileva, W. Hu, L. Hoos, G. Tetzloff, S. Yang, L. Liu, L. Kang, H. Davis, J. Hedrick, H. Lan, T. Kowalski and E. Gustafson, *J. Endocrinol.*, 2010, **205**, 225.

15. G. Brufau, M. J. Bahr, B. Staels, T. Claudel, J. Ockenga, K. H. Boker, E. J. Murphy, K. Prado, F. Stellaard, M. P. Manns, F. Kuipers and U. J. Tietge, *Nutr. Metab. (Lond)*, 2010, **7**, 73.

16. T. W. Pols, L. G. Noriega, M. Nomura, J. Auwerx and K. Schoonjans, *J. Hepatol.*, 2010, **54**, 1263.

17. (a) M. Namba, T. Matsuyama, H. Horie, K. Nonaka and S. Tarui, *Regul. Pept.*, 1983, **5**, 257; (b) M. Namba, T. Matsuyama, H. Itoh, Y. Imai, H. Horie and S. Tarui, *Regul. Pept.*, 1986, **15**, 121; (c) T. E. Adrian, G. H. Ballantyne, W. E. Longo, A. J. Bilchik, S. Graham, M. D. Basson, R. P. Tierney and I. M. Modlin, *Gut*, 1993, **34**, 1219; (d) P. Plaisancie, V. Dumoulin, J. A. Chayvialle and J. C. Cuber, *J. Endocrinol.*, 1995, **145**, 521.

18. S. Katsuma, A. Hirasawa and G. Tsujimoto, *Biochem. Biophys. Res. Commun.*, 2005, **329**, 386.

19. (a) B. Kreymann, M. A. Ghatei, G. Williams and S. R. Bloom, *Lancet*, 1987, **2**, 1300; (b) M. A. Nauck, *Am. J. Med.*, 2011, **124** (1 Suppl), S3.

20. J. J. Holst, *Physiol. Rev.*, 2007, **87**, 1409.

21. C. Thomas, A. Gioiello, L. Noriega, A. Strehle, J. Oury, G. Rizzo, A. Macchiarulo, H. Yamamoto, C. Mataki, M. Pruzanski, R. Pellicciari, J. Auwerx and K. Schoonjans, *Cell Metab.*, 2009, **10**, 167.

22. R. Pellicciari, A. Gioiello, A. Macchiarulo, C. Thomas, E. Rosatelli, B. Natalini, R. Sardella, M. Pruzanski, A. Roda, E. Pastorini, K. Schoonjans and J. Auwerx, *J. Med. Chem.*, 2009, **52**, 7958.

23. D. P. Poole, C. Godfrey, F. Cattaruzza, G. S. Cottrell, J. G. Kirkland, J. C. Pelayo, N. W. Bunnet and C. U. Corvera, *Neurogastroenterol. Motil.*, 2010, **22**, 814.

24. (a) R. Penagini, R. C. Spiller, J. J. Misiewicz and P. G. Frost, *Gut*, 1989, **30**, 609; (b) R. C. Spiller, I. F. Trotman, B. E. Higgins, M. A. Ghatei, G. K. Grimble, Y. C. Lee, S. R. Bloom, J. J. Misiewicz and D. B. Silk, *Gut*, 1984, **25**, 365.

25. T. Tolessa, M. Gutniak, J. J. Holst, S. Efendic and P. M. Hellstrom, *J. Clin. Invest.*, 1998, **102**, 764.

26. (a) G. Drivas, O. James and N. Wardle, *Br. Med. J.*, 1976, **26**, 1568; (b) R. M. Keane, T. R. Gadacz, A. M. Munster, W. Birmingham and R. A. Winchurch, *Surgery*, 1984, **95**, 439; (c) A. N. Kimmings, S. J. H. van Deventer, H. Obertop, E. A. J. Rauws and D. J. Gouma, *J. Am. Coll. Surg.*, 1995, **181**, 567.

27. (a) J. W. Greve, D. J. Gouma and W. A. Buurman, *Hepatology*, 1989, **10**, 454; (b) Y. Calmus, J. Guechot, P. Podevin, M. T. Bonnefis, J. Giboudeau and R. Poupon, *Hepatology*, 1992, **16**, 719.

28. V. Keitel, M. Donner, S. Winandy, R. Kubitz and D. Haussinger, *Biochem. Biophys. Res. Commun.*, 2008, **372**, 78.

29. G. Ramadori and T. Armbrust, *Eur. J. Gastroenterol. Hepatol.*, 2001, **13**, 777.

30. (a) M. Funaoka, M. Komatsu, I. Toyoshima, K. Mikami, T. Ono, T. Hoshino, J. Kato, T. Kuramitsu, T. Ishii and O. Masamune, *J. Gastroenterol. Hepatol.*, 1999, **14**, 652; (b) J. J. Sung and M. Y. Go, *J. Hepatol.*, 1999, **30**, 413.

31. V. Keitel, R. Reinehr, P. Gatsios, C. Rupprecht, B. Gorg, O. Selbach, D. Haussinger and R. Kubitz, *Hepatology*, 2007, **45**, 695.

32. M. Oda, H. Yokomori and J. Y. Han, *Clin. Hemorheol. Microcirc.*, 2003, **29**, 167.

33. B. Vollmar and M. D. Menger, *Physiol. Rev.*, 2009, **89**, 1269.

34. H. O. Wheeler, *Am. J. Med.*, 1971, **51**, 588.

35. D. I. Giurgiu, K. D. Saunders-Kirkwood, J. J. Roslyn and M. Z. Abedin, *Ann. Surg.*, 1997, **225**, 382.

36. V. Keitel, K. Cupisti, C. Ullmer, W. T. Knoefel, R. Kubitz and D. Haussinger, *Hepatology*, 2009, **50**, 861.

37. G. Vassileva, A. Golovko, L. Markowitz, S. J. Abbondanzo, M. Zeng, S. Yang, S. Yang, L. Hoos, G. Tetzloff, D. Levitan, N. J. Murgolo, K. Keane, H. R. Davis, J. Hedrick and E. L. Gustafson, *Biochem. J.*, 2006, **398**, 423.

38. B. Lavoie, O. B. Balemba, C. Godfrey, C. A. Watson, G. Vassileva, C. U. Corvera, M. T. Nelson and G. M. Mawe, *J. Physiol.*, 2010, **588**, 3295.

39. T. Li, S. R. Holmstrom, S. Kir, M. Umetani, D. R. Schmidt, S. A. Kliewer and D. J. Mangelsdorf, *Mol. Endocrinol.*, 2011, **25**, 1066.

40. V. Keitel, B. Görg, H. J. Bidmon, I. Zemtsova, L. Spomer, K. Zilles and D. Häussinger, *Glia*, 2010, **58**, 1794.

41. D. Häussinger and F. Schliess, *Gut*, 2008, **57**, 1156.

42. F. K. Knop, T. Vilsbøll and J. J. Holst, *Curr. Protein Pept. Sci.*, 2009, **10**, 46.

43. (a) A. M. Lambeir, C. Durinx, S. Scharpé and I. De Meester, *Crit. Rev. Clin. Lab. Sci.*, 2003, **40**, 209; (b) B. D. Green, P. R. Flatt and C. J. Bailey, *Diab. Vasc. Dis. Res.*, 2006, **3**, 159.

44. M. A. Nauck and J. J. Meier, *Nat. Rev. Endocrinol.*, 2011, **7**, 193.

45. (a) R. Burcelin, A. Da Costa, D. Drucker and B. Thorens, *Diabetes*, 2001, **50**, 1720; (b) L. Hansen, C. F. Deacon, C. Ørskov and J. J. Holst, *Endocrinology*, 1999, **140**, 5356.
46. F. K. Knop, *Am. J. Physiol. Endocrinol. Metab.*, 2010, **299**, E10.
47. H. Buchwald, R. Estok, K. Fahrbach, D. Banel, M. D. Jensen, W. J. Pories, J. P. Bantle and I. Sledge, *Am. J. Med.*, 2009, **122**, 248.
48. (a) J. Prawitt and B. Staels, *Metab. Syndr. Relat. Disord.*, 2010, **8** (Suppl. 1), S3; (b) L. Chen, J. McNulty, D. Anderson, Y. Liu, C. Nystrom, S. Bullard, J. Collins, A. L. Handlon, R. Klein, A. Grimes, D. Murray, R. Brown, D. Krull, B. Benson, E. Kleymenova, K. Remlinger, A. Young and X. Yao, *J. Pharmacol. Exp. Ther.*, 2010, **334**, 164.
49. (a) Q. Shang, M. Saumoy, J. J. Holst, G. Salen and G. Xu, *Am. J. Physiol. Gastrointest. Liver Physiol.*, 2010, **298**, G419; (b) T. Suzuki, K. Oba, Y. Igari, N. Matsumura, K. Watanabe, S. Futami-Suda, H. Yasuoka, M. Ouchi, K. Suzuki, Y. Kigawa and H. Nakano, *J. Nippon Med. Sch.*, 2007, **74**, 338.
50. (a) L. J. Bennion and S. M. Grundy, *N. Engl. J. Med.*, 1977, **296**, 1365; (b) H. Hyogo, S. Roy, B. Paigen and D. E. Cohen, *J. Biol. Chem.*, 2002, **277**, 34117; (c) R. Kohli, M. Kirby, K. D. Setchell, P. Jha, K. Klustaitis, L. A. Woollett, P. T. Pfluger, W. F. Balistreri, P. Tso, R. J. Jandacek, S. C. Woods, J. E. Heubi, M. H. Tschoep, D. A. D'Alessio, N. F. Shroyer and R. J. Seeley, *Am. J. Physiol. Gastrointest. Liver Physiol.*, 2010, **299**, G652; (d) K. Suhre, C. Meisinger, A. Döring, E. Altmaier, P. Belcredi, C. Gieger, D. Chang, M. V. Milburn, W. E. Gall, K. M. Weinberger, H. W. Mewes, M. Hrabé de Angelis, H. E. Wichmann, F. Kronenberg, J. Adamski and T. Illig, *PLoS One*, 2010, **5**, e13953; (e) J. V. Li, H. Ashrafian, M. Bueter, J. Kinross, C. Sands, C. W. le Roux, S. R. Bloom, A. Darzi, T. Athanasiou, J. R. Marchesi, J. K. Nicholson and E. Holmes, *Gut*, 2011, **60**, 1214; (f) M. Watanabe, Y. Horai, S. M. Houten, K. Morimoto, T. Sugizaki, E. Arita, C. Mataki, H. Sato, Y. Tanigawara, K. Schoonjans, H. Itoh and J. Auwerx, *J. Biol. Chem.*, 2011, **286**, 26913.
51. G. Brufau, F. Stellaard, K. Prado, V. W. Bloks, E. Jonkers, R. Boverhof, F. Kuipers and E. J. Murphy, *Hepatology*, 2010, **52**, 1455.
52. H. Herrema, M. Meissner, T. H. van Dijk, G. Brufau, R. Boverhof, M. H. Oosterveer, D. J: Reijngoud, M. Müller, F. Stellaard, A. K. Groen and F. Kuipers, *Hepatology*, 2010, **51**, 806.
53. B. W. Katona, C. L. Cummins, A. D. Ferguson, T. Li, D. R. Schmidt, D. J. Mangelsdorf and D. F. Covey, *J. Med. Chem.*, 2007, **50**, 6048.
54. R. Pellicciari, H. Sato, A. Gioiello, G. Costantino, A. Macchiarulo, B. M. Sadeghpour, G. Giorgi, K. Schoonjans and J. Auwerx, *J. Med. Chem.*, 2007, **50**, 4265.
55. R. Pellicciari, S. Fiorucci, E. Camaioni, C. Clerici, G. Costantino, P. R. Maloney, A. Morelli and T. M. Willson, *J. Med. Chem.*, 2002, **45**, 3569.
56. Y. Iguchi, T. Nishimaki-Mogami, M. Yamaguchi, F. Teraoka, T. Kaneko and M. Une, *Biol. Pharm. Bull.*, 2011, **34**, 1.

57. Y. Iguchi, M. Yamaguchi, H. Sato, K. Kihira, T. Nishimaki-Mogami and M. Une, *J. Lip. Res.*, 2011, **51**, 1432.
58. C. Genet, A. Strehle, C. Schmidt, G. Boudjelal, A. Lobstein, K. Schoonjans, M. Souchet, J. Auwerx, R. Saladin and A. Wagner, *J. Med. Chem.*, 2010, **53**, 178.
59. C. Genet, C. Schmidt, A. Strehle, K. Schoonjans, J. Auwerx, R. Saladin and A. Wagner, *Chem. Med. Chem.*, 2010, **5**, 1983.
60. Merck, US20110092582; Merck, WO2009146772.
61. O. H. Emerson, *J. Am. Chem. Soc.*, 1948, **70**, 545.
62. E. Ono, J. Inoue, T. Hashidume, M. Shimizu and R. Sato, *Biochem. Biophys. Res. Commun.*, 2011, **410**, 677.
63. (a) Takeda Pharmaceuticals, WO2004043468A1; (b) Takeda Pharmaceuticals, JP2004346059; (c) Takeda Pharmaceuticals, EP20041591120; (d) Takeda Pharmaceuticals, WO2004067008; (e) Takeda Pharmaceuticals, JP2005021151A; (f) Takeda Pharmaceuticals, US2006199795; (g) Takeda Pharmaceuticals, US20097534783; (h) Takeda Pharmaceuticals, US20097625887.
64. Arena Pharmaceuticals, WO2005116653.
65. Kalypsys, WO2008067222.
66. Kalypsys, WO2008067219.
67. (a) Kalypsys, WO2008097976; (b) Kalypsys, US20090054304; (c) Kalypsys, WO2009026241; (d) Kalypsys, WO2010014739; (e) Kalypsys, WO2010016846.
68. M. R. Herbert, D. L. Siegel, L. Staszewski, C. Cayanan, U. Banerjee, S. Dhamija, J. Anderson, A. Fan, L. Wang, P. Rix, A. K. Shiau, T. S. Rao, S. A. Noble, R. A. Heyman, E. Bischoff, M. Guha and A. Kabakibi, *Bioorg. Med. Chem. Lett.*, 2010, **20**, 5718.
69. SmithKline Beecham Corporation, WO2007127505.
70. K. A. Evans, B. W. Budzik, S. A. Ross, D. D. Wisnoski, J. Jin, R. A. Rivero, M. Vimal, G. R. Szewczyk, C. Jayawickreme, D. L. Moncol, T. J. Rimele, S. L. Armour, S. P. Weaver, R. J. Griffin, S. M. Tadepalli, M. R. Jeune, T. W. Shearer, Z. B. Chen, L. Chen, D. L. Anderson, J. D. Becherer, M. De Los Frailes and F. J. Colilla, *J. Med. Chem.*, 2009, **52**, 7962.
71. B. W. Budzik, K. A. Evans, D. D. Wisnoski, J. Jin, R. A. Rivero, G. R. Szewczyk, C. Jayawickreme, D. L. Moncol and H. Yu, *Bioorg. Med. Chem. Lett.*, 2010, **20**, 1363.
72. Exelixis, WO2010093845.
73. Exelixis WO2011071565.
74. P. D. Leeson and B. Springthorpe, *Nat. Rev. Drug Discov.*, 2007, **6**, 881.
75. S. W. Muchmore, J. J. Edmunds, K. D. Stewart and P. J. Hajduk, *J. Med. Chem.*, 2010, **53**, 4830.
76. C. A. Lipinski, F. Lombardo, B. W. Dominy and P. J. Feeney, *Adv. Drug Deliv. Rev.*, 1997, **46**, 3.
77. M. Vieth and J. J. Sutherland, *J. Med. Chem.*, 2006, **49**, 3451.

78. P. Gedeck and R. A. Lewis, *Curr. Opin. Drug Discov. Devel.*, 2008, **11**, 569.
79. A. Macchiarulo, A. Gioiello, C. Thomas, A. Massarotti, R. Nuti, E. Rosatelli, P. Sabbatini, K. Schoonjans, J. Auwerx and R. Pellicciari, *J. Chem. Inf. Model.*, 2008, **48**, 1792.
80. B. K. Kobilka and X. Deupi, *Trends Pharmacol. Sci.*, 2007, **28**, 397.
81. B. Bosier, G. G. Muccioli, E. Hermans and D. M. Lambert, *Biochem. Pharmacol.*, 2010, **80**, 1.

The Discovery and Development of MB07803, a Second-Generation Fructose-1,6-bisphosphatase Inhibitor with Improved Pharmacokinetic Properties, as a Potential Treatment of Type 2 Diabetes

QUN DANG*, PAUL D. VAN POELJE AND
MARK D. ERION

Departments of Medicinal Chemistry and Biochemistry, Metabasis
Therapeutics, Inc., 11119 North Torrey Pines Road, La Jolla, CA 92037
*E-mail: qun_dang@merck.com

11.1 Introduction

Type 2 diabetes (T2DM) is a chronic disease affecting 346 million people
worldwide, with the World Health Organization projecting that diabetes-
related deaths will double between 2005 and 2030. Hyperglycemia is one of
many hallmarks for T2DM, which is associated with an increased risk of
nephropathy, vision loss, and heart diseases.[1] Consequently, existing therapies
benefit patients by lowering blood sugar levels. Several classes of oral

RSC Drug Discovery Series No. 27
New Therapeutic Strategies for Type 2 Diabetes: Small Molecule Approaches
Edited by Robert M. Jones
© The Royal Society of Chemistry 2012
Published by the Royal Society of Chemistry, www.rsc.org

hypoglycemic agents are currently on the market to treat T2DM, including sulfonylureas (e.g. glimepiride[2]), peroxisome proliferator-activated receptor (PPAR)-γ agonists (e.g. pioglitazone[3]), biguanides (e.g. metformin[4]), and the recently approved dipeptidyl peptidase-4 (DPP4) inhibitors (e.g. sitagliptin).[5] Despite the availability of both monotherapies and combination therapies, the majority of T2DM patients fail to reach desired glucose levels and therefore are in need of new and more effective therapies.

Endogenous glucose production is a major contributor to the abnormally high blood glucose levels in T2DM patients,[6–8] and hepatic gluconeogenesis (GNG, by which 3-carbon precursors are converted enzymatically to glucose) was further identified as the major culprit for hepatic glucose overproduction in T2DM patients.[9] None of the currently marketed drugs produce their glucose-lowering effects via direct inhibition of the GNG pathway although metformin is thought to indirectly reduce GNG.

Fructose 1,6-bisphosphatase (FBPase), as the second-to-last enzyme in GNG, controls the incorporation of all 3-carbon substrates into glucose.[10] The FBPase step is not involved in the breakdown of glycogen and is well removed from the mitochondrial steps of the pathway, therefore inhibition of FBPase should theoretically have reduced risk of hypoglycemia and other mechanistic toxicities relative to other potential GNG targets.[11] This concept is supported by the existence of humans who are genetically deficient in FBPase but have near normal biochemical and clinical parameters provided they maintain an appropriate diet and avoid prolonged fasting.

Human liver FBPase is a cytosolic enzyme that consists of four identical 36.7 kD subunits each containing a substrate site and an allosteric site.[10] The intracellular activity of FBPase is regulated synergistically by fructose 2,6-bisphosphate (F2,6BP), an inhibitor that binds to the substrate site, and adenosine-5'-monophosphate (AMP), an inhibitor that binds to the allosteric site.[12,13] Both the substrate site and the AMP site of FBPase have been the target of drug discovery efforts, which have proven over the past two decades to be very challenging.[14] A structure-guided approach using AMP as a medicinal chemistry lead finally yielded the first oral FBPase inhibitor, MB06322 (**1**, also known as CS-917, Figure 11.1),[15–18] which was advanced into phase II human clinical trials and shown to lower blood glucose levels in patients with T2DM.[19] Unfortunately, a phase I study of MB06322 in combination with metformin revealed a drug–drug interaction (DDI) leading

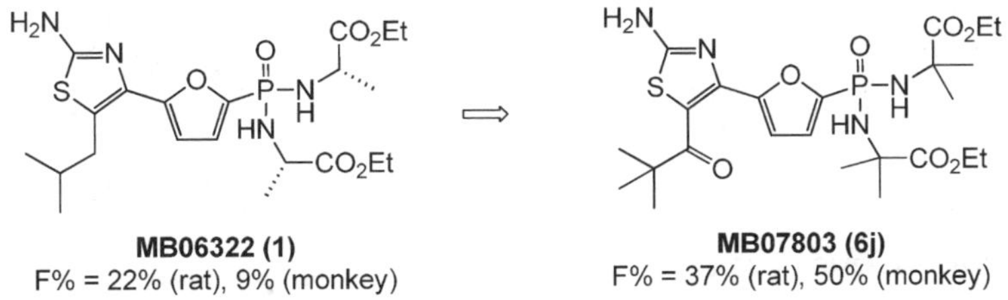

MB06322 (1)
F% = 22% (rat), 9% (monkey)

MB07803 (6j)
F% = 37% (rat), 50% (monkey)

Figure 11.1 Discovery of a second-generation FBPase inhibitor MB07803.

to elevated blood lactate levels. A second-generation oral FBPase inhibitor (MB07803) was subsequently discovered that potentially addresses the DDI uncovered for MB06322. The discovery efforts leading to MB07803 are presented here.

11.2   Discovery of First-Generation FBPase Inhibitors

11.2.1   Discovery of MB06322, First Oral FBPase Inhibitor

Using a structure-guided drug-design approach we were able to take AMP as our lead molecule and discover multiple series of potent and selective FBPase inhibitors (Figure 11.2). The program overcame many significant medicinal chemistry challenges. At the outset, our efforts focused on the discovery of non-nucleotide AMP mimics. This led to the discovery of phosphonic acid-containing purines and benzimidazoles which functioned as FBPase inhibitors *in vivo* and elicited robust glucose-lowering activity in animal models of T2DM. Advancement of these analogs was limited by their poor oral bioavailability which despite significant efforts was not solved by using phosphonic acid prodrugs. Thus, our second challenge was to search for a new scaffold that was suitable for oral delivery via a prodrug. This led to the redesign of the 5,6-fused heterocyclic nucleus in purine and benzimidazole and the discovery of phosphonic acid-containing thiazoles as potent and selective FBPase inhibitors. Efforts to achieve adequate oral bioavailability with this series using previously reported phosphonate prodrugs did not produce oral efficacy. Consequently, we initiated efforts to identify a novel prodrug class

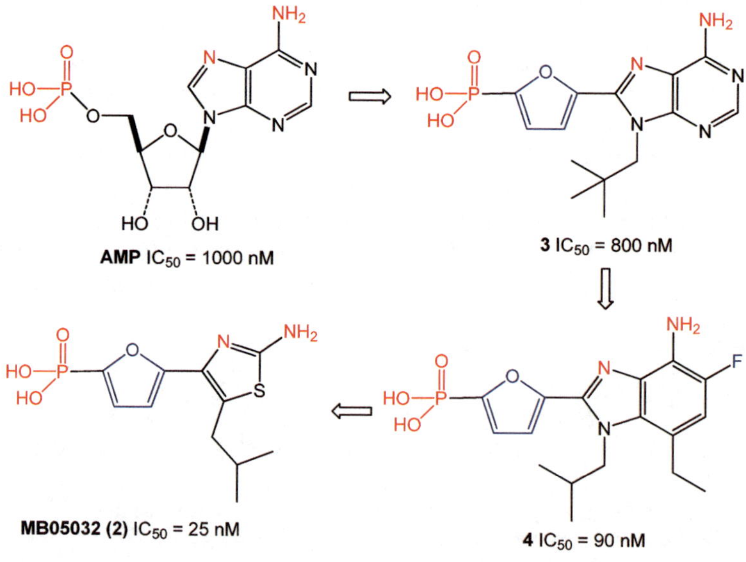

Figure 11.2   Evolution of phosphonic acid-containing FBPase inhibitors.

with the desired properties. This led to a phosphonic diamide prodrug approach, which successfully delivered our phosphonic acid-containing thiazoles with oral bioavailability (OBAV) of 20–40% and produced potent oral glucose-lowering effects in animal models of T2DM. Years of collaborative team effort culminated in the discovery of compound **1** (MB06322) as the first oral FBPase inhibitor to advance to human clinical trials in patients with T2DM.

11.2.2 Clinical Studies

MB06322 (CS-917) was evaluated in phase I and phase II clinical trials by Sankyo and reported to be safe and well tolerated. Moreover, MB06322 treatment led to meaningful reductions in fasting glucose levels in patients with T2DM.[20] Results from the clinical studies also showed that treatment with MB06322 led to modest, but non-clinically significant, elevations in lactic acid while having little to no effect on serum triglyceride levels. While monotherapy appeared safe and well tolerated, a combination study of MB06322 and metformin was subsequently reported to be associated with two cases of lactic acidosis in two patients. The reason for the DDI was not apparent and as such the finding prevented the use of MB06322 with metformin as a combination therapy. Many scenarios could be proposed to explain the observed DDI. For example, addition of MB06322 could reduce the clearance of metformin which like other members of the biguanide drug class could result in lactic acidosis in some patients. Another explanation could be that lactic acidosis arises from the more complete inhibition of GNG expected from the combination of a direct GNG inhibitor (MB6322) with an indirect GNG inhibitor (metformin).

In the oral PK study of MB06322 (**1**), there were three major metabolites observed in addition to MB06322 (**1**): MB06633 (**1a**), MB05032 (**2**), and MB05099 (**2a**) (Figure 11.3).

After oral administration, MB06322 was first metabolized by carboxyesterases leading to a mono-amidate intermediate MB06633 (**1a**), which was also observed in preclinical animal studies. A second enzymatic reaction cleaved the

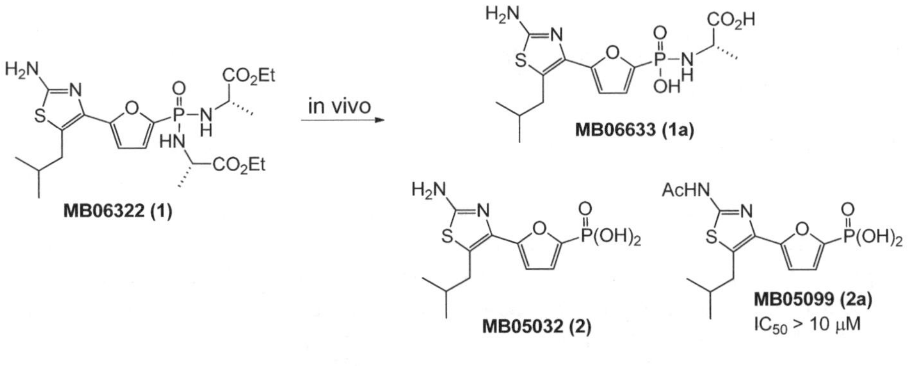

Figure 11.3 Metabolism of MB06322 (**1**) in humans.

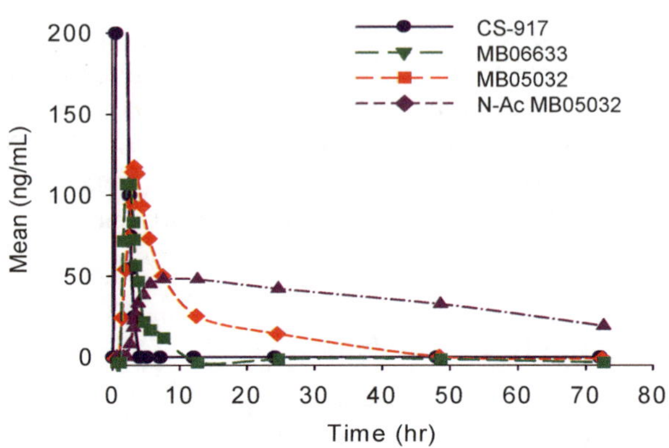

Figure 11.4 Oral human PK study of MB06322 (**1**).

P–N bond to generate the active drug MB05032. An unexpected metabolite, MB05099, was detected and its structure was determined to be the *N*-acetyl derivative of MB05032. Further investigation revealed that the enzymes responsible for the formation of MB05099 were likely arylamine *N*-acetyl transferases (NAT). There are two NAT isoforms in humans, NAT1 and NAT2. Human NAT1 protein is found in many tissues, while NAT2 protein is present predominantly in the intestine and liver.[21]

The drug levels for a PK study with healthy volunteers after a single oral dose (50 mg) of MB06322 are shown in Figure 11.4.

As shown in Figure 11.4, the conversion of MB06322 (CS-917, blue line) to the active drug MB05032 (red line) was rapid, with the mono-amidate MB06633 (green line) being also formed and cleared quickly. On the other hand, the *N*-acetyl metabolite MB05099 (pink line) exhibited a long half-life in humans and consequently accumulated to levels well above the active metabolite MB05032. Susceptibility to *N*-acetylation likely contributed to the observed low bioavailability and short half-life for MB06322, therefore we aimed to block the *N*-acetylation metabolism in the search of a second-generation FBPase inhibitor.

11.2.3 *N*-Acetylation and Possible Impact of *N*-Acetylated Metabolites on Mitochondrial Function

Another significant clinical observation during the development of MB06322 as a potential treatment for T2DM was the DDI with metformin leading to lactic acidosis. Given that lactic acidosis was also observed for metformin monotherapy and metformin was known as a chain inhibitor of mitochondrial respiration,[22] potential effects of MB05032 (**2**) and MB05099 (**2a**)[23] on mitochondrial functions were studied. Addition of MB05099 (**2a**) but not MB05032 (**2**) (data not shown) to the growth medium of fresh human hepatocytes resulted in a dose-dependent increase in the intracellular lactate/

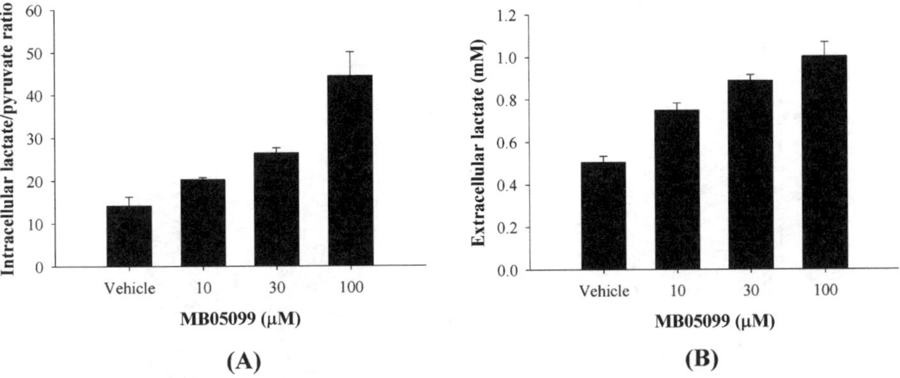

Figure 11.5 Effects of MB05099 (**2a**) on the intracellular lactate/pyruvate ratio (A) and the extracellular lactate concentration (B) of fresh human hepatocytes (mean values $\pm$ standard deviation of 2 wells/dose).

pyruvate ratio with the highest concentration of MB05099 (**2a**) increasing this ratio by $\sim$3.5-fold (Figure 11.5A). The latter effect was driven primarily by a decrease in the intracellular pyruvate concentration (data not shown). Concurrent with the increase in intracellular lactate, a marked increase in extracellular lactate levels was observed in response to incubation with MB05099 (**2a**) (Figure 11.5B), which reflects impaired oxidative capacity of the cells.

Impairment of mitochondrial function by MB05099 (**2a**) was also evident in studies with rat hepatocytes as indicated by an increased lactate/pyruvate ratio (13–63%) following incubation with MB05099 (**2a**) (3–30 μM), Figure 11.6. The effects of MB05099 (**2a**) on the lactate/pyruvate ratio in these acute studies were observed at a >100-fold lower concentration range than those with metformin, a known respiratory chain inhibitor,[22,24] which increased this ratio by 37–245% at concentrations ranging from 300 to 3000 μM but had only minor effects at concentrations <300 μM. As in the studies with human hepatocytes, the change in the lactate/pyruvate ratio in rat hepatocytes incubated with MB05099 (**2a**), or with metformin, was driven primarily by a decrease in pyruvate concentrations. Interestingly, the combined incubation of MB05099 (**2a**) and metformin led to greater perturbation of the lactate/ pyruvate ratio than observed with either agent alone (Figure 11.6).

The above studies revealed that the *N*-acetyl metabolite (MB05099) of MB05032 clearly shows surprising pharmacological effects with regards to mitochondrial function leading to altered lactate metabolism in hepatocytes. Given that metformin has been shown to exhibit similar effect on mitochondrial function and lactate levels, it is logical to hypothesize that the observed lactacidosis during the DDI trial of MB06322 and metformin combination studies could be attributed to an additive or synergistic effect of MB05099 and metformin on mitochondrial oxidative capacity. Therefore, a second-generation FBPase inhibitor that does not have the liability of *N*-acetylation should have an improved therapeutic window with regard to combination with metformin.

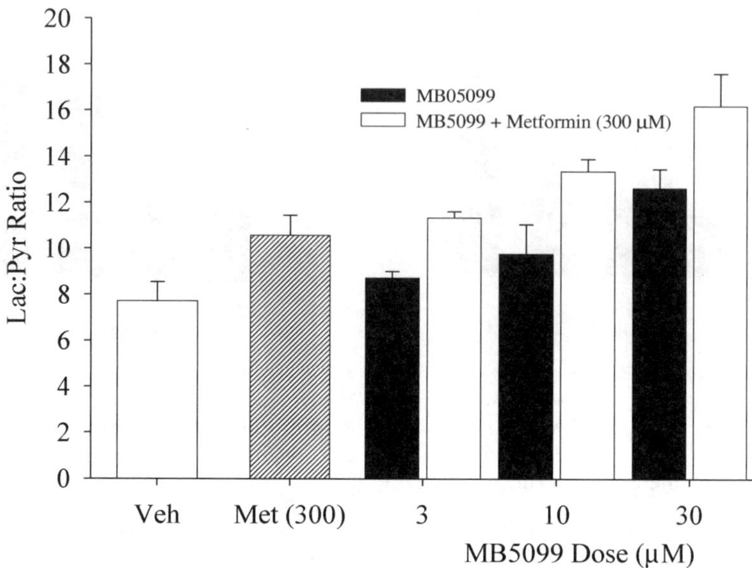

Figure 11.6 Intracellular lactate:pyruvate ratio of rat hepatocytes incubated for 16 hours with MB05099 (**2a**), metformin, or the combination of MB05099 (**2a**) and metformin (mean values $\pm$ standard deviation of 2 wells/dose).

11.3 Design Strategy for Second-Generation FBPase Inhibitors

11.3.1 SAR Summary for FBPase Inhibition by the Thiazole Scaffold

To discover a second-generation FBPase inhibitor with the desired profile, the SAR of the thiazole series of FBPase inhibitors was used to formulate strategies to reduce or eliminate *N*-acetylation of the 2-amino group. Modeling efforts and X-ray crystallography showed clearly that the 2-amino group in MB05032 forms important hydrogen bonds to Thr31 hydroxyl and the backbone carbonyl of Val17 (Figure 11.7), therefore neither substitution nor replacement of the amino group were tolerated. The furan-linker group was also fully optimized and replacement with other linkers all led to significant loss in FBPase inhibition potency. On the other hand, SAR of the C5 region of the thiazole is more permissive: analogs with C5-alkyls, esters, and aryls all inhibit FBPase potently ($IC_{50} < 100$ nM).

11.3.2 Design Strategies to Reduce or Eliminate *N*-Acetylation

To retain binding affinity, we focused our attention on the C5 position of the thiazole scaffold since the rest of the scaffold interacts with FBPase through multiple hydrogen bonds and electrostatic interactions (Figure 11.8A).

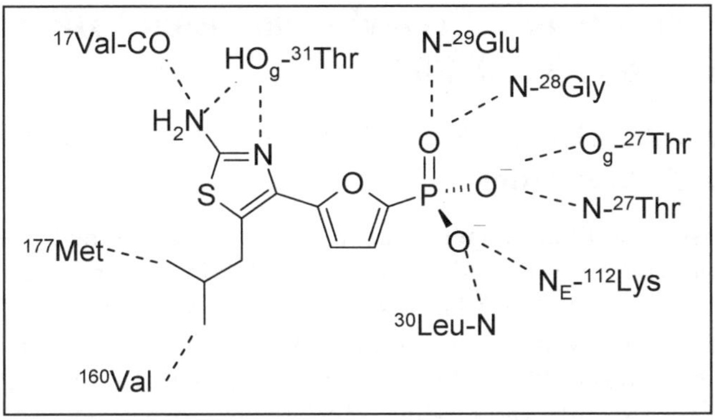

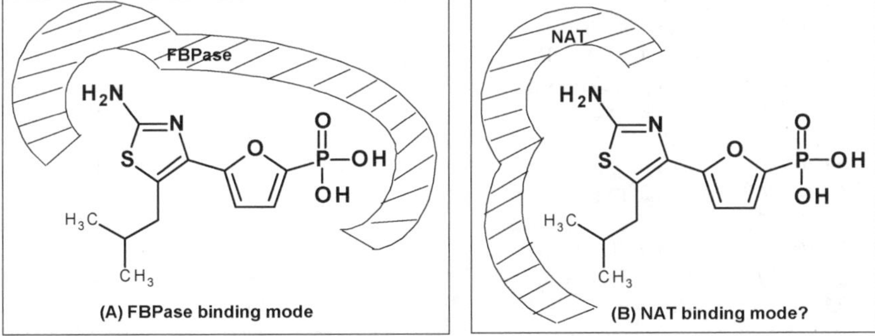

Figure 11.7   MB05032 (**2**) interactions with human FBPase.

Although substrate specificity studies of NAT1 and NAT2 for simple aryl amines have been reported,[25] there is little structural similarity between the reported arylamines and our thiazole FBPase inhibitors (MB06322 or MB05032). Consequently, we theorized a potential topological binding mode for the interaction between MB05032 and NAT enzymes (Figure 11.8B), which we used to guide our efforts to optimize the thiazole scaffold. Chemical reactions are controlled by steric and electronic effects, therefore these two aspects were investigated at the C5 position. Incorporation of C5 substituents with increased steric bulk could potentially block NAT enzyme's access to the 2-amino group (Figure 11.8B), while electron-withdrawing C5 substituents could reduce the nucleophilicity of the 2-amino group and thereby reduce its reactivity toward *N*-acetylation, or decrease the binding affinity to NAT enzyme.

Figure 11.8   Topological binding modes for MB05032 to FBPase (A) and NAT (B).

11.4   Discovery of a Second-Generation FBPase Inhibitor: MB07803

11.4.1   Investigation of C5 Steric Effects of the Thiazole Scaffold on *N*-Acetylation

Two NAT assays were developed as primary screens using the two isoforms of the human enzymes, NAT1 and NAT2. MB06322 is primarily metabolized by NAT2 (48% *N*-acetylated), while its NAT1 metabolism is only 5%. Conversely, MB05032 is primarily metabolized by NAT1 (45% *N*-acetylated) while its NAT2 metabolism is only 5%. These observations revealed that NAT1 prefers negatively charged substrates while NAT2 metabolizes neutral substrates. Therefore, our screening strategy was set to test all new phosphonic acid analogs in the human FBPase inhibition and NAT1 assays, followed by glucose lowering in normal fasted rats after intravenous (iv) administration. On the other hand, all new prodrugs were tested in the NAT2 assay followed by glucose-lowering in normal fasted rats after oral administration (p.o.).

The first concept to investigate was whether a bulky C5 group could reduce or eliminate *N*-acetylation of the 2-aminothiazole scaffold. Thus, a series of C5 analogs was investigated and the results are summarized in Table 11.1.

Results from Table 11.1 indicate that as the size of C5 groups increases then the extent of *N*-acetylation decreases; several compounds with large enough C5 groups completely avoided *N*-acetylation, while retaining FBPase inhibition

Table 11.1   Thiazole analogs with a bulky C5 group.[a]

	MB#	R^5	HL IC_{50} (nM)	G-LOW (%)	NAT1 (%)
2	MB05032	-*i*-Bu	25	65	44.9
2a	MB05099	-*i*-Bu	>10 000	ND	–
2b	MB07709	-$CH_2CH(Me)Et$	11	76	14.2
2c	MB07708	-$CH_2CH(Me)Pr$	12	73	5.5
2d	MB07745	-CH_2-cyclohexyl	59	ND[b]	5.4
2e	MB06666	neopentyl	12	80	9.3
2f	MB05317	-*i*-Pr	28	80	14
2g	MB07736	-CH_2-adamantyl	218	ND	0
2h	MB07767	-$CH_2C(Me)_2$-*i*-Pr	23	68	0
2i	MB07775	-$CH_2C(Me)_2$-*i*-Bu	12	76	2.8
2j	MB07777	-$CH_2C(Me)_2$-*n*-Pr	13	71	2.7
2k	MB07778	-$CH_2C(Me)_2Et$	28	88	4.6
2l	MB06762	-Ph-Ph	34	36	0

[a]HL, human liver FBPase; G-LOW, glucose lowering in normal fasted rats after i.v. administration at doses of 10 mg/kg; NAT1, metabolism by NAT isoform 1; [b]ND, not determined.

activity. Thus, these data validated the hypothesis that incorporating steric bulk at the C5 position could reduce *N*-acetylation of the 2-amino group. For example, thiazole **2h** showed similar potency compared to MB05032 (**2**) with regard to both human liver FBPase inhibition and glucose lowering in normal rats after i.v. administration; more importantly it does not suffer from *N*-acetylation by NAT1 unlike MB05032 (**2**).

11.4.2 Exploration of C5 Electronic Effects of the Thiazole Scaffold: Discovery of MB07729

Having confirmed that bulky C5 groups could block the *N*-acetylation of the 2-amino group of the thiazole scaffold by NAT1, the next concept to test was whether an electron-withdrawing C5 group could also reduce or eliminate *N*-acetylation of the 2-amino group. Thus, a series of C5-keto thiazoles were explored and the results are summarized in Table 11.2.

As evident from Table 11.2, smaller keto C5 groups (e.g. **5a**) showed loss in inhibitory potency against human liver FBPase compared to MB05032 (**2**), while most larger keto groups led to analogs with equal or more potent FBPase inhibitors (**5b–5i**) that completely eliminated the *N*-acetylation liability. Optimization of this series of thiazoles led to the discovery of several C5-keto thiazoles (e.g. **5b–5e, 5g–5i**) as potent inhibitors of human liver FBPase with similar degree of glucose-lowering capability compared to MB05032 (**2**). Thus, introduction of an electron-withdrawing group at the C5 position of the thiazole scaffold is another effective approach to eliminate the *N*-acetylation metabolism.

Table 11.2   Thiazole analogs with a –CO–alkyl C5 group.[a]

	MB#	R^5	HL IC_{50} (nM)	G-LOW (%)	NAT1 (%)
2	MB05032	-*i*-Bu	25	65	45
5a	MB07915	-Ac	66	ND[b]	ND
5b	MB08560	-COCH(Me)Et	11	73	0
5c	MB08508	-COCH(Et)$_2$	9	71	0
5d	MB08561	-CO-cyclobutyl	16	68	0
5e	MB08541	-CO-cyclopentyl	14	65	0
5f	MB08407	-CO-cyclohexyl	40	ND	ND
5g	MB07729	-CO-*t*-Bu	24	81	0
5h	MB07784	-COC(Me)$_2$Et	25	86	0
5i	MB07785	-COC(Et)$_2$Me	18	63	0

[a]HL, human liver FBPase; G-LOW, glucose lowering in normal fasted rats after i.v. administration at dose of 10 mg/kg; NAT1, metabolism by NAT isoform 1; [b]ND, not determined.

11.4.3 Oral Delivery of FBPase Inhibitors with no NAT Liability: Prodrug SAR

After identifying several potent thiazole FBPase inhibitors that avoided *N*-acetylation in the NAT1 assay while maintaining potent glucose-lowering activity *in vivo*, we turned our attention to the phosphonic diamide approach as a means to deliver these compounds orally. Results for selected diamide prodrugs of several thiazoles with either bulky alkyl groups or keto groups at the C5 position are summarized in Table 11.3.

Some diamide prodrugs (such as **6c**, **6d**) showed significant oral glucose lowering activity, suggesting good oral PK properties; however, these prodrugs still showed detectable *N*-acetyl metabolites. On the other hand, prodrugs from the keto series of thiazoles generated several leads (e.g. **6i**, **6j**, **6k**) that were not substrate of NAT2 enzyme, and prodrug **6j** (MB07803) also demonstrated potent glucose-lowering effects. A full pharmacokinetic study of MB07803 showed that it has good OBAV: 37% in rats, which showed 2-fold higher improvement over MB06322 (**1**). Finally, the first-generation compounds MB06322 (**1**) and its parent MB05032 (**2**) were compared with MB07803 (**6j**)

Table 11.3 Prodrug SAR of thiazoles with no *N*-acetylation liability.[a]

	R^5	X	G-LOW (%)	NAT2 (%)
1	Isobutyl	Ala-OEt	59	45
6a	-CH$_2$C(Me)$_2$Et	Ala-OEt	30	ND[b]
6b	-CH$_2$C(Me)$_2$Et	Ala-O-*i*-Pr	7	ND
6c	-CH$_2$C(Me)$_2$Et	Me-Ala-OEt	58	8
6d	-CH$_2$C(Me)$_2$Et	Me-Ala-O-*i*-Pr	49	4
6e	-CH$_2$C(Me)$_2$-*n*-Pr	Ala-O-*i*-Pr	17	ND
6f	-CH$_2$C(Me)$_2$-*n*-Pr	Me-Ala-OEt	35	ND
6g	-CH$_2$C(Me)$_2$-*n*-Pr	Me-Ala-O-*i*-Pr	17	ND
6h	-CO-*t*-Bu	Gly-OEt	22	38
6i	-CO-*t*-Bu	Ala-OEt	25 (11)[c]	0
6j	-CO-*t*-Bu	Me-Ala-OEt	68 (37)[c]	0
6k	-CO-*t*-Bu	Me-Ala-O-*i*-Pr	27	0
6l	-COC(Me)$_2$Et	Ala-OEt	12	ND
6m	-COC(Me)$_2$Et	Me-Ala-OEt	24	ND
6n	-COC(Et)$_2$Me	Ala-OEt	10	ND
6o	-COC(Et)$_2$Me	Me-Ala-OEt	15 (4)[c]	ND

[a]Gly-OEt, glycine ethyl ester; Ala-OEt, L-alanine ethyl ester; Me-Ala-OEt, 2-methylalanine ethyl ester; G-LOW, glucose lowering in normal fasted rats after oral administration at doses of 10 mg/kg; NAT2, metabolism by NAT isoform 2; [b]ND, not determined; [c]Number inside parentheses represents OBAV.

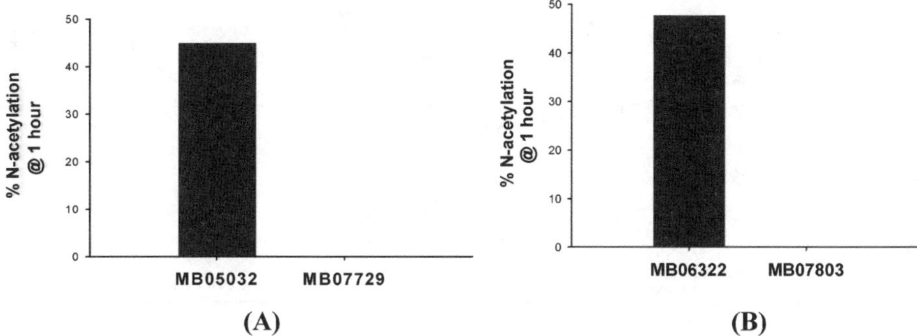

Figure 11.9   Comparison of MB07803 (**6j**)/MB07729 (**5g**) with MB06322 (**1**)/
MB05032 (**2**) in the NAT1 (A) and NAT2 (B) assays.

and its parent MB07729 (**5g**) in a head to head study in the NAT assays, and results are shown in Figure 11.9.

In the NAT assays, MB07803 and MB07729 showed distinctly different profiles compare to our first-generation compounds (MB06322 and MB05032): no *N*-acetyl metabolite was detected for either MB07729 or MB07803 in the NAT1 and NAT2 assays. Thus, the profile of MB07803 met two of the important criteria we set for a second-generation compound: no *N*-acetylation and improved OBAV.

11.5   Development of MB07803 (6j)

11.5.1   Efficacy Studies of MB07803 (6j) in Animal Models of T2DM

Similar to MB06322 (**1**), MB07803 (**6j**) elicited potent oral glucose lowering effects in various animal models of T2DM, and results from acute glucose-lowering studies in ZDF rats in the fasted (graph A) and freely feeding states (graph B) are shown in Figure 11.10.

Oral administration of MB07803 (**6j**) normalized blood glucose levels in the fasted state and significantly suppressed hyperglycemia in the freely feeding state. Multiple sub-chronic studies were also carried out with MB07803 (**6j**) and a 4-week study in ZDF rats is shown in Figure 11.11.

Treatment of 10-week-old ZDF rats with MB07803 (**6j**) as a 0.1% food admixture (approximately 100 mg per day dose) led to significant glucose lowering compared to vehicle-treated animals. The magnitude of glucose-lowering effects were maintained from week two to week four (end of the study) with no sign of tachyphylaxis. Other observations include markedly reduced urinary glucose excretion ($\sim$25% at 0.03% and $\sim$65% at 0.1% food admixture) and no effect on food intake and body weight (data not shown).

Oral efficacy in primates for MB07803 (**6j**) was studied in fasted normal cynomolgus monkeys and results are shown in Figure 11.12.

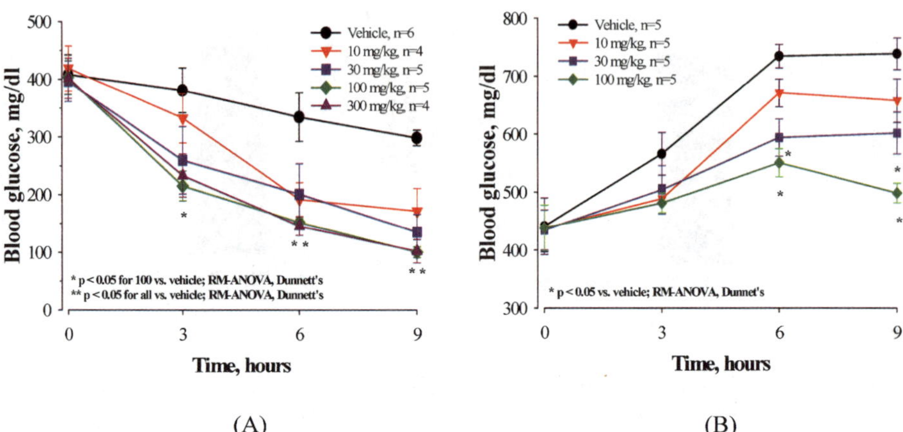

(A)   (B)

Figure 11.10   Oral efficacy of MB07803 (**6j**) in ZDF rats in the fasted (A) and freely feeding states (B).

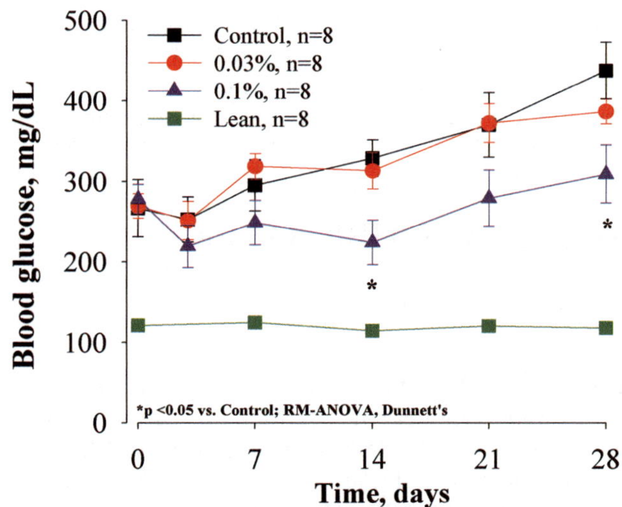

Figure 11.11   Four-week study of MB07803 (**6j**) in ZDF rats.

The study was performed using 18-hour fasted normal cynomolgus monkeys (*n* = 6 per group), which are fully dependent upon GNG to maintain glycemia. MB07803 (**6j**) was given as a solution in PEG-400 orally at the indicated doses.

A single administration of MB07803 (**6j**) to cynomolgus monkeys led to dose-dependent and significant lowering of blood glucose compared to vehicle-treated animals. There were no significant differences in levels of blood lactate and triglycerides between the MB07803 (**6j**) treated group and control animals (data not shown). Full pharmacokinetic studies of MB07803 and MB06322 in cynomolgus monkeys were also performed and the results are summarized in Table 11.4.

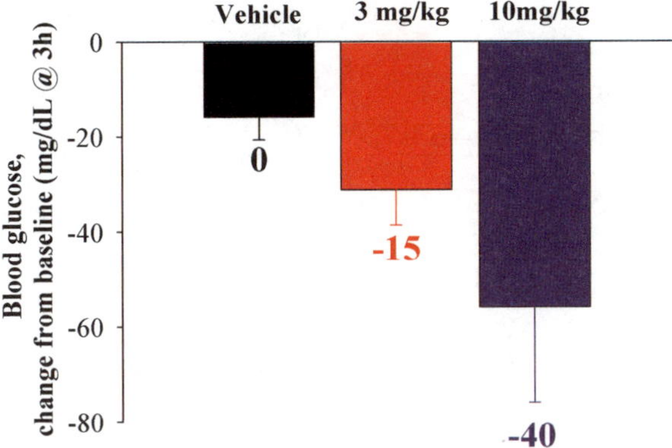

Figure 11.12 Oral efficacy of MB07803 (**6j**) in cynomolgus monkeys.

Table 11.4 Monkey PK comparison of MB06322 and MB07803.

Drugs	AUC ($\mu g \cdot h/L$)	$T_{1/2}$ (h)	C_{max} ($\mu g/mL$)	T_{max} (h)	F (%)
MB06322 (**1**)	4.73 (MB05032)	6.0	1.19	2	9
MB07803 (**6j**)	15.8 (MB07729)	8.2	2.1	3.3	50

The improvement in OBAV for MB07803 compared to MB06322 was approximately 2-fold in rats, but in monkeys the improvement is much more pronounced, a 5-fold improvement (Table 11.4)! The more significant OBAV increase in monkeys could be a direct effect of eliminating *N*-acetylation for MB07803, since the *N*-acetylation metabolite for MB06322 was observed in monkeys but not rodents.

Moreover, avoiding *N*-acetylation should maintain FBPase inhibitory activity longer (MB07729 for the case of MB07803) since the *N*-acetylated metabolite is not a FBPase inhibitor. In addition, blocking the metabolism should also reduce the clearance rate and as a consequence increase $T_{1/2}$, which may allow once daily dosing instead of bid or tid dosing required for MB06322. Indeed, longer $T_{1/2}$ was observed for MB07803 compared to MB06322 in the monkey PK study. Lastly, avoiding *N*-acetylation has the potential to improve PK variability, which for MB06322 likely arose from its poor OBAV and/or individual differences in *N*-acetylation.

11.5.2 Phase I/II Clinical Studies

11.5.2.1 Phase I Tolerability and Pharmacokinetic Studies

Five phase I clinical studies were completed to evaluate safety, tolerability, and pharmacokinetics in healthy volunteers of single doses of up to 1000 mg, and multiple doses of up to 400 mg qd for 14 days. MB07803 was found to be safe

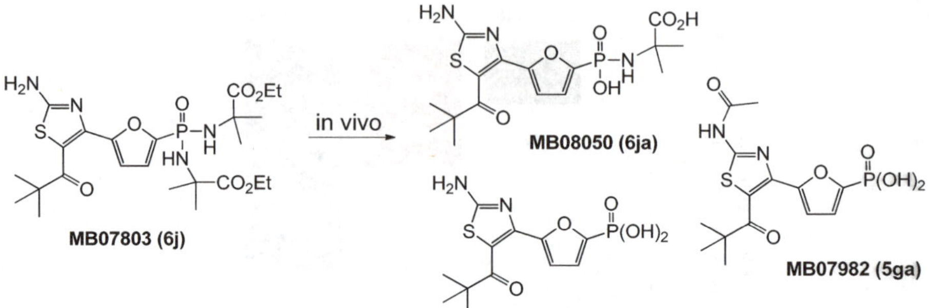

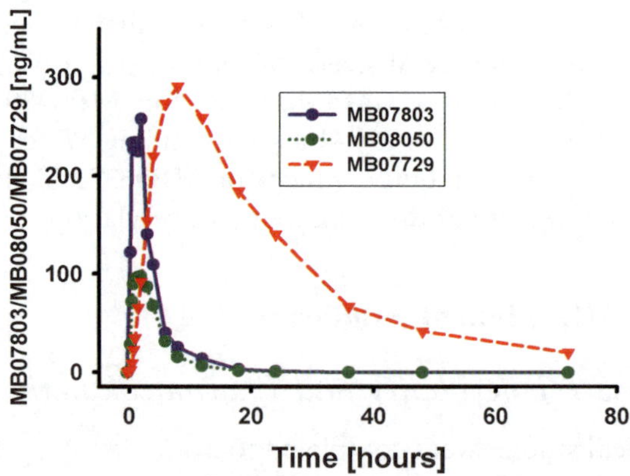

Figure 11.13   Metabolism of MB07803 (**6j**) in humans.

and well tolerated, and in the 14-day study (25, 50, 100, 200, 400 mg qd MB07803 versus placebo, $N = 40$; 8/arm; 6:2 randomization) no episodes of hypoglycemia, lactic acidosis, or lactacemia were observed.

In contrast to the single dose oral PK study of MB06322 (50 mg), there were only two major metabolites observed in a similar study with MB07803 (single 50 mg oral dose): MB08050 (**6ja**) and MB07729 (**5g**), while the *N*-acetyl derivative of MB07729 (MB07982) was not observed in this study (Figure 11.13).

To compare with the oral PK results from a MB06322 clinical study, the drug levels for a PK study with healthy volunteers after a single oral dose (50 mg) of MB07803 are shown in Figure 11.14.

In contrast to MB06322, MB07803 (blue line) was rapidly converted to the active drug MB07729 (red line) which was by far the major metabolite. The mono-amidate intermediate MB08050 (green line) was found at much lower levels, suggesting more efficient prodrug activation. Most importantly, the *N*-acetyl metabolite MB07982 was not detected in this study; although in a separate study when MB07803 was given at 600 mg dose, the *N*-acetylated

Figure 11.14   Oral human PK studies of MB07803 (**6j**).

Table 11.5   Glucose-lowering effects of MB07803 in T2DM patients.

MB07803	*FPG* $C_{max}{}^a$	*PBO-adjusted 95% CI*	$AUC_{24h}{}^b$	*PBO-adjusted 95% CI*
50 mg	−16	0.4791	−764	0.0435
200 mg	−58	0.0142	−1186	0.0026
400 mg	−55	0.0302	−1508	0.0004

[a]Fasting plasma glucose (FPG) levels were measured on day 15 and expressed as mg/dL; [b]Twenty-four hour glucose AUC were measured on day 13–14 and showed as mg·h/dL.

metabolite of MB07729 was detected but its levels were approximately 1/1000th of the levels of MB07729. Thus the goal of reducing *N*-acetylation in a second-generation FBPase inhibitor was accomplished via the discovery of MB07803.

11.5.2.2   Phase II Glucose-Lowering Activity Study

MB07803 was studied in a randomized, double-blind, placebo (PBO)-controlled, 14-day domiciled study to assess safety, tolerability, and glucose-lowering effects and acid-base balance. There were 42 T2DM patients with mean fasting plasma glucose (FPG) of 221 mg/dL and hemoglobin A1c (HbA1c) of 8.8%.

Single administration of MB07803 at 200 and 400 mg doses led to statistical and clinically significant reduction of FPG on day 15 and 24-hour glucose AUC measured on day 13–14 (Table 11.5), suggesting that MB07803 has the potential to be developed into a treatment to lower glucose in T2DM patients.

11.6   Summary

The first oral FBPase inhibitor, MB06322, was advanced into phase II human clinical trials and shown to lower blood glucose levels in patients with T2DM. However, the oral PK of MB06322 was not ideal and a long-lived metabolite, MB05099 (the *N*-acetylated derivative of the active drug MB05032), accumulated to high levels. Subsequently, a phase I study of MB06322 in combination with metformin resulted in two cases of lactic acidosis. Follow-up studies revealed that MB05099, not the active drug MB050322, impaired mitochondrial function leading to increased lactate levels in both human and rat hepatocytes. Thus, it was theorized that MB05099 may have contributed to the lactic acidosis observed with the MB06322 and metformin combination. Therefore, a key objective for a second-generation FBPase inhibitor was to greatly reduce the production of the *N*-acetylated metabolite. Exploring steric and electronic effects of substituents at the C5 position of the thiazole scaffold led to MB07729 as a potent and selective FBPase inhibitor with significantly reduced propensity to undergo *N*-acetylation. Subsequent optimization of MB07729 phosphonic diamide prodrugs led to the discovery of MB07803 as a second-generation FBPase inhibitor. MB07803 demonstrated robust glucose-lowering activity in preclinical animal models of T2DM. In addition, MB07803 exhibited good OBAV across species (rat, dog, and monkey) and a half-life consistent with once a day dosing in humans. In the clinic, MB07803 was safe and well tolerated and associated with a longer half-life and significantly lower levels of the *N*-acetylated metabolite. Robust glucose lowering was also observed for MB07803 in T2DM patients, indicating that MB07803 could represent a potential candidate for the treatment of type 2 diabetes.

References

1. G. Smushkin and A. Vella, *Medicine (Baltimore)*, 1959, **38**, 597–601.
2. S. N. Davis, *J. Diabetes Complications*, 2004, **18**, 367–376.
3. P. S. Gillies and C. J. Dunn, *Drugs*, 2000, **60**, 333–343; discussion 344–335.
4. R. S. Hundal, M. Krssak, S. Dufour, D. Laurent, V. Lebon, V. Chandramouli, S. E. Inzucchi, W. C. Schumann, K. F. Petersen, B. R. Landau and G. I. Shulman, *Diabetes*, 2000, **49**, 2063–2069.
5. N. A. Thornberry and A. E. Weber, *Curr. Top. Med. Chem.*, 2007, **7**, 557–568.
6. R. A. DeFronzo, E. Ferrannini and D. C. Simonson, *Metabolism*, 1989, **38**, 387–395.
7. C. Y. Jeng, W. H. Sheu, M. M. Fuh, Y. D. Chen and G. M. Reaven, *Diabetes*, 1994, **43**, 1440–1444.
8. G. Perriello, S. Pampanelli, P. Del Sindaco, C. Lalli, M. Ciofetta, E. Volpi, F. Santeusanio, P. Brunetti and G. B. Bolli, *Diabetes*, 1997, **46**, 1010–1016.
9. I. Magnusson, D. L. Rothman, L. D. Katz, R. G. Shulman and G. I. Shulman, *J. Clin. Invest.*, 1992, **90**, 1323–1327.
10. M. R. el-Maghrabi, M. Gidh-Jain, L. R. Austin and S. J. Pilkis, *J. Biol. Chem.*, 1993, **268**, 9466–9472.
11. B. Steinmann, S. G. van den Bergh, R. Gitzelmann et al., *Disorders of Fructose Metabolism*, McGraw-Hill, 1995.
12. S. J. Pilkis and T. H. Claus, *Annu. Rev. Nutr.*, 1991, **11**, 465–515.
13. S. L. Kelly and E. Sim, *Hum. Exp. Toxicol.*, 1991, **10**, 33–38.
14. P. D. van Poelje, Q. Dang and M. D. Erion, *Curr. Opin. Drug Discov. Devel.*, 2007, **10**, 430–437.
15. M. D. Erion, P. D. van Poelje, Q. Dang, S. R. Kasibhatla, S. C. Potter, M. R. Reddy, K. R. Reddy, T. Jiang and W. N. Lipscomb, *Proc. Natl. Acad. Sci. USA*, 2005, **102**, 7970–7975.
16. Y. Wang and B. Tomlinson, *Curr. Opin. Investig. Drugs*, 2007, **8**, 849–858.
17. M. D. Erion, Q. Dang, M. R. Reddy, S. R. Kasibhatla, J. Huang, W. N. Lipscomb and P. D. van Poelje, *J. Am. Chem. Soc.*, 2007, **129**, 15480–15490.
18. Q. Dang, S. R. Kasibhatla, K. R. Reddy, T. Jiang, M. R. Reddy, S. C. Potter, J. M. Fujitaki, P. D. van Poelje, J. Huang, W. N. Lipscomb and M. D. Erion, *J. Am. Chem. Soc.*, 2007, **129**, 15491–15502.

19. J. Triscari, J. Walker, K. Feins, B. Tao and S. R. Bruce, *The American Diabetes Association 66th Scientific Session*, Washington, DC, 2006.
20. J. Triscari, J. Walker, K. Feins, B. Tao and S. R. Bruce, *Diabetes*, 2006, **55**, (suppl 1), 444P.
21. S. Boukouvala and E. Sim, *Basic Clin. Pharmacol. Toxicol.*, 2005, **96**, 343–351.
22. M. Y. El-Mir, V. Nogueira, E. Fontaine, N. Averet, M. Rigoulet and X. Leverve, *J. Biol. Chem.*, 2000, **275**, 223–228.
23. J. R. Walker, S. Freudenthaler, C. F. Dmuchowski, T. Kaneko, N. Samata, G. Golor, S. R. Bruce and J. Triscari, *J. Clin. Pharmacol.*, 2006, **46**, 1069.
24. M. R. Owen, E. Doran and A. P. Halestrap, *Biochem. J.*, 2000, **348** Pt 3, 607–614.
25. L. Liu, A. Von Vett, N. Zhang, K. J. Walters, C. R. Wagner and P. E. Hanna, *Chem. Res. Toxicol.*, 2007, **20**, 1300–1308.

Inhibition of Glycogen Phosphorylase as a Strategy for the Treatment of Type 2 Diabetes

BRAD R. HENKE

Metabolic Pathways and Cardiovascular Therapy Area Unit, GlaxoSmithKline, 5 Moore Drive, Research Triangle Park, NC 27709, USA
E-mail: brad.r.henke@gsk.com

12.1   Introduction

Diabetes mellitus is now a worldwide public health problem with significant socioeconomic burdens. People with diabetes develop devastating chronic complications such as cardiovascular disease, retinopathy, neuropathy, and nephropathy which have negative impacts on morbidity and mortality. In 2007, diabetics cost the US in excess of \$174 billion, and recent projections indicate that total diabetes prevalence of the adult population in the US will increase from 14% in 2010 to nearly 25–28% in 2050.[1] Projections for increases in other countries are similarly sobering. The predominant form of the disease is type 2 diabetes, which accounts for approximately 90% of diabetes incidence. Type 2 diabetes is characterized by hyperglycemia, defects in pancreatic insulin secretion, and insulin resistance in skeletal muscle, adipose, and liver tissues. In addition to these derangements, the rate of endogenous glucose production is significantly elevated in type 2 diabetics in both the fed and fasted state relative to healthy individuals.[2,3] The liver produces approximately 90% of the body's endogenous glucose. In normal individuals, hepatic glucose production (HGP)

RSC Drug Discovery Series No. 27
New Therapeutic Strategies for Type 2 Diabetes: Small Molecule Approaches
Edited by Robert M. Jones

Published by the Royal Society of Chemistry, www.rsc.org

is tightly regulated by insulin and insulin's counter-regulatory hormone glucagon. However, numerous studies have shown that hepatic insulin resistance coupled with raised levels of glucagon leads to excessive HGP in the post-absorptive state in people with type 2 diabetes, which correlates well to the observed hyperglycemia.[4] Thus, drugs designed to reduce HGP would appear to be attractive approaches for the treatment of type 2 diabetes.

HGP is the sum of two metabolic pathways: glycogenolysis, which is the breakdown and release of glucose from its polymeric storage form glycogen, and gluconeogenesis, which is the *de novo* synthesis of glucose from lactate, amino acids, and glycerol. In healthy overnight-fasted subjects, the proportion of HGP attributed to glycogenolysis is 30–60% (with the remaining 40–70% attributed to gluconeogenesis) while in prolonged (>16 h) fasting, gluconeogenesis accounts for up to 90% of HGP.[5] Studies performed in type 2 diabetics estimate the glycogenolytic contribution anywhere from 12–75% of total HGP. In addition, a portion of the glucose produced by gluconeogenesis is cycled through the glycogen pool prior to efflux from the liver. Glycogen homeostasis is regulated by the key rate-limiting enzymes glycogen synthase (GS) and glycogen phosphorylase (GP), as well as a family of glycogen-targeting subunits of protein phosphatase 1 (PP1) which regulate the active and inactive states of the GS and GP enzymes. All three of these regulatory checkpoints have been the target of pharmaceutical intervention. This chapter will focus on inhibition of glycogen phosphorylase as a drug target. After a review of the characteristics of this enzyme, progress in the design, structure–activity relationships (SAR), and drug development of small molecule inhibitors at the various binding sites of this enzyme along with their potential as effective anti-hyperglycemic agents will be summarized, and the potential issues and benefits of drug therapy targeting GP will be outlined. Several excellent reviews have been previously published on this topic, so earlier work in the field will only be summarized, with this chapter focused on reports within the past 4–5 years.[5–9]

12.2 Characteristics of Glycogen Phosphorylase

12.2.1 Structure, Function, and Regulation of Glycogen Phosphorylase

The enzyme glycogen phosphorylase (GP) exists as a dimer composed of two identical 97 kDa subunits and catalyzes the phosphorolytic cleavage of the α-1,4-glycosidic bond of glycogen to produce glucose 1-phosphate (G-1-P) and a glycogen polymer shortened by one sugar residue. The liberated G-1-P is isomerized by the enzyme phosphoglucomutase to glucose-6-phosphate (G-6-P), which can then either enter the glycolytic pathway or be dephosphorylated and transported out of the cell as glucose. There are three mammalian isoforms of GP whose names are based on the tissue in which the isoform is preferentially expressed: liver, muscle, and brain. Distinct genes located on

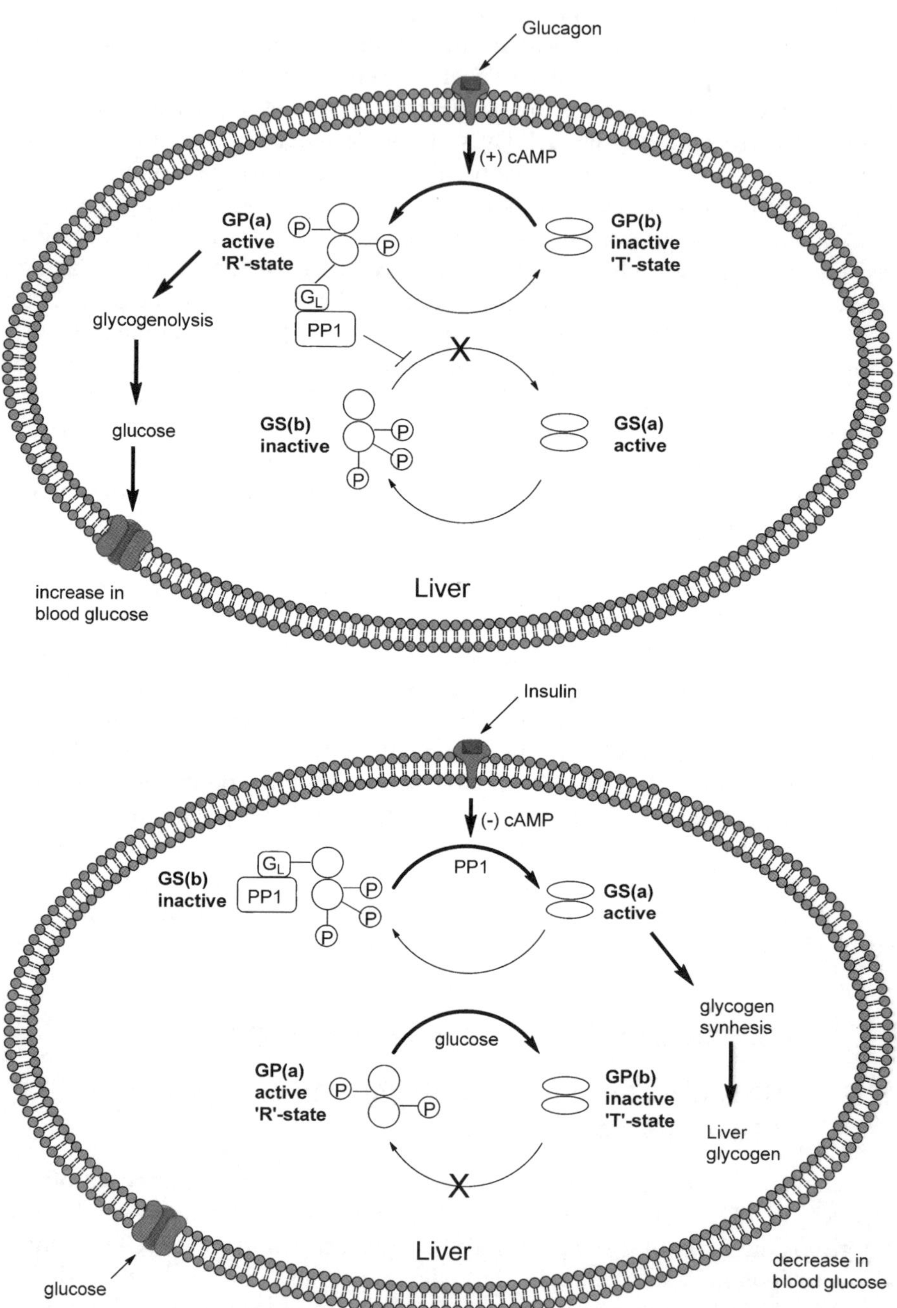

Figure 12.1   The relationship of glycogen phosphorylase in the glycolytic pathway and its modulation by glucagon and insulin.

human chromosomes 14, 11, and 20 encode these isoforms. There is a high degree of homology between the amino acid sequences of the three isoforms, particularly in residues near the catalytic and allosteric sites, with a sequence identity of approximately 80% between the human isoforms.[10] Each isoform has a distinct physiological role: the muscle isoform provides energy for muscle contraction via glucose-derived ATP generation, the brain isoform provides an emergency supply of glucose during periods of anoxia or severe hypoglycemia, and the liver isoform regulates glucose release from hepatic glycogen stores. The liver isoform is the major GP isoform involved in glucose homeostasis and thus is the focus for therapeutic intervention in type 2 diabetes.

GP cycles between two conformations: a catalytically more active GP(*a*) state (also called the "R" (relaxed) state) and a less catalytically active GP(*b*) state (also called the "T" (tensed) state). Conversion of GP(*b*) to GP(*a*) in all three GP isoforms occurs by phosphorylation of Ser14 by phosphorylase kinase as part of a cascade system initiated by cAMP, while the reverse dephosphorylation reaction is catalyzed by protein phosphatase 1 (Figure 12.1). GP activation is regulated by a variety of hormonal and neuronal signals, with insulin and glucagon being the primary hormonal regulators, and this regulation is different for the different GP isoforms. For example, in the liver phosphorylation is the major regulator of GP activation, and conversion of GP(*b*) to GP(*a*) fully activates the enzyme; the allosteric modulator AMP stimulates liver GP(*b*) by 10–20% and does not further activate liver GP(*a*).[11] However, in skeletal muscle, AMP activates GP(*b*) to 80% of maximal activity and activates muscle GP(*a*) by a further 10%. GP activity is also inhibited by glucose, G-6-P, and ATP; glycogen also asserts some regulatory control over GP via binding to a glycogen storage site. As mentioned above, a family of glycogen-targeting subunits of PP1 is also involved in modulation of GP (and GS) activity. The hepatic glycogen-targeting subunit G_L directs PP1 activity against GP and GS in the liver (Figure 12.1). The carboxy terminus of the G_L protein binds to GP(*a*) and prevents PP1 from activating GS via dephosphorylation of that enzyme;[12] thus inhibition of the binding of G_L to GP might lead to up-regulation of GS activity and a subsequent decrease in hepatic glucose output.[13] This approach to modulation of PP1 activity via inhibition of GP-G_L binding will be discussed in more detail later in the chapter.

12.2.2 Inhibitor Binding Sites of Glycogen Phosphorylase

At present, five unique sites on the GP enzyme have been identified that bind small molecule and peptide-based inhibitors (Figure 12.2). A considerable amount of detail is known about the protein–inhibitor interactions in these binding sites through numerous X-ray crystallographic studies. The general description of these inhibitor sites will be covered here, but the reader is referred to several excellent reviews for a more detailed analysis.[8,14,15]

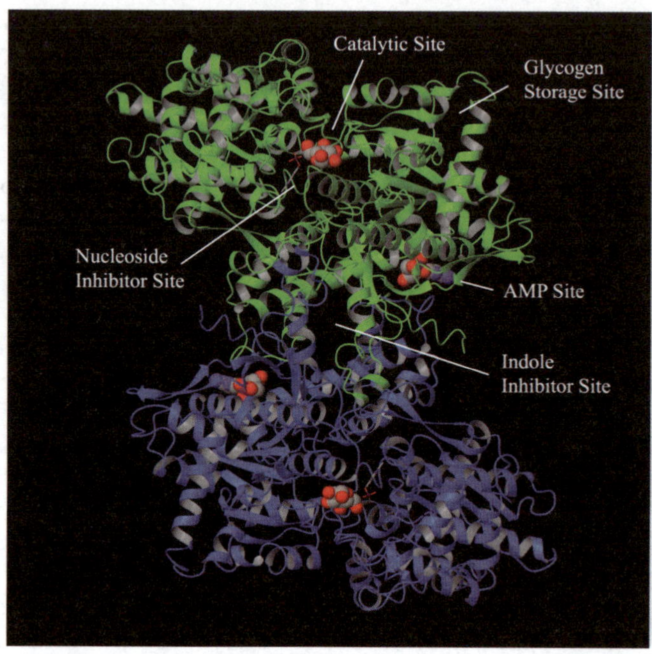

Figure 12.2   Human liver GP(*a*) homodimer complexed with AMP (magenta) and glucose (yellow). The pyridoxyl-5<pr>-phosphate cofactor is shown in light blue. Structure shown is PDB entry 1FA9. Image produced with PyMOL.

The catalytic site (Figure 12.2, with glucose bound) is buried at the center of each monomeric subunit and is accessible to solvent through a 15 Å channel, and as expected the known inhibitors are glucose-based analogs. These inhibitors promote the less active GP(*b*) state through stabilization of a conformation that blocks access of glycogen to the catalytic site. This site contains a flexible loop (residues 282–287) which makes design of very potent inhibitors fairly challenging. The glycogen storage site (Figure 12.2) is on the surface of the enzyme some 30 Å from the catalytic site. The binding region is α-helical in nature and has received much less attention than any other GP binding site as a target for inhibitor design, presumably because the protein surface in this region does not suggest that tight binding interactions can be made with small molecules having the desired drug-like properties for oral delivery. A third binding site known as the purine nucleoside, or caffeine, binding site is located on the surface of the enzyme approximately 12 Å from the entrance to the catalytic site (Figure 12.2). Binding to this site stabilizes the GP(*b*) conformation by blocking access to the catalytic site and is generally synergistic with glucose. A number of different classes of small molecules have been shown to bind this site (see below) but similar to the previous two binding sites, the potency of the known inhibitors is relatively weak.

The two binding sites that have been the most successfully exploited by drug discovery programs are the so-called AMP, or allosteric, binding site and the indole binding site. The AMP binding site is situated where the C-terminal residues of two α-helices (helices 2 and 8) come together in a V-shape, near the edge of the dimer interface, and is located about 30 Å from the catalytic site and 40 Å from the glycogen storage site (Figure 12.2). Inhibitors binding to this site deactivate the enzyme through direct inhibition of AMP binding and/or inhibition of substrate binding through stabilization of the GP(b) conformation. This site also shows considerable overlap with the binding site of carboxy terminus of the hepatic glycogen-targeting subunit G_L which suggests that blockade of the GP-G_L interaction may be part of the pharmacologic mechanism of action of AMP site inhibitors. However, an investigation with an AMP site inhibitor showed no disruption of the GP-G_L interaction as measured via a scintillation proximity assay using a labeled peptide fragment of C-terminal PP1-G_L, indicative of distinct binding sites and potentially distinct pharmacological modes of action.[16] Finally, the indole binding site is located in the central cavity region of the subunit dimerization interface and is 33 Å from the catalytic site and 50 Å from the glycogen storage site. Interestingly, an indole-site inhibitor was also shown to block the interaction of GP(a) with the C-terminus of G_L even though the indole binding site is remote from the GP(a)-G_L interface, which suggests that conformational changes induced by the binding of indole-site inhibitors are responsible for the inability of G_L to bind to GP(a).[17] Thus indole-site inhibitors may also exert their effects through lowering the affinity of GP(a) for PP1-G_L in addition to inhibition of the enzyme itself. A number of potent, efficacious, orally active, drug-like series of inhibitors that bind to each of these latter two sites have been reported.

12.3   Glycogen Phosphorylase Inhibitors

12.3.1   *In vitro* and *in vivo* Assessment of GP Inhibition

Before summarizing the recent advances in the development of glycogen phosphorylase inhibitors, a few comments should be made about the determination of *in vitro* potency and *in vivo* efficacy of GP inhibitors. It is instructive to recognize the important role that assay conditions play when trying both to compare different series of GP inhibitors from different laboratories and to correlate enzymatic potencies to cellular and *in vivo* potencies. Several different orthologs of GP (human, pig, rabbit, and rodent) as well as the three tissue isoforms, in both the phosphorylated and unphosphorylated states, have been employed as proteins in enzymatic assays. There is a high degree of structural homology between the sources of GP (tissue and species), and in some cases a comparison of the inhibitory activity of GP inhibitors has been shown to be independent of the source of enzyme (tissue or species); however, there are also cases where inhibition of a given

GPI is dramatically different across isoforms.[18,19] The activity of inhibitors is dependent upon the direction in which the assay is performed (i.e. measurement of glycogenolysis or glycogen synthesis) and particularly dependent on the concentration of physiological regulators AMP and glucose present in the screening assay. Studies have shown that glucose potentiates the inhibitory activity while AMP decreases the activity of certain GP inhibitors.[20,21] A study investigating the effects of a combination of the physiological regulators ATP, AMP, ADP, glucose, G-6-P, fructose-6-P, and UDP-glucose on GP inhibition by synthetic GP inhibitors found that addition of a cocktail containing these regulators significantly reduced the inhibitory potency of several GP inhibitors against human liver GP(*a*) relative to enzyme preparations without the added regulators.[22] An additional study with 13 GP inhibitors displaying a range of potencies against GP(*a*) provided similar results; there was a poor correlation observed between compound potencies obtained using purified liver GP(*a*) and *in vivo* minimum efficacious dose (MED) in a diabetic animal model; however, a robust correlation between enzyme K_D and MED was observed when the enzyme assay was run in the presence of liver homogenates via equilibrium dialysis.[23] This reinforces the notion that endogenous factors and free concentration at the target tissue are important variables in measuring compound potency.

Cellular potencies of GP inhibitors have routinely been measured using liver cells, generally either primary rat hepatocytes or cell lines such as HepG2 or SK-HEP-1. The cells are pre-labeled with ^{14}C-glucose by inclusion in the culture medium. The cells are treated with the test compound, and then glycogenolysis is stimulated by treatment with forskolin or glucagon. The cells are then lysed and the amount of radiolabeled glycogen is then quantified and compared to the control response to derive the activity of the compounds. As these cellular systems introduce factors such membrane permeability, endogenous modulators of GP activity, and compound metabolism, it provides valuable information regarding the potency and efficacy of potential drug candidates.

A number of preclinical *in vivo* models have been utilized to profile GP inhibitors. One simple *in vivo* assay for assessing acute efficacy of GP inhibitors involves the dosing of compound or vehicle in a genetic (e.g. ob/ob mice) or non-genetic (e.g. DIO mice) rodent model of diabetes followed by retro-orbital bleeds at measured time points post-dose to measure the decrease in blood glucose. The glucagon challenge has also been widely employed for determining acute effects of GP inhibitors, using either normal or disease strains of rats. The principle behind the test is that inhibition of the GP enzyme should attenuate the rise in blood glucose that occurs upon administration of exogenous glucagon (see Figure 12.1). Specific experimental protocols differ amongst groups; glucagon has been dosed intravenously, subcutaneously, or intraperitoneally, and test compounds have been dosed either before or after glucagon is given.[24] In some protocols the somatostatin mimetic octreotide has been co-administered with glucagon to block the counter-regulatory response of insulin to glucagon.[25] This assay provides a robust, mechanism-based readout of potency and efficacy with reasonable throughput for *in vivo* analog profiling. Oral glucose tolerance tests (OGTTs) have also been conducted in rodent disease models as an approach to determine acute efficacy.[26] Subchronic multiple dose experiments with GP inhibitors have also been performed using the same genetic and non-genetic rodent models used in acute experiments, though to date preclinical efficacy data from multiple dose experiments are fairly sparse.

12.3.2 Catalytic Site Inhibitors

α-D-Glucose inhibits GP activity by binding to the catalytic site with a $K_i = 2$ mM (rabbit muscle GP) which stabilizes the less active GP(b) conformation of the enzyme. Extensive work in identifying glucose analogs with increased affinity to GP has been reported and a detailed description of the SAR and structural information is available in recent reviews.[8,27–29] N-Acetyl-β-D-glucopyranosylamine (Table 12.1, entry 1) was one of the first glucose analog inhibitors with improved potency reported, and subsequently a large number of compounds have been prepared to explore the N-acyl substituent. Table 12.1 summarizes some key SAR findings within this series. X-Ray crystallographic studies have revealed a narrow hydrophobic channel, termed the β-channel, within this region of the catalytic site. In general, increasing the steric bulk of the acetyl group leads to modest reductions in potency[30] with the notable exception of the 2-naphthyl moiety (Table 12.1, entry 3), which does suggest that proper placement of hydrophobic groups can improve inhibitor potency. A 2-naphthyl moiety appears to provide an increase in potency over simple phenyl derivatives (Table 12.1. compare entries 2 and 3, 7 and 8) and seems to be the optimal hydrophobic substituent discovered to date. Surprisingly, short-chain aliphatic polar groups are tolerated (Table 12.1, entries 4 and 5), though extending the chain length leads to a drop in potency (data not shown).[31] Insertion of a nitrogen atom to form an aryl urea leads to a modest increase in potency (Table 12.1, compare entries 2 and 11). However, the corresponding benzyl urea (Table 12.1, entry 12) was inactive. Oxamide derivatives[8] tend to be less active than their corresponding amide analogs (Table 12.1, compare entries 3 and 6); however, acylurea derivatives[8] tend to improve potency (Table 12.1, compare entries 2 and 8), with the 2-naphthyl derivative (Table 12.1, entry 7) being the most potent N-Acyl-β-D-glucopyr-anosylamine analog reported in the literature to date. Insertion of polar functionality into the region occupied by the 2-naphthyl group diminishes activity (Table 12.1, compare entries 7, 9, and 10).[32] Further chain extensions (Table 12.1, entries 13 and 14) reduced activity, suggesting that within this series the 2-naphthoylurea provided the optimal binding interaction with the GP enzyme.

Several papers have described glucose analogs wherein the β-acyl moiety has been replaced by a variety of cyclic isosteres resulting in either spirocyclic

Table 12.1 Potencies of selected *N*-acyl-β-D-glucopyranosylamine GP inhibitors.[33]

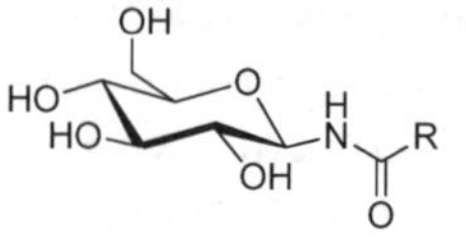

Entry	R	K_i *(μM)*	*Ref.*
1	—CH$_3$	32	8
2	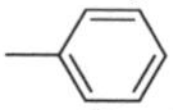 (phenyl)	81	8
3	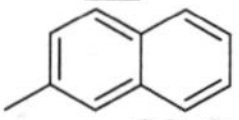 (naphthyl)	4	30
4	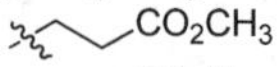 CO$_2$CH$_3$	170	31
5	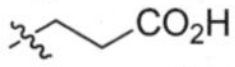 CO$_2$H	20	31
6	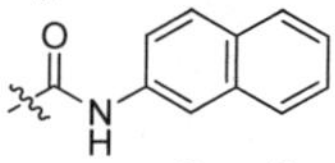 (naphthylaminocarbonyl)	56	8
7	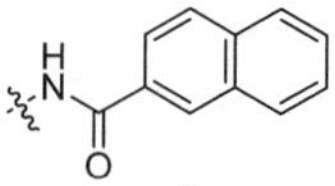 (naphthylcarboxamide)	0.35	8
8	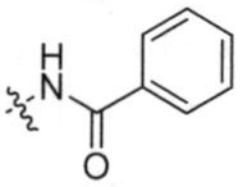 (benzamide)	5	8
9	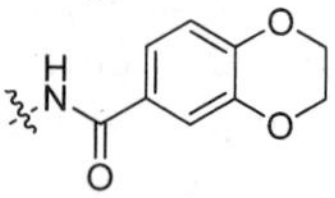 (benzodioxine carboxamide)	250	32
10	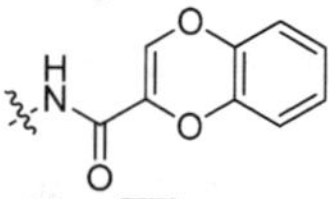 (benzodioxine carboxamide)	85	32
11	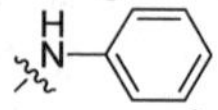 (phenylamino)	18	8
12	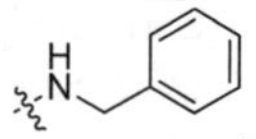 (benzylamino)	>1000	8
13	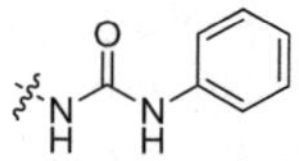 (phenylurea)	21	8
14	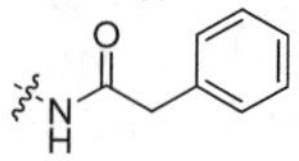 (phenylacetamide)	600	8

derivatives or *C*-glucopyranosyl derivatives as summarized in Figure 12.3. Hydantoin **1** (K_i = 3 μM) was one of the early spirocyclic analogs that showed increased enzymatic potency over that of *N*-acetyl-β-D-glucopyranosylamine,[33] while the isomeric hydantoin **2** and diketopiperazine **3** are less active. X-Ray crystallographic investigation of hydantoin **1** revealed a hydrogen bond between the amide NH and the main chain carbonyl of His377, similar to that seen with *N*-acetyl-β-D-glucopyranosylamine, suggesting that this interaction is important and that perhaps the increased rigidity of the hydantoin was responsible for the potency gain. In an effort to combine this SAR with that observed in the aforementioned acylurea series in which a large, flat hydrophobic group was directed into the β-channel, two recent papers have reported the synthesis and potency of a series of spiro-isoxazolines[34] and spiro-oxathiazoles.[35] Similar to that observed with the earlier series, the 2-naphthyl group provided the best potency, with compounds **4** and **5** (Figure 12.3) being the most potent catalytic site inhibitors reported to date. Since neither of these compounds can serve as a hydrogen bond donator to His377, the rigid nature of the spirocyclic ring overcomes the loss of that hydrogen bond. A series of oxadiazoles and triazoles have also been described which show moderate potency as GP inhibitors.[36,37] Again the 2-naphthyl derivatives were the most potent inhibitors, and there were some clear differences in enzyme potency dependent on the orientation of the heteroatoms in the five-membered ring (Figure 12.3), which suggests this ring likely makes important hydrogen bond interactions with water molecules located within the β-channel.

Finally, aza-sugar analogs, which have been designed as structural transition state mimics of glycosyl cations (the hypothesized reactive intermediate of cleavage of the α-1,4-glycosidic linkage) have also been described. The most potent of these inhibitors reported to date are 1,4-dideoxy-1,4-imino-D-arabinitol (DAB, **10**, Figure 12.3) which inhibits rabbit muscle GP(*a*) with a K_i = 0.4 μM, and isofagomine (**11**, Figure 12.3) which inhibits pig liver GP(*a*) with an IC_{50} = 0.8 μM.[38] While these compounds presumably bind at the active site, there is no X-ray crystallographic data with either DAB or isofagomine bound to GP to verify this hypothesis. Only one catalytic site inhibitor (DAB, **10**, Figure 12.3) has been reported to have *in vivo* activity in animal models of type 2 diabetes. DAB is orally bioavailable in rats (%F = 89%) with an oral half-life of 49 minutes, and metabolic studies utilizing ^{14}C-labelled DAB showed preferential distribution into the liver over muscle.[39] In glucagon-challenge studies conducted in normal rats, a 1–2 mg/kg oral dose of DAB given 15 min prior to an i.v. infusion of glucagon completely suppressed the glucagon-stimulated increase in HGP. Similarly, when dosed at 4 mg/kg 4 h prior to a bolus of glucagon, DAB reduced the glucose excursion by 50% in overnight-fasted beagle dogs. Additionally, frequent s.c. delivery of DAB was shown to acutely lower blood glucose in the ob/ob mouse model of type 2 diabetes. However, the short half-life of DAB makes it unlikely to be effective in the setting of chronic administration. Only a few additional catalytic site inhibitor reports contain data describing activity in cellular assays. Thus, the potential for catalytic site inhibitors as viable drug candidates remains largely unknown; however, given the relatively modest potency observed to date, other allosteric sites on GP appear to be more favorable drug targets.

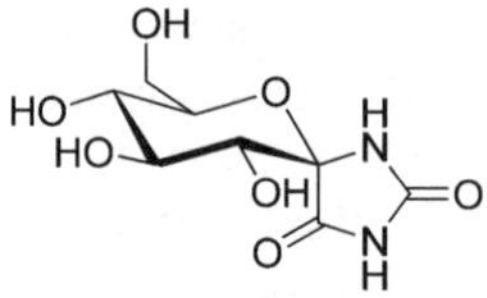

1 (K_i = 3 μM) 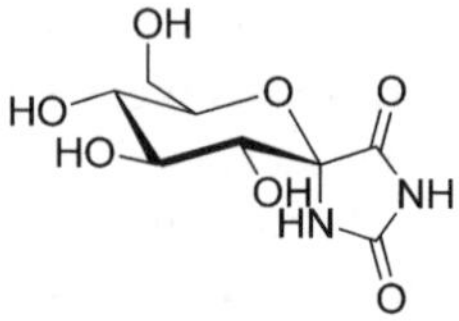 2 (K_i = 29 μM) 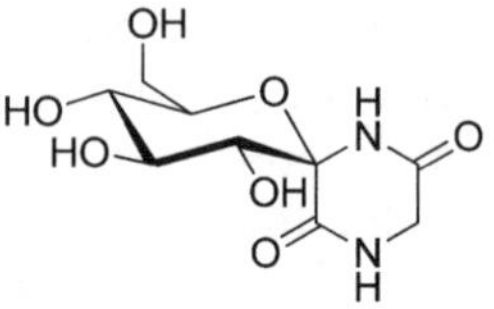 3 (K_i = 60 μM)

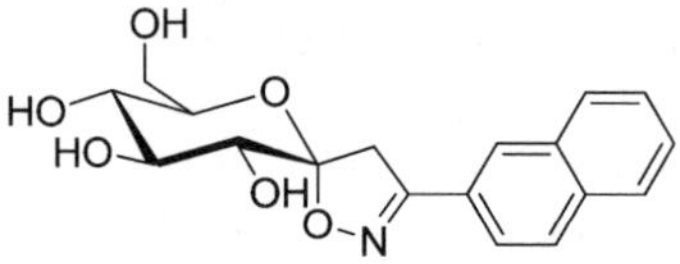

4 (K_i = 0.63 μM) 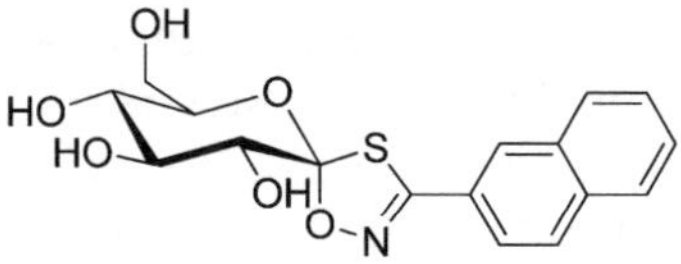 5 (K_i = 0.16 μM)

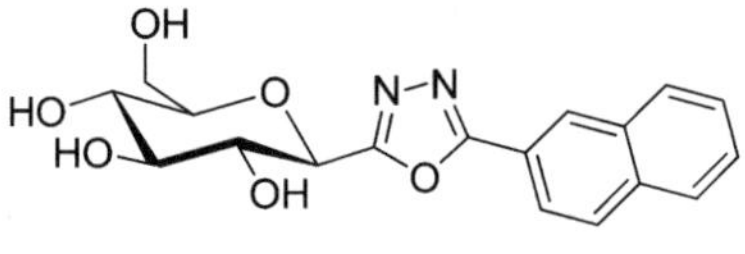

6 (K_i = >100 μM) 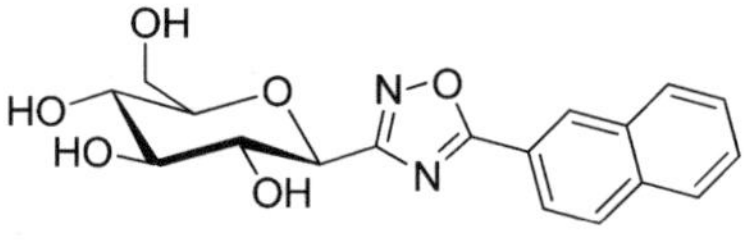 7 (K_i = 38 μM)

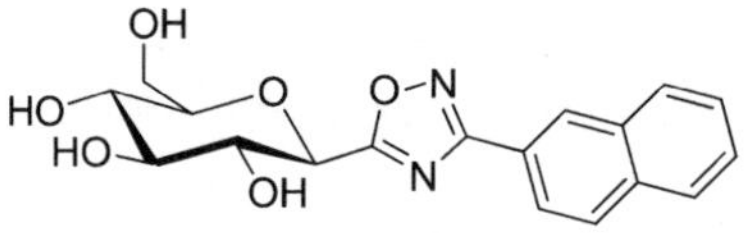

8 (K_i = 12 μM) 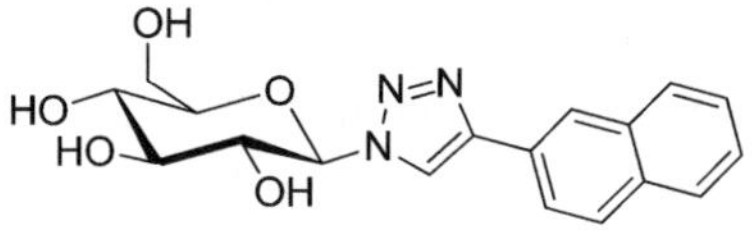 9 (K_i = 35 μM)

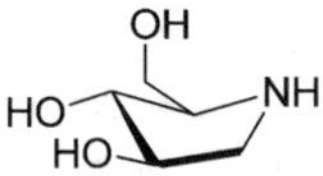

10 (K_i = 0.4 μM) 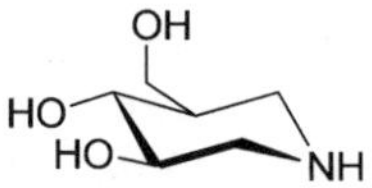 11 (K_i = 0.8 μM)

Figure 12.3 Selected catalytic site GP inhibitors and reported binding constants.

12.3.3 Glycogen Storage Site Inhibitors

Attempts at identifying potent GP inhibitors at the glycogen storage site have met with very little success. Cyclodextrins are modest inhibitors of GP(*b*), with γ-cyclodextrin and maltoheptaose inhibiting GP(*b*) with K_i values of 7.4 and 1.0 mM, respectively.[40] Since this binding site recognizes oligosaccharides, classical small molecule binding interactions are expected to be fairly weak, which most certainly accounts for the lack of progress in finding potent, drug-like inhibitors at this site via high-throughput screening.

12.3.4 Purine Nucleoside Site Inhibitors

The purine analog caffeine (Figure 12.4) is the prototypical binder at this site and binds rat liver GP with a $K_i = 100$ μM; other purine derivatives, nucleosides, and nucleotides have also been reported as weak binders.[14,41] The cyclin-dependent kinase inhibitors indirubin-5-sulfonate (**12**, Figure 12.4) and flavopiridol (**13**, Figure 12.4) both inhibit rabbit muscle GP(*b*) with IC_{50} values of 14 and 1 μM, respectively. Analogs in which the stereogenic centers of the piperidine ring in flavopiridol were replaced by an olefin (**14**, Figure 12.4) showed similar activity, suggesting that chirality in this region is unimportant in binding; however, the importance of the 2-aryl ring was revealed, as the corresponding methyl, *t*-butyl analogs were inactive.[42] Additional studies in rat hepatocytes with analogs from this series revealed that these compounds suppressed the basal rate of glycogenolysis by allosteric inhibition of GP rather than by inactivation of GP(*a*) via dephosphorylation. However, they inhibited dibutyryl cAMP-stimulated glycogenolysis by both allosteric inhibition and inactivation of GP(*a*) via dephosphorylation. No additional reports of bona fide purine nucleoside site inhibitors have appeared in the recent literature, and the potential for this site to yield robust drug candidates appears limited.

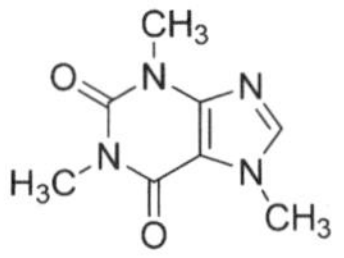

caffeine (K$_i$ = 100 μM) 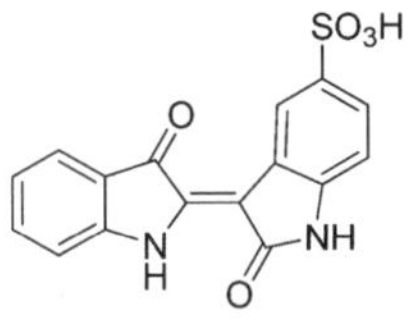   12 (K$_i$ = 14 μM) 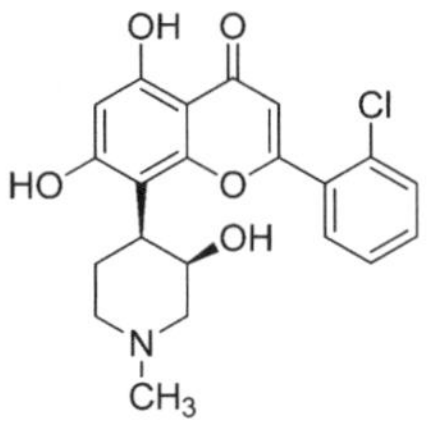   flavopyridol 13 (K$_i$ = 14 μM)

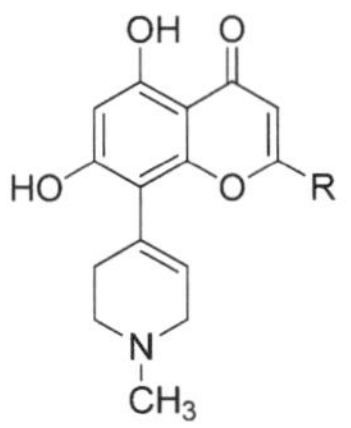

14 (R = aryl, K$_i$ = 1-2 μM; R = alkyl, K$_i$ >200 μM)

Figure 12.4 Selected purine nucleoside site GP inhibitors and reported binding constants.

12.3.5 AMP Site Inhibitors

The AMP site has been one of the two most fruitful GP inhibitor binding sites with respect to identification of potent, drug-like molecules. The first potent and efficacious AMP site inhibitors were a series of dihydropyridine dicarboxylic acids reported by scientists at Bayer. That effort culminated in the description of BAY W 1807 (Table 12.2, entry 1) as a potent inhibitor of rabbit muscle GP(*a*) (K_i = 1.6 nM).[43] The lactone prodrug BAY R 3401, which is converted via metabolic activation to BAY W 1807, was reported as a potent inhibitor of hepatic glycogenolysis in conscious dogs.[44] Scientists at Merck reported structural modifications to this series which are summarized briefly in Table 12.2.[19] Removal of the 2′-chloro substituent (Table 12.2, entry 2) decreased potency ca. 10-fold, while removal of the 2-carboxylic acid moiety led to ca. 500-fold reduction in potency (Table 12.2, entry 3). Replacement of the pendant ester functionality was also poorly tolerated (Table 12.3, entry 5), and removal of the *N*-ethyl group resulted in a 10-fold drop in potency (Table 12.2, entry 4). However, replacement of the *N*-ethyl with *N*-benzyl substituents increased both enzymatic and cellular potency (Table 12.2, entries 6–8), with the 3,4-dimethoxy analog demonstrating glucose-lowering efficacy in the db/db mouse model of type 2 diabetes.[19]

Other diacid derivatives that bind the AMP site have also been reported. A series of potent (human liver GP IC_{50} = 10 nM) phthalate-derived compounds exemplified by compound **15** (Figure 12.5) have modest selectivity for liver versus muscle GP isoforms.[45] A small set of analogs was reported; SAR

Table 12.2 Potencies of selected dihydropyridine AMP site inhibitors.

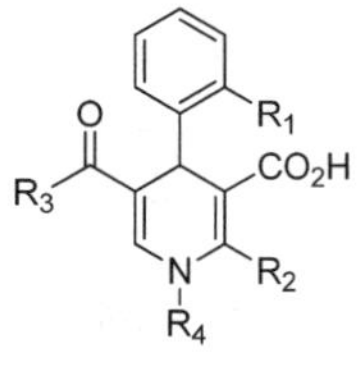

Entry	R_1	R_2	R_3	R_4	hLGP(a) K_i (µM)	Cell EC_{50} (µM)
1	Cl	CO_2H	(*i*-Pr)-O	Et	0.039	2.2
2	H	CO_2H	(*i*-Pr)-O	Et	0.395	6.5
3	Cl	H	(*i*-Pr)-O	Et	15.6	NA
4	Cl	CO_2H	(*i*-Pr)-O	H	0.692	11.0
5	Cl	CO_2H	(*i*-Pr)-NH	Et	9.1	NA
6	Cl	CO_2H	(*i*-Pr)-O	CH_2Ph	0.011	1.13
7	Cl	CO_2H	(*i*-Pr)-O	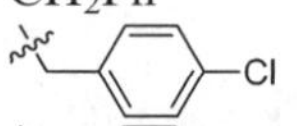 (4-chlorobenzyl)	0.002	0.48
8	Cl	CO_2H	(*i*-Pr)-O	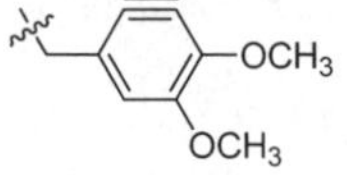 (3,4-dimethoxybenzyl)	0.004	0.27

suggests that incorporation of the 3-fluoro in the phenyl ring and 4-chloro or 4-nitro moieties in the pyridine ring of **15** provided optimal cellular potency. Compound **15** displayed *in vivo* efficacy in a glucagon-challenge assay, significantly reducing the glucagon-induced glucose excursion in normal mice at a dose of 30 mg/kg. A related series of compounds exemplified by **16** (pig liver GP IC_{50} = 74 nM) has also been described.[46] A small set of compounds was prepared from a high-throughput screening hit, and compound **16**

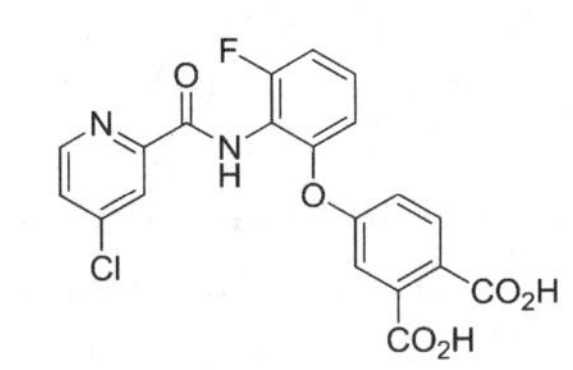

15 (IC_{50} = 10 nM)

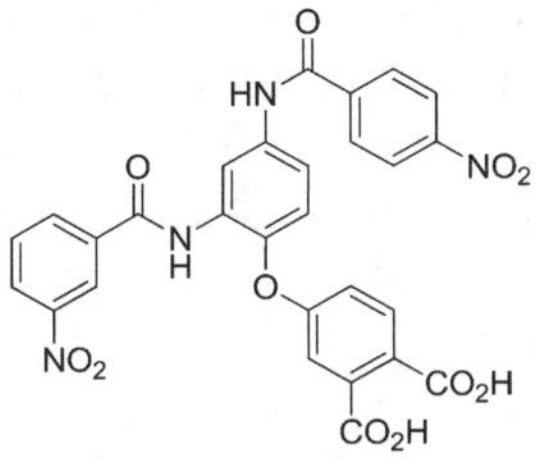

16 (IC_{50} = 74 nM)

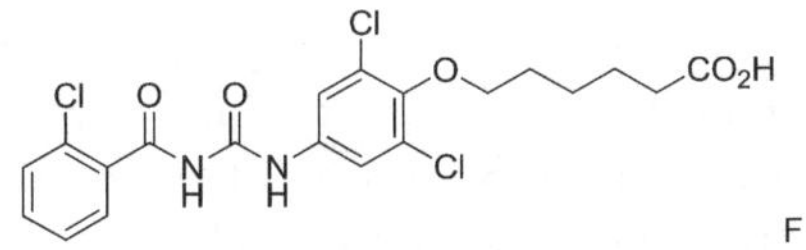

17 (IC_{50} = 2000 nM)

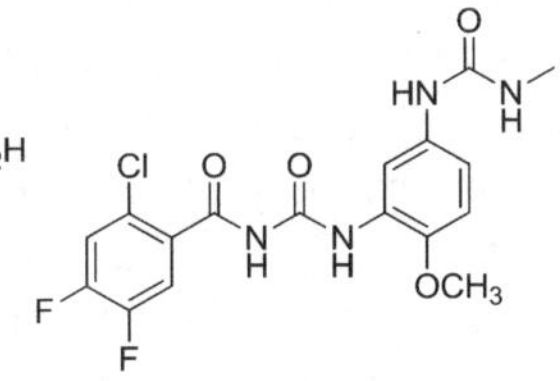

18 (IC_{50} = 53 nM)

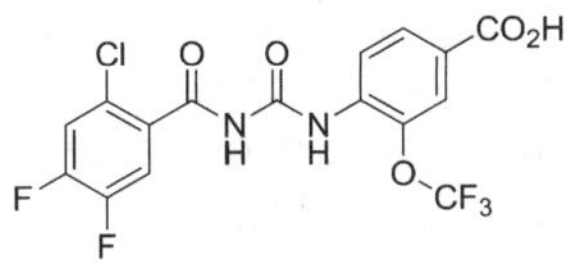

19 (AVE5688)

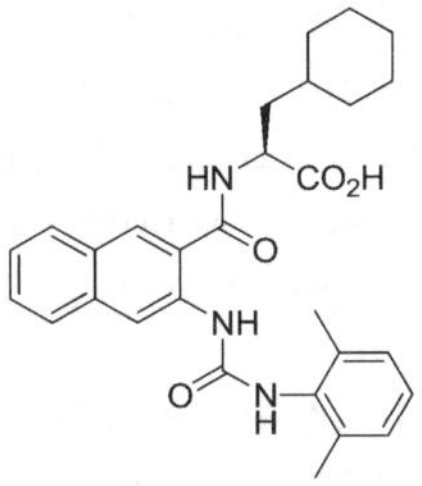

20 (IC_{50} = 2200 nM)

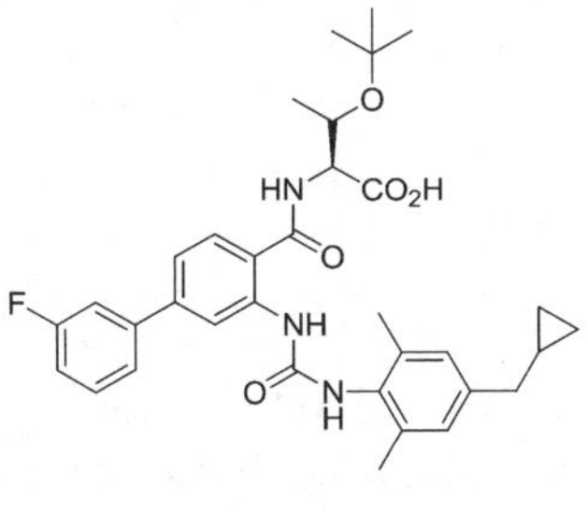

21 (IC_{50} = 3 nM)

Figure 12.5 Selected AMP site GP inhibitors and reported IC_{50} values.

emerged as the most potent inhibitor in the series. Compound **16** inhibited basal and glucagon-stimulated glycogenolysis in primary rat hepatocytes with IC_{50} values of 1.6 and 4.7 μM, respectively; no *in vivo* data with **16** were reported.

Acylurea **17** (Figure 12.5, human liver GP(*a*) IC_{50} = 2 μM, rat hepatocyte IC_{50} = 80 μM) was identified from a focused screening effort on a set of compounds with pharmacophoric similarities to known GP inhibitors. Optimization of this template through both 3D pharmacophore models and subsequently structure-based drug design resulted in compound **18**, which showed greater than 200-fold improvements in enzyme and cellular potency (Figure 12.5, human liver GP(*a*) IC_{50} = 53 nM, rat hepatocyte IC_{50} = 380 nM).[47] Intravenous administration of compound **18** to anaesthetized rats in a glucagon challenge test caused significant reduction in the glucagon-induced hyperglycemic peak. No additional *in vivo* efficacy data from this compound series has been reported; however, Sanofi-Aventis advanced a compound from this series, AVE5688 (**19**, Figure 12.5) into human clinical trials.

Recently, scientists at GlaxoSmithKline have reported a series of anthranilamide derivatives that are potent AMP site inhibitors.[25,48–50] A weakly active hit (compound **20**, Figure 12.5 and entry 1, Table 12.3) identified through a high-throughput screening campaign was systematically optimized to afford very potent and efficacious GP inhibitors, with key SAR findings reported in Tables 12.3–12.5. Initial SAR investigation of the amino acid side chain (Table 12.3) revealed that shortening of the group to an (*S*)-cyclohexyl moiety provided a significant increase in enzyme and cell potency (entry 3 versus entry 1, Table 12.3). A brief SAR survey on the aryl urea indicated that *para* substitution also boosted potency (compare entry 7 versus entry 8, Table 12.3) and also revealed the requirement of the urea moiety to achieve good potency, as the corresponding aryl carbamate (entry 10, Table 12.3), benzyl amide (entry 9, Table 12.3), phenyl amide (data not shown) and aryl sulfonamides (data not shown) were considerably less active. In addition, substitution of the urea N–H (entry 12, Table 12.3) led to inactive compounds.

Additional optimization of the amino acid side chain was then undertaken in an effort to both improve potency and remove a CYP2C9 liability identified with the cyclohexyl substituent. A number of changes to the cyclohexyl moiety failed to improve potency, with placement of polar functionality deep into this pocket detrimental to activity (entries 1–4, Table 12.4); however, the α-methylcyclohexyl derivative (entry 5, Table 12.4) was equipotent while displaying much improved metabolic stability and considerably reduced potency at CYP2C9.[50] A series of cycloalkyl derivatives (cf. entries 6–9, Table 12.4) were synthesized which demonstrated that increasing ring size/lipophilicity improved potency of GP inhibition in both the enzyme and cellular assays. A series of straight-chain alkyl derivatives were also prepared (cf. entries 10–18, Table 12.4), which demonstrated that (a) the optimal chain length was four atoms (compare entry 12 with entries 10, 11, 13), (b) δ-

Table 12.3   Potencies of selected anthranilamide AMP site inhibitors.

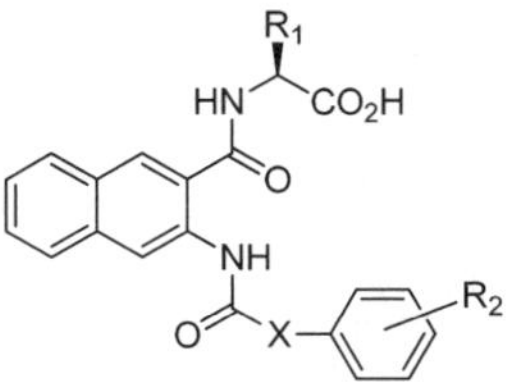

Entry	R_1	R_2	X	hLGP(a) IC_{50} (nM)	Cell IC_{50} (nM)	Ref.
1	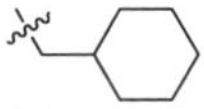 cyclohexyl	2-Cl, 6-Me	NH	2200	7400	48
2	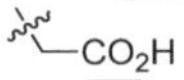 $-CO_2H$ (cyclohexyl)	2-Cl, 6-Me	NH	390	4700	48
3	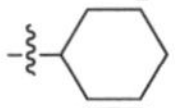 cyclohexyl	2-Cl, 6-Me	NH	73	1200	50
4	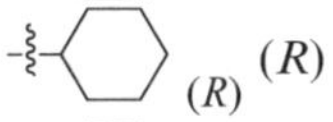 cyclohexyl (R) (R)	2-Cl, 6-Me	NH	4200	NT	48
5	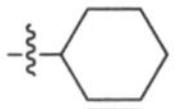 cyclohexyl	2,6- Cl	NH	21	320	50
6	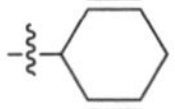 cyclohexyl	2,6-F	NH	4900	NT	48
7	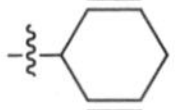 cyclohexyl	2,6-Me	NH	120	1100	50
8	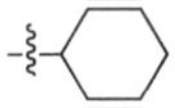 cyclohexyl	2,4,6-Me	NH	6	370	50
9	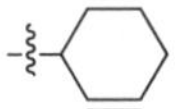 cyclohexyl	2,6-Cl	CH_2	6800	1600	48
10	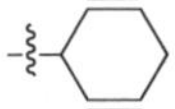 cyclohexyl	2,6-Cl	O	>10 μM	NT	48
11	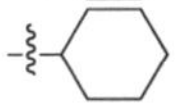 cyclohexyl	2,6-Cl	$N\text{-}CH_3$	>10 μM	NT	48

branching improved cellular potency (compare entries 12 versus 14, 15 versus 16, 17 versus 18), (c) incorporation of an ether oxygen into the side chain was well tolerated (compare entry 14 versus 16) and led to an improvement in solubility (data not shown), and (d) addition of a β-methyl group provided further improvements in enzymatic and cellular potency and additional reduction in CYP2C9 inhibition (data not shown).[49] Ultimately, the *t*-butyl threonine group (entry 18, Table 12.4) was selected as the optimal choice of substituent at this position.

Finally, with the aid of X-ray crystallography, efforts to replace the naphthalene ring system and further refine the substitution pattern of the aryl urea group were undertaken. Replacing the naphthalene was important because metabolic incubation in the presence of glutathione indicated the formation of reactive metabolites, and naphthalenes are known to form reactive metabolites via epoxidation pathways.[25] The naphthalene ring binds

Table 12.4   Potencies of selected anthranilamide AMP site inhibitors.

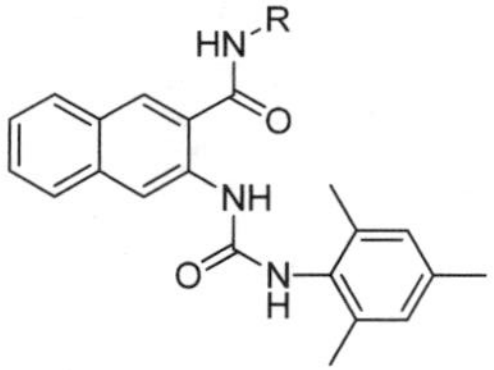

Entry	R	LGP(a) IC_{50} (nM)	Cell IC_{50} (nM)	Entry	R	LGP(a) IC_{50} (nM)	Cell IC_{50} (nM)
1	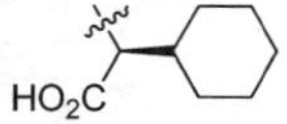	6	373	10	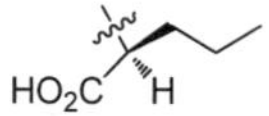	39	1019
2	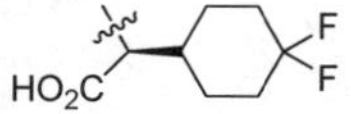	13	1000	11	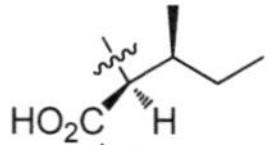	72	688
3	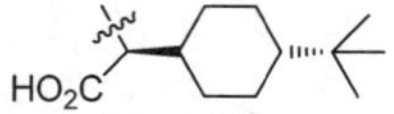	556	2400	12	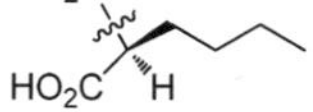	23	501
4	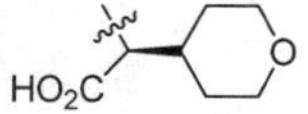	97	4873	13	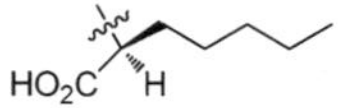	114	1533
5	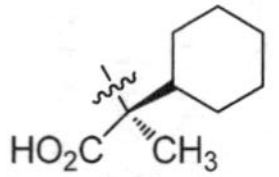	10	350	14	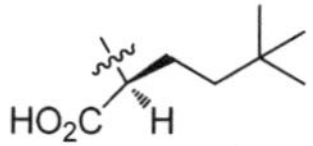	25	311
6	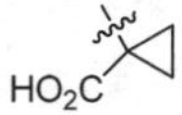	720	9023	15	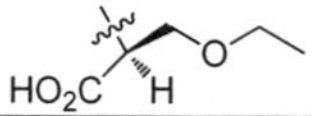	56	4467

Table 12.4 (*Continued*)

Entry	R	LGP(a) IC$_{50}$ (nM)	Cell IC$_{50}$ (nM)	Entry	R	LGP(a) IC$_{50}$ (nM)	Cell IC$_{50}$ (nM)
7		339	1390	16		44	387
8		36	585	17		298	2457
9		5	498	18		7	139

in a narrow lipophilic groove within the AMP pocket, nearly ending at the solvent front, which suggested that "out-of-plane" changes would not be tolerated, and pointed to biaryl derivatives as a reasonable alternative. Indeed, while removal of the distal ring of the naphthalene led to a considerable decrease in potency (compare entry 1 versus entry 2, Table 12.5), the biaryl replacement (entries 3–5, Table 12.5) provided analogs equipotent to the naphthalene series and importantly displayed no evidence of reactive metabolite formation. Examinations of the ligand-bound crystal structure also revealed additional space off the 4-position of the aryl ring of the urea. Replacement of the 4-methyl group with the cyclopropylmethyl group provided a small increase in enzyme potency (compare entries 2 versus 7, 4 versus 8, Table 12.5). These combined efforts culminated in compound **21** (Figure 12.5) which displayed robust effects on glucose lowering when dosed orally at 2 mg/kg in a rat glucagon challenge model and also acutely lowered glucose in an ob/ob mouse model when dosed orally at 15 mg/kg.

A series of closely related pentacyclic triterpene natural products exemplified by maslinic acid, oleanolic acid, corosolic acid, and ursolic acid (entries 1–4, Table 12.6) have been utilized in Chinese and Japanese herbal medicines and have reported positive metabolic effects in humans[51] as well as other benefits.

Table 12.5   Potencies of selected anthranilamide AMP site inhibitors.

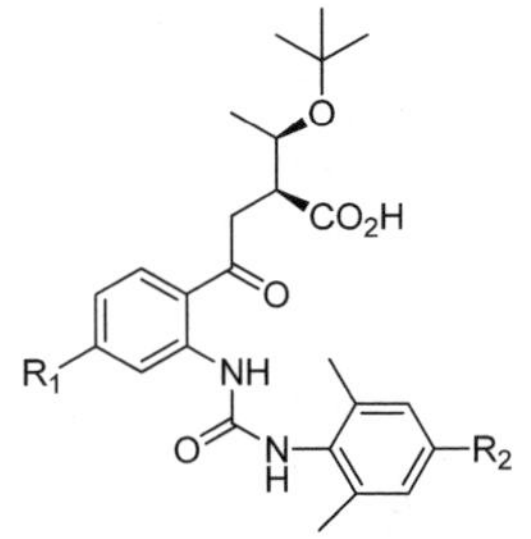

Entry	R_1	R_2	hLGP(a) IC$_{50}$ (nM)	Cell IC$_{50}$ (nM)
1	Naphthyl (fused)	CH_3	7	139
2	F	CH_3	506	1820
3	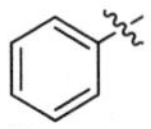 phenyl	CH_3	10	126
4	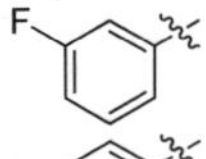 3-fluorophenyl	CH_3	15	104
5	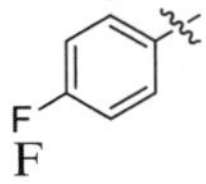 4-fluorophenyl	CH_3	7	88
6	F	 propyl	10	147
7	F	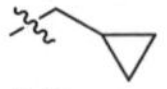 cyclopropylmethyl	4	133
8	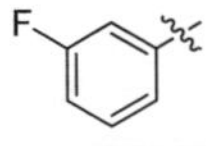 3-fluorophenyl	 cyclopropylmethyl	3	145

These compounds have demonstrated modest inhibition of GP, and the SAR of these compounds has been examined.[52–56] The parent natural products range in potency from 9 to 28 µM against rabbit muscle GP. The number and positioning of hydroxyl groups in the A ring has a modest effect on potency. The C3-monohydroxyl compounds are generally slightly more potent than the corresponding C3/C2-dihydroxylated analogs, while the C3 and C2 epimeric analogs lose approximately an order of magnitude in potency against GP (entry 3 versus entry 5, entry 3 versus entry 6, Table 12.6). The C3-hydroxyl group is not required for potency against the enzyme, as the corresponding keto analogs (entries 9, 11, and 15, Table 12.6) show equipotent or better potency than the parent compound. Interesting and somewhat puzzling data is reported on acyl derivatives at C3; for example, capping the C3-hydroxyl with a simple acetyl group provides an inactive compound (entry 13, Table 12.6) while the 4-chlorocinnamyl derivative (entry 14, Table 12.6) is the most potent triterpene inhibitor reported to date. In general, derivatives of the C28-carboxylic acid tend to be less active than the corresponding parent compound

Table 12.6 Potencies of selected triterpene AMP site inhibitors.

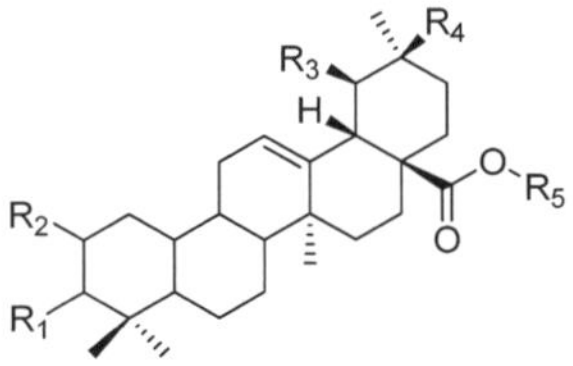

Entry	R_1	R_2	R_3	R_4	R_5	*RM GPa* IC_{50} *(µM)*	*Ref.*
1 (maslinic acid)	β-OH	α-OH	H	CH_3	H	28	56
2 (oleanolic acid)	β-OH	H	H	CH_3	H	14	52
3 (corosolic acid)	β-OH	α-OH	CH_3	H	H	20	53
4 (ursolic acid)	β-OH	H	CH_3	H	H	9	8
5	β-OH	β-OH	CH_3	H	H	115	53
6	α-OH	α-OH	CH_3	H	H	212	53
7	β-OH	α-OH	CH_3	H	CH_2Ph	41	53
8	β-OH	α-OH	CH_3	H	CH_2CO_2H	NI	53
9	=O	=O	CH_3	H	CH_2Ph	7.3	53
10	β-OH	α-OBz	CH_3	H	H	5.1	53
11	=O	H	H	CH_3	H	18	52
12	β-OH	H	H	CH_3	CH_2CO_2H	63	52
13	β-OAc	H	H	CH_3	H	NI	52
14	β-O(4-Cl)cinnamyl	H	H	CH_3	H	3.3	52
15	=O	CH(OH)	H	CH_3	CH_2Ph	6.3	52

(cf. entries 7 and 8, Table 12.6), and the SAR in this portion of the molecule is inconsistent (compare entries 8 and 12, Table 12.6), with a few compounds showing slightly improved potency while close structural analogs do not. Overall the improvements in potency over the naturally occurring compounds have been modest, and a systematic evaluation of the SAR of this class of compounds is no doubt hindered by the poor synthetic tractability of this class of compound.

More recently, glucoconjugates, nucleoside conjugates, and several dimers at the C3 and C28 positions of oleanolic acid have been prepared utilizing Click chemistry.[57–59] All of the C3 glucoconjugates and nucleoside conjugates and most of the C28 glucoconjugates and nucleoside conjugates profiled as considerably less active than oleanolic acid; however, glucoconjugate **22** (Figure 12.6) was approximately 10-fold more potent than oleanolic acid. The origin of this increase in potency is not clear; molecular docking of compound **22** using the crystal structure of asiatic and maslinic acids suggest the compound binds in the AMP site, but the orientation of oleanolic acid is reversed relative to that of asiatic and maslinic acid due to additional stabilizing interactions of the hydroxyls of the sugar moiety. Two oleanolic acid dimers also showed approximately 10-fold increase in potency at GP, the

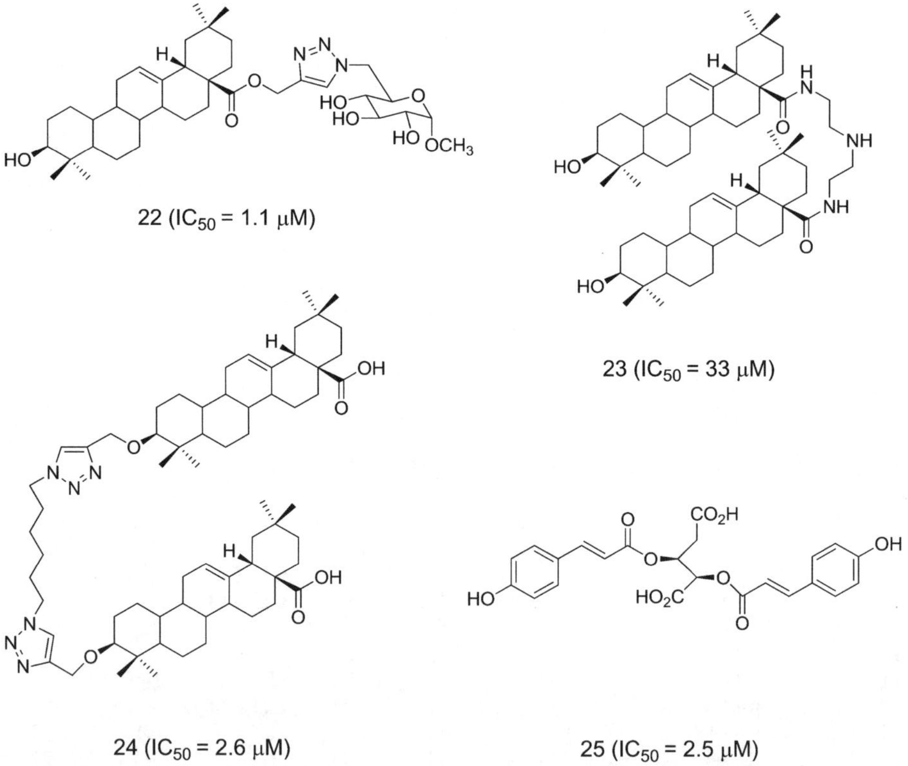

22 (IC$_{50}$ = 1.1 μM)

23 (IC$_{50}$ = 33 μM)

24 (IC$_{50}$ = 2.6 μM)

25 (IC$_{50}$ = 2.5 μM)

Figure 12.6   Selected AMP site GP inhibitors and reported IC$_{50}$ values.

C28 dimer **23** (Figure 12.6) and the C3 dimer **24** (Figure 12.6). While these compounds are presumed to bind at the AMP site this has not been confirmed. Finally, another natural product, FR258900 (**25**, Figure 12.6), isolated from a fungal broth, has been shown to inhibit human liver GP(a) with an IC_{50} = 2.5 µM and bind to the AMP site.[60] FR258900 effectively lowered glucose acutely in two animal models of type 2 diabetes,[61] though no further work on this compound has been reported.

12.3.6 Indole Site Inhibitors

In 1998, while conducting a high-throughput screen against human liver GP, scientists at Pfizer discovered a structurally novel series of GP inhibitors[62] that was later reported to bind at the interface of the GP homodimer.[63] This class of molecules contains an indole-2-carboxamide moiety, as exemplified by CP-91,149 (entry 1, Table 12.7) and thus the binding site was termed the indole site. CP-91,149 inhibits human liver GP(a) with an IC_{50} = 110 nM in the presence of 7.5 mM glucose, and was 5–10 times less potent in the absence of glucose. A cell-based assay measuring ^{14}C-labeled glycogen in SK-HEP-1 cells activated by addition of forskolin (which induces phosphorylation of liver GP(b) to GP(a)) was utilized to evaluate inhibitory activity in a human liver-derived tissue culture cell line. CP-91,149 blocked the forskolin effect with an IC_{50} = 1.5 µM in this assay. The SAR of this new series of compounds was explored in some detail,[64,65] and is summarized below and in Table 12.7.

Within the series where R_2 in Table 12.7 is 2-hydroxy-4-phenylbutyric amides, examination of the stereochemistry confirmed that the (3S,2R) configuration of CP-91,149 (entry 1) was preferred by 15–80 fold over the three other possible stereoisomers (data not shown). Replacement of the phenyl moiety with a cyclohexyl group (entry 2) led to considerable drop in potency against GP, so the phenyl group was held constant during further SAR examination. Structurally simplified phenylalanine-derived analogs were found to be equipotent to the 2-hydroxy-4-phenylbutyric amides (compare entries 1 and 4, Table 12.7). Examination of the 5-chloro substituent revealed this to be the optimal group, as replacement with other groups led to decreases in potency (entries 5–7, Table 12.7). The importance of the 1H-indole to enzyme binding was revealed by the corresponding N-methyl and benzofuran analogs (entries 8 and 9, Table 12.7) which were inactive. Further modifications to the 2-carboxamide side chain were made in an effort to reduce molecular weight and hydrophobicity and resulted in the discovery of glycine amide derivatives (entry 10, Table 12.7) that were equipotent *in vitro* to the previous series. Optimization of the groups on the tertiary amide provided a number of potent analogs with improved solubility and cell activity (cf. entry 11, Table 12.7). A number of compounds from this series were profiled *in vivo* for their ability to lower glucose in diabetic (ob/ob) mice in the fed state; interestingly this revealed a poor correlation between *in vivo* potency and potency in the enzyme and cellular assays amongst structurally related

Table 12.7 Potencies of selected indole site inhibitors.

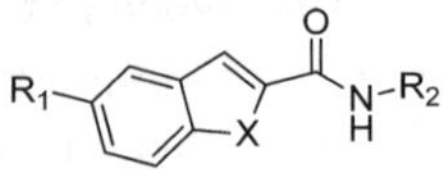

$$R_1\text{—}\underset{X}{\text{(benzo-fused ring)}}\text{—}C(=O)\text{—}NH\text{—}R_2$$

Entry	X	R_1	R_2	hLGP(a) IC$_{50}$ (nM)	Cell IC$_{50}$ (μM)	Ref.
1	NH	Cl	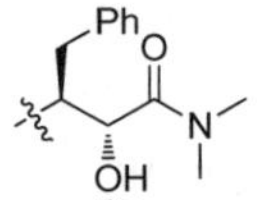 (Ph, OH; C(=O)N(Me)Me)	110	1.5	64
2	NH	Cl	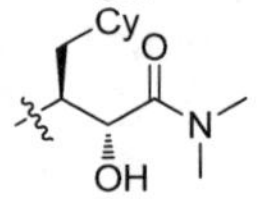 (Cy, OH; C(=O)N(Me)Me)	6500	NT	64
3	NH	Cl	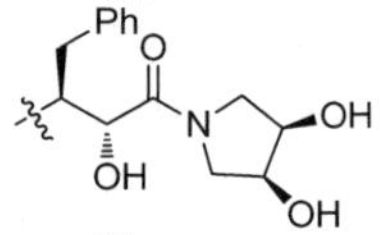 (Ph, OH; C(=O)–pyrrolidine-diol)	52	NR	23
4	NH	Cl	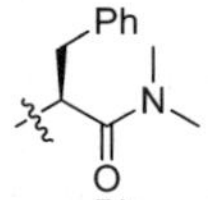 (Ph; C(=O)N(Me)Me)	82	1.9	64
5	NH	F	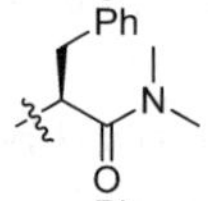 (Ph; C(=O)N(Me)Me)	430	2.3	64
6	NH	H	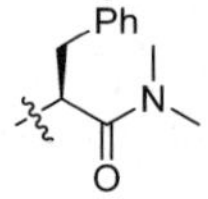 (Ph; C(=O)N(Me)Me)	400	7.8	64
7	NH	OMe	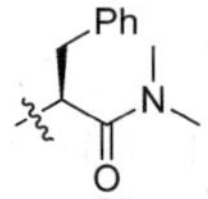 (Ph; C(=O)N(Me)Me)	4700	NT	64
8	O	Cl	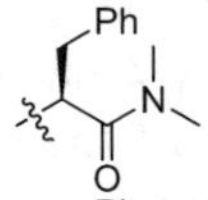 (Ph; C(=O)N(Me)Me)	>10000	NT	64
9	NMe	Cl	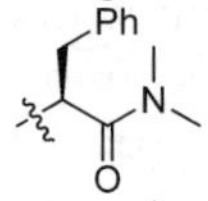 (Ph; C(=O)N(Me)Me)	>10000	NT	64
10	NH	Cl	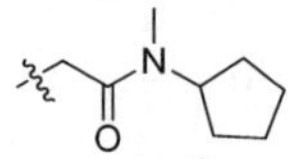 (C(=O)N(Me)–cyclopentyl)	55	1.7	65
11	NH	Cl	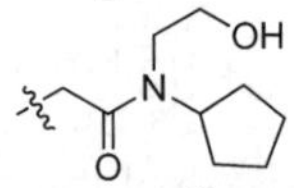 (C(=O)N(CH$_2$CH$_2$OH)–cyclopentyl)	57	0.14	65
12	NH	Cl	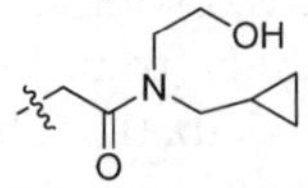 (C(=O)N(CH$_2$CH$_2$OH)–CH$_2$-cyclopropyl)	160	1.5	65

Table 12.7 (*Continued*)

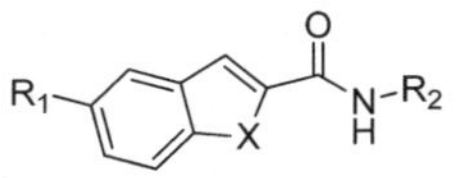

R₁—[indole/benzothiophene ring, X]—C(=O)—NH—R₂

Entry	X	R_1	R_2	$hLGP(a)$ IC_{50} (nM)	Cell IC_{50} (μM)	Ref.
13	NH	Cl	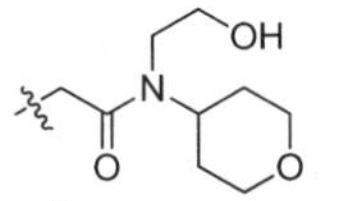 [structure]	250	1.4	65
14	NH	Cl	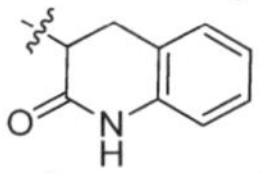 [structure]	25	1.8	66
15	NH	Cl	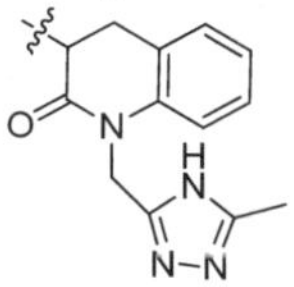 [structure]	140	1.2	66

compounds. For example, the cyclopentyl ethanolamine analog (entry 11, Table 12.7) showed excellent potency *in vivo*, while the cyclopropylmethyl derivate (entry 12, Table 12.7) was inactive, despite having good *in vitro* potency, while a 4-tetrahydropyanyl analog (entry 13, Table 12.7) that was considerably weaker in enzyme potency was equipotent to the cyclopentyl analog *in vivo*. Nevertheless, a number of compounds from these series displayed good rodent pharmacokinetics and glucose-lowering efficacy. Ultimately a compound from the 2-hydroxy-4-phenylbutyric amide series, CP-368,296 (ingliforib, entry 3, Table 12.7) was selected for advancement into human clinical trials (see below). The paucity of preclinical information published on CP-368,296 makes it difficult to analyze Pfizer's rationale for selection of this particular molecule for clinical development.

Subsequent to the publications from Pfizer scientists, other groups have reported structurally similar series of compounds which also bind at the indole inhibitor site. Merck scientists reported a series of indole analogs containing 3,4-dihydroquinolin-2-one derivatives at the C-2 carboxamide position (entry 14, Table 12.7).[66] Interestingly, the GP enzyme shows little stereochemical discrimination at the C3 quinolinone center, as both (*R*)- and (*S*)-isomers have similar potency (data not shown). A thorough examination of the substitution pattern on the indole ring again revealed that the 5-chloro substituent is preferred, while substitutions at the 4-, 6-, and 7-positions lead to decreases in potency (data not shown). Substitution on the lactam nitrogen were reasonably well tolerated (cf entry 15, Table 12.7) but did not improve activity over the parent compound. A number of these compounds were active in a glucagon-stimulated glycogenolysis assay in primary rat hepatocytes, but no *in vivo* data were reported.

Scientists at AstraZeneca have reported a series of thieno[2,3-*b*]pyrrole and thieno[3,2-*b*]pyrrole carboxamides that serve as indole bioisosteres.[67–69] The parent thienopyrrole carboxamides (entries 1 and 3, Table 12.8) are considerably less active than the 5-chloroindole analogs reported by Pfizer (entry 4, Table 12.7); however, potency can be restored by the introduction of a 2-chloro (entries 2 and 4, Table 12.8) or 2-bromo (data not shown) substituent. The 2,3-dichloro thieno[3,2-*b*]pyrrole analog (entry 5, Table 12.8) was the most potent reported. Similar to that reported for the 4-chloroindole series, methylation of either the pyrrole nitrogen or the secondary carboxamide led to a loss of activity. Subsequent exploration of the carboxamide substituent revealed the 3,4-dihydro-2-quinoline moiety (entry 2, Table 12.9) previously reported by Merck to provide optimal potency, along with novel 2-indanyl derivatives (entries 3 and 6, Table 12.9). Optimization of solubility, oral bioavailability, and free drug fraction led to the *N*-2,3-dihyroxypropyl-3,4-dihydro-2-quinoline compound GPi688 (entry 5, Table 12.9) which was

Table 12.8   Potencies of selected indole site inhibitors.

Entry	R_1	*hLGP(a) IC$_{50}$ (nM)*	*Hepatocyte Cell IC$_{50}$ (µM)*
1	thieno[2,3-*b*]pyrrole carboxamide	2310	NT
2	2-chloro-thieno[2,3-*b*]pyrrole carboxamide	17	1.4
3	thieno[3,2-*b*]pyrrole carboxamide	>10000	NT
4	2-chloro-thieno[3,2-*b*]pyrrole carboxamide	577	NT
5	2,3-dichloro-thieno[3,2-*b*]pyrrole carboxamide	5	1.0
6	2,3-dibromo-*N*-methyl-thieno[3,2-*b*]pyrrole carboxamide	>10000	NT
7	2,3-dibromo-thieno[3,2-*b*]pyrrole *N*-methyl carboxamide	>10000	NT

evaluated for *in vivo* efficacy in a glucagon challenge model and via an oral glucose tolerance test in obese Zucker rats.[26] Oral administration of GPi688 dose-dependently inhibited glucagon-induced hyperglycemia in Zucker rats with an ED_{50} of approximately 8 mg/kg, with the top dose of 50 mg/kg inhibiting 100% of the glucagon response after 45 minutes and showing ca. 50% inhibition out to 24 h. However, GPi688 showed only minimal efficacy (7% reduction in $AUC_{glucose}$) in an OGTT in obese Zucker rats despite increasing hepatic glycogen content by 50% at $t = 4$ hours post-dose.

The 4-chloroindole carboxamide also served as a template for scientists at Astellas Pharma, who have reported potent inhibitors of GP by making additional changes at the carboxamide substituents.[70,71] The simple 4-hydroxymethylphenyl and 4-hydroxyethylphenyl carboxamides (entries 1 and 2, Table 12.10) are low micromolar inhibitors of human liver GP(a). Extending the carbon chain decreases potency (data not shown), while the 1,2-diol shows

Table 12.9   Potencies of selected indole site inhibitors.

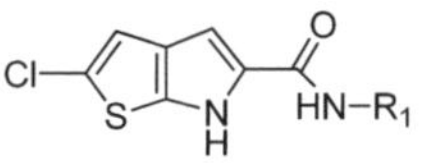

Entry	R_1	$hLGP$(a) IC_{50} (nM)	Hepatocyte Cell IC_{50} (μM)	Ref.
1	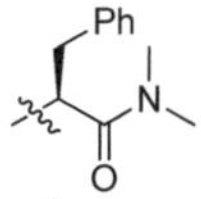	17	1.4	69
2	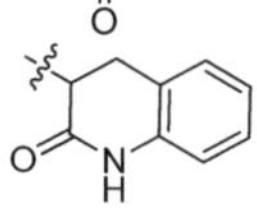 (+/−) (+/−)	41	0.69	69
3	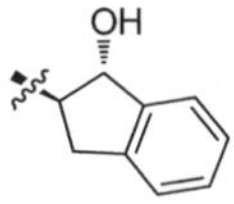 (+/−) (+/−)	319	3.6	69
4	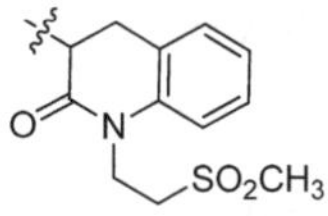 (+/−) SO₂CH₃ (+/−)	63	2.7	68
5	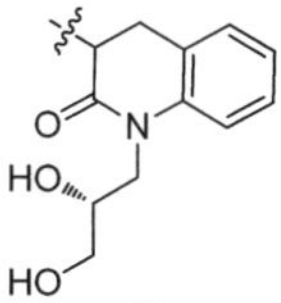	29	1.1	68
6	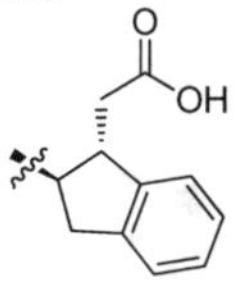	40	NR	67

a modest improvement (entry 3, Table 12.10). In general, additional substitution in the 2- and 3-positions of the phenyl ring slightly decreased potency – with the exception of fluorine, which increased potency (cf. entry 4, Table 12.10). Conversion of the benzene ring into a 2-pyridyl derivative showed a further increase in enzyme potency to 250 nM (entry 5, Table 12.10). This compound was also quite potent in inhibiting glucagon-induced glucose

Table 12.10 Potencies of selected indole site inhibitors.

Entry	R_1	hLGP(a) IC_{50} (nM)	Cell IC_{50} (μM)	Ref.
1	*(4-hydroxymethylphenyl)*	1600	NA	70
2	*(4-(2-hydroxyethyl)phenyl)*	1200	NA	71
3	*(4-(1,2-dihydroxyethyl)phenyl)*	900	NA	71
4	*(4-fluoro-3-(1,2-dihydroxyethyl)phenyl)*	420	NA	71
5	*(pyridyl-1,2-dihydroxyethyl)*	250	0.62	71
6	*(1-hydroxy-indanyl)*	440	NA	70
7	*(1-hydroxy-tetralinyl)*	320	NA	70
8	*(2,2-difluoro-1-hydroxy-tetralinyl)*	68	NA	70
9	*(fluoro-difluoro-hydroxy-tetralinyl)*	36	NA	70
10	*(fluoro-difluoro-hydroxy-tetralinyl)*	20	690	70

output in cultured human primary hepatocytes (IC_{50} = 620 nM) and acutely lowered glucose (26% reduction in AUC) when dosed orally at 50 mg/kg to db/db mice. Additional investigation revealed that conformational constraint of the 4-hydroxy methyl via 5- and 6-membered rings also provided potent GP inhibitors (entries 6 and 7, Table 12.10). Incorporation of fluorine atoms adjacent to the hydroxyl group provided an additional 5-fold increase in potency (entry 8, Table 12.10), and introduction of fluorine atoms into the phenyl ring provided a further potency increase (entry 9, Table 12.10). The α-enantiomer of this tetra-fluorohydroxy analog (entry 10, Table 12.10) was potent *in vitro* (IC_{50} = 20 nM against hLGP(a), IC_{50} = 690 nM in cultured hepatocytes), showed excellent oral pharmacokinetic properties in rats (%F = 100, $t_{1/2}$ = 12 h) and showed excellent acute *in vivo* glucose-lowering ability in db/db mice (26% reduction in glucose at 10 mg/kg dose).

12.3.7 Inhibitors of the GP-G_L Interaction

While no reports describing structure–activity relationships of small molecules that bind to the GP-G_L site have appeared in peer-reviewed journals at the time of this review, scientists at Boehringer Ingelheim have filed several patent applications describing a series of arylsulfonyl-based compounds that inhibit the binding of G_L to GP(a), exemplified by compound **26** (Figure 12.7).[72–75] The mechanism of action of one of these inhibitors has been characterized in comparison to known indole site and AMP site inhibitors.[16] The GP-G_L site inhibitor suppressed the interaction between a labeled G_L peptide and GP(a) with an IC_{50} = 150 nM but importantly did not inhibit the enzymatic activity of GP. In primary rat hepatocytes at elevated glucose levels (10 mM glucose), the GP-G_L site inhibitor increased glycogen synthesis and glycogen accumulation similarly to the indole site and AMP site inhibitors. However, prolonged treatment with the GP-G_L site inhibitor did not lead to a further increase in glycogen levels, unlike that seen with the indole site and AMP site inhibitors. In addition, at low glucose levels (2.5 mM glucose), only the indole site and AMP site inhibitors increased glycogen amounts, while the GP-G_L site inhibitor allowed the mobilization of glycogen because it did not inhibit the enzymatic activity of GP. The pharmacological profile of a GP-G_L site inhibitor may offer advantages over that of direct GP inhibitors with respect to potential side effects (see below), and additional data on the *in vivo* and clinical effects of GP-G_L site inhibitors are anxiously awaited.

12.3.8 Inhibitors with an Unknown Binding Mode

A number of GP inhibitors have been reported in which the binding site has not been confirmed by X-ray crystallography. Many of these compounds are weak inhibitors of GP, inactive in cellular assays, and their SAR is not well characterized, making them unlikely to hold any real therapeutic potential. The most potent of these inhibitors will be reviewed briefly here.

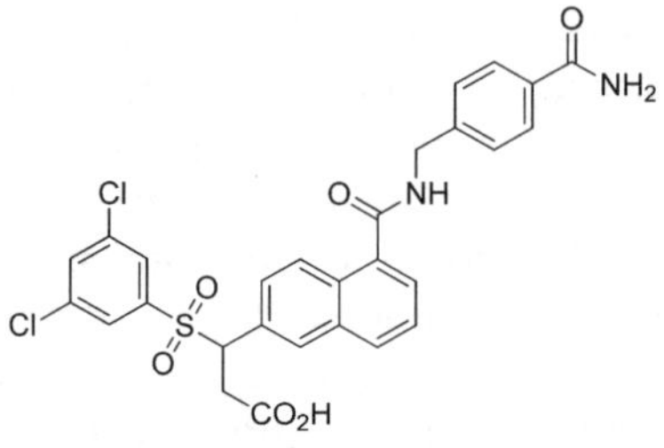

27 (IC$_{50}$ = 0.11 μM)

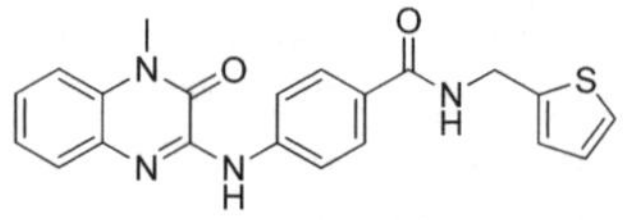

26

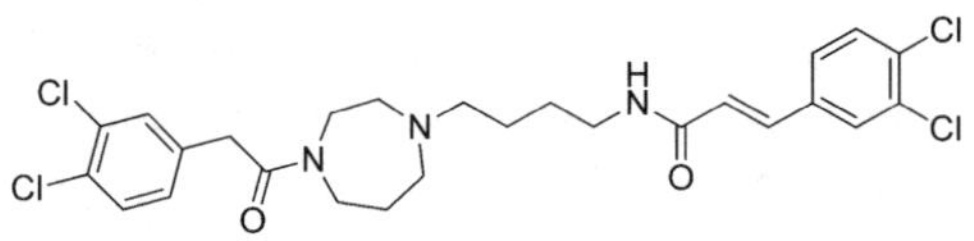

28 (IC$_{50}$ = 0.17 μM)

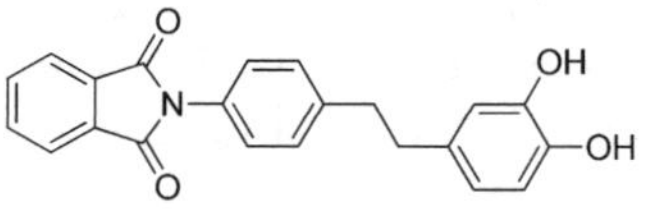

29 (IC$_{50}$ = 2.7 μM)

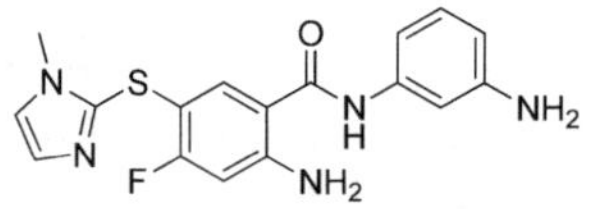

30 (IC$_{50}$ = 2.7 μM)

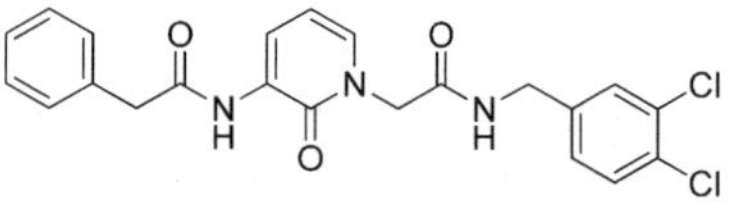

31 (IC$_{50}$ = 6.3 μM)

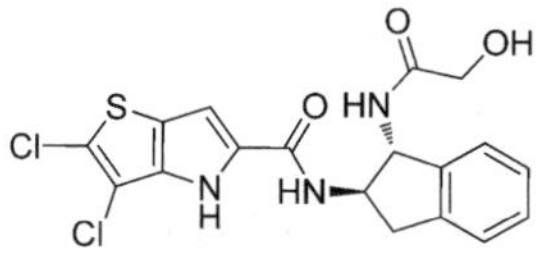

32 (GPi921)

Figure 12.7 Selected GP inhibitors with an unknown binding mode and reported IC$_{50}$ values.

Scientists from Johnson & Johnson[76] have described a series of anilinoquinoxalinone GP inhibitors exemplified by compound **27** (Figure 12.7). Structural changes in the aniline moiety had the most impact on enzyme potency in this series, while changes to the quinoxalinone portion of the template had little effect. Compound **27** inhibited rabbit muscle GP(*a*) with an IC$_{50}$ = 110 nM. Unfortunately, none of these compounds were active in cellular assays or in diabetic animal models, and the authors suggest that poor physicochemical properties of the series are likely to be the source of the problem.

A series of propenamide GP inhibitors exemplified by compound **28** (Figure 12.7) were reported by scientists at GlaxoSmithKline.[77] This series was derived from a high-throughput screen. Most changes to the aryl substituents led to inactive compounds, while some degree of tolerance to changes in the central portion of the molecule was observed. Compound **28** was the most potent inhibitor in the series with an IC_{50} = 170 nM against human liver GP(*a*) in the presence of 10 mM glucose. Unfortunately, none of these compounds displayed cellular activity, and this series was abandoned in favor of the aforementioned anthranilic acid series (see Section 12.3.5).

A series of phthalimide-based compounds were recently described as GP inhibitors.[78] The series of compounds originated from work on thalidomide analogs that possessed LXR and α-glucosidase activity. The structure–activity relationships within this series are generally flat and unremarkable, and compound **29** (Figure 12.7), with an IC_{50} = 2.7 μM against GP(*a*) (species not disclosed) is the most potent compound reported.

A series of benzamides that display both modest inhibition of GP and simultaneous activation of glucokinase (GK) have also been reported. The series was derived from a known potent GK activator previously described by scientists from Banyu that also displayed weak GP inhbition.[79,80] Exploration of this template revealed that potency of GP inhibition could be improved by incorporation of a *m*-substituted phenyl group off the amide, with an amino substituent appearing to be optimal.[81] Changes to the heteroaryl group attached to the thiol had little effect on GP inhibition but clearly modulated the ability of these compounds to activate GK.[82] Within the series of analogs, compound **30** (Figure 12.7) was the most potent inhibitor of rabbit muscle GP(*a*), with an IC_{50} = 2.7 μM; this compound also caused a 2-fold activation of GK with an EC_{50} = 12.5 μM. Molecular docking simulations were utilized to probe possible binding sites and suggest that these compounds bind at the indole binding site, but this has not been confirmed by crystallographic studies. Unfortunately no cell-based or *in vivo* activity of the compounds was reported. There is data to suggest that combined GP inhibition and GK activation is an effective strategy for controlling hepatic glycogen metabolism. Graded over-expression of glucokinase in rat hepatocytes caused a progressive increase in glycogen synthesis, and pharmacologic blockade of GP by the indole GP inhibitor CP-91149 further increased glycogen synthesis to a greater extent than the sum of either treatment alone.[83] This result is hypothesized to be due in part to the synergy of G-6-P and the indole GP inhibitor to promote inactivation of GP and in part to the complementary roles of elevation of G-6-P and depletion of GP(*a*) in activating glycogen synthase activity. Whether this synergism is evident in an *in vivo* setting remains to be determined.

Finally, a series of 2-pyridone inhibitors exemplified by compound **31** (Figure 12.7) have been discovered using a chemogenomics approach targeting the GP-G_L interface.[84] The 2-pyridone scaffold was selected based on its ability to mimic the Leu-Gly residues contained within the conserved C-terminus of the PP1-G_L protein that binds to GP(*a*). A series of analogs were prepared and characterized based on their ability to inhibit rabbit muscle GP(*a*). Generally weak enzyme inhibition was observed; however, inhibition of

GP-G_L binding was not measured, and inhibition of enzyme activity is arguably not the best measure of activity for compounds designed to target the GP-G_L site (see above). No cellular or *in vivo* data have been reported for this series.

12.4 GP Inhibitors as Therapeutic Agents for Type 2 Diabetes

12.4.1 Clinical Results with GP Inhibitors

Despite the wide range of small molecule GP inhibitors that have been reported in the literature, very few GP inhibitors have been administered in humans, and there is a paucity of literature that has been published on the clinical effects of those compounds.

Pfizer scientists first reported clinical effects in 2001 on their first-generation indole site GP inhibitor, CP-316,819, on glycemia in response to a glucagon challenge in healthy individuals.[85] Overnight-fasted subjects ($n = 4$) were administered 6 mg/kg CP-316,819 or vehicle followed by an i.v. bolus of 0.5 mg glucagon 1 h later. Administration of CP-316,819 induced a significant reduction (32–86%) in peak glucose excursion compared to placebo. Importantly, CP-316,819 did not affect baseline glycemia relative to placebo. Ingliforib (CP-368,296, entry 3, Table 12.7) subsequently replaced CP-316,819 in Pfizer's clinical development program for reasons that have not been elucidated. An overview of the results with ingliforib in three separate clinical trials in Type 2 diabetics has been reported in oral sessions, but no details have appeared to date in the peer-reviewed literature.[86] In a 15-day multiple-dose escalation study in type 2 diabetics, ingliforib was dosed orally at 50, 200, and 400 mg once daily on days 1–4, twice daily on days 5–9, and three times daily on days 10–15. Fasting plasma glucose (FPG) levels in the placebo-treated group increased by an average of 27 mg/dL over the course of the study; in contrast, ingliforib caused a dose-dependent decrease in FPG levels, with the 400 mg dose lowering glucose an average of 37 mg/dL relative to baseline values. However, in a subsequent 28-day phase II trial in type 2 diabetics, ingliforib at doses of 75 and 200 mg twice daily showed no significant difference in FPG levels compared to the placebo group. Ingliforib was also studied in a 12-week study in type 2 diabetics refractory to sulfonylurea treatment. Doses of 50, 100, and 200 mg were given twice daily and both FPG and HbA$_{1c}$ were measured. Ingliforib failed to show any significant difference from placebo in either FPG or HbA$_{1c}$ levels, though there was a significant and unexpected decrease in FPG in the placebo group. Poor pharmacokinetic properties in humans are unlikely to explain the lack of chronic efficacy in the latter study with ingliforib, as plasma drug levels at the end of the study were reported to be at the expected level.

No additional information from clinical studies on glycogen phosphorylase inhibitors in type 2 diabetics has appeared in peer-reviewed publications, though several pharmaceutical companies have placed GP inhibitors into clinical development. Sanofi-Aventis conducted a phase I trial of the GP inhibitor AVE56588 (**19**, Figure 12.5) but development was discontinued in 2005 for reasons not disclosed. Prosidion (now Astellas Pharma) successfully completed a phase I clinical trial in 80 healthy volunteers of the GP inhibitor PSN-357 in 2006 and was reported to be in phase II clinical development, but no further information on the compound has been reported. The chemical structure and accompanying pharmacological data on PSN-357 have not been reported in the literature. GlaxoSmithKline successfully completed a phase I clinical trial of their GP inhibitor GSK1362885 and have reported interim results.[87] In this study, GSK1362885 was safe and generally well-tolerated across all groups. A clear dose-dependent suppression of glucose profiles was observed in response to a glucagon challenge, with nearly complete suppression of the baseline-corrected glucose AUC(0–20 min) at doses of 100 and 300 mg compared to placebo.

12.4.2 Challenges with GP Inhibitors

Based on the mechanistic rationale and preclinical evidence summarized above, inhibition of GP is a promising strategy for the attenuation of hyperglycemia associated with type 2 diabetes; however, as with any novel pharmacotherapy there are a number of development challenges and potential concerns regarding the safety, tolerability, and efficacy of pharmacologic inhibition of GP in humans.

One obvious potential concern based on the GP inhibitor mechanism of glucose lowering is increased risk of hypoglycemia. However, most GP inhibitors show an *in vitro* glucose dependency on GP inhibition, being more potent in the presence of high glucose concentrations. This may be a "self-regulating" mechanism in which hyperglycemia enhances liver phosphorylase inhibition whereas a decrease in glucose concentrations towards normoglycemic levels attenuates liver phosphorylase inhibition, thus avoiding rebound hypoglycemia. Hypoglycemic episodes have not been reported after administration of GP inhibitors in preclinical animal models, and non-diabetic mice remain normoglycemic upon treatment with potent GP inhibitors at doses up to 100 mg/kg.[5] However, the lack of ability to mobilize glycogen due to inhibition of GP mimics certain aspects of fasting-induced starvation, where glycogen stores are depleted and thus glycogenolysis is nearly absent, and mild hypoglycemia is commonly observed in the setting of starvation in humans. In addition, mild fasting hypoglycemia is observed in people with glycogen storage disease (GSD) type VI, which is characterized by defects in the liver GP enzyme (see below). The ability of GP inhibitors to induce hypoglycemia will need to be carefully evaluated in the clinical setting, particularly with respect to their potential to exacerbate hypoglycemia when dosed in combination with insulin or insulin secretagogues.

A second potential safety concern is the effect of GP inhibition on muscle functional capacity. Skeletal muscle glycogen stores represent an essential energy source during exercise, and depletion of muscle glycogen stores promotes muscle fatigue.[88] Since many type 2 diabetics are overweight and advised to increase their level of exercise as part of a treatment regimen, adverse effects on skeletal muscle function during exercise would severely limit the utility of a GP inhibitor.

Some information on the effect of blocking muscle GP activity can be gleaned by studying individuals with a genetic deficiency in their muscle GP enzyme. These individuals have an inactivating mutation that results in a complete absence of functional muscle GP activity and a condition called GSD type V or McArdle's disease. GSD type V is characterized by myalgia, severe cramps, muscle stiffness, and muscle weakness coupled with early fatigue after exercise.[89] The severity and age of onset of these symptoms varies widely. The defect is most notable during either high intensity exercise of short duration or during less intense but sustained activity. The symptoms are often mild enough during childhood that clinical diagnosis does not occur until adulthood. Liver and brain phosphorylase activity is normal. Dietary management, weight control, and exercise maintenance are utilized in McArdle's patients to manage this disease.

Effects of pharmacologic GP inhibition on skeletal muscle function after both acute and prolonged muscle contraction in a perfused rat hindlimb model have been reported.[90,91] The potent indole-site GP inhibitor CP-316,819 did not affect muscle force production or muscle energy metabolism during 20 seconds of maximal contraction or 10 minutes of sub-maximal contraction in an *in situ* pump-perfused rat gastrocnemius-plantaris-soleus muscle model.[90] CP-316,819 was able to reduce GP activation at rest and during contraction in this model. However, inhibition of GP with CP-316,819 during prolonged (60 min) muscle contraction resulted in a 35% greater muscle fatigue than the control group.[91] This was accompanied by lower muscle lactate efflux and glucose yet a higher rate of oxygen consumption indicative of impaired carbohydrate utilization which suggests pharmacologic GP inhibition could negatively affect endurance capacity. As mentioned above, the sequence homology between the liver and muscle GP isoforms is very high, and the GP inhibitors reported to date do not demonstrate appreciable selectivity for the liver versus muscle isoform. One potential strategy to mitigate this risk would be to develop compounds, prodrugs, or other conjugates that selectively distribute into liver tissue over skeletal muscle tissue. Such strategies have been employed with other anti-diabetic targets.[92–94]

Perhaps the most significant safety concern with GP inhibitors is the potential for excessive glycogen accumulation in the liver. As in muscle, there is a GSD associated with genetic defects in the GP liver isoform: GSD type VI or Hers' disease, in which hepatomegaly and growth retardation are the main clinical symptoms, with hypertriglyceridemia, hypercholesterolemia, and elevated liver enzymes also present in some cases.[95] GSD type VI is rare and fairly benign, with clinical symptoms mainly present during childhood. Several mutations in the liver GP gene have been linked to GSD type V. Enzymological studies of patients with these mutations indicates that their liver GP still retains residual enzyme activity (ca. 25% of normal) which may account for the relatively mild phenotype of this disease.

There is one published report on the effects of prolonged administration of a GP inhibitor on the biochemical and histological features of the liver. Floettmann *et al.*[96] dosed ZDF rats with the indole site inhibitor GPi921 (**32**, Figure 12.7) at a dose of 53 mg/kg/day for 28 days and assessed both glycemic control and liver function and histology. Pharmacokinetic analysis revealed that plasma drug levels reached ca. 25-fold above the IC_{50} value of GPi921 at C_{max} and were at least 15-fold above the IC_{50} throughout the exposure period, and liver levels were approximately twice the levels measured in plasma, indicating that robust and sustained levels of GP inhibition occurred throughout the study. As expected, a reduction in both plasma glucose and glycated hemoglobin levels was observed in GPi921-treated animals relative to vehicle controls. A significant increase ($>2\times$) in liver glycogen levels was observed after 10 days of treatment and further increased after 28 days. Liver weights increased approximately 30% relative to vehicle controls, and histological analysis showed moderate to severe levels of hepatocellular vacuolation and lipid accumulation. In addition, increased evidence of inflammation, fibrosis, and necrosis was observed, along with an increase in ALT and AST levels. These observations are similar to those features seen with human GSDs. The authors reported that similar observations were made following administration of a chemically different class of GP inhibitor (data not shown), which suggests the findings are not due to off-target activities of GPi921.

These observations allow some hypotheses to be drawn about the metabolic adaptation to chronic dosing of a GP inhibitor, and the fate of hepatic glucose once it can no longer be converted to glycogen. There are three options for disposing of the excess glucose under saturating glycogen storage conditions, with the assumption that GP inhibition does not cause any insulin-sensitizing effects in extraheptaic tissues which lead to increased glucose uptake. First, glucose could be released into the plasma via activation of glucose-6-phosphatase; if this situation occurs one might predict a reduction in the durability of efficacy of a GP inhibitor. Interestingly a lack of durability of glucose lowering is precisely what was reported clinically with ingliforib (see above). However, other factors such as incomplete enzyme inhibition at the clinical doses could also explain this effect, and more data are needed to determine the cause of ingliforib's lack of efficacy. Second, glucose could be converted to lactate via glycolysis. This is unlikely a major pathway upon GP inhibition, as lactic acidosis is not a common feature of GSD type VI. However, this possibility should not be discounted and careful monitoring of acidosis would need to be done in the clinical setting, particularly if a GP inhibitor were used in combination with metformin. The third possibility is that glucose could be converted into triglycerides via *de novo* lipogenesis. This appears to be the result with chronic dosing of GPi921 in the ZDF rat. Whether or not activation of the same *de novo* lipogenesis pathway would be as prominent in type 2 diabetics with pharmacologic GP inhibition is unclear and awaits further clinical data. However, the data suggests that both the drug exposure profile and dosing regimens with GP inhibitors should be carefully considered to avoid a sustained saturating level of hepatic glycogen.

12.4.3 Opportunities for GP Inhibitors

Assuming that GP inhibitors are shown to be safe and well-tolerated clinically, what are the opportunities for developing these agents as value-added medicines for the treatment of type 2 diabetes given the data thus far? To compete favorably in a broad market with the current armament of anti-diabetic medicines, which includes metformin, DPP-4 inhibitors, GLP-1 agonists, PPARγ agonists, insulin, sulfonylureas, etc. a GP inhibitor would likely need to demonstrate either superior glycemic efficacy to current therapies or demonstrate equivalent glycemic efficacy plus a beneficial effect in at least one other diabetic co-morbid condition (e.g. weight loss, reduction in cardiovascular disease, decrease in LDL cholesterol, etc.).

Though some classes of anti-diabetic agents have shown modest differences in overall glycemic efficacy as measured by HbA1c, the March 2011 U.S. Department of Health and Human Services *Comparative Effectiveness and Safety of Oral Diabetes Medications for Adults with Type 2 Diabetes* (www.effectivehealthcare.ahrq.gov/reports/final.cfm) concludes that "…most diabetes medications (metformin, thiazolidinediones, sulfonylureas, and repaglinide) reduced HbA1c to a similar degree, by about 1 absolute percentage point when compared with baseline values, after 3 or more months of treatment", with the caveat that GLP-1 agonists had not been on the market long enough to produce a full comparison for that drug class. Thus while it is reasonable to expect that a GP inhibitor can produce efficacy similar to current "gold standard" therapy such as metformin, the likelihood of producing clinically superior glycemic control via this mechanism is probably low.

Much more research is needed to understand the effects of GP inhibitors on the various co-morbidities associated with type 2 diabetes. Preclinical data suggests GP inhibitors are "weight-neutral" in animal models of type 2 diabetes, and no clinical data have been published on the effects of GP inhibition on body weight in diabetic patients. No reports of GP inhibitors having beneficial effects on cholesterol levels in animal models of diabetes have appeared; as mentioned above, the one observation thus far reported is increased hepatic triglyceride levels with chronic GP inhibitor administration in rats, which would be considered detrimental rather than beneficial. GP inhibitors also appear to have no significant effects on blood pressure, whole body insulin resistance, and inflammation. The effects of GP inhibition on vascular tone, kidney function, beta cell function, neurobiology, etc. in the diabetic setting are simply not known.

Despite these challenges, robust opportunities for GP inhibitors may exist. Inhibition of GP offers a novel mechanism of glucose lowering that could be useful in combination with the currently available agents to provide additive or possibly synergistic results. Combination therapy has clearly demonstrated advantages over monotherapy with respect to glycemic control in type 2 diabetics. The hepatic-based mechanism of GP inhibition may prove very useful in combination with other anti-diabetic agents, most of which do not have a substantial hepatic component to their glucose-lowering mechanism of action. Unfortunately no clinical or preclinical studies have been published in which a GP inhibitor has been used in combination with another anti-diabetic agent. In addition, GP inhibitors may be particularly effective in certain subpopulations of diabetics. One intriguing possibility is type 2 diabetics with very high nocturnal glucose excursions that are poorly controlled, resulting in high morning fasting blood glucose. In this setting hepatic glucose output is the major determinant of elevated glucose levels, and a short-acting GP inhibitor given before bedtime may be particularly effective in controlling overnight plasma glucose levels without sustained accumulation of hepatic glycogen. No epidemiological studies have been done to evaluate the size of this subpopulation of diabetics. Clinical data from these types of combination studies and subpopulation studies will be important in determining the most appropriate clinical setting in which to use a GP inhibitor.

12.5   Conclusions

Inhibition of glycogen phosphorylase represents a novel mechanism for modulation of hepatic glucose production. A number of labs have produced potent, structurally novel GP inhibitors, and X-ray crystallography has established that these inhibitors interact with several different allosteric binding sites on the protein. Meticulous application of medicinal chemistry has resulted in the development of a number of compounds with good drug-like properties that lower plasma glucose in diabetic rodent models. Despite these advances, the future of GP inhibitors as a medicine for type 2 diabetes remains unclear. There are safety and tolerability issues related to chronic GP inhibition that remain to be discharged. The clinical data to date with GP inhibitors are sparse and not particularly encouraging, and most large pharmaceutical companies are no longer listing GP inhibitors in their development pipelines. Furthermore, the development of any new anti-diabetic agent is an extremely daunting task given the current regulatory and medicine reimbursement climate in many countries. Inhibition of glycogen phosphor-ylase may yet represent a safe, effective, and novel therapy for treatment of type 2 diabetes; it will rest on the willingness of a pharmaceutical company to accept the risks and cost of a full clinical development program in the face of substantial challenges to obtain the answer.

Acknowledgment

The author wishes to thank Dr Stephen Thomson for his critical review of the manuscript.

References

1. J. P. Boyle, T. J. Thompson, E. W. Gregg, L. E. Barker and D. F. Williamson, *Popul. Health Metr.*, 2010, **8**, 29.
2. P. J. Lefebvre and A. J. Scheen, *Eur. J. Clin. Invest.*, 1999, **29**, 1.
3. M. Roden and E. Bernroider, *Best Pract. Res., Clin. Endocrinol. Metab.*, 2003, **17**, 365.
4. J. Radziuk and S. Pye, *Diabetes. Metab. Res. Rev.*, 2001, **17**, 250.
5. J. L. Treadway, P. Mendys and D. J. Hoover, *Expert Opin. Invest. Drugs*, 2001, **10**, 439.
6. B. R. Henke and S. M. Sparks, *Mini-Rev. Med. Chem.*, 2006, **6**, 845.
7. J. T. Link, *Curr. Opin. Invest. Drugs*, 2003, **4**, 421.
8. L. Somsak, K. Czifrak, M. Toth, E. Bokor, E. D. Chrysina, K. M. Alexacou, J. M. Hayes, C. Tiraidis, E. Lazoura, D. D. Leonidas, S. E. Zographos and N. G. Oikonomakos, *Curr. Med. Chem.*, 2008, **15**, 2933.
9. W. A. Loughlin, *Mini-Rev. Med. Chem.*, 2010, **10**, 1139.
10. C. B. Newgard, P. K. Hwang and R. J. Fletterick, *Crit. Rev. Biochem. Mol. Biol.*, 1989, **24**, 69.
11. A. Pautsch, N. Stadler, O. Wissdorf, E. Langkopf, W. Moreth and R. Streicher, *J. Biol. Chem.*, 2008, **283**, 8913.
12. C. G. Armstrong, M. J. Doherty and P. T. W. Cohen, *Biochem. J.*, 1998, **336**, 699.
13. I. R. Kelsall, D. Rosenzweig and P. T. W. Cohen, *Cell. Signalling*, 2009, **21**, 1123.
14. N. G. Oikonomakos, *Curr. Protein Pept. Sci.*, 2002, **3**, 561.
15. N. G. Oikonomakos and L. Somsak, *Curr. Opin. Invest. Drugs*, 2008, **9**, 379.
16. D. Zibrova, R. Grempler, R. Streicher and S. G. Kauschke, *Biochem. J.*, 2008, **412**, 359.
17. I. R. Kelsall, S. Munro, I. Hallyburton, J. L. Treadway and P. T. W. Cohen, *FEBS Lett.*, 2007, **581**, 4749.
18. K. Fosgerau, N. Westergaard, B. Quistorff, N. Grunnet, M. Kristiansen and K. Lundgren, *Arch. Biochem. Biophys.*, 2000, **380**, 274.
19. A. K. Ogawa, C. A. Willoughby, R. Bergeron, K. P. Ellsworth, W. M. Geissler, R. W. Myers, J. Yao, G. Harris and K. T. Chapman, *Bioorg. Med. Chem. Lett.*, 2003, **13**, 3405.
20. B. Andersen and N. Westergaard, *Biochem. J.*, 2002, **367**, 443.
21. N. Ercan-Fang, M. C. Gannon, V. L. Rath, J. L. Treadway, M. R. Taylor and F. Q. Nuttall, *Am. J. Physiol.*, 2002, **283**, E29–E37.

22. N. Ercan-Fang, M. R. Taylor, J. L. Treadway, C. B. Levy, P. E. Genereux, E. M. Gibbs, V. L. Rath, Y. Kwon, M. C. Gannon and F. Q. Nuttall, *Am. J. Physiol.*, 2005, **289**, E366–E372.

23. L. J. Yu, Y. Chen, J. L. Treadway, R. K. McPherson, S. C. McCoid, E. M. Gibbs and D. J. Hoover, *J. Pharmacol. Exp. Ther.*, 2006, **317**, 1230.

24. S. J. G. Loxham, J. Teague, S. M. Poucher, De School, A. V. Turnbull and F. Carey, *J. Pharmacol. Toxicol. Methods*, 2007, **55**, 71.

25. S. A. Thomson, P. Banker, D. M. Bickett, J. A. Boucheron, H. L. Carter, D. C. Clancy, J. P. Cooper, S. H. Dickerson, D. M. Garrido, R. T. Nolte, A. J. Peat, L. R. Sheckler, S. M. Sparks, F. X. Tavares, L. Wang, T. Y. Wang and J. E. Weiel, *Bioorg. Med. Chem. Lett.*, 2009, **19**, 1177.

26. S. M. Poucher, S. Freeman, S. J. G. Loxham, G. Convey, J. B. Bartlett, De School, J. Teague, M. Walker, A. V. Turnbull, A. D. Charles, F. Carey and S. Berg, *Br. J. Pharmacol.*, 2007, **152**, 1239.

27. E. D. Chrysina, *Mini-Rev. Med. Chem.*, 2010, **10**, 1093.

28. J. P. Praly and S. Vidal, *Mini-Rev. Med. Chem.*, 2010, **10**, 1102.

29. T. Gimisis, *Mini-Rev. Med. Chem.*, 2010, **10**, 1127.

30. Z. Gyorgydeak, Z. Hadady, N. Felfoldi, A. Krakomperger, V. Nagy, M. Toth, A. Brunyanszki, T. Docsa, P. Gergely and L. Somsak, *Bioorg. Med. Chem.*, 2004, **12**, 4861.

31. K. Czifrak, Z. Hadady, T. Docsa, P. Gergely, J. Schmidt, L. Wessjohann and L. Somsak, *Carbohydr. Res.*, 2006, **341**, 947.

32. Z. Czako, L. Juhasz, A. Kenez, K. Czifrak, L. Somsak, T. Docsa, P. Gergely and S. Antus, *Bioorg. Med. Chem.*, 2009, **17**, 6738.

33. M. Gregoriou, M. E. M. Noble, K. A. Watson, E. F. Garman, T. M. Krulle, C. De La Fuente, G. W. J. Fleet, N. G. Oikonomakos and L. N. Johnson, *Protein Sci.*, 1998, **7**, 915.

34. M. Benltifa, J. M. Hayes, S. Vidal, D. Gueyrard, P. G. Goekjian, J. P. Praly, G. Kizilis, C. Tiraidis, K. M. Alexacou, E. D. Chrysina, S. E. Zographos, D. D. Leonidas, G. Archontis and N. G. Oikonomakos, *Bioorg. Med. Chem.*, 2009, **17**, 7368.

35. V. Nagy, M. Benltifa, S. Vidal, E. Berzsenyi, C. Teilhet, K. Czifrak, G. Batta, T. Docsa, P. Gergely, L. Somsak and J. P. Praly, *Bioorg. Med. Chem.*, 2009, **17**, 5696.

36. E. Bokor, T. Docsa, P. Gergely and L. Somsak, *Bioorg. Med. Chem.*, 2010, **18**, 1171.

37. M. Toth, S. Kun, E. Bokor, M. Benltifa, G. Tallec, S. Vidal, T. Docsa, P. Gergely, L. Somsak and J. P. Praly, *Bioorg. Med. Chem.*, 2009, **17**, 4773.

38. N. G. Oikonomakos, C. Tiraidis, D. D. Leonidas, S. E. Zographos, M. Kristiansen, C. U. Jessen, L. Norskov-Lauritsen and L. Agius, *J. Med. Chem.*, 2006, **49**, 5687.

39. P. Mackay, L. Ynddal, J. V. Andersen and J. G. McCormack, *Diabetes, Obes. Metab.*, 2003, **5**, 397.

40. N. Pinotsis, D. D. Leonidas, E. D. Chrysina, N. G. Oikonomakos and I. M. Mavridis, *Protein Sci.*, 2003, **12**, 1914.

41. J. L. Ekstrom, T. A. Pauly, M. D. Carty, W. C. Soeller, J. Culp, D. E. Danley, D. J. Hoover, J. L. Treadway, E. M. Gibbs, R. J. Fletterick, Y. S. N. Day, D. G. Myszka and V. L. Rath, *Chem. Biol.*, 2002, **9**, 915.

42. L. J. Hampson, C. Arden, L. Agius, M. Ganotidis, M. N. Kosmopoulou, C. Tiraidis, Y. Elemes, C. Sakarellos, D. D. Leonidas and N. G. Oikonomakos, *Bioorg. Med. Chem.*, 2006, **14**, 7835.

43. S. E. Zographos, N. G. Oikonomakos, K. E. Tsitsanou, D. D. Leonidas, E. D. Chrysina, V. T. Skamnaki, H. Bischoff, S. Goldmann, K. A. Watson and L. N. Johnson, *Structure*, 1997, **5**, 1413.

44. M. Shiota, P. A. Jackson, H. Bischoff, M. Mccaleb, M. Scott, M. Monohan, D. W. Neal and A. D. Cherrington, *Am. J. Physiol.*, 1997, **273**, E868–E879.

45. Z. Lu, J. Bohn, R. Bergeron, Q. Deng, K. P. Ellsworth, W. M. Geissler, G. Harris, P. E. McCann, B. McKeever, R. W. Myers, R. Saperstein, C. A. Willoughby, J. Yao and K. Chapman, *Bioorg. Med. Chem. Lett.*, 2003, **13**, 4125.

46. M. Kristiansen, B. Andersen, L. F. Iversen and N. Westergaard, *J. Med. Chem.*, 2004, **47**, 3537.

47. T. Klabunde, K. U. Wendt, D. Kadereit, V. Brachvogel, H. J. Burger, A. W. Herling, N. G. Oikonomakos, M. N. Kosmopoulou, D. Schmoll, E. Sarubbi, R. E. von K. Schonafinger and E. Defossa, *J. Med. Chem.*, 2005, **48**, 6178.

48. K. A. Evans, Y. H. Li, F. T. Coppo, T. L. Graybill, M. Cichy-Knight, M. Patel, J. Gale, H. Li, S. H. Thrall, D. Tew, F. Tavares, S. A. Thomson, J. E. Weiel, J. A. Boucheron, D. C. Clancy, A. H. Epperly and P. L. Golden, *Bioorg. Med. Chem. Lett.*, 2008, **18**, 4068.

49. S. M. Sparks, P. Banker, D. M. Bickett, D. C. Clancy, S. H. Dickerson, D. M. Garrido, P. L. Golden, A. J. Peat, L. R. Sheckler, F. X. Tavares, S. A. Thomson and J. E. Weiel, *Bioorg. Med. Chem. Lett.*, 2009, **19**, 981.

50. S. M. Sparks, P. Banker, D. M. Bickett, H. L. Carter, D. C. Clancy, S. H. Dickerson, K. A. Dwornik, D. M. Garrido, P. L. Golden, R. T. Nolte, A. J. Peat, L. R. Sheckler, F. X. Tavares, S. A. Thomson, L. Wang and J. E. Weiel, *Bioorg. Med. Chem. Lett.*, 2009, **19**, 976.

51. M. Fukushima, F. Matsuyama, N. Ueda, K. Egawa, J. Takemoto, Y. Kajimoto, N. Yonaha, T. Miura, T. Kaneko, Y. Nishi, R. Mitsui, Y. Fujita, Y. Yamada and Y. Seino, *Diabetes Res. Clin. Pract.*, 2006, **73**, 174.

52. J. Chen, J. Liu, L. Zhang, G. Wu, W. Hua, X. Wu and H. Sun, *Bioorg. Med. Chem. Lett.*, 2006, **16**, 2915.

53. X. Wen, J. Xia, K. Cheng, L. Zhang, P. Zhang, J. Liu, L. Zhang, P. Ni and H. Sun, *Bioorg. Med. Chem. Lett.*, 2007, **17**, 5777.

54. X. Wen, H. Sun, J. Liu, K. Cheng, P. Zhang, L. Zhang, J. Hao, L. Zhang, P. Ni, S. E. Zographos, D. D. Leonidas, K. M. Alexacou, T. Gimisis, J. M. Hayes and N. G. Oikonomakos, *J. Med. Chem.*, 2008, **51**, 3540.

55. X. Wen, H. Sun, J. Liu, G. Wu, L. Zhang, X. Wu and P. Ni, *Bioorg. Med. Chem. Lett.*, 2005, **15**, 4944.

56. X. Wen, P. Zhang, J. Liu, L. Zhang, X. Wu, P. Ni and H. Sun, *Bioorg. Med. Chem. Lett.*, 2006, **16**, 722.

57. K. Cheng, J. Liu, X. Liu, H. Li, H. Sun and J. Xie, *Carbohydr. Res.*, 2009, **344**, 841.

58. K. Cheng, J. Liu, H. Sun and J. Xie, *Chem. Biodiversity*, 2010, **7**, 690.

59. K. Cheng, J. Liu, H. Sun and J. Xie, *Synthesis*, 2010, 1046.

60. C. Tiraidis, K. M. Alexacou, S. E. Zographos, D. D. Leonidas, T. Gimisis and N. G. Oikonomakos, *Protein Sci.*, 2007, **16**, 1773.

61. S. Furukawa, K. Murakami, M. Nishikawa, O. Nakayama and M. Hino, *J. Antibiot.*, 2005, **58**, 503.

62. W. H. Martin, D. J. Hoover, S. J. Armento, I. A. Stock, R. K. McPherson, D. E. Danley, R. W. Stevenson, E. J. Barrett and J. L. Treadway, *Proc. Natl. Acad. Sci. USA*, 1998, **95**, 1776.

63. V. L. Rath, M. Ammirati, D. E. Danley, J. L. Ekstrom, E. M. Gibbs, T. R. Hynes, A. M. Mathiowetz, R. K. McPherson, T. Olson, V. J. L. Treadway and D. J. Hoover, *Chem. Biol.*, 2000, **7**, 677.

64. D. J. Hoover, S. Lefkowitz-Snow, J. L. Burgess-Henry, W. H. Martin, S. J. Armento, I. A. Stock, R. K. McPherson, P. E. Genereux, E. M. Gibbs and J. L. Treadway, *J. Med. Chem.*, 1998, **41**, 2934.

65. S. W. Wright, V. L. Rath, P. E. Genereux, D. L. Hageman, C. B. Levy, L. D. McClure, S. C. McCoid, R. K. McPherson, T. M. Schelhorn, D. E. Wilder, W. J. Zavadoski, E. M. Gibbs and J. L. Treadway, *Bioorg. Med. Chem. Lett.*, 2005, **15**, 459.

66. K. G. Rosauer, A. K. Ogawa, C. A. Willoughby, K. P. Ellsworth, W. M. Geissler, R. W. Myers, Q. Deng, K. T. Chapman, G. Harris and D. E. Moller, *Bioorg. Med. Chem. Lett.*, 2003, **13**, 4385.

67. S. N. L. Bennett, A. D. Campbell, A. Hancock, C. Johnstone, P. W. Kenny, A. Pickup, A. T. Plowright, N. Selmi, I. Simpson, A. Stocker, D. P. Whalley and P. R. O. Whittamore, *Bioorg. Med. Chem. Lett.*, 2010, **20**, 3511.

68. A. M. Birch, P. W. Kenny, N. G. Oikonomakos, L. Otterbein, P. Schofield, P. R. O. Whittamore and D. P. Whalley, *Bioorg. Med. Chem. Lett.*, 2007, **17**, 394.

69. P. R. O. Whittamore, M. S. Addie, S. N. L. Bennett, A. M. Birch, M. Butters, L. Godfrey, P. W. Kenny, A. D. Morley, P. M. Murray, N. G. Oikonomakos, L. R. Otterbein, A. D. Pannifer, J. S. Parker, K. Readman, P. S. Siedlecki, P. Schofield, A. Stocker, M. J. Taylor, L. A. Townsend, D. P. Whalley and J. Whitehouse, *Bioorg. Med. Chem. Lett.*, 2006, **16**, 5567.

70. K. Onda, R. Shiraki, T. Ogiyama, K. Yokoyama, K. Momose, N. Katayama, M. Orita, T. Yamaguchi, M. Furutani, N. Hamada, M. Takeuchi, M. Okada, M. Ohta and S. Tsukamoto, *Bioorg. Med. Chem.*, 2008, **16**, 10001.

71. K. Onda, T. Suzuki, R. Shiraki, Y. Yonetoku, K. Negoro, K. Momose, N. Katayama, M. Orita, T. Yamaguchi, M. Ohta and S. Tsukamoto, *Bioorg. Med. Chem.*, 2008, **16**, 5452.

72. H. Wagner, E. Langkopf, R. Streicher, M. Eckhardt, A. Schuler-Metz, A. Pautsch and C. Schoelch, *WO Patent*, 2008099000 (2008).

73. H. Wagner, E. Langkopf, M. Eckhardt, R. Streicher, C. Schoelch, A. Schuler-Metz and A. Pautsch, *WO Patent*, 2008113760 (2008).

74. H. Wagner, E. Langkopf, M. Eckhardt, R. Streicher, C. Schoelch, A. Schuler-Metz and A. Pautsch, *WO Patent*, 2009016119 (2009).

75. E. Langkopf, F. Himmelsbach, J. Mack, A. Pautsch, C. Schoelch, A. Schuler-Metz, R. Streicher and H. Wagner, *WO Patent*, 2009127723 (2009).

76. J. Dudash, Y. Zhang, J. B. Moore, R. Look, Y. Liang, M. P. Beavers, B. R. Conway, P. J. Rybczynski and K. T. Demarest, *Bioorg. Med. Chem. Lett.*, 2005, **15**, 4790.

77. Y. H. Li, F. T. Coppo, K. A. Evans, T. L. Graybill, M. Patel, J. Gale, H. Li, F. Tavares and S. A. Thomson, *Bioorg. Med. Chem. Lett.*, 2006, **16**, 5892.

78. K. Motoshima, M. Ishikawa, K. Sugita and Y. Hashimoto, *Biol. Pharm. Bull.*, 2009, **32**, 1618.

79. K. Kamata, M. Mitsuya, T. Nishimura, J. i. Eiki and Y. Nagata, *Structure*, 2004, **12**, 429.

80. T. Nishimura, T. Iino, M. Mitsuya, M. Bamba, H. Watanabe, D. Tsukahara, K. Kamata, K. Sasaki, S. Ohyama, H. Hosaka, M. Futamura, Y. Nagata and J. i. Eiki, *Bioorg. Med. Chem. Lett.*, 2009, **19**, 1357.

81. L. Chen, H. Li, J. Liu, L. Zhang, H. Liu and H. Jiang, *Bioorg. Med. Chem.*, 2007, **15**, 6763.

82. L. Zhang, H. Li, Q. Zhu, J. Liu, L. Chen, Y. Leng, H. Jiang and H. Liu, *Bioorg. Med. Chem.*, 2009, **17**, 7301.

83. L. J. Hampson and L. Agius, *Diabetes*, 2005, **54**, 617.

84. N. D. Karis, W. A. Loughlin, I. D. Jenkins and P. C. Healy, *Bioorg. Med. Chem.*, 2009, **17**, 4724.

85. J. L. Treadway, C. B. Levy, D. J. Hoover, R. W. Stevenson, E. M. Gibbs, R. A. Gelfand, O. Kuye, *Diabetes*, 2001, **50** (Suppl. 2), 536.

86. J. L. Treadway, *Symposia on New Targets for Glycemic Control*, 65th American Diabetes Association National Meeting, San Diego, CA, 2005.

87. L. V. Vasist, J. Lin, J. S. Stuart, R. L. Byerly, S. K. Swan, S. A. Thomson, F. Terschan, C. Cannon, R. V. Clark, Abstract OR24-5, 92nd Annual Meeting of the Endocrine Society, 2010.

88. A. Casey, A. H. Short, S. Curtis and P. L. Greenhaff, *Eur. J. Appl. Physiol. Occup. Physiol.*, 1996, **72**, 249.

89. C. Bartram, R. H. T. Edwards and R. J. Beynon, *Biochim. Biophys. Acta, Mol. Basis Dis.*, 1995, **1272**, 1.

90. D. J. Baker, J. A. Timmons and P. L. Greenhaff, *Diabetes*, 2005, **54**, 2453.

91. D. J. Baker, P. L. Greenhaff, A. MacInnes and J. A. Timmons, *Diabetes*, 2006, **55**, 1855.
92. Q. Dang, Y. Liu, D. K. Cashion, S. R. Kasibhatla, T. Jiang, F. Taplin, J. D. Jacintho, H. Li, Z. Sun, Y. Fan, J. DaRe, F. Tian, W. Li, T. Gibson, R. Lemus, P. D. van Poelje, S. C. Potter and M. D. Erion, *J. Med. Chem.*, 2011, **54**, 153.
93. J. M. Fujitaki, E. E. Cable, B. R. Ito, B. H. Zhang, J. Hou, C. Yang, D. A. Bullough, J. L. Ferrero, P. D. van Poelje, D. L. Linemeyer and M. D. Erion, *Drug Metab. Dispos.*, 2008, **36**, 2393.
94. T. W. von Geldern, N. Tu, P. R. Kym, J. T. Link, H. S. Jae, C. Lai, T. Apelqvist, P. Rhonnstad, L. Hagberg, K. Koehler, M. Grynfarb, A. Goos-Nilsson, J. Sandberg, M. Oesterlund, T. Barkhem, M. Hoeglund, J. Wang, S. Fung, D. Wilcox, P. Nguyen, C. Jakob, C. Hutchins, M. Faernegrdh, B. Kauppi, L. Oehman and P. B. Jacobson, *J. Med. Chem.*, 2004, **47**, 4213.
95. J. Hendrickx and P. J. Willems, *Hum. Genet.*, 1996, **97**, 551.
96. E. Floettmann, L. Gregory, J. Teague, J. Myatt, C. Hammond, S. M. Poucher and H. B. Jones, *Toxicol. Pathol.*, 2010, **38**, 393.

SIRT1 Activators in Development

ROBERT B. PERNI, VIPIN SURI, THOMAS V. RIERA, JOSEPH WU, CHARLES A. BLUM, GEORGE P. VLASUK AND JAMES L. ELLIS*

Sirtris, a GSK Company, 200 Technology Square, Cambridge, MA 02139, USA
*E-mail: James.5.Ellis@gsk.com

13.1 Introduction

The sirtuins are a family of seven evolutionarily conserved enzymes that require the co-factor nicotinamide adenine dinucleotide (NAD^+) and primarily catalyze the deacetylation and ADP-ribosylation of ε-acetyl-Lys residues in protein substrates.[1–3] Interest in these enzymes stems from the roles they are thought to play in human disease including metabolic, inflammatory, and neurodegenerative disease. Of particular interest is SIRT1, the most well-characterized of the mammalian sirtuins, which has been implicated in a number of age-related diseases and biological functions involving cell survival, apoptosis, stress resistance, fat storage, insulin production, and glucose and lipid homeostasis. Involvement in these diverse biologies is thought to occur through deacetylation of its many known *in vivo* protein substrates, including histones H1, H3, and H4, p53, p300, FOXOs 1, 3a, and 4, p65, HIVTat, PGC-1α, PCAF, MyoD, PPARγ, Ku70, and others.[3,4]

Studies in which SIRT1 protein and activity levels have been manipulated, through gene-deletion or over-expression in mice, have validated the beneficial impact of increased SIRT1 activity in several models of disease including those

RSC Drug Discovery Series No. 27
New Therapeutic Strategies for Type 2 Diabetes: Small Molecule Approaches
Edited by Robert M. Jones
© The Royal Society of Chemistry 2012
Published by the Royal Society of Chemistry, www.rsc.org

involving metabolic stress.[5,6] This has recently been observed in humans where reduced SIRT1 expression in insulin-sensitive tissues was associated with reduced energy expenditure.[7] Therefore, for many of the diseases in which SIRT1 is thought to play a role, therapeutic benefits are predicted to follow from the administration of activators of this enzyme's deacetylase activity. Over the past several years, SIRT1-activating compounds (STACs), including resveratrol and more specific, chemically distinct molecules, have been developed.[8–10] When tested in cell-based and animal models of disease, STACs produce effects consistent with direct activation of this enzyme.[8,11–17]

At the molecular level, much remains to be learned concerning the mechanism by which these compounds accelerate SIRT1-catalyzed deacetylation. This has been a controversial area, with recent studies supporting an allosteric mechanism of activation which forms the basis for ongoing and future investigations.

This chapter attempts to cover the broad recent literature relating to the biology of SIRT1 and the efforts in the design and characterization of pharmacological activators of this enzyme. It is clear that much remains to be learned about SIRT1. However, significant progress has been made in understanding the fundamental aspects of SIRT1 biology and in designing drug-like compounds to modulate its activity *in vivo* with the hope that these molecules will be the basis for new medicines for a broad range of age-related diseases.

13.2 Role of SIRT1 in Metabolic Regulation

The role of SIRT1 in the regulation of metabolic homeostasis has been extensively explored.[18–20] The discovery of the central role of *Sir2*, the yeast homolog of SIRT1, as the link between caloric restriction and lifespan extension in the budding yeast *Saccharomyces cerevisiae*, immediately suggested a role for SIRT1 in metabolic regulation.[21] While subsequent studies have suggested that the effect of SIRT1 in caloric restriction associated lifespan extension may have greater relevance in specific organisms and under specific nutritional conditions,[22] the role of SIRT1 in regulating several aspects of metabolism has been very well established in a number of organisms.[23] Although this chapter focuses on the current understanding of the role of SIRT1 in metabolic regulation in mice and humans, there are a number of excellent studies exploring the effect of SIRT1 homologs in metabolic control in yeast,[24] *Caenorhabitis elegans*,[25] *Drosophila melanogaster*,[26] and others.

As discussed below, many physiological and biochemical systems appear to be influenced by SIRT1. SIRT1 has been suggested to deacetylate a broad array of metabolic regulators such as peroxisome proliferator and activated receptor gamma coactivator 1 alpha (PGC1α),[27] nuclear hormone receptors such as liver X receptor (LXR),[28] forkhead box subgroup O (FOXO) transcription factors,[29,30] cAMP response element binding protein (CREB),[14] CREB regulated transcriptional coactivator 2 (CRTC2),[14] and sterol response element binding protein (SREBP).[31] In addition, SIRT1 also regulates key cellular pathways such as those involved in autophagy and mammalian target of rapamycin (mTOR)[32,33] signaling through mechanisms that are only beginning to be understood.

SIRT1 also directly regulates activity of metabolic enzymes such as acetyl-CoA synthase I (ACS1),[34] 3-methyl-3-hydroxyglutaryl coenzyme A synthase (HMGCS1),[35] and others.

In addition to the regulation of activity through reversible deacetylation, the SIRT1 deacetylation reaction consumes NAD^+ and generates nicotinamide. NAD^+/NADH serves as a cofactor for a number of metabolic redox reactions such as those involved in harvesting energy from various substrates and coupling the tricarboxylic acid (TCA) cycle to adenosine-5′-triphosphate (ATP) generation. NAD^+ also serves as a substrate for adenosine-5′-diphosphate (ADP) ribosylation reaction as well as for the generation of cyclic ADP ribose for calcium signaling. Nicotinamide generated by SIRT1 deacetylation and other cellular enzymes is converted back to NAD through the actions of the enzymes nicotinamide mononucleotide phosphotransferase (NAMPT) and nicotinamide mononucleotide adenyltransferase (NMNAT).[36] In several tissues, NAMPT has been suggested to be regulated by the circadian transcription factor complex CLOCK/BMAL1.[37,38] The observation that SIRT1 can repress CLOCK/BMAL1 activity directly through the deacetylation of BMAL1 completes a circadian–metabolic NAD oscillatory loop that could be critical to metabolic regulation.[39] SIRT1 repression of CLOCK/BMAL1 also appears to impact a number of other circadian regulators and since metabolic regulation is a key circadian output, the circadian systems provides another potential mechanism linking SIRT1 to metabolic control.[40]

The SIRT1 deacetylation reaction also produces 2′*O*-acetyl ADP ribose (OAADPR), although the function of this reaction product remains unclear at this time. However, a number of enzymes have been identified that can metabolize OAADPR either to ADPR by OAADPR deacetylases or to AMP and *O*-acetyl ribose phosphate by ADP-ribose pyrophosphatases.[41] Deletion of yeast *ysa1*, a nudix family ADP–ribose pyrophosphatase, confers resistance to oxidative insults.[42] Although as yet unsubstantiated, a signaling role for OAADPR appears to be an attractive proposition.

The understanding of the role of SIRT1 in metabolic regulation draws heavily on studies utilizing genetic manipulation of SIRT1 levels as well as pharmacological agents. Section 13.5 of this chapter discusses the metabolic effects of pharmacological SIRT1 modulators. Genetic strategies to mechanistically dissect the metabolic effects of SIRT1 include whole body knockouts, tissue specific knockouts, conditional knockouts as well as various transgenic over-expression models.[18,43] At least three different knockout strategies have been reported. McBurney *et al.*[44] deleted exon 5 and 6 of murine *sirt1* that resulted in no detectable SIRT1 protein in the homozygous animals. Absence of SIRT1 resulted in approximately 50% reduction in viability and the surviving animals exhibited severe developmental defects. These animals were generated on a CD1/129/SV mixed background. Survival of these animals is reported to be better on an FVB background, although the developmental defects persist.[45] More recently, Wang *et al.*[45] generated alleles of *sirt1* with a loxP site flanking exon 5 and 6 that allows for generation of tissue specific as well as conditional knockouts. Using ELL-cre to create a whole body *sirt1* knockout, Wang *et al.*[45] reported postnatal survival rates of about 1% in 129SVEV/FVB background and 9.3% in a 129/SVEV/FVB/Black Swiss background. Cheng *et al.*[46] generated alleles of *sirt1* with loxP sites flanking exon 4 resulting in the deletion of a 51-amino acid region in the catalytic domain of SIRT1. The mutation produced a truncated SIRT1 protein of unknown function that has not been completely characterized. In a 129/SV/ C57BL6 background, these authors reported less than 10% viability of the homozygous progeny, 67% died postnatally within the first week. The floxed *sirt1*$^{\Delta ex4/\Delta ex4}$ animal has been used extensively to generate tissue-specific and conditional *sirt1*$^{\Delta ex4/\Delta ex4}$ animals and has contributed greatly to the knowledge of SIRT1 function. While the tissue-specific knockouts have generally shown normal viability, the potential exists for tissue-specific developmental alterations in *sirt1*$^{\Delta ex4/\Delta ex4}$.[47] Additionally, the potential for the truncated protein in these animals to carry residual functionality or to act as a dominant negative has not been rigorously evaluated and therefore care needs to be taken in the interpretation of those results.[45] Finally, genetic background effects can be particularly profound in metabolic settings and needs to be considered.[48]

A number of transgenic lines have also been reported for SIRT1. Pfluger *et al.*[49] used a bacterial artificial chromosome (BAC) containing a 174 kilobase (kb) fragment to over-express *sirt1*, which included *sirt1* promoter as well as an adjacent gene *Dnajc*. The mice were on a 97% C57/BL6/ 3% CBA background and over-expressed SIRT1 by 2–4-fold in various tissues. Herranz *et al.*[50] used a BAC containing a shorter fragment that did not include the *Dnajc* locus but included endogenous *sirt1* promoter to generate SIRT1 transgenic animals as well as *herc4*, the gene to the left of *sirt1* in the murine genome. These animals were backcrossed eight generations to a 99% C57/BL6 / 1% CBA background and over-expressed SIRT1 2–3-fold. Banks *et al.*[51] also used a BAC strategy that included endogenous *sirt1* promoter but also included *herc4*. The animals were backcrossed ten generations to an essentially pure C57/BL6 background and over-expressed SIRT1 2–3-fold in most tissues except spleen where ~7-fold over-expression was observed. Importantly, Pfluger *et al.*[49] as well as Banks *et al.*[51] demonstrated rescue of perinatal lethality in *sirt1* null animals upon crossing the knockouts with the transgenics.

A number of pharmacological tools have been used in the literature to dissect mechanisms underlying the effects of SIRT1 on metabolic regulation. These are discussed in Section 13.5 of this chapter. It needs to be noted that the selectivity, specificity, dose dependence, and pharmacokinetic-pharmacodynamic correlations of several SIRT1 modulators reported in the literature have not been carefully evaluated.

13.2.1 SIRT1 and Regulation of Energy Balance

Body weight regulation in mammals requires tight control of food intake and energy expenditure.[52] A number of centers in the central nervous system participate in these control mechanisms, primarily in the hypothalamus.[53] Neurons in the arcuate nucleus are critically required to integrate a number of hormonal signals that regulate feeding.[53] These neurons include the pro-opiomelanocortin (POMC) expressing neurons that are activated in response to anorexigenic signals and the neuropeptide Y/Agouti-related peptide (NPY/AgRP) expressing neurons. SIRT1 is abundantly expressed in a number of hypothalamic centers including arcuate, venteromedial, dorsomedial, paraventricular, supraoptic, and suprachiasmatic nuclei.[54] Alterations in energy balance in SIRT1 knockouts and transgenics have been reported in many studies and have been extensively reviewed elsewhere.[55,56] A number of recent studies have systematically dissected the role of SIRT1 in specific neuronal subsets and signaling pathways.[57,58]

Cakir *et al.*[59] used the selective SIRT1 inhibitor EX-527 as well as SIRT1-specific siRNAs to show that reduction in SIRT1 activity resulted in reduced food intake in fasted rats through the activation of the melanocortin system. These authors suggested SIRT1 activation reduced POMC neuron expression and increased AgRP neuron expression through deacetylation of FOXO1 transcription factor.

Dietrich *et al.*[60] also reported a reduction in food intake following administration of EX-527 in mice, as well as in mice lacking catalytically active SIRT1 in AgRP specific neurons (*AgRP-sirt1$^{\Delta ex4/\Delta ex4}$*). EX-527 reduced inhibitory but not excitatory synapse recruitment into POMC neurons and reduced mitochondrial density. Interestingly, UCP2 knockout animals were resistant to EX-527-induced changes in synaptic plasticity and mitochondrial density as well as EX-527-induced reduction in food intake. Furthermore, both EX-527 dosed as well as the *AgRP-sirt1$^{\Delta ex4/\Delta ex4}$* animals exhibited a muted response to the orexigenic hormone ghrelin. Velasquez *et al.*[61] also reported reduced ghrelin-induced food intake upon central SIRT1 inhibition by EX-527 in rats. Ghrelin induced AMPK phosphorylation as well as p53 deacetylation and both these effects were blunted by central administration of EX-527. Importantly, EX-527 did not affect ghrelin-induced growth hormone secretion which is known to be mediated by a distinct set of neurons. Consistent with these findings, Satoh *et al.*[62] reported increased ghrelin-induced activation of neurons in the dorsomedial hypothalamus and lateral hypothalamus in brain-specific *SIRT1* transgenic mice (BRASTO).

AgRP-sirt1$^{\Delta ex4/\Delta ex4}$ females showed reduced body weight with significant reductions in both lean and fat mass but energy expenditure in ad lib fed *AgRP-sirt1$^{\Delta ex4/\Delta ex4}$* females was unchanged. However, food deprivation for 2 h prior to and during the dark phase significantly increased oxygen consumption due to increased locomotor activity in these animals leading the authors to propose a role for SIRT1 in promoting a negative energy balance.

Coppari and colleagues have systematically dissected the role of SIRT1 in neuronal subsets by utilizing deletion of catalytically active SIRT1 in specific neurons. Ramadori *et al.*[58] reported that *POMC-sirt1*$^{\Delta ex4/\Delta ex4}$ mutant females demonstrated increased susceptibility to diet-induced obesity on a high fat (58% kcal fat with sucrose) diet, without any effects on glucose or lipid parameters or insulin sensitivity. Increased body weight was due to a reduction in basal metabolic rate in *POMC-sirt1*$^{\Delta ex4/\Delta ex4}$ females on a high fat diet (HFD), without a change in substrate preference. These authors attributed the reduced energy expenditure to a reduction in brown adipocyte content in the white adipose tissue of the *POMC-sirt1*$^{\Delta ex4/\Delta ex4}$ mice due to a selective reduction in sympathetic nerve activity in the perigonadal white adipose tissue. Consistent with this mechanism, the selective β3-adrenergic receptor agonist, CL316,243 reversed the changes in body weight, energy expenditure, and brown adipose content of white adipose tissue in *POMC-sirt1*$^{\Delta ex4/\Delta ex4}$ females. Leptin signaling in the POMC neurons, as assessed by nuclear exclusion of FOXO1 upon leptin administration, appeared to be blunted in *POMC-sirt1*$^{\Delta ex4/\Delta ex4}$ females suggesting an important role for SIRT1 in the regulation of leptin sensitivity of POMC neurons.

Very recently, Ramadori *et al.*[57] have also generated animals that lack catalytically active SIRT1 specifically in the steroidogenic factor 1 (SF1) neurons of the ventromedial hypothalamus. SF1 neuronal activity is sensitive to a number of hormonal and nutritional signals including leptin, orexin, and glucose and SF1 neurons are believed to be important in body weight regulation as well as in glucose homeostasis. *SF1-sirt1*$^{\Delta ex4/\Delta ex4}$ showed greater susceptibility to obesity on a HFD due to a reduction in locomotor activity and energy expenditure. In addition, *SF1-sirt1*$^{\Delta ex4/\Delta ex4}$ mice also showed remarkable sensitivity to HFD-induced skeletal muscle insulin resistance. *SF1-sirt1*$^{\Delta ex4/\Delta ex4}$ animals also showed defects in leptin and orexin signaling. Conversely, mice over-expressing SIRT1 in SF1 specific neurons were protected from HFD-induced obesity and muscle insulin resistance and exhibited increased leptin sensitivity suggesting an important role of SIRT1 in the SF1 neurons in the maintenance of metabolic homeostasis in calorie-rich conditions.

13.2.2 SIRT1 and Carbohydrate Metabolism

The opposing processes of glycolysis and gluconeogenesis provide metabolic buffering to maintain glucose homeostasis in a wide range of nutritional conditions. During nutrient deprivation, the liver channels substrates into glucose production through gluconeogenesis.[63] Glucagon signals through the hepatic glucagon receptor to stimulate the gluconeogenic program via CRTC2 dephosphorylation, nuclear translocation, and recruitment to the promoters of gluconeogenic genes such as *phosphoenolpyruvate carboxykinase* (*PEPCK*) and *glucose-6-phoshatase* (*G6Pase*).[63,64] Insulin, on the other hand, suppresses gluconeogenesis through phosphorylation and cytoplasmic sequestration of the gluconeogenic transcription factor FOXO1. In addition, PGC1α promotes gluconeogenesis by co-activating FOXO1 whereas signal transducer and activator of transcription 3 (STAT3) suppresses gluconeogenesis through inhibition of PGC1α expression.[63]

Rodgers *et al.*[65,66] reported that SIRT1-dependent deacetylation of PGC1α during fasting conditions increased the activity of PGC1α at the promoters of gluconeogenic genes *pepck* and *g6pase* and reduced the activity of PGC1α at the promoters of glycolytic genes *glucokinase* (*Gck*) and *liver pyruvate kinase* (*LPK*).

Liu *et al.*[14] reported that SIRT1 deacetylates CRTC2, a major regulator of glucagon-stimulated gluconeogenic gene expression. After an 8 h fast, glucagon signaling resulted in a reduction in salt inducible kinase 2 (SIK2)-dependent phosphorylation p300 histone acetyltransferase at Ser89. Dephosphorylated p300 then acetylated CRTC2 at Lys682 and increased the activity of CRTC2 at *g6pase* promoter as well as prevented CRTC2 ubiquitination and subsequent degradation. After 18 h, however, CRTC2 was deacetylated by SIRT1 leading to ubiquitin-dependent degradation. During prolonged fasting, the deacetylation-dependent activation of FOXO1 on the promoter of gluconeogenic enzymes appeared to be the dominant mechanism.

Nie *et al.*[67] observed increased STAT3 acetylation after prolonged fasting in animals treated with the SIRT1 inhibitor EX-527 or antisense oligonucleotides (ASO) specific for SIRT1. Moreover, SIRT1 ASO administration reduced glucose output in wild-type but not in STAT3 liver-specific knockouts suggesting that STAT3 is an important mediator of the effects of SIRT1 on the gluconeogenic program.

Erion *et al.*[68] also used an ASO approach to explore the effect of SIRT1 knockdown on metabolic regulation in a diabetic rat model. Diabetes was induced by a low dose of streptozotocin followed by a combination of high fructose and HFD. ASO administration reduced SIRT1 mRNA by 77% and 91% in liver and white adipose tissue, respectively. Intriguingly, ASO administration reduced food intake by 17% in rats on lean diet and 12% in diabetic rats. ASO-treated diabetic rats also showed increased hepatic insulin sensitivity. Acetylation of STAT3, FOXO1, and PGC1α were reduced while transcription of gluconeogenic genes was reduced, consistent with a role of SIRT1 in the activation of hepatic gluconeogenesis.

The role of SIRT1 in other pathways of carbohydrate metabolism has been less well explored, although a potential role for SIRT1 in the regulation of glycolysis through HIF1α deacetylation and inactivation has been suggested by Lim and colleagues.[69]

13.2.3   SIRT1 and Lipid Metabolism

The initial observations of Picard *et al.*[70] suggested a role for SIRT1 in the differentiation of white adipose tissue through repression of PPARγ activity. In addition, *SIRT1*[+/−] mice showed a reduction in fasting-induced fatty acid release from white adipose tissues suggesting a role for SIRT1 in lipid mobilization from adipose tissue under nutrient deprivation.

The role of SIRT1 in lipid metabolism in the liver has been extensively explored. Chen *et al.*[71] reported protection from HFD-induced hepatic steatosis as well as reduced HFD-induced weight gain and glucose intolerance in mice lacking catalytically active SIRT1 in the liver (*Liver-sirt1$^{ex4/ex4}$*, LKO-A) due to a reduction in liver X receptor (LXR) and SREBP activity. LXR has been shown to be deacetylated and activated by SIRT1, although deacetylation eventually leads to ubiquitination and degradation of LXR. SREBP has also been shown to be deacetylated by SIRT1 leading to ubiquitin-dependent degradation of SREBP.

Purshotham *et al.*[72] also generated *Liver-sirt1$^{ex4/ex4}$* (LKO-B) on a 98% C57/BL6 background and discovered a significant reduction in PPARα regulated genes in the LKO-B animals. Furthermore, SIRT1 directly interacted with PPARα and SIRT1-dependent deacetylation was required for transcriptional activation by PGC1α at PPARα response elements (PPREs). On a high fat (40% Kcal fat / 0.21% cholesterol) diet, LKO-B animals showed significantly greater weight gain, increased hepatic steatosis, and serum free fatty acids. In addition, fasting-induced β-hydroxybutyrate was significantly lower in the LKO-B animals suggesting impaired β-oxidation.

The differences between LKO-A and LKO-B mice are remarkable since similar knockout strategies and diets were used in the two studies. Genetic backgrounds were also similar between the two studies, although it is unclear if the background was identical. However, the LKO-A studies were performed in animals that had been on the HFD for a shorter duration (6 weeks) compared to LKO-B studies. Nevertheless, the opposing results do suggest the sensitivity of the SIRT1 mechanism to slight changes in experimental conditions.

Wang *et al.*[73] recently reported a third liver-specific SIRT1 knockout study. Instead of exon 4 deleted *sirt1*, these authors used a null allele of *sirt1* to generate *Liver-sirt1$^{-/-}$* mice (LKO-C). A significant fraction (7/9) of LKO-C mice developed hepatic steatosis even on a chow diet compared to only 2/12 control animals at 12 months of age. Serum triglycerides were also significantly elevated in 12-month-old LKO-C mice. Interestingly, analysis of gene expression in 2-month-old animals suggested little change in glycolytic, β-oxidation, triglyceride synthesis or uptake pathways. However, lipogenic genes such as fatty acid synthase (FAS), acetyl-coenzyme A carboxylase 1 (ACC1), elongase of long chain fatty acids 6 (ELOVL6), and stearoyl coenzyme A desaturase (SCD1) were significantly increased in 2-month-old as well as 6-month-old LKO-C animals relative to control animals. The increase in lipogenic genes was due to an increase in expression of carbohydrate response element binding protein (ChREBP). In LKO-C livers, an increase in acetylation of histone 3 at Lys9 (H3K9) and histone 4 at Lys16 (H4K16) was observed suggesting chromatin changes consistent with increased transcription activity at ChREBP promoter. Unfortunately, these authors did not study the LKO-C mice on a HFD to allow for a direct comparison with

LKO-A and LKO-B animals. However, an effect of reduced SIRT1 activity on lipogenesis was also reported by Xu *et al.*,[74] who observed increased hepatic steatosis in *SIRT1*$^{+/-}$ animals on medium and high fat diets.

Reduced hepatic steatosis upon feeding a HFD has also been reported in mice lacking *Deleted in Breast Cancer 1* (*DBC1*) gene.[75,76] DBC1 directly associates with and inhibits SIRT1 deacetylase activity and the dynamics of DBC1-SIRT1 complex have been suggested to underlie the changes in SIRT1 deacetylase activity in conditions of differing nutrient availability.[75] The mechanisms underlying reduced steatosis in the *DBC1* knockout animal are unclear, although AMPK phosphorylation maybe involved.[75] Mice lacking *DBC1* also showed reduced hepatic inflammation. Endotoxin-induced inflammatory responses in Kupffer cells from *DBC1*-deleted mice were significantly reduced, likely due to a reduction in NF-κB activation.[75]

13.2.4 SIRT1 and Insulin Secretion and Sensitivity

SIRT1 is expressed at moderate levels in the pancreatic islets but not in exocrine pancreas. Moynihan *et al.*[77] over-expressed SIRT1 under the control of human insulin promoter in C57/BL6 mice. β-Cell-specific SIRT1 transgenic (BESTO) mice over-expressed SIRT1 by >10-fold endogenous levels specifically in the insulin producing β-cells. Three-month-old male and female BESTO mice showed improved glucose tolerance, although insulin secretion during glucose tolerance test appeared to be higher only in males. Additionally, there were modest improvements in glucose tolerance in 8-month-old animals, suggesting a reduction in age-associated functional decline in β-cells in BESTO mice. The expression of a number of islet genes was altered in BESTO mice, including uncoupling protein 2 (UCP2), an important regulator of glucose-stimulated insulin secretion (GSIS). The observed reduction in UCP2 was consistent with increased glucose-stimulated insulin secretion as reduced UCP2 increases the coupling efficiency of oxidative phosphorylation, driving up the ATP/ADP ratio and promoting insulin secretion.

Bordone *et al.*[78] reported reduced GSIS upon SIRT1 knockdown in rat INS-1 cells and mouse MIN6 cells. SIRT1 knockdown increased UCP2 levels in these cells and reduction of siRNA-mediated reduction in UCP2 in INS-1 cells already knocked down for SIRT1 restored glucose-stimulated insulin secretion. In addition, SIRT1 was found to directly bind to and repress transcriptional activity at the UCP promoter.

SIRT1 has also been reported to be protective against cytokine-induced toxicity in rat insulinoma (RIN) cells as well as in rat primary islets. Lee *et al.*[79] showed improved survival as well as GSIS after IL-1β and IFN-γ treatment in primary rat islets infected with adenovirus-SIRT1. Mechanisms underlying the improvement were suggested to be a reduction in inducible nitric oxide synthase (iNOS) expression and activity as well as reduction in NF-κB signaling. The effect of SIRT1 on β-cell preservation *in vivo* has not been fully explored and remains an exciting avenue for further research.

An effect of SIRT1 on insulin sensitivity has been observed in many cellular and *in vivo* models. SIRT1 over-expression has been shown to improve glucose tolerance in mice fed a HFD in a number of studies. Banks *et al.*[51] reported improved glucose tolerance in HFD fed or aged mice over-expressing SIRT1 2–3-fold in various tissues. SIRT1 over-expressing mice showed improvement in both hepatic insulin sensitivity as well as whole body glucose uptake. Adiponectin was increased by 30–40% in the SIRT1 transgenics. In addition, several markers of adiponectin signaling such as phospho-AMPK, PPARα, carnitine palmitoyl transferase I (CPT1), and adiponectin receptor 2 (AdipoR2) were increased in the liver of SIRT1 transgenics suggesting increased adiponectin signaling underlies the improvement in insulin sensitivity.

Pfluger *et al.*[49] also reported improved glucose tolerance and reduction in glucose production during pyruvate challenge in SIRT1 transgenic animals. However, these authors did not observe a significant effect of SIRT1 over-expression on insulin sensitivity by an insulin tolerance test suggesting that the metabolic improvements were primarily due to improved hepatic glucose metabolism.

Of note is the apparent divergence between the role of SIRT1 in promoting gluconeogenesis discussed above and the reduction in hepatic glucose output in the livers of SIRT1 over-expressing mice on a HFD. The most likely explanation for the difference is that the role of SIRT1 in maintaining glucose output under fasting conditions has been primarily studied in insulin-sensitive, non-steatotic animals on a chow diet. In a HFD setting, the gluconeogenic program is adversely affected by hepatic insulin resistance and increased steatosis and the benefits of SIRT1 on improving hepatic insulin sensitivity as well as reducing hepatic steatosis underlie the overall reduction in hepatic glucose output. However, a detailed study of the effect of SIRT1 on hepatic gluconeogenesis program in HFD-fed animals has not been reported and such a study would be of great utility in bridging the gap in understanding the role of SIRT1 in hepatic glucose metabolism under various nutritional conditions.

Sun *et al.*[80] reported transcriptional repression of protein tyrosine phosphatase-1B (PTP1B) upon SIRT1 over-expression in C2C12 myotubes through H3K9 deacetylation at the PTP1B promoter. PTP1B terminates insulin action through insulin receptor dephosphorylation and a reduction in PTP1B activity has been reported to improve insulin as well as leptin sensitivity. Reduced expression of PTP1B upon SIRT1 over-expression in chondrocytes has also been reported by Gagarina *et al.*,[81] although in the context of chondrocytes, a reduction in PTPT1B signaling improved cell survival through insulin-like growth factor (IGF1) pathway.

13.2.5 SIRT1 and Other Hormones

In addition to insulin, SIRT1 has been shown to impact the production, secretion, and signaling of a number of other hormonal signals as well, although the mechanisms are only beginning to be understood. As noted above, SIRT1 has been shown to be involved in leptin, ghrelin, orexin, adiponectin, and IGF1 signaling.

Akieda-Asai *et al.*[82] recently reported the stimulation of thyroid stimulated hormone (TSH) secretion by SIRT1. Over-expression of SIRT1 in primary rat anterior pituitary cells increased TSH secretion by about ~2-fold while SIRT1 knockdown in these cells reduced TSH secretion. Interestingly, SIRT1 primarily affected exocytosis of TSH secretory granules, a process known to require phosphatidylinositol (4,5) diphosphate (PI(4,5)P2). A search for SIRT1 substrates led to the identification of PIP5Kγ as a mediator of SIRT1 effects on exocytosis of TSH granules as knockdown of PIP5Kγ abolished the effect of SIRT1 over-expression on TSH secretion. PIP5Kγ was readily deacetylated by SIRT1 and the deacetylation increased the activity of PIP5Kγ by about 2-fold. PIP5Kγ acetylation and PIP2 levels were elevated in brains from *SIRT1*$^{-/-}$ mice. In addition, TSH secretion was reduced in *SIRT1*$^{-/-}$ mice despite higher TSH content in the pituitary. TSH regulates thyroid hormone T3/T4 levels, which are closely tied to energy expenditure. Alterations in T3/T4 levels have been observed in SIRT1 knockouts and transgenics and may underlie the observed changes in energy expenditure in these animals.

13.3   Biochemistry of SIRT1 Activation

13.3.1   General Characterization of Enzyme Activation

Activation of an enzyme-catalyzed reaction can be described in a mechanism-independent way by Equation 13.1.

$$\frac{v_x}{v_o} = 1 + \frac{RV_{max} - 1}{1 + \dfrac{EC_{50}}{[X]_o}} \tag{13.1}$$

v_x/v_0 is the relative rate of reaction in the presence (v_x) versus absence (v_0) of activator (X), RV_{max} is the maximal relative velocity at infinite activator concentration, and EC_{50} is the concentration of activator producing half of the maximal activation. The EC_{50} is an apparent binding constant relevant to the experimental conditions and is a useful measure for ranking compounds. Measurement of the EC_{50}, however, depends upon a good determination of the maximum activation. This is not always possible as factors such as compound solubility or potency may be limiting. In these cases the $EC_{1.5}$, or concentration of activator required to produce 1.5-fold activation can be utilized. The $EC_{1.5}$ and EC_{50} are related by Equation 13.2.

$$EC_{1.5} = \frac{EC_{50}}{2RV_{max} - 3} \tag{13.2}$$

It can be seen that the $EC_{1.5}$ is dependent upon both the EC_{50} and RV_{max}. The reciprocal of $EC_{1.5}$ (Equation 13.3) is analogous to the steady-state parameter V_{max}/K_M for substrate specificity and is thus a measure of activator efficacy.

$$(EC_{1.2})^{-1} = 2\left(\frac{RV_{max} - 1.5}{EC_{50}}\right) \qquad (13.3)$$

13.3.2 Demonstration of Direct Activation of SIRT1 by STACs

In 2003, Howitz *et al.* reported the first screen for STACs identifying several plant polyphenols, the most potent of which was resveratrol (3,4′,5-trihydroxy-trans-stilbene, Figure 13.1).[83] Activation of SIRT1 was measured by following deacetylation of the fluorogenic substrate Ac-Arg-His-Lys-Lys^{Ac}-AMC using the Fluor de Lys assay (Enzo Life Sciences, Farmingdale, NY). Steady-state kinetic characterization revealed that resveratrol activates by lowering the K_M for the peptide substrate.[8,83] Investigations of the molecular details of SIRT1 activation have emphasized the dependence on substrate structural features. This was highlighted by two reports in 2005 demonstrating that resveratrol enhances the SIRT1-catalyzed deacetylation of Ac-Arg-His-Lys-Lys^{Ac}-AMC but not the corresponding peptide lacking the AMC group, Ac-Arg-His-Lys-Lys^{Ac}-NH$_2$.[84,85] The observation that activation requires a fluorophore covalently attached to the peptide substrate was later confirmed by Beher *et al.*[86] and extended by Pacholec *et al.*[87] to include the STACs, SRT1460, SRT1720, and SRT2183 (Figure 13.1) developed by Sirtris Pharmaceuticals Inc. and the TAMRA-labeled peptide, Ac-Glu-Glu-Lys^{biotin}-Gly-Gln-Ser-Thr-Ser-Ser-His-Ser-Lys^{Ac}-Nle-Ser-Thr-Glu-Gly-Lys^{TAMRA}-Glu-Glu-NH$_2$.[8] Pacholec *et al.* further reported that these STACs do not activate two full-length protein substrates and that SRT1460 and SRT1720 bind to the TAMRA peptide but, not to the same peptide lacking the TAMRA group suggesting that they activate SIRT1 indirectly.[87] These results raised concerns that activation of SIRT1 was only relevant to fluorescently labeled substrates although a specific mechanism for how STAC binding to the fluorophore could result in activation of SIRT1 was not proposed.

The simplest model to describe these data is one of substrate enhancement where substrate and activator form an activated complex (*X-S*) which is better utilized by enzyme. Compared to free substrate, the *X-S* complex displays enhanced binding and/or turnover by enzyme. A key feature of this mechanism is that activator potency is directly related to the K_d for formation of the *X-S* complex. This was the basis for the report by Dai and Kustigian *et al.* that first demonstrated that there is no correlation between the $EC_{1.5}$ values for activation and K_d values for binding to the TAMRA peptide for a set of 20 STACs from distinct chemical classes with varied potency (Figure 13.2)[88] which was in contrast to the generalization proposed by Pacholec *et al.* This set included SRT1460 but excluded resveratrol, SRT1720, and SRT2183 because low solubility prohibited K_d determination for the TAMRA peptide. Many

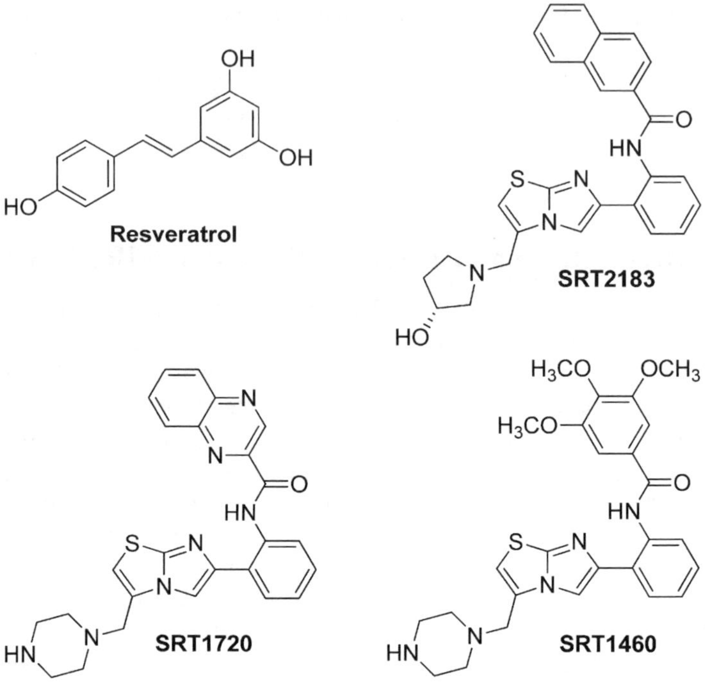

Figure 13.1 Structures of resveratrol and SRT1460, SRT1720, and SRT2183.

potent STACs ($EC_{1.5} < 500$ nM) were identified which displayed no detectable binding to the TAMRA peptide (>100 μM K_d) leading the authors to conclude that STAC binding to TAMRA peptide is unrelated to SIRT1 activation. Also significant was the identification of two pairs of structural isomers (**8** and **19** as well as **11** and **16** from Figure 13.2) which exhibit large changes in $EC_{1.5}$ values (>300-fold and 25-fold, respectively). Large changes in activity resulting from seemingly small differences in structure are well-documented for ligand interactions with geometrically defined protein binding pockets. However, this is harder to explain for compounds binding to aromatic ring systems, such as TAMRA and AMC, which are presumably dominated by $\pi-\pi$ interactions. In accordance, these compound pairs exhibit less than 4-fold differences in K_d values for TAMRA peptide.[88] Together, this led Dai and Kustigian *et al.* to propose that STACs directly interact with SIRT1 as allosteric modulators.

In this view, the hydrophobic, aromatic fluorophores serve as activation cofactors for SIRT1. Dai and Kustigian *et al.* investigated whether these properties could be generalized to peptide substrates composed only of natural amino acids by testing for activation of the SIRT1-catalyzed deacetylation of peptides with the p53-based sequence Ac-Arg-His-Lys-Lys^{Ac}-X-NH$_2$ where X was either Ala, Phe, Trp, NH$_2$, or AMC. While no activators were found when

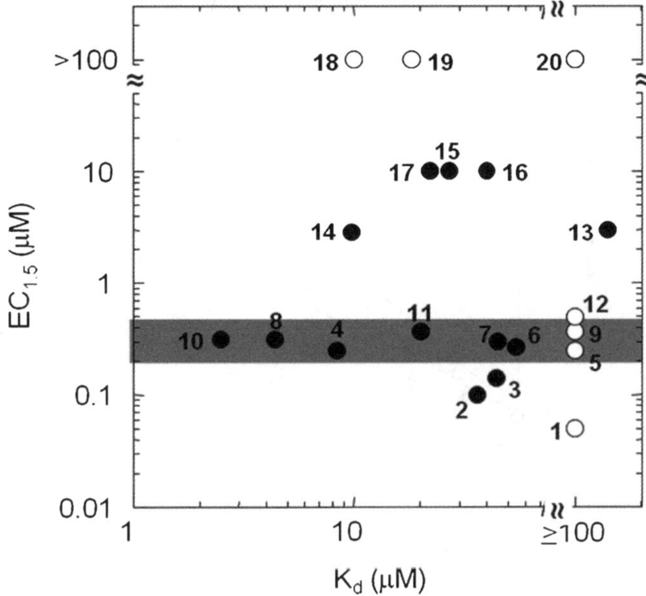

Figure 13.2 $EC_{1.5}$ values from the activation of SIRT1 using the TAMRA peptide plotted versus the corresponding K_d values for STAC binding to the TAMRA peptide measured by ITC. Open circles represent compounds with $EC_{1.5}$ or K_d values > 100 μM. The gray bar highlights STACs with $EC_{1.5}$ values around 0.3 μM, but whose K_d values range from 2.5 μM to >100 μM (>40-fold range). The data labels are the compound numbers from reference 88. This research was originally published in the *Journal of Biological Chemistry*. Dai, H., Kustigian, L., Carney, D., Considine, T., Hubbard, B. P., Perni, R. B., Riera, T. V., Szczepankiewicz, B., Vlasuk, G. P., Stein, R. L. SIRT1 activation by small molecules: kinetic and biophysical evidence for direct interaction of enzyme and activator. *J. Biol. Chem.*, 2010, **285**(43), 32695–32703. © The American Society for Biochemistry and Molecular Biology.

X was Ala or NH_2, two compounds activated when X was one of the hydrophobic, aromatic amino acids Phe or Trp (Figure 13.3), demonstrating that SIRT1 deacetylation can be activated with substrates composed solely of natural amino acids.[88]

A key remaining question is the identification of the enzyme form to which STACs bind. Most of the STACs tested do not bind to apo-SIRT1,[8,88] suggesting that they interact with an enzyme form after substrate has bound. Consistent with this, it has been demonstrated that SRT1460 binds to the SIRT1-TAMRA peptide complex by ITC.[8,87] This is exactly analogous to the mechanism of uncompetitive inhibition. Thus STACs may be uncompetitive activators which exert K_M-lowering effects through the stabilization of enzyme–substrate complexes.[8,88,89] While questions remain concerning the detailed mechanism of activation of SIRT1 by small molecules it is clear that

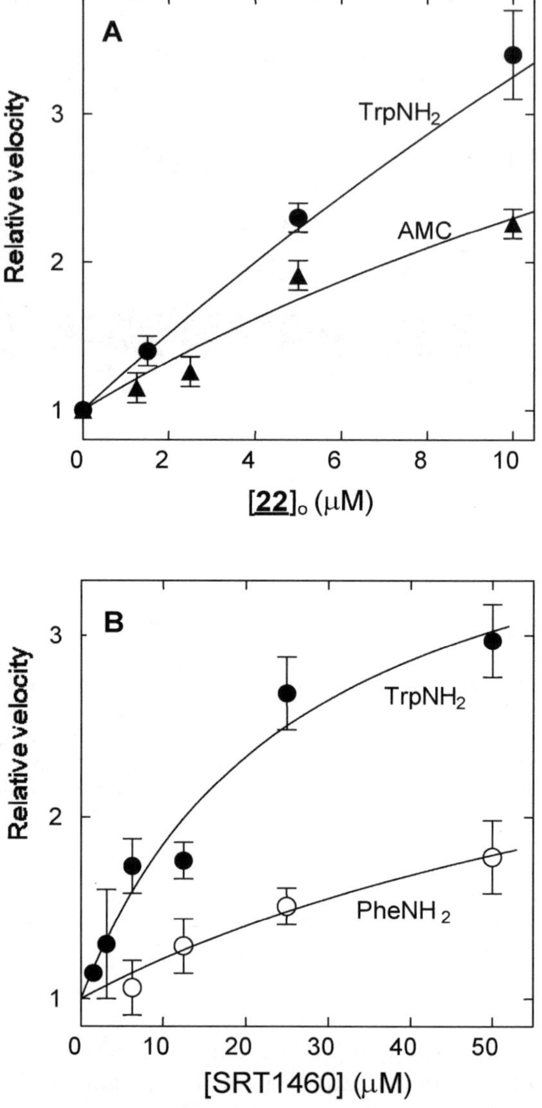

Figure 13.3 Activation of the SIRT1-catalyzed deacetylation of peptides of the sequence Ac-Arg-His-Lys-Lys^{Ac}-X-NH$_2$ by compound **22** from reference 88 (A) or SRT1460 (B). The residue at position X is specified on the graph. This research was originally published in the *Journal of Biological Chemistry*. Dai, H., Kustigian, L., Carney, D., Considine, T., Hubbard, B. P., Perni, R. B., Riera, T. V., Szczepankiewicz, B., Vlasuk, G. P., Stein, R. L. SIRT1 activation by small molecules: kinetic and biophysical evidence for direct interaction of enzyme and activator. *J. Biol. Chem.*, 2010, **285**(43), 32695–32703. © The American Society for Biochemistry and Molecular Biology.

STACs can directly activate the enzyme and that the likely mechanistic explanation involves allosteric modulation which has been described for several other classes of small molecule enzyme activators.[90]

13.4   The Medicinal Chemistry of SIRT1 Activators

Medicinal chemistry intervention on enzymatic pathways usually revolves around the inhibition of enzyme function. There are exceptions such as small molecule activation of glucokinase[91] and of protein kinase PDK1, but such examples are not common.[92] How activation of enzymes can be accomplished with small molecules is far less obvious than inhibiting an enzyme, though it is highly likely that allosteric mechanisms are at play. Unfortunately for SIRT1 there is no crystal structure available for assisting in mechanistic and drug design efforts, although there has been at least one attempt at modeling the protein.[93] Nevertheless, there are an increasing number of reports identifying STACs and these are summarized here. SIRT1 activation data is presented in a two ways, EC_{50} and $EC_{1.5}$. These terms are defined in Section 13.3.1. To date little cellular data has been reported for SIRT1 activators so no correlation of enzymatic activity to any cellular readout currently exists.

13.4.1   Polyphenols

The first small molecule activator of SIRT1 described in the literature was the polyphenol resveratrol (**13.1**).[94,95] This *trans*-stilbene analog has been demonstrated to modulate numerous biological pathways.[96,97] The large number of targets with which **13.1** appears to interact make discerning the true contribution of SIRT1 activation to its purported health benefits difficult to ascertain. The reactive, polyphenolic nature of this class of compound is problematic from an SAR perspective and the likelihood of developing potent SIRT1 activators with adequate drug-like properties from this starting point is small. Nevertheless, a limited body of work exists in this area. Resveratrol analogs **13.2** (pinosylvin) and **13.3** exhibit similar activating properties to resveratrol itself (fold-activation at 100 μM = 2.2, 3.3, and 2.1 for **13.1**, **13.2**, and **13.3**, respectively).[98]

13.1 R = OH; Resveratrol
13.2 R = H; Pinosylvin
13.3 R = OCH_3

13.4.2 Isoflavones

Isoflavones[99] have been identified as having activating properties toward SIRT1. Some members of the family appear to activate SIRT1 as well as up-regulate mitochondrial biogenesis. SIRT1 activity was determined using the fluorometric Fluor de Lys assay[100] and basic SAR has been established. For example in the, albeit limited, series studied, a free 7-hydroxyl substituent is required for SIRT1 activation. Of note from a pharmacophore perspective is the fact that both benzene rings of compounds **13.4–13.9** can overlay exactly with the phenyl rings of **13.1** (resveratrol). In addition both C3 and C4 of the isoflavone are sp^2 as is the olefinic linker of **13.1** (Figure 13.4). To date direct resveratrol analogs on the flavone scaffold have not been prepared. Compounds **13.4–13.6** display approximately a five-fold maximal activation of SIRT1. In contrast activators **13.7–13.9** do not activate SIRT1 at the maximum concentrated tested. There is no obvious structural delineation of active compounds from inactive ones.

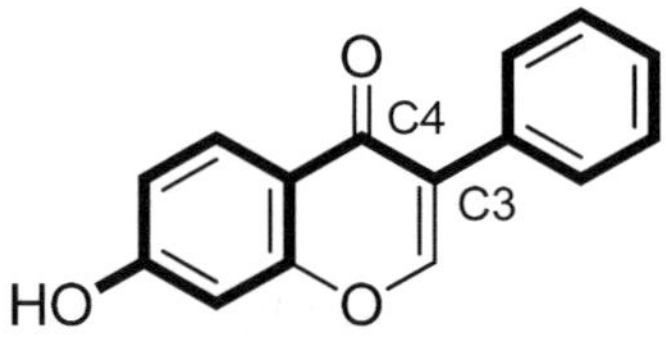

Figure 13.4   Resveratrol core (bolded line) overlay with 3-phenyl-isoflavone core.

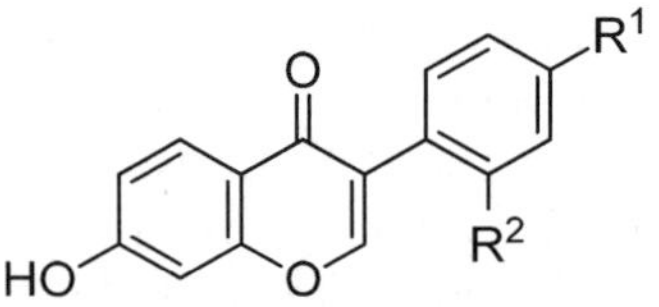

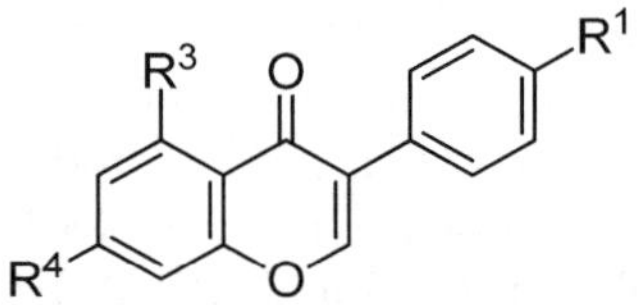

13.4 R^1 = OH, R^2 = H
13.5 R^1 = OCH$_3$, R^2 = H
13.6 R^1 = Cl, R^2 = Cl

13.7 R^1 = OH, R^3 = OH, R^4 = OH
13.8 R^1 = OCH$_3$, R^3 = OH, R^4 = OH
13.9 R^1 = OCH$_3$, R^3 = H, R^4 = OCH$_3$

13.4.3 Dihydropyridines

Calcium channel blockers derived from the 3,5-diacetyl-1,4-dihydropyridine scaffold comprise a broadly marketed class of anti-hypertensive agents. Closely related analogs, **13.10–13.15**, have been shown to possess the ability to both activate and inhibit SIRT1.[101] The SIRT1 modulators differ from that vast array of Ca^{2+} blockers only by the inclusion of an N-substituent. Curiously, the investigators imply that this series is derived from the observation of SIRT1 inhibitory activity of nicotinamide and the activating activity of *iso*-nicotinamide. Structurally, the relationships of *iso*-nicotinamide to the dihydropyridines is minimal and difficult to connect *a priori* on the basis of published SAR data. In general compounds with R = benzyl were activators (**13.10–13.12**) while compounds with other N-substituents were found to inhibit SIRT1 (**13.13–13.15**). Results were obtained with the Fluor de Lys assay.

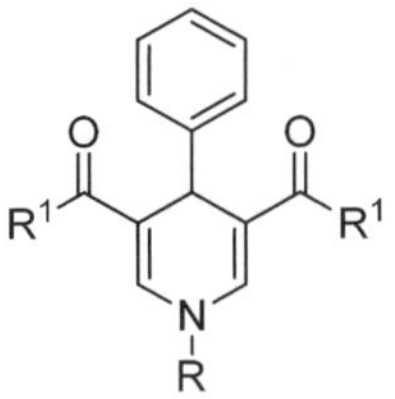

Activators

13.10 R = CH₂Ph R¹ = OEt
13.11 R = CH₂Ph R¹ = OH
13.12 R = CH₂Ph R¹ = NH₂

Inhibitors

13.13 R = cyclopropy R¹ = OEt
13.14 R = Ph R¹ = OEt
13.15 R = CH₂CH₂Ph R¹ = OEt

13.4.4 Dihydroquinolones

Annulated dihydroquinolones[102] are another recently described series of SIRT1 activators. These compounds were identified via the screening of a library of indole derivatives (**13.16–13.18**) which were prepared for general pharmacological screening. *In vitro* enzyme activation was determined with a SIRT1 fluorescence assay the origin of which was not disclosed. Though **13.18** appeared to cause a clear increase in SIRT1 activity as indicated by increased fluorescence, **13.16** or **13.17** displayed marginal, if any, activity.

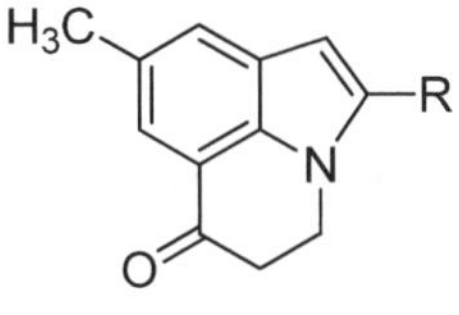

13.16 R = 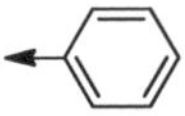
13.17 R 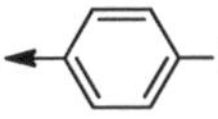 = —CH₃
13.18 R = t-Bu

13.4.5 Oxazolopyridines and Related Analogs

Oxazolopyridines (OAPs, e.g. **13.19**) were originally identified via high-throughput screening of a diversified library using a fluorescent polarization assay. Analogous azabenzimidazoles (ABIs, e.g. **13.22–13.24**) were subsequently created from SAR studies of core modifications.[9] More detailed SAR of these series was determined using a high-throughput mass spectrometry assay (HTMS).[8,9] All activators subsequently discussed in this chapter were evaluated using the HTMS assay.

13.19

SIRT1 activation is reported as an $EC_{1.5}$ value which is the concentration of the test article required to increase the enzyme activity by 50%. The *ortho* positioning of the aryl amide substituent (**13.20**, $EC_{1.5}$ = 0.9 μM) on the phenyl ring was found to be preferred relative to the *meta* position (**13.19**, $EC_{1.5}$ = 6 μM). Placement of the amide *para* to the core obliterated measureable activity (**13.21**, $EC_{1.5}$ >100 μM). Overall the SAR of this series is relatively insensitive to small changes. Modification of core ring atoms of OAP **13.20** ($EC_{1.5}$ = 0.9 μM) to give azabenzimidazole **13.22** ($EC_{1.5}$ = 0.5 μM) produces a marginal effect. Removal of one of the ring nitrogens gives benzimidazoles such as **13.23** and **13.24** ($EC_{1.5}$ = 0.4 μM and 0.3 μM, respectively) with activity little changed relative to **13.22**.

13.20 R^1 = R^2 = H

13.21 R^1 = H R^2 =

13.22 X = N, R^1 =

13.23 X = CH, R^1 =

13.24 X = CH, R^1 =

13.4.6   Imidazothiazoles, Thiazaolopyridines, and Related Analogs

Among the first synthetic SIRT1 activators reported following the OAPs/ABIs were imidazothiazoles (IATs).[8,10] This series represents an extension of the OAP/ABI series SAR and resulted in novel series exemplified by compounds such as **13.25–13.28**. Incorporation of substituents such as the methylaminopiperazinyl group off the core improved water solubility for this series relative to the unsubstituted IATs.[10] The *in vitro* and *in vivo* properties of several members of this series including **13.26** ($EC_{1.5}$ = 2.9 µM) and **13.28** ($EC_{1.5}$ = 0.16 µM) have also been described.[8] Furthermore it has been reported that activation appears to depend on the presence of an internal hydrogen bond acceptor for the amide NH on the activator core ring system. Isothermal calorimetry performed on **13.26** showed 1:1 binding stoichiometry with SIRT1 with K_d = 16.2 µM.

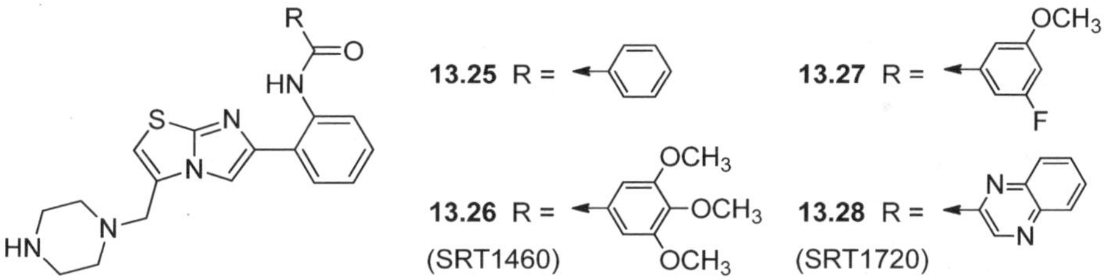

13.25 R =

13.26 R =   (SRT1460)

13.27 R =

13.28 R =   (SRT1720)

Thiazolopyiridines[103] (TAPs) are related to the IATs via modification of the bicyclic core while maintaining the relative positions of methyl amino substituents as well as the *o*-amidophenyl group. Given the apparent SAR of the OAPs, ABIs, and IATs, the heterobicyclic cores of SIRT1 activators are apparently interchangeable with respect to their ability to activate SIRT1. The retention of SIRT1 activation across multiple cores suggests that the core represents little more than structural scaffolding holding other key substituents in the proper geometric orientation while the cores themselves provide little high energy contact with the target. Activators incorporating a variety of biaryl ring systems appended via the amide linker maintain the ability to activate SIRT1 (**13.29–13.33**). $EC_{1.5}$ values for **13.29**, **13.30**, **13.32**, and **13.33** range from 0.23 µM to 0.66 µM while **13.31** is the most potent compound ($EC_{1.5}$ = 65 nM). These results indicate that small changes in polarity and charge distribution on the terminal region of the molecule can have significant effects on the ability of a compound in this series to activate SIRT1.

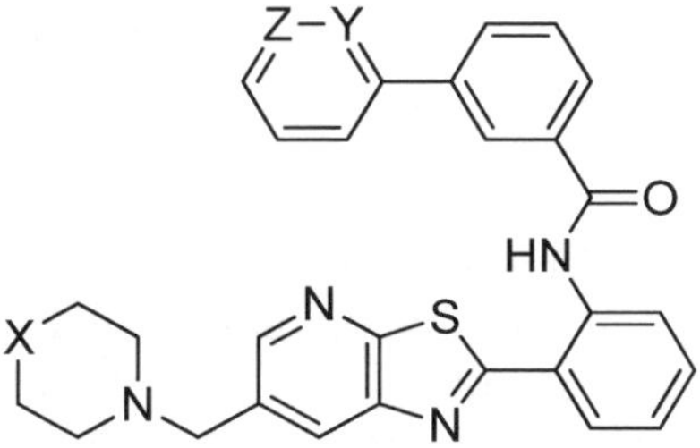

13.29   X = O,   Y = CH,   Z = CH
13.30   X = O,   Y = N,   Z = CH
13.31   X = O,   Y = CH,   Z = N
13.32   X = NH,   Y = CH,   Z = CH
13.33   X = NH,   Y = N,   Z = CH

Modification of the TAP scaffold by rearrangement of the methyl amino group, referred to as a water solubilizing group, away from the core to a pendant group[10] results in a new series of activators (**13.34–13.38**).[104,105] It was subsequently determined that the TAP, or any cyclic structure, was not required as the "core" to induce activation. Thiadiazoles **13.36** and **13.37** exhibit $EC_{1.5}$ = 0.67 µM and 0.10 µM, respectively, similar to that of **13.34** ($EC_{1.5}$ = 0.042 µM). Indeed when the TAP was replaced by a simple acetyl group (e.g. **13.38**), SIRT1 activation was still observed ($EC_{1.5}$ = 0.71 µM). Of particular note is compound **13.35** which displays significantly reduced SIRT1 activity. This compound lacks the remote basic nitrogen on the R-group that the other active examples possess.

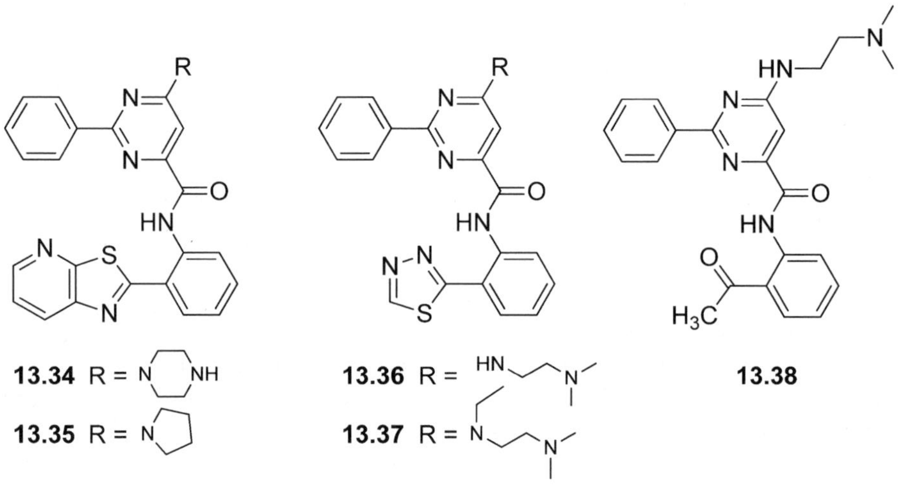

13.34 R =

13.35 R =

13.36 R =

13.37 R =

13.38

13.4.7 Newly Disclosed Activators

A number of recently published patent applications disclose a variety of series that differ primarily in the bicyclic core structures while maintaining very similar substituents appended from those cores. Consequently, these series may be reasonably expected to possess similar SAR. The following discussion summarizes the differences and similarities across the newest SIRT1 activators.

One recently disclosed series is based on a rearranged (relative to **13.21**, **13.22**) benzimidazole scaffold.[106,107] This particular series is related to the original benzimidazole series in that it possesses an aryl ring at the benzimidazole 2-position. Unlike the previously described compounds this series has the amide side chain appended at the 7-position of the benzimidazole instead of the *ortho*-position of the 2-phenyl group. Compounds **13.39–13.46** were chosen to exemplify the effects of a diversity of substituents. Examples such as **13.39** with a fully saturated amide substituent displayed no activating activity at the highest concentration tested while addition of a heteroatom, as in **13.41**, restores activity ($EC_{1.5}$ = 7 µM). Aromatic substituents exhibit the greatest activating ability (**13.40** and **13.42–13.45**) although pyridine derivative **13.43** ($EC_{1.5}$ = 22 µM) is a substantially weaker activator than **13.42**, **13.44**, or **13.45** ($EC_{1.5}$ = 2.1 µM, 0.35 µM, 0.09 µM, respectively). Further SAR analysis demonstrates that the reversal of the amide linkage (e.g. **13.46** $EC_{1.5}$ < 10 µM) generally maintains activity.

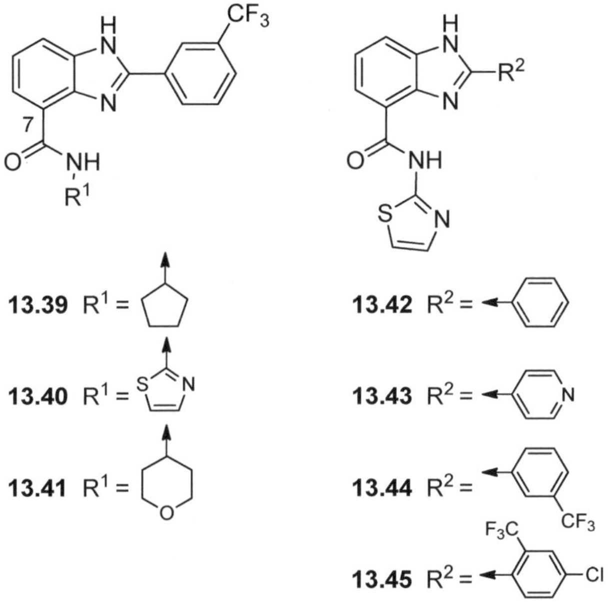

Another series closely related to benzimidazoles **13.39–13.46** is a series of quinoline derivatives exemplified by **13.47–13.51**.[108] The overall spatial distribution of the phenyl and amide substituents is similar to that of the benzimidazoles with the only significant change being the ring expansion from 5 to 6 in the right-hand ring of the core affecting the relative positions of the two terminal rings. Activity in this series overall appears similar to that of the benzimidazoles (**13.39–13.46**) though exact values for activation have not been reported.

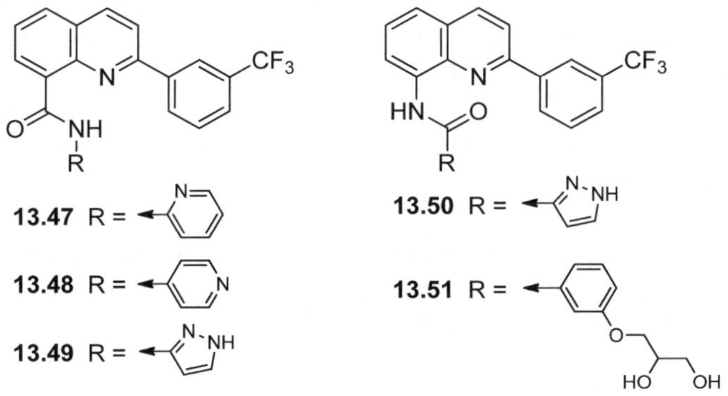

Pyridine activators[109] were evolved from the quinoline series by reduction the benzene ring of the quinoline. Tetrahydroquinoline derivatives such **13.52**, **13.54**, and **13.55**, exhibit $EC_{1.5}$ values of $<10\ \mu M$. Incorporation of an oxygen in the saturated ring as in **13.53** reduced activity at least 2.5-fold. Relative to **13.55**, compound **13.56** was less active. Further it was found that SIRT1 activation remained ($EC_{1.5} < 10\ \mu M$) if the reduced ring was removed leaving a pendant ethyl side chain (**13.57–13.58**) although reversal of the amide again significantly attenuated activation (**13.59**).

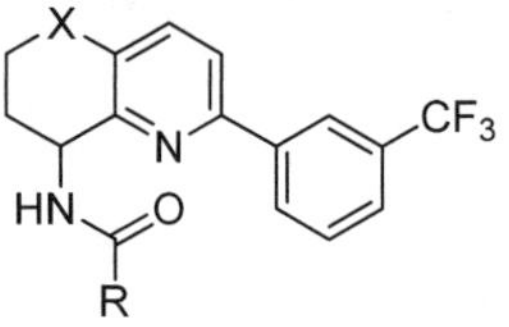

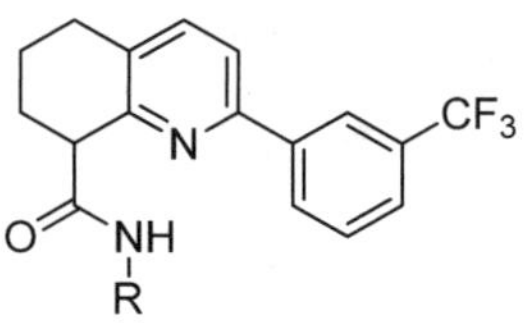

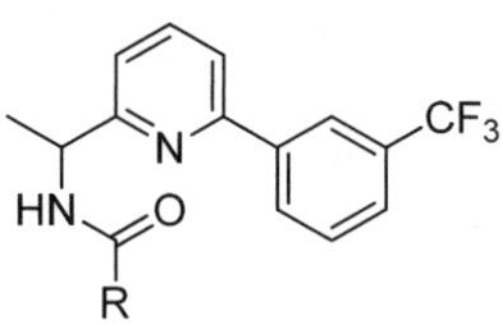

13.52 X = CH₂ R =

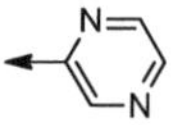

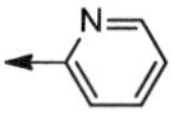

13.53 X = O   R =

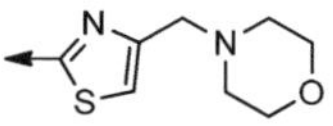

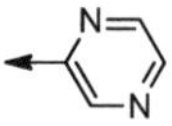

13.54 X = CH₂ R =

13.55 R =

13.56 R =

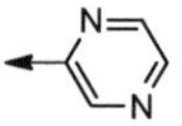

13.57 R =

13.58 R =

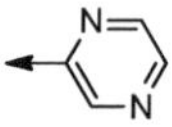

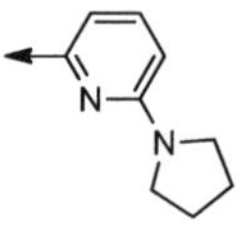

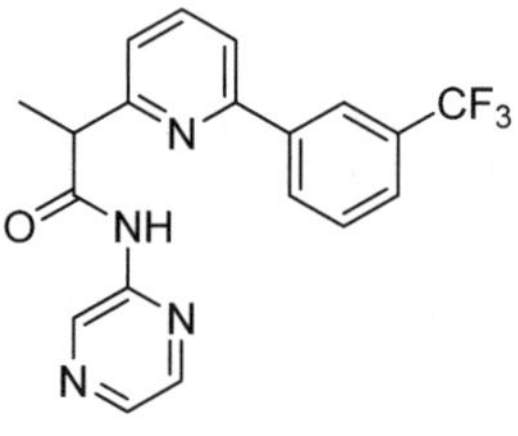

13.59

Quinolones and quinoxazolines, are closely related series recently disclosed in a published patent application.[110] In these series activation is sensitive to substitution on the 3-position nitrogen given that **13.60** displays $EC_{1.5} < 1\ \mu M$ while **13.61** is at least 25-fold less active. On the other hand, the directionality of the amide linker appears to have minimal effect (**13.62**, $EC_{1.5} < 1\ \mu M$). Similarly, removal of the 3-position nitrogen to provide the simple quinolones, **13.63** and **13.64**, also has little impact ($EC_{1.5} < 1\ \mu M$ for both) on the ability of these compounds to activate SIRT1.

In contrast to the quinolone and quinoxazilone series the directionality of the amide does appear to play a role for the closely related chromenone series,[111] where the 1-position nitrogen has been replaced by an oxygen atom. Chromenone **13.65** displays potent activity ($EC_{1.5} < 1\ \mu M$) while **13.67** exhibits greatly reduced activating ability ($EC_{1.5} > 25\ \mu M$). In this series the SAR is also sensitive to the position of substituents on the phenyl ring (**13.65** $EC_{1.5} < 1\ \mu M$ versus **13.66** $EC_{1.5} > 25\ \mu M$).[111]

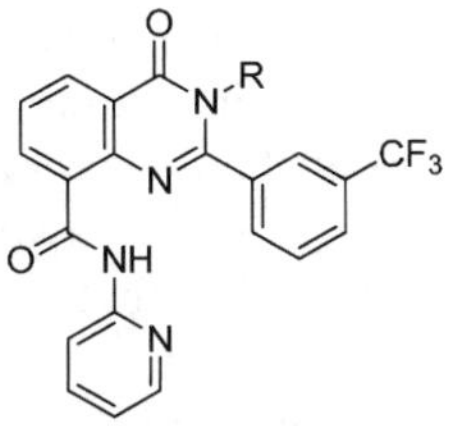

13.60 R = CH$_3$
13.61 R = CH$_2$CH$_2$CH$_3$

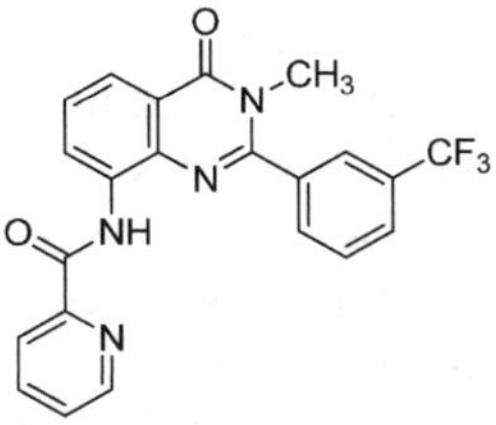

13.62

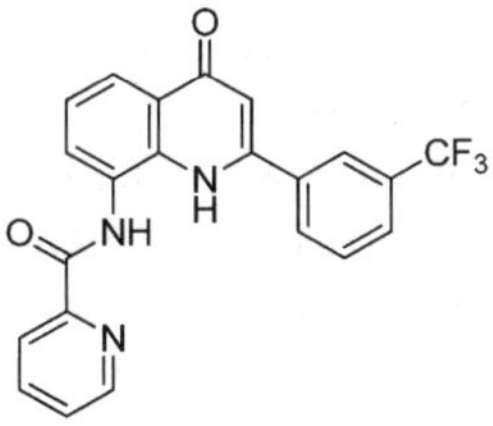

13.63

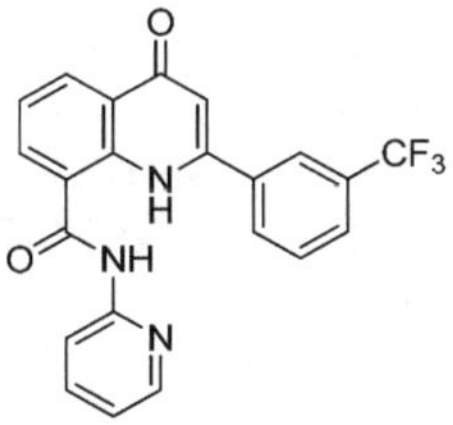

13.64

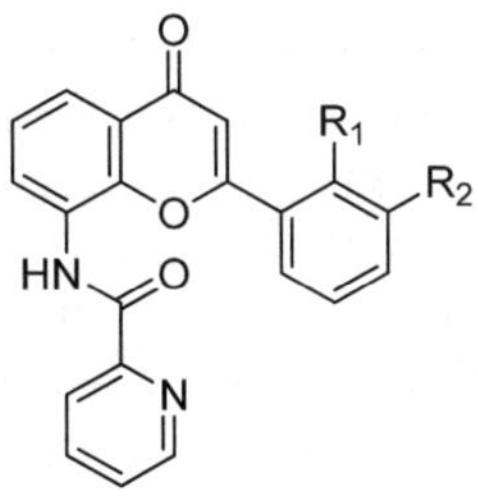

13.65 R^1 = CF$_3$, R^2 = H
13.66 R^1 = H, R^2 = CF$_3$

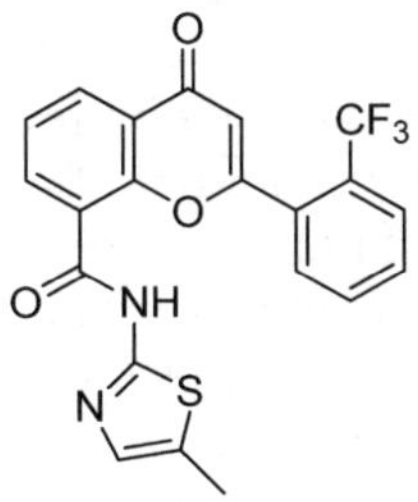

13.67

13.5   Preclinical Studies with SIRT1 Activators

There is little clinical data available for the selective SIRT1 activators. SRT2104 is being studied in multiple trials including trials in type 2 diabetes and psoriasis and in an acute inflammation study involving LPS challenge of normal volunteers. Preliminary data demonstrating an anti-inflammatory effect of SRT2104 from the latter study was presented recently at the World Inflammation Congress.[112] In addition First Time in Human Trials have been carried out for SRT2379 and SRT3025 (unpublished). Preclinically, SRT1720 has been the most extensively tested compound with data available from many labs in several disease models, including models of diabetes,[8,11,113] inflammation,[16,114] and fatty liver disease.[17,31]

In the original paper on SRT1720,[8] the compound was tested in three different models of type 2 diabetes at a dose of 100 mg/kg. In the diet-induced obesity (DIO) model, SRT1720 significantly improved most metabolic parameters, including a reduction in blood glucose that was maintained over 10 weeks of dosing, a reduction in the hyperinsulinemia, and a reduction in glucose excursion in the intraperitoneal glucose tolerance test that was comparable to rosiglitazone. SRT1720 also increased mitochondrial biogenesis in DIO mice as measured by an increase in citrate synthase levels in the gastrocnemius muscle. In the leptin deficient ob/ob mouse, SRT1720 significantly decreased blood glucose after 1 week of treatment with no effect on body weight. In the Zucker fa/fa model of diabetes, SRT1720 improved insulin sensitivity as evidenced by a reduction in glucose and insulin levels during an oral glucose tolerance test. In a hyperinsulinemic-euglycemic clamp study in Zucker fa/fa rats, SRT1720 increased the glucose infusion rate, the glucose disposal rate, and the insulin-stimulated glucose disposal rate after 4 weeks of dosing.

In the study by Feige *et al.*,[115] SRT1720 was added to the chow at equivalents of 100 and 500 mg/kg/day in mice fed a HFD. The compound was found to protect from DIO by preventing fat accumulation and by increasing energy expenditure. Lipid profiles (triglycerides and cholesterol) were also improved by SRT1720. In agreement with the results from Milne *et al.*,[8] SRT1720 reduced glucose and insulin levels and increased metabolic parameters in the hyperinsulinemic-euglycemic clamp paradigm. SRT1720 significantly increased muscle performance as measured by treadmill testing, grip strength, and rotarod performance. There was a switch in the contractile phenotype of gastrocnemius muscle from fast to slow twitch accompanied by an increase in genes controlling fatty acid oxidation (FAO). The livers from SRT1720-treated animals had reduced fat and increased expression of genes involved in mitochondrial metabolism and FAO. SRT1720 also stimulated energy expenditure in brown fat and again increased genes involved in mitochondrial function and FAO. In this study, SRT1720 was shown to increase deacetylation of known SIRT1 targets, including p53, PGC-1α, and FOXO1.

These studies have been significantly extended by Minor *et al.*[116] who administered SRT1720 in food at doses of 30 and 100 mg/kg in 1-year-old mice fed a HFD for the remainder of their lifespan. SRT1720 significantly increased the mean and maximum lifespan of the mice while conferring significant health benefits including reduced liver steatosis, increased insulin sensitivity, and a reduction in inflammation and apoptotic markers. Moreover, the altered gene expression profile of the HFD diet fed mice was almost normalized to that of normal chow mice in the HFD animals receiving SRT1720. The positive benefit of SRT1720 on mitochondrial respiration was shown to be SIRT1-dependent by using SIRT1 conditional knockout animals. Additionally, the ability of SRT1720 to increase resistance to oxidant stress was absent in mouse embryonic fibroblasts (MEFs) from SIRT1 knockout mice.

In addition to the positive metabolic effects of SIRT1 activation produced by compounds like SRT1720, it is also possible that SIRT1 activators could affect insulin sensitivity by inhibiting inflammation.[16,113,116] The treatment of Zucker fa/fa rats with the SIRT1 activator SRT2379 (100 mg/kg) (structure not disclosed) improved glucose tolerance, reduced hyperinsulinemia, and enhanced glucose uptake during a hyperinsulinemic-euglycemic clamp study.[113] The metabolic improvement was accompanied by a reduction in inflammatory cytokines (TNFα, IL-6) and chemokines (MCP-1) in adipose tissue and a reduction in the inflammatory state of macrophages in adipose tissue. In this same study, it was shown that knockdown of SIRT1 increased the activation of the JNK and IKK inflammatory pathways and increased LPS-induced TNFα release in primary macrophages and in macrophage cell lines. Activation of SIRT1 by SRT1720 had the opposite effect to SIRT1 knockdown and inhibited LPS-stimulated inflammatory pathways and TNFα release. The SIRT1 dependence of the effect of SRT1720 was demonstrated by the observation that SRT1720 no longer inhibited LPS-induced effects on JNK, IKK, and TNFα release when SIRT1 was knocked down using siRNA.[113]

Further anti-inflammatory effects of SIRT1 activation with SRT1720 have been demonstrated in adipocytes[16] and in human gestational tissues.[114] SIRT1 knockdown in adipocytes inhibited insulin-stimulated glucose uptake and increased activation of inflammatory pathways as evidenced by increased phosphorylation of JNK and insulin receptor substrate 1 (IRS-1). The opposite effect was seen with the SIRT1 activators SRT1720 and SRT2530.[116] In mice fed a high-fat diet, SRT1720 (100 mg/kg) decreased the amount of acetylated NF-κB and reduced the expression levels of the NF-κB target genes JNK, iNOS, and TNFα receptor-associated factor 2. In LPS-stimulated human placental villous tissue, SRT1720 inhibited IL-6 and IL-8 release as well as that of prostaglandins E$_2$ and F$_{2\alpha}$.[114]

Consistent with the effects on liver seen by Feige *et al.*[115] and Minor *et al.*,[116] SRT1720 has also been shown to ameliorate fatty liver in monosodium glutamate (MSG) mice, a model of nonalcoholic fatty liver disease (NAFLD).[17] SRT1720 (200 mg/kg) inhibited triglyceride accumulation in the livers of MSG mice and reduced the elevated levels of aminotransferase seen in control mice. This positive impact on NAFLD by SRT1720 was accompanied by a decrease in the hepatic expression of lipogenic genes as well as genes involved in inflammation (inflammatory cytokines, macrophage infiltration) and oxidative stress. Serum levels of triglycerides, free fatty acids, and cholesterol were also reduced by SRT1720. SIRT1 has been shown to be involved in lipid and cholesterol regulation through deacetylation of the key transcription factor sterol regulatory element-binding protein (SREBP) leading to decreased stability of SREBP and a decrease in lipid synthesis and fat deposition.[31] *In vitro*, SRT2183 decreased SREBP as well as expression of SREBP target genes. SRT1720 (100 mg/kg) given to ob/ob mice for 4 weeks decreases levels of SREBP target genes and inhibits liver steatosis.[31]

SIRT1 activation may also produce beneficial metabolic effects through changes in energy balance. Energy balance during fasting is controlled by a fasting-inducible switch consisting of the protein lysine acetyltransferase p300 and the deacetylase SIRT1 which controls energy balance through the regulation of TORC2 and FOXO1. Liver-specific knockout of SIRT1 or a SIRT1 inhibitor increases TORC2 activity and glucose output.[14] Over-expression of SIRT1 or administration of SRT1720 to hepatocytes decreased TORC2 activity. SRT1720 (100 mg/kg) also decreased the amount of acetylated TORC2 in Zucker fa/fa rats.

The positive effect of SRT1720 on mitochondrial biogenesis seen by Milne *et al.*,[8] Feige *et al.*,[115] and Minor *et al.*[116] has also been seen in renal proximal tubule cells (RPTC).[117,118] The compound increased mitochondrial biogenesis through deacetylation of the SIRT1 target PGC1α and this increase in mitochondrial biogenesis was able to prevent oxidant-induced injury of the RPTC. SRT1720 has also been shown to prevent renal fibrosis in the mouse model of unilateral ureteral obstruction (UUO).[119] SRT1720 (100 mg/kg) reduced apoptosis and fibrosis in the UUO model, potentially through an increase in COX2 expression. The effect of SRT1720 is in direct contrast to what is observed in mice heterozygous for SIRT1 which show increased apoptosis and fibrosis in the UUO model.[119]

13.5.1 SIRT1 Dependence of SIRT1 Modulators

The SIRT1 dependence of SRT1720 has recently been called into question by Pacholec *et al.*, who were unable to show activation of non-fluorometric SIRT1 substrates by SRT1720 in enzymology studies and demonstrated a lack of selectivity against a panel of non-related molecular targets.[87] As discussed earlier, it may not be possible to replicate the cellular milieu of the SIRT1 enzyme in a test tube. Therefore, the SIRT1 dependence can only be studied in cell-based systems using appropriate inhibitors and/or siRNA and ultimately in animals in which SIRT1 has been conditionally knocked out. Thus, there have been multiple studies with SRT1720 that have consistently been able to demonstrate the SIRT1 dependence of this compound and other SIRT1 activators in cell-based assays[16,113,115–117] as well as *in vivo*[116] thereby confirming the primary molecular target as SIRT1.

In the study by Feige *et al.*, the ability of SRT1720 to stimulate oxygen consumption in muscle cells was abolished in cells in which SIRT1 had been knocked down by shRNA.[115] In the study of the anti-inflammatory effects of SRT1720 in macrophages,[113] the ability of SRT1720 to inhibit LPS-induced effects on JNK and IKK phosphorylation and TNFα release was abolished in cells in which SIRT1 had been knocked down using siRNA targeting. In addition CD11c expression, which represents the activated pro-inflammatory phenotype of RAW cells, was inhibited by SRT1720 in contrast to the effects seen following SIRT1 knockdown which had the opposite effect as well as eliminated the impact of SRT1720.[113] Minor *et al.* have demonstrated the SIRT1-dependency of SRT1720 *in vitro* and *in vivo*.[116] The ability of SRT1720 to confer stress resistance to H_2O_2 in wild-type MEFs was lost in MEFs from SIRT1 knockout mice. SRT1720 also rescued respiratory capacity in mitochondria and wild-type mice treated with a HFD but not in SIRT1 knockout mice. In the paper by Funk *et al.*, the SIRT1 dependence of SRT1720 was demonstrated using the putative SIRT1 inhibitors sirtinol and nicotinamide.[117] Sirtinol and nicotinamide are weak, relatively non-specific SIRT1 inhibitors so other biological processes may be at play but the inhibitor data in conjunction with the above-mentioned knockdown data provide strong evidence of SIRT1-dependence for SRT1720. This SIRT1 dependence shown for SRT1720[16,113,115,117] suggests that the off-target binding activity reported by Pacholec *et al.*[87] does not influence the biological effects seen by the different investigators and highlights the importance of checking any pharmacological tool for on-target effects by using appropriate knockdown or knockout techniques.

Pacholec *et al.* were unable to show efficacy with SRT1720 (30 and 100 mg/kg) in ob/ob mice aside from a reduction in insulin levels.[87] This is in contrast to numerous studies from multiple laboratories which have shown *in vivo* efficacy with SRT1720 in multiple models of diabetes[8,113,115] inflammation,[16] and fatty liver disease.[17] The lack of efficacy observed by Pacholec *et al.* may have been due to the toxicity that was observed in their studies. This toxicity is puzzling, as other laboratories[8,17] have seen no impact on body weights or food intake or other obvious toxicity. Most striking is the study carried out at the National Institute on Aging where no toxicity was observed in mice administered SRT1720 at 100 mg/kg/day in chow for 80 weeks. Indeed, in this study SRT1720 increased lifespan and had significant health benefits in mice fed a HFD.[116]

A systems biology approach has also been used to look at the signaling pathways which are activated by calorie restriction, resveratrol, and SRT1720.[15] Signaling pathways involved in mitochondrial biogenesis, metabolic pathways, and inflammatory pathways were all similarly affected by resveratrol, SRT1720, or calorie restriction. This provides strong evidence of a common molecular target involved with the three interventions, with SIRT1

being the most likely candidate based on the known biology associated with
SRT1720, resveratrol, and calorie restriction.

Consistent with the systems biology data described above, the SIRT1
dependency of resveratrol has been shown in a number of different contexts. In
yeast, resveratrol extends lifespan but this effect is lost in Sir2 null mutant
yeast.[120] Kaeberlein *et al.*[121] reported that resveratrol was unable to activate Sir2
or increase lifespan in yeast. This inability to see lifespan extension by resveratrol
is in contrast to the original paper by Howitz *et al.*[120] and subsequent work in
yeast with resveratrol[122,123] and resveratrol derivatives.[123] In addition,
resveratrol has been shown to produce lifespan extension in multiple species
including worms,[124,125] flies,[125,126] and mice on a HFD.[127] In mouse embryonic
fibroblasts (MEFs) isolated from wild-type mice, resveratrol increases expression of PGC1α, and cytochrome c, an effect which is absent in MEFS from
SIRT1 knockout mice.[128] In 3T3-L1 adipocytes, resveratrol inhibits adipogenesis, an effect which is lost in SIRT1 RNAi-treated cells.[70] Further studies in
adipocytes have shown that resveratrol's effects on glucose uptake, *de novo*
lipogenesis and cytokine expression are also SIRT1-dependent.[129] In an *in vivo*
setting, Boily *et al.*[130] show that the anti-tumorigenic effect of resveratrol in a
classical two-stage carcinogenesis model is significantly reduced in SIRT1
knockout mice. Resveratrol, like SRT1720, has been shown to inhibit cytokine
and prostaglandin release in human gestational tissues stimulated by LPS. This
effect is absent in cells in which SIRT1 has been knocked down by siRNA.[114]
Recently Price *et al.*[131] showed that mice treated with a moderate dose of
resveratrol had increased mitochondrial biogenesis and function along with
AMPK activation and increased NAD^+ levels in skeletal muscle. This effect was
reproduced by over-expression of SIRT1 and was absent in SIRT1 conditional
knockout mice thus providing convincing evidence of the SIRT1-dependence of
resveratrol's effects.

13.6   Conclusions

SIRT1 is a fascinating enzyme with a broad and varied biology across multiple
organ systems. This breadth of biology has been the basis for the focus on the
development of modulators of this enzyme, particularly on small molecule
activators which can mimic the beneficial effects seen in transgenic animals
over-expressing SIRT1. The chemistry for these compounds has focused on a
clear set of core structural motifs with distinct structure–activity relationships.
What is needed is the molecular detail of how these compounds bind to SIRT1,
which can only come from crystallographic data that so far has been elusive.
The early controversy regarding whether small molecules could directly
activate SIRT1 *in vitro* appears to have been clarified with the proposal of an
allosteric mechanism of activation which is consistent with the data using
fluorophore-tagged and un-tagged peptide substrates containing natural
amino acids. What remains is elucidating the mechanism of activation with
macromolecular substrates *in vitro* and extending this to cells and *in vivo*

systems. With the rapid development of technologies for determination of quantitative acetylation profiles of protein substrates, it is hoped that a specific signature for SIRT1 activity could be developed that would provide an avenue for monitoring SIRT1 modulation *in vivo* as well as clinical studies. The pharmacology of SIRT1 activators has provided tantalizing clues to the potential for these compounds as future broad spectrum medicines for the treatment of disease associated with aging. Time will tell if this ultimate goal of targeting SIRT1 will result in a new generation of medicines; however, the significant progress that has taken place to date points to this becoming a reality in the near future.

References

1. A. A. Sauve, I. Celic, J. Acalos, H. Deng, J. D. Boeke and V. L. Schramm, *Biochemistry*, 2001, **40**, 15456–15463.
2. A. A. Sauve and V. L. Schramm, *Biochemistry*, 2003, **42**, 9249–9256.
3. A. A. Sauve, C. Wolberger, V. L. Schramm and J. D. Boeke, *Annu. Rev. Biochem.*, 2006, **75**, 435–465.
4. L. Guarente, *Cold Spring Harb. Symp. Quant. Biol.*, 2007, **72**, 483–488.
5. M. C. Haigis and D. Sinclair, *Annu. Rev. Pathol.*, 2010, **5**, 253–259.
6. J. A. Baur, *Mech. Ageing Dev.*, 2010, **131**, 261–269.
7. J. Rutanen, N. Yaluri, S. Modi, J. Pihlajamaki, M. Vanttinen, P. Itkonen, S. Kainulainen, H. Yamamoto, M. Lagouge, D. A. Sinclair, P. Elliott, C. Westphal, J. Auwerx and M. Laakso, *Diabetes*, 2010, **59**, 829–835.
8. J. C. Milne, P. D. Lambert, S. Schenk, D. P. Carney, J. J. Smith, D. J. Gagne, L. Jin, O. Boss, R. B. Perni, C. B. Vu, J. E. Bemis, R. Xie, J. S. Disch, P. Y. Ng, J. J. Nunes, A. V. Lynch, H. Yang, H. Galonek, K. Israelian, W. Choy, A. Iffland, S. Lavu, O. Medvedik, D. A. Sinclair, J. M. Olefsky, M. R. Jirousek, P. J. Elliott and C. H. Westphal, *Nature*, 2007, **450**, 712–716.
9. J. E. Bemis, C. B. Vu, R. Xie, J. J. Nunes, P. Y. Ng, J. S. Disch, J. C. Milne, D. P. Carney, A. V. Lynch, L. Jin, J. J. Smith, S. Lavu, A. Iffland, M. R. Jirousek and R. B. Perni, *Bioorg. Med. Chem. Lett.*, 2009, **19**, 2350–2353.
10. C. B. Vu, J. E. Bemis, J. S. Disch, P. Y. Ng, J. J. Nunes, J. C. Milne, D. P. Carney, A. V. Lynch, J. J. Smith, S. Lavu, P. D. Lambert, D. J. Gagne, M. R. Jirousek, S. Schenk, J. M. Olefsky and R. B. Perni, *J. Med. Chem.*, 2009, **52**, 1275–1283.
11. J. N. Feige, M. Lagouge, C. Canto, A. Strehle, S. M. Houten, J. C. Milne, P. D. Lambert, C. Mataki, P. J. Elliott and J. Auwerx, *Cell. Metab.*, 2008, **8**, 347–358.
12. J. A. Funk, S. Odejinmi and R. G. Schnellmann, *J. Pharmacol. Exp. Ther.*, 2010, **333**, 593–601.
13. L. Jin, H. Galonek, K. Israelian, W. Choy, M. Morrison, Y. Xia, X. Wang, Y. Xu, Y. Yang, J. J. Smith, E. Hoffmann, D. P. Carney, R. B.

Perni, M. R. Jirousek, J. E. Bemis, J. C. Milne, D. A. Sinclair and C. H. Westphal, *Protein Sci.*, 2009, **18**, 514–525.

14. Y. Liu, R. Dentin, D. Chen, S. Hedrick, K. Ravnskjaer, S. Schenk, J. Milne, D. J. Meyers, P. Cole, J. Yates, *Nature*, 2008, **456**, 269–273.

15. J. J. Smith, R. D. Kenney, D. J. Gagne, B. P. Frushour, W. Ladd, H. L. Galonek, K. Israelian, J. Song, G. Razvadauskaite, A. V. Lynch, D. P. Carney, R. J. Johnson, S. Lavu, A. Iffland, P. J. Elliott, P. D. Lambert, K. O. Elliston, M. R. Jirousek, J. C. Milne and O. Boss, *BMC Syst. Biol.*, 2009, **3**, 31.

16. T. Yoshizaki, J. C. Milne, T. Imamura, S. Schenk, N. Sonoda, J. L. Babendure, J. C. Lu, J. J. Smith, M. R. Jirousek and J. M. Olefsky, *Mol. Cell. Biol.*, 2009, **29**, 1363–1374.

17. Y. Yamazaki, I. Usui, Y. Kanatani, Y. Matsuya, K. Tsuneyama, S. Fujisaka, A. Bukhari, H. Suzuki, S. Senda, S. Imanishi, K. Hirata, M. Ishiki, R. Hayashi, M. Urakaze, H. Nemoto, M. Kobayashi and K. Tobe, *Am. J. Physiol. Endocrinol. Metab.*, 2009, **297**, E1179–E1186.

18. J. P. Silva and C. Wahlestedt, *Drug Discov. Today*, **15**, 781–791.

19. D. Herranz and M. Serrano, *Nat. Rev. Cancer*, **10**, 819–823.

20. J. Yu and J. Auwerx, *Ann. NY Acad. Sci.*, 2009, **1173** (Suppl. 1), E10–19.

21. S. J. Lin, P. A. Defossez and L. Guarente, *Science*, 2000, **289**, 2126–2128.

22. D. B. Lombard, S. D. Pletcher, C. Canto and J. Auwerx, *Nature*, 2011, **477**, 410–411.

23. L. Guarente, *N. Engl. J. Med.*, 2011, **364**, 2235–2244.

24. Y. Y. Lin, J. Y. Lu, J. Zhang, W. Walter, W. Dang, J. Wan, S. C. Tao, J. Qian, Y. Zhao, J. D. Boeke, S. L. Berger and H. Zhu, *Cell*, 2009, **136**, 1073–1084.

25. S. Bamps, J. Wirtz, F. R. Savory, D. Lake and I. A. Hope, *Mech. Ageing Dev.*, 2009, **130**, 762–770.

26. R. Luthi-Carter, D. M. Taylor, J. Pallos, E. Lambert, A. Amore, A. Parker, H. Moffitt, D. L. Smith, H. Runne, O. Gokce, A. Kuhn, Z. Xiang, M. M. Maxwell, S. A. Reeves, G. P. Bates, C. Neri, L. M. Thompson, J. L. Marsh and A. G. Kazantsev, *Proc. Natl. Acad. Sci. USA*, 2011, **107**, 7927–7932.

27. J. T. Rodgers, C. Lerin, Z. Gerhart-Hines and P. Puigserver, *FEBS Lett.*, 2008, **582**, 46–53.

28. X. Li, S. Zhang, G. Blander, J. G. Tse, M. Krieger and L. Guarente, *Mol. Cell*, 2007, **28**, 91–106.

29. F. Wang, M. Nguyen, F. X. Qin and Q. Tong, *Aging Cell*, 2007, **6**, 505–514.

30. J. Nakae, Y. Cao, H. Daitoku, A. Fukamizu, W. Ogawa, Y. Yano and Y. Hayashi, *J. Clin. Invest.*, 2006, **116**, 2473–2483.

31. A. K. Walker, F. Yang, K. Jiang, J. Y. Ji, J. L. Watts, A. Purushotham, O. Boss, M. L. Hirsch, S. Ribich, J. J. Smith, K. Israelian, C. H. Westphal, J. T. Rodgers, T. Shioda, S. L. Elson, P. Mulligan, H. Najafi-Shoushtari, J. C. Black, J. K. Thakur, L. C. Kadyk, J. R. Whetstine, R.

Mostoslavsky, P. Puigserver, X. Li, N. J. Dyson, A. C. Hart and A. M. Naar, *Genes Dev.*, 2010, **24**, 1403–1417.

32. A. Salminen and K. Kaarniranta, *Cell. Signal.*, 2009, **21**, 1356–1360.

33. H. S. Ghosh, M. McBurney and P. D. Robbins, *PLoS One*, **5**, e9199.

34. W. C. Hallows, S. Lee and J. M. Denu, *Proc. Natl. Acad. Sci. USA*, 2006, **103**, 10230–10235.

35. M. D. Hirschey, T. Shimazu, J. A. Capra, K. S. Pollard and E. Verdin, *Aging (Albany NY)*, 2011, **3**, 635–642.

36. A. Nikiforov, C. Dolle, M. Niere and M. Ziegler, *J. Biol. Chem.*, 2011, **286**, 21767–21778.

37. Y. Nakahata, S. Sahar, G. Astarita, M. Kaluzova and P. Sassone-Corsi, *Science*, 2009, **324**, 654–657.

38. K. M. Ramsey, J. Yoshino, C. S. Brace, D. Abrassart, Y. Kobayashi, B. Marcheva, H. K. Hong, J. L. Chong, E. D. Buhr, C. Lee, J. S. Takahashi, S. Imai and J. Bass, *Science*, 2009, **324**, 651–654.

39. S. Imai, *Biochim. Biophys. Acta*, 2010, **1804**, 1584–1590.

40. W. Huang, K. M. Ramsey, B. Marcheva and J. Bass, *J. Clin. Invest.*, 2011, **121**, 2133–2141.

41. L. Tong and J. M. Denu, *Biochim. Biophys. Acta*, 2010, **1804**, 1617–1625.

42. L. Tong, S. Lee and J. M. Denu, *J. Biol. Chem.*, 2009, **284**, 11256–11266.

43. M. C. Haigis and D. A. Sinclair, *Annu. Rev. Pathol.*, 2010, **5**, 253–295.

44. M. W. McBurney, X. Yang, K. Jardine, M. Hixon, K. Boekelheide, J. R. Webb, P. M. Lansdorp and M. Lemieux, *Mol. Cell Biol.*, 2003, **23**, 38–54.

45. R. H. Wang, K. Sengupta, C. Li, H. S. Kim, L. Cao, C. Xiao, S. Kim, X. Xu, Y. Zheng, B. Chilton, R. Jia, Z. M. Zheng, E. Appella, X. W. Wang, T. Ried and C. X. Deng, *Cancer Cell*, 2008, **14**, 312–323.

46. H. L. Cheng, R. Mostoslavsky, S. Saito, J. P. Manis, Y. Gu, P. Patel, R. Bronson, E. Appella, F. W. Alt and K. F. Chua, *Proc. Natl. Acad. Sci. USA*, 2003, **100**, 10794–10799.

47. J. Gao, W. Y. Wang, Y. W. Mao, J. Graff, J. S. Guan, L. Pan, G. Mak, D. Kim, S. C. Su and L. H. Tsai, *Nature*, 2010, **466**, 1105–1109.

48. M. F. Champy, M. Selloum, V. Zeitler, C. Caradec, B. Jung, S. Rousseau, L. Pouilly, T. Sorg and J. Auwerx, *Mamm. Genome*, 2008, **19**, 318–331.

49. P. T. Pfluger, D. Herranz, S. Velasco-Miguel, M. Serrano and M. H. Tschop, *Proc. Natl. Acad. Sci. USA*, 2008, **105**, 9793–9798.

50. D. Herranz, M. Munoz-Martin, M. Canamero, F. Mulero, B. Martinez-Pastor, O. Fernandez-Capetillo and M. Serrano, *Nat. Commun.*, 2010, **1**, 3.

51. A. S. Banks, N. Kon, C. Knight, M. Matsumoto, R. Gutierrez-Juarez, L. Rossetti, W. Gu and D. Accili, *Cell. Metab.*, 2008, **8**, 333–341.

52. B. M. Spiegelman and J. S. Flier, *Cell*, 2001, **104**, 531–543.

53. G. J. Morton and M. W. Schwartz, *Physiol. Rev.*, 2011, **91**, 389–411.

54. G. Ramadori, C. E. Lee, A. L. Bookout, S. Lee, K. W. Williams, J. Anderson, J. K. Elmquist and R. Coppari, *J. Neurosci.*, 2008, **28**, 9989–9996.

55. G. Ramadori and R. Coppari, *Pharmacol. Res.*, 2010, **62**, 48–54.

56. X. Li and N. Kazgan, *Int. J. Biol. Sci.*, 2011, **7**, 575–587.

57. G. Ramadori, T. Fujikawa, J. Anderson, E. D. Berglund, R. Frazao, S. Michan, C. R. Vianna, D. A. Sinclair, C. F. Elias and R. Coppari, *Cell. Metab.*, 2011, **14**, 301–312.

58. G. Ramadori, T. Fujikawa, M. Fukuda, J. Anderson, D. A. Morgan, R. Mostoslavsky, R. C. Stuart, M. Perello, C. R. Vianna, E. A. Nillni, K. Rahmouni and R. Coppari, *Cell. Metab.*, 2010, **12**, 78–87.

59. I. Cakir, M. Perello, O. Lansari, N. J. Messier, C. A. Vaslet and E. A. Nillni, *PLoS One*, 2009, **4**, e8322.

60. M. O. Dietrich, C. Antunes, G. Geliang, Z. W. Liu, E. Borok, Y. Nie, A. W. Xu, D. O. Souza, Q. Gao, S. Diano, X. B. Gao and T. L. Horvath, *J. Neurosci.*, 2010, **30**, 11815–11825.

61. D. A. Velasquez, G. Martinez, A. Romero, M. J. Vazquez, K. D. Boit, I. G. Dopeso-Reyes, M. Lopez, A. Vidal, R. Nogueiras and C. Dieguez, *Diabetes*, 2011, **60**, 1177–1185.

62. A. Satoh, C. S. Brace, G. Ben-Josef, T. West, D. F. Wozniak, D. M. Holtzman, E. D. Herzog and S. Imai, *J. Neurosci.*, 2010, **30**, 10220–10232.

63. H. V. Lin and D. Accili, *Cell. Metab.*, 2011, **14**, 9–19.

64. J. Y. Altarejos and M. Montminy, *Nat. Rev. Mol. Cell Biol.*, 2011, **12**, 141–151.

65. J. T. Rodgers, C. Lerin, W. Haas, S. P. Gygi, B. M. Spiegelman and P. Puigserver, *Nature*, 2005, **434**, 113–118.

66. J. T. Rodgers and P. Puigserver, *Proc. Natl. Acad. Sci. USA*, 2007, **104**, 12861–12866.

67. Y. Nie, D. M. Erion, Z. Yuan, M. Dietrich, G. I. Shulman, T. L. Horvath and Q. Gao, *Nat. Cell Biol.*, 2009, **11**, 492–500.

68. D. M. Erion, S. Yonemitsu, Y. Nie, Y. Nagai, M. P. Gillum, J. J. Hsiao, T. Iwasaki, R. Stark, D. Weismann, X. X. Yu, S. F. Murray, S. Bhanot, B. P. Monia, T. L. Horvath, Q. Gao, V. T. Samuel and G. I. Shulman, *Proc. Natl. Acad. Sci. USA*, 2009, **106**, 11288–11293.

69. J. H. Lim, Y. M. Lee, Y. S. Chun, J. Chen, J. E. Kim and J. W. Park, *Mol. Cell*, 2010, **38**, 864–878.

70. F. Picard, M. Kurtev, N. Chung, A. Topark-Ngarm, T. Senawong, R. Machado De Oliveira, M. Leid, M. W. McBurney and L. Guarente, *Nature*, 2004, **429**, 771–776.

71. D. Chen, J. Bruno, E. Easlon, S. J. Lin, H. L. Cheng, F. W. Alt and L. Guarente, *Genes Dev.*, 2008, **22**, 1753–1757.

72. A. Purushotham, T. T. Schug, Q. Xu, S. Surapureddi, X. Guo and X. Li, *Cell Metab.*, 2009, **9**, 327–338.

73. R. H. Wang, C. Li and C. X. Deng, *Int. J. Biol. Sci.*, 2010, **6**, 682–690.

74. F. Xu, Z. Gao, J. Zhang, C. A. Rivera, J. Yin, J. Weng and J. Ye, *Endocrinology*, 2010, **151**, 2504–2514.

75. C. Escande, C. C. Chini, V. Nin, K. M. Dykhouse, C. M. Novak, J. Levine, J. van Deursen, G. J. Gores, J. Chen, Z. Lou and E. N. Chini, *J. Clin. Invest.*, 2010, **120**, 545–558.

76. W. Zhao, J. P. Kruse, Y. Tang, S. Y. Jung, J. Qin and W. Gu, *Nature*, 2008, **451**, 587–590.

77. K. A. Moynihan, A. A. Grimm, M. M. Plueger, E. Bernal-Mizrachi, E. Ford, C. Cras-Meneur, M. A. Permutt and S. Imai, *Cell Metab.*, 2005, **2**, 105–117.

78. L. Bordone, M. C. Motta, F. Picard, A. Robinson, U. S. Jhala, J. Apfeld, T. McDonagh, M. Lemieux, M. McBurney, A. Szilvasi, E. J. Easlon, S. J. Lin and L. Guarente, *PLoS Biol.*, 2006, **4**, e31.

79. J. H. Lee, M. Y. Song, E. K. Song, E. K. Kim, W. S. Moon, M. K. Han, J. W. Park, K. B. Kwon and B. H. Park, *Diabetes*, 2009, **58**, 344–351.

80. C. Sun, F. Zhang, X. Ge, T. Yan, X. Chen, X. Shi and Q. Zhai, *Cell Metab.*, 2007, **6**, 307–319.

81. V. Gagarina, O. Gabay, M. Dvir-Ginzberg, E. J. Lee, J. K. Brady, M. J. Quon and D. J. Hall, *Arthritis Rheum.*, 2010, **62**, 1383–1392.

82. S. Akieda-Asai, N. Zaima, K. Ikegami, T. Kahyo, I. Yao, T. Hatanaka, S. Iemura, R. Sugiyama, T. Yokozeki, Y. Eishi, M. Koike, K. Ikeda, T. Chiba, H. Yamaza, I. Shimokawa, S. Y. Song, A. Matsuno, A. Mizutani, M. Sawabe, M. V. Chao, M. Tanaka, Y. Kanaho, T. Natsume, H. Sugimura, Y. Date, M. W. McBurney, L. Guarente and M. Setou, *PLoS One*, 2010, **5**, e11755.

83. K. T. Howitz, K. J. Bitterman, H. Y. Cohen, D. W. Lamming, S. Lavu, J. G. Wood, R. E. Zipkin, P. Chung, A. Kisielewski, L. L. Zhang, B. Scherer and D. A. Sinclair, *Nature*, 2003, **425**, 191–196.

84. M. T. Borra, B. C. Smith and J. M. Denu, *J. Biol. Chem.*, 2005, **280**, 17187–17195.

85. M. Kaeberlein, T. McDonagh, B. Heltweg, J. Hixon, E. A. Westman, S. D. Caldwell, A. Napper, R. Curtis, P. S. DiStefano, S. Fields, A. Bedalov and B. K. Kennedy, *J. Biol. Chem.*, 2005, **280**, 17038–17045.

86. D. Beher, J. Wu, S. Cumine, K. W. Kim, S. C. Lu, L. Atangan and M. Wang, *Chem. Biol. Drug Des.*, 2009, **74**, 619–624.

87. M. Pacholec, J. E. Bleasdale, B. Chrunyk, D. Cunningham, D. Flynn, R. S. Garofalo, D. Griffith, M. Griffor, P. Loulakis, B. Pabst, X. Qiu, B. Stockman, V. Thanabal, A. Varghese, J. Ward, J. Withka and K. Ahn, *J. Biol. Chem.*, 2010, **285**, 8340–8351.

88. H. Dai, L. Kustigian, D. Carney, A. Case, T. Considine, B. P. Hubbard, R. B. Perni, T. V. Riera, B. Szczepankiewicz, G. P. Vlasuk and R. L. Stein, *J. Biol. Chem.*, 2010, **285**, 32695–32703.

89. A. A. Sauve, *Curr. Pharm. Des.*, 2009, **15**, 45–56.

90. J. A. Zorn and J. A. Wells, *Nat. Chem. Biol.*, 2010, **6**, 179–188.

91. F. M. Matschinsky, *Nat. Rev. Drug Discov.*, 2009, **8**, 399–416.

92. V. Hindie, A. Stroba, H. Zhang, L. A. Lopez-Garcia, L. Idrissova, S. Zeuzem, D. Hirschberg, F. Schaeffer, T. J. D. Jorgensen, M. Engel, P. M. Alzari and R. C. Biondi, *Nat. Chem. Biol.*, 2009, **5**, 758–764.

93. I. Auterio, S. Costantini and G. Colonna, *PLoS ONE*, 2009, **4**, 1–10.

94. K. T. Howitz, K. J. Bitterman, H. Y. Cohen, D. W. Lamming, S. Lavu, J. G. Wood, R. E. Zipkin, P. Chung, A. Kisielewski, L. L. Zhang, B. Scherer and D. A. Sinclair, *Nature*, 2003, **425**, 191–196.

95. M. T. Borra, B. C. Smith and J. M. Denu, *J. Biol. Chem.*, 2005, **280**, 17187–17195.

96. S. Pervaiz and A. L. Holme, *Antioxid. Redox Signaling*, 2009, **11**, 2851–2897.

97. J. A. Baur, K. J. Pearson, N. L. Price, H. A. Jamieson, C. Lerin, A. Kalra, V. V. Prabhu, J. S. Allard, G. Lopez-Lluch, K. Lewis, P. J. Pistell, S. Poosala, K. G. Becker, O. Boss, D. Gwinn, M. Wang, S. Ramaswamy, K. W. Fishbein, R. G. Spencer, E. G. Lakatta, D. Le Couteur, R. J. Shaw, P. Navas, P. Puigserver, D. K. Ingram, R. de Cabo and D. A. Sinclair, *Nature*, 2006, **444**, 337–342.

98. H. Y. Cohen, C. Miller, K. J. Bitterman, N. R. Wall, B. Hekking, B. Kessler, K. T. Howitz, M. Gorospe, R. de Cabo and D. A. Sinclair, *Science*, 2004, **305**, 390–392.

99. K. A. Rasbach and R. G. Schnellmann, *J. Pharmacol. Exp. Ther.*, 2008, **325**, 536–543.

100. The SIRT1 fluorometric activity assay (BioMol) is also known as the Fluor de Lys assay.

101. A. V. Mai, S. Meade, V. Carafa, M. Tadugno, A. Nebbioso, A. Galmozzi, N. Mitro, E. De Fabiani, L. Altucci and A. Kazantsev, *J. Med. Chem.*, 2009, **52**, 5496–5504.

102. M. Layek, M. Appi Reddy, A. V. Dhanunjaya Rao, M. Alvala, M. K. Arunasree, A. Islam, K. Mukkanti, J. Iqbal and M. Pal, *Org. Biomol. Chem.*, 2011, **9**, 1004–1007.

103. J. E. Bemis, J. S. Disch, P. Y. Ng, C. Oalmann, R. B. Perni and C. B. Vu, *US Patent*, 7,893,086 (2011).

104. C. Oalmann, R. B. Perni, J. S. Disch, C. B. Vu, B. Szczepankiewicz, G. Gualtieri, R. L. Casaubon and K. Koppetsch, *Patent Appl.*, PCT WO 2009/00058348.

105. K. J. Koppetsch, C. J. Oalmann, B. Szczepankiewicz, C. B. Vu, G. Gualtieri, R. Casaubon, J. S. Disch, D. J. Gagne, A. Cote, M. Davis, E. Lainez and R. B. Perni, *240th National Meeting of the American Chemical Society*, Boston, MA, 2010.

106. J. S. Disch, C. B. Vu, L. McPherson, P. Y. Ng, J. E. Bemis, D. P. Carney, A. V. Lynch, C. Loh, S. Ribich, P. Romero, J. J. Smith, D. J. Gagne, A. Cote, M. Davis, E. Lainez and R. B. Perni, *240th National Meeting of the American Chemical Society*, Boston, MA, 2010.

107. C. B. Vu, J. S. Disch, P. Y. Ng, C. A. Blum and R. B. Perni, *Patent Appl.*, PCT WO 2010/003048.

108. C. B. Vu, J. S. Disch, S. K. Springer, C. A. Blum and R. B. Perni, *Patent Appl.*, PCT WO 2009/134973.

109. R. D. Narayan, J. E. Disch, R. B. Perni and C. B. Vu, *Patent Appl.*, PCT WO 2010/056549.

110. C. B. Vu, C. Oalmann, J. E. Disch, R. B. Perni, and B. White, *Patent Appl.*, PCT WO 2010/037129.
111. C. Oalmann, R. B. Perni and C. B. Vu, *Patent Appl.*, PCT WO 2010/037127.
112. A. J. van der Meer, B. Scicluna, J. Lin, E. W. Jacobson, G. P. Vlasuk and T. Van der Poll, *Inflam. Res.*, 2011, **60**, S1–S321.
113. T. Yoshizaki, S. Schenk, T. Imamura, J. L. Babendure, N. Sonoda, E. J. Bae, Y. Oh da, M. Lu, J. C. Milne, C. Westphal, G. Bandyopadhyay and J. M. Olefsky, *Am. J. Physiol. Endocrinol. Metab.*, 2010, **298**, E419–428.
114. M. Lappas, A. Mitton, R. Lim, G. Barker, C. Riley and M. Permezel, *Biol. Reprod.*, 2011, **84**, 167–178.
115. J. N. Feige, M. Lagouge, C. Canto, A. Strehle, S. M. Houten, J. C. Milne, P. D. Lambert, C. Mataki, P. J. Elliott and J. Auwerx, *Cell. Metab.*, 2008, **8**, 347–358.
116. R. K. Minor, J. A. Baur, A. P. Gomes, T. M. Ward, A. Csiszar, E. M. Mercken, K. Abdelmohsen, Y.-K. Shin, C. Canto, M. Scheibye-Knudsen, M. Krawczyk, P. M. Irusta, A. Martín-Montalvo, B. P. Hubbard, Y. Zhang, E. Lehrmann, A. A. White, N. L. L. Price, W. R. Swindell, K. J. Pearson, K. G. Becker, V. A. Bohr, M. Gorospe, J. M. Egan, M. I. Talan, J. Auwerx, C. H. Westphal, J. L. Ellis, Z. Ungvari, G. P. Vlasuk, P. J. Elliott, D. A. Sinclair and R. de Cabo, *Scientific Reports*, 2011, vol. 1.
117. J. A. Funk, S. Odejinmi and R. G. Schnellmann, *J. Pharmacol. Exp. Ther.*, 2010, **333**, 593–601.
118. C. C. Beeson, G. C. Beeson and R. G. Schnellmann, *Anal. Biochem.*, 2010, **404**, 75–81.
119. W. He, Y. Wang, M. Z. Zhang, L. You, L. S. Davis, H. Fan, H. C. Yang, A. B. Fogo, R. Zent, R. C. Harris, M. D. Breyer and C. M. Hao, *J. Clin. Invest.*, 2010, 1056–1068.
120. K. T. Howitz, K. J. Bitterman, H. Y. Cohen, D. W. Lamming, S. Lavu, J. G. Wood, R. E. Zipkin, P. Chung, A. Kisielewski, L. L. Zhang, B. Scherer and D. A. Sinclair, *Nature*, 2003, **425**, 191–196.
121. M. Kaeberlein, T. McDonagh, B. Heltweg, J. Hixon, E. A. Westman, S. D. Caldwell, A. Napper, R. Curtis, P. S. DiStefano, S. Fields, A. Beldalov and B. K. Kennedy, *J. Biol. Chem.*, 2005, **280**, 17038–17045.
122. S. Jarolim, J. Millen, G. Heeren, P. Laun, D. S. Goldfarb and M. Breitenbach, *FEMS Yeast Res.*, 2004, **5**, 169–177.
123. H. Yang, J. A. Baur, A. Chen, C. Miller, J. K. Adams, A. Kisielewski, K. T. Howitz, R. E. Zipkin and D. A. Sinclair, *Aging Cell*, 2007, **6**, 35–43.
124. M. Viswanathan, S. K. Kim, A. Berdichevsky and L. Guarente, *Dev. Cell*, 2005, **9**, 605–615.
125. T. M. Bass, D. Weinkove, K. Houthoofd, D. Gems and L. Partridge, *Mech. Ageing Dev.*, 2007, **128**, 546–552.
126. J. G. Wood, B. Rogina, S. Lavu, K. Howitz, S. L. Helfand, M. Tatar and D. Sinclair, *Nature*, 2004, **430**, 686–689.

127. J. A. Baur, K. J. Pearson, N. L. Price, H. A. Jamieson, C. Lerin, A. Kalra, V. V. Prabhu, J. S. Allard, G. Lopez-Lluch, K. Lewis, P. J. Pistell, S. Poosala, K. G. Becker, O. Boss, D. Gwinn, M. Wang, S. Ramaswamy, K. W. Fishbein, R. G. Spencer, E. G. Lakatta, D. Le Couteur, R. J. Shaw, P. Navas, P. Puigserver, D. K. Ingram, R. de Cabo and D. A. Sinclair, *Nature*, 2006, **444**, 337–342.
128. M. Lagouge, C. Argmann, Z. Gerhart-Hines, H. Meziane, C. Lerin, F. Daussin, N. Messadeq, J. Milne, P. Lambert, P. Elliott, B. Geny, M. Laakso, P. Puigserver and J. Auwerx, *Cell*, 2006, **127**, 1109–1122.
129. P. Fischer-Posovszky, V. Kukulus, D. Tews, T. Unterkircher, K. M. Debatin, S. Fulda and M. Wabitsch, *Am. J. Clin. Nutr.*, 2010, **92**, 5–15.
130. G. Boily, X. H. He, B. Pearce, K. Jardine and M. W. McBurney, *Oncogene*, 2009, **28**, 2882–2893.
131. N. Price, A. Gomes, A. Ling, F. Duarte, A. Martin-Montalvo, B. North, B. Agarwal, L. Ye, G. Ramadori, J. Teodoro, B. Hubbard, A. Varela, J. Davis, B. Varamini, A. Hafner, R. Moaddel, A. Rolo, R. Coppari, C. Palmeira, R. de Cabo, J. Baur and D. Sinclair, *Cell Metabolism*, 2012, **15**, 675–690.

Long-Chain Free Fatty Acid Receptor Agonists

JONATHAN B. HOUZE

Amgen Inc., 1120 Veterans Blvd., South San Francisco, CA 94080, USA
E-mail: jhouze@amgen.com

14.1 Introduction: Diabetes and Free Fatty Acids

With the prevalence of diabetes expected to grow from 171 million patients in the year 2000 to 366 million patients by 2030,[1] the burden of an epidemic of type 2 diabetes (T2DM) is straining healthcare systems worldwide. Despite an impressive array of pharmacotherapies for T2DM, the need for new agents and approaches remains acute.[2,3] Type 2 diabetes manifests as an "ominous octet" of metabolic defects including decreased insulin secretion, decreased glucose uptake by bodily tissues, decreased incretin effects, increased hepatic glucose output, increased lipolysis, increased renal glucose reabsorption, increased glucagon secretion, and neurotransmitter dysfunction.[4] Pre-eminent among these defects is the marked decrease in the insulin secretory response by the β-cells of the pancreas to glucose loads. β-Cell glucose sensitivity declines significantly (50–70%) as patients progress from normal to impaired glucose tolerance. Once patients become overtly diabetic, glucose sensitivity declines further until >90% of the original healthy β-cell insulin secretory response is lost.[5]

The ability of free fatty acids (FFA) to acutely increase glucose stimulated insulin secretion (GSIS) has been extensively studied both *in vitro* and *in vivo*.[6] However, chronic exposure to high levels of FFAs blunts insulin secretion.[7] The magnitude of the FFA stimulatory effect was found to be quite large in fasted rats where infusion of nicotinic acid (which inhibits lipolysis leading to

RSC Drug Discovery Series No. 27
New Therapeutic Strategies for Type 2 Diabetes: Small Molecule Approaches
Edited by Robert M. Jones

Published by the Royal Society of Chemistry, www.rsc.org

very low plasma FFA concentrations) resulted in virtual abolition of GSIS.[8] In fasted humans, a nicotinic acid infusion lowered insulin AUC by approximately 50% after glucose stimulus compared to controls.[9] In both species, co-infusing a lipid source of FFAs along with nicotinic acid restored and even augmented GSIS compared to controls. In order to assess the effect of individual FFAs in augmenting GSIS, further studies were carried out in perfused rat pancreas.[10] Long-chain (C_{18}) fatty acids were more effective than medium-chain (C_8) fatty acids at enhancing GSIS, whereas unsaturation led to decreased GSIS at a given chain length.

The mechanism by which FFAs exert their effects on GSIS in the β-cell is multifaceted and not yet completely delineated. One facet of FFA signaling involves their conversion to long-chain acyl CoA esters (LC-CoA).[11] Metabolism of glucose results in increased levels of malonyl-CoA which is an allosteric inhibitor of carnitine palmitoyl-transferase (CPT)-1. With the β-oxidation pathway closed off by inhibition of CPT-1, LC-CoA esters accumulate in the cytosol and are available for generation of complex lipids such as di-acyl glycerols (DAG), triglycerides, and phospholipids that are important in intracellular signaling pathways. Another facet involves the direct binding to and activation of one or more G-protein coupled receptors (GPCRs) that specifically respond to FFAs.[11] It is these free fatty acid receptors that are the subject of this chapter.

14.2   Free Fatty Acid Receptors

The GPCR family accounts for nearly 30% of the targets of current therapeutic drugs.[12] Considering only the non-sensory GPCRs, 367 family members that respond to endogenous ligands have been identified.[13] Of the known GPCRs, less than 10% are the target of a drug.[14] The GPCR family therefore provides a goldmine of potential drug targets. Those GPCRs which were identified by gene sequence, but with unknown biological ligands or function, are termed orphan GPCRs. Free fatty acids have been shown to bind and activate several formerly orphan GPCRs including FFA1 (GPR40),[15–17] FFA2 (GPR43),[18–20] FFA3 (GPR41),[18,19] GPR84,[21] and GPR120.[22,23] Table 14.1 provides a summary of the salient features of these receptors. Several reviews covering the biology and pharmacology of all or various combinations of the free fatty acid receptors have appeared.[24–30] Of these five potential targets, the long-chain FFA receptors FFA1 and GPR120 are most closely implicated in the regulation of insulin secretion and other processes relevant to T2DM. Thus they will be the focus of this chapter.

14.2.1   Biology of FFA1

The gene encoding FFA1 was identified along with its family members FFA2 and FFA3 by Sawzdargo and colleagues in 1997.[31] Since their ligand specificity and function were unknown, these orphan receptors were termed GPR40,

Table 14.1 Free fatty acid receptors.

IUPHAR Recommended Name[a]	*FFA1*	*GPR120*	*GPR84*	*FFA2*	*FFA3*
Previous Names	GPR40, FFAR1	PGR4	EX33	GPR43, FFAR2	GPR41, FFAR3
Primary Transduction Mechanism(s)	$G_{q/11}$ G_s	$G_{q/11}$, β-arrestin 2	$G_{i/o}$	$G_{i/o}$, $G_{q/11}$	$G_{i/o}$
Most Preferred FFA Ligands	Long-chain, unsaturated ($\geq C_{10}$)	Long-chain, unsaturated ($\geq C_{18}$)	Medium-chain, saturated (C_{10}–C_{12})	Short-chain (C_3)	Short-chain (C_3–C_5)
Highest Expression Tissue(s)	Pancreatic islets, enteroendocrine cells	Enteroendocrine cells, macrophages, adipocytes	Bone marrow, leukocytes, lung	Enteroendocrine, leukocytes, adipocytes	Enteroendocrine, adipocytes
Tissue Function(s)	Islet insulin secretion, gut hormone release	Gut hormone release, anti-inflammatory effects in macrophages, adipocyte function	Cytokine release	Lipolysis inhibition, adipogenesis	PYY release, leptin release

[a]D. B. Bylund, R. A. Bond, D. C. Eikenburg, J. P. Hieble, R. Hills, K. P. Minneman, S. Parra, IUPHAR database (IUPHAR-DB), http://www.iuphar-db.org/DATABASE/GPCRListForward

GPR43, and GPR41 respectively. The sequence similarity between FFA1 and its other family members is somewhat limited, showing 33% and 34% similarity compared to FFA2 and FFA3 respectively.[24] In 2003, great interest was focused on FFA1 by the reports that it is highly expressed in pancreatic β-cells and at least partially mediates the FFA amplification of GSIS.[15–17,32] A key difference between FFA1-mediated insulin secretion and that brought about by the established sulfonylurea insulin secretagogues is that activation of FFA1 does not cause insulin secretion from β-cells when low concentrations of glucose are present.[15,32,33] Insulin secretion stimulated by administration of sulfonylureas occurs through blockade of ATP-sensitive potassium channels (K_{ATP}) and is independent of glucose concentration.[34] The glucose-dependent nature of FFA1-mediated insulin secretion decreases the potential of FFA1 agonists to induce hypoglycemia (see Section 14.3.3.1), a common side effect of the sulfonylureas.[35]

The role of FFA1 in insulin secretion from islet β-cells is supported by expression data. An examination of the 5′-flanking region of the *FFAR1* gene revealed binding sites for the β-cell specific transcription factors PDX1 and BETA2.[36] High expression of *FFA1* in human pancreatic islets was verified from samples taken from patients undergoing pancreatectomy, with *FFA1* expression levels found to be comparable to that of the K_{ATP} channel and the GLP-1 receptor – both targets of anti-diabetic agents.[37] In addition to its presence in islet β-cells, *FFA1* has been reported to be expressed in islet α-cells and to mediate FFA-stimulated glucagon release; however, there is conflicting evidence on this point.[14,38]

Outside pancreatic islets, *FFAR1* expression has been reported in the CNS, liver, heart, skeletal muscle, and immune cells, again with some conflicting data.[15–17,39] The role of FFA1 in these tissues has not yet been identified. However, recent reports that *FFAR1* is expressed in enteroendocrine cells where it contributes to FFA-induced secretion of the gut hormones glucagon-like peptide 1 (GLP-1), glucose-dependent insulinotropic peptide (GIP), and cholecystokinin (CCK) have added an exciting new dimension to the activity of FFA1.[40–43] GLP-1 is known to amplify GSIS as part of the so-called incretin effect, and serves as a drug target in its own right with both direct mimics (synthetic peptides with much greater half-life) and inhibitors of its degradation currently in use as anti-diabetic agents.[44] Therefore, agonists of FFA1 may not only directly amplify GSIS in islet β-cells but may also amplify GSIS indirectly via GLP-1.

Aspects of FFA1 signaling in the islet β-cell are summarized in Figure 14.1. FFA1 was found to be coupled to the $G\alpha_{q/11}$ signaling pathway in all the initial reports.[15–17] When treated with long-chain fatty acids, mouse insulinoma (MIN6), Chinese hamster ovary (CHO), and human embryonic kidney (HEK293) cells over-expressing GPR40 all showed an increase in intracellular calcium ion concentration ($[Ca^{2+}]_i$).[15,17,45] Studies by Shapiro and colleagues in the β-cell line INS-1E demonstrated that a $G\alpha_{q/11}$-specific inhibitor partially blocked the expected increase in $[Ca^{2+}]_i$ after treatment with palmitic acid.[46]

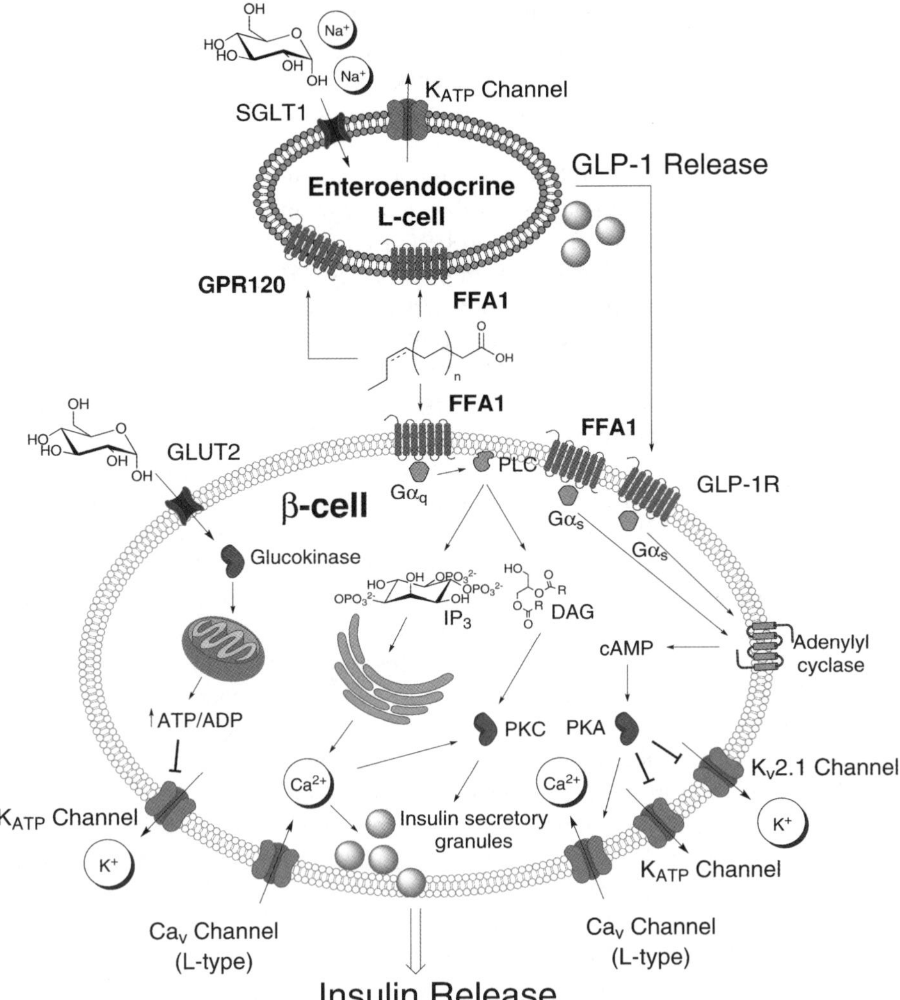

Figure 14.1 FFA1 mechanism of action.

The $G\alpha_{q/11}$ signaling pathway involves the activation of phospholipase C (PLC) which generates inositol-1,4,5-triphosphate (IP_3) and DAG. The IP_3 thus generated can open a gated Ca^{2+} channel in the ER membrane that allows Ca^{2+} to flow into the cytoplasm and raise $[Ca^{2+}]_i$. The increased $[Ca^{2+}]_i$ induces protein kinase C (PKC) to translocate to the inner surface of the plasma membrane and become activated by DAG. The role of PKC in GSIS in β-cells is complex and not well understood;[47] however, multiple reports indicate that PKC inhibitors partially inhibit fatty acid enhancement of GSIS.[48]

In addition to $G\alpha_{q/11}$-coupling, FFA1 has also been shown to signal through $G\alpha_s$.[49] Activation of $G\alpha_s$ in turn activates adenylyl cyclase which leads to increased intracellular concentrations of cyclic AMP (cAMP). Cyclic AMP activates the cAMP-dependent protein kinase (PKA) which can phosphorylate several ion channels involved in GSIS including the ATP-sensitive potassium channel (K_{ATP}), the L-type Ca channel (Ca_v), and voltage-dependent potassium channels (K_v).[50] The effects of phosphorylating these channels will be discussed below.

The consequences of FFA1 activation can be integrated into the established pathway for glucose-stimulated insulin secretion in the β-cell[51] to form a picture of how FFA1 amplifies this process (Figure 14.1). Glucose enters the β-cell through the low-affinity but high-capacity GLUT2 transporter. It is phosphorylated by glucokinase (which serves as a proximal glucose sensor for the β-cell) and enters glycolysis and eventually oxidative phosphorylation to generate ATP. The increasing ATP/ADP ratio in the cell leads to closure of the K_{ATP} channel which depolarizes the cell, initiating the upstroke of an action potential.[52] Phosphorylation of the regulatory subunit of the K_{ATP} channel (also known as the sulfonylurea receptor, SUR1) by PKA promotes its closure and this is one mechanism by which GLP-1 is known to amplify GSIS.[50]

The action potential is propagated by opening of L-type Ca_v channels. The influx of Ca^{2+} through L-type Ca_v channels serves as the master signal triggering insulin release,[53] and PKC activation can sensitize insulin secretory vesicles to the Ca^{2+} signal.[47] Both PKA and PKC can phosphorylate the L-type Ca_v channel.[53] In the case of PKA, this phosphorylation enhances Ca^{2+} flux.[53] However, the case with PKC is more complex. Acute stimulation with PKC activators does not always increase Ca^{2+} flux through the channel, but deprivation of PKC activity can reduce Ca_v currents by $\sim 50\%$.[53]

As the β-cell becomes increasingly depolarized with Ca^{2+} influx, K_v channels open to allow potassium ions to exit the cell and mediate repolarization.[54] Phosphorylation of K_v channels by PKA disfavors the opening of these channels, thus extending the action potential and insulin release.[55] Treatment of MIN6 cells with the FFA1 ligand linoleic acid reduced voltage-gated K^+ current in a FFA1 and PKA-dependent manner.[49]

Taken together, the changes in phosphorylation state of the K_{ATP}, Ca_v, and K_v channels of the β-cell as a consequence of FFA1 (and GLP-1) signaling serve to substantially amplify the amount of insulin released for a given glucose stimulus. However, unless an initial surge of glucose sets the stage by increasing the ATP/ADP ratio in the β-cell, these changes remain latent. While this picture is certainly incomplete, and it ignores any potential long-term consequences in the β-cell from FFA1 signaling such as effects on insulin stores and β-cell survival, it provides at least a framework to understand the multiple pathways by which FFA1 activation can potentiate GSIS.

Because it has been very difficult until relatively recently to obtain pure populations of isolated enteroendocrine cells,[56] signal transduction in the L-cell is less well understood than in β-cells. A recent report by Reimann and co-workers shows that many of the components used by the β-cell to sense glucose and other nutrients are also present in the L-cell.[42] These components include the K_{ATP} channel and its associated regulatory subunit SUR1, glucokinase,

FFA1, GPR119, and GPR120. Despite their presence in the L-cell, these proteins do not necessarily play exactly the same role that they do in the β-cell. While primary L-cells respond to sulfonylureas like tolbutamide with action potential firing,[42] humans dosed with sulfonylureas showed no enhancement of GLP-1 or GIP plasma levels during an oral glucose tolerance test (OGTT).[57] An additional trigger, the sodium-coupled glucose transporter-1 (SGLT1) may modulate initiation of action potentials that lead to release of GLP-1.[58]

The dual mechanism of action of FFA1 involving both direct potentiation of GSIS in the islet β-cells as well as stimulating GLP-1 release from L-cells provides a compelling rationale to discover and develop agonists of FFA1 as glucose-dependent insulin secretagogues for the treatment of T2DM. As will be covered in Section 14.3, many groups have identified synthetic FFA1 activators for this purpose.

14.2.2 Biology of GPR120

The gene encoding GPR120 was identified in 2003 along with several other orphan GPCRs that lack close homology to any other receptors in the family.[59] In 2005, Hirasawa and co-workers reported that GPR120 was activated by long-chain fatty acids with a substantial preference for polyunsaturated fatty acids.[22] These observations were later confirmed by Oh and colleagues.[23] As with FFA1, the carboxylic acid functionality is crucial for GPR120 activation since the methyl ester of a fatty acid was inactive.[22] Despite the similar ligand preference of GPR120 and FFA1, there is only low homology ($\sim 10\%$) between the two receptors.[27]

GPR120 has been shown to be primarily expressed in the large intestine, lung, mature adipocytes, and macrophages.[22,23] The function of GPR120 in the lung is unknown. Within the large intestine, GPR120 was localized using a specific antibody to cells also staining with an anti-GLP-1 antibody.[60] This observation is further supported by the detection of GPR120 expression in isolated primary murine L-cells as well as the murine enteroendocrine cell line STC-1.[22,42] In STC-1 cells, treatment with α-linolenic acid (a potent FFA agonist of GPR120) has been shown to stimulate secretion of the hormones GLP-1 and CCK.[22,61] Thus, while not directly found in β-cells, GPR120 may indirectly mediate effects on GSIS via the incretin pathway.

Activation of GPR120 stimulates a rise in $[Ca^{2+}]_i$ in cells over-expressing the receptor, suggesting that it is coupled to the $G\alpha_{q/11}$ signaling pathway.[22] However, Oh and colleagues demonstrated tissue-specific signaling pathways wherein enhancement of glucose uptake by adipocytes was $G\alpha_{q/11}$ mediated; however, anti-inflammatory effects in macrophages were found to be β-arrestin 2 mediated.[23] Figure 14.2 summarizes the different signaling pathways active in the two different tissues. Activation of GPR120 leads to its phosphorylation.[62] Typically, activated GPCRs are phosphorylated by GPCR kinases (GRK) that lead to the binding of the β-arrestins. β-arrestin then mediates receptor internalization via clathrin-coated pits.[63] In macro-

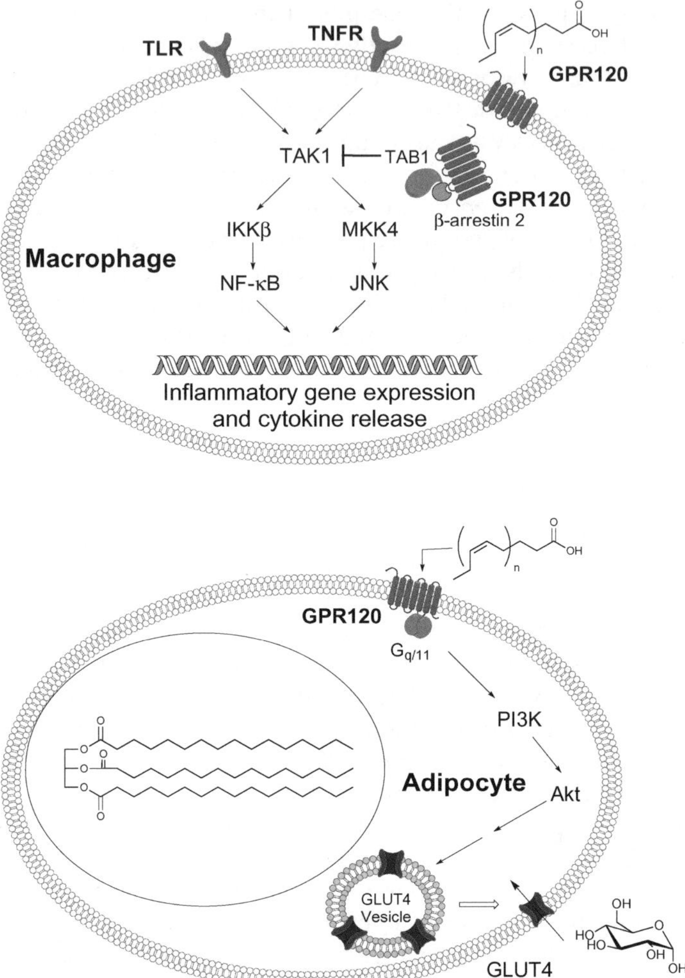

Figure 14.2 GPR120 mechanism of action.

phages, the internalized GPR120-β-arrestin 2 complex binds the transforming growth factor-β activated kinase 1 (TAK1) binding protein 1 (TAB1) and prevents it from activating TAK1, effectively inhibiting pro-inflammatory signaling through the IKKβ/NFκB and JNK/AP1 pathways.[22,23] The pathway in adipocytes, however, involves Gα$_q$-dependent activation of PI3K and Akt, leading to the translocation of GLUT4-containing vesicles to the plasma membrane to allow glucose transport into the cell. This insulinomimetic effect is 30–50% as great as that brought about by treatment with insulin itself and was additive to sub-maximal doses of insulin.[23]

In macrophages, GPR120 has been found to mediate the anti-inflammatory effects of ω-3 fatty acids on these cells.[23] In obese humans, macrophages are recruited into adipose tissue (where they can comprise up to 40% of cells) and secrete pro-inflammatory cytokines such as TNFα that promote insulin resistance not only in the adjacent adipocytes but in other tissues as well.[64–66] The combined effects of GPR120 in both macrophages and adipocytes should lead to whole body insulin sensitization. While adipose tissue only accounts for ~10% of whole body glucose disposal by itself, it does play an important role in modulating insulin resistance in other metabolically important tissues such as liver and skeletal muscle through the secretion of FFAs and adipokines.[65,67] By relieving TNFα-mediated insulin resistance in adipocytes (through its effects in macrophages), the effects of GPR120 agonists in adipose tissue can extend beyond just promoting GLUT4 translocation to include inhibition of hormone-sensitive lipase activity. The resulting drop in plasma FFAs can lead to greater insulin sensitivity in skeletal muscle and other beneficial effects on the diabetic condition.[65]

GPR120 KO mice show effects consistent with the above hypothesis. On a normal chow diet, GPR120 KO mice are less glucose tolerant, hyperinsulinemic, more insulin resistant, and show greater macrophage infiltration into adipose tissue than their wild-type counterparts.[23] While both wild-type and GPR120 KO mice grow equally obese and insulin resistant on a high fat diet that is low in ω-3 FA, ω-3 FA supplementation led to decreased macrophage infiltration into adipose tissue, reduced inflammatory markers, and improved insulin sensitivity in wild-type mice but was without effect in the knockouts.

As with FFA1, GPR120 can potentially positively affect the diabetic condition through multiple pathways. While initially reported as a target for modulating GLP-1 secretion and thus potentially affecting GSIS,[22] the recent report by Oh and colleagues on the effects of GPR120 activation in adipocytes and macrophages[23] has raised the possibility of GPR120 agonists acting as insulin sensitizers. As will be covered in Section 14.4, several groups have identified potent, synthetic GPR120 activators, but none have yet been reported to have entered clinical development.

14.3 FFA1 Receptor Agonists

While FFA1 has a fairly wide specificity showing activation with fatty acids (FA) from 6–22 carbons in length, the most potent known endogenous FA ligands are the highly unsaturated ω-3 fatty acids eicosapentaenoic acid (EPA, C20:5n-3) and docosahexaenoic acid (DHA, C22:6n-3).[15,17] The carboxylic acid functionality is crucial for FFA1 activation since the methyl ester of one of the more potent endogenous free fatty acids, linoleic acid (C18:2n-6), is inactive.[15] Disclosed synthetic agonists of FFA1 mimic the fatty acid structure by consisting of an acidic moiety (which is usually, but not, necessarily a carboxylic acid) and a lipophilic tail. The general pharmacophore that most synthetic FFA1 agonists conform to is shown in Figure 14.3. The acidic

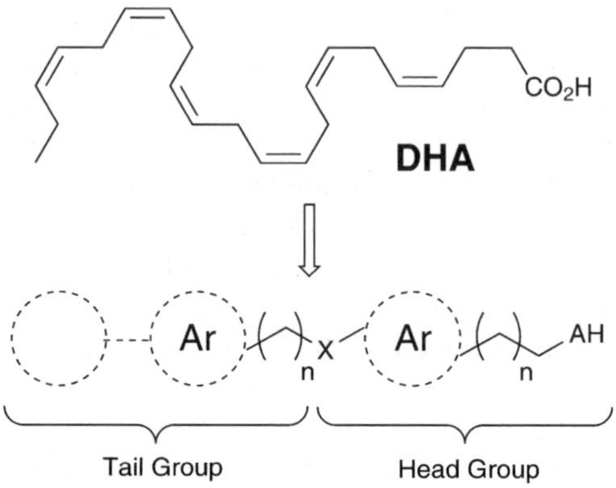

DHA

Tail Group   Head Group

Figure 14.3   FFA1 pharmacophore.

moiety is attached by a short linker (1 or 2 carbon atoms) to an aromatic ring. This combination will be referred to as the "head group" in this and subsequent sections. The head group is attached to a second ring system by another short linker that usually contains at least one heteroatom. As will be shown, there is substantial diversity in this second ring system; however, there is almost always at least one aromatic ring present. Presumably the aromatic rings mimic the multiple double bonds found in the most potent endogenous FFA1 ligands EPA and DHA. The portion of the FFA1 agonist starting from the second linker and encompassing the second ring system will be referred to as the "tail group".

FFA1 has attracted considerable interest as a target for anti-diabetic agents, with synthetic agonists reported by many groups. These efforts have been previously reviewed.[68,69] This section will focus on those FFA1 agonists reported since 2009 when the later of the reviews appeared. Reported FFA1 agonists have been grouped by the degree of conformational constraint present in the head group (open chain, bicyclic, or tricyclic) with each class considered in its own sub-section. In addition, because of the large number of reports on open chain FFA1 agonists, this class has been further divided into those head groups containing a carboxylic acid and those employing a bioisosteric replacement.

14.3.1   Open Chain Carboxylic Acids

Head groups containing open chain carboxylic acids comprise the most commonly reported class of synthetic FFA1 agonists. Table 14.2 collects representative examples from recent reports on this class. As Table 14.2 shows, members of this class are generally derivatives of dihydrocinnamic acid. However, there is substantial diversity in the tail group portion of the molecules.

The most straightforward examples in this class are the unsubstituted linkers found in two reports by Christiansen and colleagues.[70,71] Optimization of a hit based on high-throughput screening (HTS) resulted in the identification of compound **14.1**, designated TUG-424, as potent FFA1 agonist with an EC_{50} of 32 nM. TUG-424 was found to amplify GSIS in an FFA1-dependent manner in mouse islets.[70] Starting from a different hit in their HTS, the same group reported compound **14.2**, designated TUG-469.[71] TUG-469 showed high potency against FFA1 (EC_{50} = 19 nM) and amplified GSIS in the rat β-cell line INS-1E.

Mono-substitution at the β-position of the head group linker has been shown to be tolerated by FFA1. A methylacetylenyl substituent at the β-position has been found to be particularly interesting, as shown by compounds **14.3**–**14.6** shown in Table 14.2. Compound **14.3** was first disclosed in 2005 by Akerman and colleagues at Amgen;[72,73] however, it was only recently that the compound, designated AMG 837, was reported to be selected for clinical development.[74]

Shimada and colleagues have disclosed unique spirocyclic tail groups that combine well with various open chain carboxylic acid head groups.[75] Compound **14.4** was selected from among the most potent examples included; however, only a potency range was disclosed. An extended tail group incorporating a spiro[indene-1,4'-piperidine] system was disclosed by Hamdouchi and colleagues.[76,77] Compound **14.5** displayed an EC_{50} of 186 nM in a Ca^{2+} FLIPR assay and was found to amplify GSIS in both MIN6 cells and intact rat islets. Compound **14.5** also improved glucose excursion in an i.p. glucose tolerance test (ipGTT) carried out in male Balb/c mice with an ED_{50} of only 0.09 mg/kg.[76] Finally, relatively small tail groups such as the aryl ether found in compound **14.6** reported by Walsh and colleagues are also able to potently activate FFA1 in combination with a substituted open chain carboxylic acid head group.[78] In addition to high potency (EC_{50} = 71 nM), compound **14.6** showed an excellent PKDM profile in mouse with a long half-life (7.8 h) and ~100% oral bioavailability. These properties translated into robust activity after a 10 mg/kg dose in an ipGTT carried out in normal lean mice with 66% inhibition of $AUC_{glucose}$ following a glucose challenge.[78]

Some groups have elected to enclose the linker from the carboxylic acid to the first aromatic ring into another ring. Defossa and colleagues have disclosed aryl-substituted cyclopropanecarboxylic acids as head groups combined with a unique oxalamide tail group such as compound **14.7**, which shows an EC_{50} of 0.7 μM against FFA1.[79] Another example of this approach was disclosed by Ellsworth and colleagues wherein the carboxylic acid is appended to a pyrrolidine ring, as shown by compound **14.8**.[80]

14.3.2 Open Chain Carboxylate Bioisosteres

Bioisosteric replacement of specific functional groups is a well-established strategy in drug discovery.[81,82] Given the importance of the carboxylic acid functionality in FFAs for binding and activation of FFA1, bioisosteric replacement of the carboxylate would be expected to lead to novel FFA1

Table 14.2   Open chain FFA1 agonists.

Cmpd.	Structure	FFA1 EC$_{50}$ (µM)	Ref.
14.1 TUG-424		0.032 [a]	[70]
14.2 TUG-469		0.019 [b]	[71]
14.3 AMG 837		0.013 [c]	[73]
14.4		0.01 - 0.1 [b]	[75]
14.5		0.186 [b]	[76]
14.6		0.071 [d]	[78]
14.7		0.7 [b]	[79]
14.8		0.065 [e]	[80]

[a] Ca^{2+} fluorescence assay in 1321N1 cells stably transfected with FFA1; [b] Ca^{2+} FLIPR in HEK293 cells stably transfected with FFA$_1$; [c] Ca^{2+} chemiluminescence assay in CHO cells transiently transfected with FFA1; [d] Ca^{2+} FLIPR in CHO cells stably transfected with FFA1; [e] Ca^{2+} FLIPR in unspecified cells expressing FFA1

agonists. In one of the initial reports on FFA1, thiazoledinediones, a well-known carboxylate bioisostere, were shown by Kotarsky and colleagues to activate FFA1.[16] Table 14.3 summarizes representative examples from recent reports of FFA1 agonists containing carboxylate bioisosteres.

Zhou and colleagues have reported optimizing a series of thiazolidinedione-containing FFA1 agonists to arrive at the high potency (EC_{50} = 10 nM) compound **14.9a**.[83] In a separate report, the pharmacology of compound **14.9a** and a related analog **14.9b** that possessed an improved PKDM profile was extensively explored.[84] Compound **14.9a** amplified GSIS in mouse islets in a FFA1-dependent manner. The effect of compound **14.9a** was found to be strongly glucose dependent in an islet perifusion experiment with minimal effects seen in islets exposed to 8 mM glucose, but significant amplification was seen in the presence of 16 mM glucose.[84] The ability of compound **14.9a** to improve glucose homeostasis in a diet-induced obese (DIO) mouse model was also examined in a sub-chronic setting (10 days dosing). Compound **14.9a** was able to significantly reduce $AUC_{glucose}$ ($\sim$50% after a 10 mg/kg dose) after ipGTT. This activity was maintained after 10 days of dosing.[84]

Other carboxylate bioisosteres are active on FFA1. Oxadiazolidinedione-containing FFA1 agonist compound **14.10** displayed an EC_{50} of 69 nM on FFA1 as disclosed by Negoro and colleagues.[85] Finally, hydroxyisoxazoles (**14.11**) have been disclosed as FFA1 agonists by Okano and colleagues.[86]

Table 14.3 Open chain carboxylate bioisosteres.

Cmpd.	Structure	FFA1 EC_{50} (μM)	Ref.
14.9a (R = CH$_3$) **14.9b** (R = H)		0.010 [a] 0.069 [a]	[83]
14.10		0.069 [a]	[85]
14.11		ND	[86]

[a] Ca^{2+} FLIPR in CHO cells stably transfected with FFA1

14.3.3 Bicyclic Carboxylates and Carboxylate Bioisosteres

Head groups containing bicyclic structures are relatively less common than open chain variants, but may benefit from improved potency, selectivity, and PKDM properties compared to the corresponding unconstrained structures. Notably, the most clinically advanced FFA1 agonist, TAK-875, belongs to this class.

14.3.3.1 TAK-875

Reported in 2010 by Negoro and colleagues at Takeda, compound **14.14** is a highly potent FFA1 agonist with $EC_{50} = 14$ nM in a Ca^{2+} FLIPR assay.[87] Compound **14.14** was derived from an even more potent parent compound **14.13** that in turn could trace its origin to a relatively simple lead compound **14.12**.[88] Despite the high potency of compound **14.13**, the unsubstituted β-position of the dihydrocinnamic acid head group left the compound vulnerable to metabolism via β-oxidation.[87] Therefore, the Takeda group examined a number of fused, bicyclic headgroups with the aim of preventing β-oxidation and ensuring selectivity against other receptors recognizing free fatty acids. The dihydrobenzofuran structure of compound **14.14** fulfilled these design aims and was selected for clinical development, designated TAK-875.[87]

The preclinical pharmacology of TAK-875 was examined extensively by Tsujihata and colleagues.[89] TAK-875 amplified GSIS in INS-1 β-cell line as well as intact rat islets. Importantly, extended incubation with TAK-875 in INS-1 cells did not lead to the inhibition of GSIS ("lipotoxicity") seen with endogenous fatty acids.[7] *In vivo* studies in normal SD rats showed that TAK-875 did not significantly lower plasma glucose concentration in the absence of a glucose challenge whereas sulfonylurea insulin secretagogues showed the

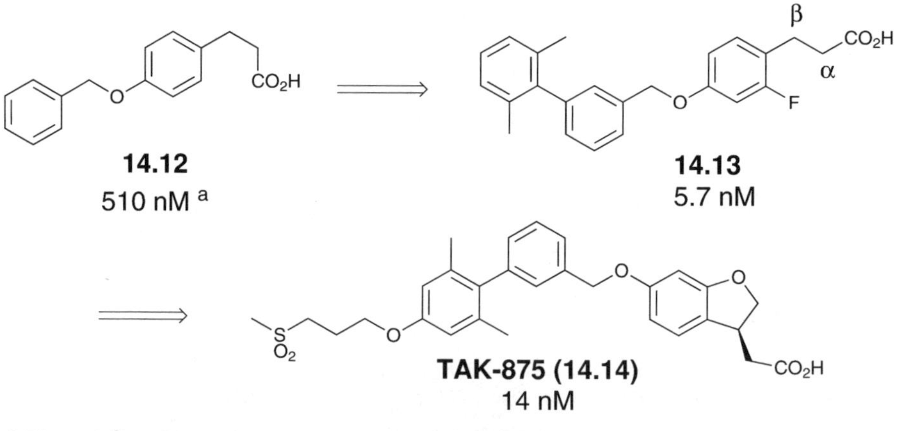

14.12

510 nM [a]

14.13

5.7 nM

TAK-875 (14.14)

14 nM

[a] EC_{50} in Ca^{2+} FLIPR in CHO cells stably transfected with FFA1

Scheme 14.1 TAK-875.

expected lowering. However, TAK-875 did significantly lower fasting plasma glucose levels in the hyperglycemic Zucker diabetic fatty rat model of T2DM.

In a single ascending dose study carried out in healthy volunteers,[90] TAK-875 proved to have a very favorable pharmacokinetic profile. Plasma $t_{1/2}$ (28–36 h) was more than sufficient to allow once-daily dosing. AUC and C_{max} showed good linearity up to a 200 mg dose, with higher doses showing a slightly greater than dose proportional exposure. Adverse events were few, mild, and transient. No glucose-lowering was seen in the healthy volunteers, but this is consistent with the mechanism of FFA1 requiring elevated plasma glucose concentrations (such as those seen in diabetic patients) in order to see effects.

In contrast to the study in healthy volunteers, significant fasting plasma glucose (FPG) lowering by TAK-875 was observed in a 2-week study carried out in diabetic patients.[91] FPG lowering was evident by the second day of dosing and had reached ~ 60 mg/dL ($\sim 27\%$) by day 13 in the top dose (400 mg) group. In addition, plasma glucose concentrations were also significantly lowered by up to 90 mg/dL 2-h post OGTT at doses of 100 mg and above. These are impressive reductions in plasma glucose concentrations in diabetic patients, and they would be expected to lead to significant lowering of glycated hemoglobin (HbA_{1c}), the "gold-standard" biomarker used to track long-term plasma glucose control in diabetic patients.[92] Indeed this turned out to be the case when Burant and colleagues reported $\sim 1\%$ lowering of HbA_{1c} in diabetic patients treated with 100 mg of TAK-875 once-daily for 3 months.[93] The extent of HbA_{1c} lowering with TAK-875 was similar to that seen with the sulfonylurea glimepiride; however, the incidence of hypoglycemia in the glimepiride-treated patients was 7-fold higher than TAK-875 treated patients (16.1% versus 2.3%).

The clinical efficacy of TAK-875 in diabetic patients provides proof-of-concept for the use of FFA1 agonists in the treatment of diabetes, and is a milestone for the class. The translation of the preclinical observations on the glucose-dependent nature of FFA1 activity into the clinic is of particular interest. While many hurdles remain before an FFA1 agonist can be approved as an anti-diabetic therapeutic, the efficacy of TAK-875 is sure to spark great interest in the field.

14.3.3.2 Other Bicyclic FFA1 Agonists (Table 14.4)

Negoro and colleagues have disclosed conformational constraints in addition to the dihydrobenzofuran head group employed in TAK-875.[94] In compound **14.15** a dihydrobenzofuran is also employed in the tail group portion of the molecule. While the exact potency of compound **14.15** was not disclosed, it was shown to lower $AUC_{glucose}$ and increase $AUC_{insulin}$ after an OGTT in streptozotocin (STZ) treated rats. Humphries and colleagues have taken a very different approach in their report of tetrahydroisoquinolinones as FFA1 agonists.[95] Unlike other FFA1 agonists, compound **14.15** does not comfortably fit into the pharmacophore described in Section **14.3**. The carboxylic acid is situated in the center of the molecule with no long hydrophobic tail evident.

Table 14.4   Bicyclic FFA1 agonists.

Cmpd.	Structure	FFA1 EC$_{50}$ (μM)	Ref.
14.15	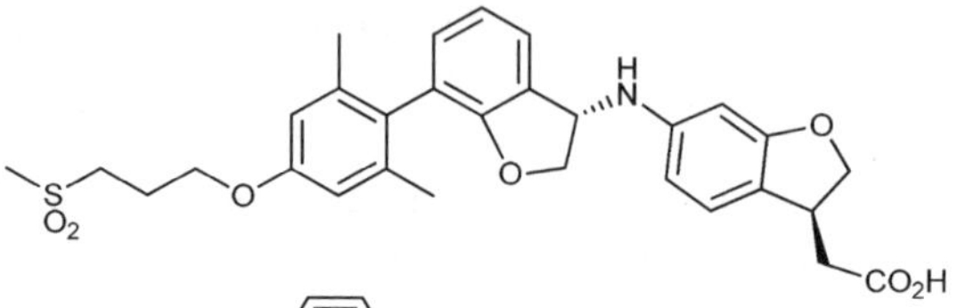	ND	[94]
14.16	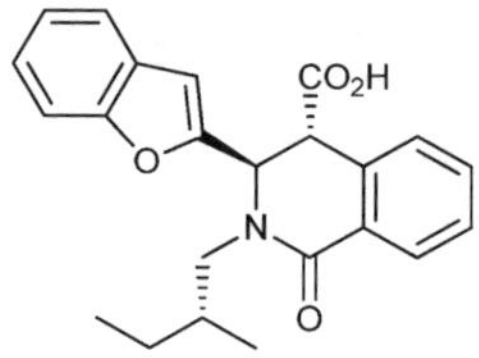	0.020 [a]	[95]
14.17	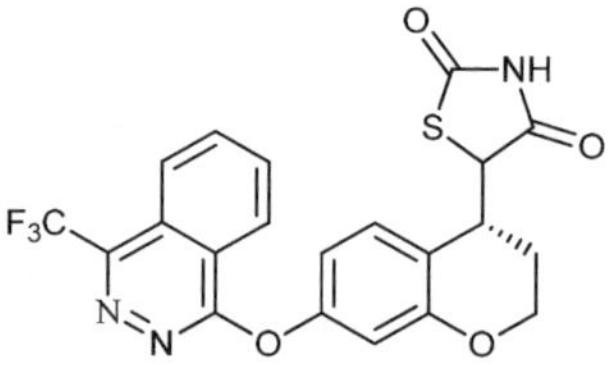	ND	[96]

[a] Ca^{2+} FLIPR in unspecified cells expressing FFA1

Regardless, compound **14.16** is potent on FFA1 (EC$_{50}$ = 20 nM) and displayed a satisfactory PKDM profile in rats with $t_{1/2}$ = 5.4 h. Lastly in this section, Davis and colleagues have disclosed conformationally constrained thiazolidinediones as FFA1 agonists; however, no additional details were provided.[96]

14.3.4   Tricyclic Carboxylates and Carboxylate Bioisosteres

Due to their synthetic complexity and multiple stereocenters, it is not surprising that tricyclic head groups are even less common than their bicyclic counterparts. Nevertheless, the possible advantages of improved potency, selectivity, and PKDM properties compared to the corresponding unconstrained structures are operative in this class as well (Table 14.5).

Thiazolidinediones displayed from a bridged tetrahydronaphthalene have been combined with a variety of tail groups in structures disclosed as FFA1 agonists by Josien and colleagues (compound **14.18**); however, no further details were provided.[97–99] Liang and colleagues disclosed fused cyclopropanecarboxylic acids such as compound **14.19** as FFA1 agonists with no additional details.[100] Finally, Brown and colleagues disclosed that spirocyclic head groups such as compound **14.20** can act as FFA1 agonists, with the

Table 14.5 Tricyclic FFA1 agonists.

Cmpd.	Structure	FFA1 EC$_{50}$ (µM)	Ref.
14.18	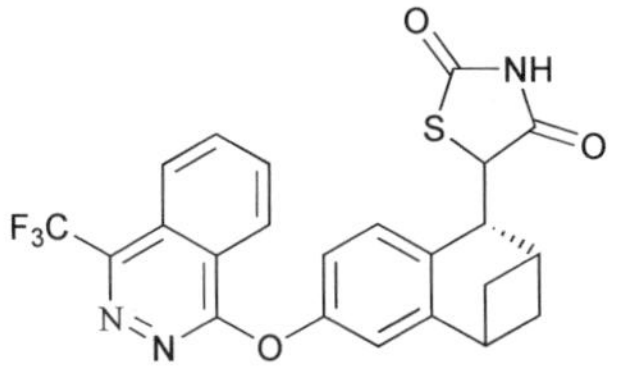	ND	[99]
14.19	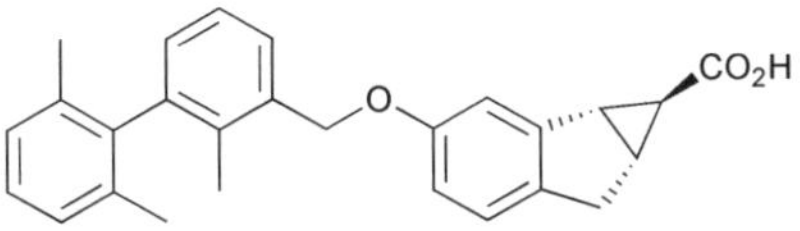	< 0.1^a	[100]
14.20	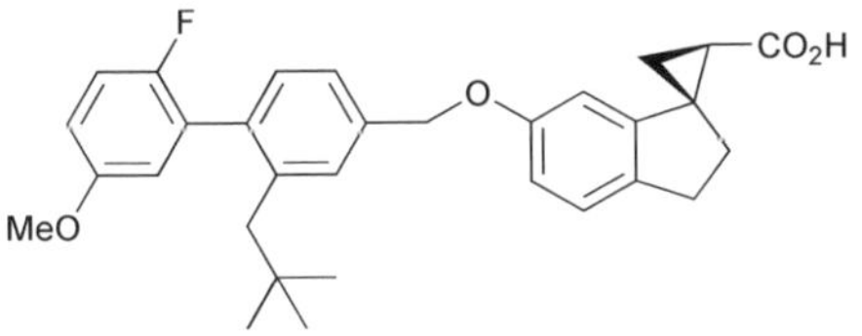	0.01 - 0.1^b	[101]

^{a} EC$_{50}$ in Ca^{2+} FLIPR assay in CHO cells stably transfected with FFA1 ^{b} EC$_{50}$ in Ca^{2+} aequorin assay in CHO cells stably transfected with FFA1

example shown possessing an EC$_{50}$ between 10 and 100 nM in a Ca^{2+} flux assay carried out in CHO cells transfected with FFA1.[101]

14.4 GPR120 Receptor Agonists

GPR120 possesses a somewhat narrower specificity for endogenous free fatty acids than FFA1. While FFA1 will respond (albeit with low potency) to FFAs down to 6 carbons in length, GPR120 requires at least 14 carbons before significant activation occurs.[22] Both receptors prefer highly unsaturated fatty acids, but GPR120 responds most potently to the triply unsaturated α-linolenic acid (C18:3n-3) and γ-linolenic acid (C18:3n-6) in contrast to the preference of FFA1 for the more highly unsaturated EPA and DHA.[22] Due to the later discovery of GPR120 compared to FFA1, relatively fewer synthetic GPR120 agonists have been disclosed. However, the same basic pharmacophore seen in synthetic FFA1 agonists carries over to GPR120 agonists.

14.4.1 Carboxylic Acids

Among the disclosures of synthetic GPR120 agonists, reports are roughly evenly divided between those agonists incorporating a carboxylic acid and those incorporating a bioisosteric replacement for the carboxylate. Table 14.6 collects representative examples from the disclosures of carboxylic acid-containing synthetic GPR120 agonists.

In an early report, Suzuki and colleagues reported the modification of a known PPAR-γ agonist into a selective GPR120 agonist (compound **14.21**) with an EC_{50} of 1.2 μM in a Ca^{2+} mobilization assay.[102] Compound **14.21**, designated NCG21, shows 16-fold selectivity over FFA1 and negligible remaining PPAR-γ activity. NCG21 stimulated GLP-1 secretion in the murine enteroendocrine cell line STC-1.[103] Finally, when NCG21 was delivered directly into the colon of fasting C57/Bl6 mice, significant elevation of plasma GLP-1 concentrations was found.[103]

Several additional classes of carboxylic acid GPR120 agonists have been disclosed. Phenoxyacetic acids such as compound **14.22** have been disclosed by Epple and colleagues as GPR120 agonists, but with few further details.[104] Ma and colleagues have disclosed GPR120 agonists with a variety of linkers and substitutions between the carboxylic acid functionality and the first aromatic ring. A representative compound (**14.23**) was dosed in C57/Bl6 mice and

Table 14.6 Carboxylic acid-containing GPR120 agonists.

Cmpd.	Structure	GPR120 EC_{50} (μM)	Ref.
14.21 **NCG21**		1.2 [a]	[102]
14.22		ND	[104]
14.23		< 1 [a]	[105]
14.24		ND	[106]

[a] Ca^{2+} FLIPR in HEK293 cells stably transfected with GPR120

reduced $AUC_{glucose}$ by 45% during an OGTT.[105] A related compound series incorporating various substituted benzofurans in the tail group was disclosed by Shi and colleagues, with representative compound **14.24** reducing $AUC_{glucose}$ by 44% during an ipGTT in mice.[106]

14.4.2 Carboxylate Bioisosteres

As shown in Table 14.7, carboxylate bioisostere-containing compounds disclosed as GPR120 agonists include not only well-known replacements such as tetrazoles, sulfonic acids, and hydroxyisoxazoles, but also an unusual pyridine ring carboxylate replacement.

Represented by compound **14.25**, Epple and colleagues have disclosed a series of tetrazole-containing GPR120 agonists related to their carboxylic acid-containing series (exemplified by compound **14.22**).[107] He and colleagues have disclosed the sulfonic acids exemplified by compound **14.26**.[108] A series of compounds incorporating a hydroxyisoxazole carboxylate bioisostere, as shown in compound **14.27**, was disclosed by Hashimoto and colleagues and

Table 14.7 Carboxylate bioisostere-containing GPR120 agonists

Cmpd.	Structure	GPR120 EC$_{50}$ (μM)	Ref.
14.25		ND	[107]
14.26		0.078 [a]	[108]
14.27		0.07 [a]	[109]
14.28		0.18 [a]	[112]

[a] Ca^{2+} FLIPR in CHO cells stably transfected with GPR120

displays potent activity on GPR120 with an EC_{50} of 70 nM in a Ca^{2+} mobilization assay.[109,110] While uncommon, the employment of a pyridine ring as a replacement for a carboxylate is precedented.[111] As disclosed by Arakawa and colleagues, the pyridine-containing compound **14.28** with a reasonably potent EC_{50} of 180 nM in Ca^{2+} mobilization assay proves that an acidic functionality is not absolutely necessary for GPR120 activity.[112]

14.5 Conclusions

Despite over 50 years of anti-diabetic drug discovery and development leading to a diverse array of therapeutic agents, type 2 diabetes remains a grievous illness and was the seventh leading cause of death in the United States in 2007.[113] The long-chain FFA receptors FFA1 and GPR120 are expressed in several tissues relevant to diabetes and can mediate multiple beneficial effects. FFA1 has been demonstrated to amplify glucose-stimulated insulin secretion from islet β-cells and also stimulates incretin secretion from intestinal enteroendocrine cells. GPR120 is also reported to stimulate incretin secretion from enteroendocrine cells, but in addition mediates the anti-inflammatory effects of ω-3 fatty acids in macrophages and increases glucose uptake by adipocytes. Both receptors have attracted significant interest and several groups have reported synthetic agonists to one or the other receptor that are up to 100-fold more potent than the endogenously occurring long-chain fatty acids originally identified as ligands. Since the identification of GPR120 as a potential target for anti-diabetic drug discovery is relatively recent, there are as of yet no reports of GPR120 agonists entering clinical development. In contrast, the somewhat more mature field of FFA1 agonists has yielded multiple clinical compounds. The recent proof-of-concept data showing that the FFA1 agonist TAK-875 can lower FPG and HbA_{1c} levels in diabetic patients with fewer incidents of hypoglycemia than established insulin secretagogues highlight the potential of this target and will ensure continuing interest in this field.

References

1. S. Wild, G. Roglic, A. Green, R. Sicree and H. King, *Diabetes Care*, 2004, **27**, 1047–1053.
2. H. E. Lebovitz, *Nat. Rev. Endocrinol.*, 2011, **7**, 408–419.
3. A. A. Tahrani, C. J. Bailey, S. Del Prato and A. H. Barnett, *Lancet*, 2011, **378**, 182–197.
4. R. A. DeFronzo, *Diabetes*, 2009, **58**, 773–795.
5. E. Ferrannini, A. Gastaldelli, Y. Miyazaki, M. Matsuda, A. Mari and R. A. DeFronzo, *J. Clin. Endocrinol. Metab.*, 2005, **90**, 493–500.
6. J. D. McGarry and R. L. Dobbins, *Diabetologia*, 1999, **42**, 128–138.
7. E. P. Haber, H. M. A. Ximenes, J. Procópio, C. R. O. Carvalho, R. Curi and A. R. Carpinelli, *J. Cell. Physiol.*, 2003, **194**, 1–12.

8. D. T. Stein, V. Esser, B. E. Stevenson, K. E. Lane, J. H. Whiteside, M. B. Daniels, S. Chen and J. D. McGarry, *J. Clin. Invest.*, 1996, **97**, 2728–2735.

9. R. L. Dobbins, M. W. Chester, M. B. Daniels, J. D. McGarry and D. T. Stein, *Diabetes*, 1998, **47**, 1613.

10. D. T. Stein, B. E. Stevenson, M. W. Chester, M. Basit, M. B. Daniels, S. D. Turley and J. D. McGarry, *J. Clin. Invest.*, 1997, **100**, 398–403.

11. C. J. Nolan, M. S. R. Madiraju, V. Delghingaro-Augusto, M.-L. Peyot and M. Prentki, *Diabetes*, 2006, **55**, S16–S23.

12. J. P. Overington, B. Al-Lazikani and A. L. Hopkins, *Nat. Rev. Drug Discov.*, 2006, **5**, 993–996.

13. D. K. Vassilatis, J. G. Hohmann, H. Zeng, F. Li, J. E. Ranchalis, M. T. Mortrud, A. Brown, S. S. Rodriguez, J. R. Weller, A. C. Wright, J. E. Bergmann and G. A. Gaitanaris, *Proc. Natl. Acad. Sci. USA*, 2003, **100**, 4903–4908.

14. A. Wise, K. Gearing and S. Rees, *Drug Discov. Today*, 2002, **7**, 235–246.

15. Y. Itoh, Y. Kawamata, M. Harada, M. Kobayashi, R. Fujii, S. Fukusumi, K. Ogi, M. Hosoya, Y. Tanaka, H. Uejima, H. Tanaka, M. Maruyama, R. Satoh, S. Okubo, H. Kizawa, H. Komatsu, F. Matsumura, Y. Noguchi, T. Shinohara, S. Hinuma, Y. Fujisawa and M. Fujino, *Nature*, 2003, **422**, 173–176.

16. K. Kotarsky, N. E. Nilsson, E. Flodgren, C. Owman and B. Olde, *Biochem. Biophys. Res. Commun.*, 2003, **301**, 406–410.

17. C. P. Briscoe, M. Tadayyon, J. L. Andrews, W. G. Benson, J. K. Chambers, M. M. Eilert, C. Ellis, N. A. Elshourbagy, A. S. Goetz, D. T. Minnick, P. R. Murdock, H. R. Sauls, Jr., U. Shabon, L. D. Spinage, J. C. Strum, P. G. Szekeres, K. B. Tan, J. M. Way, D. M. Ignar, S. Wilson and A. I. Muir, *J. Biol. Chem.*, 2003, **278**, 11303–11311.

18. A. J. Brown, S. M. Goldsworthy, A. A. Barnes, M. M. Eilert, L. Tcheang, D. Daniels, A. I. Muir, M. J. Wigglesworth, I. Kinghorn, N. J. Fraser, N. B. Pike, J. C. Strum, K. M. Steplewski, P. R. Murdock, J. C. Holder, F. H. Marshall, P. G. Szekeres, S. Wilson, D. M. Ignar, S. M. Foord, A. Wise and S. J. Dowell, *J. Biol. Chem.*, 2003, **278**, 11312–11319.

19. E. Le Poul, C. Loison, S. Struyf, J.-Y. Springael, V. Lannoy, M.-E. Decobecq, S. Brezillon, V. Dupriez, G. Vassart, J. Van Damme, M. Parmentier and M. Detheux, *J. Biol. Chem.*, 2003, **278**, 25481–25489.

20. N. E. Nilsson, K. Kotarsky, C. Owman and B. Olde, *Biochem. Biophys. Res. Commun.*, 2003, **303**, 1047–1052.

21. J. Wang, X. Wu, N. Simonavicius, H. Tian and L. Ling, *J. Biol. Chem.*, 2006, **281**, 34457–34464.

22. A. Hirasawa, K. Tsumaya, T. Awaji, S. Katsuma, T. Adachi, M. Yamada, Y. Sugimoto, S. Miyazaki and G. Tsujimoto, *Nat. Med.*, 2005, **11**, 90–94.

23. D. Y. Oh, S. Talukdar, E. J. Bae, T. Imamura, H. Morinaga, W. Q. Fan, P. Li, W. J. Lu, S. M. Watkins and J. M. Olefsky, *Cell*, 2010, **142**, 687–698.

24. L. A. Stoddart, N. J. Smith and G. Milligan, *Pharmacol. Rev.*, 2008, **60**, 405–417.

25. P. Wellendorph, D. Johansen Lars and H. Brauner-Osborne, *Mol. Pharmacol.*, 2009, **76**, 453–465.

26. S. Talukdar, J. M. Olefsky and O. Osborn, *Trends Pharmacol. Sci.*, 2011, **32**, 543–550.

27. T. Hara, A. Hirasawa, A. Ichimura, I. Kimura and G. Tsujimoto, *J. Pharm. Sci.*, 2011, **100**, 3594–3601.

28. D. Y. Oh and W. S. Lagakos, *Curr. Opin. Clin. Nutr. Metab. Care*, 2011, **14**, 322–327.

29. S. Miyauchi, A. Hirasawa, A. Ichimura, T. Hara and G. Tsujimoto, *J. Pharmacol. Sci.*, 2010, **112**, 19–24.

30. B. Ahren, *Nat. Rev. Drug Discov.*, 2009, **8**, 369–385.

31. M. Sawzdargo, S. R. George, T. Nguyen, S. Xu, L. F. Kolakowski, Jr. and B. F. O'Dowd, *Biochem. Biophys. Res. Commun.*, 1997, **239**, 543–547.

32. A. Salehi, E. Flodgren, N. E. Nilsson, J. Jimenez-Feltstrom, J. Miyazaki, C. Owman and B. Olde, *Cell Tissue Res.*, 2005, **322**, 207–215.

33. C. P. Briscoe, A. J. Peat, S. C. McKeown, D. F. Corbett, A. S. Goetz, T. R. Littleton, D. C. McCoy, T. P. Kenakin, J. L. Andrews, C. Ammala, J. A. Fornwald, D. M. Ignar and S. Jenkinson, *Br. J. Pharm.*, 2006, **148**, 619–628.

34. P. Proks, F. Reimann, N. Green, F. Gribble and F. Ashcroft, *Diabetes*, 2002, **51**, S368–S376.

35. S. A. Amiel, T. Dixon, R. Mann and K. Jameson, *Diabet. Med.*, 2008, **25**, 245–254.

36. R. Bartoov-Shifman, G. Ridner, K. Bahar, N. Rubins and M. D. Walker, *J. Biol. Chem.*, 2007, **282**, 23561–23571.

37. T. Tomita, H. Masuzaki, H. Iwakura, J. Fujikura, M. Noguchi, T. Tanaka, K. Ebihara, J. Kawamura, I. Komoto, Y. Kawaguchi, K. Fujimoto, R. Doi, Y. Shimada, K. Hosoda, M. Imamura and K. Nakao, *Diabetologia*, 2006, **49**, 962–968.

38. E. Flodgren, B. Olde, S. Meidute-Abaraviciene, M. S. Winzell, B. Ahren and A. Salehi, *Biochem. Biophys. Res. Commun.*, 2007, **354**, 240–245.

39. D. Ma, B. Tao, S. Warashina, S. Kotani, L. Lu, D. B. Kaplamadzhiev, Y. Mori, A. B. Tonchev and T. Yamashima, *Neurosci. Res.*, 2007, **58**, 394–401.

40. S. Edfalk, P. Steneberg and H. Edlund, *Diabetes*, 2008, **57**, 2280–2287.

41. H. E. Parker, A. M. Habib, G. J. Rogers, F. M. Gribble and F. Reimann, *Diabetologia*, 2009, **52**, 289–298.

42. F. Reimann, A. M. Habib, G. Tolhurst, H. E. Parker, G. J. Rogers and F. M. Gribble, *Cell Metab.*, 2008, **8**, 532–539.

43. A. P. Liou, X. Lu, Y. Sei, X. Zhao, S. Pechhold, R. J. Carrero, H. E. Raybould and S. Wank, *Gastroenterology*, 2011, **140**, 903–912, e903/901–e903/904.

44. D. J. Drucker and M. A. Nauck, *Lancet*, 2006, **368**, 1696–1705.

45. K. Fujiwara, F. Maekawa and T. Yada, *Am. J. Physiol. Endocrinol. Metab.*, 2005, **289**, E670–E677.

46. H. Shapiro, S. Shachar, I. Sekler, M. Hershfinkel and M. D. Walker, *Biochem. Biophys. Res. Commun.*, 2005, **335**, 97–104.

47. Q.-F. Wan, Y. Dong, H. Yang, X. Lou, J. Ding and T. Xu, *J. Gen. Physiol.*, 2004, **124**, 653–662.

48. C. Schmitz-Peiffer and T. J. Biden, *Diabetes*, 2008, **57**, 1774–1783.

49. D. D. Feng, Z. Luo, S.-g. Roh, M. Hernandez, N. Tawadros, D. J. Keating and C. Chen, *Endocrinology*, 2006, **147**, 674–682.

50. W. Kim and J. M. Egan, *Pharmacol. Rev.*, 2008, **60**, 470–512.

51. P. E. MacDonald, J. W. Joseph and P. Rorsman, *Phil. Trans. R. Soc. B*, 2005, **360**, 2211–2225.

52. S. J. Ashcroft, *J. Membrane Biol.*, 2000, **176**, 187–206.

53. S.-N. Yang and P.-O. Berggren, *Endocrinol. Rev.*, 2006, **27**, 621–676.

54. P. E. MacDonald and M. B. Wheeler, *Diabetologia*, 2003, **46**, 1046–1062.

55. P. E. MacDonald, X. Wang, F. Xia, W. El-kholy, E. D. Targonsky, R. G. Tsushima and M. B. Wheeler, *J. Biol. Chem.*, 2003, **278**, 52446–52453.

56. S. Engelstoft Maja, L. Egerod Kristoffer, B. Holst and W. Schwartz Thue, *Cell Metabolism*, 2008, **8**, 447–449.

57. A. El-Ouaghlidi, E. Rehring, J. J. Holst, A. Schweizer, J. Foley, D. Holmes and M. A. Nauck, *J. Clin. Endocrinol. Metab.*, 2007, **92**, 4165–4171.

58. F. M. Gribble, L. Williams, A. K. Simpson and F. Reimann, *Diabetes*, 2003, **52**, 1147–1154.

59. R. Fredriksson, J. Hoglund Par, E. I. Gloriam David, C. Lagerstrom Malin and B. Schioth Helgi, *FEBS Lett.*, 2003, **554**, 381–388.

60. S. Miyauchi, A. Hirasawa, T. Iga, N. Liu, C. Itsubo, K. Sadakane, T. Hara and G. Tsujimoto, *Naunyn-Schmiedeberg's Arch. Pharmacol.*, 2009, **379**, 427–434.

61. T. Tanaka, S. Katsuma, T. Adachi, T.-a. Koshimizu, A. Hirasawa and G. Tsujimoto, *Naunyn-Schmiedeberg's Arch. Pharmacol.*, 2008, **377**, 523–527.

62. R. N. Burns and N. H. Moniri, *Biochem. Biophys. Res. Commun.*, 2010, **396**, 1030–1035.

63. A. Marchese, C. Chen, Y. M. Kim and J. L. Benovic, *Trends Biochem. Sci.*, 2003, **28**, 369–376.

64. S. P. Weisberg, D. McCann, M. Desai, M. Rosenbaum, R. L. Leibel and A. W. Ferrante, *J. Clin. Invest.*, 2003, **112**, 1796–1808.

65. A. Guilherme, J. V. Virbasius, V. Puri and M. P. Czech, *Nat. Rev. Mol. Cell Biol.*, 2008, **9**, 367–377.

66. G. S. Hotamisligil, D. L. Murray, L. N. Choy and B. M. Spiegelman, *Proc. Natl. Acad. Sci. USA*, 1994, **91**, 4854–4858.

67. V. Catalán, J. Gómez-Ambrosi, A. Rodríguez, J. Salvador and G. Frühbeck, *Expert Opin. Pharmacother.*, 2009, **10**, 239–254.

68. J. C. Medina and J. B. Houze, *Annu. Rep. Med. Chem.*, 2008, **43**, 75–85.

69. S. B. Bharate, K. V. S. Nemmani and R. A. Vishwakarma, *Expert Opin. Ther. Pat.*, 2009, **19**, 237–264.

70. E. Christiansen, C. Urban, N. Merten, K. Liebscher, K. K. Karlsen, A. Hamacher, A. Spinrath, A. D. Bond, C. Drewke, S. Ullrich, M. U. Kassack, E. Kostenis and T. Ulven, *J. Med. Chem.*, 2008, **51**, 7061–7064.

71. E. Christiansen, M. E. Due-Hansen, C. Urban, N. Merten, M. Pfleiderer, K. K. Karlsen, S. S. Rasmussen, M. Steensgaard, A. Hamacher, J. Schmidt, C. Drewke, R. K. Petersen, K. Kristiansen, S. Ullrich, E. Kostenis, M. U. Kassack and T. Ulven, *ACS Med. Chem. Lett.*, 2010, **1**, 345–349.

72. J. B. Houze, L. Zhu, Y. Sun, M. Akerman, W. Qiu, A. J. Zhang, R. Sharma, M. Schmitt, Y. Wang, J. Liu, J. Liu, J. C. Medina, J. D. Reagan, J. Luo, G. Tonn, J. Zhang, J. Y.-L. Lu, M. Chen, E. Lopez, K. Nguyen, L. Yang, L. Tang, H. Tian, S. J. Shuttleworth and D. C.-H. Lin, *Bioorg. Med. Chem. Lett.*, 2012, **22**, 1267–1270.

73. M. Akerman, J. Houze, D. C. H. Lin, J. Liu, J. Luo, J. C. Medina, W. Qiu, J. D. Reagan, R. Sharma, S. J. Shuttleworth, Y. Sun, J. Zhang and L. Zhu, *PCT Application*, WO 2005/086661.

74. D. C.-H. Lin, J. Zhang, R. Zhuang, F. Li, K. Nguyen, M. Chen, T. Tran, E. Lopez, J. Y. L. Lu, X. N. Li, L. Tang, G. R. Tonn, G. Swaminath, J. D. Reagan, J.-L. Chen, H. Tian, Y.-J. Lin, J. B. Houze and J. Luo, *PLoS One*, 2011, **6**, e27270.

75. T. Shimada, H. Ueno, K. Tsutsumi, K. Aoyagi, T. Manabe, S.-Y. Sasaki and S. Katoh, *US Patent Appl.*, 2009/0170908.

76. C. Hamdouchi, J. P. Lineswala and P. Maiti, *PCT Application*, WO 2011/046851.

77. C. Hamdouchi, J. P. Lineswala and P. Maiti, *PCT Application*, WO 2011/066183.

78. S. P. Walsh, A. Severino, C. Zhou, J. He, G.-B. Liang, C. P. Tan, J. Cao, G. J. Eiermann, L. Xu, G. Salituro, A. D. Howard, S. G. Mills and L. Yang, *Bioorg. Med. Chem. Lett.*, 2011, **21**, 3390–3394.

79. E. Defossa, T. Klabunde, V. Dietrich, S. Stengelin, G. Haschke, A. Herling, J. Kuhlmann and S. Bartoschek, *PCT Application*, WO 2009/039942.

80. B. A. Ellsworth, W. R. Ewing and E. Jurica, *PCT Application*, WO 2011/044073.

81. G. A. Patani and E. J. LaVoie, *Chem. Rev.*, 1996, **96**, 3147–3176.

82. N. A. Meanwell, *J. Med. Chem.*, 2011, **54**, 2529–2591.

83. C. Zhou, C. Tang, E. Chang, M. Ge, S. Lin, E. Cline, C. P. Tan, Y. Feng, Y.-P. Zhou, G. J. Eiermann, A. Petrov, G. Salituro, P. Meinke, R. Mosley, T. E. Akiyama, M. Einstein, S. Kumar, J. Berger, A. D. Howard, N. Thornberry, S. G. Mills and L. Yang, *Bioorg. Med. Chem. Lett.*, 2010, **20**, 1298–1301.

84. C. P. Tan, Y. Feng, Y.-P. Zhou, G. J. Eiermann, A. Petrov, C. Zhou, S. Lin, G. Salituro, P. Meinke, R. Mosley, T. E. Akiyama, M. Einstein, S.

Kumar, J. P. Berger, S. G. Mills, N. A. Thornberry, L. Yang and A. D. Howard, *Diabetes*, 2008, **57**, 2211–2219.

85. K. Negoro, F. Iwasaki, K. Ohnuki, T. Kurosaki, K. Tsuchiya, K. Kuramoto, S. Yoshida and T. Soga, *US Patent Appl.*, 2010/0267775.

86. A. Okano and M. Okochi, *PCT Application*, WO 2011/052756.

87. N. Negoro, S. Sasaki, S. Mikami, M. Ito, M. Suzuki, Y. Tsujihata, R. Ito, A. Harada, K. Takeuchi, N. Suzuki, J. Miyazaki, T. Santou, T. Odani, N. Kanzaki, M. Funami, T. Tanaka, A. Kogame, S. Matsunaga, T. Yasuma and Y. Momose, *ACS Med. Chem. Lett.*, 2010, **1**, 290–294.

88. S. Sasaki, S. Kitamura, N. Negoro, M. Suzuki, Y. Tsujihata, N. Suzuki, T. Santou, N. Kanzaki, M. Harada, Y. Tanaka, M. Kobayashi, N. Tada, M. Funami, T. Tanaka, Y. Yamamoto, K. Fukatsu, T. Yasuma and Y. Momose, *J. Med. Chem.*, 2011, **54**, 1365–1378.

89. Y. Tsujihata, R. Ito, M. Suzuki, A. Harada, N. Negoro, T. Yasuma, Y. Momose and K. Takeuchi, *J. Pharmacol. Exp. Ther.*, **339**, 228–237.

90. H. Naik, M. Vakilynejad, J. Wu, P. Viswanathan, N. Dote, T. Higuchi and E. Leifke, *J. Clin. Pharmacol.*, 2011, doi: 10.1177/0091270011409230.

91. E. Leifke, J. Wu, P. Viswamathan, M. S. Kipnes and M. Vakilynejad, *Diabetes*, 2011, **60** (Suppl. 1), A114–A115.

92. E. J. Caveney and O. J. Cohen, *J. Diabetes Sci. Technol.*, 2011, **5**, 192–197.

93. C. F. Burant, P. Viswanathan, J. Marcinak, C. Cao, M. Vakilynejad, B. Xie and E. Leifke, *The Lancet*, 2012, **379**, 1403–1411.

94. N. Negoro, Y. Terao, S. Mikami and T. Yukawa, *PCT Application*, WO 2010/143733.

95. P. S. Humphries, J. W. Benbow, P. D. Bonin, D. Boyer, S. D. Doran, R. K. Frisbie, D. W. Piotrowski, G. Balan, B. M. Bechle, E. L. Conn, K. J. Dirico, R. M. Oliver, W. C. Soeller, J. A. Southers and X. Yang, *Bioorg. Med. Chem. Lett.*, 2009, **19**, 2400–2403.

96. J. L. Davis, M. J. Mayer and H. B. Josien, *PCT Application*, WO 2010/091176.

97. H. B. Josien, J. W. Clader, W. J. Greenlee, M. J. Mayer, R. J. Herr, J. L. Davis, K. Deng, M. M. Hsia and S. Wan, *PCT Application*, WO 2010/085525.

98. H. B. Josien, J. W. Clader, A. Stamford, W. J. Greenlee, M. J. Mayer, J. L. Davis, M. M. Hsia and S. Wan, *PCT Application*, WO 2010/085522.

99. H. B. Josien, J. W. Clader, W. J. Greenlee, M. J. Mayer, J. L. Davis and S. Wan, *PCT Application*, WO 2010/085528.

100. G.-B. Liang, X. Liao, W. Liu, P. E. Finke, D. Kim, L. Yang and S. Lin, *PCT Application*, WO 2009/058237.

101. S. P. Brown, P. Dransfield, X. Du, Z. Fu, J. Houze, X. Jiao, S. Lai, A.-R. Li, J. Liu, Z. Ma, J. C. Medina, V. Pattaropong, W. Shen, M. Vimolratana, Y. Wang, Z. Wang, M. Yu and L. Zhu, *PCT Application*, WO 2010/045258.

102. T. Suzuki, S.-i. Igari, A. Hirasawa, M. Hata, M. Ishiguro, H. Fujieda, Y. Itoh, T. Hirano, H. Nakagawa, M. Ogura, M. Makishima, G. Tsujimoto and N. Miyata, *J. Med. Chem.*, 2008, **51**, 7640–7644.
103. Q. Sun, A. Hirasawa, T. Hara, I. Kimura, T. Adachi, T. Awaji, M. Ishiguro, T. Suzuki, N. Miyata and G. Tsujimoto, *Mol. Pharmacol.*, 2010, **78**, 804–810.
104. R. Epple, M. Azimiora, C. Cow, R. Russo, V. Nikulin and G. Lelais, *PCT Application*, WO 2008/103501.
105. J. Ma, A. Novack, I. Nashashibi, P. Pham, C. J. Rabbat, J. Song, D. F. Shi, Z. Zhao, Y.-J. Choi and X. Chen, *PCT Application*, WO 2010/048207.
106. D. F. Shi, J. Song, J. Ma, A. Novack, P. Pham, I. Nashashibi, C. J. Rabbat and X. Chen, *PCT Application*, WO 2010/080537.
107. R. Epple, M. Azimiora, C. Cow, R. Russo, V. Nikulin and G. Lelais, *PCT Application*, WO 2008/103500.
108. X. He, X. Zhu, K. Yang, R. Epple and H. Liu, *PCT Application*, WO 2010/008831.
109. N. Hashimoto, Y. Sasaki, C. Nakama and M. Ishikawa, *US Patent Appl.*, 2010/0130559.
110. M. Ishikawa, T. Haketa, C. Nakama, T. Nishimura, J. Shibata, T. Shimamura and T. Yamakawa, *US Patent Appl.*, 2011/0065739.
111. L. Zhuang, J. S. Wai, M. W. Embrey, T. E. Fisher, M. S. Egbertson, L. S. Payne, J. P. Guare, J. P. Vacca, D. J. Hazuda, P. J. Felock, A. L. Wolfe, K. A. Stillmock, M. V. Witmer, G. Moyer, W. A. Schleif, L. J. Gabryelski, Y. M. Leonard, J. J. Lynch, S. R. Michelson and S. D. Young, *J. Med. Chem.*, 2003, **46**, 453–456.
112. K. Arakawa, T. Nishimura, Y. Sugimoto, H. Takahashi and T. Shimamura, *PCT Application*, WO 2010/104195.
113. J. Xu, K. D. Kochanek, S. L. Murphy and B. Tejada-Vera, *National Vital Stat. Rep.*, 2010, **58**, 1–135.

Glucagon Receptor Antagonists in Development

DUANE E. DEMONG* AND M. W. MILLER

Department of Medicinal Chemistry, Merck Research Laboratories, 126 E. Lincoln Avenue, Mailstop RY50G-341, Rahway, NJ 07065, USA
*E-mail: duane.demong@merck.com

15.1 Introduction

In 1923, while attempting to improve the extraction and purification of insulin from beef pancreas, Kimball and Murlin discovered an unknown substance which, when administered to depancreatized dogs, afforded a hyperglycemic response.[1] The authors named this substance glucagon. In 1953, Staub, Sinn, and Behrens at Lilly Research Laboratories reported the isolation and crystallization of pure glucagon.[2,3] The same group at Lilly determined glucagon to be a 29 amino acid linear peptide with the sequence H_2N-His-Ser-Gln-Gly-Thr-Phe-Thr-Ser-Asp-Tyr-Ser-Lys-Tyr-Leu-Asp-Ser-Arg-Arg-Ala-Gln-Asp-Phe-Val-Gln-Trp-Leu-Met-Asn-Thr-COOH.[4–7]

Produced by α-cells of the islets of Langerhans in the pancreas via selective cleavage of the proglucagon polypeptide, the peptide hormone glucagon is counter-regulatory to insulin in the process of glucose homeostasis.[8] Binding of glucagon to the glucagon receptor (GCGR), present in the liver, stimulates a hyperglycemic event caused by both glycogenolysis and gluconeogenesis.[9–11] The glucagon receptor is a class B GPCR in the B1 subfamily. Upon formation of the complex of glucagon with its receptor, a G_s-type G-protein is released which, in turn, activates adenylate cyclase for further signaling.[12,13] Diabetics

RSC Drug Discovery Series No. 27
New Therapeutic Strategies for Type 2 Diabetes: Small Molecule Approaches
Edited by Robert M. Jones

Published by the Royal Society of Chemistry, www.rsc.org

with elevated blood glucose levels have been shown to possess elevated glucagon levels when compared with non-diabetics.[14]

Isolation of cDNA that encodes a human glucagon receptor (hGCGR) has been accomplished by probing a liver tissue cDNA library.[15] Separate reports described the cloning of a functional human glucagon receptor from human liver tissue and subsequent expression in COS-7 and *Drosophila* Schneider 2 (S2) cells.[16,17]

Mice with a null mutation of the glucagon receptor (mGCGR$^{-/-}$) demonstrate reduced plasma glucose levels in both fed and fasted states.[18,19] These receptor knockout mice also possessed improved glucose tolerance versus wild-type. Glucagon levels and glucagon-like peptide-1 (GLP-1) levels were significantly elevated relative to wild-type and α-cell hyperplasia was also observed. The knockout animals also displayed a statistically significant increase in serum levels of low density lipoprotein (LDL) when compared with wild-type animals. Genomic, proteomic, and metabolomic profiling of the receptor knockout mice demonstrated significant differences relative to wild-type.[20] Up-regulation of glycolysis, fatty acid synthesis, and cholesterol synthesis pathways was observed. Bile acid production also increased with the mGCGR$^{-/-}$ animals. Also observed was a significant inhibition of gluconeogenesis and amino acid degradation. Transgenic mice which express the hGCGR have also been developed.[21] The impetus for generating these animals was for the *in vivo* testing of hGCGR antagonists that possessed inferior potency at the rodent receptor.

15.2 Peptide Glucagon Receptor Antagonists

The first tool for probing the effect of glucagon receptor antagonism was reported by Sundby *et al.*[22] Removal of the *N*-terminal histidine from glucagon produced a peptide (des-His-glucagon) that competed with radiolabeled glucagon for binding to rat liver membranes, but did not activate adenylyl cyclase. At an intraperitoneal (IP) dose of 2 mg/kg in transgenic mice containing hGCGR, des-His-glucagon was shown to occupy 78% of the hepatic glucagon receptors and to efficiently block an exogenous glucagon-stimulated increase in plasma glucose levels. Subsequently, Hruby and co-workers prepared the glucagon receptor antagonist [1-N^{α}-trinitrophenylhisti-dine-12-homoarginine]glucagon (THG).[23] THG was determined to not only inhibit glucagon-stimulated activation of adenylyl cyclase in rat liver membranes, but also demonstrated a significant decrease in plasma glucose levels in a streptozotocin-induced rat model of type 2 diabetes mellitus (T2DM).[24] These results were important in that they demonstrated, for the first time, a preclinical proof of concept that antagonists of the glucagon receptor could potentially be used as a treatment for diabetes.

15.3 Monoclonal Antibodies

The first monoclonal antibodies to the glucagon receptor were generated by Iwanij and co-workers.[25] The four antibodies were generated against partially purified rat glucagon receptor (rGCGR) which was generated from rat liver membranes. Chronic administration of a mouse glucagon receptor (mGCGR) monoclonal antibody to male C57BL/6 mice with diet-induced obesity (DIO) demonstrated decreased fasting glucose levels relative to control.[26]

The first monoclonal antibodies to the hGCGR were prepared at Bayer (CIV395.7A) and Novo Nordisk (hGR-2 F6).[27,28] Both antibodies were shown to bind competitively to the human glucagon receptor. Additionally, CIV395.7A was shown to be a functional antagonist of hGCGR and selective over hGLP-1R. A crystal structure of hGR-2 F6 was also disclosed.

Recently, researchers at Amgen described the production and profiling of a number of fully human monoclonal antibodies to the hGCGR.[29] One antibody in particular (mAb B) was demonstrated to displace [^{125}I]-glucagon from its receptor (hGCGR binding IC_{50} = 0.47 $\pm$ 0.06 nM, cynomologus monkey GCGR binding IC_{50} = 0.43 $\pm$ 0.08 nM, mGCGR binding IC_{50} = 0.21 $\pm$ 0.03 nM). Additionally, mAb B was shown to be a functional antagonist of the glucagon receptor (hGCGR cAMP pIC_{50} = 8.29 $\pm$ 0.19, cynomologus monkey GCGR cAMP pIC_{50} = 7.87 $\pm$ 0.26, mGCGR cAMP pIC_{50} = 8.74), but was not a functional antagonist of hGLP-1R. Subcutaneous (SQ) administration of mAb B to C57BL/6 mice at 1 and 10 mg/kg afforded a dose-dependent decrease in fasting plasma glucose levels without hypoglycemia. An intraperitoneal glucose tolerance test (IPGTT) also demonstrated improved glucose tolerance at both the 1 and 10 mg/kg SQ doses. Male cynomolgus monkeys were treated with a single dose of 3 mg/kg mAb B, 30 mg/kg mAb B or vehicle. Oral glucose tolerance tests (OGTT) were then performed on days 3, 8, and 17 after administration of mAb B. Animals treated with mAb B at both doses showed an improvement in glucose tolerance at days 3 and 8. Plasma insulin levels were unchanged between the mAb B-treated and vehicle-treated groups, but glucagon levels and GLP-1 levels were elevated in the mAb B-treated groups up to day 8. Further profiling of mAb B in wild-type and GLP-1R knockout mice demonstrated that the glucose regulating effects of mAb B are dependent upon the existence of a functional GLP-1 receptor.[30]

15.4   Small Molecule Glucagon Receptor Antagonists

Since the early 1990s, an increasing number of reports of small molecule glucagon receptor antagonists have surfaced in both the patent and peer-reviewed literature. The chemotypes described cover a broad chemical space. A number of these chemotypes will be summarized in the proceeding sections.

15.4.1   β-Alanine Benzamides and their Isosteres

15.4.1.1   Discovery and Preclinical Development

The utility of β-alanine benzamides as glucagon receptor antagonists was first disclosed by Lau and co-workers in 2000.[31] Compound **15.1** (Figure 15.1) was initially found in an HTS screen as a weak antagonist of the human glucagon receptor (hGCGR binding IC_{50} = 7000 nM).[32] A follow-up compound library identified the β-alanine benzamide-containing compound **15.2** as a potent antagonist of the human glucagon receptor (hGCGR binding IC_{50} = 23–55 nM, ratGCGR binding IC_{50} = 56 nM) with a modest pharmacokinetic (PK) profile (**dog**: F_{po} = 10–20%, $t_{1/2}$ = 101 min (IV); **rat**: C_{max} = 106 ng/mL, T_{max} = 60 min, AUC = 13045 ng·min/mL, $t_{1/2}$ = 90 min IV, F_{po} = 17%). Further profiling of **15.2** demonstrated its ability to inhibit a glucagon-stimulated rise in blood glucose at a dose of 3 mg/kg i.v. in a Sprague Dawley rat-based glucagon challenge model. Compound **15.2** was also screened against the human gastric inhibitory polypeptide receptor (hGIPR) and the human glucagon-like peptide-1 receptor (hGLP-1R), both of which belong to the same family of GPCRs as the glucagon receptor.[33] Gastric inhibitory polypeptide (GIP) is known to stimulate pancreatic insulin secretion, therefore making antagonism of hGIPR undesirable. Furthermore, agonism of the GLP-1R via long acting GLP-1 analogs or inhibition of dipeptidylpeptidase IV (DPP-IV) have shown clinical effect in ameliorating T2DM, therefore necessitating hGCGR antagonists which are devoid of antagonistic effects at the hGLP-1R. Compound **15.2** possessed 100-fold selectivity over the hGLP-1R (hGLP-1R binding IC_{50} = 2536 nM), but no selectivity over the hGIPR (hGIPR binding IC_{50} = 28.3 nM).

Additional SAR studies led to the discovery of compound **15.3** (hGCGR IC_{50} = 12–27 nM, ratGCGR IC_{50} = 122 nM), which possessed a significantly improved PK profile over that of **15.2** (**dog**: F_{po} = 65%, $t_{1/2}$ = 92 min; **rat**: C_{max} = 275 ng/mL, T_{max} = 60 min, AUC = 178 226 ng·min/mL, $t_{1/2}$ = 82 min IV, F_{po} = 69%) and a modest 5-fold improvement in selectivity over hGIPR (hGIPR binding IC_{50} = 64 nM). Compound **15.3** exhibited statistically significant inhibition of glucagon-induced rise in blood glucose at a dose of 10 mg/kg IV, but not at 3 mg/kg IV. Upon dosing **15.3** at 100 mg/kg p.o. in a leptin deficient ob/ob mouse model of T2DM, a significant decrease in blood glucose concentration was observed.

Further SAR studies around the β-alanine moiety revealed that the *R*-isoserine benzamide **15.4** (NNC 25-0926) (hGCGR IC_{50} = 3.9 nM, hGIPR binding IC_{50} = 359 nM, hGLP-1R IC_{50} = 17 045 nM) possessed an improved selectivity profile over both hGIPR and hGLP-1R compared to its β-alanine homolog **15.3** with no loss in hGCGR binding potency. The rat pharmaco-kinetic data for **15.4**, although not as robust as **15.3**, was acceptable ($t_{1/2}$ = 53 min IV, F_{po} = 32%).

Insulin-resistant high-fat fed (HFF) mice were treated chronically (up to 30 days) with compound **15.4** at 300 mg/kg via gavage.[34] High-fat fed animals treated with compound **15.4** for 17–18 days demonstrated a statistically significant reduction in fasting plasma glucose increase versus the HFF vehicle-treated group. Plasma glucagon levels in the antagonist-treated group were also elevated compared to the vehicle-treated lean mice. An intravenous glucose tolerance test (IVGTT) was also performed at day 18 of antagonist

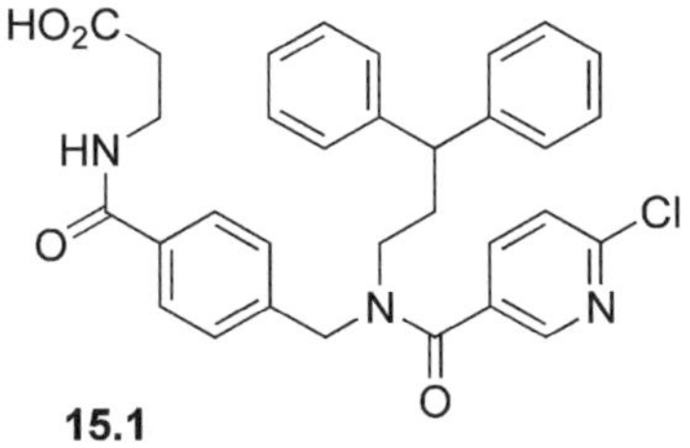

15.1
hGCGR Binding IC$_{50}$: 7000 nM

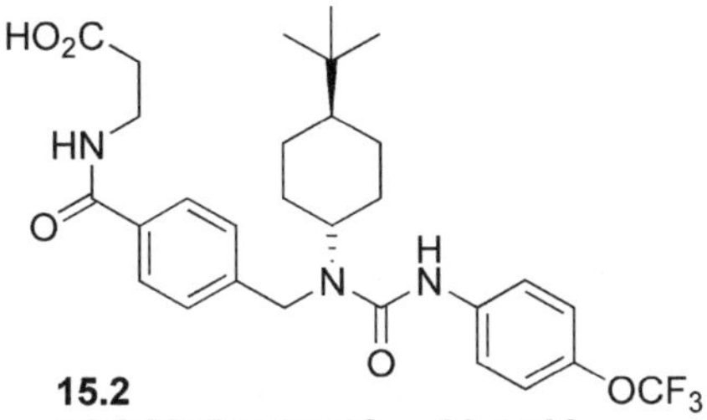

15.2
hGCGR Binding IC$_{50}$: 23.4 nM
hGIPR Binding IC$_{50}$: 28.3 nM
hGLP-1R Binding IC$_{50}$: 2536 nM

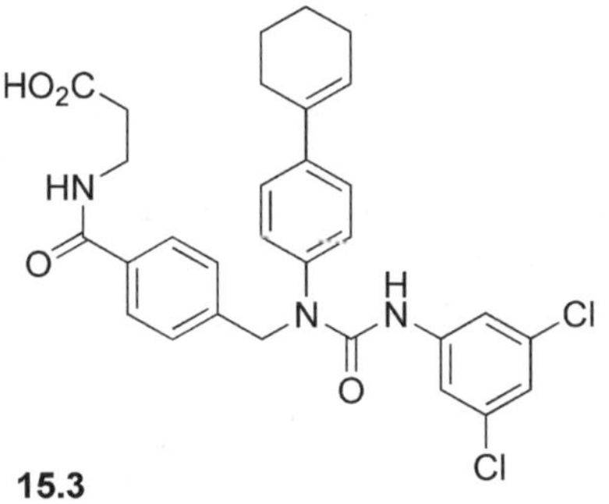

15.3
hGCGR Binding IC$_{50}$: 12 nM
hGIPR Binding IC$_{50}$: 64 nM
hGLP-1R Binding IC$_{50}$: 4995 nM

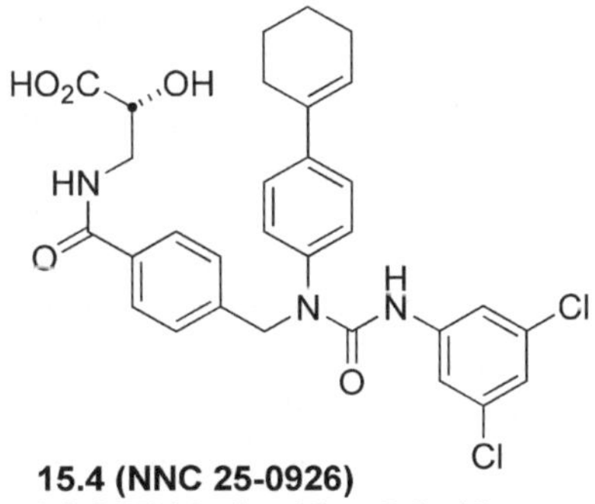

15.4 (NNC 25-0926)
hGCGR binding IC$_{50}$: 3.9 nM
hGIPR binding IC$_{50}$: 359 nM
hGLP-1R binding IC$_{50}$: 17045 nM

Figure 15.1 Early β-alanine benzamide-based glucagon receptor antagonists.

treatment. In the IVGTT, the group treated with **15.4** showed an increase in glucose clearance compared to untreated HFF mice. An oral glucose tolerance test (OGTT) demonstrated an improvement in glucose tolerance and a decrease in plasma insulin levels in the HFF mice treated with **15.4** versus those that were treatment naïve. Treatment with **15.4** resulted in an increase in islet α-cell mass, but no cellular damage was observed.

Compound **15.4** at 20 mg/kg (intragastric infusion) was also shown to inhibit the glucagon-stimulated increase of plasma glucose levels in a conscious dog hepatic glucose production model.[35] The resulting inhibition of hepatic glucose production was positively correlated with the plasma levels of **15.4** sampled from the hepatic portal vein.

A series of fluorinated β-alanine benzamides, exemplified by compounds such as **15.5** and **15.6**, have also been described as potent glucagon receptor antagonists with selectivity over GIP (Figure 15.2).[36] While no compound-specific potency data were disclosed, all compounds exemplified were stated to possess hGCGR binding IC$_{50}$ values of <1000 nM, with some compounds registering IC$_{50}$ values <100 nM.

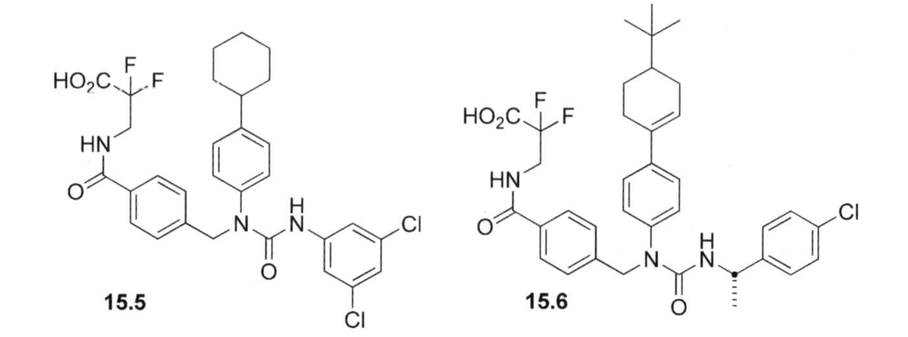

15.5

15.6

Figure 15.2 α,α-Difluoro β-alanine benzamides.

In addition to β-alanine benzamides, various aminotetrazole and aminoalkyltetrazole benzamides have been described as viable β-alanine bioisosteres (Figure 15.3).[37] For example, the aminotetrazole benzamide **15.7**, a close analog of β-alanine benzamide **15.3**, has been disclosed as a glucagon receptor antagonist. Analogs of **15.2**, such as tetrazoles **15.8** and **15.9**, have also been described as glucagon receptor antagonists.

Acyclic urea core structures such as **15.10** and **15.11** have also been described in the patent literature (Figure 15.4).[38] In these instances, the β-

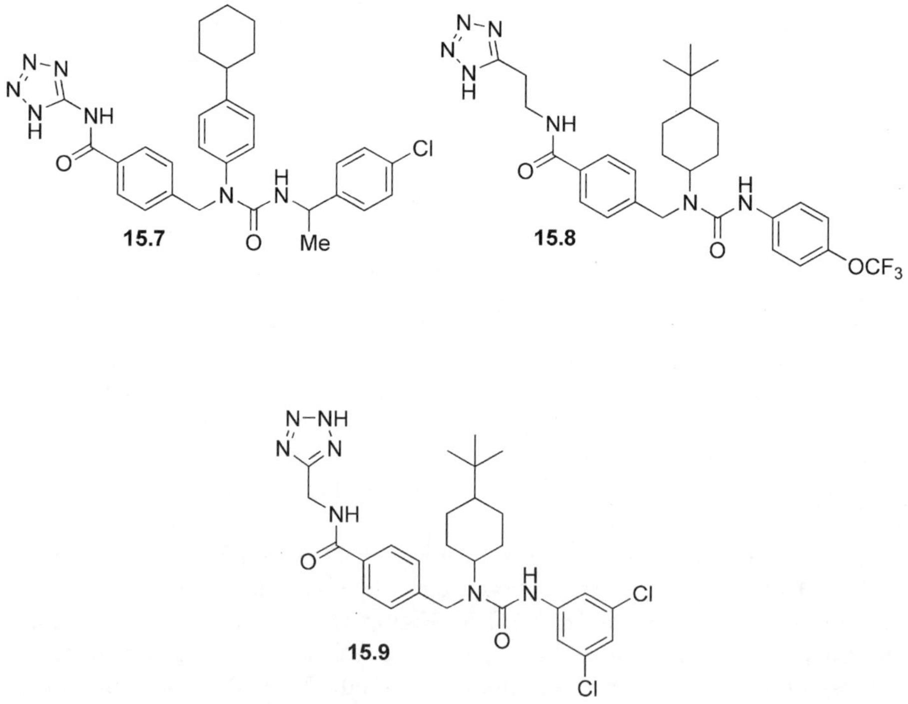

15.7

15.8

15.9

Figure 15.3 Tetrazole-based β-alanine bioisosteres.

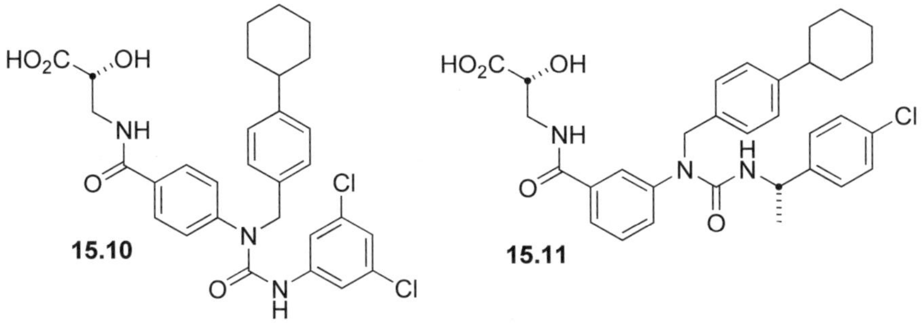

Figure 15.4 (*R*)-Isoserine benzamides.

alanine benzamide is connected to the urea nitrogen directly, rather than by a benzylic methylene group as described in the previous examples. While biological data was not provided for specific compounds, the most preferred compounds of the invention were stated to have an hGCGR binding IC$_{50}$ of <100 nM.

Upon discovery of the β-alanine benzamide and isosteres thereof as the key structural motif for imparting binding affinity to the glucagon receptor, a number of non-urea containing acyclic core structures were explored (Figure 15.5). Butanedione benzamides **15.12**, dibenzylamino benzamides **15.13**, and *N*-benzyl carbamoylbenzamides **15.14** have all been described as viable core structures for the effective display of the β-alanine residue and its

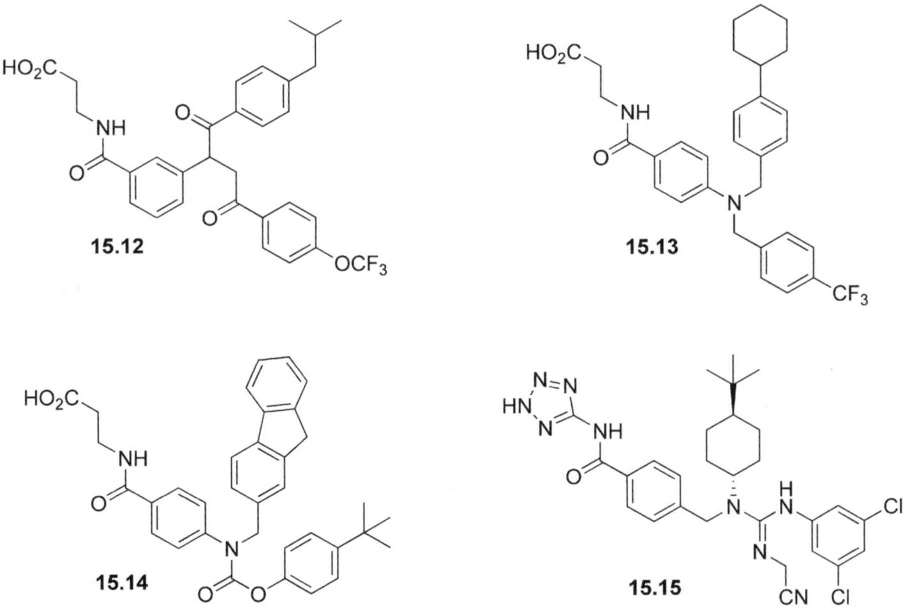

Figure 15.5 Acyclic, non-urea-based cores.

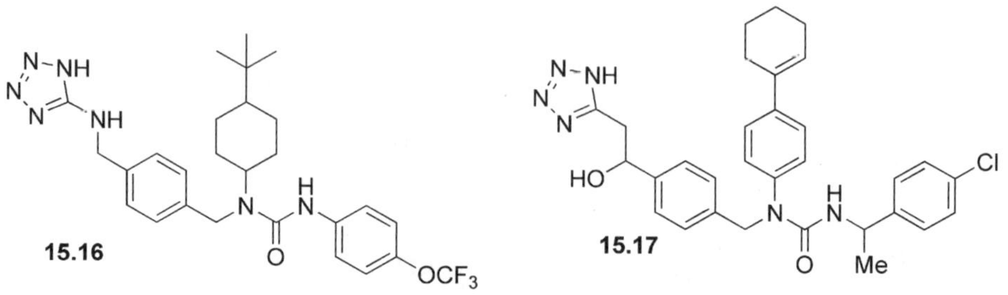

15.16

15.17

Figure 15.6   Non-benzamide-containing tetrazole-based antagonists.

bioisosteres.[38] While specific biological data were not provided for individual compounds, preferred compounds were said to have a hGCGR binding IC$_{50}$ of <100 nM. Additionally, a substituted guanidine core exemplified by cyanoguanidine **15.15** has also been reported in the patent literature.[39]

Additional disclosures have shown that the benzamide functionality present in the previous examples may not be necessary (Figure 15.6).[31,40] Tetrazole-terminated compounds such as **15.16** and **15.17** are devoid of the amide linker, yet bind to the hGCGR with an IC$_{50}$ <1000 nM.

A number of conformationally constrained urea analogs have also been investigated. The indanyl urea analogs **15.18** and **15.19** sought to restrict rotation around the benzylic portion of the benzamide (Figure 15.7).[41,42] The more potent enantiomers of both structures (absolute configuration was not disclosed) displayed excellent hGCGR binding affinity and robust hGCGR functional activity along with modest selectivity over hGIPR (**15.18**: hGCGR IC$_{50}$ = 4.1 nM, hGCGR cAMP IC$_{50}$ = 33 nM, hGIPR cAMP IC$_{50}$ = 132 nM; **15.19**: hGCGR IC$_{50}$ = 2.3 nM, hGCGR cAMP IC$_{50}$ = 15 nM, hGIPR cAMP IC$_{50}$ = 296 nM). All compounds in the report were stated to possess excellent selectivity over hGLP-1R (hGLP-1R cAMP IC$_{50}$ > 10000 nM). Compound **15.18** demonstrated modest PK in mice (F_{po} = 17%, $t_{1/2}$ = 5.6 h, C_{max} = 130 nM, AUC (dose normalized, p.o.) = 560 nM·h/dose, V_{dss} = 1.6 L/kg, Cl = 9 mL/min·kg). The β-alanine analog **15.18** was dosed chronically as an admixture in chow to high-fat fed hGCGR-containing transgenic mice that displayed moderate hyperglycemia. A statistically significant correction of non-fasting plasma glucose levels was observed by day 3 when dosed at 3 mg/kg.

The closely related spiro-urea analogs **15.20** and **15.21** displayed moderate hGCGR binding and functional potency (Figure 15.8) (**15.20**: hGCGR IC$_{50}$ =

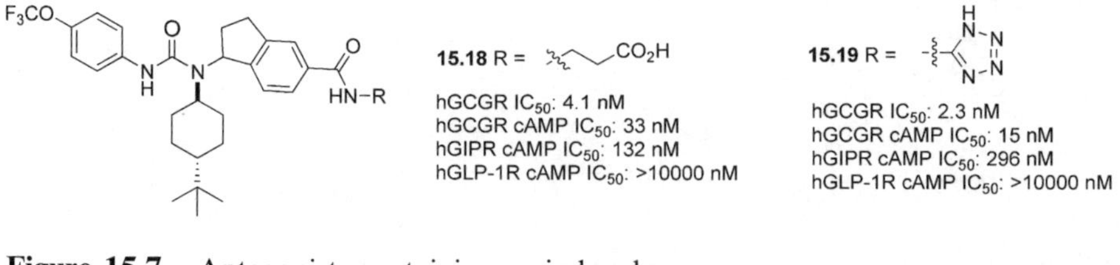

15.18 R = ⟶CO$_2$H

hGCGR IC$_{50}$: 4.1 nM
hGCGR cAMP IC$_{50}$: 33 nM
hGIPR cAMP IC$_{50}$: 132 nM
hGLP-1R cAMP IC$_{50}$: >10000 nM

15.19 R =

hGCGR IC$_{50}$: 2.3 nM
hGCGR cAMP IC$_{50}$: 15 nM
hGIPR cAMP IC$_{50}$: 296 nM
hGLP-1R cAMP IC$_{50}$: >10000 nM

Figure 15.7   Antagonists containing an indanylurea core.

15.20 R = (CH₂)₂CO₂H equivalent structure

15.20 R = ~CO_2H

hGCGR IC_{50}: 182 nM
hGCGR cAMP IC_{50}: 282 nM
hGIPR cAMP IC_{50}: ND
hGLP-1R cAMP IC_{50}: ND

15.21 R = (aminotetrazole)

hGCGR IC_{50}: 12 nM
hGCGR cAMP IC_{50}: 92 nM
hGIPR cAMP IC_{50}: 430+/-600 nM
hGLP-1R cAMP IC_{50}: 4600+/-1900 nM

Figure 15.8 Spirocyclic urea benzamides.

182 nM, hGCGR cAMP IC_{50} = 282 nM; **15.21**: hGCGR IC_{50} = 12 nM, hGCGR cAMP IC_{50} = 92 nM).[43,44] Compound **15.21** was also shown to possess modest selectivity over the hGIPR (hGIPR IC_{50} = 430 $\pm$ 600 nM) and good selectivity over hGLP-1R (hGLP-1R IC_{50} = 4600 $\pm$ 1900 nM). The pharmacokinetic profile of **15.21** in a variety of species is modest (Table 15.1). The aminotetrazole benzamide **15.21** also effectively blunted glucagon-stimulated blood glucose elevation when administered at 10 mg/kg p.o. to hGCGR transgenic mice. It was also noted that the *in vivo* glucagon receptor occupancy at 1h post dose in the 10 mg/kg arm was 52%.

Compound **15.21** was also dosed chronically (11 days) at 1, 3, and 10 mg/kg in hGCGR transgenic mice with high-fat diet-induced obesity.[45] The 10 mg/kg group demonstrated oral glucose tolerance when an oral glucose tolerance test (OGTT) was performed on day 11. Additionally, the 10 mg/kg group showed improved fasting plasma glucose as well. There was a dose-dependent increase in plasma glucagon levels observed with increasing doses of **15.21**. Upon dosing of **15.21** for up to 82 days, glucagon levels remained elevated relative to control, as did GLP-1 levels. Additionally, unlike observations made in mGCGR$^{-/-}$ mice, chronic treatment with **15.21** did not afford significant changes in the morphology of pancreatic islet cells. The sum of these results added significant support to the possible therapeutic value of glucagon receptor antagonism in T2DM.

Table 15.1 Pharmacokinetic profile of **15.21**.

Species	*Mouse*	*hGCGR* *Mouse*	*Rat*	*Dog*
Normalized oral AUC (µM·h/dose)	0.25	0.47	0.26	1.6
C_{max} (µM)	0.07	0.13	0.15	0.66
F (%)	20	24	13	38
$t_{1/2}$ (h)	3.5	4.6	1.8	2.4
Clp (ml/min/kg)	24	16	15	6.8
V_{dss} (L/kg)	4.3	4.2	1.3	0.48
Dose	IV: 1 mg/kg PO: 2 mg/kg	IV: 1 mg/kg PO: 2 mg/kg	IV: 1 mg/kg PO: 2 mg/kg	IV: 0.5 mg/kg PO: 1.0 mg/kg
Vehicle	5:10:85 DMSO: polysorbate 80: water	NS	NS	NS

*NS: Not specified.

Additional efforts to discover potent and selective hGCGR antagonists have led to cyclic core structures exemplified by imidazolone **15.22**, cyclic urea **15.23**, and iminooxazolidine **15.24** (Figure 15.9).[46,47] While no specific *in vitro* data was provided for **15.22**, preferred compounds from the disclosure that it is a part of possess an hGCGR binding IC_{50} of 100 nM or lower. Compounds **15.23** and **15.24** were reported to inhibit glucagon-stimulated cAMP production at a level between 50% and 100% at 20 μM.

A series of spiro-imidazolone hGCGR antagonists have recently been disclosed (Figure 15.10).[48] Functional antagonism data for compounds **15.25** (hGCGR cAMP IC_{50} = 452 nM) and **15.26** (hGCGR cAMP IC_{50} = 57 nM) were provided. Additional compounds such as **15.27** and **15.28** were singled out as preferred, but no biological data was provided.

A wide variety of acyclic non-urea-containing cores have been described in the literature. Amido analogs, exemplified by (+)-**15.29** and (+)-**15.30**, have been reported to exhibit potent binding affinities along with moderate functional potency (Figure 15.11) ((+)-**15.29**: hGCGR binding K_i = 11 nM, hGCGR cAMP K_i = 300 nM; (+)-**15.30**: hGCGR binding K_i = 9 nM, hGCGR cAMP K_i = 144 nM).[49] Using a similar core structure to (+)-**15.29** and a taurine benzamide bioisostere, compound **15.31** was prepared (hGCGR binding IC_{50} = 5 nM, human hepatocyte cAMP EC_{50} = 4 nM).[50] Compound **15.31** also possessed oral bioavailability (F = 73%) in rats and reduced non-fasting glucose by 57% over 24 h in a db/db mouse model of T2DM at an oral dose of 30 mg/kg. The structurally similar compound **15.32**

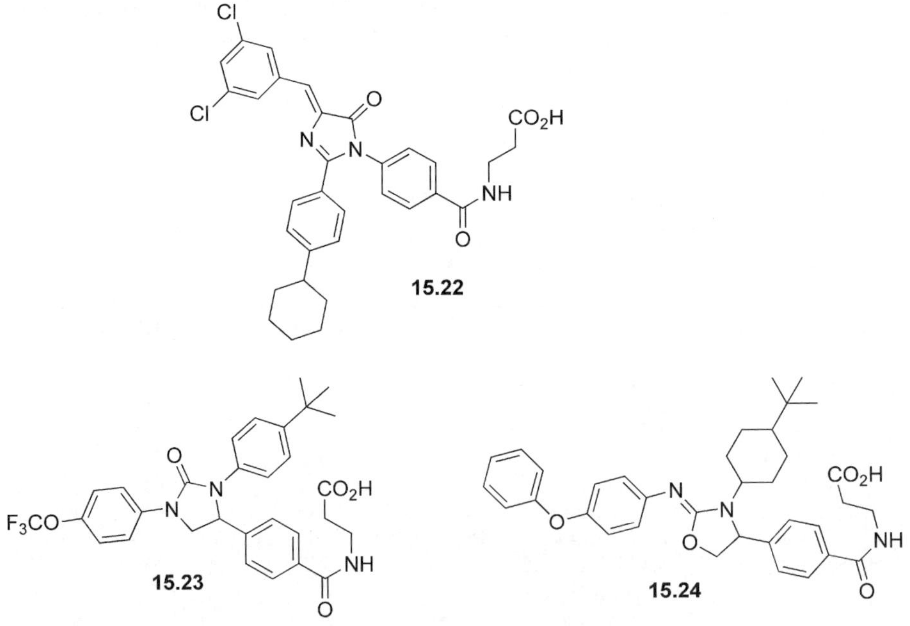

Figure 15.9 Additional partially saturated heterocyclic scaffolds.

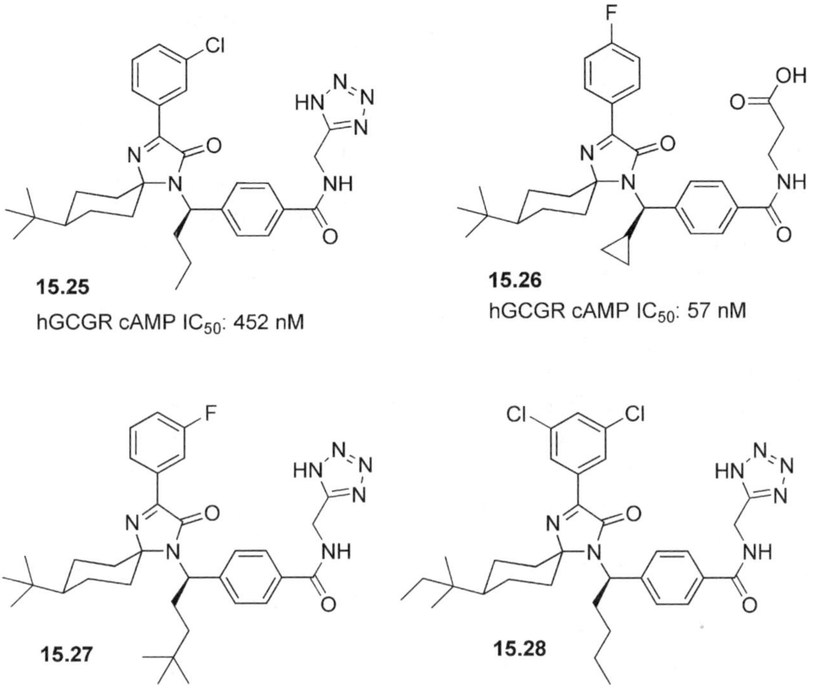

15.25

hGCGR cAMP IC_{50}: 452 nM

15.26

hGCGR cAMP IC_{50}: 57 nM

15.27

15.28

Figure 15.10 Spiroimidazolone-based antagonists.

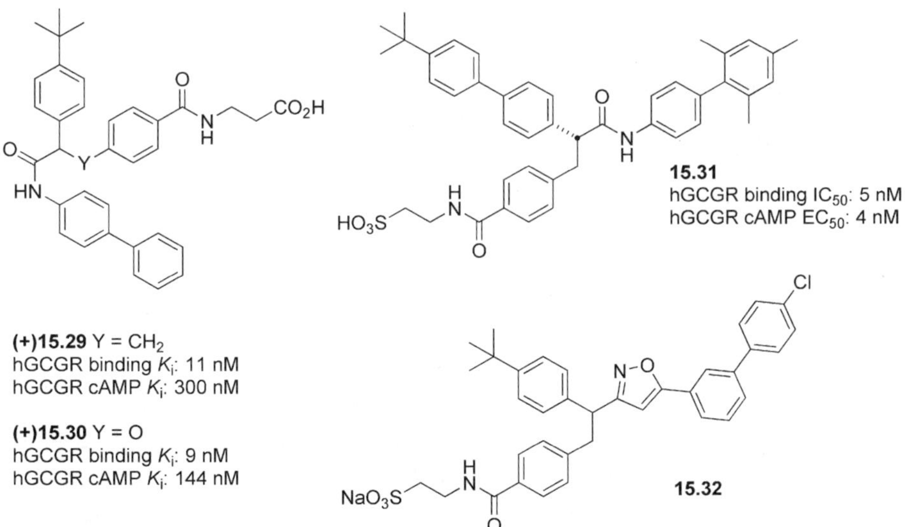

(+)15.29 Y = CH_2
hGCGR binding K_i: 11 nM
hGCGR cAMP K_i: 300 nM

(+)15.30 Y = O
hGCGR binding K_i: 9 nM
hGCGR cAMP K_i: 144 nM

15.31
hGCGR binding IC_{50}: 5 nM
hGCGR cAMP EC_{50}: 4 nM

15.32

Figure 15.11 Amide and isoxazole scaffolds.

was investigated in a glucagon challenge PD assay in beagles.[51] At an oral dose of 38 mg/kg, compound **15.32** inhibited an exogenous glucagon-stimulated glucose excursion by 40%.

Lower molecular weight structures such as the methylene-linked benzamides **15.33** and **15.34** have also been described in the patent literature (Figure 15.12).[52] The binding affinities of these compounds have been described as moderate to potent (hGCGR binding IC_{50} = 1–500 nM), with no specific data having been disclosed for individual structures. In this series, the quinoline and naphthalene have also been replaced with benzimidazole and benzisoxazole. Structurally constrained analogs such as the spiro-cycloalkyl compounds **15.35** and **15.36** have also been disclosed which possess hGCGR binding IC_{50} values in the range of 1 to 500 nM.[53] Structures **15.37** and **15.38**, in which the β-alanine benzamide has been linked to a triaryl moiety via a methylene group or an oxygen atom have been shown to be potent hGCGR antagonists with binding IC_{50} values in the 1 to 500 nM range.[54]

A variety of antagonists in which the β-alanine benzamide is connected to the remaining pharmacophores via an aminomethyl group have also been described (Figure 15.13). Compound (±)-**15.39** was shown to have good affinity for the glucagon receptor (hGCGR binding IC_{50} = 40 nM).[55] Inversion of the aminomethyl linker afforded a series of *para*-aminobenzamide antagonists exemplified by (±)-**15.40** (hGCGR binding IC_{50} = 57 nM).[56] Compound **15.41** (hGCGR binding IC_{50} = 84 nM) was shown to attenuate

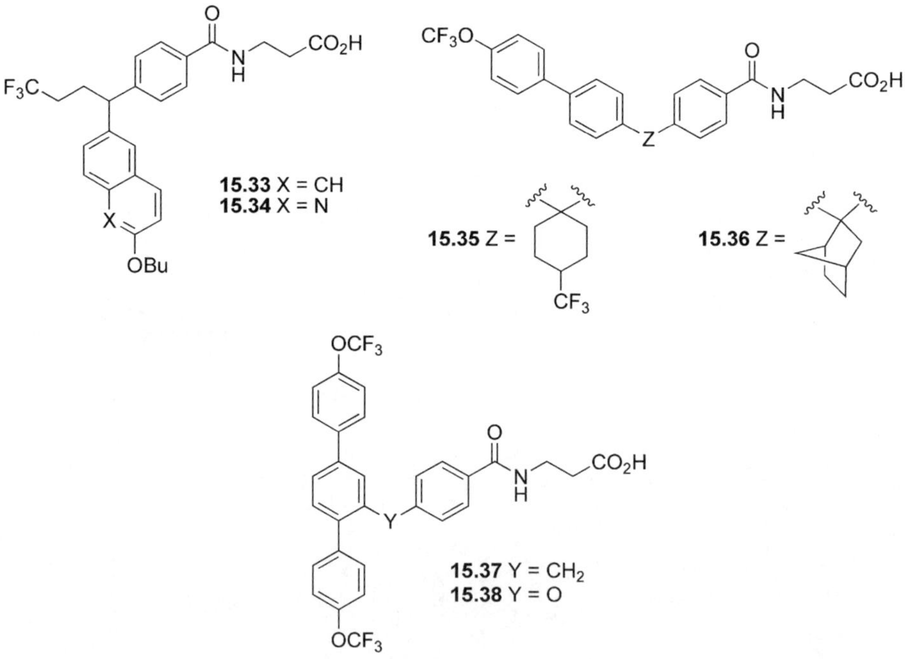

Figure 15.12 Additional approaches to β-alanine benzamide display.

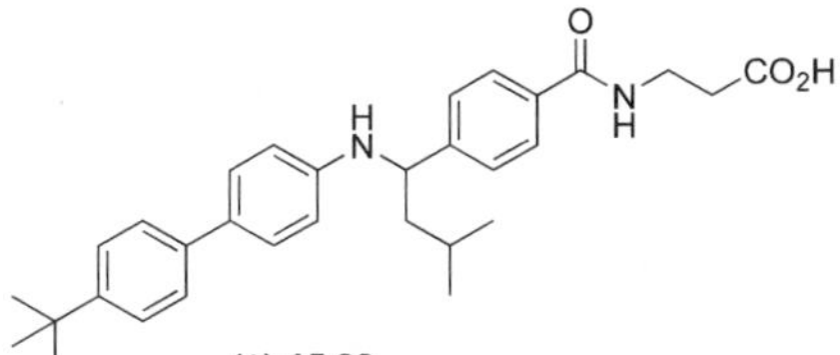

(±)-15.39
hGCGR binding IC_{50}: 40 nM

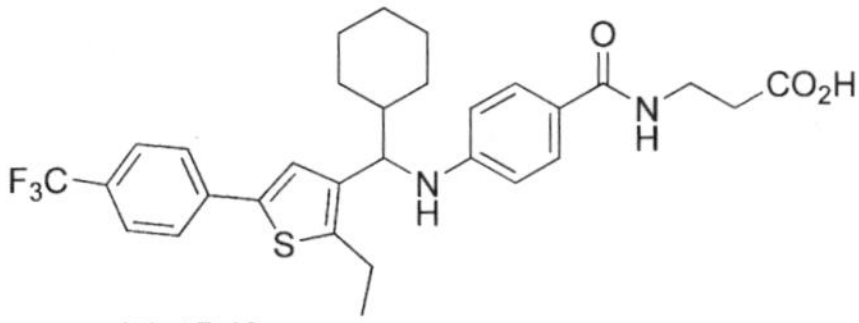

(±)-15.40
hGCGR binding IC_{50}: 57 nM

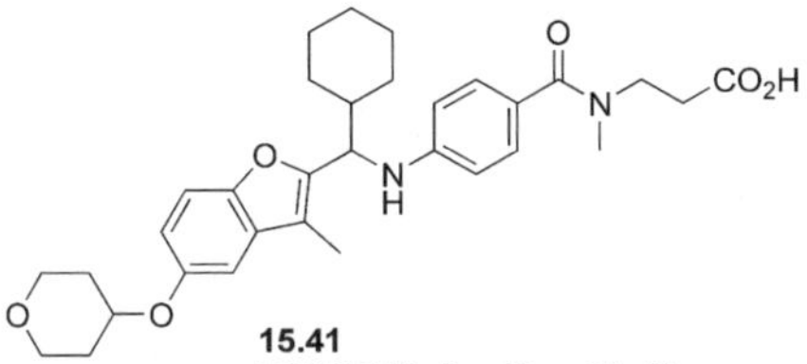

15.41
hGCGR binding IC_{50}: 84 nM

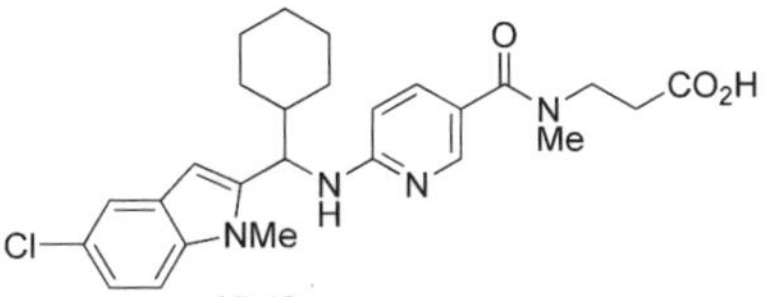

15.42
hGCGR binding: 90% @ 10000 nM

Figure 15.13   Aminomethyl-linked benzamides.

exogenous glucagon-stimulated glucose excursion (43% of control) at 10 mg/kg p.o. in a Sprague Dawley rat glucagon challenge model.[57] At 3 mg/kg p.o. in a Sprague Dawley rat glucagon challenge model, compound **15.42** (hGCGR binding: 90% inhibition at 10000 nM) partially blocked glucagon-stimulated glucose increase (45% of control).

Antagonists containing a pyrrolidine core (exemplified by compounds **15.43–15.47**) have been recently described in the patent literature (Figure 15.14).[58] The β-alanine benzamide is connected to the core via the pyrrolidine nitrogen. Most of the compounds prepared were stated to have hGCGR functional potency (cAMP IC_{50}) in the range of 100 to 3000 nM. One unspecified compound registered an hGCGR cAMP IC_{50} of less than 100 nM.

Attachment of the β-alanine benzamide to additional pharmacophore fragments has also been accomplished via a methylenoxy group (Figure 15.15). Compounds (±)-**15.48** and (±)-**15.49**, in which the methylenoxy point of attachment to the benzamide are regioisomeric, demonstrated modest binding

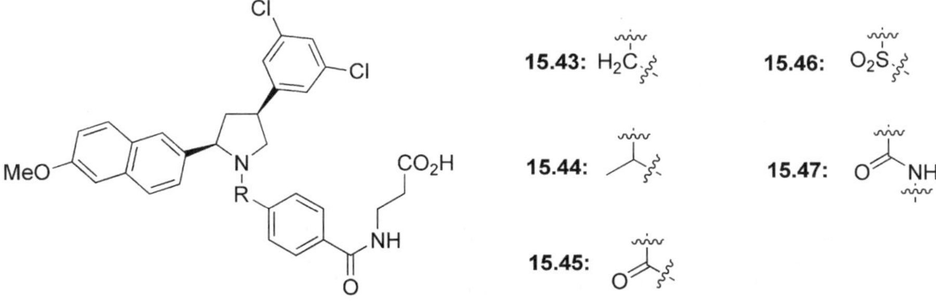

15.43: H_2C

15.46: O_2S

15.44:

15.47:

15.45:

Figure 15.14   Antagonists utilizing a pyrrolidine core.

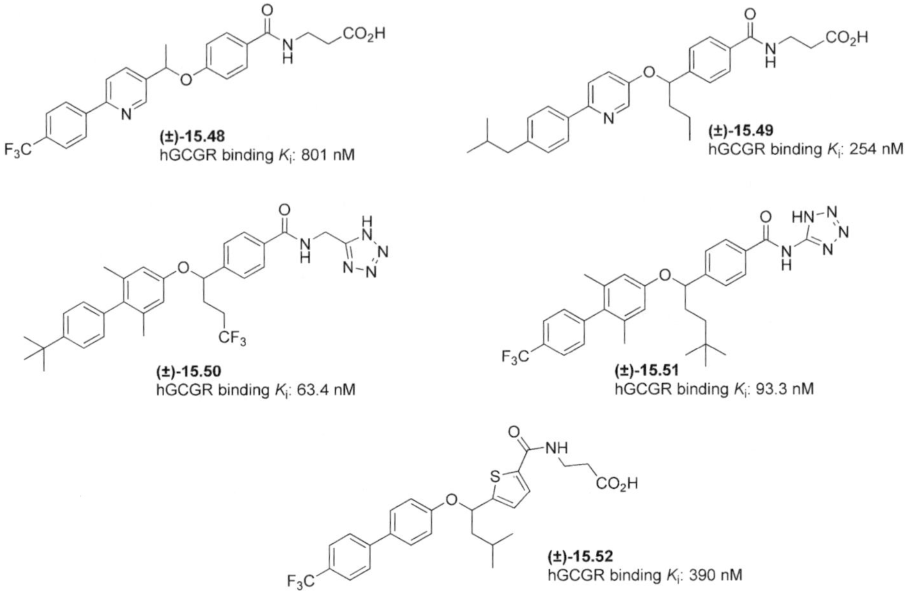

(±)-15.48
hGCGR binding K_i: 801 nM

(±)-15.49
hGCGR binding K_i: 254 nM

(±)-15.50
hGCGR binding K_i: 63.4 nM

(±)-15.51
hGCGR binding K_i: 93.3 nM

(±)-15.52
hGCGR binding K_i: 390 nM

Figure 15.15   Methylenoxy-linked benzamides.

affinity ((±)-**15.48**: hGCGR binding K_i = 801 nM; (±)-**15.49**: hGCGR binding K_i = 254 nM).[59,60] Aminomethyltetrazole benzamides such as (±)-**15.50** (hGCGR binding K_i = 63.4 nM) and aminotetrazole benzamides such as (±)-**15.51** (hGCGR binding K_i = 93.3 nM) have both been used as bioisosteres for the β-alanine benzamide in this structural series.[61] Replacement of the benzamide with a thiophene amide afforded compounds such as (±)-**15.52** (hGCGR binding K_i = 390 nM).[62] The absolute stereochemistry of these compounds was not disclosed.

Thiomethylene-linked thiophene amides (±)-**15.53** (hGCGR binding K_i = 18.4 nM) and (±)-**15.54** (hGCGR binding K_i = 36.0 nM) both possess potent affinity for the human glucagon receptor (Figure 15.16).[63]

Extension of the benzamide linker to an ethylenoxy group afforded compounds such as (±)-**15.55**, which demonstrated an improved binding

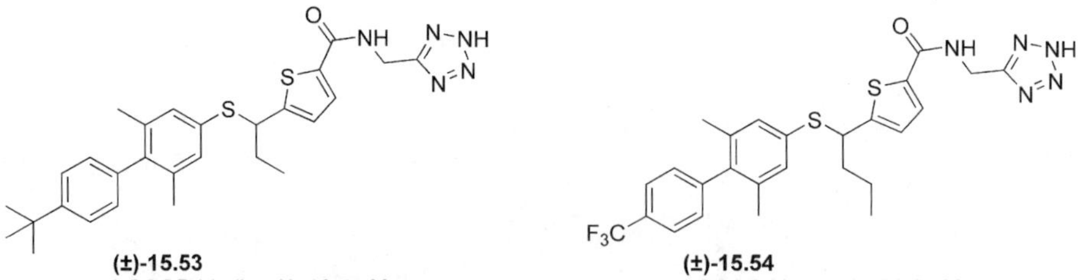

(±)-15.53
hGCGR binding K_i: 18.4 nM

(±)-15.54
hGCGR binding K_i: 36.0 nM

Figure 15.16   Thiomethylene-linked thiophene amides.

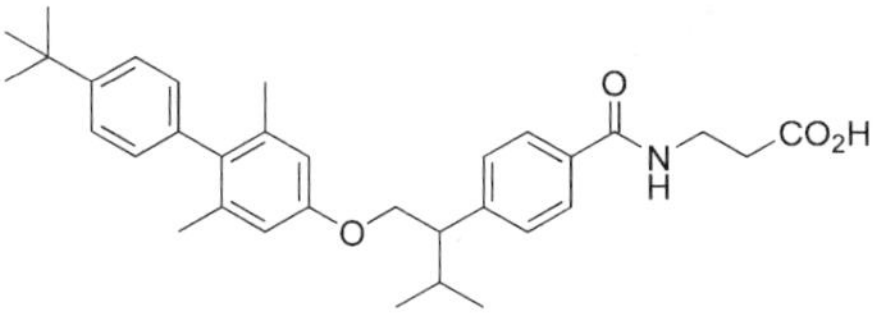

(±)-15.55
hGCGR binding K_i: 61 nM

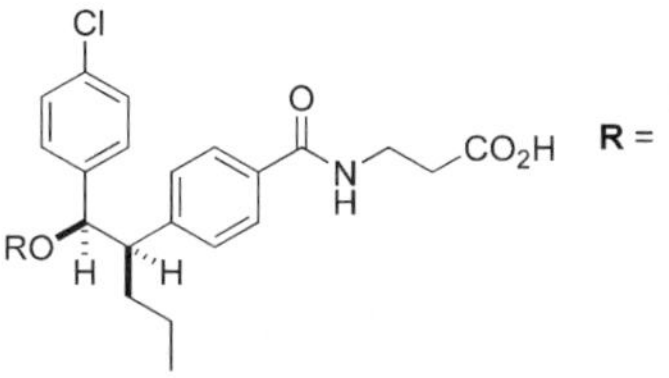

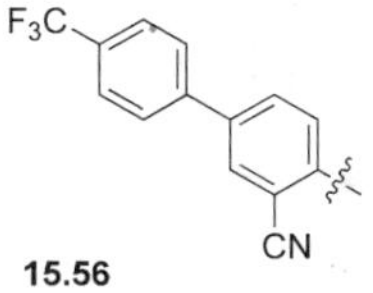

15.56
hGCGR binding IC_{50}: 0.3 nM

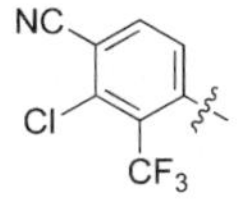

15.57
hGCGR binding IC_{50}: 1.4 nM

Figure 15.17 Antagonists containing an ethylenoxy-linked benzamide.

affinity (hGCGR binding K_i = 61 nM) over the methylenoxy-linked compounds (Figure 15.17).[64] Compounds such as **15.56** (hGCGR binding IC_{50} = 0.3 nM) and **15.57** (hGCGR binding IC_{50} = 1.4 nM), in which there is substitution on both carbons of the ethylenoxy linker, showed a marked improvement in affinity for the human glucagon receptor.[65]

Acylindole analogs such as **15.58** have also been disclosed as glucagon receptor antagonists (Figure 15.18).[66] Compound **15.58** has been described as a potent human glucagon receptor antagonist that displays good PK properties and efficacy in rodent diabetes models.[67] Structurally similar isoindoline amide antagonists such as the exquisitely potent **15.59** (hGCGR binding IC_{50} = 0.1 nM) and the aminotetrazole benzamide **15.60** (hGCGR binding IC_{50} = 2.2 nM) have also been reported in the patent literature.[68]

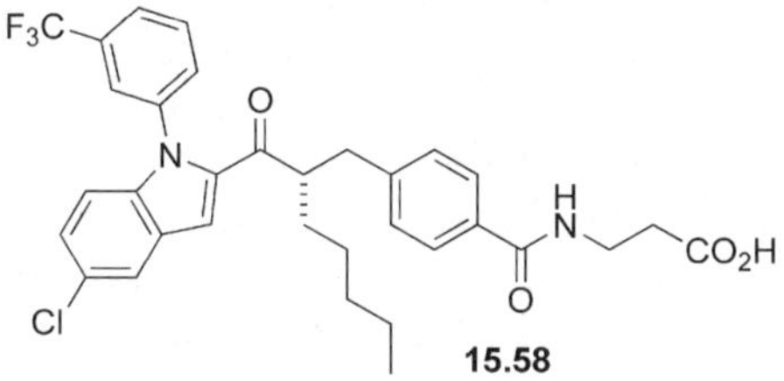

15.58

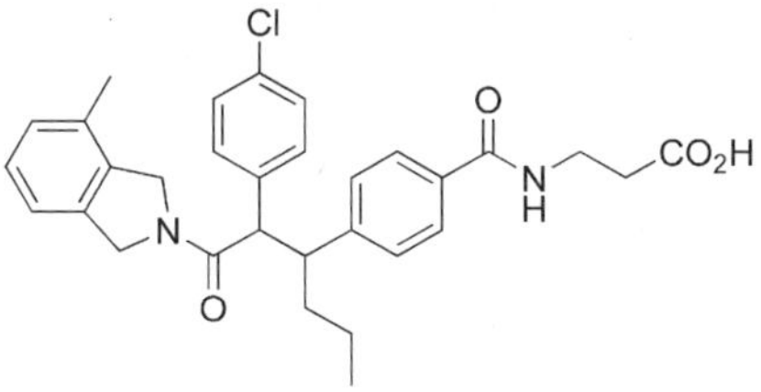

15.59
hGCGR binding IC_{50}: 0.1nM

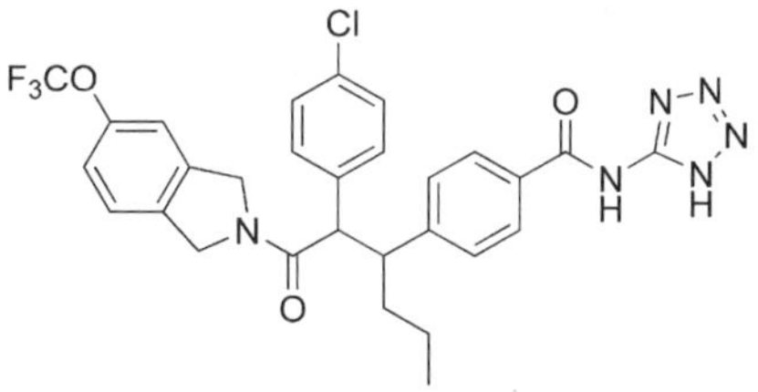

15.60
hGCGR binding IC_{50}: 2.2nM

Figure 15.18 2-Acylindoles and amidoisoindolines.

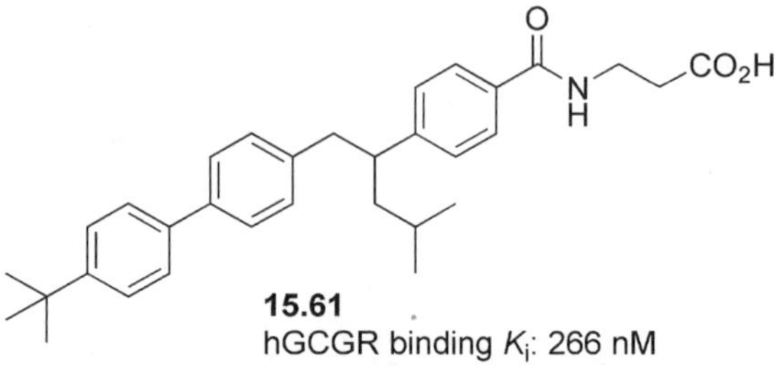

15.61
hGCGR binding K_i: 266 nM

Figure 15.19   Ethylene-linked β-alanine benzamides.

A significant amount of research has been devoted to glucagon receptor antagonists where an ethylene linker is used to display the key binding elements. Compound **15.61** has demonstrated moderate affinity for the glucagon receptor (Figure 15.19, hGCGR binding K_i = 266 nM).[69]

Ethylene-linked antagonists in which one ethylene carbon is substituted with an aryl group and the second ethylene carbon is substituted with an alkyl group have also been described. Compound **15.62** was described as having an hGCGR binding IC_{50} between 1 and 500 nM (Figure 15.20).[70,71] While no further detail regarding this compound has been published, it is also the subject of a patent application regarding crystalline polymorphic forms.[72] The structurally related indazole **15.63** (hGCGR binding IC_{50} = 0.8 nM) has also been reported.[73]

A series of additional patent applications expanding the chemical space around structures **15.62** and **15.63** were published in 2010 (Figure 15.21). The hGCGR binding affinities of many of the compounds prepared were reported to be in the single-digit nanomolar or picomolar IC_{50} range.

Quinolines such as **15.64** (hGCGR binding IC_{50} = 0.3 nM) and quniazolines such as **15.65** (hGCGR binding IC_{50} = 1.7 nM) serve as suitable isosteres for the indole present in **15.62**.[74] Substituted naphthalenes **15.66** (hGCGR binding IC_{50} = 0.37 nM) and **15.67** (hGCGR binding IC_{50} = 3.3 nM) have also been reported.[75] Finally, benzothiophenes represented by the exceptionally potent **15.68** (hGCGR binding IC_{50} = 0.1 nM) were also prepared.[76]

As an extension of their initial efforts with urea-based β-alanine benzamide hGCGR antagonists, Madsen and co-workers also probed heteroaromatic replacements for the urea functionality.[77,78] Their efforts afforded a series of potent and selective hGCGR antagonists which contained an aminothiazole

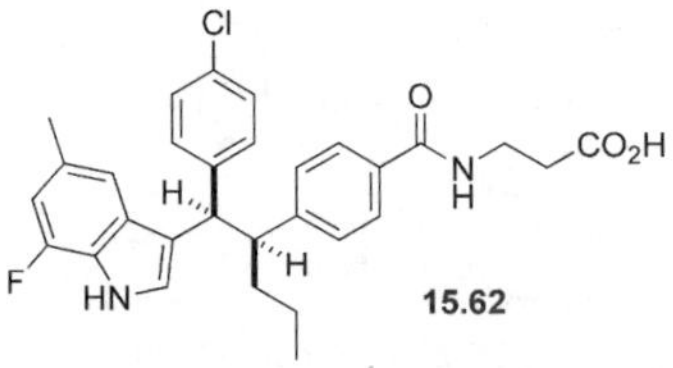

15.62

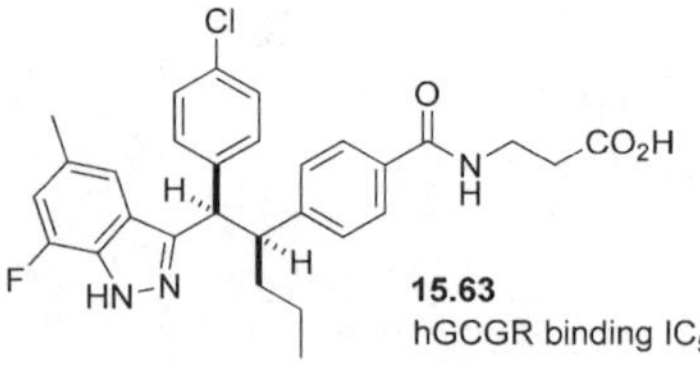

15.63
hGCGR binding IC_{50}: 0.8 nM

Figure 15.20   Indole- and indazole-containing ethylene-linked β-alanine benzamides.

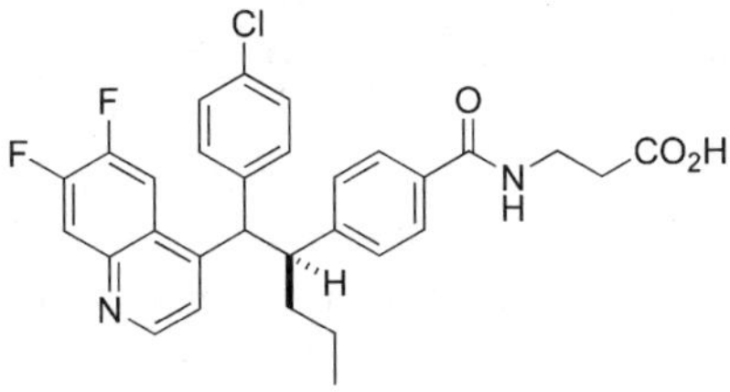

15.64
hGCGR binding IC_{50}: 0.3 nM

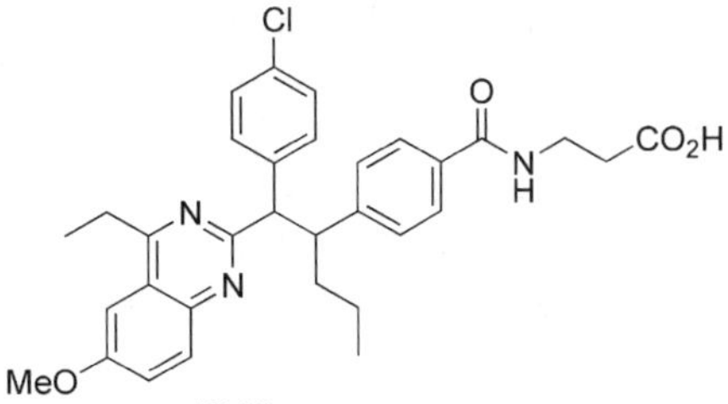

15.65
hGCGR binding IC_{50}: 1.7 nM

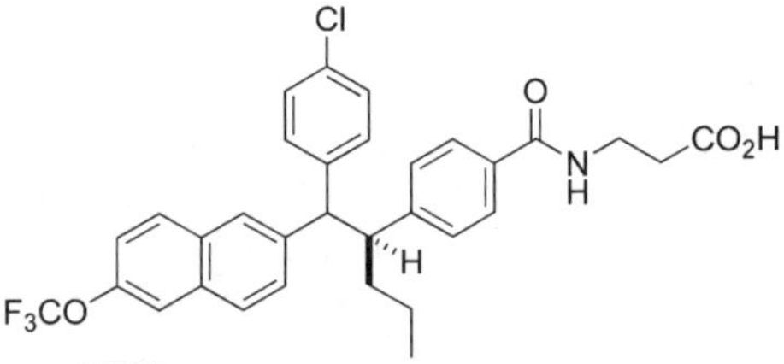

15.66
hGCGR binding IC_{50}: 0.37 nM

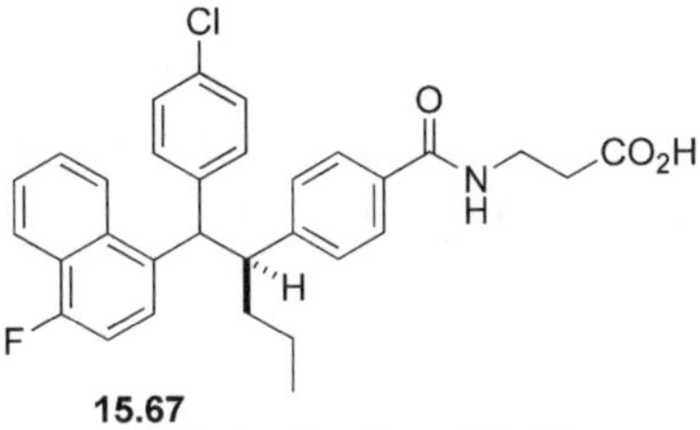

15.67
hGCGR binding IC_{50}: 3.3 nM

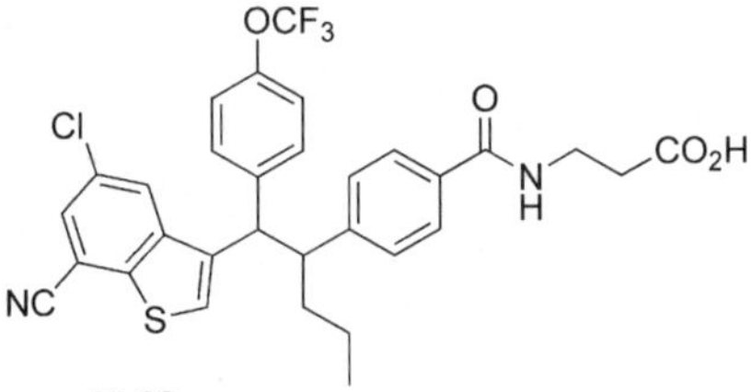

15.68
hGCGR binding IC_{50}: 0.1 nM

Figure 15.21 Additional ethylene-linked β-alanine benzamides.

core. Of particular interest was compound **15.69** (Figure 15.22) (hGCGR binding IC_{50} = 93 nM, hGIPR binding IC_{50} = 1100 nM). Compound **15.69** was found to have acceptable oral exposure in both rats and dogs (PK rat: F_{po}

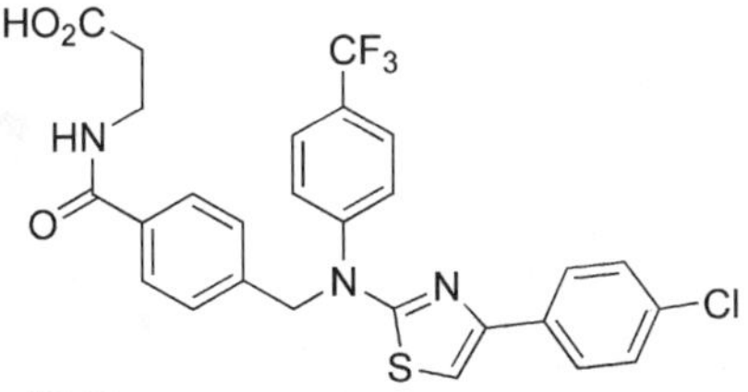

15.69
hGCGR binding IC_{50}: 93 nM
hGIPR binding IC_{50}: 1100 nM

Figure 15.22 Aminothiazole core.

$= 58\%$, $t_{1/2} = 228$ min, $C_{max} = 2100$ ng/mL, $T_{max} = 85$ min, Cl $= 1$ mL/min·kg; PK dog: $F_{po} = 141\%$, $t_{1/2} = 104$ min, $C_{max} = 1253$ ng/mL, $T_{max} = 90$ min, Cl $= 2.23$ mL/min·kg). Additionally, compound **15.69** was shown to inhibit glucose production in primary human hepatocytes (IC$_{50}$ $= 6700$ nM). In macaca mulatta monkeys, at 3 mg/kg and 1 mg/kg IV, **15.69** was shown to completely block hyperglycemia induced by administration of exogenous glucagon.

A series of antagonists containing an aminobenzimidazole core have also been prepared (Figure 15.23).[79–81] Compound **15.70** is a potent glucagon receptor antagonist (hGCGR binding IC$_{50}$ $= 11$ nM, hGCGR cAMP IC$_{50}$ $= 62$ nM, dogGCGR cAMP IC$_{50}$ $= 473 \pm 134$ nM, rhesusGCGR IC$_{50}$ $= 111$ nM) and is selective (hGIP cAMP IC$_{50}$ $= 6000$ nM, hGLP-1R cAMP IC$_{50}$ $= 6800$ nM). Modest oral bioavailability was observed for **15.70** in a variety of species (**mouse**: oral AUC$_{norm}$ $= 0.40$ μM·h/dose, $F = 5\%$, Cl $= 4$ mL/min·kg; **rat**: oral AUC$_{norm}$ $= 0.40$ μM·h/dose, $F = 15\%$, Cl $= 12$ mL/min·kg; **dog**: oral AUC$_{norm}$ $= 2.1$ μM·h/dose, $F = 27\%$, Cl $= 3.8$ mL/min·kg; **rhesus monkey**: oral AUC$_{norm}$ $= 0.45$ μM·h/dose, $F = 23\%$, Cl $= 16$ mL/min·kg). In transgenic hGCGR mice, **15.70** was shown to inhibit the exogenous glucagon-induced excursion of glucose at 3, 10, and 30 mg/kg p.o. A similar pharmacodynamic (PD) experiment in rhesus monkeys at oral doses of **15.70** at 3 and 10 mg/kg resulted in similar inhibition of glucagon-mediated glucose production. Oral administration of **15.70** over 10 days at 30 mg/kg per day to transgenic hGCGR mice on a high-fat diet resulted in almost complete correction of plasma glucose levels compared to the lean control mice. A similar series of phenylenediamine-derived guanidine core structures, exemplified by **15.71**, have also been reported in the patent literature.[82]

The benzimidazole core has been utilized in other series of glucagon receptor antagonists as well. An *N*-dehydroabietylbenzimidazole-based glucagon receptor antagonist **15.72**, which incorporated the β-alanine benzamide at the benzimidazole 2-position, has been reported in the patent literature (Figure 15.24).[83] No biological data was provided for this compound.

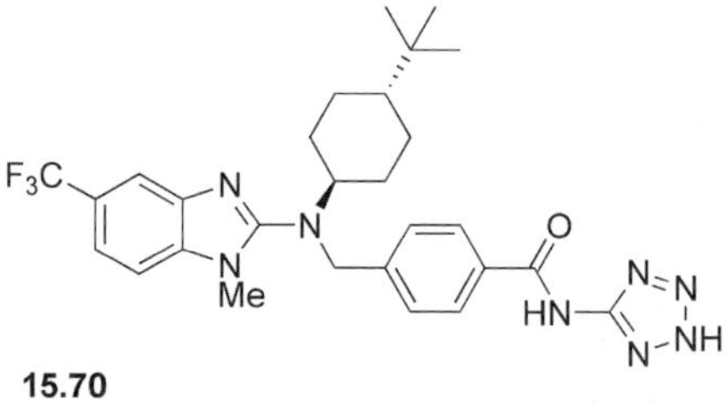

15.70
hGCGR binding IC$_{50}$: 11 nM
hGCGR cAMP IC$_{50}$: 62 nM
hGIP cAMP IC$_{50}$: 6000 nM
hGLP-1R cAMP IC$_{50}$: 6800 nM

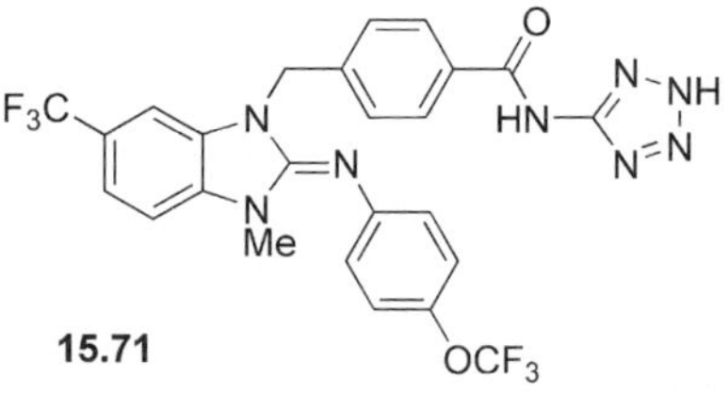

15.71

Figure 15.23 2-aminobenzimidazole antagonists.

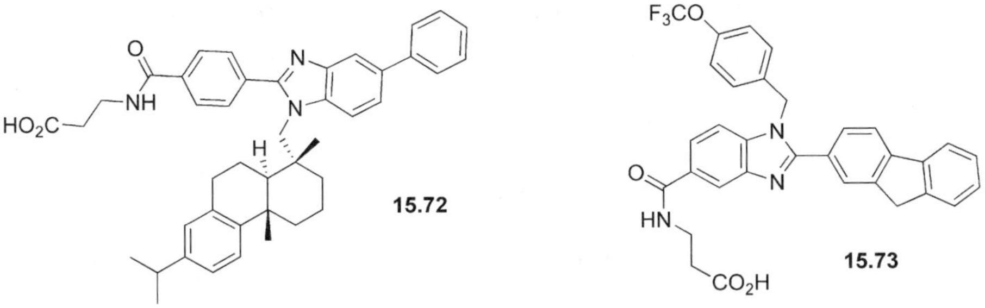

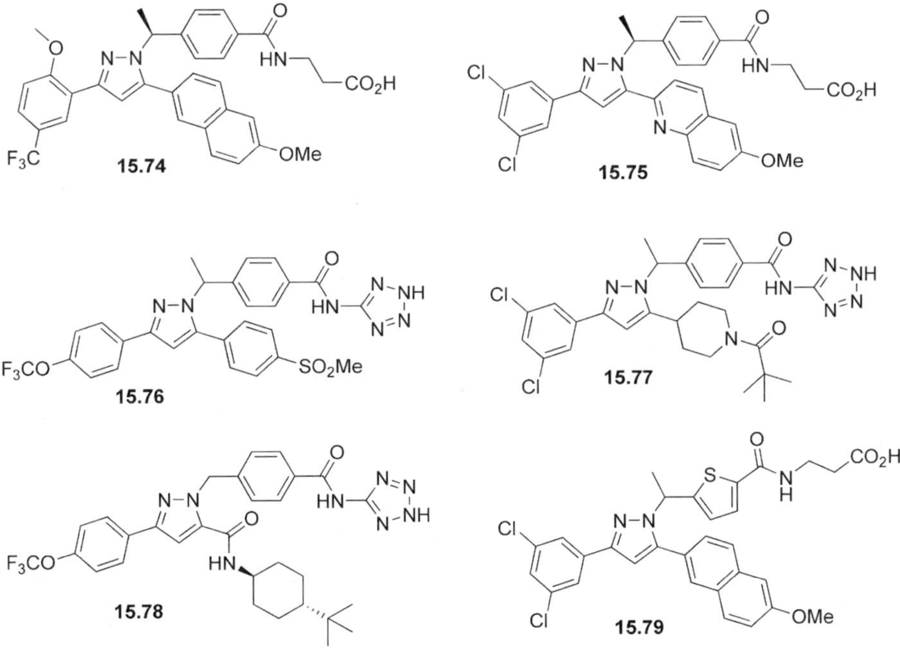

Figure 15.24 Benzimidazole scaffolds.

Compound **15.73** is unique in that the β-alanine amide is connected directly to the benzimidazole core, rather than through a benzamide.[84]

Over the last several years, a number of pyrazole-based hGCGR antagonists have been described in the patent and peer-reviewed literature. The trifluoromethyl-alkoxyphenyl substituted pyrazoles exemplified by **15.74** were reported to exhibit potent binding and functional activity (Figure 15.25) (hGCGR binding IC_{50} = 1–10 nM, hGCGR cAMP IC_{50} = 5–50 nM).[85,86] Related pyrazoles substituted with a bicyclic heteroaryl group, such as **15.75**, have also been disclosed.[87] Additionally, bis-aryl substituted pyrazoles, exemplified by **15.76**, and cycloheteroalkyl substituted pyrazoles, such as **15.77**, have also been reported.[88] No specific potency data was disclosed in the

Figure 15.25 β-Alanine benzamide antagonists containing a pyrazole core.

applications describing analogs encompassing **15.75**, **15.76**, and **15.77**. Amido-pyrazoles exemplified by **15.78** were described, and it was reported that these exhibited binding IC_{50} values from 1 to 500 nM.[89] Related thiophene amides such as **15.79** were recently reported in the patent literature.[90] While no compound-specific biological data were provided for compounds in this patent application, the most potent compounds were described as having an hGCGR cAMP IC_{50} of less than 500 nM.

Another series of pyrazole-based antagonists exemplified by **15.80**–**15.82** were reported wherein the benzamide was linked to the pyrazole via a heteroatom rather than carbon (Figure 15.26).[91] Biological data for the individual compounds was not provided, but the most potent compounds were stated as having an hGCGR cAMP IC_{50} of less than 500 nM.

Recently, the *in vitro* and *in vivo* profiles of structures exemplified by **15.83** and **15.84** were summarized (Figure 15.27).[92] These compounds exhibited

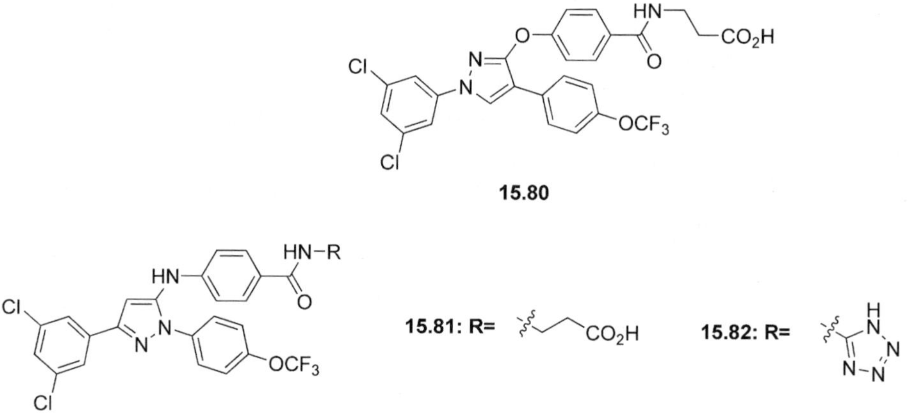

Figure 15.26 β-Alanine benzamides linked to a pyrazole core via oxygen or nitrogen.

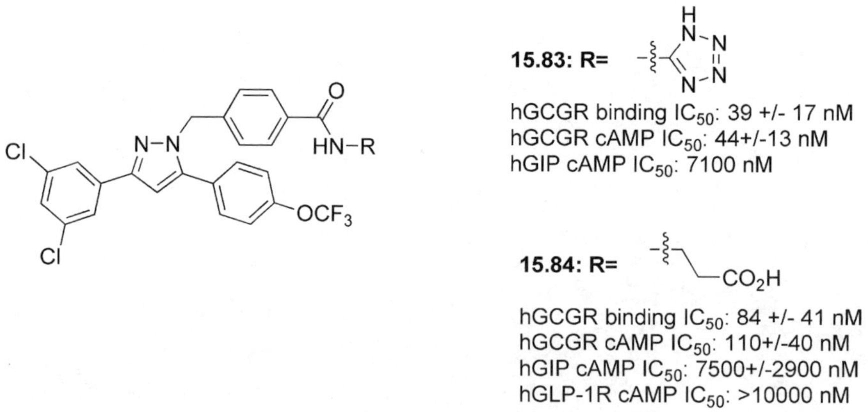

Figure 15.27 β-Alanine and aminotetrazole benzamides displayed via a pyrazole core.

Table 15.2 Pharmacokinetic profile of **15.84**.

Species	*Mouse*	*Rat*	*Dog*	*Monkey (Rhesus)*
Normalized oral AUC (μM·h/dose)	0.73 $\pm$ 0.09	11 $\pm$ 4	27 $\pm$ 9	30.0 $\pm$ 0.5
C_{max} (μM)	0.23 $\pm$ 0.05	1.6 $\pm$ 0.3	7.7 $\pm$ 1.5	4.2 $\pm$ 0.6
F (%)	15 $\pm$ 2	57 $\pm$ 20	46 $\pm$ 16	68 $\pm$ 1
$t_{1/2}$ (h)	2.3 $\pm$ 0.4	6.5	3.6 $\pm$ 0.1	10 $\pm$ 2
Clp (ml/min/kg)	6.1 $\pm$ 1.1	1.6	0.49 $\pm$ 0.11	0.68 $\pm$ 0.21
V_{dss} (L/kg)	1.1 $\pm$ 0.3	0.75	0.14 $\pm$ 0.03	0.55 $\pm$ 0.05
Dose	i.v.: 1 mg/kg; p.o.: 2 mg/kg		i.v.: 0.5 mg/kg; p.o.: 2.0 mg/kg	
Vehicle	5:10:85		i.v.: 2:5:3 EtOH:PEG400:water	
	DMSO: polysorbate 80: water		p.o.: 0.5% methylcellulose in water	

acceptable potency against the glucagon receptor (**15.83**: hGCGR binding = 39 $\pm$ 17 nM, hGCGR cAMP = 44 $\pm$ 13 nM; **15.84**: hGCGR binding = 84 $\pm$ 41 nM, hGCGR cAMP = 110 $\pm$ 40 nM) with excellent selectivity over the hGIP receptor (**15.83**: hGIP cAMP = 7100 nM; **15.84**: hGIP cAMP = 7500 $\pm$ 2900 nM). Compound **15.84** was also found to be selective over the hGLP-1R (cAMP IC_{50} > 10000 nM) and the pituitary adenylate cyclase-activating polypeptide (PACAP, cAMP IC_{50} > 7000 nM) receptor. As has often been seen with small molecule GCGR antagonists, the functional and binding potency can vary greatly between species. Compound **15.84** possessed modest functional potency at the dog (dGCGR cAMP IC_{50} = 400 $\pm$ 150 nM) and rhesus monkey (rhesusGCGR cAMP IC_{50} = 140 $\pm$ 20 nM) receptors, but poor functional potency at the rat (rGCGR cAMP IC_{50} = 2600 $\pm$ 200 nM) and mouse (mouseGCGR cAMP IC_{50} = 200 nM) receptors.

Compound **15.84** underwent extensive *in vivo* profiling. Acceptable oral PK was observed for **15.84** across species (Table 15.2). In a glucagon challenge PD assay in transgenic hGCGR mice, **15.84** blocked exogenous glucagon-stimulated glucose excursion when dosed at both 10 mg/kg and 30 mg/kg p.o., 1h prior to glucagon administration. The receptor occupancy was determined to be 51% at the 10 mg/kg dose and 87% at the 30 mg/kg dose. Compound **15.84** was also profiled in a glucagon challenge PD assay in rhesus monkeys. Compound **15.84** was able to completely block glucagon-stimulated plasma glucose increase at both 1 and 10 mg/kg when dosed 4 h prior to the glucagon challenge.

15.85 (MK-0893)
hGCGR binding IC_{50}: 6 nM
hGCGR cAMP IC_{50}: 17 nM
hGIP cAMP IC_{50}: 1020 nM
hGLP-1R cAMP IC_{50}: >10000 nM

Figure 15.28 MK-0893.

15.4.1.2   Clinical Development

Merck recently described the discovery and clinical development of the potent and selective glucagon receptor antagonist MK-0893 (**15.85**: hGCGR binding IC_{50} = 6 nM, hGCGR cAMP IC_{50} = 17 nM, hGIPR IC_{50} = 1020 nM, hGLP-1R IC_{50} > 10 000 nM) (Figure 15.28).[93–97] Compound **15.85** was studied in a trial of 342 individuals having T2DM for at least 4 years.[98] The study lasted 12 weeks, during which time an equal number of individuals were given **15.85** at 20, 40, 60, or 80 mg once daily (qd), metformin at 1000 mg twice daily (bid) or placebo. Compound **15.85** demonstrated a statistically significant, dose-dependent decrease in fasting plasma glucose levels and HbA1c, with an overall improvement in glycemic control at doses ≥ 40 mg qd when compared to metformin. Unfortunately, compared to placebo **15.85** was shown to increase LDL-C, ALT, and body weight.

Compound **15.85** was also profiled in a human glucagon challenge model.[99] At a 200 mg qd oral dose, **15.85** decreased exogenous glucagon-stimulated glucose excursion by 59% and completely blunted glucagon-stimulated glucose excursion at the 1000 mg qd dose.

Based on these results, **15.85** was explored as a monotherapy in T2DM patients for 4 weeks at either 40 mg qd or 120 mg qd and the results were compared to metformin administered at 1000 mg twice daily (bid) for 4 weeks. A total of 74 patients were included in the study. Using change in 24-h weighted mean glucose (WMG) as the primary endpoint, change in WMG for **15.85** at 40 mg qd was 25.9 mg/dL, for **15.85** at 120 mg qd was 53.6 mg/dL, and for metformin at 1000 mg bid was 26.0 mg/dL. At the 120 mg dose, a trend was observed for increases in LDL-cholesterol, blood pressure, and liver function test values (LFTs).

In order to assess the effect of treatment with **15.85** on the ability of individuals to recover from hypoglycemia, individuals were first subjected to two out of three possible single dose treatments: placebo, 200 mg qd **15.85**, and 1000 mg qd **15.85**, with a 3-week washout between treatments.[100] After each of the two doses, plasma glucose levels were maintained at 50 mg/dL via hypoglycemic clamp for 30 minutes. Following the glycemic clamp, the time to >70 mg/dL plasma glucose levels were measured (placebo: 33.3 min; 200 mg **15.85**: 45.0 min; 1000 mg **15.85**: 59.1 min). Minimal increase in time to recovery for the 200 mg dose of **15.85** ($\sim$60% functional hGCGR blockade) when compared to placebo. This level of hGCGR blockade had previously been demonstrated to be sufficient for the lowering of plasma glucose, therefore leading to the suggestion that hGCGR antagonists can be used at doses that improve glucose homeostasis with minimal risk of delayed recovery from hypoglycemia. A second related study in patients with T2DM demonstrated that **15.85** at an oral dose corresponding to near-complete hGCGR blockade (1000 mg qd) in conjunction with administration of propranolol at 80 mg three times daily (tid) also delayed recovery (estimated mean difference: 32 minutes) from an induced hypoglycemic state.[101]

Eli Lilly has also recently reported preclinical and phase I human trial results for their hGCGR antagonist LY2409021 (hGCGCR K_i = 6.66 ± 0.64 nM;

mouse GCGR K_i = 75.3 $\pm$ 11.4 nM).[102] The structure of this compound has not been disclosed. In hyperglycemic ob/ob mice, acute dosing of LY2409021 lowered plasma glucose levels with an ED_{50} of 7.4 mg/kg. When dosed at 50 mg/kg per day for 14 days in ob/ob mice, LY2409021 significantly lowered plasma fructosamine levels from 335 $\pm$ 9.1 μM on day 1 of treatment to 271 $\pm$ 9.8 μM on day 14 of treatment. During this study, an increase in plasma glucagon levels was observed and a decrease in plasma insulin levels was observed. In a transgenic hGCGR C57BL/6 mouse model with streptozotocin (STZ)-induced hyperglycemia, LY2409021 demonstrated reduced plasma glucose levels at 6h post dose (ED_{50} = 1.39 mg/kg). In a liver tissue *ex vivo* binding assay in this model, at 30 mg/kg, LY2409021 hGCGR receptor occupancy was determined to be 81.7% at 6h post dose.

LY2409021 was administered in a rising single dose human subject study to both healthy volunteers (23 males) and individuals with T2DM (5 males and 4 females).[103] The T2DM patients had fasting blood glucose (FBG) levels ranging from 93.6 to 207.0 mg/dL. For the healthy volunteers, the doses administered were 2.5, 10, 30, 100, 250, and 500 mg and for the T2DM volunteers, the doses were 75, 200, and 500 mg. The t_{max} in this study ranged from 4 to 8h, the $t_{1/2}$ from 50.8 to 58.6h and the apparent clearance (clearance/ bioavailability: (CL/F)) from 0.232 to 0.396 L/h. The changes in FBG in healthy volunteers ranged from +2.9 to -11.5 mg/dL and in T2DM subjects from -21.9 to -33.3 mg/dL.

To determine if the glucose lowering observed following treatment with LY2409021 was glucagon receptor-mediated, 21 healthy volunteers were administered LY2409021 or placebo at various doses, followed by an infusion of 6,6-[^{2}H$_2$]glucose (9h post LY2409021 administration), and 3h later a simultaneous infusion of somatostatin, insulin, and glucagon. Exogenous glucagon-stimulated hepatic glucose output was blocked in a dose-dependent manner by LY2409021.[104]

A third study with LY2409021 looked at doses of 5, 30, 60, and 90 mg in T2DM subjects over a period of 28 days.[105] The t_{max} in this study ranged from 6 to 8 h, the $t_{1/2}$ from 56.1 to 61.9 h and the apparent clearance (CL/F) from 0.263 to 0.345 L/h. There was also a dose-proportional increase in C_{max} observed. At day 28, statistically significant reductions in HbA1c were observed at the 60mg (-0.53%, p = 0.0117) and 90 mg doses (-0.43%, p = 0.0391). Fasting glucagon levels rose in all dose levels versus baseline and GLP-1 levels rose by 59% at the 90 mg dose. Four hypoglycemic events were observed at the 90 mg dose with the minimum glucose level being 62 mg/dL. In the 90 mg dose group, hepatic transaminases increased by greater than 3-fold the upper limit of normal in 5 out of 9 patients.

These clinical results provide significant proof of concept for the use of glucagon receptor antagonists for glycemic control in humans with T2DM. Hypoglycemia and lipid metabolism appear to be issues that need to be monitored in subsequent clinical investigations.

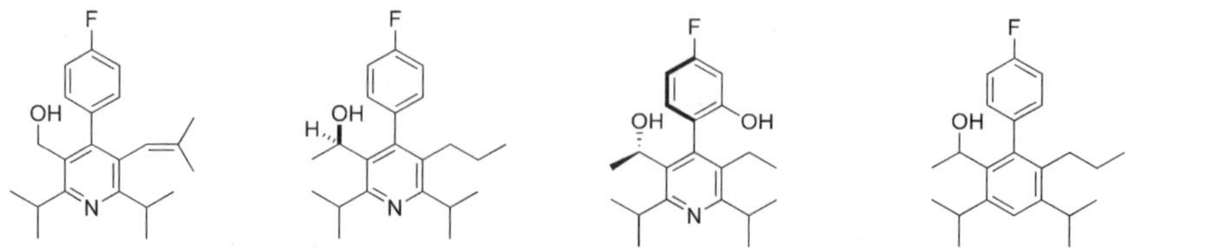

15.86
hGCGR binding IC$_{50}$: 7000 nM
hGCGR cAMP IC$_{50}$: 2000 nM

15.87
hGCGR binding IC$_{50}$: 110 nM
hGCGR cAMP IC$_{50}$: 65 nM

15.88
hGCGR binding IC$_{50}$: 16 nM

(+)-15.89
hGCGR binding IC$_{50}$: 110 nM

Figure 15.29   Substituted biaryl glucagon receptor antagonists.

15.4.2   Biaryl Glucagon Receptor Antagonists

Hindered biaryls are another class of small molecule hGCGR antagonists that have seen a significant amount of research. Initially identified through a high-throughput screening (HTS) campaign, compound **15.86** demonstrated modest binding and functional antagonism of the hGCGR (Figure 15.29) (hGCGR binding IC$_{50}$ = 7 µM, hGCGR cAMP IC$_{50}$ = 2 µM).[106] Potency optimization studies afforded the chiral alcohol **15.87** (hGCGR binding IC$_{50}$ = 110 nM, hGCGR cAMP IC$_{50}$ = 65 nM). Introduction of a 2′-hydroxy substituent on the phenyl ring yielded atropisomeric analogs with high affinity for the glucagon receptor such as compound **15.88** (hGCGR binding IC$_{50}$ = 16 nM).[107,108]

Substitution of a phenyl ring for the pyridyl ring present in the previous analogs afforded (+)-**15.89** (BAY 27-9955).[109,110] Compound (+)-**15.89** competitively inhibits the binding of glucagon to the hGCGR (hGCGR binding IC$_{50}$ = 110 nM).[111] A receptor occupancy study was performed in a transgenic hGCGR mouse model with (+)-**15.89**.[112] The results indicated that a minimum of 50–60% coverage is necessary to effectively block hyperglycemia induced by exogenous glucagon administration. Compound (+)-**15.89** has also been administered to healthy, non-diabetic humans as part of a glucagon challenge pharmacodynamic assay.[105] An infusion of somatostatin clamped the endogenous production of insulin and glucagon. Exogenous insulin was infused in order to maintain normal insulin levels. Hyperglucagonemia (2 times normal) was then induced via infusion of exogenous glucagon (3 ng/kg·min) over 3h. The individuals who received a placebo experienced a greater than 100% increase in glucose production, while those treated with 70 mg (+)-**15.89** experienced a 72% increase and those treated with 200 mg (+)-**15.89** experienced only a 25% increase in glucose production. These results demonstrated for the first time that a glucagon receptor antagonist could inhibit glucagon-stimulated hepatic glucose production in humans.

15.4.3   Phenol-Based hGCGR Antagonists

A number of glucagon receptor antagonists possessing catechols and substituted phenols have also been reported. Mercaptobenzimidazoles contain-

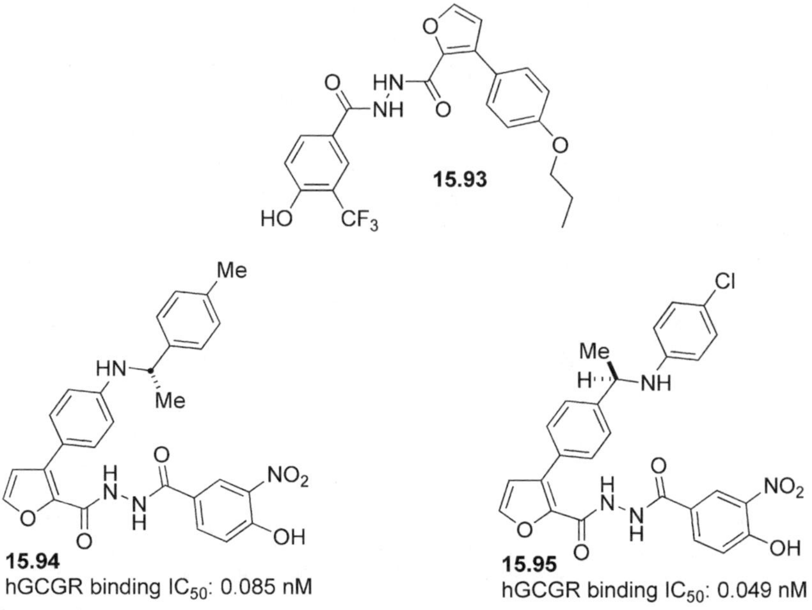

15.90 (NNC 92-1687)
hGCGR binding IC_{50}: 20000+/-2800 nM

15.91
hGCGR binding IC_{50}: 20 nM

15.92 (NNC 25-2504)
hGCGR binding IC_{50}: 2.3 nM

Figure 15.30   Catechol- and cyanophenol-based glucagon receptor antagonists.

ing a catechol exemplified by compound **15.90** (Figure 15.30: NNC 92-1687, hGCGR binding IC_{50} = 20 $\pm$ 2.8 μM) were initially described.[113] Additional alkylidene hydrazides such as compounds **15.91** (hGCGR binding IC_{50} = 20

15.93

15.94
hGCGR binding IC_{50}: 0.085 nM

15.95
hGCGR binding IC_{50}: 0.049 nM

Figure 15.31   Hydrazide-linked phenol-based glucagon receptor antagonists.

nM) and **15.92** (NNC 25-2504, hGCGR binding IC_{50} = 2.3 nM, rGCGR binding IC_{50} = 0.43 nM) have also been reported.[114–116] Compound **15.92** was determined to be a true non-competitive antagonist of the rat glucagon receptor, but a mixed competitive and non-competitive antagonist of the human glucagon receptor. Compound **15.92** also competitively inhibited glucose production in isolated rat primary hepatocytes (K_i = 14 nM). At 10 mg/kg i.v. **15.92** demonstrated the ability to prevent exogenous glucagon-stimulated plasma glucose increase in Sprague-Dawley rats. The oral bioavailability of **15.92** in dogs was determined to be modest (F = 15%).

More recently, a structurally similar series of furanyl hydrazides exemplified by compounds **15.93** and **15.94** (hGCGR binding IC_{50} = 0.085 nM) have also been reported (Figure 15.31).[117,118] Compound **15.94** was shown to lower plasma glucose in ob/ob mice at doses as low as 0.1 mg/kg. Compound **15.95** (hGCGR binding IC_{50} = 0.049 nM) was also described as a potent hGCGR antagonist.[119]

15.4.4   Additional Small Molecule hGCGR Antagonists

Quinoxaline **15.96** (CP-99,711) was the first non-peptidic glucagon receptor antagonist reported in the literature (Figure 15.32).[120] Compound **15.96** inhibited binding of $[^{125}I]$-glucagon to Sprague Dawley rat liver membranes (rat liver membrane binding IC_{50} = 4000 $\pm$ 1000 nM). Additionally, **15.96** was determined to be a functional antagonist of the rat GCGR (rat liver membrane cAMP IC_{50} = 7000 $\pm$ 1000 nM). It was also noted in this report that **15.96** is a non-selective antagonist for the glucagon receptor. The development of a series of structurally related pyrrolo[1,2-*a*]quinoxaline glucagon receptor antagonists exemplified by compound **15.97** followed this initial report.[121] Like quinoxaline **15.96**, compound **15.97** exhibited micromolar binding affinity for the rat glucagon receptor (rat membrane binding IC_{50} = 5000 nM). Compound **15.97** also exhibited micromolar affinity for the rat truncated GLP-1 receptor (tGLP-1R binding IC_{50} = 2500 nM). Compound **15.96** was also analyzed in this rat publication and was found to be more potent than originally reported (rat liver membrane binding IC_{50} = 100 nM, tGLP-1R IC_{50} = 300 nM).

The fungal metabolite skyrin **15.98** and oxyskyrin **15.99** have been shown to block the glucagon-stimulated production of cAMP in rat hepatocytes (Figure 15.33: for **15.98**, 50% at 30 µM), primary cultures of human

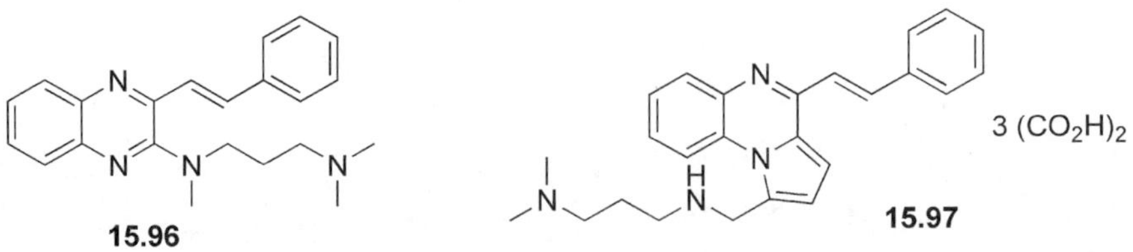

Figure 15.32   Quinoxaline and pyrrolo[1,2-*a*]quinoxaline antagonists.

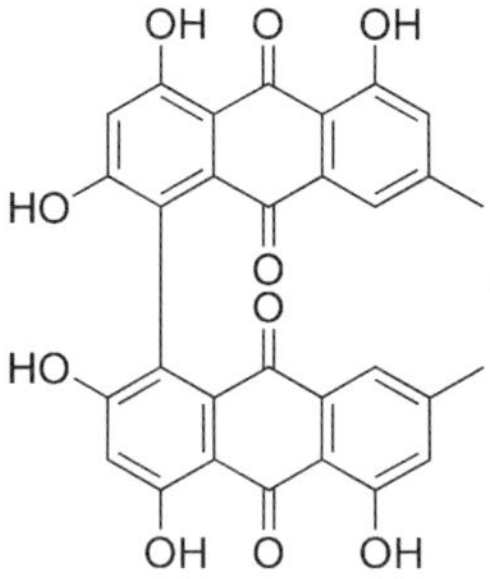

Figure 15.33 Skyrin and oxyskyrin.

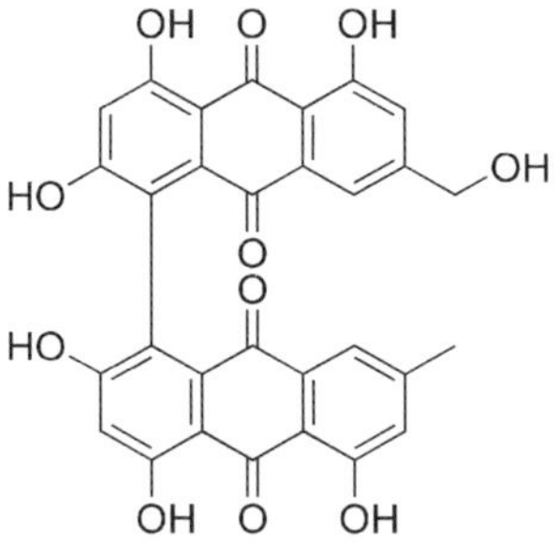

hepatocytes (>50% at 10 μM), and CHO cells transfected with hGCGR (for **15.98**, 30% at 30 μM).[122,123] Interestingly, it was observed that **15.98** (at 50 μM) had no effect on the ability of unlabeled glucagon to displace [^{125}I]-glucagon from membranes prepared from CHO cells transfected with hGCGR. Additionally, **15.98** (at 30 μM) had no effect on the ability of [^{125}I]-glucagon to bind to these same membranes. Compound **15.98** does not affect cAMP production in CHO cells transfected with the GLP-1 receptor. This suggests that **15.98** is a selective non-competitive, functional antagonist of rGCGR.

Disubstituted pyridyl imidazoles and disubstituted pyridyl pyrroles (known p38 MAP kinase inhibitors) have also been previously identified as glucagon receptor antagonists (Figure 15.34). Compared to the initial hits, pyridyl imidazole **15.100** possessed improved glucagon receptor affinity (hGCGR binding IC$_{50}$ = 6.5 nM) and minimal p38 inhibition (20% inhibition at 40 000 nM).[124] Compound **15.101** has been reported as a potent glucagon receptor antagonist (hGCGR binding IC$_{50}$ = 6 nM) with oral bioavailability in both rats and mice.[125,126] In the course of these investigations it was discovered that there was a significant decrease in glucagon receptor binding affinity for compounds **15.100** and **15.101** when the receptor binding assay was performed in the presence of physiological concentrations of Mg^{2+} (hGCGR binding IC$_{50}$ (5 mM MgCl$_2$ present) **15.100**: 53 nM; **15.101**: 170 nM). Additional studies were performed with **15.101** in order to determine its mechanism of glucagon receptor antagonism.[127] In an hGCGR cAMP functional assay, increasing concentrations of compound **15.101** concurrently increased the apparent EC$_{50}$ for glucagon-stimulated production of cAMP while decreasing the maximum stimulation of cAMP production by glucagon. The rate of dissociation of [^{125}I]-glucagon from the hGCGR was increased 4-fold in the presence of compound **15.101** at 1000 nM. These results suggested that **15.101** is a non-competitive antagonist of the glucagon receptor. Site-directed mutagenesis of selected residues in the trans-membrane region of the human glucagon receptor, suggest that **15.101** binds in the trans-membrane domain. A rationale for the effect of divalent cations such as Mg^{2+} on the potency of **15.100** and **15.101** has yet to be determined.

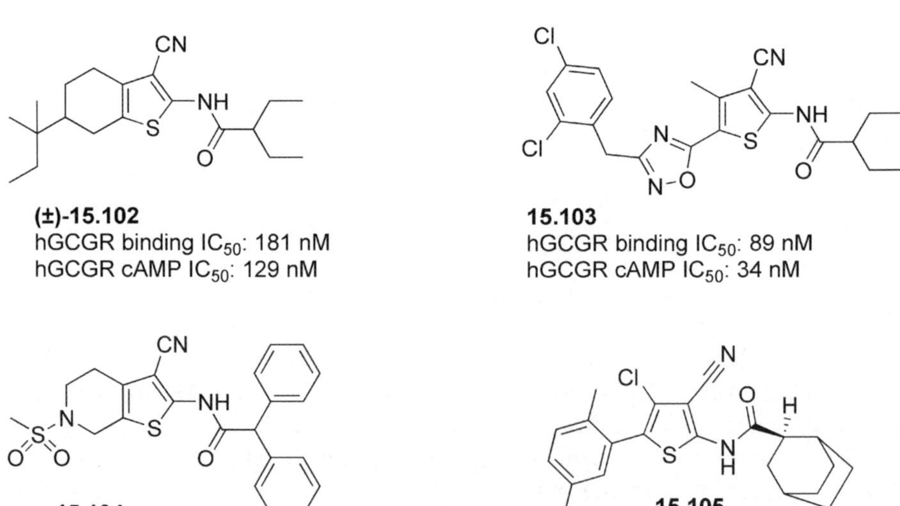

15.100
hGCGR binding IC$_{50}$: 6.5 nM

15.101
hGCGR binding IC$_{50}$: 6 nM

Figure 15.34 Pyridylimidazoles and pyridylpyrroles.

Bicyclic cyanothiophene ($\pm$)-**15.102** has been reported to be a competitive antagonist of the glucagon receptor (Figure 15.35) (hGCGR binding IC$_{50}$ = 181 nM, hGCGR cAMP IC$_{50}$ = 129 nM).[128] When administered at 50 mg/kg intraperitoneally ($\pm$)-**15.102** blocked exogenous glucagon-stimulated glucose excursion in hGCGR-containing transgenic mice. Lead optimization studies resulted in the discovery of the ring-opened cyanothiophene **15.103** (hGCGR binding IC$_{50}$ = 89 nM, hGCGR cAMP IC$_{50}$ = 34 nM).[129] The mouse oral PK profile of **15.103** was significantly improved over that of ($\pm$)-**15.102**. Compounds ($\pm$)-**15.102** and **15.103** were both profiled in the hGCGR-containing transgenic mouse glucagon challenge model at 100 mg/kg p.o. No reduction in glycogenolysis was observed with oral dosing of these compounds. The observation that there is a 15-fold decrease in the potency of **15.103** in the hGCGR cAMP assay when performed in the presence of 5% mouse plasma has been offered as a possible explanation for the lack of *in vivo* efficacy. Structurally similar compounds **15.104** and **15.105** have also been disclosed as glucagon receptor antagonists.[130,131]

(±)-15.102
hGCGR binding IC$_{50}$: 181 nM
hGCGR cAMP IC$_{50}$: 129 nM

15.103
hGCGR binding IC$_{50}$: 89 nM
hGCGR cAMP IC$_{50}$: 34 nM

15.104

15.105

Figure 15.35 2-Amino-3-cyanothiophenes and their derivatives.

15.5 Conclusions

Beginning with peptide antagonists, followed by monoclonal antibodies and orally bioavailable small molecules, a wide variety of glucagon receptor antagonists have demonstrated their ability to improve glucose homeostasis in preclinical animal models of T2DM. Recently, select small molecule hGCGR antagonists have demonstrated the same effect in human clinical trials in patients with T2DM. Although encouraging, these results are tempered by undesired changes in body weight, lipid profile, and liver enzyme levels upon administration of an hGCGR antagonist in humans. Additionally, the potential risk of hypoglycemia and α-cell hyperplasia upon administration of an hGCGR antagonist must also be considered. Ongoing research in the glucagon receptor antagonist arena is likely to shed additional light on the viability of this target for the treatment of T2DM.

References

1. C. P. Kimball and J. R. Murlin, *J. Biol. Chem.*, 1923, **58**, 337.
2. A. Staub, L. Sinn and O. K. Behrens, *Science*, 1953, **117**, 628.
3. A. Staub, L. Sinn and O. K. Behrens, *J. Biol. Chem.*, 1955, **214**, 619.
4. V. W. Bromer, A. Staub, E. R. Diller, H. L. Bird, L. G. Sinn and O. K. Behrens, *J. Am. Chem. Soc.*, 1957, **79**, 2794.
5. W. W. Bromer, L. G. Sinn and O. K . Behrens, *J. Am. Chem. Soc.*, 1957, **79**, 2798.
6. W. W. Bromer, A. Staub, L. G. Sinn and O. K . Behrens, *J. Am. Chem. Soc.*, 1957, **79**, 2801.
7. L. G. Sinn, O. K . Behrens and W. W. Bromer, *J. Am. Chem. Soc.*, 1957, **79**, 2805.
8. H. Ferner, *Am. J. Digestive Dis.*, 1953, **20**, 301.
9. S. L. Pohl, L. Birnbaumer and M. Rodbell, *Science*, 1969, **164**, 566.
10. E. Struck, J. Ashmore and O. Wieland, *Biochem Z.*, 1965, **343**, 107.
11. J. H. Exton and C. R. Park, *Pharmacol. Rev.*, 1966, **18**, 181.
12. L. J. Jelinik, S. Lok, G. B. Rosenberg, R. A. Smith, F. J. Grant, S. Biggs, P. A. Bensch, J. L. Kuijper, P. O. Sheppard, C. A. Sprecher, P. J. O'Hara, D. Foster, K. M. Walker, L. H. J. Chen, P. A. McKernan and W. Kindsvogel, *Science*, 1993, **259**, 1614.
13. K. E. Mayo, L. J. Miller, D. Bataille, S. Dalle, B. Goke, B. Thorens and D. J. Drucker, *Pharmacol. Rev.*, 2003, **55**, 167.
14. A. D. Baron, L. Schaeffer, P. Shragg and O. G. Kolterman, *Diabetes*, 1987, **36**, 274.
15. S. Lok, J. L. Kuijper, L. J. Jelinek, J. M. Kramer, T. E. Whitmore, C. A. Sprecher, S. Matthewes, F. J. Grant, S. H. Biggs, G. B. Rosenberg, P. O. Sheppard, P. J. O'Hara, D. C. Foster and W. Kindsvogel, *Gene*, 1994, **140**, 203.

16. D. J. MacNeil, J. L. Occi, P. J. Hey, C. D. Strader and M. P. Graziano, *Biochem. Biophys. Res. Commun.*, 1994, **198**, 328.

17. M. R. Tota, L. Xu, A. Sirotina, C. D. Strader and M. P. Graziano, *J. Biol. Chem.*, 1995, **270**, 26466.

18. J. C. Parker, K. M. Andrews, M. R. Allen, J. L. Stock and J. D. McNeish, *Biochem. Biophys. Res. Commun.*, 2002, **290**, 839.

19. R. W. Gelling, X. Q. Du, D. S. Dichmann, J. Romer, H. Huang, L. Cui, S. Obici, B. Tang, J. J. Holst, C. Fledelius, P. B. Johansen, L. Rossetti, L. A. Jelicks, P. Serup, E. Nishimura and M. J. Charron, *Proc. Natl. Acad. Sci.*, 2003, **100**, 1438.

20. J. Yang, M. L. MacDougall, M. T. McDowell, L. Xi, R. Wei, W. J. Zavadoski, M. P. Molloy, J. D. Baker, M. Kuhn, O. Cabrera and J. L. Treadway, *BMC Genomics*, 2011, **12**, 281.

21. L.-L. Shiao, M. A. Cascieri, M. Trumbauer, H. Chen and K. A. Sullivan, *Transgenic Res.*, 1999, **8**, 295.

22. M. Rodbell, L. Birnbaumer, S. L. Pohl and F. Sundby, *Proc. Natl. Acad. Sci.*, 1971, **68**, 909.

23. M. D. Bregman, D. Trivedi and V. Hruby, *J. Biol. Chem.*, 1980, **255**, 11725.

24. V. Hruby, M. D. Bregman and D. Trivedi, *Science*, 1982, **215**, 115.

25. V. Iwanij and A. C. Vincent, *J. Biol. Chem.*, 1990, **265**, 21302.

26. W. Gu, H. Yan, K. A. Winters, R. Komorowski, S. Vonderfecht, L. Atangan, G. Sivits, D. Hill, J. Yang, V. Bi, Y. Shen, S. Hu, T. Boone, R. A. Lindberg and M. M. Veniant, *J. Pharmacol. Exp. Ther.*, 2009, **331**, 871.

27. J. Buggy, A. Rossomando, M. MacDougall, D. Mierz, D. Wunderlich and H. Yoo-Warren, *Horm. Metab. Res.*, 1996, **28**, 215.

28. L. M. Wright, A. M. Brzozowski, R. E. Hubbard, A. C. W. Pike, S. M. Roberts, R. N. Skovgaard, I. Svendsen, H. Vissing and R. P. Bywater, *Acta Crystallogr.*, 2000, **D56**, 573.

29. H. Yan, W. Gu, J. Yang, V. Bi, Y. Shen, E. Lee, K. A. Winters, R. Komorowski, C. Zhang, J. J. Patel, D. Caughey, G. S. Elliott, Y. Y. Lau, J. Wang, Y.-S. Li, T. Boone, R. A. Lindberg, S. Hu and M. M. Veniant, *J. Pharmacol. Exp. Ther.*, 2009, **329**, 102.

30. W. Gu, K. A. Winters, A. S. Motani, R. Komorowski, Y. Zhang, Q. Liu, X. Wu, I. C. Rulifson, G. Sivitis Jr., M. Graham, H. Yan, P. Wang, S. Moore, T. Meng, R. A. Lindberg and M. M. Veniant, *Am. J. Physiol. Endocrinol. Metab.*, 2010, **299**, E624.

31. Novo Nordisk A/S and Agouron Pharmaceuticals Inc., *WO Pat. Appl.*, 69810, 2000.

32. J. Lau, C. Behrens, U. G. Sidelmann, L. B. Knudsen, B. Lundt, C. Sams, L. Ynddal, C. L. Brand, L. Pridal, A. Ling, D. Kiel, M. Plewe, S. Shi and P. Madsen, *J. Med. Chem.*, 2007, **50**, 113.

33. J. T. Kodra, A. S. Jorgensen, B. Andersen, C. Behrens, C. L. Brand, I. T. Christensen, M. Guldbrandt, C. B. Jeppesen, L. B. Knudsen, P. Madsen,

E. Nishamura, C. Sams, U. G. Sidelmann, R. A. Pedersen, F. C. Lynn and J. Lau, *J. Med. Chem.*, 2008, **51**, 5387.

34. M. Winzell, C. L. Brand, N. Wierup, U. G. Sidelmann, F. Sundler, E. Nishimura and B. Ahren, *Diabetologia*, 2007, **50**, 1453.

35. N. Rivera, C. A. Everett-Grueter, D. S. Edgerton, T. Rodewald, D. W. Neal, E. Nishimura, M. O. Larsen, L. O. Jacobsen, K. Kristensen, C. L. Brand and A. D. Cherrington, *J. Pharmacol. Exp. Ther.*, 2007, **321**, 743.

36. Novo Nordisk A/S, *WO Pat. Appl.*, 040446 A1, 2002.

37. Novo Nordisk A/S and Agouron Pharmaceuticals Inc., *WO Pat. Appl.*, 069810 A1, 2000.

38. Novo Nordisk A/S, *WO Pat. Appl.*, 056763 A2, 2004.

39. Novo Nordisk A/S, *WO Pat. Appl.*, 051357 A1, 2003.

40. Novo Nordisk A/S, *US Patent App.*, 203946 A1, 2003.

41. R. Liang, L. Abrardo, E. J. Brady, M. R. Candelore, V. Ding, R. Saperstein, L. M. Tota, M. Wright, S. Mock, C. Tamvakopolous, S. Tong, S. Zheng, B. B. Zhang, J. R. Tata and E. R. Parmee Bioorg. *Med. Chem. Lett.*, 2007, **17**, 587.

42. Merck & Co, *WO Pat. Appl.*, 104826 A2, 2006.

43. D.-M. Shen, F. Zhang, E. J. Brady, M. R. Candelore, Q. Dallas-Yang, V. D.-H. Ding, J. Dragovic, W. P. Feeney, G. Jiang, P. E. McCann, S. Mock, S. A. Qureshi, R. Saperstein, X. Shen, C. Tamvakopoulos, X. Tong, L. M. Tota, M. J. Wright, X. Yang, S. Zheng, K. T. Chapman, B. B. Zhang, J. R. Tata and E. R. Parmee, *Bioorg. Med. Chem. Lett.*, 2005, **15**, 4564.

44. Merck & Co, *WO Pat. Appl.*, 050039, 2004.

45. J. Mu, G. Jiang, E. Brady, Q. Dallas-Yang, F. Liu, J. Woods, E. Zycband, M. Wright, Z. Li, K. Lu, L. Zhu, X. Shen, R. SinhaRoy, M. L. Candelore, S. A. Qureshi, D.-M. Shen, F. Zhang, E. R. Parmee and B. B. Zhang, *Diabetologia*, 2011, **54**, 2381.

46. Novo Nordisk A/S, *WO Pat. Appl.*, 058845 A2, 2005.

47. Abbott Laboratories, *US Pat. Appl.*, 0209928 A1, 2004.

48. Schering Corporation, *WO Pat. Appl.*, 039789 A1, 2010.

49. R. Kurukulasuriya, B. K. Sorensen, J. T. Link, J. R. Patel, H.-W. Jae, M. X. Winn, J. R. Rohde, N. D. Grihalde, C. W. Lin, C. A. Ogiela, A. L. Adler and C. A. Collins, *Bioorg. Med. Chem. Lett.*, 2004,**14**, 2047.

50. Metabasis Therapeutics Inc., *WO Pat. Appl.*, WO 019830 A1, 2010.

51. Metabasis Therapeutics Inc., *WO Pat. Appl.*, 098244 A1, 2008.

52. Merck & Co, *WO Pat. Appl.*, 136577, 2007.

53. Merck & Co, *WO Pat. Appl.*, 111864, 2007.

54. Merck & Co, *WO Pat. Appl.*, 102067, 2006.

55. Eli Lilly and Company, *WO Pat. Appl.*, 106181, 2007.

56. Takeda Pharmaceutical Company Ltd., *WO Pat. Appl.*, 057784 A1, 2009.

57. Takeda Pharmaceutical Company Ltd., *WO Pat. Appl.*, 110520 A1, 2009.

58. Schering Corporation, *WO Pat. Appl.*, 037815, 2011.

59. Eli Lilly and Company, *WO Pat. Appl.*, 118542, 2005.

60. Eli Lilly and Company, *WO Pat. Appl.*, 123668, 2005.

61. Eli Lilly and Company, *WO Pat. Appl.*, 120270 A2, 2007.

62. Eli Lilly and Company, *WO Pat. Appl.*, 086488, 2006.

63. Eli Lilly and Company, *WO Pat. Appl.*, 120284 A2, 2007.

64. Eli Lilly and Company, *WO Pat. Appl.*, 114855, 2007.

65. Merck & Co, *WO Pat. Appl.*, 098948, 2010.

66. Merck & Co, *US Pat. Appl.*, 0088071, 2007.

67. C. J. Sinz, A. Bittner, R. M. Kim, E. Brady, M. R. Candelore, V. D.-H. Ding, G. Jiang, Z. Lin, A. R. Lins, P. McCann, C. Miller, K. Nam, S. A. Qureshi, F. Salituro, R. Saperstein, J. Shang, D. Szalkowski, L. Tota, M. Wright, R. Wang, S. Xu, X. Yang, B. Zhang, M. Hammond, J. Tata and E. Parmee, MEDI 16, *235th ACS National Meeting*, New Orleans, LA, April, 2008.

68. Merck & Co, *WO Pat. Appl.*, 093535, 2010.

69. Eli Lilly and Company, *WO Pat. Appl.*, 123581 A1, 2007.

70. Merck & Co, *WO Pat. Appl.*, 042223, 2008.

71. Merck & Co, *US Pat.*, 7 687 534 B2, 2010.

72. Merck Sharp and Dohme Corp, *WO Pat. Appl.*, 080971 A1, 2010.

73. Merck Sharp and Dohme Corp, *WO Pat. Appl.*, 098994 A1, 2010.

74. Merck Sharp and Dohme Corp, *WO Pat. Appl.*, 030722 A1, 2010.

75. Merck Sharp and Dohme Corp, *WO Pat. Appl.*, 071750 A1, 2010.

76. Merck Sharp and Dohme Corp, *WO Pat. Appl.*, 088061 A1, 2010.

77. Novo Nordisk A/S, *WO Pat. Appl.*, 002480 A1, 2004.

78. P. Madsen, J. T. Kodra, C. Behrens, E. Nishimura, C. B. Jeppesen, L. Pridal, B. Andersen, L. B. Knudsen, C. Valcarce-Aspegren, M. Guldbrandt, I. T. Christensen, A. S. Jorgensen, L. Ynddal, C. L. Brand, M. A. Bagger and J. Lau, *J. Med. Chem.*, 2009, **52**, 2989.

79. Merck & Co., Inc., *WO Pat. Appl.*, 100875 A2, 2004.

80. X. Yang, M. L. Yates, M. R. Candelore, W. Feeney, D. Hora, R. M. Kim, E. R. Parmee, J. P. Berger, B. B. Zhang and S. A. Qureshi, *Eur. J. Pharm.*, 2007, **555**, 8.

81. R. M. Kim, J. Chang, A. R. Lins, E. Brady, M. R. Candelore, Q. Dallas-Tang, V. Ding, J. Dragovic, S. Iliff, G. Jiang, S. Mock, S. Qureshi, R. Saperstein, D. Szalkowski, C. Tamvakopoulos, L. Tota, M. Wright, X. Yang, J. R. Tata, K. Chapman, B. B. Zhang and E. R. Parmee, *Bioorg. Med. Chem. Lett.*, 2008, **18**, 3701.

82. Merck & Co., Inc., *WO Pat. Appl.*, 065680 A1, 2005.

83. Boehringer Ingelheim International GmbH, *US Pat.*, 7 151 114 B2, 2006.

84. Novo Nordisk A/S, *WO Pat. Appl.*, 053938 A1, 2003.

85. Merck & Co., Inc., *US Pat. Appl.*, 0088070 A1, 2007.

86. Merck & Co., Inc., *WO Pat. Appl.*, 121097 A2, 2005.

87. Merck & Co., Inc., *WO Pat. Appl.*, 014618 A2, 2006.

88. Merck & Co., Inc., *WO Pat. Appl.*, 069158 A2, 2004.

89. Merck & Co., Inc., *WO Pat. Appl.*, 017055, 2006.

90. Schering Corp, *WO Pat. Appl.*, 144664 A1, 2010.

91. Schering Corp, *WO Pat. Appl.*, 140342 A1, 2009.
92. D.-M. Shen, E. J. Brady, M. R. Candelore, Q. Dallas-Yang, V. D.-H. Ding, W. P. Feeney, G. Jiang, M. E. McCann, S. Mock, S. A. Qureshi, R. Saperstein, X. Shen, X. Tong, L. M. Tota, M. J. Wright, X. Yang, S. Zheng, K. T. Chapman, B. B. Zhang, J. R. Tata and E. R. Parmee, *Bioorg. Med. Chem. Lett.*, 2011, **21**, 78.
93. E. R. Parmee, Abstracts of Papers, *241st ACS National Meeting & Exposition*, Anaheim, CA, United States, March 27–31, 2011, MEDI-31.
94. C. Drahl, *Chem. Eng. News*, April 18, 2011, **89**, Number 16, 37.
95. Merck & Co., Inc., *US Pat. Appl.*, 0272794 A1, 2005.
96. Merck & Co., Inc., *WO Pat. Appl.*, 015999 A2, 2007.
97. Merck Sharp and Dohme Corp, *WO Pat. Appl.*, 035558, 2009.
98. S. S. Engel, L. Xu, P. J. Andryuk, M. J. Davies, J. Amatruda, K. Kaufman and B. J. Goldstein, *Abstracts of 71st Scientific Sessions of the American Diabetes Association*, June 24–28, 2011, Abstract number 0309-OR.
99. M. Ruddy, B. Pramanik, J. Lunceford, S. Li, C. Cilissen, S. A. Stoch, J. Wagner, J. Amatruda, G. Herman and K. Kaufman, *Abstracts of 71st Scientific Sessions of the American Diabetes Association*, June 24–28, 2011, Abstract number 0311-OR.
100. M. D. Troyer, M. Hompesch, B. Pramanik, W. Zheng, K. Win, S. Dunbar, S. Li, M. Ruddy, J. Amatruda, K. Kaufman, J. Wagner and S. A. Stoch, *Abstracts of 71st Scientific Sessions of the American Diabetes Association*, June 24–28, 2011, Abstract number 0494P.
101. M. D. Troyer, M. Hompesch, T. Jax, B. Pramanik, F. Liu, L. Morrow, K. Win, T. Reynders, C. Liu, S. Engel, T. Heise, E. Maliwat, W. S. Denney, D. E. Kelley, J. A. Wagner and S. A. Stoch, *Abstracts of 71st Scientific Sessions of the American Diabetes Association*, June 24–28, 2011, Abstract number 0495P.
102. T. B. Farb, W. Ma, J. V. Ficorilli, E. D. Hawkins, J. T. Brozinick, K. W. Sloop, R. A. Owens, M. B. Brenner, P. A. Hipskind and J. S. Moyers, *Abstracts of 71st Scientific Sessions of the American Diabetes Association*, June 24–28, 2011, Abstract number 1673-P.
103. R. P. Kelley, E. J. Abu-Raddad, L. S. Tham, H. Fu, J. A. Pinaire and M. A. Deeg, *Abstracts of 71st Scientific Sessions of the American Diabetes Association*, June 24–28, 2011, Abstract number 1004P.
104. L. S. Tham, E. J. Abu-Raddad, C. N. Lim, M. T. Loh, W. T. Ng, J. A. Pinaire and R. P. Kelley, *Abstracts of 71st Scientific Sessions of the American Diabetes Association*, June 24–28, 2011, Abstract number 0416PP.
105. R. P. Kelley, P. Garhyan, E. J. Abu-Raddad, H. Fu, C. N. Lim, M. J. Prince, J. A. Pinaire, M. T. Loh and M. A. Deeg, *Abstracts of 71st Scientific Sessions of the American Diabetes Association*, June 24–28, 2011, Abstract number 0305OR.

106. G. H. Ladouceur, J. H. Cook, E. M. Doherty, W. R. Shoen, M. L. MacDougall and J. N. Livingston, *Bioorg. Med. Chem. Lett.*, 2002, **12**, 461.

107. G. H. Ladouceur, J. H. Cook, E. M. Doherty, W. R. Shoen, M. L. MacDougall and J. N. Livingston, *Bioorg. Med. Chem. Lett.*, 2002, **12**, 1303.

108. G. H. Ladouceur, J. H. Cook, D. L. Hertzog, J. H. Jones, T. Hundertmark, M. Korpusik, T. G. Lease, J. N. Livingston, M. L. MacDougall, M. H. Osterhout, K. Phelan, R. H. Romero, W. R. Schoen, C. Shao and R. A. Smith, *Bioorg. Med. Chem. Lett.*, 2002, **12**, 3421.

109. Bayer Corp, *WO Pat. Appl.*, 04528, 1998.

110. Bayer Corp, *US Pat.*, 6 218 431, 2001.

111. K. F. Peterson and J. T. Sullivan, *Diabetologia*, 2001, **44**, 2018.

112. Q. Dallas-Yang, X. Shen, M. Strowski, E. Brady, R. Saperstein, R. E. Gibson, D. Szalkowski, S. A. Qureshi, M. R. Candelore, J. E. Fenyk-Melody, E. R. Parmee, B. B. Zhang and G. Jiang, *Eur. J. Pharm.*, 2004, **501**, 225.

113. P. Madsen, L. B. Knudsen, F. C. Wiberg and R. D. Carr, *J. Med. Chem.*, 1998, **41**, 5150.

114. A. Ling, Y. Hong, J. Gonzalez, V. Gregor, A. Polinsky, A. Kuki, S. Shi, K. Teston, D. Murphy, J. Porter, D. Kiel, J. Lakis, K. Anderes and J. May, *J. Med. Chem.*, 2001, **44**, 3141.

115. A. Ling, M. Plewe, J. Gonzalez, P. Madsen, C. K. Sams, J. Lau, V. Gregor, D. Murphy, K. Teston, A. Kuki, S. Shi, L. Truesdale, D. Kiel, J. May, J. Lakis, K. Anderes, E. Iatsimirskaia, U. G. Sidelmann, L. B. Knudsen, C. L. Brand and A. Polinsky, *Bioorg. Med. Chem. Lett.*, 2002, **12**, 663.

116. P. Madsen, A. Ling, M. Plewe, C. K. Sams, L. B. Knudsen, U. G. Sidelmann, L. Ynddal, C. L. Brand, B. Andersen, D. Murphy, M. Teng, L. Truesdale, D. Kiel, J. May, A. Kuki, S. Shi, M. D. Johnson, K. A. Teston, J. Feng, J. Lakis, K. Anderes, V. Gregor and J. Lau, *J. Med. Chem.*, 2002, **45**, 5755.

117. Dainippon Pharmaceutical Co, Ltd., *WO Pat. Appl.*, 064404 A1, 2003.

118. Dainippon Sumitomo Pharma Co, Ltd., *WO Pat. Appl.*, 131669 A1, 2010.

119. Dainippon Sumitomo Pharma Co, Ltd., *WO Pat. Appl.*, 007722 A1, 2011.

120. J. L. Collins, P. J. Dambek, S. W. Goldstein and W. S. Faraci, *Bioorg. Med. Chem. Lett.*, 1992, **2**, 915.

121. J. Guillon, P. Dallemagne, B. Pfeiffer, P. Renard, D. Manechez, A. Kervran and S. Rault, *Eur. J. Med. Chem.*, 1998, **33**, 293.

122. Zymogenetics Inc., *WO Pat. Appl.*, 9414427 A2, 1994.

123. J. C. Parker, R. K. McPherson, K. M. Andrews, C. B. Levy, J. S. Dubins, J. E. Chin, P. V. Perry, B. Hulin, D. A. Perry, T. Inagaki, K. A. Dekker, K. Tachikawa, Y. Sugie and J. L. Treadway, *Diabetes*, 2000, **49**, 2079.

124. L. L. Chang, K. L. Sidler, M. A. Cascieri, S. de Laszlo, G. Koch, B. Li, M. MacCoss, N. Mantlo, S. O'Keefe, M. Pang, A. Rolando and W. K. Hagmann, *Bioorg. Med. Chem. Lett.*, 2001, **11**, 2549.
125. Merck & Co., Inc., *US Pat.*, 5 776 954, 1998.
126. S. E. De Laszlo, C. Hacker, B. Li, D. Kim, M. MacCoss, N. Mantlo, J. V. Pivnichny, L. Colwell, G. E. Koch, M. A. Cascieri and W. K. Hagmann, *Bioorg. Med. Chem. Lett.*, 1999, **9**, 641.
127. M. A. Cascieri, G. E. Koch, E. Ber, S. J. Sadowski, D. Louizides, S. E. de Laszlo, C. Hacker, W. K. Hagmann, M. MacCoss, G. G. Chicchi and P. P. Vicario, *J. Biol. Chem.*, 1999, **274**, 8694.
128. S. A. Qureshi, M. R. Candelore, D. Xie, X. Yang, L. M. Tota, V. D.-H. Ding, Z. Li, A. Bansal, C. Miller, S. M. Cohen, G. Jiang, E. Brady, R. Saperstein, J. L. Duffy, J. R. Tata, K. T. Chapman, D. E. Moller and B. B. Zhang, *Diabetes*, 2004, **53**, 3267.
129. J. L. Duffy, B. A. Kirk, Z. Konteatis, E. L. Campbell, R. Liang, E. J. Brady, M. R. Candelore, V. D. H. Ding, G. Jiang, F. Liu, S. A. Qureshi, R. Saperstein, D. Szalkowski, S. Tong, L. M. Tota, D. Xie, X. Yang, P. Zafian, S. Zheng, K. T. Chapman, B. B. Zhang and J. R. Tata, *Bioorg. Med. Chem. Lett.*, 2005, **15**, 1401.
130. Hoffmann-La Roche Inc., *US Pat. Appl.*, 209943 A1, 2004.
131. Boehringer Ingelheim Pharma GMBH & Co. KG, *WO Pat. Appl.*, 042850 A1, 2006.

ACC Inhibitors in Development

MATTHEW P. BOURBEAU

Amgen Inc, 1 Amgen Center Dr, Thousand Oaks, CA 91320, USA
E-mail: Bourbeau@amgen.com

16.1   Introduction

Metabolic syndrome is an increasingly worrisome condition affecting up to 25% of the American adult population.[1] As currently defined, a patient suffering from metabolic syndrome must exhibit central obesity as well as two or more of the following symptoms: elevated triglyceride levels (>150 mg/dL), reduced HDL levels (<40 mL/dL in males, <50 mg/dL in females), increased blood pressure (>130 diastolic, or >85 diastolic), and elevated fasting glucose (>110 mg/dL, or previously diagnosed diabetes). Abnormal fatty acid metabolism is thought to be a key player in the development of metabolic syndrome, and studies have demonstrated a correlation between insulin resistance and intracellular accumulation of fatty acid metabolites and triglycerides in tissues such as muscle and liver.[2] Accumulation of fatty acid metabolites is controlled by fatty acid uptake, as well as fatty acid synthesis and oxidation. One strategy to potentially treat metabolic syndrome is to perturb one or more of these pathways (block fatty acid update, reduce fatty acid synthesis, or increase fatty acid oxidation).

The acetyl-CoA carboxylases (ACCs) are key enzymes that have emerged as attractive targets to mitigate fatty acid synthesis and increase fatty acid oxidation. The ACCs consist of three functional domains: the biotin carboxylase (BC) domain, the biotin carboxyl carrier protein (BCCP) domain, and the carboxy transferase (CT) domain.[3] There are two known isoforms of mammalian ACC – ACC1 and ACC2 (Figure 16.1).[4,5] Both isoforms catalyze

---

RSC Drug Discovery Series No. 27
New Therapeutic Strategies for Type 2 Diabetes: Small Molecule Approaches
Edited by Robert M. Jones

Published by the Royal Society of Chemistry, www.rsc.org

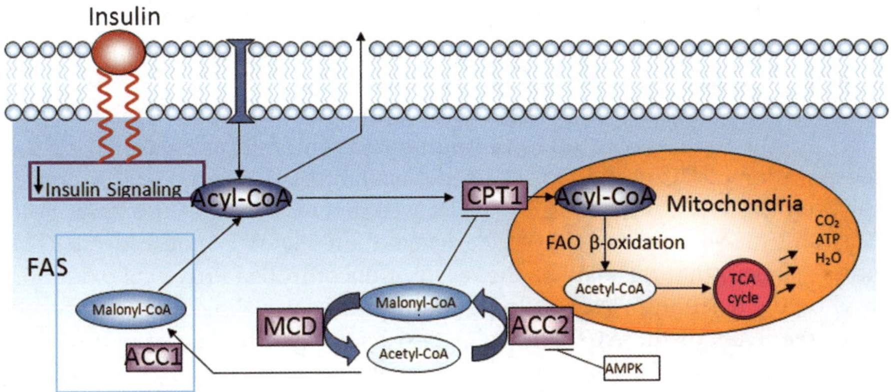

Figure 16.1 ACC1 is located in the cytosol and controls the rate-limiting step of fatty acid synthesis (FAS) by conversion of acetyl-CoA to malonyl-CoA. Excess free fatty acids (acyl-CoAs) have been shown to impair insulin signaling. ACC2 is associated with the mitochondrial membrane. Malonyl-CoA produced by ACC2 is thought to serve as a negative regulator of carityl palmitate transfer protein 1 (CPT1), which shuttles free fatty acyls across the mitochondrial membrane for subsequent fatty acid oxidation (FAO). ACC2 activity is inhibited by AMPK via phosphorylation at Ser 218. The supply of malonyl-CoA is also regulated by malonyl-CoA decarboxylase (MCD), which converts malonyl-CoA back to acetyl-CoA.

the conversion of acetyl-CoA to malonyl-CoA. ACC1 is primarily located in lipogenic tissue, where it is localized in the cytosol. ACC1 serves as the rate-limiting promoter of fatty acid synthesis by the production of malonyl-CoA.[4] ACC2, on the other hand, is primarily located in energy-expending tissue (muscle) and is associated with mitochondrial membrane, where it serves as a negative regulator of fatty acid uptake by the mitochondria by inhibition of carnitine palmitoyltransferase I (CPT1).[6] CPT1 is responsible for conjugation of free fatty acyls to carnitine for transfer across the mitochondrial membrane and subsequent β-oxidation.[7] Therefore, one could imagine that inhibition of the ACCs would significantly reduce the level of fatty acid metabolites *in vivo* by reducing the amount of fatty acids that are synthesized, as well as increasing the amount of fatty acids that are consumed by β-oxidation.

16.2   ACC2 Mouse Knockout Studies

16.2.1   Wakil's Studies

Initial interest in the ACCs as a therapeutic target class was stimulated by a series of publications from Wakil studying the phenotype of ACC2 knockout mice (ACC2$^{-/-}$ mice).[8] The biotin binding domain of mouse ACC2 was replaced with HTRP cassette and the resulting ACC2$^{-/-}$ mice were found to be viable. Malonyl-CoA levels in liver, heart, and muscle tissues were then measured. While there was no change in the malonyl-CoA level observed in the liver (consistent with the notion that ACC1 is the dominate isoform in liver), the malonyl-CoA levels observed in heart and skeletal muscle were reduced by $10\times$ and $30\times$ respectively when compared to wild-type (WT) mice. Upon closer examination, it was found that fatty acid synthesis, as measured by the incorporation of ^{14}C-acetyl-CoA, was identical in the WT and ACC2$^{-/-}$ mice. However, the livers of the ACC2$^{-/-}$ mice were lighter and had 20% lower lipid levels and 80–90% lower triglyceride levels than the WT mouse livers. The authors suggest that this may be the result of uncontrolled fatty acid oxidation cause by the deletion of ACC2. In addition to the lower lipid and triglyceride levels, the livers of the ACC2$^{-/-}$ mice also had a 20% reduction in glycogen content. There was a 20% reduction in blood glucose levels seen in the ACC2$^{-/-}$ mice relative to the WT mice, as well as reduced plasma fatty acid levels and higher triglyceride levels. The authors postulated that this may be due to increased triglyceride mobilization from the liver to supply increased oxidation in muscle. Fasted ACC2$^{-/-}$ mice also showed $4\times$ higher levels of β-hydroxybutyrate relative to WT mice, indicative of increase fatty acid oxidation. Indeed, when muscle samples were examined for fatty acid oxidation, measured by oxidation of ^{3}H-palmitate, there was a 30% increase in basal fatty acid oxidation seen in the ACC2$^{-/-}$ mice. Additionally, while treatment with insulin was shown to reduce fatty acid oxidation in WT mice by 45%, there was no such change observed in the ACC2$^{-/-}$ mice when treated with insulin. Wakil's group then studied the ACC2$^{-/-}$ mice in a 27-week feeding/body weight study. Over the course of the experiment, the ACC2$^{-/-}$ mice consumed 20–30% more food than the WT mice. Remarkably, despite the increase in food intake, the ACC2$^{-/}$$^{-}$ mice exhibited 10% lower body weight than the WT mice at the end of the study. There were also lower fat levels measured in the adipose tissue of the ACC2$^{-/-}$ mice than the WT mice.

In order to further profile the metabolic effect of ACC2 deletion, the ACC2$^{-/-}$ mice were placed on a high fat/ high carbohydrate diet (32% fat/38% carbohydrate) at 6–8 weeks of age and compared to WT mice placed on the same diet.[9] After 4 months on this diet, both male and female ACC2$^{-/-}$ mice were lower in weight than the corresponding WT mice (30 g versus 44 g for males, 20 g versus 26 g for females). This could largely be attributed to an increase in fat mass observed in the WT mice relative to the ACC2$^{-/-}$ mice. The mice were then subjected to a 12-hour fast followed by blood glucose measurement, which was determined to be 122 $\pm$ 28 mg/dL and 73 $\pm$ 20 mg/dL respectively for the WT and ACC2$^{-/-}$ mice. The insulin levels of the WT mice were also 5 times higher than those observed in the ACC2$^{-/-}$ mice (4 $\pm$ 1.6 ng/mL versus 0.8 $\pm$ 0.24 ng/mL), suggesting that the ACC2$^{-/-}$ mice were more insulin sensitive compared to the WT mice. This was confirmed with a glucose tolerance test, where the WT mice exhibited higher blood glucose levels than the ACC2$^{-/-}$ mice prior to glucose dosing (WT 173 $\pm$ 26 mg/dL versus ACC2$^{-/-}$ 106 $\pm$ 11 mg /dL) and relative glucose intolerance (WT versus ACC GTT AUC comparison) compared to the ACC2$^{-/-}$ mice. To investigate the mechanism of action of the improved metabolic phenotype from $ACC2^{-/-}$ deletion, soleus muscle samples were also taken from the animals and ^{3}H-palmitate oxidation was measured. Consistent with the previous study, there was a 20% increase in palmitate oxidation in the $ACC2^{-/-}$ mice relative to WT and the amount of oxidation observed in the $ACC2^{-/-}$ mice was not affected by the addition of insulin. A 2- to 3-fold increase in ^{3}H-palmitate oxidation relative to WT was also observed in hepatocyte samples from the $ACC2^{-/-}$ mice. Consistent with this observation, CPT1 activity was 40% higher in $ACC2^{-/-}$ mice versus WT mice. Interestingly, when the hepatocytes were treated with exogenous malonyl-CoA, the level of CPT1 activity of the $ACC2^{-/-}$ mice returned to the same level as in the WT animals.

Subsequent studies by Wakil also looked at the effect of ACC2 knockout in adipose tissue.[10] Samples of white adipose tissue from $ACC2^{-/-}$ mice and WT mice were found to have comparable levels of malonyl-CoA (1.00 $\pm$ 0.34 and 1.04 $\pm$ 0.30 ng/mmol respectively) The total ACC activity measured from the epididymal fat pads of $ACC2^{-/-}$ and WT mice were also similar (4.63 $\pm$ 1.09 versus 4.20 $\pm$ 1.23 nmol/min per mg of protein, respectively) and, taken together, these two pieces of data suggest that the majority of ACC activity in adipose tissue comes from ACC1. When cultured adipocytes from $ACC2^{-/-}$ mice and WT mice were assayed for ^{14}C-palmitate oxidation, however, the $ACC2^{-/-}$ mice showed an 80% increase in the amount of fatty acid oxidation relative to WT mice (3.12 $\pm$ 0.46 versus 1.71 $\pm$ 0.12 nmol/2 $\times$ 10^5 cells, respectively). Treatment of these adipocytes with insulin decreased the oxidation level for both the $ACC2^{-/-}$ mice and WT mice, but the $ACC2^{-/-}$ mice still showed a higher level of palmitate oxidation than the WT mice (2.20 $\pm$ 0.23 versus 1.47 $\pm$ 0.14 nmol/2 $\times$ 10^5 cells, respectively). Given this result, it is not surprising that when lipolysis was measure in adipocytes cultured from the epididymal fat pad, the rate of lipolysis was 2.4-fold higher in the $ACC2^{-/-}$ mice adipocytes than in the corresponding WT adipocytes (2.38 $\pm$ 0.38 versus 1.00 $\pm$ 0.33 nmol/2 $\times$ 10^5 cells, respectively). What is perhaps more remarkable, however, is that when ^{14}C-glucose oxidation was measured in adipocytes, it was found to be $\sim$ 50% higher in the $ACC2^{-/-}$ versus WT (16.75 $\pm$ 3.75 versus 11.23 $\pm$ 1.26 nmol/2 $\times$ 10^5 cells respectively). This would seem to be in violation of the Randle cycle hypothesis, which suggests that the observed increase in fatty acid oxidation should cause a corresponding decrease in glucose oxidation.[11]

The results of the previously discussed studies suggest that ACC2 knockout results in the increase of both fat and carbohydrate oxidation, which would in principle lead to an increase in whole body energy utilization. In a subsequent report, Wakil discussed this phenomenon in more detail.[12] First, the total energy utilization of $ACC2^{-/-}$ mice and WT mice on regular or high fat diets was measured, and it was determined that the $ACC2^{-/-}$ mice expended 15% more energy on regular chow (60% carbohydrate, 10% fat, 50% protein) and 19% more energy on high fat chow (24% carbohydrates, 55% fat, 21% protein) than the corresponding WT mice. Despite this difference on energy utilization, when the respiratory quotients (RQ) of $ACC2^{-/-}$ and WT of mice on regular chow were determined, they were found to be identical in both the light and dark phase. This suggests that any increase in fatty acid oxidation is occurring in conjunction with a similar increase in carbohydrate oxidation, as was suggested in the prior paper. A similar effect was observed when the animals were placed on high fat chow. Plasma ketones were also found to be elevated in the $ACC2^{-/-}$ mice, confirming that increased fatty acid oxidation was in fact occurring. To better determine whole body and tissue specific glucose uptake, a hyperinsulinemic euglycemic clamp study was performed using a radio-labeled glucose infusion. The $ACC2^{-/-}$ mice required a 2-fold increase in the rate of glucose infusion to maintain euglycemia, indicative of an increased insulin response relative to WT mice. The $ACC2^{-/-}$ mice also exhibited a 78% increase in insulin-induced suppression of hepatic glucose production, a 42% increase in whole body glucose uptake, and whole body increases in glycolysis (26%) and glycogen synthesis (75%). Glucose uptake by skeletal muscle and heart in the $ACC2^{-/-}$ mice was also increased by 66% and 100%, respectively. Interestingly, the $ACC2^{-/-}$ mice also showed a 26% reduction in plasma insulin levels, despite the fact that the insulin infusion rates were constant for both the $ACC2^{-/-}$ mice and WT mice. Long chain fatty acyl-CoA levels and triglyceride levels were observed to be lower in $ACC2^{-/-}$ mice than WT mice. Concentrations of diacylglycerol and the membrane/cytosol ratio of diacylgly-cerol were also decreased by $\sim 50\%$ in high fat fed $ACC2^{-/-}$ mice versus WT mice, although there was no change observed in ceramide levels. Long chain acyl-CoAs, diacylglycerol, and ceramides have been proposed to activate the serine/threonine kinase cascade (through kinases such as PKCΘ and PKCε), which may contribute to insulin resistance.[13] Supporting this hypothesis, PKCΘ and PKCε activities were 30–40% lower in muscle and liver, respectively, in the $ACC2^{-/-}$ mice versus WT mice.

In summary, the $ACC2^{-/-}$ mice generated by Wakil show a profoundly beneficial metabolic phenotype. Basel levels of malonyl-CoA in heart and muscle were significantly reduced, and fatty acid oxidation was markedly increased in these tissues. This suggests that any production of malonyl-CoA in these tissues by ACC1 was not able to compensate for the deletion of ACC2 activity, possible due to local levels of malonyl-CoA decarboxylase in the mitochondria.[14] When placed on a high fat diet, the $ACC2^{-/-}$ mice showed improvement versus WT in virtually all metabolic parameters studied, suggesting that pharmacological inhibition of ACC2 might lead to a beneficial effect in diabetic patients.

16.2.2 Cooney's Studies

Subsequent to the publications from Wakil's group, Cooney and co-workers have reported an $ACC2^{-/-}$ mice model with a markedly different phenotype.[15] The $ACC2^{-/-}$ mice were generated in a different manner from the mice used in Wakil's lab. Exon 12 of the ACC2 gene (which contains the biotin carboxylase domain) was flanked with loxP sites and introduced to mice with the C57BL/6 background. These mice were crossed into a cre-deleter strain, which removed exon 12 and introduced an early stop codon. The cre was then bred out of the mice, and the subsequent offspring were confirmed to lack ACC2 protein. Malonyl-CoA levels in the $ACC2^{-/-}$ mice were found to be lower in heart and muscle. Isolated soleus muscle also showed a 57% increase in fatty acid oxidation, which is higher than the increase in fatty acid oxidation reported by Wakil (20%). $ACC2^{-/-}$ mice showed an increase in RQ versus WT mice in the dark phase (0.92 $\pm$ 0.02 versus 0.87 $\pm$ 0.02, respectively) and a trend towards an increase in RQ in the light phase (0.88 $\pm$ 0.02 versus 0.85 $\pm$ 0.02, respectively), which differs from Wakil's experiments that showed no difference in RQ value between $ACC2^{-/-}$ mice and WT mice. Additionally, Cooney's studies showed no change in overall energy expenditure between $ACC2^{-/-}$ mice and WT mice, again contradicting the results from Wakil's study. When placed on a high fat diet, there were no differences observed in food intake, body weight, fat deposition, insulin resistance, glucose and insulin tolerance, and glucose deposition into skeletal muscle during a glucose tolerance test. Taken as a whole, these results suggest that there is little metabolic benefit to ACC2 knockout. Cooney does not offer a definitive explanation for why there are such pronounced differences between the animals generated in his labs and Wakil's, but he suggests that it may be due to differences in the breeding protocols. For example, Cooney's KO mice were generated on pure C57BL6 background while Wakil's KO mice are in a mixed background. Cooney also points out that of the ~2000 mouse knockout models that have been examined, 30% exhibit a lean phenotype, suggesting that there may be other genetic components that may play a role in Wakil's animals.[16]

16.2.3 Lowell's Studies

Shortly after Cooney's report was published, a third $ACC2^{-/-}$ mouse was reported by Lowell and co-workers.[17] The strategy used to generate the knockout was similar to that employed by Cooney. The exon containing the biotin-binding domain as well the upstream exon were flanked by loxP inserts. The upstream exon was included to induce a frameshift mutation and nonsense mutation in any translated protein. In addition to a whole body knockout, the loxP-flanked animals were bred with a strain containing a muscle specific cre-recombinase, allowing them to generate muscle-specific ACC2 deletion. The muscle-specific $ACC2^{-/-}$ mice showed no change versus WT mice in body weight, food intake, body composition, insulin levels, or glucose levels. The global $ACC2^{-/-}$ mice showed a similar phenotype to the muscle-specific knockouts, with the exception that there was a slight increase in insulin levels relative to WT mice. Samples of soleus muscle from the global $ACC2^{-/-}$ mice showed a 40% reduction in total ACC activity relative to WT, but there was no difference in fatty acid oxidation levels or malonyl-CoA levels relative to the

WT mice. When the hearts of the global $ACC2^{-/-}$ mice were examined, there was a modest (30%) reduction in malonyl-CoA levels relative to WT mice. Serum triglycerides, non-esterified fatty acids, and β-hydroxybutyrate levels were all unchanged. A comparison of the expression levels of a subset of genes involved in metabolism between the $ACC2^{-/-}$ and WT mice was performed. There was no difference observed in the expression levels of pPARα, ACC1, or Pgc1α, but there was a decrease in the expression level of malonyl-CoA decarboxylase in the $ACC2^{-/-}$ mice. The $ACC2^{-/-}$ mice showed a lowering of RQ in both the light and dark phases, as well as a slight but non-significant increase on oxygen consumption relative to WT mice. However, when placed on a high fat diet (45% fat) for 11 weeks, the $ACC2^{-/-}$ mice did not show any difference from WT mice in body weight, insulin levels, or blood glucose levels. A separate cohort of animals placed on a high fat/high carbohydrate diet (45% fat, 35% sucrose) for 36 weeks showed a similar lack of differentiation between $ACC2^{-/-}$ and WT mice. As a final test, the $ACC2^{-/-}$ allele was bred into leptin-deficient mice that achieve non-diet induced obesity. Much the same as the other examples, these $ACC2^{-/-}$ mice show not measurable difference in body weight gain from the WT mice. Lowell postulates that since the Wakil knockout retained the majority of the ACC2 protein due to the fact that no stop codon was inserted into the construct, it may be the case that the resulting catalytically inactive, mutant ACC2 protein is exerting a dominant negative effect on ACC1. In addition, the reduction in expression of the malonyl-CoA decarboxylase, an enzyme involved in degradation of malonyl-CoA, could also contribute to the phenotype.

16.2.4 Summary of ACC2 Knockout Data

Three different strains of $ACC2^{-/-}$ mice have been generated (Figure 16.2). Given the remarkable difference in the phenotypes observed between Wakil's group and the groups of Cooney and Lowell, it is difficult from a genetic perspective to accurately determine the potential effect of ACC2 inhibition on metabolic parameters. Cooney's hypothesis on differences in the methodology used to generate the knockouts and Lowell's hypothesis that the Wakil animals may be exhibiting a dominant negative effect are potential explanations for the observed differences. Another consideration is that because the majority of the ACC2 protein is intact in Wakil's $ACC2^{-/-}$ mice, other uncharacterized

Group	Malonyl-CoA Levels	Fatty Acid Oxidation	Body Weight/Food Intake	Adiposity	Energy Expenditure	Insulin Sensitivity	Glucose Levels/GTT
Wakil	Muscle: 90% ↓ Heart: 90% ↓	Muscle↑ Liver ↑ Whole Body ↑	Body Wt ↓ Food Intake ↑	Fat mass ↓ Fatty liver ↓	Increased	Improved	Glucose: 20%↓ Improved GTT
Cooney	Muscle: 50% ↓ Heart: 60% ↓	Muscle↑ Whole Body ↑	No Changes	No Changes	No Change	No Change	No Changes
Lowell	Heart: 30% ↓	No changes	No Changes	No Changes	No Change	No Change	No Changes

Figure 16.2 Comparison of published ACC2 knockout phenotypes.

functions of the ACC2 protein may still be present in those $ACC2^{-/-}$ mice, but absent in the mice where the majority of the protein has been deleted. An interesting follow up experiment to these studies would be to do a more broad gene profiling experiment between all three of the knockouts and look for differences in expression of both genes known to effect metabolism as well as any other genes whose expression profile changed significantly.

16.3 ACC1 Knockout Studies

16.3.1 Total ACC1 Knockout

Full knockout of ACC1 in mice has been shown to be embryonically lethal.[18] The $ACC1^{+/-}$ heterozygotes have 50% less ACC1 mRNA, but this did not translate into any measurable difference in metabolic phenotype between the $ACC1^{+/-}$ and WT mice. The authors postulate that the embryonic lethality is due to a requirement for *de novo* fatty acid synthesis by the embryo in the early stages of development.

16.3.2 Tissue-Specific ACC1 Knockout: Wakil

After determining a global ACC1 knockout was embryonically lethal, Wakil and co-workers used loxP technology to create a liver-specific ACC1 knockout $(LACC1^{-/-})$.[19] C57BL/6 mice containing an ACC1 gene with loxP sites flanking exon 22 (which contains the biotin-binding domain) were crossed with mice containing cre-recombinase under the control of rat albumin promoter (liver specific cre), resulting in $LACC1^{-/-}$ mice. RT-PCR analysis confirmed >95% deletion exon 22, and the mutant protein was observed by Western blotting. When compared to WT mice, the $LACC1^{-/-}$ mice had a 75% reduction in total liver ACC activity and a 70% reduction in malonyl-CoA. Under non fasted conditions, there were no differences observed in the levels of blood glucose, insulin, triglycerides, ketone bodies, or cholesterol in $LACC1^{-/-}$ mice and WT mice. There was, however, a reduction in non-esterified fatty acids observed in the $LACC1^{-/-}$ mice versus WT mice (0.83 ± 0.07 versus 0.61 ± 0.05 mEq/L, respectively), as well as a reduction in liver triglycerides (7.8 ± 0.8 versus 0.48 ± 0.4 mg/g, respectively). When the mice were fasted for 24 h, there were no differences observed in any of the parameters measured. In order to better assess the differences between $LACC1^{-/-}$ mice and WT under conditions where fatty acid synthesis would be expected to be enhanced, the mice were placed on a fat-free diet for 10 days. After this feeding period, lipid accumulations in the livers of the $LACC1^{-/-}$ mice were 66% lower than in WT mice, and non-esterified fatty acid levels were 30% lower . However, there were no differences observed in blood glucose levels, liver glycogen levels, or fat pad weight. Blood insulin levels were slightly elevated in $LACC1^{-/-}$ mice compared to WT mice. Hepatocyte samples were collected from animals placed on the fat-free diet and the rate of fatty acid synthesis was measured by incorporation of ^{14}C-acetate. As expected, the rate of fatty acid synthesis was 50% lower in the LACC1$^{-/-}$ than in WT mice. This study concludes that tissue-specific ACC1 knockout in the liver reduces fatty acid synthesis, suggesting that malonyl-CoA synthesized by ACC2 in the liver cannot compensate for the deletion of ACC1.

16.3.3 Tissue-Specific ACC1 Knockout: Kusunoki

A second report characterizing a liver specific ACC1$^{-/-}$ mouse appeared from the Kusunoki lab[20] shortly after Wakil's publication. The knockout was generated using loxP flanking in a similar manner to the approach used by Wakil, with the exception that exon 46, located in the CT domain of ACC1, was targeted for deletion rather than exon 22. Interestingly, neither wild-type ACC1 nor the supposed mutant ACC1 were observed in the livers of the LACC1$^{-/-}$ mice, suggesting that the exon 46 deletion effects mRNA translation. The metabolic phenotype of this second LACC1$^{-/-}$ mouse model was observed to be significantly different than the phenotype reported by Wakil. There was no difference between the observed liver malonyl-CoA levels in the LACC1$^{-/-}$ or WT mice, in either a fed or fasted state. There was also no change observed in *de novo* fatty acid synthesis between the LACC1$^{-/-}$ or WT mice. However, ACC2 levels were increased in the liver samples from LACC1$^{-/-}$ mice relative to WT mice (2.6-fold increase in ACC2 mRNA and 1.4-fold increase in ACC2 protein). ACC2 activity was also increased 2.2-fold in the LACC1$^{-/-}$ livers. These observations suggested that ACC2 is being up-regulated to compensate for the deletion of ACC1. To confirm that the malonyl-CoA observed in the liver samples is in fact the result of ACC2 activity, isolated LACC1$^{-/-}$ hepatocytes were treated with 5-(tetradecycloxyl)-2-furancaboxylic acid (TOFA), a pan-ACC inhibitor. TOFA was found to inhibit *de novo* fatty acid synthesis in a dose-dependent manner, confirming that the observed fatty acid synthesis is in fact ACC2 dependent. This data argues against the hypothesis that there are separate malonyl-CoA pools produced by ACC1 and ACC2 and that up-regulation of ACC2 can, in fact, compensate for the absence ACC1.

16.3.4 Summary of ACC1 Knockout Data

While global knockout of ACC1 in mice proved to be lethal *in utero*, two liver-specific ACC1 knockout mouse models have been developed. The phenotypes of the two knockout strains differ markedly in their liver malonyl-CoA levels, ACC activity, and *de novo* fatty acid oxidation. The origin of this difference is not clear, although, as the two knockouts were generated in a different manner, it is possible that this is the cause of the discrepancy. Nevertheless, the knockout published by Kusunoki strongly suggests that there are not discrete malonyl-CoA pools generated by ACC1 and ACC2. Thus, it might be expected that in order to illicit a broad pharmacological effect with an ACC inhibitor, it would be required to inhibit both ACC1 and ACC2 rather than one of the individual isoforms.

16.4 ACC as a Target for Cancer Treatment

While the majority of therapeutic interest in ACC inhibition has focused on metabolic disease, there is also some indication that ACC inhibition could be useful in the treatment of cancer.[21] ACC1 has been shown to be up-regulated in a number of cancer types. Additionally, knockdown of ACC1 expression with siRNA in cancer cells has been shown to lead to oxidative stress and apoptosis in cancer cells, while having little effect on normal cells.[22,23] Known ACC inhibitors have been shown to be cytotoxic to LNCap, PC-3M, NCI-H460, HCT-8, and HCT-15 cells while having little effect on a non-cancerous cell line, strengthening the hypothesis that some cancer cells may be dependent on ACC1 activity for proliferation.[24,25] To date, no *in vivo* studies involving ACC inhibitors as cancer therapeutics have been reported.

16.5 ACC Inhibitors for the Treatment of Metabolic Syndrome

There has been a great deal of interest in the medicinal chemistry community, particularly in response to the initial work published by Wakil and co-workers,[8,9] in the development of ACC inhibitors as potential clinical agents for the treatment of metabolic syndrome and related disorders.[26–28] Rather than exhaustively detailing all of the reported ACC inhibitors from the primary literature, this chapter will focus primarily on those programs that have generated *in vivo* data with the ACC inhibitors that have been developed. In the aggregate, this approach should give the reader the most useful information in evaluating the potential therapeutic utility of ACC inhibition.

16.5.1 Pfizer's ACC Inhibitors

16.5.1.1 Anthracene Series

The first well-documented example of a drug discovery program targeting ACC inhibitors was reported by Harwood and co-workers in 2003.[29] A high-throughput screen was conducted using partially purified rat ACC1 (ACC1:ACC2 ratio 85:15). The screen was biased towards finding compounds that inhibited either citrate-mediated ACC activation or blocked the acetyl-CoA binding site by utilizing concentrations of the substrates at half maximal activation or activity (citrate = 2 mM, K_a = 2.7 mM; acetyl-CoA = 30 μM, K_m = 33 μM) while saturating concentrations of ATP and $KHCO_3$ were utilized (ATP = 4 mM, K_m = 43 μM; $KHCO_3$ = 1.5 mM, K_m = 2.5 mM). Presumably this was done to avoid generating ATP competitive hits, which might prove to have issues of selectivity over other ATP utilizing enzymes, such as kinases. The most "attractive" hit from the screening campaign was CP-610431 (**1**) (Figure 16.3). Compound **1** inhibited both ACC1 and ACC2 in a non-selective manner. When the enantiomers of **1** were separated, it was found that essentially all of the ACC inhibition was coming from the *R* enantiomer (compound **2**). Kinetic analysis of **2** showed that the compound bound reversibly to the protein and was non-competitive for acetyl-CoA, bicarbonate, and citrate, while the compound was uncompetitive for ATP. In addition to enzymatic activity, compound **2** also inhibited fatty acid synthesis (IC_{50} = 1.6 μM), triglyceride synthesis (IC_{50} = 1.8 μM), triglyceride secretion (IC_{50} = 3.0 μM), and ApoB secretion (IC_{50} = 5.7 μM) in HepG2 cells. While compound **2** proved to be a useful *in vitro* tool compound, it had a short half-life when incubated with liver microsomes from a variety of species (8.5 min human, 7.2 min rat, 7.2 min CD1 mouse). *N*-De-ethylation was found to be the major metabolic product (the *N*-de-ethylated product was significantly less effective as an ACC inhibitor), so an SAR campaign was undertaken to find a suitable replacement for the diethylamide. While a full account of this work has not been published, what has been disclosed is that the corresponding morpholino analog CP-640186 (compound **3**) achieved similar levels of enzymatic ACC inhibition as **2**, and it also possessed somewhat better microsomal stability, particularly in the case of rat (half-life data: 13.8 min human, >45 min rat, 21.8 min mouse). Compound **3** was shown (Table 16.1A) to lower fatty acid synthesis (IC_{50} = 0.62 μM) and triglyceride synthesis (IC_{50} = 1.8 μM) in HepG2 cells. In addition, **3** was also shown to lower *in vivo* fatty acid oxidation in Sprague Dawley rats, CD1 mice, and ob/ob mice. Tissue biopsies obtained from Sprague Dawley rats dosed with **3** also showed lowered malonyl-CoA levels, consistent with ACC inhibition (Table 16.1B).

Consistent with increased fatty acid oxidation, Sprague Dawley rats fed high carbohydrate chow showed a 64% reduction in RQ when dosed with 100 mg/kg of **3**, indicating a complete shift from carbohydrate to fat oxidation as the primary mode of metabolism. Taken in totality, this paper suggests treatment with **3** reduces fatty acid synthesis and increases fatty oxidation *in vivo*, consistent with the predicted effects of a dual ACC1/ACC2 inhibitor.

	R Group	Stereochemistry	Rat liver isozyme (ACC1)	Rat Muscle Isozyme (ACC2)
1	Ethyl	*R/S*	265	460
2	Ethyl	*R*	36	55
3	Morpholino (cyclized)	*R*	53	61

Figure 16.3 Inhibition of ACC activity in lead molecules (values reported as IC_{50}).

Table 16.1   (A) Inhibition of fatty acid synthesis by **3**; Dawley rats. (B) Inhibition of malonyl-CoA production in Sprague Dawley rat tissue biopsies.

(A)

Fatty acid Synthesis Inhibition (Species)	*ED_{50} (mg/kg)*
Sprague Dawley rat	13
CD1 Mouse	11
ob/ob mouse	4

(B)

Tissue Type	*ED_{50} (mg/kg)*
Liver	55
Heart	8
Soleus muscle	6
Quadriceps	15

Despite the encouraging effects seen with compound **3** in the previously described studies, there have been no additional reports in the primary literature describing broader metabolic effects (i.e. insulin lowering, improved glucose tolerance) of treatment with **3** in animals exhibiting a metabolically diseased state.[30] There have been two posters presented at scientific meetings, however, which detail the effect of chronic dosing of **3** in mice.

The first study was initiated by dosing male ob/ob mice for 6 weeks with 150 mg/kg/day of **3**.[31] Somewhat surprisingly, the mice dosed with **3** showed an increase in blood glucose levels and a decrease in insulin levels relative to control animals. Coupled with an increase in triglyceride and cholesterol levels, these results can be interpreted as a worsening of the diabetic phenotype. An observed increase in ketone bodies, as well as decreases in liver lipids and body weight, suggest that the ACCs are, in fact, being inhibited. It was postulated that the high compound dose may have been responsible for the undesired effect, so two additional studies were performed dosing ob/ob mice at 50 and 100 mg/kg/day respectively. However, similar effects of blood glucose elevation, insulin lowering, and triglyceride elevation were seen in both dose groups. After 6 weeks of dosing, the 100 mg/kg/day group had a dose reduction to 20 mg/kg. This resulted in normalization of the blood glucose and triglyceride levels, but the insulin levels remained lower than the control group. The animals in the 100 mg/kg dose group showed a reduction in body weight relative to control, even after the dose was lowered to 20 mg/kg. There was no difference observed in body weight in the 50 mg/kg dose group relative to control. Perhaps not surprisingly, given the observed effects on blood glucose and insulin levels, when the mice dosed with **3** at 50 and 100 mg/kg/day were subjected to an oral glucose tolerance test the observed glucose intolerance was greater than in the control animals. The poster concludes that the utility of ACC inhibition in diabetes is unclear.

Rather than using ob/ob mice, the second reported study with chronic dosing of **3** utilized diet-induced obese (DIO) C57BL6 mice.[32] In this case, there was a significant reduction in body weight observed in the mice dosed with **3** at 50 mg/kg/day. There were no significant changes in blood or hepatic lipid parameters in the animals dosed with **3**. However, in a hyperinsulinemic euglycemic clamp study, there was a lower glucose infusion rate observed in the compound **3** treated group with no change observed in insulin levels. The authors conclude that **3** showed a significant effect in preventing obesity with marginal improvements in diabetes.

16.5.1.2 Spirochromanone Series

Subsequent to the pioneering studies on **3**, continued medicinal chemistry efforts at Pfizer resulted in the development of another series of ACC inhibitors based on a spirochromanone core (Figure 16.4).[33] High-throughput screening identified compound **4** as a promising hit compound, with an IC_{50} of 8950 nM against rat ACC1 (rACC1). Compound **4** was docked into a co-crystal structure of **3** complexed with the CT domain of yeast ACC.[34] It appeared from the overlay of **3** and **4** that the quinoline amide of **4** was interacting with the protein in a suboptimal manner. Indeed, when the quinoline was replaced with an anthracene ring, similar to **3**, the resulting compound **5** showed a significant improvement in rACC1 inhibition (IC_{50} = 325 nM). Wishing to move beyond the anthracene ring system, which was deemed to be unattractive, an extensive SAR investigation identified an indazole as a potential replacement (**6**, rACC1 IC_{50} = 3380 nM), albeit with a significant loss in potency that could not be rescued while retaining the methylquinoline ring system. It was postulated that the major shortcoming of **6**

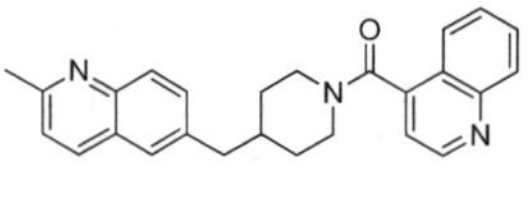

4
rACC1 IC_{50} = 8950 nM

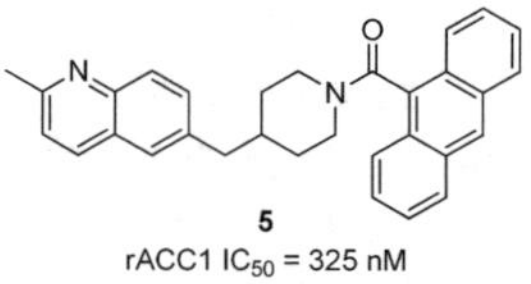

5
rACC1 IC_{50} = 325 nM

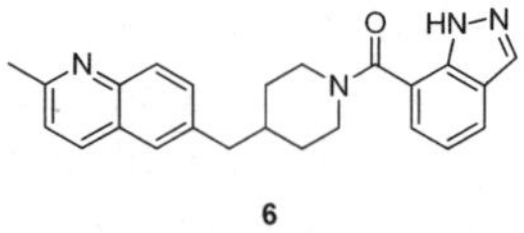

6
rACC1 IC_{50} = 3380 nM

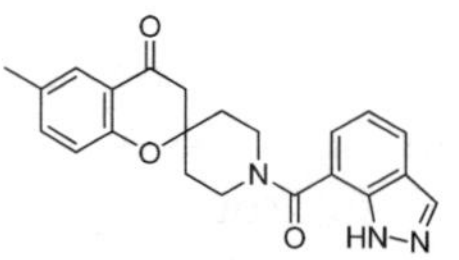

7
rACC1 IC_{50} = 634 nM
hACC2 IC_{50} = 641 nM

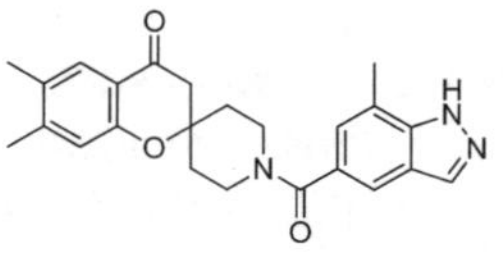

8
rACC1 IC_{50} = 22 nM
hACC2 IC_{50} = 48 nM

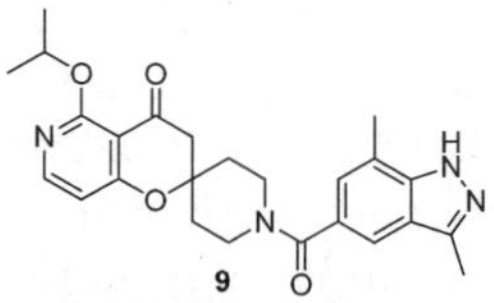

9
rACC1 IC_{50} = 12 nM
hACC2 IC_{50} = 20 nM

Figure 16.4 Development of spirochromanone **9**.

was that it was only able to form one hydrogen bond to the ACC protein (between the amide carbonyl and Glu-B2026), whereas compound **3** was able to form two hydrogen bonds (one between the anthracene amide and Glu-B2026 and one between the morpholine amide carbonyl and Gly-B1958). A *de novo* design effort was then undertaken to identify potential replacements for the methylquinoline ring of **6** which would be disposed to forming this second hydrogen bond. This resulted in the synthesis of spirochromanone **7**, which showed a significant improvement in ACC activity (rACC1 IC_{50} = 634 nM, hACC2 IC_{50} = 641 nM). A co-crystal structure of **7** with yeast ACC showed that **7** did in fact form the two key hydrogen bonds with Gly-B1958 and Glu-B2026, while also forming a third hydrogen bond between the indazole NH and Glu-B2026. While an interesting development, the benefit of this third hydrogen bond is likely negligible, as an internal hydrogen bond between Glu-B2026 and Arg-B1954 had to be broken to accommodate this new hydrogen bond. An extensive SAR investigation was then conducted around the spirochromanone core. It was found that indazole regioisomers such as **8** showed significant improvements in ACC inhibition relative to **7** (rACC1 IC_{50} = 22 nM, hACC2 IC_{50} = 48 nM). Further studies resulted in even more potent compounds, such as pyridine spirochromanone **9** (rACC1 IC_{50} = 12 nM, hACC2 IC_{50} = 20 nM). Compound **9** was further profiled in *in vivo* pharmacokinetic (PK) experiments in rat and dog. Compound **9** showed moderate to low CL in both species (31 $\pm$ 5 mL/min/kg in rat, 3.6 $\pm$ 0.4 mL/min/kg in dog). When assessing oral PK of **9** in mice, it was found that there was a high variability in bioavailability depending upon the vehicle used for formulation (Table 16.2). No further *in vivo* data was reported for **9**.

Work in the spirochromanone series was further detailed in two patent applications filed in 2009.[35,36] In particular, the second patent exemplified only one compound (**10**: rACC1 IC_{50} = 17.2 nM, hACC2 IC_{50} = 6.7 nM). In addition to enzymatic data, the patent included PD data showing inhibition of malonyl-CoA in liver and muscle at various doses (Table 16.3). A greater impact in malonyl-CoA levels is seen at a lower dose in the liver than in muscle; however, it is not clear what is causing this effect. A process chemistry route for the synthesis of multigram quantities of **10** was subsequently disclosed, and it was stated that **10** had been nominated as an early development candidate.[37] Subsequently, two additional applications exemplifying pyrazolospiroketone ACC inhibitors structurally related to **10** have been published.[38,39] The first application did not include any biological data, but the there were only four compounds specifically exemplified (Figure 16.5, **11–14**). The second application included hACC1 and hACC2 enzymatic IC_{50} values for the included compounds. Of the exemplified compounds, **15** and **16** showed the greatest enzymatic potency (**15**: hACC1 IC_{50} = 24.1 nM, hACC2 IC_{50} = 9.8 nM; **16**: hACC1 IC_{50} = 24.8 nM, hACC2 IC_{50} = 8.4 nM). While not as potent as **15** and **16**, compound **17** (hACC1 IC_{50} = 96.1 nM, hACC2 IC_{50} = 48.6 nM) was prepared in large scale (20.5 g) suggesting that this compound was studied in some detail.

Table 16.2 (A) i.v. dosing of **9** in rat (i.v. dose 1 mg/kg, p.o. dose 3 mg/kg). (B) i.v. dosing in male beagle (1 mg/kg).

(A)

i.v. Dosing

$T_{1/2}$ *(h)*	*CL (mL/min/kg)*	V_{dss} *(L/kg)*
1.4 ± 0.2	31 ± 5	3.3 ± 0.2

p.o. Dosing

Vehicle	C_{max}	*%F*
Methylcellulose	79 ± 19	32 ± 5
Lipid emulsion	13 ± 2	8 ± 0
Spray-dried dispersion	63 ± 21	52 ± 5

(B)

$T_{1/2}$ *(h)*	*CL (mL/min/kg)*	V_{dss} *(L/kg)*
6.1 ± 0.5	3.6 ± 0.4	1.6 ± 0

There have been no reports of any compound from Pfizer's program reaching clinical studies. However, Pfizer did initiate a clinical study to develop a malonyl-CoA biomarker assay.[40] The study took muscle biopsies from volunteers to assay malonyl-CoA levels. This study was discontinued due to adverse reactions to the biopsies (pain). Presumably, had this study been successful this assay would have been useful for monitoring the PD efficacy of ACC inhibitors in the clinic. The current status of Pfizer's ACC program is unknown.

Table 16.3 Percent inhibition of malonyl-CoA by **10** relative to control.

10

rACC1 IC_{50} = 17.2 nM

hACC2 IC_{50} = 6.7 nM

Dose *(mg/kg)*	*Muscle Malonyl-CoA (quadriceps)*	*Liver Malonyl-CoA*
1	2 ± 5	35 ± 2.6
3	24 ± 6.8	54 ± 1.2
10	49 ± 2.4	71 ± 3.5
30	57 ± 1.0	64 ± 3.6

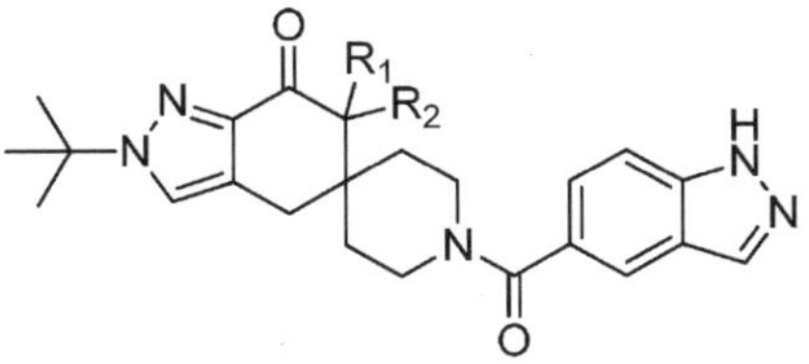

11 R$_1$ = H, R$_2$ = H
12 R$_1$ = H, R$_2$ = Me
13 R$_1$ = Me, R$_2$ = H
14 R$_1$ = Me, R$_2$ = Me

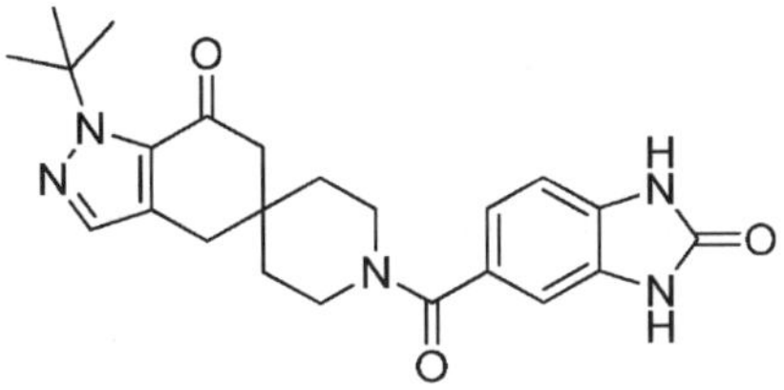

15
IC$_{50}$ hACC1 = 24.1 nM
IC$_{50}$ hACC2 = 9.8 nM

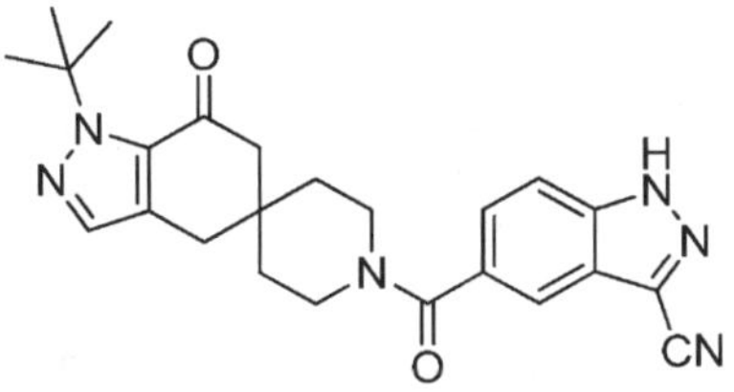

16
IC$_{50}$ hACC1 = 24.8 nM
IC$_{50}$ hACC2 = 8.4 nM

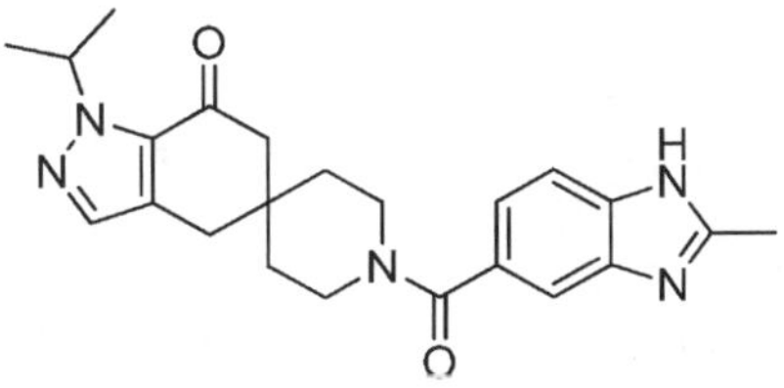

17
IC$_{50}$ hACC1 = 96.1 nM
IC$_{50}$ hACC2 = 48.6 nM

Figure 16.5 Pyrazolospirotetone ACC inhibitors.

16.5.2 Abbott's ACC Inhibitors

Researchers from Abbott have published a series of papers disclosing a set of alkynyl-heterocyclic ACC inhibitors with significant selectivity for ACC2 versus ACC1. The initial high-throughput screening hit was hydroxyurea **18** (hACC1 IC$_{50}$ = 1.0 μM, hACC2 IC$_{50}$ = 80 nM) (Figure 16.6), which was originally synthesized as part of Abbott's 5-lipoxygenase inhibitor program.[41] This was the first ACC inhibitor reported with significant selectivity for ACC2 versus ACC1. Owing to concerns surrounding the embryonic lethality of ACC1 knockout in mice,[18] the Abbott team chose to focus on further improving the ACC2/ACC1 selectivity window. Replacement of the hydro-xyurea with a urea (**19**: hACC1 IC$_{50}$ ≥ 30.0 μM, hACC2 IC$_{50}$ = 80 nM) resulted in increased ACC2/ACC1 selectivity relative to **18**, while replacement of the urea with an acetamide (**20**: hACC1 IC$_{50}$ = 260 nM, hACC2 IC$_{50}$ = 11 nM) showed similar ACC2/ACC1 selectivity to **18**, albeit with increased relative potency. Truncation of the phenoxy group of **20** to an isopropoxy group improved the ACC2/ACC1 selectivity (**21**: hACC1 IC$_{50}$ ≥ 30.0 μM, hACC2 IC$_{50}$ = 19 nM), while, interestingly, further truncation to an isopropyl group resulted in **22**, which was the only compound in this report with preferential selectivity for ACC1 versus ACC2 (hACC1 IC$_{50}$ = 1.2 μM, hACC2 IC$_{50}$ ≥ 30.0 μM). Replacement of the acetamide with a methylcarbo-

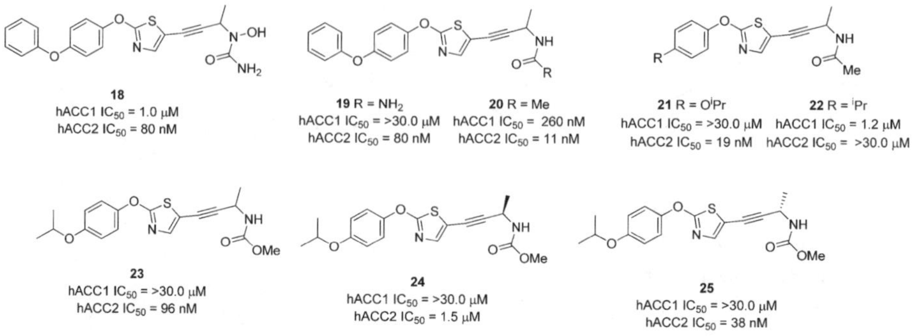

18
hACC1 IC_{50} = 1.0 µM
hACC2 IC_{50} = 80 nM

19 R = NH_2
hACC1 IC_{50} = >30.0 µM
hACC2 IC_{50} = 80 nM

20 R = Me
hACC1 IC_{50} = 260 nM
hACC2 IC_{50} = 11 nM

21 R = O^{i}Pr
hACC1 IC_{50} = >30.0 µM
hACC2 IC_{50} = 19 nM

22 R = ^{i}Pr
hACC1 IC_{50} = 1.2 µM
hACC2 IC_{50} = >30.0 µM

23
hACC1 IC_{50} = >30.0 µM
hACC2 IC_{50} = 96 nM

24
hACC1 IC_{50} = >30.0 µM
hACC2 IC_{50} = 1.5 µM

25
hACC1 IC_{50} = >30.0 µM
hACC2 IC_{50} = 38 nM

Figure 16.6 Alkynylthiazole ACC2 selective inhibitors.

nate resulted in **23** (hACC1 $IC_{50} \geq 30.0$ µM, hACC2 $IC_{50} = 96$ nM), which possessed good ACC2/ACC1 selectivity, in addition to having promising *in vivo* rat PK properties (5 mg/kg oral dose: $\%F = 80$, $C_{\max} = 0.90$ µg/mL, AUC = 18.9 µg·h/mL). The enantiomers of **23** were separated (**24** and **25**), and it was found that the *S* enantiomer **25** was responsible for the majority of ACC2 inhibition (**24**: hACC1 $IC_{50} \geq 30.0$ µM, hACC2 $IC_{50} = 1.5$ µM; **25**: hACC1 $IC_{50} \geq 30.0$ µM, hACC2 $IC_{50} = 38$ nM). Both **24** and **25** were advanced to an *in vivo* pharmacodynamic study in Sprague Dawley rats. Treatment with **25** at 10 and 50 mg/kg resulted in a 36% and 54% reduction in muscle malonyl-CoA respectively (measurements taken 3-h post dose). There was a slight, but significant, reduction in liver malonyl-CoA levels at the 50 mg/kg dose (26%), but no change observed at 10 mg/kg dose. As ACC2 is the dominant isoform expressed in muscle, while ACC1 is the dominant isoform expressed in liver, the observed *in vitro* selectivity translated to the *in vivo* experiment. Not surprisingly, the less active enantiomer **24** showed no effect in the *in vivo* study.

In a follow-up paper, it was disclosed that **25** was advanced to an anesthetized rat cardiovascular safety study to gauge the potential of **25** for further development.[42] During the course of the study, it was found that treatment with **25** led to severe cardiovascular events, as well as seizure. Other alkynyl thiazoles profiled in this assay showed similar effects. Through a series of SAR investigations, it was found that replacement of the alkyne functional group with a five-membered heterocycle such as thiazole (**26**) or isoxazole (**27**) could ameliorate the observed toxicity (Figure 16.7). Similar plasma levels and brain to plasma ratios were observed for **25–27**, so it was deemed unlikely that the observed toxicities were simply due to CNS effects. A gene expression profiling experiment was then undertaken to examine the effect of 3 days of dosing with **25** or **27** on cardiac gene expression pattern (**25** and **27** possess nearly identical pharmacokinetic properties) where it was found that treatment with **25** and **27** result in distinctly different cardiac gene expression patterns (data not shown). In fact, the gene expression pattern of the animals treated with **25** was observed to be similar to the gene expression pattern induced by treatment with doxorubicin, a chemotherapeutic agent with known cardio-

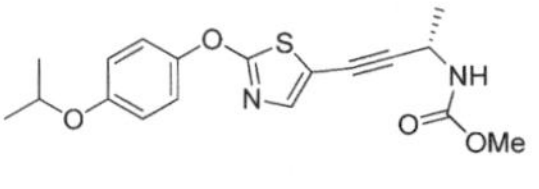

25
hACC1 IC_{50} = >30.0 μM
hACC2 IC_{50} = 38 nM
Brain to plasma ratio: 1.7
plasma conc: 75 μg/mL

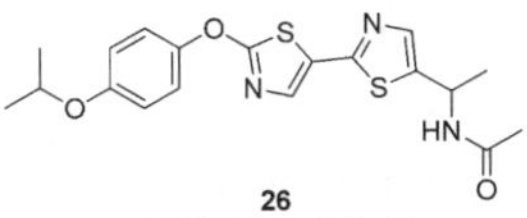

26
hACC1 IC_{50} = 210 nM
hACC2 IC_{50} = 28 nM
Brain to plasma ratio: 2.5
plasma conc: 62 μg/mL

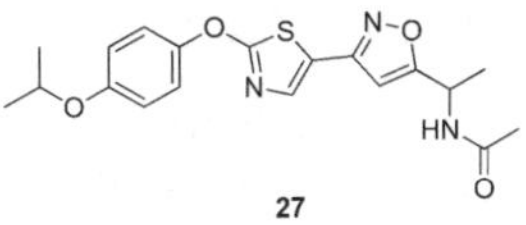

27
hACC1 IC_{50} = 93 nM
hACC2 IC_{50} = 8 nM
Brain to plasma ratio: 0.54
plasma conc: 80 μg/mL

Figure 16.7 Heterocyclic analogs of alkynyl thiazoles.

vascular liabilities. In a more broad look at gene expression pattern change, it was found that **25** impacted mitochondrial oxidative phosphorylation pathway, such as NADH dehydrogenase (ubiquinone), ubiquinol-cytochrome c reductase, cytochrome c oxidase, and ATP synthase. Again, these are observations that are similar to those seen upon treatment with doxorubicin, but no such effects were seen when animals were treated with **27**. The authors concluded that the observed toxicities were likely due to off mechanism effects, as **25** and **27** possess similar levels of ACC inhibition, but have profoundly different gene expression effects.

In a subsequent experiment, both **24** and **25** were advanced to a 2-week *in vivo* efficacy study in ob/ob mice.[43] It was expected that **25** might show an improvement in metabolic properties due to ACC 2 inhibition, and that **24** would not do so, thus serving as a negative control. However, much to the surprise of the authors, treatment with both **24** and **25** led to significant reductions in both glucose and triglyceride levels. In order to better understand this unexpected observation, a gene profiling experiment was undertaken. Livers from Sprague Dawley rats treated with **24** and **25** were assayed for gene expression changes, and the observations were compared to the DrugMatrix database, which contains expression profiles for rats treated with approx. 1000 different compounds.[44] The observed gene expression changes were most consistent with those seen from treatment with peroxisome proliferator-activated receptor (PPAR)-α activators, such as bezafibrate, clofibrate, fenofibrate, and nafenopin. As PPARα activators are known to improve diabetic phenotypes in rodents, such activity would clearly confound the ability to accurately assess these compounds *in vivo*.[45] In a follow-up experiment using immunohistochemistry staining to detect peroxisome proliferation, it was found that treatment with **24** and **25** did in fact result in increased peroxisome levels. Thus, the efficacy seen by dosing **24** and **25** in ob/ob mice is likely the result of PPARα activation. Interestingly, when human hepatocytes were treated with **24** and **25**, no evidence of PPARα activation was observed, suggesting that the observed induction may be rodent specific.

Three additional papers report more extensive SAR investigations in this series, but they do not include any additional *in vivo* data.[46–48] The current status of Abbott's ACC program is unknown.

16.5.3 Tashio's ACC Inhibitors

Tashio's efforts to develop a dual ACC1/ACC2 inhibitor were initiated by analysis of CP-640186 (Compound **3**) reported by Pfizer.[49] After screening their scaffold collection, the researchers found that compound **28** appeared to be an attractive starting point for further SAR investigations (Figure 16.8). Compound **28** shows reasonable hACC1/2 inhibition (IC_{50} = 239 nM, data is reported as human ACC1/2 inhibition, presumably partially purified tissue extract was being used as the ACC source); however, the compounds suffered from poor human microsomal stability (hMS) (19% remaining after 15 min incubation). Efforts were then undertaken to improve both the ACC potency and microsomal stability of this series (Figure 16.9). It was found that the morpholino group could be replaced with a variety of function groups, such as phenyl (**29**: hACC1/2 IC_{50} = 636 nM, hMS not determined). The anthracene group could also be replaced with a 2,6-diphenylpyridyl group (**30**: hACC1/2 IC_{50} = 693 nM, hMS = 96% remaining after 15 min), which retained much of the ACC enzymatic potency, but also showed a significant improvement in microsomal stability. Replacement of the phenyl amide with a 4-acylpiperidine group (**31**: hACC1/2 IC_{50} = 126 nM, hMS = 79% remaining after 15 min) resulted an improvement in ACC potency with a slight loss in microsomal stability. Most of the microsomal stability was rescued by 4-methylation of the phenyl rings of the 2,6-diphenylpyridine (**32**: hACC1/2 IC_{50} = 76 nM, hMS = 87% remaining after 15 min). Compound **32** was profiled more fully in a variety of assays, and was found to potently inhibit both recombinant hACC1 and hACC2 (IC_{50} = 101 nM and 23 nM, respectively). In a rat hepatocyte cellular assay, **32** also inhibited fatty acid synthesis (FAS) (IC_{50} = 340 nM) and increased fatty acid oxidation (EC_{50} = 580 nM). When administered to Sprague Dawley rats, **32** was also shown to have reasonable PK properties (10 mg/kg p.o. dosing: C_{max} = 107 ng/mL, T_{max} = 4 h, AUC = 1000 ng·h/mL).

In a subsequent disclosure, further efforts to improve the potency of **32** were reported.[50] Molecular modeling studies with **32** suggested that the one of the phenyl rings of the 2,6-diphenyl piperidine ring was not engaging the ACC protein in an optimal fashion. Therefore, an effort was undertaken to replace

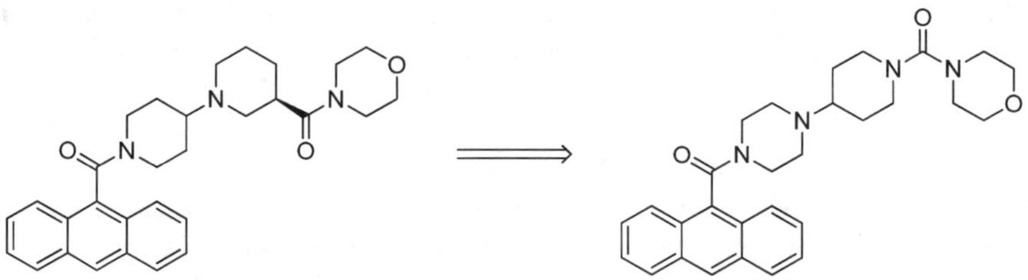

3
rACC1 IC_{50} = 53 nM
rACC2 IC_{50} = 61 nM

28
hACC1/2 IC_{50} = 239 nM
hMS = 19% remaining after 15 min

Figure 16.8 Rational design of **28**.

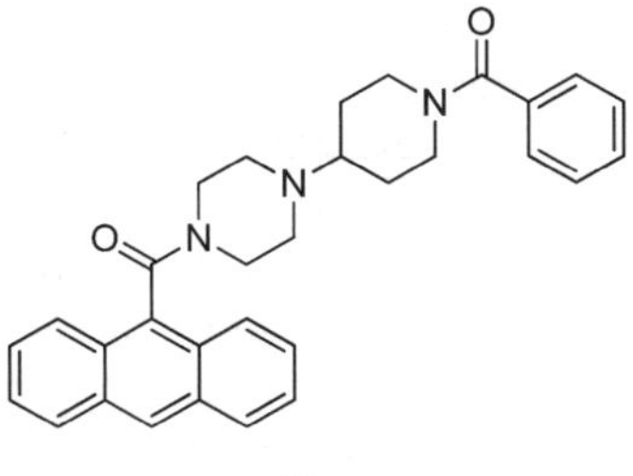

29
hACC1/2 IC$_{50}$ = 636 nM
hMS = not determined

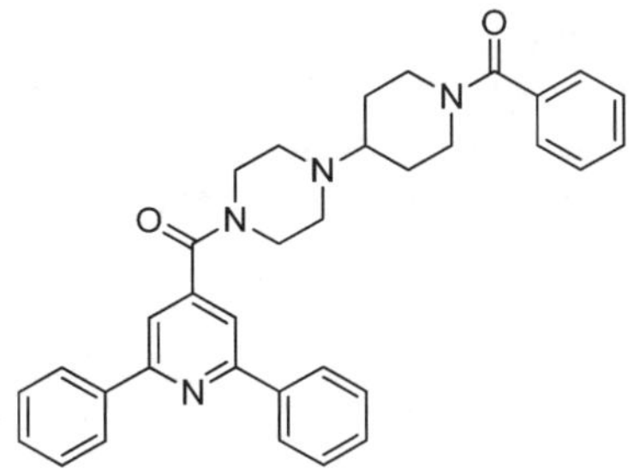

30
hACC1/2 IC$_{50}$ = 693 nM
hMS = 96% remaining after 15 min

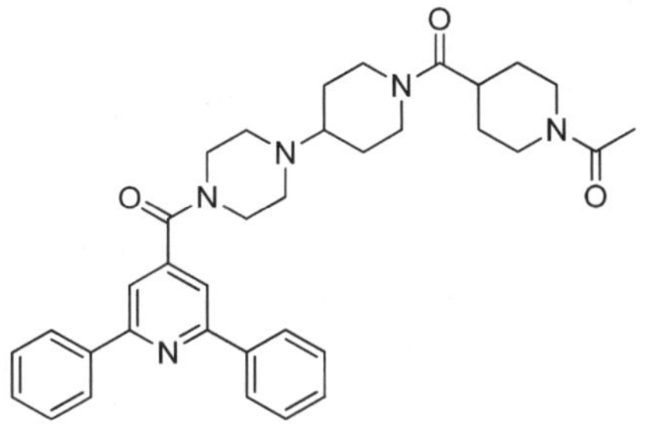

31
hACC1/2 IC$_{50}$ = 126 nM
hMS = 79% remaining after 15 min

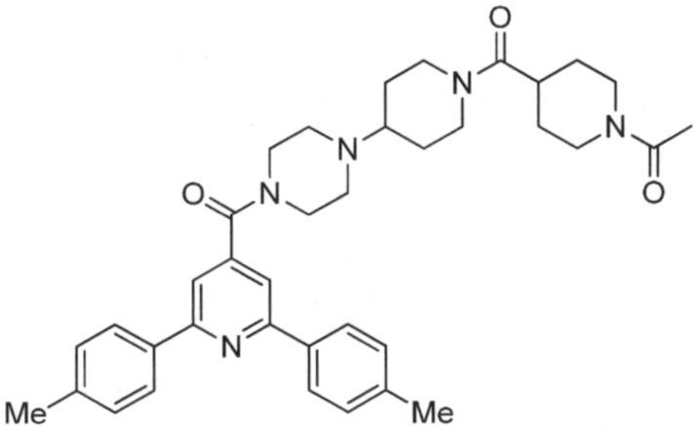

32
hACC1/2 IC$_{50}$ = 76 nM
hMS = 87% remaining after 15 min

Figure 16.9 Development of piperazine amide **32**.

or modify one of the phenyl ring (Figure 16.10). Substituting the 4-position of the phenyl ring with a 2-hydroxy ethyl group (**33**) resulted in an improvement in enzymatic potency relative to **32** (**33**: hACC1/2 IC$_{50}$ = 27 nM). However, when **33** was evaluated in its ability to inhibit FAS, it demonstrated a poor enzyme to cell shift (FAS IC$_{50}$ = 1622 nM). The cause for this shift is unclear, but it may be related to the relatively poor passive permeability of **33** relative to **32** (PAMPA permeability at pH 6.2 **32**: 61.5 $\times$ 10^{-6} cm/s, **33**: 6.7 $\times$ 10^{-6} cm/s). Further examination of changes to the phenyl ring led to the

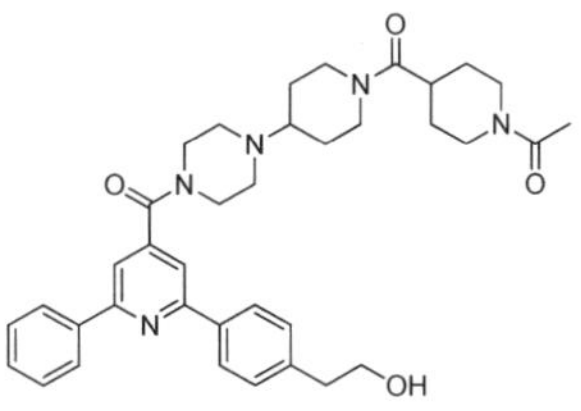

33
hACC1/2 IC$_{50}$ = 27 nM
FAS IC$_{50}$ = 1622 nM

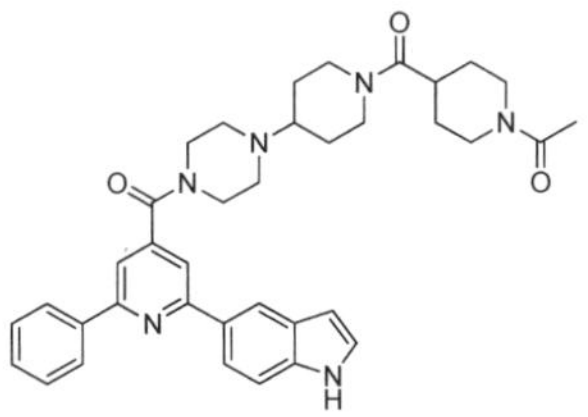

34
hACC1/2 IC$_{50}$ = 55 nM
FAS IC$_{50}$ = 467 nM

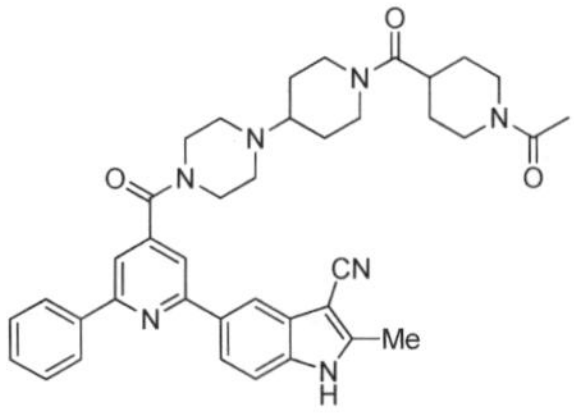

35
hACC1/2 IC$_{50}$ = 9.7 nM
FAS IC$_{50}$ = ND

Figure 16.10 Additional piperazine amide analogs.

identification of indole **34**. While there was a slight loss in enzymatic potency relative to **33** (**34** hACC1/2 IC_{50} = 55 nM), there was a much smaller cellular shift observed (FAS IC_{50} = 467 nM). The passive permeability of **34** was greater than that of **33** (PAMPA permeability at pH 6.2 = 46.5 $\times$ 10^{-6} cm/s), again suggesting that the cellular shift may be related to permeability. Additional SAR investigations around the indole ring were undertaken, and some compounds, such as **35**, showed improvements in enzymatic potency (hACC1/2 IC_{50} = 9.7 nM). However, no *in vivo* data was included in this report.

A more recent publication describes further modifications to this series of ACC inhibitors and includes *in vivo* efficacy data.[51] Further efforts to optimize/replace the 2,6-diphenylpiperazine group of **32** (possibly inspired by related work conducted at Takeda, see below) resulted in the replacement of the 2,6-diphenylpyridine group with a benzothiophene urea, giving compound **36** (hACC1/2 IC_{50} = 74 nM) which retained ACC activity while possessing a lower molecular weight (Figure 16.11). Efforts were then made to re-optimize the acylpiperazine amide portion of the molecule. Alkyl amides such as **37** retained good enzymatic potency; however, there was a significant cell shift seen in the hepatocyte FAS assay (hACC1/2 IC_{50} =49 nM, FAS IC_{50} = 309 nM). Somewhat surprisingly, of all the inputs reported, *t*-butoxycarbamate **38** possessed the best potency in both the enzymatic and cellular FAS assays (hACC1/2 IC_{50} = 24 nM, FAS IC_{50} = 79 nM respectively). Not surprisingly, given the classical use of a *t*-butoxycarbamate as an acid-labile protecting group, **38** suffered from poor acid stability (29% of compound remaining after 3 h incubation in aqueous solution at pH 1.2). An effort was undertaken to find acid-stable replacements for the *t*-butoxycarbamate, and the mono-fluoro

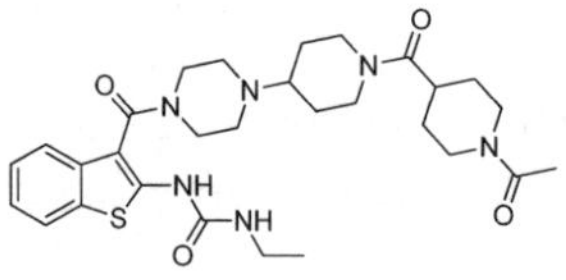

36
hACC1/2 IC_{50} = 74 nM

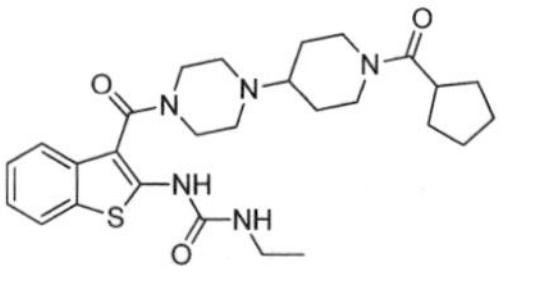

37
hACC1/2 IC_{50} = 49 nM
FAS IC_{50} = 309 nM

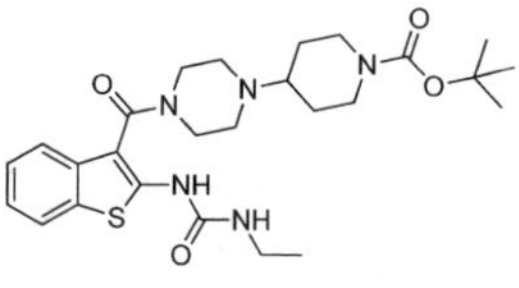

38
hACC1/2 IC_{50} = 24 nM
FAS IC_{50} = 79 nM
Acid stability = 29%

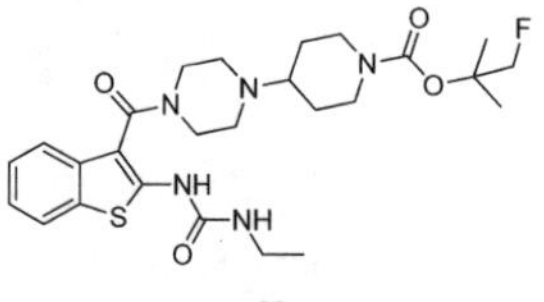

39
hACC1/2 IC_{50} = 17 nM
FAS IC_{50} = 61 nM
Acid stability = 98%

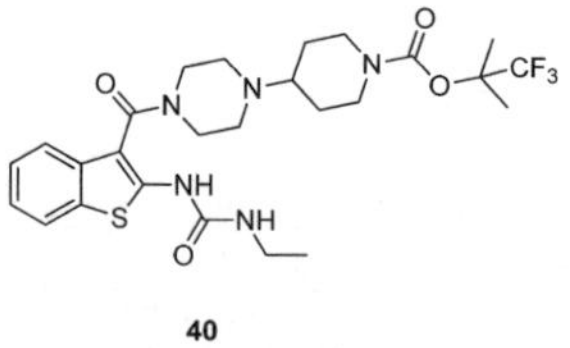

40
hACC1/2 IC_{50} = 31 nM
FAS IC_{50} = 81 nM
Acid stability = 100%

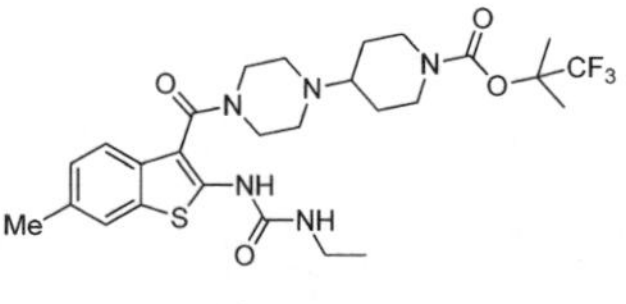

41
hACC1/2 IC_{50} = 58 nM
FAS IC_{50} = 58 nM

Figure 16.11 Benzthiazole analogs.

t-butoxy and trifluoromethyl *t*-butoxy analogs **39** and **40** were found to possess similar enzymatic and cellular potency to **38** (**39**: hACC1/2 IC$_{50}$ = 17 nM, FAS IC$_{50}$ = 61 nM; **40**: hACC1/2 IC$_{50}$ = 31 nM, FAS IC$_{50}$ = 81 nM) with essentially total acid stability. Changes to the substitution pattern of the benzothiophene ring were then examined. This resulted in the synthesis of compound **41**, which had a 1:1 cellular potency to enzymatic potency ratio, and was advanced for further profiling (**41**: hACC1/2 IC$_{50}$ = 58 nM, FAS IC$_{50}$ = 58 nM). Compound **41** was as a potent inhibitor of recombinant hACC1 and hACC2 (IC$_{50}$ = 192 nM and 95 nM respectively), as well as rACC1/2 (IC$_{50}$ = 32 nM). In addition to inhibiting FAS in hepatocytes, **41** also increased fatty acid oxidation (EC$_{50}$ = 370 nM). Compound **41** was found to have favorable oral PK properties in Sprague Dawley rats (10 mg/kg dose, C_{max} = 603 ng/mL, T_{max} = 5.33 h, AUC = 7480 ng·h/mL). Interestingly, the compound concentration observed in liver samples was significantly higher than the concentration observed in plasma (10,300 ng/g of tissue versus 435 ng/mL). Compound **41** was then advanced to an acute *in vivo* efficacy study in Sprague Dawley rats, where it was found to lower *de novo* fatty acid synthesis in liver by 74.6% 1-h post dose (10 mg/kg oral dosing). Based on those results, **41** was advanced to a long-term efficacy study in fructose drinking rats, a dietary model of hypertriglycemia. After 12 days of dosing at 10 mg/kg, plasma triglyceride levels were lowered by ~75% relative to control, and liver triglyceride levels were lowered by ~80% relative to control. Unfortunately, animals dosed with **41** exhibited considerable skin irritation around the nose and toes. The authors point out that lipid synthesis is required for normal skin development, and that due the suppression of fatty acid synthesis seen with **41**, it is possible that this is a mechanism-based effect. Further examination of the origin of this observed skin effect would be required to gauge its ramifications on the human therapeutic potential of ACC inhibition.

16.5.4 Torrent's ACC Inhibitors

Torrent has reported as series of spirochromanone ACC inhibitors.[52] These compounds were design as hybrids of **3** and known spirochromanone ACC inhibitors (reported by Merck-Banyu, see below),[53–55] and are also similar to compounds reported by Tashio (for example, **30**). The initial reported spirochromanone amide possessed modest inhibitor activity against rACC tissue extract (**42**, IC$_{50}$ = 120 nM) (Figure 16.12). However, select substitutions at the 6-position (**43**, **45**) or the 7-position (**44**) of the spirochromanone was shown to significantly improve enzymatic potency (**43**: rACC IC$_{50}$ = 6 nM, **44**: rACC IC$_{50}$ = 26 nM, **45**: rACC IC$_{50}$ = 14 nM). Replacement of the 2,6-diphenylpyridine amide with either a xanthene or thioxanthene amide led to reduced potency (**46**: r ACC IC$_{50}$ = 99 nM, **47**: rACC 64% inhibition at 10 μM).

Several of the most potent compounds synthesized were assayed for permeability to assess their utility for *in vivo* experiments. Compound **45** was found to have the best permeability (PAMPA, Pe 26.76 × 10^{-6} cm/s at pH

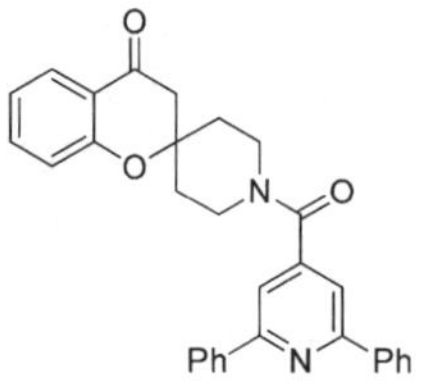

42
rACC IC$_{50}$ = 1200 nM

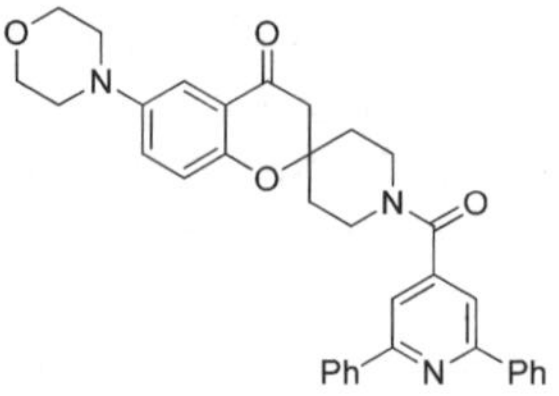

43
rACC IC$_{50}$ = 6 nM

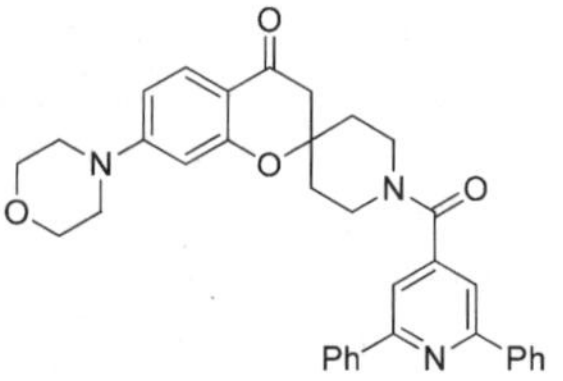

44
rACC IC$_{50}$ = 26 nM

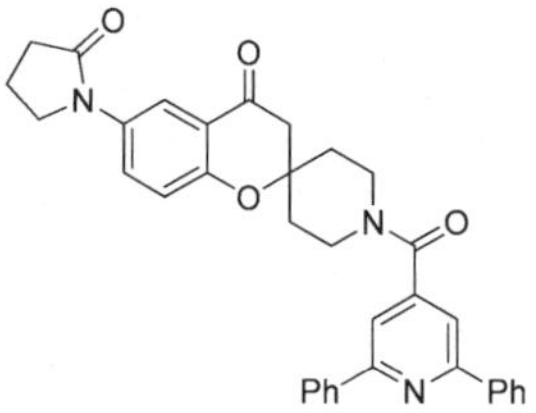

45
rACC IC$_{50}$ = 14 nM

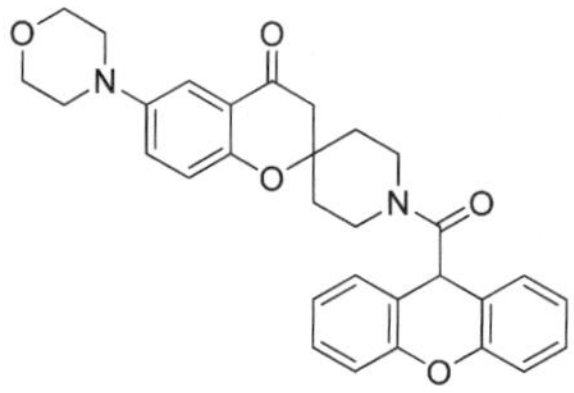

46
rACC IC$_{50}$ = 99 nM

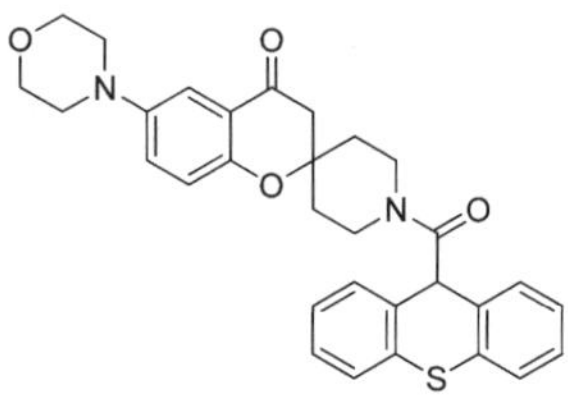

47
rACC 64% inhibition
at 10 μM

Figure 16.12 Torrent's ACC inhibitors.

6.8) was thus advanced to a RQ study in C57BL6 mice. Compound **45** was observed to cause a dose-dependent reduction in RQ relative to control mice (control RQ $\approx$ 0.92, **45**: 30 mg/kg RQ $\approx$ 0.85, 60 mg/kg RQ $\approx$ 0.83), indicating a shift in metabolic preference from carbohydrate oxidation to fatty acid oxidation, as would be expected from ACC 2 inhibition.

16.5.5 Cropsolutions' ACC Inhibitors

Cropsolutions has filed a patent application claiming compounds with a 4-oxo-3,4-dihydrospiro[benzo[e][1,3]oxazine-2,4'-piperidine] core as ACC inhibitors.[56] These compounds are clearly related to Pfizer's spirochromanone compounds (e.g. **9**) that have been previously discussed. Although a large number of compounds are included in the application, only six compounds have *in vivo* data included (Figure 16.13, **48–53**). Compound **51** was the most potent ACC enzymatic inhibitor included (IC$_{50}$ = 2 nM for both hACC1 and hACC2) and it also showed the greatest percent inhibition of fatty acid synthesis in rat (55% at 10 mg/kg dose, 78% at 30 mg/kg dose). Interestingly, compound **52**, while possessing somewhat lower ACC enzymatic activity than **51** (hACC1 IC$_{50}$ = 200 nM, hACC2 IC$_{50}$ = 21 nM), shows improved fatty acid synthesis inhibition at the 10 mg/kg dose (67%). No data for a 30 mg/kg dose was included. It is possible that the morpholine is improving the properties of **52** (solubility, permeability, etc.) resulting in a higher effective *in vivo* concentra-

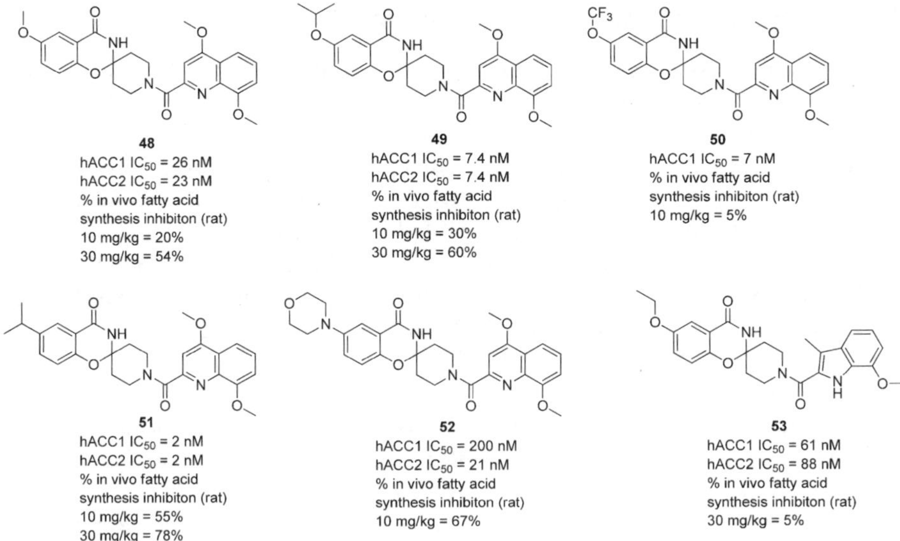

Figure 16.13   Cropsolutions ACC inhibitors.

tion. It is also interesting that **52** is the only compound included that shows appreciable ACC1/ACC2 selectivity.

In addition to ACC inhibition data, the patent also includes the effect of the treatment of four different cancer cell lines with ACC inhibitors (Table 16.4). All compounds included showed some inhibition of proliferation in all four cell lines at two different concentrations. It does not appear that any of the cell lines examined are more or less sensitive to ACC inhibition. None of the other active medicinal chemistry programs covered in this chapter have disclosed anti-proliferative data with ACC inhibitors.

Table 16.4   Percentage inhibition of proliferation of four cancer cell types relative to control upon treatment with two difference concentrations of ACC inhibitor for 72 h.

Compound	Concentration (μM)	A2780	HCT-116	MDA-MB-231	PC-3
48	10	57.6	59.4	50.2	47.7
	3.3	86.5	70.2	76.1	60.7
49	10	56.5	45	47.8	52.1
	3.3	80.5	57.2	81.4	54.7
50	10	65.5	41.6	29.5	55.7
	3.3	97.4	61.4	54.1	63.0
51	10	52.1	41.8	39.2	33.2
	3.3	59.5	53.6	74.2	48.9
52	10	56.4	58.5	56.1	39.0
	3.3	81.6	63.6	73.7	50.7
53	10	25.1	39.1	34.5	67.2
	3.3	65.0	64.0	88.9	76.5

16.5.6   Sanofi-Aventis' ACC Inhibitors

Sanofi-Aventis' ACC program was initiated by the discovery of acetal **54** in a high-throughput screening campaign targeting hACC2 (Figure 16.14).[57] While **54** inhibited hACC2 with an IC_{50} of 630 nM, it had no activity on hACC1, rACC1, or rACC2. The subsequent medicinal chemistry efforts were focused on developing dual ACC1/2 inhibitor with activity across different species. It was found that replacement of the naphthyl ring of **54** with a phenyl ring was tolerated, and, with transposition of the ether linkage on the alkyl side chain, resulted in **55**, which retained potency on hACC2 while showing activity, albeit somewhat weakly, on hACC1, rACC2, and rACC1 (**55**: hACC2 IC_{50} = 270 nM, hACC1 IC_{50} = 2.7 μM, rACC1 IC_{50} = 17 μM, rACC1 IC_{50} = 25 μM). Transposition of the ether from the *meta* to the *para* position, relative to the acetal, and the incorporation of an *ortho* pyridine resulted in **56**, which showed improved hACC2 activity with no significant improvement of hACC1, rACC2, and rACC1 potency (**56**: hACC2 IC_{50} = 70 nM, hACC1 IC_{50} = 14.6 μM, rACC1 IC_{50} ≥ 30 μM, rACC1 IC_{50} = 14 μM). Indeed, none of the reported acetal-containing compounds achieved IC_{50} values of <1 μM for hACC1, rACC2, or rACC1. However, replacement of the acetal with a phenyl ether, as well as the addition of a stereogenic methyl group alpha to the acetamide, resulted in compound **57**, which retained potency for hACC2 and showed improved potency for hACC1 and rat ACC1(**57**: hACC2 IC_{50} = 210 nM, hACC1 IC_{50} = 750 μM, rACC1 IC_{50} ≥ 30 μM, rACC1 IC_{50} = 2.1 μM). Truncation of the benzyl ether to isopropyl ether led to **58**, which inhibited all four ACC isoforms with IC_{50} values of <5 μM (**58**: hACC2 IC_{50} = 200 nM, hACC1 IC_{50} = 1.8 μM, rACC1 IC_{50} = 4.5 μM, rACC1 IC_{50} = 2.6 μM). Transposition of the pyridine nitrogen to the position *ortho* to the isopropyl ether, as well as isolating the *S*-enantiomer, which showed greater activity, resulted in compound **59**, which showed reasonable enzymatic inhibition for all

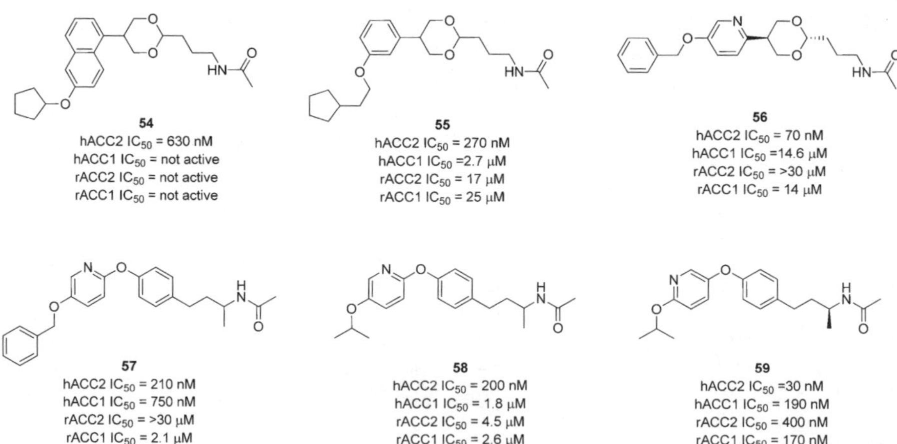

Figure 16.14   Sanofi-Aventis' ACC inhibitors.

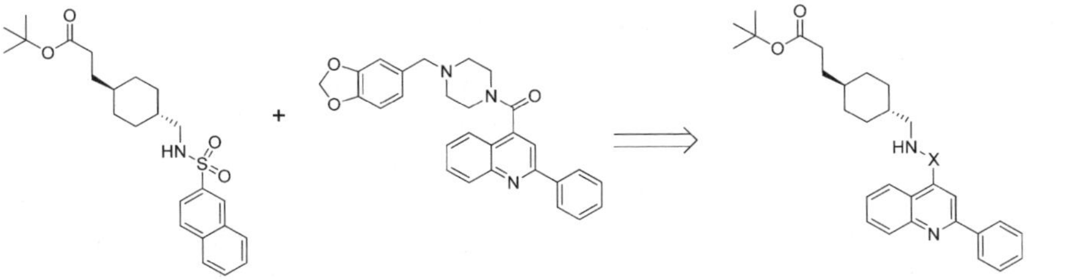

Figure 16.15 Mapping of binding features of hits **60** and **62** to achieve greater enzymatic potency.

four ACC isoforms assayed (**59**: hACC2 IC_{50} = 30 nM, hACC1 IC_{50} = 190 nM, rACC1 IC_{50} = 400 nM, rACC1 IC_{50} = 170 nM).

Compound **59** was assayed in a pharmacological safety panel and found to be inactive against 30 unrelated transporters, receptors, and enzymes. It was also inactive against PARR α, γ, and δ, as well as pyruvate carboxylase. Based on the ACC potency and selectivity against other targets, **59** was advanced to i.v. and p.o. rat PK experiments (Table 16.5). In the i.v. study, **59** displayed moderate clearance and V_{dss}. When dosed in an oral study, the C_{max} observed in liver samples was significantly higher than that observed in plasma. This was also the case when samples were taken 6-hours post dose (C_{6h}). Compound **59** was found to stimulate fatty acid oxidation in isolate human hepatocytes in a dose-dependent manner at 1, 3, and 10 μM concentrations. Dosing with 50 mg/kg of **59** in rats resulted in a decrease in RQ, suggesting an increase in fatty acid oxidation. Moreover, when administered to obese female ZDF rats at 30 mg/kg for 3 days, **59** resulted in an ~40% reduction in plasma triglyceride levels relative to vehicle, indicating an improvement in dyslipidemia.

Compound **47** has undoubtedly been a useful proof of concept molecule for Sanofi-Aventis' ACC program. The current status of the program is uncertain.

Table 16.5 i.v. and p.o. PK data for **59** in Wister rates.

Parameter	Plasma	Liver
i.v. dose (3 mg/kg)		
C (L/h/kg)	1.6	
V_{dss} (L/kg)	2.6	
$T_{1/2}$ (h)	1.5	
p.o. dose (10 mg/kg)		
C_{max} (μmol/L or kg)	3.50	21.8
t_{max} (h)	2.0	2.0
C_{6h} (μmol/L or kg)	1.59	7.59
$AUC_{(0-inf)}$ (μmol·h/L or kg)	19	120
$T_{1/2}$ (h)	1.8	3
%F	98	

16.5.7   Astra Zeneca's ACC Program

A series of cyclohexyl-based ACC1/2 dual inhibitors was recently reported by Astra Zeneca.[58] High-throughput screening seeking hACC2 inhibitors identified two chemical series, represented by **60** (hACC2 IC_{50} = 2.5 µM) and **61** (hACC2 IC_{50} = 3.6 µM), that were followed up on (Figure 16.15). It was postulated that **60** and **61** were occupying overlapping chemical space, and thus two hybrid molecules were prepared in an attempt to combine the key interactions from both parent molecules. The resulting compounds **62** and **63** both showed a significant improvement in enzymatic potency (hACC2 IC_{50} = 690 nM and 210 nM, respectively), particular in the case of amide **63**. Despite this promising enzymatic potency, **63** suffered from high lipophilicity (log D = 5.9) and high human plasma protein binding (unbound fraction = 0.1%), as well as low aqueous solubility (< 1.0 µM). While **63** exhibited good permeability in a caco-2 assay (P_{app} A-B 22 × 10^{-6} cm/s), it suffered from high human microsomal clearance (HLM Clint = 124 µL/min/mg) and high rat *in vivo* clearance (3.6 L/h/kg). Sulfonamide **62** possessed similar liabilities. In an effort to lower the lipophilicity of the series, with the hope of improving other properties in the process, the Astra-Zeneca team used calculated ligand lipophilicity efficiency (LLE) as a tracking tool for new compounds that were synthesized.[59,60] LLE is a composite value taking into account both potency and lipophilicity (in this case, LLE = pIC_{50} – log D). Thus, compounds with a larger LLE would be thought to be more ideal. Based on LLE, amide **63** was therefore prioritized over sulfonamide **62** (**62** LLE = −0.5, **63** LLE = 0.8).

As the phenylquinoline portion of **63** was a large contributor to lipophilicity, early efforts centered on attempting to modify/replace this portion of the molecule. Attempts to replace the quinoline, such as pyridine **64** (Figure 16.16), resulted in a significant loss in ACC potency (**64**: hACC2 IC_{50} = 4.5 µM, hACC1 IC_{50} = 6.6 µM). Replacement of the phenyl ring with more polar heterocycles, however, was more fruitful. In particular, pyridine **65** and pyrazole **66** both have comparable potency to **63**, but due to lower log D values, the LLE of these two compounds was significantly higher than **63** (**65**: hACC2 IC_{50} = 210 nM, hACC1 IC_{50} = 1.1 µM, LLE = 2.3; **66**: hACC2 IC_{50} = 570 nM, hACC1 IC_{50} = 2.3 µM, LLE = 2.7). Piperazine **67** is also noteworthy. Even though the ACC potency is somewhat reduced relative to **63**, the LLE is again considerably higher due to an almost three log difference in log D (**67**: hACC2 IC_{50} = 1.3 µM, hACC1 IC_{50} = 13 µM, LLE = 3.0). Despite the improvement in LLE, these compounds did not show any marked improvement in stability (for example, **65** HLM Clint = 146 µL/min/mg) and none of the initial phenylquinoline replacements showed improved potency for hACC2. At this point, competitive binding NMR experiments were run showing that **63** bound to the same portion of the ACC CT domain as **3**. It was hypothesized that the phenyl quinoline of **63** was occupying the sample space as the anthracene of **3**, which is relatively close to the hydrophilic surface of the protein. Therefore, it was believed that increased potency might be achieved by adding relatively small, polar groups to the phenylquinoline system that might

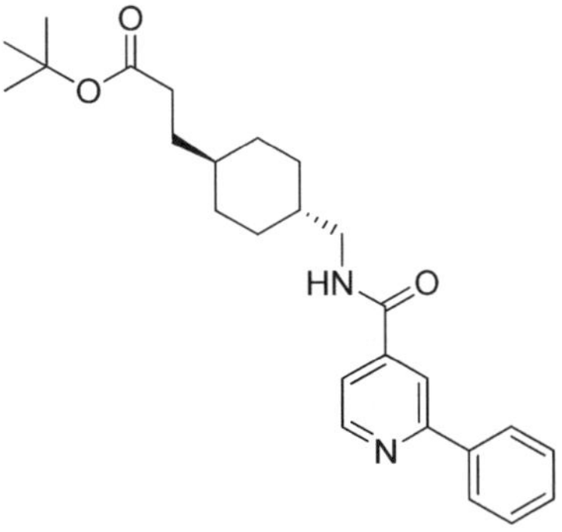

64
hACC2 IC_{50} = 4.5 μM
hACC1 IC_{50} = 6.6 μM
LogD 4.7
LLE 0.6

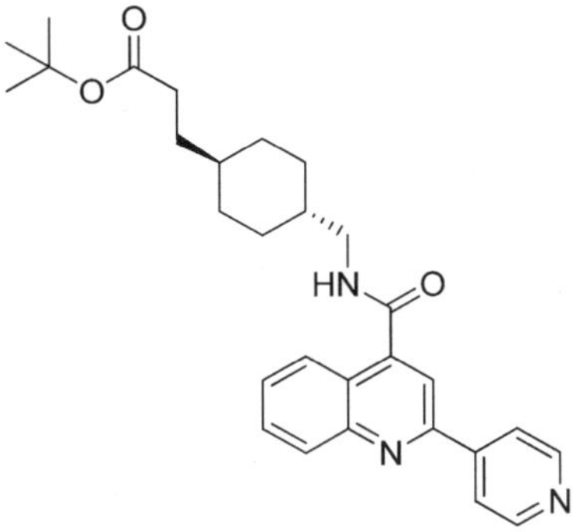

65
hACC2 IC_{50} =210 nM
hACC1 IC_{50} = 1.1 μM
LogD 4.4
LLE 2.3

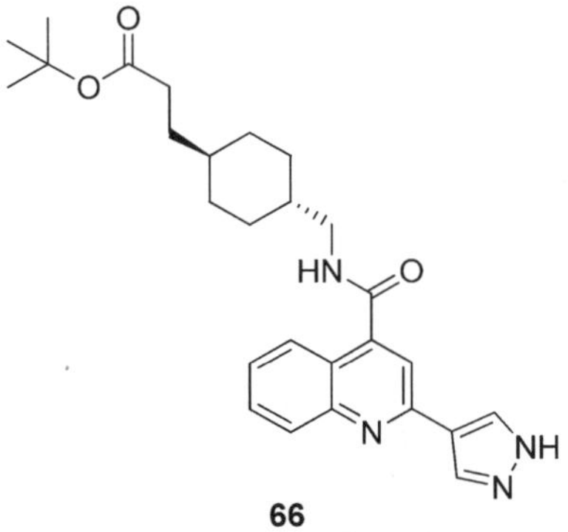

66
hACC2 IC_{50} =570 nM
hACC1 IC_{50} = 2.3 μM
LogD 3.5
LLE 2.7

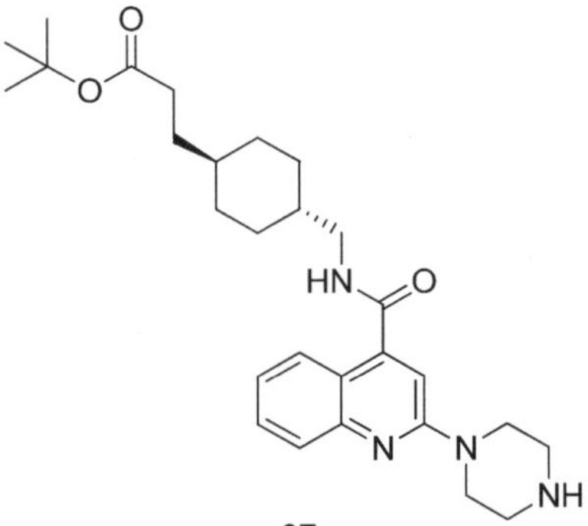

67
hACC2 IC_{50} = 1.3 μM
hACC1 IC_{50} = 13 μM
LogD 2.9
LLE 3.0

Figure 16.16 Astra-Zeneca's initial ACC inhibitors.

favorably interact with hydrophilic residues (Figure 16.17). This resulted in the benzyl amines **68**, which showed as significant improvement in hACC2 inhibition, LLE, and, for the first time in this series, inhibited hACC1 with an IC_{50} of <1 μM (**68**: hACC2 IC_{50} = 52 nM, hACC1 IC_{50} = 240 nM, LLE = 3.4). Gratifyingly, **68** also showed significant improvements in microsomal stability (HLM Clint = 14 μL/min/mg) and plasma protein binding (unbound fraction = 1.4%). However, the aqueous solubility of **68** was still modest (8 μM) and the compound also demonstrated significant inhibition of cytochrome p450 isoforms 2C9, 3A4, and 1A2 (IC_{50} = 0.8, 2.4, and 4.0 μM, respectively). Recalling the improved LLE of pyrazole **66** and piperazine **67**, it was postulated that addition of a basic amine side chain into each of the

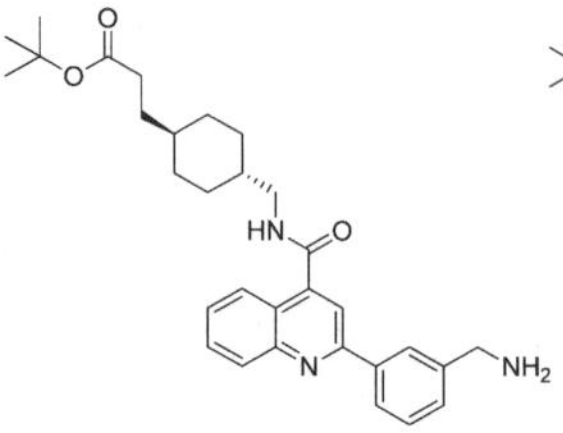

68
hACC2 IC$_{50}$ = 52 nM
hACC1 IC$_{50}$ = 240 nM
LogD 3.4
LLE 3.9

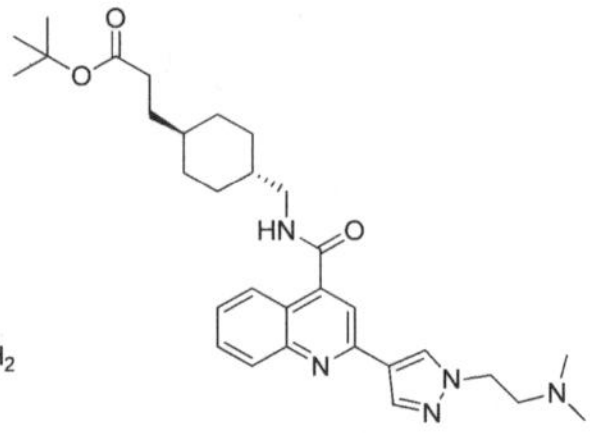

69
hACC2 IC$_{50}$ = 240 nM
hACC1 IC$_{50}$ = 1.3 μM
LogD 3.6
LLE 3.0

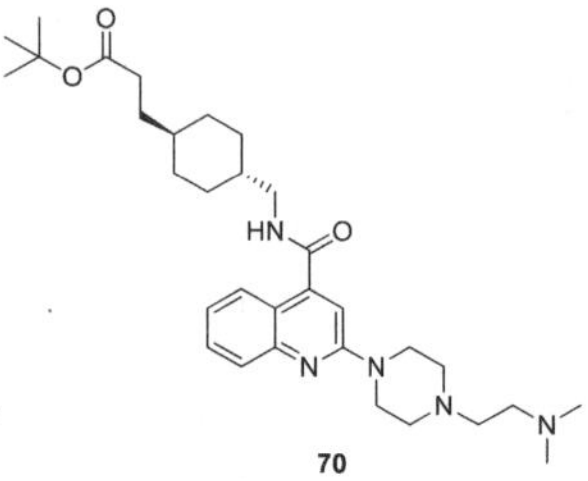

70
hACC2 IC$_{50}$ = 730 nM
hACC1 IC$_{50}$ = 5.3 μM
LogD 3.4
LLE 3.9

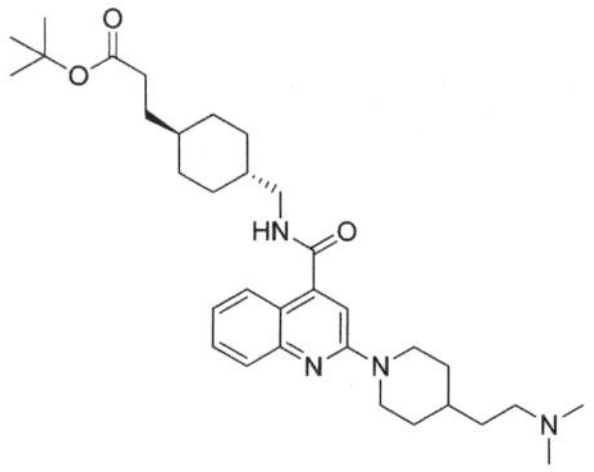

71
hACC2 IC$_{50}$ = 110 nM
hACC1 IC$_{50}$ = 1.2 μM
LogD 3.6
LLE 3.3

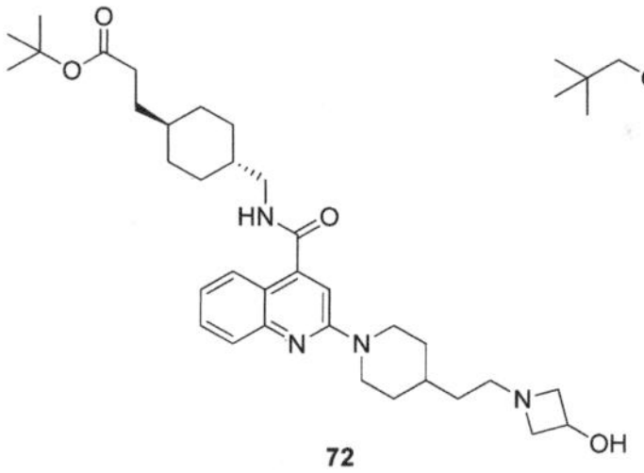

72
hACC2 IC$_{50}$ = 99 nM
hACC1 IC$_{50}$ = 1.0 μM
LogD 3.6
LLE 3.5

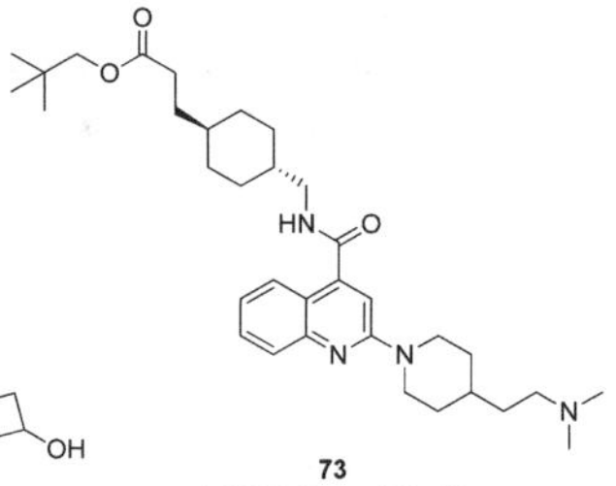

73
hACC2 IC$_{50}$ = 130 nM
hACC1 IC$_{50}$ = 1.0 μM
LogD 4.0
LLE 2.9

Figure 16.17   Astra-Zeneca's more potent ACC inhibitors.

heterocycles would lead to an increase in hACC2 potency, as was seen in **68**. This resulted in **69** and **70** which both showed reasonable hACC2 potency (**69**: hACC2 IC$_{50}$ = 240 nM, hACC1 IC$_{50}$ = 1.3 μM, LLE = 3.0; **70**: hACC2 IC$_{50}$ = 730 nM, hACC1 IC$_{50}$ = 5.3 μM, LLE = 3.9). Moreover, **70** showed significantly improved aqueous solubility (73 μM) and microsomal stability (HLM Clint < 12 μL/min/mg), and had no inhibitory activity against a panel of cytochrome P450 isoforms. Replacement of the piperazine with a piperidine (**71**) resulted in a further improvement in hACC2 potency (**71**: hACC2 IC$_{50}$ = 110 nM, hACC1 IC$_{50}$ = 1.2 μM, LLE = 3.3). The dimethyl amine of **71** could also be replaced with a azetiden-2-ol (**72**) without loss of potency (**72**: hACC2 IC$_{50}$ = 99 nM, hACC1 IC$_{50}$ = 1.0 μM, LLE = 3.5). A considerable effort was expended attempting to replace the acid labile *t*-butoxy carbonyl group. While neopentyl carbonate **73** retained most of the enzymatic potency, all other replacements that were examined resulted in a considerable loss of activity (**73**: hACC2 IC$_{50}$ = 130 nM, hACC1 IC$_{50}$ = 1.0 μM, LLE = 2.9). Although it was not discussed, given Tashio's success in replacing a *t*-butyoxy carbamate with either a mono-fluoro *t*-butoxy or a trifluoromethyl *t*-butoxy carbamates, it would be interesting to see if substitution with these functional groups would have resulted in enzymatic potency being retained.

Compounds **71** and **72** were advanced to *in vivo* PK experiments in C57BL6 mice(Table 16.6). Both compounds had moderate clearance, moderate to high

Table 16.6 *In vitro* PK for **71** (dosed 20 μmol/kg i.v., 20 μmol/kg p.o.)and **72** (dosed 5 μmol/kg i.v., 10 μmol /kg p.o.).

Parameter	71	72
CL (i.v.) mL/min/kg	46	33
V_{dss} (i.v.) (L/kg)	6.7	3.6
Oral $t_{1/2}$ (h)	4.5	4.3
%F	131	135
Free fraction (human) (% unbound)	3.4	3.3

volumes of distribution, reasonable oral half-life, and excellent oral bioavailability. They were inactive in a hERG binding assay, and they possessed reasonable unbound free fractions. The compounds were then assayed for enzymatic potency against rACC2 and rACC1 and found to have similar activity on both enzymes (**71**: rACC2 IC_{50} = 190 nM, rACC1 IC_{50} = 460 nM; **72**: rACC2 IC_{50} = 150 nM, rACC1 IC_{50} = 250 nM). Compounds **71** and **72** were then assayed in a liver PD experiment in obese Zucker rats. Obese Zucker rats exhibit elevated levels of hepatic malonyl-CoA relative to lean animals (obese 10 nmol/g, lean 1.0 nmol/g). After a 2-h continuous i.v. infusion, both **71** and **72** lowered hepatic malonyl-CoA levels in obese Zucker rats by $\sim$50% at concentrations 2–3 fold higher than the *in vitro* rACC2 IC_{50} when adjusted for free fraction.

By using LLE as a tracking measure, the Astra-Zeneca team was able to develop compounds like **71** and **72** which lowered hepatic malonyl-CoA levels in obese rats. Unfortunately, no data were included showing any improvement in metabolic parameters, such as plasma triglyceride lowering, making it impossible to frame the context of the observed PD effect in terms of efficacy.

16.5.8 Other ACC Inhibitors with *in vivo* Data: Soraphen A

In addition to the previously discussed programs targeting the development of ACC inhibitors, the natural product soraphen A (**74**, Figure 16.18) has also been found to be a useful ACC inhibitor. Compound **74** has been reported to a be a 1–5 nM inhibitor of hACC1 and hACC2.[30,61] It has also been shown to increase fatty acid oxidation in HepG2 cells. When dosed in Wistar rats, treatment with **74** results in increased palmitate oxidation, as well as increase lipid metabolism as measured by RQ. Compound **74** has been shown to bind to the BC domain of ACC crystallographically, making it the only reported ACC inhibitor to bind in this manner.[62] Recently, **74** has been dosed in a 6-week efficacy study in high-fat (HF) diet-induced obese C57BL6 mice.[63] Two dose groups of 50 and 100 mg/kg/day were utilized in the study. In addition to the mice dosed with **74**, there were two control groups; a lean, regular chow (RC) group and a HF diet obese group. The **74** dosed mice showed reduced weight gain relative to the HF group at both dose levels, with a more

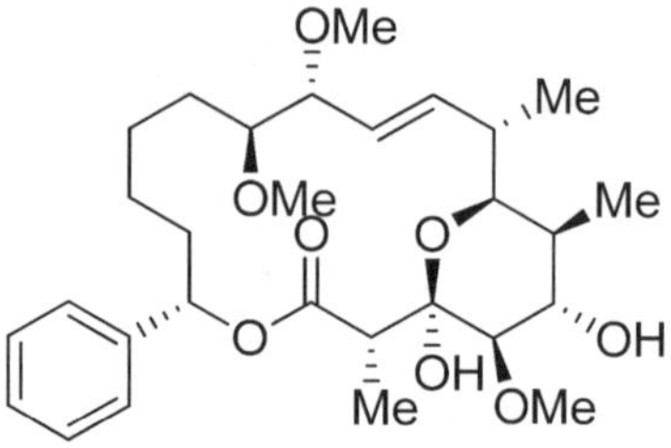

74

Figure 16.18   Soraphen A.

pronounced effect observed at the 100 mg/kd/day dose. While elevated insulin levels were observed in the HF control group, mice dosed with either dose of **74** had insulin levels that were on par with the RC mice. β-Hydroxybutyrate levels (a marker of fat metabolism) were reduced in the HF mice relative to the RC mice, and the animals in the two dose groups had β-hydroxybutyrate on par with the RC group, indicating that treatment with **74** shifted the metabolic preference of the animals towards fat utilization. When put in a euglycemic clamp study, the glucose infusion rates (GIR) of the both groups of the **74**-dosed animals were significantly higher than the HF animals, indicating that there was improved glucose tolerance. Somewhat surprisingly, though, during the clamp study, hepatic glucose production relative to basal was reduced by 50% in the group dosed with 50 mg/kg/day of **74** and the HF mice, where as the animals dosed with 100 mg/kg/day of **74** only exhibited a 10% reduction in hepatic glucose production. Furthermore, the level of hepatic malonyl-CoA was observed to be identical between the HF mice and the mice receiving 50 mg/kg/day of **74**, both of which were slightly lower that the levels of malonyl-CoA observed in the RC group. The mice in the 100 mg/kg/day dose group showed identical malonyl-CoA levels to the RC mice.

While there is clearly a favorable metabolic profile achieved by dosing HF mice with **74**, some of the observations warrant further examination. It is difficult to understand the observed differences in hepatic glucose production between the 50 and 100 mg/kg/day of **74** dose groups. Moreover, it is hard to imagine how an ACC inhibitor would result in an improvement in metabolic phenotype without resulting in an appreciable lowering of malonyl-CoA levels. One thing that is not clear from the paper is how much time elapsed between the administration of the last dose of **74** and when the mice were euthanized for tissue collection. It is possible that during this period of time there was a transient lowering of malonyl-CoA following the dosing of **74**, but that **74** had cleared from the animals systems and malonyl-CoA levels had thus returned to baseline levels. In order to better understand this, it would be helpful to run a time-course PD study using **74** to see if in fact there is some window of time after dosing when malonyl-CoA levels are reduced.

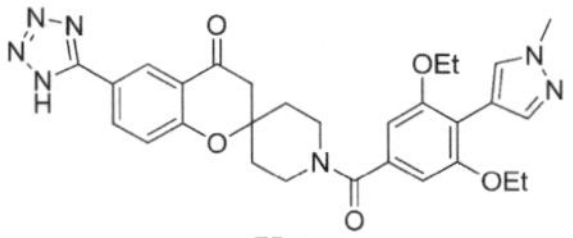

75
100% inhibition of ACC1
and ACC2 at 1.0 μM

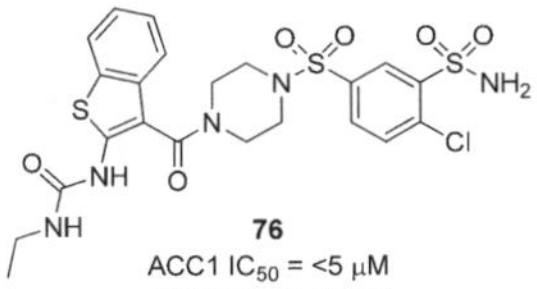

76
ACC1 IC_{50} = <5 μM
ACC2 IC_{50} = <5 μM

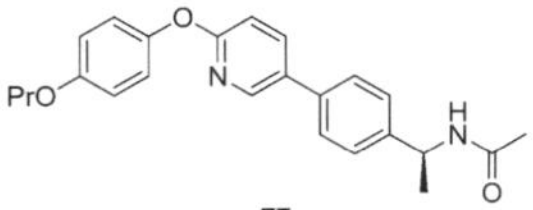

77
hACC1 IC_{50} = 35 nM
hACC2 IC_{50} = 8 nM

Figure 16.19 Additional ACC inhibitors.

16.5.9 Other ACC Inhibitors of Note

In addition to the previously discussed programs, there have been several other active programs directed at developing ACC inhibitors, albeit with no reported *in vivo* efficacy (Figure 16.19). Merck and Banyu have reported a series of spirochromanone based ACC inhibitors, exemplified by **75**, that are reported to inhibit enzymatic ACC activity.[53–55] However, percent enzyme inhibition is the only reported data (**75** is reported to inhibit 100% of ACC1 and ACC2 activity at 1 μM). A series of benzothiophene ACC inhibitors has been reported by Takeda.[64,65] Enzyme inhibition data is only reported in ranges (<1 μM, 1–10 μM, >10 μM), which makes it difficult to determine SAR trends (representative example **76**, for example, is reported to inhibit ACC1 and ACC2 with IC50s < 5 μM). However, there is a noticeable similarity between the exemplified compounds and benzothiophene piperazines (such as **36**) reported by Tashio,[51] so it is possible information about the SAR of the Takeda series could be extrapolated by comparing compounds with similar examples from Tashio's publications. BMS has reported a series of phenoxy amides, exemplified by **77**, that in the best cases inhibit hACC activity with IC_{50} values in the low nanomolar range (**77**: hACC1 IC_{50} = 35 nM, hACC2 IC_{50} = 8 nM).[66] These compounds are structurally similar to those published by Abbott and Sanofi-Aventis (see **39** and **59** for representative examples), so it is plausible that the compounds share the same binding pocket.

16.6 Conclusions

A considerable amount of effort has been devoted to the development of ACC inhibitors for the treatment of metabolic disease. Early efforts were driven by a favorable metabolic profile observed in ACC2 knockout animals, although subsequent knockout studies have called into question the benefit of ACC inhibition on improving metabolic endpoints. Based on results reported by the groups discussed above, a few general conclusions about the effectiveness of ACC inhibitors can be reached. First, ACC inhibitors have been shown to lower malonyl-CoA levels *in vivo*, using either an ACC1/2 dual inhibitor (Pfizer, Astra-Zeneca) or an ACC2 selective inhibitor (Abbott). ACC inhibitors have been shown to lower RQ *in vivo*, indicating a shift in metabolic preference from carbohydrates to fatty acids (Pfizer, Abbott, Torrent, Sanofi-Aventis). ACC inhibitors can decrease fatty acid synthesis (Pfizer,

Cropsolutions) and, in multidose experiments, it has been shown that plasma and liver triglyceride levels can be lowered by treatment with ACC inhibitors (Tashio, Sanofi-Aventis), with the caveat that in Tashio's study, skin toxicity was observed that may be related to decreased fatty acid synthesis. What has not been clearly demonstrated in the studies measuring plasma triglyceride levels is whether or not there is a clear correlation between pharmacodynamic coverage (malonyl-CoA lowering) and efficacy (triglyceride lowering).

While the effect of ACC inhibition on plasma triglyceride levels has been fairly clearly demonstrated, the effect of ACC inhibition of more downstream metabolic parameters, such as glucose and insulin levels, is less clear. In two long-term studies, it appeared that treatment with **3** resulted in either a modest benefit or possibly a detrimental effect. Long-term treatment with **74** resulted in improved glucose/insulin response; however, there was no evidence for significant pharmacodynamic coverage (malonyl-CoA lowering), making it difficult to gauge whether the observed insulin effects were ACC mediated or not.

The future of ACC inhibitors in the clinical treatment of metabolic disease is uncertain. Other than Pfizer's aborted study seeking to develop a malonyl-CoA biomarker in humans, no other clinical activity related to ACC inhibitors has been reported. Nevertheless, the interesting results showing *in vivo* triglyceride lowering that have appeared over the last 2 years are encouraging. Further work to better correlate pharmacodynamic coverage with efficacy, and additional studies looking at the effects of long-term dosing with ACC inhibitors on broader metabolic parameters may thus be warranted.

References

1. E. S. Ford, W. H. Giles and W. H. Dietz, *J. Am. Med. Assoc.*, 2002, **287**, 356.
2. D. B. Savage, K. F. Petersen and G. I. Shulman, *Physiol. Rev.*, 2007, **87**, 507.
3. S. E. Ploakis, R. B. Guchhait, E. E. Zwergel, M. D. Lane and T. G. Cooper, *J. Biol. Chem.*, 1974, **249**, 6657.
4. J. D. McGarry and D. W. Foster, *Annu. Rev. Biochem.*, 1980, **49**, 395.
5. K.-H. Kim, *Annu. Rev. Nutr.*, 1997, **17**, 77.
6. A. Bianchi, J. L. Evans, A. J. Iverson, A.-C. Nordlund, T. D. Watts and L. A. Witters, *J. Biol. Chem.*, 1990, **265**, 1502.
7. B. B. Rasmussen, U. C. Holmback, E. Volpi, B. Morio-Liondore, D. Paddon-Jones and R. R. Wolfe, *J. Clin. Invest.*, 2002, **110**, 1687.
8. L. Abu-Elheiga, M. M. Matzuk, K. A. H. Abo-Hashema and S. J. Wakil, *Science*, 2001, **291**, 2613.
9. L. Abu-Elheiga, W. Oh, P. Kordari and S. J. Wakil, *Proc. Natl. Acad. Sci.*, 2003, **100**, 10207.
10. W. Oh, L. Abu-Elheiga, P. Kordari, Z. Gu, T. Shaikenov, S. S. Chirala and S. J. Wakil, *Proc. Natl. Acad. Sci.*, 2005, **102**, 1384.

11. P. J. Randel, P. B. Garland, E. A. Newsholme and C. N. Hales, *Ann. N.Y. Acad. Sci.*, 1965, **131**, 324.

12. C. S. Choi, D. B. Savage, L. Abu-Elheiga, Z.-X. Liu, S. Kim, A. Kulkarni, A. Distefano, Y.-J. Hwang, R. M. Reznik, R. Codella, D. Zhang, G. W. Cline, S. Wakil and G. I. Shulman, *Proc. Natl. Acad. Sci.*, 2007, **104**, 16480.

13. G. I. Shulman, *J. Clin. Invest.*, 2000, **106**, 171.

14. G. W. Goodwin and H. Taegtmeyer, *J. Am. Physiol. Endocrinol.*, 1999, **277**, E772.

15. K. L. Hoehn, N. Turner, M. M. Swarbrick, D. Wilks, E. Preston, Y. Phua, H. Joshi, S. M. Furler, M. Larance, B. D. Hegarty, S. J. Leslie, R. Pickford, A. J. Hoy, E. W. Kraegen, D. E. James and G. J. Cooney, *Cell Metab.*, 2010, **11**, 70.

16. D. R. Reed, M. P. Lawler and M. G. Tordoff, *BMC Genet.*, 2008, **9**, 4.

17. D. P. Olson, T. Pulinilkunnil, G. W. Cline, G. I. Shulman and B. B. Lowell, *Proc. Natl. Acad. Sci.*, 2010, **107**, 7598.

18. L. Abu-Elheiga, M. M. Matzuk, P. Kordari, W. Oh, T. Shalkenov, Z. Gu and S. J. Wakil, *Proc. Natl. Acad. Sci.*, 2005, **102**, 12011.

19. J. Mao, F. J. DeMayo, H. Li, L. Abu-Elheiga, Z. Gu, T. Shaikenov, P. Kordari, S. S. Chirala, W. C. Heird and S. J. Wakil, *Proc. Natl. Acad. Sci.*, 2006, **103**, 8552.

20. N. Harada, Z. Oda, Y. Hara, K. Fujinama, M. Okawa, K. Ohbuchi, M. Yonemoto, Y. Ikeda, K. Ohwaki, K. Aragane, Y. Tamai and J. Kununoki, *Mol. Cell. Biol.*, 2007, **27**, 1881.

21. C. Wang, S. Rajput, K. Watabe, D.-F. Liao and D. Cao, *Frontiers in Bioscience*, 2010, 515.

22. V. Chages, M. Cambot, K. Moreau, G. M. Lenoir and V. Joulin, *Cancer Res.*, 2006, **66**, 5287.

23. K. Brusselmans, E. DeSchrijver, G. Verhoeven and J. V. Swinnen, *Cancer Res.*, 2005, **65**, 6719.

24. A. Beckers, S. Organe, L. Timmermans, K. Scheys, A. Peeters, K. Brusselmans, G. Verhoeven and J. V. Swinnen, *Cancer Res.*, 2007, **67**, 8180.

25. C. Wang, C. Xu, M. Sun, D. Luo, D. F. Liao and D. CaO, *Biochem. Biophys. Res. Commun.*, 2009, **385**, 302.

26. J. W. Corbett and H. J. Harwood, *Recent Pat. Cardiol.*, 2007, **2**, 162.

27. J. W. Corbett, *Expert Opin. Ther. Pat.*, 2009, **19**, 943.

28. M. P. Bourbeau, J. G. Allen. and W. Gu, *Annu. Rep. Med. Chem.*, 2010, **45**, 95.

29. H. J. Harwood, S. F. Petras, L. D. Shelly, L. M. Zaccaro, D. A. Perry, M. R. Makowski, D. M. Hargrove, K. A. Martin, W. R. Tracey, J. G. Chapman, W. P. McGee, D. K. Dalvie, V. F. Soliman, W. H. Martin, C. J. Mularski and S. A. Eisenbeis, *J. Biol. Chem.*, 2003, **278**, 37099.

30. H. J. Harwood, *Expert Opin. Ther. Targets*, 2005, **9**, 267.

31. J. L. Treadway, R. K. McPherson, S. F. Petras, L. D. Shelly, K. S. Fredrick, K. Sagawa, D. A. Perry and H. J. Harwood, presented at the 64th Scientific Sessions Meeting, June 4–8, 2004, Orlando, FL. Abstract Number 679-P. The abstract and the poster can be viewed at http://professional.diabetes.org

32. M. Schreurs, M. H. Oosterveer, T. H. v. Dikj, R. Havinga, D.-J. Reijngoud and F. Kuipers, presented at 43rd EASD Annual Meeting, September 17–21, 2007. Amsterdam. The abstract can be viewd at htpp://easd.org

33. J. W. Corbett, K. D. Freeman-Cook, R. E. F. Vajdos, F. Rajamohan, D. Kohls, E. Marr, H. Zhang, L. Tong, M. Tu, S. Murdande, S. D. Doran, J. A. Houser, W. Song, C. J. Jones, S. B. Coffey, L. Buzon, M. L. Minich, K. J. Dirico, S. Tapley, R. K. McPherson, E. Sugarman, H. J. Harwood and W. Esler, *Bioorg. Med. Chem. Lett.*, 2010, **20**, 2383.

34. H. Zhang, B. Tweel, J. Li and L. Tong, *Structure*, 2004, **12**, 1683.

35. J. W. Corbett, R. L. Elliott, K. D. Freeman-Cook, D. A. Griffith and D. P. Phillion, *Patent Application*, WO 2009/144554, 2009.

36. K. D. Freeman-Cook and B. M. Samas, *Patent Application*, WO 2009/144555, 2009.

37. C. Limberakis, S. Bader, L. M. Buzon, S. B. Coffey, B. G. C. J. W. Corbett, K. D. Freeman-Cook, K. E. Henegar, C. S. Jones, K. L. Nelson, D. P. Phillon, W. Song, J. D. Weaver and Z. Zhao, presented at the 240th ACS National Meeting, Boston, MA, United States, August 22–26, 2010.

38. S. W. Bagley, D. A. Griffith and D. W.-S. Kung, *Patent Application*, WO 2011/058437 A1, 2011.

39. S. W. Bagley, A. Griffith and D. W.-S. Kung, *Patent Application*, US 2011/0111046 A1, 2011.

40. http://www.clinicaltrials.gov/ct2/show/NCT01003444?term=B1501004&rank=1

41. Y. G. Gu, M. Weitzburg, R. F. Clark, X. Xu, Q. Li, T. Zhang, T. M. Hansen, G. Liu, Z. Xin, X. Wang, R. Wang, T. McNally, H. Camp, B. A. Beutel and H. L. Sham, *J. Med. Chem.*, 2006, **49**, 3770.

42. Y. G. Gu, M. Weitzberg, R. F. Clark, X. Xu, Q. Li, N. L. Lubbers, Y. Yang, D. W. A. Beno, D. L. Widomski, T. Zhang, T. M. Hansen, R. F. Keyes, J. W. Waring, S. L. Carroll, X. Wang, R. Wang, C. H. Healan-Greenberg, E. A. Blomme, B. A. Beutel, H. L. Sham and H. S. Camp, *J. Med. Chem.*, 2007, **50**, 1078.

43. J. F. Waring, Y. Yang, C. H. Healan-Hreenberg, A. L. Alder, R. Dickinson, T. McNally, X. Wang, M. Weitzberg, X. Xu, A. Lisowski, S. E. Warder, W. G. Gu, B. A. Zinker, E. A. Blomme and H. S. Camp, *J. Pharmacol. Exp. Ther.*, 2008, **324**, 507.

44. B. Ganter, S. Tugendreich, C. I. Pearson, E. Ayanoglu, S. Baumhueter, K. A. Bostian, L. Brady, L. J. Browne, J. T. Calvin, G.-J. Day, N. Breckenridge, S. Dunlea, B. P. Eynon, L. M. Furness, J. Ferng, M. R. Fielden, S. Y. Fujimoto, L. Gong, C. Hu, R. Idury, M. S. B. Judo, K. L. Kolaja, M. D. Lee, C. McSorley, J. M. Minor, R. V. Nair, G. Natsoulis, P.

Nguyen, S. M. Nicholson, H. Pham, A. H. Roter, D. Sun, S. Tan, S. Thode, A. M. Tolley, A. Vladimirova, J. Yang, Z. Zhou and K. Jarnagin, *J. Biotechnol.*, 2005, **119**, 219.

45. E. H. Koh, M. S. Kim, J. Y. Park, H. S. Kim, J. Y. Youn, H. S. Park, J. H. Youn and K. U. Lee, *Diabetes*, 2003, **52**, 2331.

46. R. F. Clark, T. Zhang, Z. Xin, G. Liu, Y. Wang, T. M. Hansen, X. Wang, R. Wang, X. Zhang, E. Frevert, H. S. Camp, B. A. Beutel, H. L. Sham and Y. G. Gu, *Bioorg. Med. Chem. Lett.*, 2006, **16**, 6078.

47. R. F. Clark, T. Zhang, X. Wang, R. Wang, X. Zhang, H. S. Camp, B. A. Beutel, H. L. Sham and Y. G. Gu, *Bioorg. Med. Chem. Lett.*, 2007, **17**, 1961.

48. X. Xu, M. Weitzberg, R. F. Keyes, Q. Li, R. Wang, X. Wang, X. Zhang, E. U. Frevert, H. S. Camp, B. A. Beutel, H. L. Sham and W. G. Gu, *Bioorg. Med. Chem. Lett.*, 2007, **17**, 1803.

49. T. Chonan, T. Oi, D. Yamamoto, M. Yashiro, D. Wakasug, H. Tanaka, A. Ohoka-Sugita, F. Io, H. Koretsune and A. Hiratate, *Bioorg. Med. Chem. Lett.*, 2009, **19**, 6645.

50. T. Chonan, H. Tanaka, D. Yamamoto, M. Yashiro, T. Oi, D. Wakasugi, A. Ohoka-Sugita, F. Io, H. Koretsune and A. Hiratate, *Bioorg. Med. Chem. Lett.*, 2010, **20**, 3965.

51. T. Chonan, D. Wakasugi, D. Yamamoto, M. Yashiro, T. Oi, H. Tanaka, A. Ohoka-Sugita, F. Io, H. Koretsune and A. Hiratate, *Bioorg. Med. Chem.*, 2011, **19**, 1580.

52. P. Shinde, S. K. Srivastava, R. Odedara, D. Tuli, S. Munshi, J. Patel, S. P. Zambad, R. Sonawane, R. C. Gupta, V. Chauthaiwale and C. Dutt, *Bioorg. Med. Chem. Lett.*, 2009, **19**, 949.

53. T. Yamakawa, H. Jona, K. Niiyama, K. Yamada, T. Iino, M. Ohkubo, H. Imamura, J. Shibata, J. Kusunoki and L. Yang, *Patent Application*, WO 2007/011809 A1, 2007.

54. T. Yamakawa, H. Jona, K. Niiyama, K. Yamada, T. Iino, M. Ohkubo, H. Imamura, J. Kusunoki and L. Yang, *Patent Application*, WO 2007/011811 A1, 2007.

55. T. Iino, H. Jona, H. Kurihara, M. Nakamura, K. Niiyama, J. Shibata, T. Shimamura, H. Watanabe, T. Yamakawa and L. Yang, *Patent Application*, WO 2008/088692 A2, 2008.

56. R. Anderson, S. Breazeale, T. Elich and S.-F. Lee, *Patent Application*, US 2010/0009982 A1, 2010.

57. S. Keil, M. Mller, G. Zoller, G. Haschke, K. Schroeter, M. Glien, S. Ruf, I. Focken, A. W. Herling and D. Schmoll, *J. Med. Chem.*, 2010, **53**, 8679.

58. C. Bengtsson, S. Blaho, D. Blomberg Saitton, K. Brickmann, J. Broddefalk, Ö. Davidsson, T. Drmota, R. Folmer, K. Hallberg, S. Hallén, R. Hovland, E. Isin, P. Johannesson, B. Kull, L.-O. Larsson, L. Löfgren, K. E. Nilsson, T. Noeske, N. Oakes, A. T. Plowright, V. Schnecke, P. Ståhlberg, P. Sörme, H. Wan, E. Wellner and L. Öster, *Bioorg. Med. Chem.*, 2011, **19**, 3039.

59. P. D. Leeson and B. Springthorpe, *Nat. Rev. Drug Disc.*, 2007, **6**, 881.
60. M. P. Edwards and D. A. Price, *Annu. Rep. Med. Chem.*, 2010, **45**, 381.
61. M. Gubeler and J. Mizrahi, *Patent Application*, WO 2003/011867 A1, 2003.
62. Y. Shen, S. L. Volrath, S. C. Weatherly and T. D. Elich, *Mol. Cell*, 2004, **16**, 881.
63. M. Schreurs, *Diabetes Obes. Metab.*, 2009, **11**, 987.
64. E. Chang and M. H. McNeill, *Patent Application*, US 2009/0005375 A1, 2009.
65. E. Chang, T. Duong and A. Vassar, *Patent Application*, US 2009/0253725 A1, 2009.
66. T. S. Haque, N. Liang, R. Golla, R. Seethala, Z. Ma, W. R. Ewing, C. B. Cooper, M. A. Pelleymounter, M. A. Poss and D. Cheng, *Bioorg. Med. Chem. Lett.*, 2009, **19**, 5872.

Subject Index